T0320792

Topics in Graph Theory

The interplay continues to grow between graph theory and a wide variety of models and applications in mathematics, computer science, operations research, and the natural and social sciences.

Topics in Graph Theory is geared toward the more mathematically mature student. The first three chapters provide the basic definitions and theorems of graph theory and the remaining chapters introduce a variety of topics and directions for research. These topics draw on numerous areas of theoretical and applied mathematics, including combinatorics, probability, linear algebra, group theory, topology, operations research, and computer science. This makes the book appropriate for a first course at the graduate level or as a second course at the undergraduate level.

The authors build upon material previously published in *Graph Theory and Its Applications, Third Edition,* by the same authors. That text covers material for both an undergraduate and graduate course, while this book builds on and expands the graduate-level material.

Features

- Extensive exercises and applications.
- Flexibility: appropriate for either a first course at the graduate level or an advanced course at the undergraduate level.
- Opens avenues to a variety of research areas in graph theory.
- Emphasis on topological and algebraic graph theory.

Jonathan L. Gross is a professor of computer science at Columbia University. His research interests include topology and graph theory.

Jay Yellen is a professor of mathematics at Rollins College. His current areas of research include graph theory, combinatorics, and algorithms.

Mark Anderson is a professor of mathematics and computer science at Rollins College. His research interests in graph theory center on the topological or algebraic side.

Discrete Mathematics and Its Applications
Series editors:
Miklos Bona, Donald L. Kreher, Douglas B. West

Algorithmics of Nonuniformity
Tools and Paradigms
Micha Hofri, Hosam Mahmoud

Handbook of Geometric Constraint Systems Principles
Edited by Meera Sitharam, Audrey St. John, Jessica Sidman

Introduction to Chemical Graph Theory
Stephan Wagner, Hua Wang

Extremal Finite Set Theory
Daniel Gerbner, Balazs Patkos

The Mathematics of Chip-Firing
Caroline J. Klivans

Computational Complexity of Counting and Sampling
Istvan Miklos

Volumetric Discrete Geometry
Karoly Bezdek, Zsolt Langi

The Art of Proving Binomial Identities
Michael Z. Spivey

Combinatorics and Number Theory of Counting Sequences
István Mező

Applied Mathematical Modeling
A Multidisciplinary Approach
Douglas R. Shier, K.T. Wallenius

Analytic Combinatorics
A Multidimensional Approach
Marni Mishna

50 years of Combinatorics, Graph Theory, and Computing
Edited By Fan Chung, Ron Graham, Frederick Hoffman, Ronald C. Mullin, Leslie Hogben, Douglas B. West

Fundamentals of Ramsey Theory
Aaron Robertson

Methods for the Summation of Series
Tian-Xiao He

The Lambert W Function and its Generalizations and Applications
István Mező

Combinatorics of Permutations, Third Edition
Miklos Bona

Topics in Graph Theory
Jonathan L. Gross, Jay Yellen, Mark Anderson

https://www.routledge.com/Discrete-Mathematics-and-Its-Applications/book-series/CHDISMTHAPP

Topics in Graph Theory

Jonathan L. Gross
Jay Yellen
Mark Anderson

CRC Press is an imprint of the
Taylor & Francis Group, an **informa** business
A CHAPMAN & HALL BOOK

First edition published 2023
by CRC Press
6000 Broken Sound Parkway NW, Suite 300, Boca Raton, FL 33487-2742

and by CRC Press
4 Park Square, Milton Park, Abingdon, Oxon, OX14 4RN

CRC Press is an imprint of Taylor & Francis Group, LLC

© 2023 Taylor & Francis Group, LLC

Reasonable efforts have been made to publish reliable data and information, but the author and publisher cannot assume responsibility for the validity of all materials or the consequences of their use. The authors and publishers have attempted to trace the copyright holders of all material reproduced in this publication and apologize to copyright holders if permission to publish in this form has not been obtained. If any copyright material has not been acknowledged please write and let us know so we may rectify in any future reprint.

Except as permitted under U.S. Copyright Law, no part of this book may be reprinted, reproduced, transmitted, or utilized in any form by any electronic, mechanical, or other means, now known or hereafter invented, including photocopying, microfilming, and recording, or in any information storage or retrieval system, without written permission from the publishers.

For permission to photocopy or use material electronically from this work, access www.copyright.com or contact the Copyright Clearance Center, Inc. (CCC), 222 Rosewood Drive, Danvers, MA 01923, 978-750-8400. For works that are not available on CCC please contact mpkbookspermissions@tandf.co.uk

Trademark notice: Product or corporate names may be trademarks or registered trademarks and are used only for identification and explanation without intent to infringe.

ISBN: 978-0-367-50787-9 (hbk)
ISBN: 978-1-032-49239-1 (pbk)
ISBN: 978-1-003-05123-7 (ebk)

DOI: 10.1201/9781003051237

Typeset in CMR10 font
by KnowledgeWorks Global Ltd.

Publisher's note: This book has been prepared from camera-ready copy provided by the authors.

CONTENTS

PREFACE

This graduate-level text seeks to unify a wide variety of concepts, both abstract and concrete. Establishing connections between seemingly diverse topics strengthens one's understanding and appreciation of the power and beauty of graph theory. Many of the topics draw on numerous areas of theoretical and applied mathematics. These include combinatorics, probability, linear algebra, group theory, topology, operations research, and computer science.

Mark Anderson, Jonathan L. Gross, and Jay Yellen

PREFACE

Chapter 1

FOUNDATIONS

INTRODUCTION

A graph is a mathematical structure, representing a symmetric relation on a set. The elements of the set are called vertices or nodes, and the edges of a graph are pairs of vertices that are related. Graphs can be used to model a variety of situations, both concrete and abstract. Examples include electrical circuits, roadways, molecular structure, interactions within ecosystems, sociological relationships, and databases. The chapter begins with a formal definition of graph, basic terminology that allows us to describe and differentiate graphs, and examples of different families of graphs.

One prominent aspect of the structure is the system of smaller graphs inside a graph, which are called its *subgraphs*. Subgraphs arise explicitly or implicitly in almost any discussion of graphs. For instance, in the process of building a graph G from scratch, vertex by vertex, and edge by edge, the entire sequence of graphs formed along the way is a nested chain of subgraphs in G.

Relations that are not symmetric are modeled by directed graphs. In this case, edges are *ordered* pairs of vertices. Directed graphs are used to model roadways that include one-way streets, interactions within ecosystems (predator to prey) the flow of control in a computer program, and the subset relations of the members of the power set of a set (superset to subset). The concepts of undirected graphs are extended to concepts for directed graphs.

1.1 BASIC DEFINITIONS AND TERMINOLOGY

Graphs are highly versatile models for analyzing a wide range of practical problems in which points and connections between them have some physical or conceptual interpretation. Placing such analysis on solid footing requires precise definitions, terminology, and notation.

DEFINITION: A ***graph*** $G = (V, E)$ is a mathematical structure consisting of two finite sets V and E. The elements of V are called ***vertices*** (or ***nodes***), and the elements of E are called ***edges***. Each edge has a set of one or two vertices associated to it, which are called its ***endpoints***.

TERMINOLOGY: An edge is said to ***join*** its endpoints.

DEFINITION: ***Adjacent vertices*** are two vertices that are joined by an edge.

DEFINITION: ***Adjacent edges*** are two distinct edges that have an endpoint in common.

DEFINITION: If vertex v is an endpoint of edge e, then v is said to be ***incident*** on e, and e is incident on v.

DEFINITION: The (***open***) ***neighborhood*** of a vertex v in a graph G, denoted $N(v)$, all vertices adjacent to v. The ***closed neighborhood*** of v is given by $N[v] = N(v) \cup \{v\}$. The vertices in $N(v)$ are said to be ***neighbors*** of v.

NOTATION: When G is not the only graph under consideration, the notations V_G and E_G (or $V(G)$ and $E(G)$) are used for the vertex- and edge-sets of G, and the notations $N_G(v)$ and $N_G[v]$ are used for the neighborhoods of v.

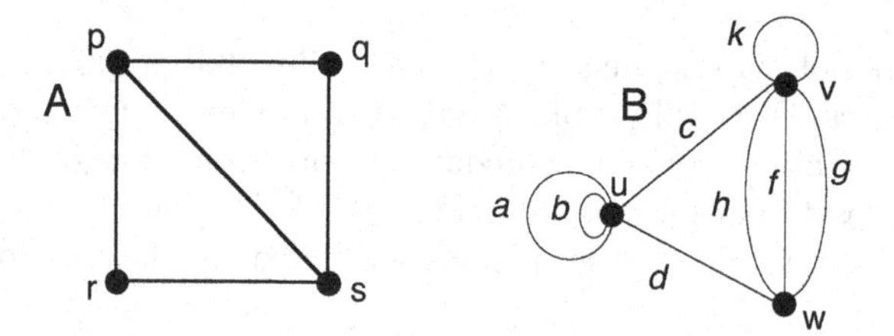

Figure 1.1.1 Line drawings of a graph A and a graph B.

Example 1.1.1: The vertex- and edge-sets of graph A in Figure 1.1.1 are given by

$$V_A = \{p, q, r, s\} \quad \text{and} \quad E_A = \{pq, pr, ps, rs, qs\}$$

and the vertex- and edge-sets of graph B are given by

$$V_B = \{u, v, w\} \quad \text{and} \quad E_B = \{a, b, c, d, f, g, h, k\}$$

Notice that in graph A, we are able to denote each edge simply by juxtaposing its endpoints, because those endpoints are unique to that edge. On the other hand, in graph B, where some edges have the same set of endpoints, we use explicit names to denote the edges.

Simple Graphs and General Graphs

In certain applications of graph theory and in some theoretical contexts, there are frequent instances in which an edge joins a vertex to itself or multiple edges have the same set of endpoints. In other applications or theoretical contexts, such instances are absent.

DEFINITION: A ***proper edge*** is an edge that joins two distinct vertices.

DEFINITION: A ***self-loop*** is an edge that joins a single endpoint to itself.[†]

DEFINITION: A ***multi-edge*** is a collection of two or more edges having identical endpoints. The ***edge multiplicity*** is the number of edges within the multi-edge.

DEFINITION: A ***simple graph*** has neither self-loops nor multi-edges.

DEFINITION: A ***loopless graph*** (or ***multi-graph***) may have multi-edges but no self-loops.

DEFINITION: A (***general***) ***graph*** may have self-loops and/or multi-edges.

Example 1.1.1 continued: Graph A in Figure 1.1.1 is simple. Graph B is not simple; the edges a, b, and k are self-loops, and the edge-sets $\{f, g, h\}$ and $\{a, b\}$ are multi-edges.

TERMINOLOGY: When we use the term *graph* without a modifier, we mean a *general graph*. An exception to this convention occurs when an entire section concerns simple graphs only, in which case, we make an explicit declaration at the beginning of that section.

TERMINOLOGY NOTE: Some authors use the term *graph* without a modifier to mean simple graph, and they use *pseudograph* to mean general graph.

Degree of a Vertex

DEFINITION: The ***degree*** (or ***valence***) of a vertex v in a graph G, denoted $deg(v)$, is the number of proper edges incident on v plus twice the number of self-loops.[‡]

TERMINOLOGY: A vertex of degree d is also called a ***d-valent vertex***.

NOTATION: The smallest and largest degrees in a graph G are denoted $\delta_{\min}$ and $\delta_{\max}$ (or $\delta_{\min}(G)$ and $\delta_{\max}(G)$ when there is more than one graph under discussion). Some authors use δ instead of $\delta_{\min}$ and Δ instead of $\delta_{\max}$.

DEFINITION: The ***degree sequence*** of a graph is the sequence formed by arranging the vertex degrees in non-increasing order.

Example 1.1.2: Figure 1.1.2 shows a graph and its degree sequence.

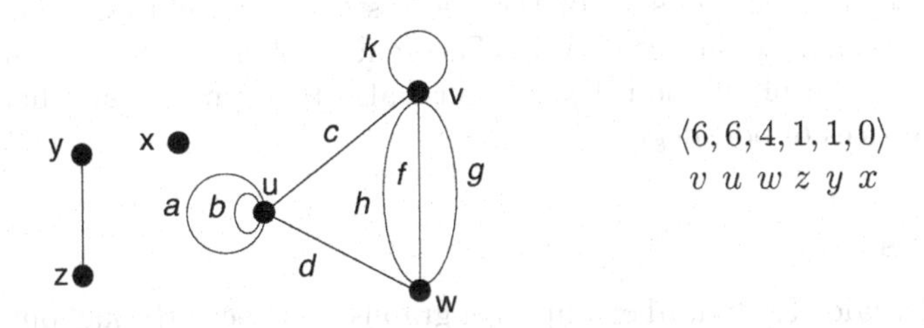

Figure 1.1.2 A graph and its degree sequence.

Although each graph has a unique degree sequence, two structurally different graphs can have identical degree sequences.

[†] We use the term "self-loop" instead of the more commonly used term "loop" because loop means something else in many applications.

[‡] Applications of graph theory to physical chemistry motivate the use of the term *valence*.

Example 1.1.3: Figure 1.1.3 shows two different graphs, G and H, with the same degree sequence.

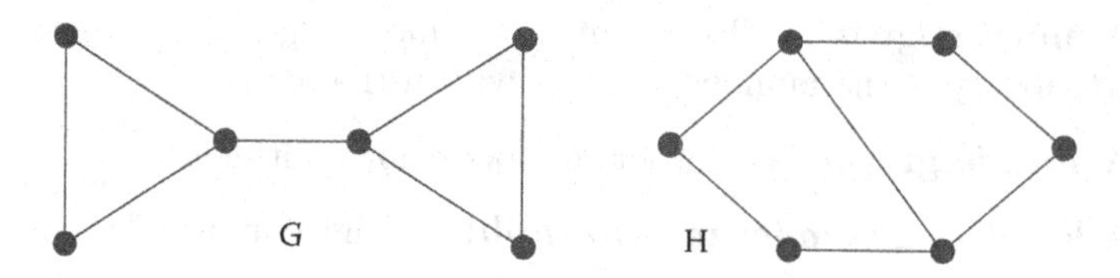

Figure 1.1.3 Two graphs whose degree sequences are both $\langle 3,3,2,2,2,2\rangle$.

The following result shows that the degree sequence of a simple graph must have at least two equal terms.

Proposition 1.1.1: *A simple graph G on two or more vertices must have at least one pair of vertices whose degrees are equal.*

Proof: Suppose that the graph G has n vertices. Then there are n possible degree values, namely $0,\ldots,n-1$. However, there cannot be both a vertex of degree 0 and a vertex of degree $n-1$, since the presence of a vertex of degree 0 implies that each of the remaining $n-1$ vertices is adjacent to at most $n-2$ other vertices. Hence, the n vertices of G can realize at most $n-1$ possible values for their degrees. Thus, the *pigeonhole principle* implies that at least two of the n vertices have equal degree. ♢

The work of Leonhard Euler (1707–1783) is regarded as the beginning of graph theory as a mathematical discipline. The following theorem of Euler establishes a fundamental relationship between the vertices and edges of a graph.

Theorem 1.1.2: [***Euler's Degree-Sum Theorem***] *The sum of the degrees of the vertices of a graph is twice the number of edges.*

Proof: Each edge contributes two to the degree sum. ♢

Corollary 1.1.3: *In a graph, there is an even number of vertices having odd degree.*

Proof: Consider separately, the sum of the degrees that are odd and the sum of those that are even. The combined sum is even by Theorem 1.1.2, and since the sum of the even degrees is even, the sum of the odd degrees must also be even. Hence, there must be an even number of vertices of odd degree. ♢

Regular Graphs

There is a multitude of standard examples of graphs that recur throughout graph theory. Many of them are regular graphs.

DEFINITION: A ***regular*** graph is a graph whose vertices all have equal degree. A ***k-regular*** graph is a regular graph whose common degree is k. That is, the degree sequence of a k-regular graph is $\langle k,k,k,\ldots,k\rangle$.

DEFINITION: A ***complete graph*** is a simple graph such that every pair of vertices is joined by an edge. Any complete graph on n vertices is denoted K_n; it is an $(n-1)$-regular graph.

DEFINITION: The complete graph on one vertex, K_1, is referred to as a ***trivial graph***.

Example 1.1.4: Complete graphs on one, two, three, four, and five vertices are shown in Figure 1.1.4.

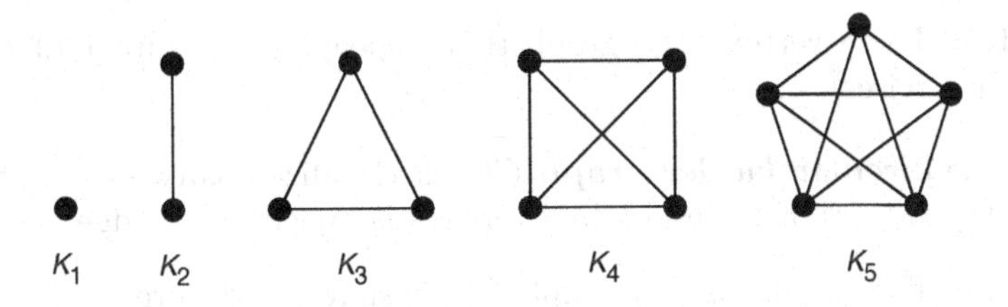

Figure 1.1.4 The first five complete graphs.

DEFINITION: The five regular polyhedra illustrated in Figure 1.1.5 are known as the ***platonic solids***. Their vertex and edge configurations form regular graphs called the ***platonic graphs***.

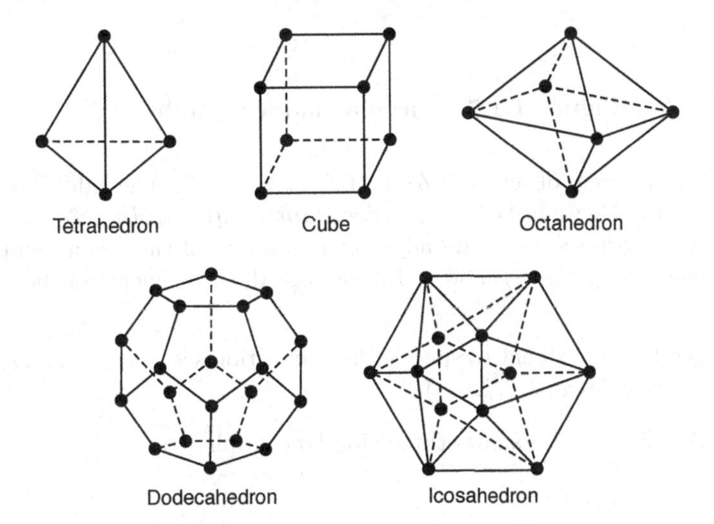

Figure 1.1.5 The five platonic graphs.

DEFINITION: The ***Petersen graph*** is the 3-regular graph represented by the line drawing in Figure 1.1.6. Because it possesses a number of interesting graph-theoretic properties, the Petersen graph is frequently used both to illustrate established theorems and to test conjectures.

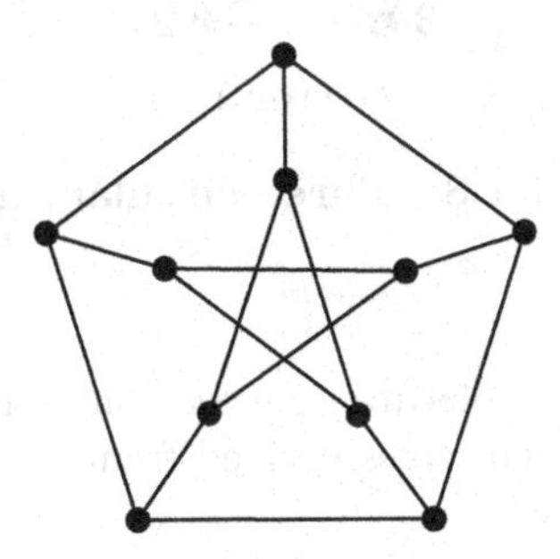

Figure 1.1.6 The Petersen graph.

DEFINITION: The ***hypercube graph*** Q_n is the n-regular graph whose vertex-set is the set of bitstrings of length n, such that there is an edge between two vertices if and only if they differ in exactly one bit.

Example 1.1.5: The 8-vertex cube graph that appeared in Figure 1.1.5 is a hypercube graph Q_3 (see Exercises).

DEFINITION: The ***circular ladder graph*** CL_n is visualized as two concentric n-cycles in which each of the n pairs of corresponding vertices is joined by an edge.

Example 1.1.6: The circular ladder graph CL_4 is shown in Figure 1.1.7.

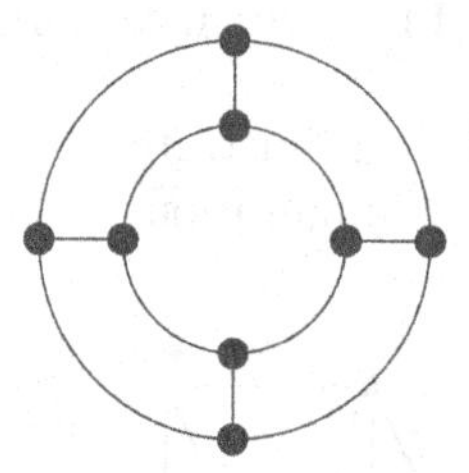

Figure 1.1.7 Circular ladder graph CL_4.

DEFINITION: To the group of integers $Z_n = \{0, 1, \ldots, n-1\}$ under addition modulo n and a set $S \subseteq \{1, \ldots, n-1\}$, we associate the ***circulant graph*** $circ(n : S)$ whose vertex-set is Z_n, such that two vertices i and j are adjacent if and only if there is a number $s \in S$ such that $i + s = j \bmod n$ or $j + s = i \bmod n$. In this regard, the elements of the set S are called ***connections***.

NOTATION: It is often convenient to specify the connection set $S = \{s_1, \ldots, s_r\}$ without the braces, and to write $circ\,(n : s_1, \ldots, s_r)$.

Example 1.1.7: Figure 1.1.8 shows three circulant graphs.

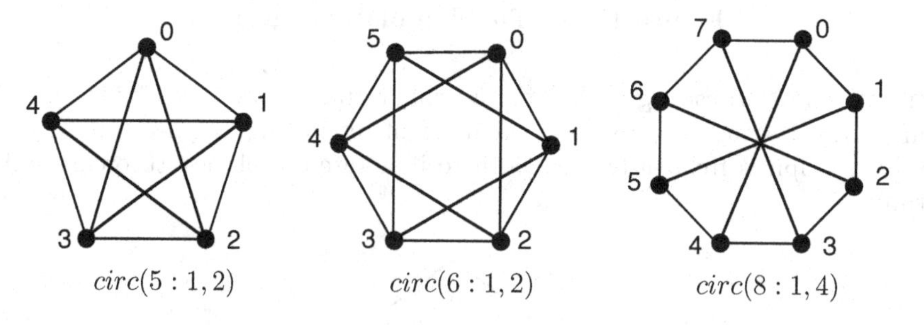

Figure 1.1.8 Three circulant graphs.

Remark: Notice that circulant graphs are simple graphs. Circulant graphs are a special case of ***Cayley graphs***, which are themselves derived from a special case of **voltage graphs**. These are discussed in Chapter 10.

Bipartite Graphs

DEFINITION: A ***bipartite*** graph G is a graph whose vertex-set V can be partitioned into two subsets U and W, such that each edge of G has one endpoint in U and one endpoint in W. The pair U, W is called a ***(vertex) bipartition*** of G, and U and W are called the ***bipartition subsets***.

Example 1.1.8: Two bipartite graphs are shown in Figure 1.1.9. The bipartition subsets are indicated by the solid and hollow vertices.

Figure 1.1.9 Two bipartite graphs.

Proposition 1.1.4: *A bipartite graph cannot have any self-loops.*

Proof: This is an immediate consequence of the definition. ◇

DEFINITION: A ***complete bipartite graph*** is a simple bipartite graph such that every vertex in one of the bipartition subsets is joined to every vertex in the other bipartition subset. Any complete bipartite graph that has m vertices in one of its bipartition subsets and n vertices in the other is denoted $K_{m,n}$.†

Example 1.1.9: The complete bipartite graph $K_{3,4}$ is shown in Figure 1.1.10.

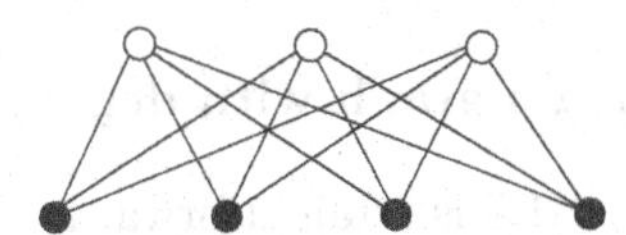

Figure 1.1.10 The complete bipartite graph $K_{3,4}$.

Line Graphs

DEFINITION: The ***line graph*** $L(G)$ of a graph G has a vertex for each edge of G, and two vertices in $L(G)$ are adjacent if and only if the corresponding edges in G are adjacent.

Example 1.1.10: Figure 1.1.11 shows a graph G and its line graph $L(G)$.

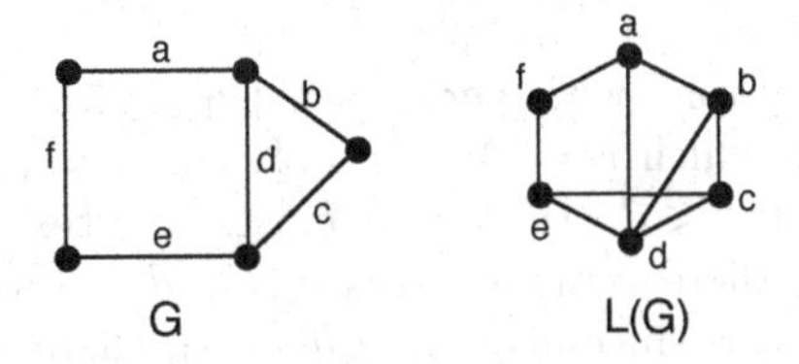

Figure 1.1.11 A graph and its line graph.

†The sense in which $K_{m,n}$ is a unique object is described in §2.1.

Graphic Sequences

Remark: By Theorem 1.1.2, the sum of the terms of a degree sequence of a graph is even. Theorem 1.1.5 establishes the following converse: any non-increasing, nonnegative sequence of integers whose sum is even is the degree sequence of some graph. Example 1.1.11 illustrates the construction used in its proof.

Theorem 1.1.5: *Suppose that $\langle d_1, d_2, \ldots, d_n \rangle$ is a sequence of nonnegative integers whose sum is even. Then there exists a graph with vertices $v_1, v_2, \ldots, v_n$ such that $deg(v_i) = d_i$, for $i = 1, \ldots, n$.*

Proof: Start with n isolated vertices $v_1, v_2, \ldots, v_n$. For each i, if d_i is even, draw $\frac{d_i}{2}$ self-loops on vertex v_i, and if d_i is odd, draw $\frac{d_i-1}{2}$ self-loops. By Corollary 1.1.3, there is an even number of odd d_i's. Thus, the construction can be completed by grouping the vertices associated with the odd terms into pairs and then joining each pair by a single edge. ◇

Example 1.1.11: To construct a graph whose degree sequence is $\langle 5, 4, 3, 3, 2, 1, 0 \rangle$, start with seven isolated vertices $v_1, v_2, \ldots, v_7$. For the even-valued terms of the sequence, draw the appropriate number of self-loops on the corresponding vertices. Thus, v_2 gets two self-loops, v_5 gets one self-loop, and v_7 remains isolated. For the four remaining odd-valued terms, group the corresponding vertices into any two pairs, for instance, v_1, v_3 and v_4, v_6. Then join each pair by a single edge and add to each vertex the appropriate number of self-loops. The resulting graph is shown in Figure 1.1.12.

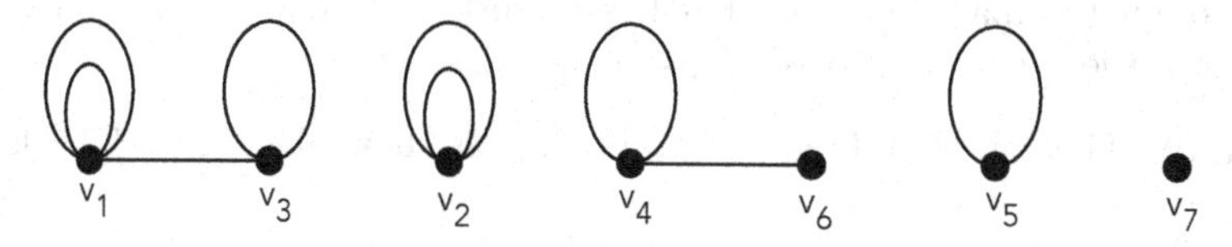

Figure 1.1.12 Constructing a graph with degree sequence $\langle 5, 4, 3, 3, 2, 1, 0 \rangle$.

The construction in Theorem 1.1.5 is straightforward but hinges on allowing the graph to be non-simple. A more interesting problem is determining when a sequence is the degree sequence of a *simple* graph.

DEFINITION: A non-increasing sequence $\langle d_1, d_2, \ldots, d_n \rangle$ is said to be ***graphic*** if it is the degree sequence of some simple graph. That simple graph is said to ***realize*** the sequence.

Remark: If $\langle d_1, d_2, \ldots, d_n \rangle$ is the degree sequence of a simple graph, then, clearly, $d_1 \leq n - 1$.

Theorem 1.1.6: *Let $\langle d_1, d_2, \ldots, d_n \rangle$ be a graphic sequence, with $d_1 \geq d_2 \geq \ldots \geq d_n$. Then there is a simple graph with vertex-set $\{v_1, \ldots, v_n\}$ satisfying deg $(v_i) = d_i$ for $i = 1, 2, \ldots, n$, such that v_1 is adjacent to vertices $v_2, \ldots, v_{d_1+1}$.*

Proof: Among all simple graphs with vertex-set $\{v_1, v_2, \ldots, v_n\}$ and $deg(v_i) = d_i$, $i = 1, 2, \ldots, n$, let G be one for which $r = |N_G(v_1) \cap \{v_2, \ldots, v_{d_1+1}\}|$ is maximum. If $r = d_1$, then the conclusion follows. If $r < d_1$, then there exists a vertex $v_s, 2 \leq s \leq d_1 + 1$, such that v_1 is not adjacent to v_s, and there exists a vertex $v_t, t > d_1 + 1$ such that v_1 is adjacent to v_t (since $deg(v_1) = d_1$). Moreover, since $deg(v_s) \geq deg(v_t)$, there exists a vertex v_k such that v_k is adjacent to v_s but not to v_t. Let $\tilde{G}$ be the graph obtained from G by replacing the edges v_1v_t and v_sv_k with the edges v_1v_s and v_tv_k (as shown in Figure 1.1.13). Then the degrees are all preserved and $v_s \in N_{\tilde{G}}(v_1) \cap \{v_3, \ldots, v_{d_1+1}\}$. Thus, $|N_{\tilde{G}}(v_1) \cap \{v_2, \ldots, v_{d_1+1}\}| = r + 1$, which contradicts the choice of G and completes the proof. ◇

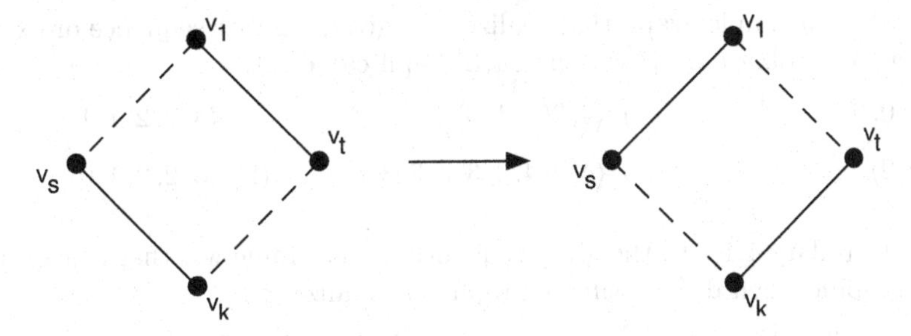

Figure 1.1.13 Switching adjacencies while preserving all degrees.

Corollary 1.1.7: [***Havel (1955) and Hakimi (1961)***] *A sequence $\langle d_1, d_2, \ldots, d_n \rangle$ of nonnegative integers such that $d_1 \leq n-1$ and $d_1 \geq d_2 \geq \ldots \geq d_n$ is graphic if and only if the sequence $\langle d_2 - 1, \ldots, d_{d_1+1} - 1, d_{d_1+2}, \ldots, d_n \rangle$ is graphic.* ♢ *(Exercises)*

Remark: Corollary 1.1.7 yields a recursive algorithm that decides whether a non-increasing sequence is graphic.

Example 1.1.12: We start with the sequence $\langle 3, 3, 2, 2, 1, 1 \rangle$. Figure 1.1.14 illustrates the use of Corollary 1.1.7 to iteratively determine if the sequence is graphic and then backtracks, leading to a graph that realizes the original sequence. The hollow vertex shown in each backtracking step is the new vertex added at that step.

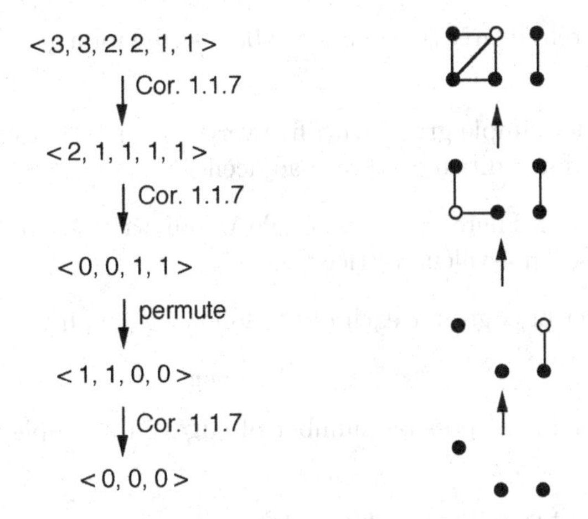

Figure 1.1.14 Testing and realizing the sequence $\langle 3, 3, 2, 2, 1, 1 \rangle$.

EXERCISES for Section 1.1

1.1.1 Draw a graph with the given degree sequence.

(a) $\langle 8, 7, 3 \rangle$ (b) $\langle 9, 8, 8, 6, 5, 3, 1 \rangle$

1.1.2 Draw a simple graph with the given degree sequence.

(a) $\langle 6, 4, 4, 3, 3, 2, 1, 1 \rangle$ (b) $\langle 5, 5, 5, 3, 3, 3, 3, 3 \rangle$

1.1.3 Either draw a simple graph that realizes the given number sequence or explain, without resorting to Corollary 1.1.7, why no such graph can exist.

(a) $\langle 2,2,1,0,0\rangle$ (b) $\langle 4,3,2,1,0\rangle$ (c) $\langle 4,2,2,1,1\rangle$

(d) $\langle 2,2,2,2\rangle$ (e) $\langle 4,4,4,4,3,3,3,3\rangle$ (f) $\langle 3,2,2,1,0\rangle$

1.1.4 Apply Corollary 1.1.7 to the given sequence to determine whether it is graphic. If the sequence is graphic, then draw a simple graph that realizes it.

(a) $\langle 7,6,6,5,4,3,2,1\rangle$ (b) $\langle 5,5,5,4,2,1,1,1\rangle$

(c) $\langle 7,7,6,5,4,4,3,2\rangle$ (d) $\langle 5,5,4,4,2,2,1,1\rangle$

1.1.5 Use Theorem 1.1.6 to prove Corollary 1.1.7.

1.1.6 Write an algorithm that applies Corollary 1.1.7 repeatedly until a sequence of all zeros or a sequence with a negative term results.

1.1.7 Given a group of nine people, is it possible for each person to shake hands with exactly three other people?

1.1.8 Draw a graph whose degree sequence has no duplicate terms.

1.1.9 What special property of a function must the *endpts* function have for a graph to have no multi-edges?

1.1.10 How many different degree sequences can be realized for a graph having three vertices and three edges?

1.1.11 Does there exist a simple graph with five vertices, such that every vertex is incident with at least one edge, but no two edges are adjacent?

1.1.12 Prove or disprove: There exists a simple graph with 13 vertices, 31 edges, three 1-valent vertices, and seven 4-valent vertices.

1.1.13 Find the number of edges for each of the following graphs.

(a) K_n (b) $K_{m,n}$

1.1.14 What is the maximum possible number of edges in a simple bipartite graph on m vertices?

1.1.15 Draw the smallest possible non-bipartite graph.

1.1.16 Determine the values of n for which the given graph is bipartite.

(a) K_n (b) C_n (c) P_n

1.1.17 Draw a 3-regular bipartite graph that is not $K_{3,3}$.

1.1.18 Determine whether the given graph is bipartite. In each case, give a vertex bipartition or explain why the graph is not bipartite.

(a)

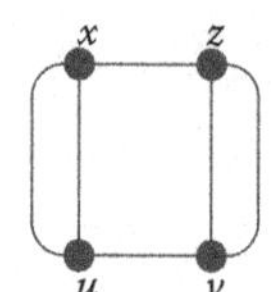

(b)

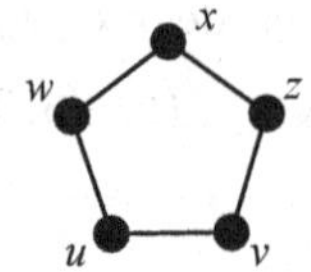

(c)

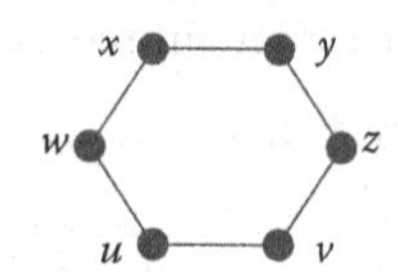

1.1.19 Label the vertices of the cube graph in Figure 1.1.5 with 3-bit binary strings so that the labels on adjacent vertices differ in exactly one bit.

1.1.20 Prove that the graph Q_n is bipartite.

1.1.21 For each of the platonic graphs, is it possible to trace a *tour* of all the vertices by starting at one vertex, traveling only along edges, never revisiting a vertex, and never lifting the pen off the paper? Is it possible to make the tour return to the starting vertex?

1.1.22 Prove or disprove: There does not exist a 5-regular graph on 11 vertices.

1.1.23 Chartrand and Lesniak [ChLe04] define a pair of sequences of nonnegative integers $\langle a_1, a_2, \ldots a_r \rangle$ and $\langle b_1, b_2, \ldots b_t \rangle$ to be ***bigraphical*** if there exists a bipartite graph G with bipartition subsets $U = \{u_1, u_2, \ldots u_r\}$ and $W = \{w_1, w_2, \ldots w_t\}$ such that $deg\,(u_i) = a_i, i = 1, 2, \ldots, r$, and $deg(w_i) = b_i, i = 1, 2, \ldots, t$. Prove that a pair of nonincreasing sequences of nonnegative integers $\langle a_1, a_2, \ldots, a_r \rangle$ and $\langle b_1, b_2, \ldots, b_t \rangle$ with $r \geq 2$, $0 < a_1 \leq t$, and $0 < a_1 \leq r$ is bigraphical if and only if the pair $\langle a_2, \ldots, a_r \rangle$ and $\langle b_1 - 1, b_2 - 1, \ldots, b_{a_1} - 1, b_{a_1+1}, b_{a_1+2}, \ldots, b_t \rangle$ is bigraphical.

1.1.24 Find all the 4-vertex circulant graphs.

1.1.25 Show that each of the following graphs is a circulant graph.

(a)

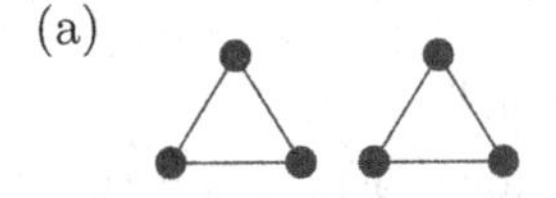

(b)

(c)

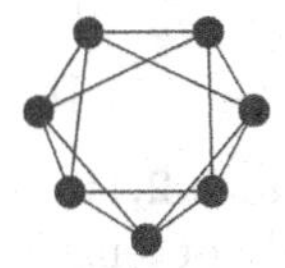

1.1.26 State a necessary and sufficient condition on the positive integers n and k for $circ(\,n : k)$ to be the cycle graph C_n.

1.1.27 Find necessary and sufficient conditions on the positive integers n and k for $circ(n : k)$ to be the graph consisting of $n/2$ mutually nonadjacent edges.

1.2 WALKS AND CONNECTIVITY

Many applications call for graph models that can represent traversal and distance. For instance, the number of node-links traversed by an email message on its route from sender to recipient is a form of distance. Beyond physical distance, another example is that a sequence of tasks in an activity-scheduling network forms a *critical path* if a delay in any one of the tasks would cause a delay in the overall project completion. This section and the following one clarify the notion of walk and related terminology.

Walks

In proceeding continuously from a starting vertex to a destination vertex of a physical representation of a graph, one would alternately encounter vertices and edges. Accordingly, a *walk* in a graph is modeled by such a sequence.

DEFINITION: In a graph G, a ***walk*** from vertex v_0 to vertex v_n is an alternating sequence

$$W = \langle v_0, e_1, v_1, e_2, \ldots, v_{n-1}, e_n, v_n \rangle$$

of vertices and edges, such that $endpts(e_i) = \{v_{i-1}, v_i\}$, for $i = 1, \ldots, n$. One might abbreviate the representation as an edge sequence from the starting vertex to the destination vertex

$$W = \langle v_0, e_1, e_2, \ldots, e_n, v_n \rangle$$

In a simple graph, there is only one edge between two consecutive vertices of a walk, so one could abbreviate the representation as a vertex sequence

$$W = \langle v_0, v_1, \ldots, v_n \rangle$$

TERMINOLOGY: A walk from a vertex x to a vertex y is also called an **x-y walk**.

DEFINITION: The ***length*** of a walk is the number of edge-steps in the walk sequence.

DEFINITION: A walk of length 0, i.e., with one vertex and no edges, is called a ***trivial walk***.

DEFINITION: A ***closed walk*** is a walk that begins and ends at the same vertex. An ***open walk*** begins and ends at different vertices.

DEFINITION: The ***concatenation*** of two walks $W_1 = \langle v_0, e_1, \ldots, v_{k-1}, e_k, v_k \rangle$ and $W_2 = \langle v_k, e_{k+1}, v_{k+1}, e_{k+2}, \ldots, v_{n-1}, e_n, v_n \rangle$ such that walk W_2 begins where walk W_1 ends, is the walk

$$W_1 \circ W_2 = \langle v_0, e_1, \ldots, v_{k-1}, e_k, v_k, e_{k+1}, \ldots, v_{n-1}, e_n, v_n \rangle$$

Example 1.2.1: Figure 1.2.1 shows the concatenation of an open walk of length 3 with an open walk of length 2, resulting in a closed walk of length 5. Notice that W_1 and W_2 have no repeated vertices or edges (and hence are *paths* as defined in the next section), but their concatenation does have repeated edges.

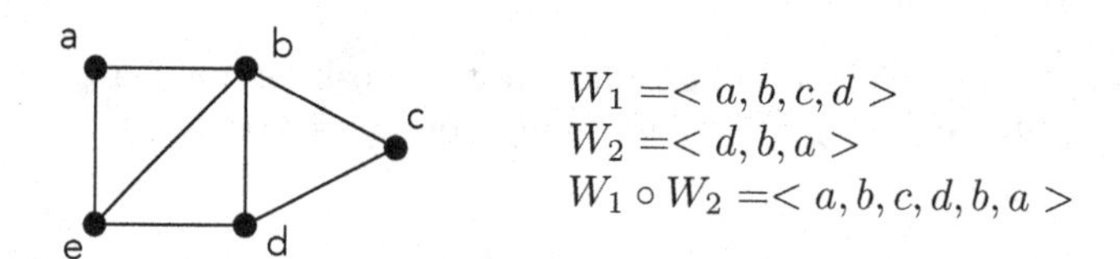

Figure 1.2.1 Concatenation of two walks.

DEFINITION: A ***subwalk*** of a walk $W = \langle v_0, e_1, v_1, e_2, \ldots, v_{n-1}, e_n, v_n \rangle$ is a subsequence of consecutive entries $S = \langle v_j, e_{j+1}, v_{j+1}, \ldots, e_k, v_k \rangle$ such that $0 \le j \le k \le n$, that begins and ends at a vertex. Thus, the subwalk is itself a walk.

Distance

DEFINITION: The ***distance*** $d(s,t)$ from a vertex s to a vertex t in a graph G is the length of a shortest $s - t$ walk if one exists; otherwise, $d(s,t) = \infty$.

Example 1.2.2: In the following graph, $d(u, w) = 4$ and $d(x, y) = 1$.

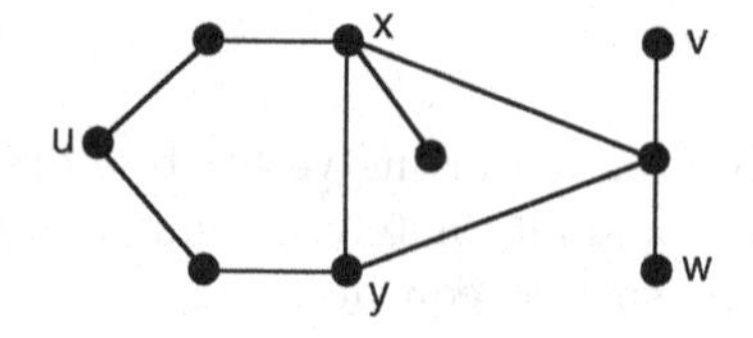

A shortest walk contains no repeated vertices or edges (see Exercises). It is instructive to think about how one might find a shortest walk. Ad hoc approaches are adequate for small graphs, but a systematic algorithm is essential for larger graphs. See Figure 1.2.2.

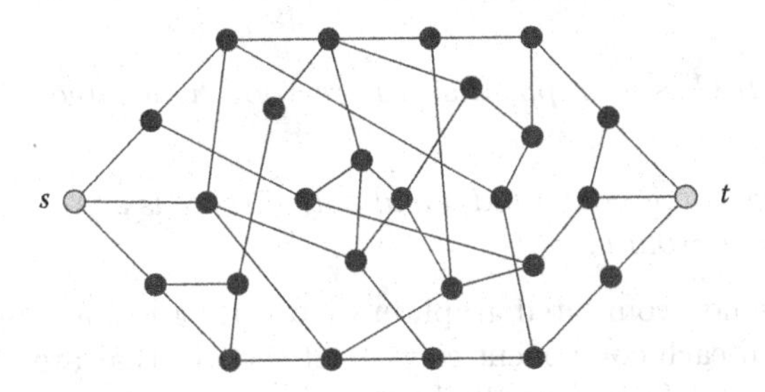

Figure 1.2.2 How might you find a shortest walk from s to t in this graph?

Eccentricity, Diameter, and Radius

DEFINITION: The ***diameter*** of a graph G, denoted $diam(G)$, is the maximum distance between two vertices in G. That is,

$$diam\,(G) = \max_{x,y \in V_G} \{d\,(x,y)\}$$

DEFINITION: The ***eccentricity*** of a vertex v in a graph G, denoted $ecc(\,v)$, is the distance from v to a vertex farthest from v. That is,

$$ecc\,(v) = \max_{x \in V_G} \{d\,(v,x)\}$$

DEFINITION: The ***radius*** of a graph G, denoted $rad(G)$, is the minimum of the vertex eccentricities. That is,

$$rad\,(G) = \min_{x \in V_G} \{ecc\,(x)\}$$

DEFINITION: A ***central vertex*** v of a graph G is a vertex with minimum eccentricity, i.e., $ecc(v) = rad(G)$.

Example 1.2.3: The graph of Figure 1.2.3 below has diameter 4, achieved by the vertex pairs u, v and u, w. The vertices x and y have eccentricity 2 and all other vertices have greater eccentricity. Thus, the graph has radius 2 and central vertices x and y.

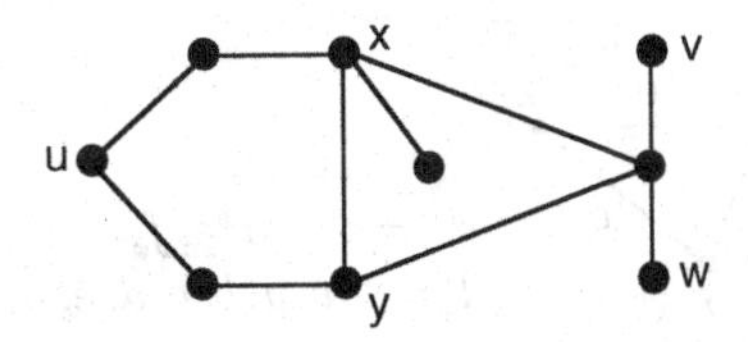

Figure 1.2.3 A graph with diameter 4 and radius 2.

Example 1.2.4: Let G be the graph whose vertex-set is the set of all people on the planet and whose edges correspond to the pairs of people who are acquainted. Then according to the *six degrees of separation* conjecture, the graph G has diameter 6.

Connectedness

DEFINITION: Vertex v is ***reachable from*** vertex u if there is a walk from u to v.

DEFINITION: A graph is ***connected*** if for every pair of vertices u and v, there is a walk from u to v.

Proposition 1.2.1: *If G is a graph, then the relation 'reachable from' is an equivalence relation on V_G.* ◇ *(Exercises)*

Corollary 1.2.2: *A graph is connected if and only if there is a vertex v such that for every other vertex is reachable from v.*

Example 1.2.5: The non-connected graph in Figure 1.2.4 is made up of connected pieces called *components*, and each component consists of vertices that are all reachable from one another. This concept is defined formally in §2.1.

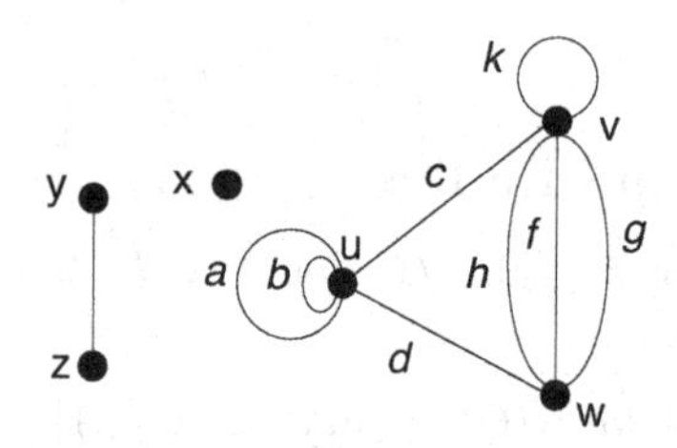

Figure 1.2.4 A non-connected graph with three components.

Paths

DEFINITION: A ***trail*** is a walk with no repeated edges.

DEFINITION: A ***path*** is a trail with no repeated vertices (except possibly the initial and final vertices).

DEFINITION: A walk, trail, or path is ***trivial*** if it has only one vertex and no edges.

TERMINOLOGY NOTE: Unfortunately, there is no universally agreed-upon terminology for walks, trails, and paths. For instance, some authors use the term *path* instead of walk and *simple path* instead of path. Others use the term *path* instead of trail. It is best to check the terminology when reading articles about this material.

Example 1.2.6: For the graph shown in Figure 1.2.5, $W = \langle v, a, e, f, a, d, z\rangle$ is the edge sequence of a walk but not a trail, because edge a is repeated, and $T = \langle v, a, b, c, d, e, u\rangle$ is a trail but not a path, because vertex x is repeated.

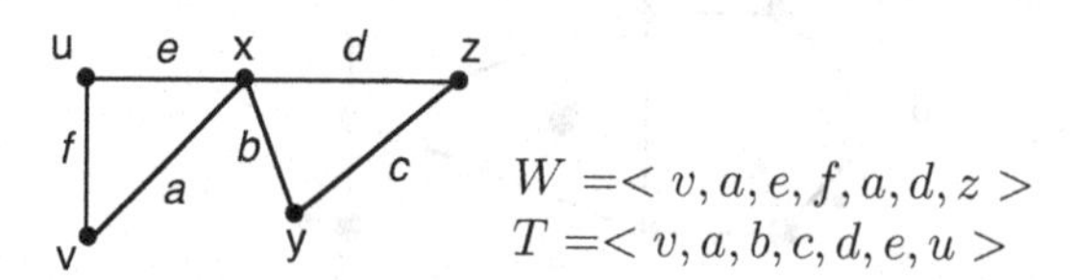

Figure 1.2.5 Walk W is not a trail, and trail T is not a path.

DEFINITION: A graph is a ***path graph*** if there exists a non-closed path that contains all of its vertices and edges. A path graph with n vertices has $n - 1$ edges and is denoted P_n. Note that P_1 is considered to be a path graph.

Example 1.2.7: Path graphs P_2 and P_4 are shown in Figure 1.2.6.

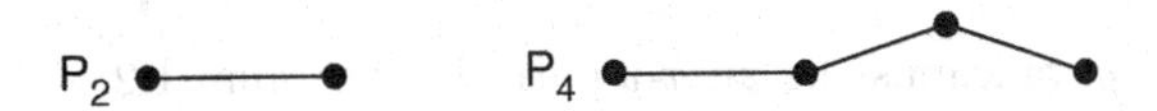

Figure 1.2.6 Path graphs P_2 and P_4.

Deleting Closed Subwalks from Walks

DEFINITION: Given a walk $W = \langle v_0, e_1, \ldots, v_{n-1}, e_n, v_n \rangle$ that contains a nontrivial, closed subwalk $W' = \langle v_k, e_{k+1}, \ldots, v_{m-1}, e_m, v_k \rangle$, the ***reduction of walk W by subwalk W'***, denoted $W - W'$, is the walk

$$W - W' = \langle v_0, e_1, \ldots, v_{k-1}, e_k, v_k, e_{m+1}, v_{m+1}, \ldots, v_{n-1}, e_n \rangle$$

Thus, $W - W'$ is obtained by deleting from W all of the vertices and edges of W' except v_k. Or, less formally, $W - W'$ is obtained by ***deleting W' from W***.

Example 1.2.8: Figure 1.2.7 shows a simple graph and the edge sequences of three walks, W, W', and $W - W'$. Observe that $W - W'$ is traversed by starting a traversal of walk W at vertex u and avoiding the detour W' at vertex v by continuing directly to vertex w.

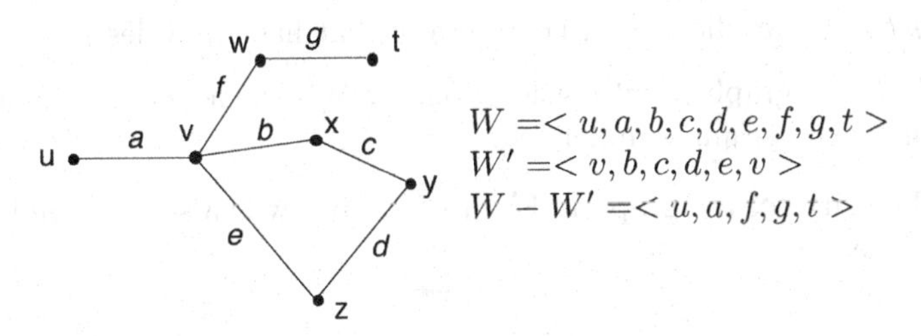

Figure 1.2.7 The reduction walk $W - W'$.

Remark: The reduction of a walk by a (nontrivial) closed subwalk yields a walk that is obviously shorter than the original walk.

DEFINITION: Given two walks A and B, the walk B is said to be a ***reduced walk*** of A if there exists a sequence of walks $A = W_1, W_2, \ldots, W_r = B$ such that for each $i = 1, \ldots, r-1$, walk W_{i+1} is the reduction of W_i by some closed subwalk of W_i.

The next three results show that the concepts of distance and reachability in a graph can be defined in terms of paths instead of walks. This often simplifies arguments in which the detours of a walk can be ignored.

Lemma 1.2.3: *Every open x-y walk W is either an x-y path or contains a closed subwalk.*

Proof: If W is not an x-y path, then the subsequence of W between repeated vertices defines a closed subwalk of W. ◇

Theorem 1.2.4: *Let W be an open x-y walk. Then either W is an x-y path or there is an x-y path that is a reduced walk of W.*

Proof: If W is not an x-y path, then delete a closed subwalk from W to obtain a shorter x-y walk. Repeat the process until the resulting x-y walk contains no closed subwalks and, hence, is an x-y path, by Lemma 1.2.3. ◇

Corollary 1.2.5: *The distance from a vertex x to a reachable vertex y is always realizable by an x-y path.*

Proof: A shortest x-y walk must be an x-y path, by Theorem 1.2.4. 

Cycles

DEFINITION: A nontrivial closed path is called a ***cycle***.

DEFINITION: A graph is a ***cycle graph*** if there exists a cycle that contains all of its vertices and edges. A cycle graph with n vertices has n edges and is denoted C_n.

Example 1.2.9: The cycle graphs C_1, C_2, and C_4 are shown in Figure 1.2.8.

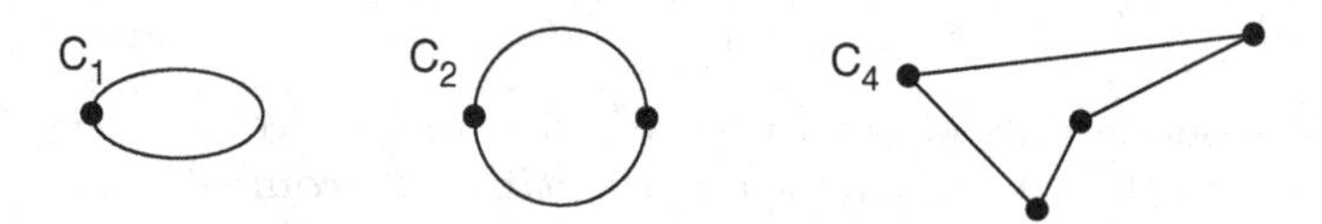

Figure 1.2.8 Cycle graphs C_1, C_2, and C_4.

DEFINITION: A ***tree*** is a connected graph that has no cycles.

DEFINITION: A ***forest*** is an acyclic graph, i.e., a graph that has no cycles.

DEFINITION: The ***girth*** of a graph G with at least one cycle is the length of a shortest cycle in G. The girth of an acyclic graph is undefined.

Example 1.2.10: The girth of the graph in Figure 1.2.9 below is 3 since there is a 3-cycle but no 2-cycle or 1-cycle.

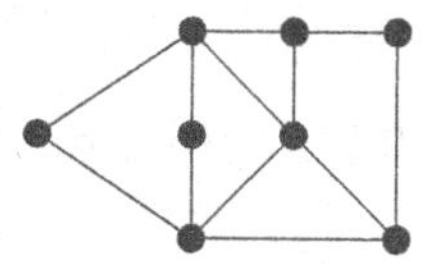

Figure 1.2.9 A graph with girth 3.

DEFINITION: A cycle that includes every vertex of a graph is called a ***Hamiltonian cycle***.†

DEFINITION: A ***Hamiltonian graph*** is a graph that has a Hamiltonian cycle.

Example 1.2.11: The graph in Figure 1.2.10 is Hamiltonian. The edges of the Hamiltonian cycle $\langle u, z, y, x, w, t, v, u \rangle$ are shown in bold.

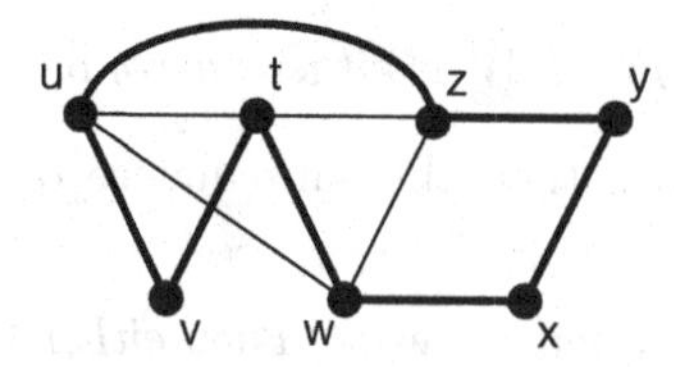

Figure 1.2.10 A Hamiltonian graph.

†The term "Hamiltonian" is in honor of the Irish mathematician William Rowan Hamilton.

Example 1.2.12: The graphs in Figure 1.2.11 are not Hamiltonian.

Figure 1.2.11 Two graphs that are not Hamiltonian.

The following result gives an important characterization of bipartite graphs.

Theorem 1.2.6: *A graph G is bipartite if and only if it has no cycles of odd length.*

Proof: *Necessity* ($\Rightarrow$): Suppose that G is bipartite. Since traversing each edge in a walk switches sides of the bipartition, it requires an even number of steps for a walk to return to the side from which it started. Thus, a cycle must have even length.

Sufficiency ($\Leftarrow$): Let G be a graph with $n \geq 2$ vertices and no cycles of odd length. Without loss of generality, assume that G is connected. Pick any vertex u of G, and define a partition (X, Y) of V as follows:

$$X = \{x \mid d(u, x) \text{ is even}\}$$
$$Y = \{y \mid d(u, y) \text{ is odd}\}$$

If (X, Y) is not a bipartition of G, then there are two vertices in one of the sets, say v and w, that are joined by an edge, say e. Let P_1 be a shortest $u - v$ path, and let P_2 be a shortest $u - w$ path. By definition of the sets X and Y, the lengths of these paths are both even or both odd. Starting from vertex u, let z be the last vertex common to both paths (see Figure 1.2.12).

Since P_1 and P_2 are both shortest paths, their $u \to z$ sections have equal length. Thus, the lengths of the $z \to v$ section of P_1 and the $z \to w$ section of P_2 are either both even or both odd. But then the concatenation of those two sections with edge e forms a cycle of odd length, contradicting the hypothesis. Hence, (X, Y) is a bipartition of G. ♢

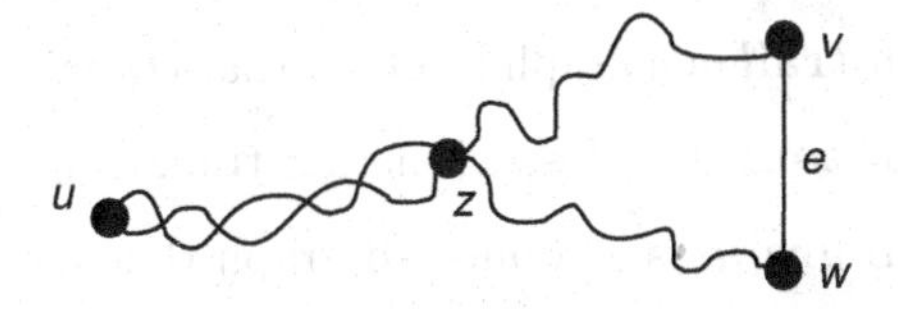

Figure 1.2.12 Figure for sufficiency part of Theorem 1.2.6 proof.

Trails

The following proposition is similar to Theorem 1.2.4.

Proposition 1.2.7: *Every nontrivial, closed trail T contains a subwalk that is a cycle.*

Proof: Let T' be a minimum-length, nontrivial, closed subwalk of T. Its minimum length implies that T' has no proper closed subwalks, and, hence, its only repeated vertices are its first and last. Thus, T' is a cycle. ♢

Remark: The assertion of Proposition 1.2.7 is no longer true if T is merely a closed walk (see Exercises).

DEFINITION: A collection of edge-disjoint cycles, $C_1, C_2, \ldots, C_m$, is called a ***decomposition*** of a closed trail T if each cycle C_i is either a subwalk or a reduced walk of T and the edge-sets of the cycles *partition* the edge-set of trail T.

Theorem 1.2.8: *A closed trail can be decomposed into edge-disjoint cycles.*

Proof: We use a proof by induction on the number of edges. A closed trail having only one edge is itself a cycle. Assume that the theorem is true for all closed trails having m or fewer edges. Next, let T be a closed trail with $m + 1$ edges. By Proposition 1.2.7, the trail T contains a cycle, say C, and $T - C$ is a closed trail having m or fewer edges. By the induction hypothesis, the trail $T - C$ can be decomposed into edge-disjoint cycles. Therefore, these cycles together with C form an edge-decomposition of trail T. ♢

Example 1.2.13: Consider the closed trail $T = < u, a, b, c, d, e, f, g, h, i, j, k, u >$ in Figure 1.2.13. The three cycles C_1, C_2, and C_3, depicted with bold, dashed, and solid lines, respectively, form a decomposition of T.

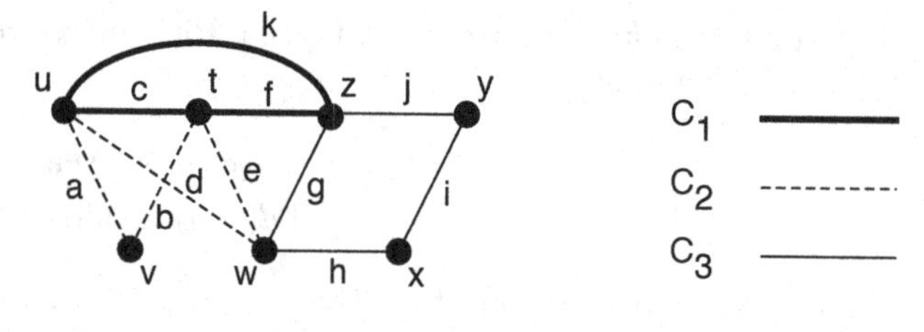

$$T = < u, a, b, c, d, e, f, g, h, i, j, k, u >$$

$$C_1 = < z, k, c, f, z >; \quad C_2 = < u, a, b, e, d, u >; \quad C_3 = < w, h, i, j, g, w >$$

Figure 1.2.13 A closed trail T decomposed into three cycles.

Eulerian Trails

DEFINITION: An ***Eulerian trail*** in a graph is a trail that contains every edge of that graph.

DEFINITION: An ***Eulerian tour*** is a closed Eulerian trail.

DEFINITION: An ***Eulerian graph*** is a connected graph that has an Eulerian tour.†

Example 1.2.14: The graph in Figure 1.2.14 is an Eulerian graph illustrated by the Eulerian tour T, shown in Figure 1.2.13.

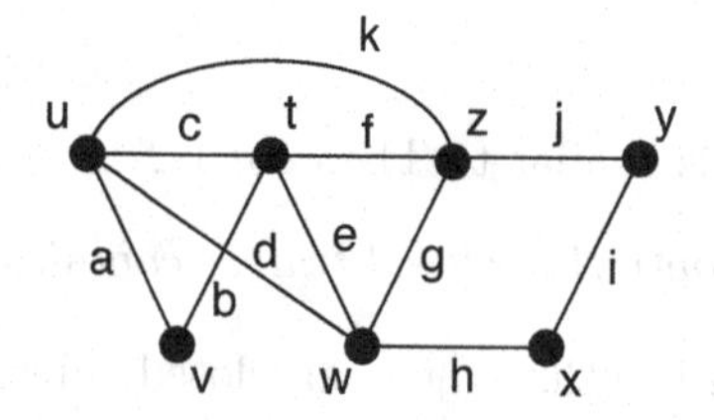

Figure 1.2.14 An Eulerian graph.

†The term "Eulerian" is in honor of the Swiss mathematician Leonhard Euler.

Application 1.2.1: *Traversing the Edges of a Network* Suppose that the graph shown in Figure 1.2.14 represents a network of railroad tracks. A special cart equipped with a sensing device will traverse every section of the network, checking for imperfections. Can the cart be routed so that it traverses each section of track exactly once and then returns to its starting point?

The problem is equivalent to determining whether the graph is Eulerian, and as seen from Example 1.2.14, trail T does indeed provide the desired routing.

By Theorem 1.2.8, every Eulerian tour decomposes into edge-disjoint cycles, which establishes the following corollary.

Corollary 1.2.9: *Every vertex in an Eulerian graph has even degree.*

Establishing the converse of Corollary 1.2.9 results in the following classical characterization of Eulerian graphs, which appears in Chapter 3.

Theorem 3.2.11: [Eulerian-Graph Characterization] The following statements are equivalent for a connected graph G.

1. G is Eulerian.
2. The degree of every vertex in G is even.
3. E_G is the union of the edge-sets of a set of edge-disjoint cycles of G.

EXERCISES for Section 1.2

1.2.1 Determine which of the following vertex sequences represent walks in the graph below.

(a) $\langle u, v\rangle$ (b) $\langle v\rangle$ (c) $\langle u, z, v\rangle$ (d) $\langle u, v, w, v\rangle$

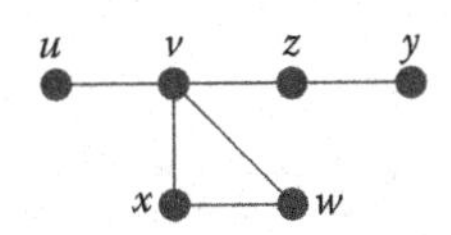

1.2.2 Find all walks of length 4 or 5 from vertex w to vertex r in the following graph.

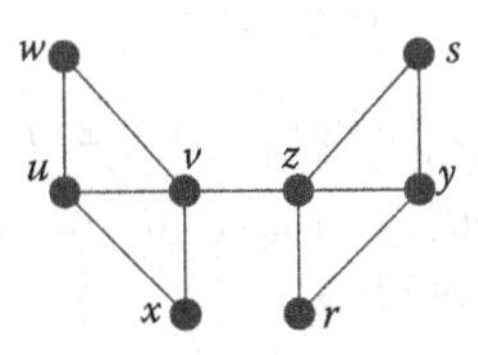

1.2.3 Find the distance between vertices x and y in the following graph.

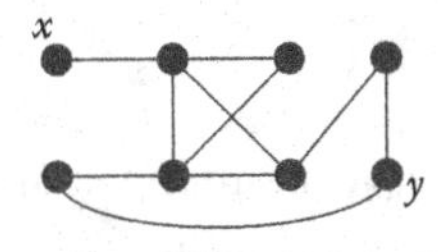

1.2.4 Is there an even-length walk between two antipodal vertices (i.e., endpoints of a long diagonal) of a cube?

1.2.5 Draw an 8-vertex connected graph with as few edges as possible.

1.2.6 Draw a 7-vertex connected graph such that the removal of any one edge results in a non-connected graph.

1.2.7 Draw an 8-vertex connected graph with no closed walks, except those that retrace at least one edge.

1.2.8 Draw a 5-vertex connected graph that remains connected after the removal of any two of its vertices.

1.2.9 A 3×3 mesh graph and a 3×4 mesh graph are shown below. For what values of m and n is the $m \times n$ grid graph Hamiltonian?

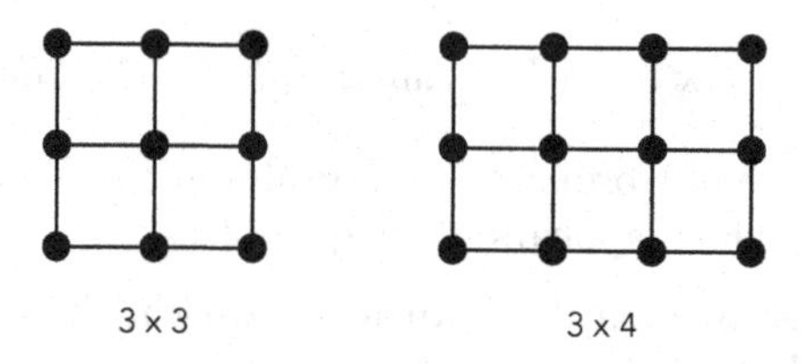

1.2.10 Determine the diameter, radius, and central vertices of the given graph.

(a) The graph in Exercise 1.2.1.

(b) The graph in Exercise 1.2.2.

(c) The path graph P_n, $n \geq 3$.

(d) The cycle graph C_n, $n \geq 4$.

(e) The complete graph K_n, $n \geq 3$.

(f) The complete bipartite graph $K_{m,n}$, $m \geq n \geq 3$.

(g) The Petersen graph.

(h) The hypercube graph Q_n?

(i) The circular ladder graph CL_n , $n \geq 4$.

(j) The circulant graph $circ(n : m)$.

(k) The circulant graph $circ(n : a, b)$.

1.2.11 Let G be a connected graph. Prove that

$$rad\,(G) \leq diam\,(G) \leq 2 \cdot rad\,(G)$$

1.2.12 Let u and v be any two vertices of a connected graph G. Prove that there exists a $u - v$ walk containing all the vertices of G.

1.2.13 Prove that a shortest walk between two vertices cannot repeat a vertex or an edge.

1.2.14 Let x and y be two different vertices in the given graph. Find the number of x-y walks of length 2 and of length 3.

(a) The complete graph K_4.

(b) The complete bipartite graph $K_{3,3}$, assuming x and y are adjacent.

(c) The complete bipartite graph $K_{3,3}$, assuming x and y are not adjacent.

1.2.15 Prove that the distance function d on a graph G satisfies the *triangle inequality*. That is,

$$\text{For all } x, y, z \in V_G, \quad d(x, z) \leq d(x, y) + d(y, z).$$

1.2.16 Prove Proposition 1.2.1.

1.2.17 Let G be a connected graph. Prove that if $d(x, y) \geq 2$, then there exists a vertex w such that $d(x, y) = d(x, w) + d(w, y)$.

1.2.18 Let G be an n-vertex simple graph such that $\deg(v) \geq \frac{n-1}{2}$ for every vertex $v \in V_G$. Prove that graph G is connected.

1.2.19 Explain how Corollary 1.2.9 follows from Theorem 1.2.8.

1.2.20 Determine the girth of the given graph.

(a) The hypercube graph Q_n.
(b) The circular ladder graph CL_n.
(c) The Petersen graph.
(d) The circulant graph $circ(n : m)$.
(e) The circulant graph $circ(n : a, b)$.

1.2.21 Determine whether the given graph is Hamiltonian.

(a) The hypercube graph Q_n.
(b) The circular ladder graph CL_n.
(c) The Petersen graph.
(d) The circulant graph $circ(n : a, b)$.

1.2.22 Find an Eulerian trail, if possible, for the given graph.

(a) The hypercube graph Q_3.
(b) The circular ladder graphs CL_4.
(c) The Petersen graph.

1.2.23 Construct an Eulerian trail, if possible, for the given graph.

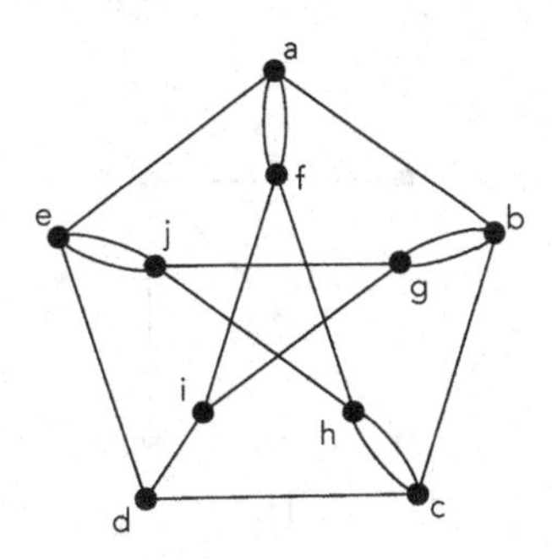

1.2.24 Prove that if v is a vertex on a nontrivial, closed trail, then v lies on a cycle.

1.2.25 Prove or disprove: Every graph that has a closed walk of odd length has a cycle.

1.2.26 Prove that in a digraph, a shortest directed walk from a vertex x to a vertex y is a directed path from x to y.

1.2.27 Suppose G is a simple graph whose vertices all have degree at least k.

a. Prove that G contains a path of length k.

b. Prove that if $k \geq 2$, then G contains a cycle of length at least k.

1.2.28 Prove that if v is a vertex on a nontrivial, closed trail, then v lies on a cycle.

1.2.29 Prove or disprove: Every graph that has a closed walk of odd length has a cycle.

1.2.30 Prove that in a digraph, a shortest directed walk from a vertex x to a vertex y is a directed path from x to y.

1.2.31 Suppose G is a simple graph whose vertices all have degree at least k.

a. Prove that G contains a path of length k.

b. Prove that if $k \geq 2$, then G contains a cycle of length at least k.

1.3 Subgraphs

DEFINITION: A graph $G' = (V', E')$ is a ***subgraph*** of graph $G = (V, E)$ if $V' \subseteq V$ and $E' \subseteq E$. It is a ***proper subgraph*** of G if V' is a proper subset of V or E' is a proper subset of E.

Example 1.3.1: *Figure 1.3.1 shows a graph and three proper subgraphs.*

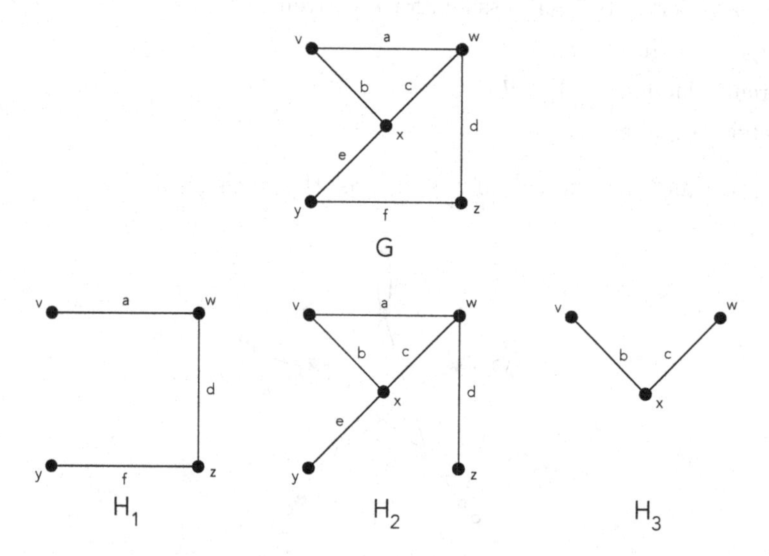

Figure 1.3.1 A graph G and three (proper) subgraphs H_1, H_2, and H_3.

Properties of a given graph are often determined by the existence or non-existence of certain types of subgraphs. For example, bipartite graphs never have an odd cycle graph as a subgraph.

Induced Subgraphs

DEFINITION: For a given graph G, the ***subgraph induced on a vertex subset*** U of V_G, denoted $G(U)$, is the subgraph of G whose vertex-set is U and whose edge-set consists of all edges in G that have both endpoints in U. That is,

$$V_{G(U)} = U \quad \text{and} \quad E_{G(U)} = \{e \in E_G | \ endpts(e) \subseteq U\}$$

Example 1.3.2: Figure 1.3.2 shows a 10-vertex graph G and two induced subgraphs (shown in bold), one on a 5-vertex subset, and the other on a 7-vertex subset.

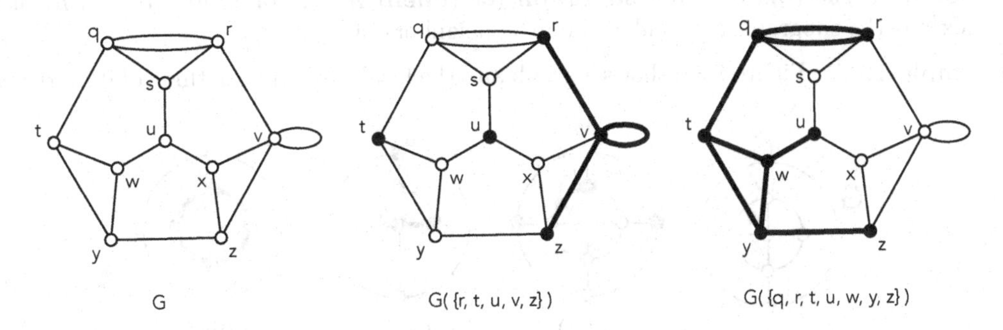

Figure 1.3.2 Two subgraphs induced on vertex subsets.

DEFINITION: For a given graph G, the ***subgraph induced on an edge subset*** D of E_G, denoted $G(D)$, is the subgraph of G whose edge-set is D and whose vertex-set consists of all vertices that are incident with at least one edge in D. That is,

$$V_{G(D)} = \{v \in V_G \mid v \in endpts(e), \text{ for some } e \in D\} \text{ and } E_{G(D)} = D$$

Example 1.3.3: Figure 1.3.3 shows a graph G and two induced subgraphs (shown in bold), one on a 8-edge subset, and the other on a 6-edge subset.

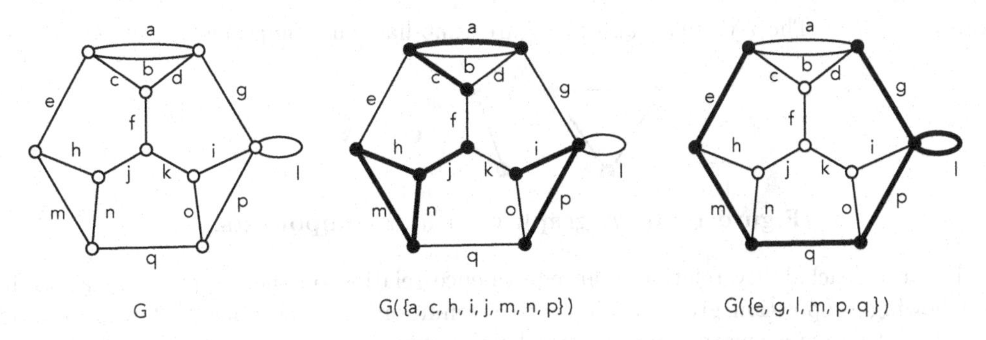

Figure 1.3.3 Two subgraphs induced on edge subsets.

DEFINITION: The ***center of a graph*** G, denoted $Z(G)$, is the subgraph induced on the set of central vertices of G.

Example 1.3.4: The vertices of the graph in Figure 1.3.4 are labeled by their eccentricities. Since the minimum eccentricity is 3, the vertices of eccentricity 3 lie in the center.

Remark: In §1.4, we show that any graph can be the center of a graph.

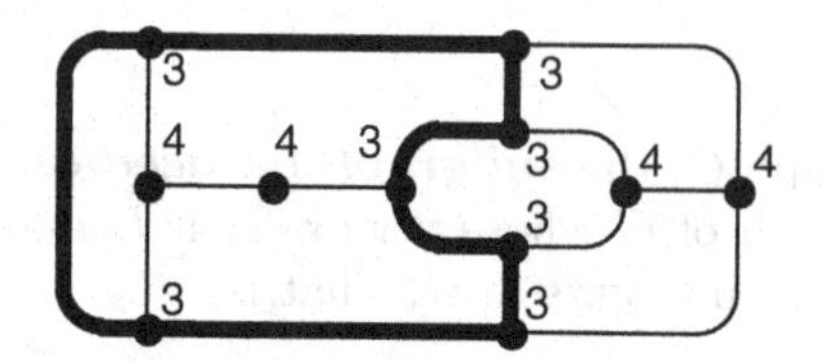

Figure 1.3.4 A graph whose center is a 7-cycle.

Local Subgraphs

DEFINITION: The (***open***) ***local subgraph*** (or (***open***) ***neighborhood subgraph***) of a vertex v is the subgraph $L(v)$ induced on the neighbors of v.

Example 1.3.5: Figure 1.3.5 shows a graph and the local subgraphs of three of its vertices.

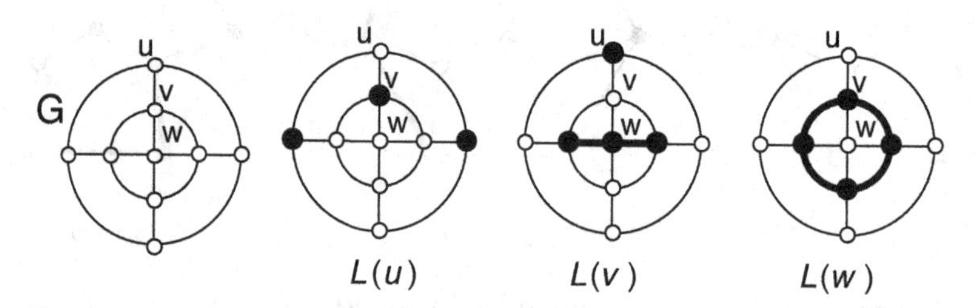

Figure 1.3.5 A graph G and three of its local subgraphs.

Components

DEFINITION: A ***component*** of a graph G is a *maximal* connected subgraph of G. In other words, a connected subgraph H is a component of G if H is not a proper subgraph of any connected subgraph of G.

The only component of a connected graph is the entire graph. Intuitively, the components of a non-connected graph are the "whole pieces" it comprises.

Example 1.3.6: The 7-vertex graph in Figure 1.3.6 has four components.

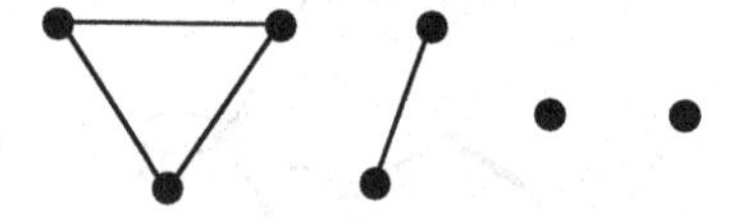

Figure 1.3.6 A graph with four components.

Since the reachability relation is an equivalence relation on the vertex-set of a graph, the following proposition gives an alternative definition of a component of a graph and shows that there is a one-to-one correspondence between the components of a graph and the equivalence classes of the reachability relation.

Proposition 1.3.1: *A subgraph H of a graph G is a component of G if and only if V_H is an equivalence class of the reachability relation on V_G and H is the subgraph induced on its vertex-set, i.e., $H = G(V_H)$.* ♢ *(Exercises)*

NOTATION: The number of components of a graph G is denoted $c(G)$.

DEFINITION: In a graph G, the ***component of a vertex*** v, denoted $C(v)$, is the subgraph induced on the subset of all vertices reachable from v.

Cliques and Independent Sets

DEFINITION: A subset S of V_G is called a ***clique*** if every pair of vertices in S is joined by at least one edge, and no proper superset of S has this property.

Thus, a clique of a graph G is a maximal subset of mutually adjacent vertices in G.

TERMINOLOGY NOTE: The graph-theory community seems to be split on whether to require that a clique be maximal with respect to the pairwise-adjacency property. The prose sense of "clique" as an *exclusive* group of people motivates our decision to include maximality in the mathematical definition.

DEFINITION: The ***clique number*** of a graph G is the number $\omega(G)$ of vertices in a largest clique in G.

Example 1.3.7: There are three cliques in the graph shown in Figure 1.3.7. In particular, each of the vertex subsets $\{u, v, y\}$, $\{u, x, y\}$, and $\{y, z\}$ corresponds to a complete subgraph, and no proper superset of any of them corresponds to a complete subgraph. The clique number is 3, the cardinality of the largest clique.

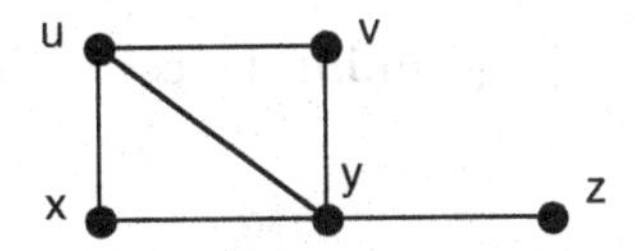

Figure 1.3.7 A graph with three cliques.

Example 1.3.8: If G is a simple graph, then it follows from the definition of a clique that the subgraph induced on a clique is a complete graph. Some authors refer to this complete graph as a clique.

DEFINITION: A subset S of V_G is said to be an ***independent set*** if no pair of vertices in S is joined by an edge. That is, S is a subset of mutually nonadjacent vertices of G.

DEFINITION: The ***independence number*** of a graph G is the number $\alpha(G)$ of vertices in a largest independent set in G.

Remark: The clique number $\omega(G)$ and the independence number $\alpha(G)$ of a graph may be regarded as complementary concepts, as will be shown in §1.4, by Theorem 1.4.1.

Spanning Subgraphs

DEFINITION: A ***spanning subgraph*** H of a graph G is a subgraph such that $V_H = V_G$.

DEFINITION: A ***spanning tree*** of a graph is a spanning subgraph that is a tree.

Example 1.3.9: Three spanning subgraphs of a non-simple graph are shown in Figure 1.3.8 with their edges in bold. The subgraph in (c) is a spanning tree.

A number of important properties related to spanning trees are established in Chapter 3. Two of these properties are previewed here.

Proposition 1.3.2: *A graph is connected if and only if it contains a spanning tree.*

♢ *(Exercises)*

Proposition 1.3.3: *Every acyclic subgraph of a connected graph G is contained in at least one spanning tree of G.* ♢ *(Exercises or Proposition 3.2.2)*

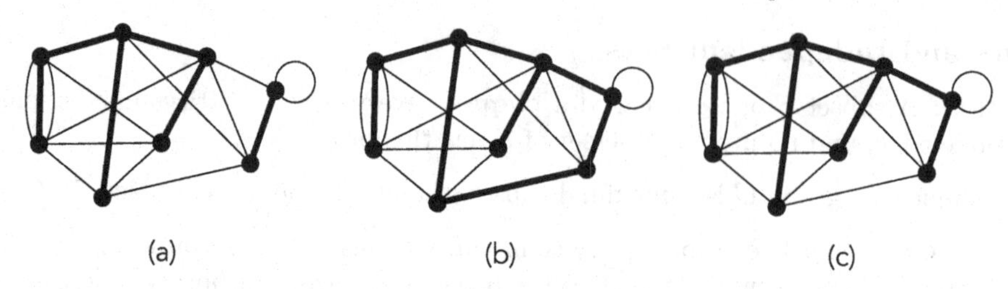

Figure 1.3.8 Three spanning subgraphs.

Example 1.3.10: In Figure 1.3.9, the isolated vertex and the bold edges with their end-points form a forest that spans a 12-vertex graph G. Each component of the forest is a spanning tree of the corresponding component of G.

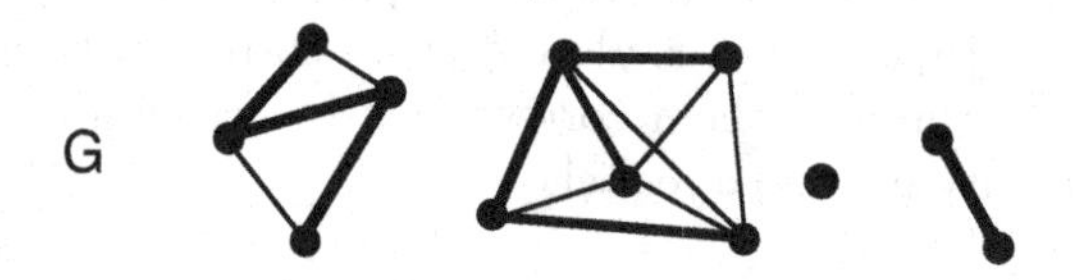

Figure 1.3.9 A spanning forest H of graph G.

Vertex-Deletion Subgraphs

DEFINITION: If v is a vertex of a graph G, then the ***vertex-deletion subgraph*** $G - v$ is the subgraph with vertex-set $V_G - \{v\}$ and edge-set $E_{G-v} = \{e \in E_G \mid e \text{ is not incident on } v\}$. More generally, if $U \subseteq V_G$, then the result of iteratively deleting all the vertices in U is denoted $G - U$.

DEFINITION: If e is an edge of a graph G, then the ***edge-deletion subgraph*** $G - e$ is the subgraph with vertex-set V_G and edge-set $E_G - \{e\}$. More generally, if $D \subseteq E_G$, then the result of iteratively deleting all the edges in D is denoted $G - D$.

Example 1.3.11: The next Figure 1.3.10 shows a vertex-deletion subgraph and an edge-deletion subgraph of a graph G.

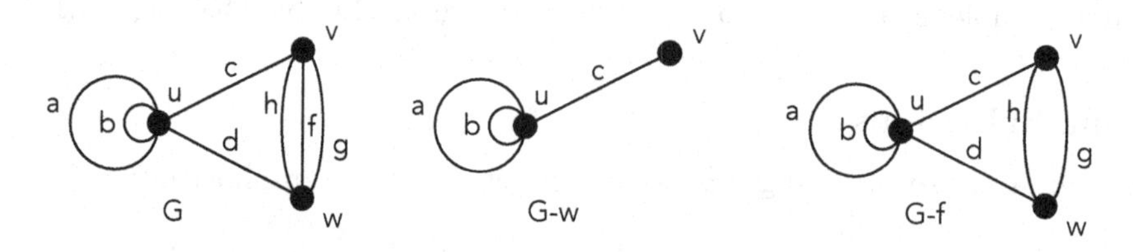

Figure 1.3.10 The result of deleting the vertex w or edge f from graph G.

Proposition 1.3.4: *For any graph G and any vertex subset U,*

$$G(U) = G - (V_G - U) \qquad \diamondsuit \text{ (Exercises)}$$

Remark: Note that the Graph $G-w$ in Figure 1.3.10 is also the graph induced on the edge set $\{a, b, c\}$. It is not an edge-deletion subgraph of G since it does not have the vertex w. This shows that the assertion analogous to Proposition 1.3.4 does not hold for edge-subsets.

Network Vulnerability

The following definitions identify the most vulnerable parts of a network. These definitions lead to different characterizations of a graph's connectivity, which appear in Chapter 2.

DEFINITION: A ***vertex-cut*** in a graph G is a vertex-set U such that $G - U$ has more components than G.

DEFINITION: A vertex v is a ***cut-vertex*** (or ***cutpoint***) in a graph G if $\{v\}$ is a vertex-cut.

Example 1.3.12: If the graph in Figure 1.3.11 represents a communications network, then a breakdown at any of the four cut-vertices w, x, y, or z would destroy the connectedness of the network. Also, $\{u, v\}$ and $\{w, x\}$ are both vertex-cuts, and $\{u, v\}$ is a *minimal* vertex-cut (i.e., no proper subset of $\{u, v\}$ is a vertex-cut).

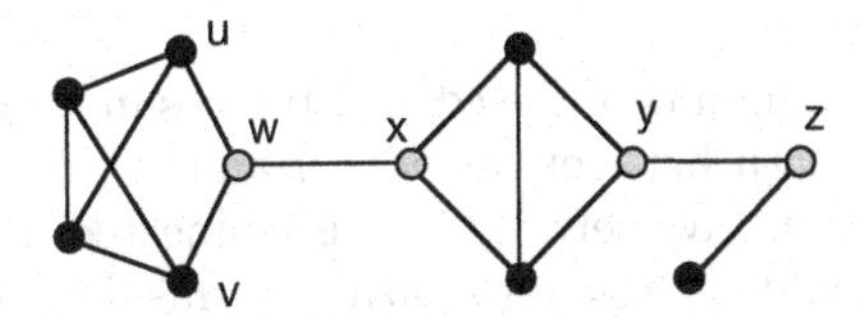

Figure 1.3.11 A graph with four cut-vertices.

DEFINITION: An ***edge-cut*** in a graph G is a set of edges D such that $G - D$ has more components than G.

DEFINITION: An edge e is a ***cut-edge*** (or ***bridge***) in a graph G if $\{e\}$ is an edge-cut.

Example 1.3.13: For the graph shown in Figure 1.3.12, edges a, b, and c are cut-edges; $\{r, s, t\}$ is an edge-cut; and $\{r, s\}$ is a *minimal* edge-cut.

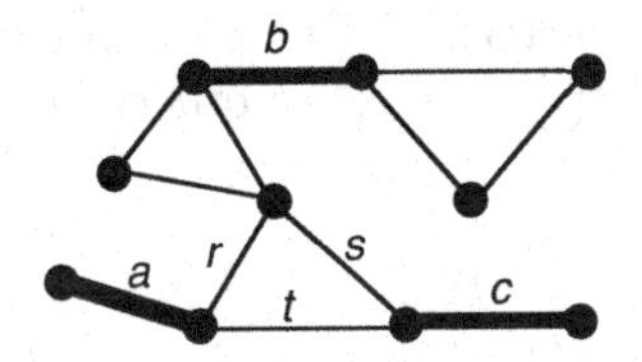

Figure 1.3.12 A graph with three cut-edges.

Observation: In a connected graph G, let e be any edge, with endpoints u and v. Then in the edge-deletion graph $G - e$, every vertex is reachable either from u or from v (or possibly from both). Thus, the only possible components of $G - e$ are the component $C_{G-e}(u)$ that contains the vertex u and the component $C_{G-e}(v)$. These two components coincide if and only if there is a path in G from u to v that *avoids* the edge e (in which case $G - e$ is connected). For example, in Figure 1.3.13, the graph $G - e$ has two components, $C_{G-e}(u)$ and $C_{G-e}(v)$. Notice that had there been an edge in the original graph joining the vertices w and z, then $G - e$ would have been connected.

DEFINITION: An edge e of a graph is called a ***cycle-edge*** if e lies in some cycle of that graph.

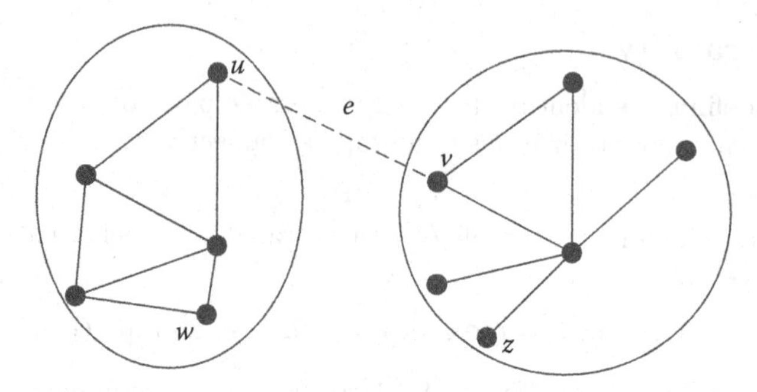

Figure 1.3.13 The components $C_{G-e}(u)$ and $C_{G-e}(v)$.

Proposition 1.3.5: *Let e be an edge of a connected graph G. Then $G - e$ is connected if and only if e is a cycle-edge of G.*

Proof: Let u and v be the endpoints of the edge e. If e lies in some cycle of G, then the long way around that cycle is a path between vertices u and v, and by the observation above, the graph $G - e$ is connected. Conversely, if $G - e$ is connected, then there is a path P from u to v that avoids e. Thus, adding edge e to path P forms a cycle in G containing e. ♢

The first of the following two corollaries restates Proposition 1.3.5 as a characterization of cut-edges. The second corollary combines the result of the proposition with the observation above to establish that deleting a cut-edge increases the number of components by at most one.

Corollary 1.3.6: *An edge of a graph is a cut-edge if and only if it is not a cycle-edge.*

Proof: Apply Proposition 1.3.5 to the component that contains that edge. ♢

Corollary 1.3.7: *Let e be any edge of a graph G. Then*

$$c(G-e) = \begin{cases} c(G), & \textit{if } e \textit{ is a cycle-edge} \\ c(G)+1, & \textit{otherwise} \end{cases}$$

The Graph-Reconstruction Problem

DEFINITION: Let G be a graph with $V_G = \{v_1, v_2, \ldots, v_n\}$. The ***vertex-deletion subgraph list*** of G is the list of the n subgraphs $H_1, \ldots, H_n$, where $H_k = G - v_k$, $k = 1, \ldots, n$.

Remark: The word "list" is being used informally here to mean the *multiset*† of vertex-deletion subgraphs of G. That is, the unordered collection of the n vertex-deletion subgraphs, some of which may be isomorphic.

Example 1.3.14: A graph G and its labeled vertex-deletion subgraph list is shown in Figure 1.3.14. Notice that the first, third, and fourth vertex-deletion subgraphs are essentially the same graph, as are the second and fifth. A more precise definition of what it means for two graphs to be the same (i.e., *isomorphic*) will be given in Chapter 2.

---

†A set has no order and no repetitions, while a list has order and can have repetitions. A multiset is a hybrid, having no order, but may have repetitions.

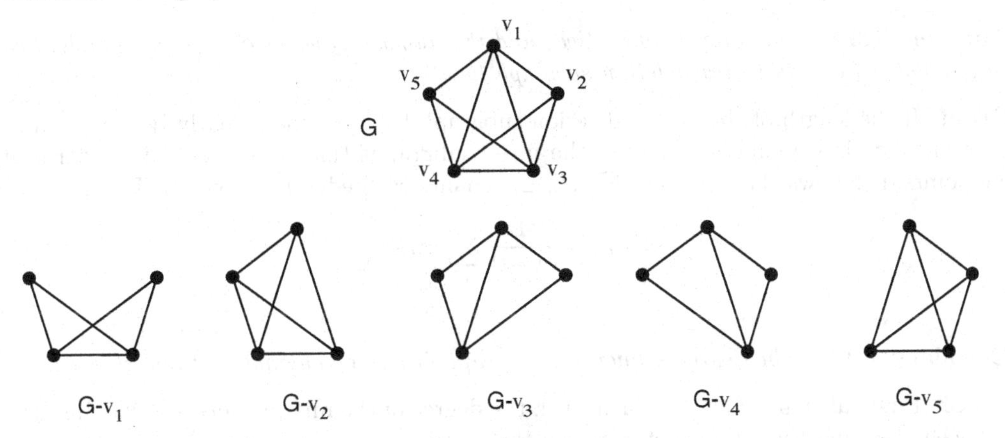

Figure 1.3.14 A graph and its vertex-deletion subgraph list.

Remark: Observe that the two different 2-vertex graphs shown below have the same vertex-deletion subgraph list.

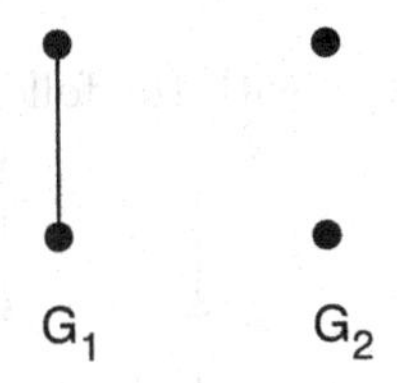

DEFINITION: The ***deck*** of a graph is its vertex-deletion subgraph list, with no labels on the vertices. We regard each individual vertex-deletion subgraph as being a ***card*** in the deck.

Example 1.3.14 continued: The deck for the graph G is shown in Figure 1.3.15.

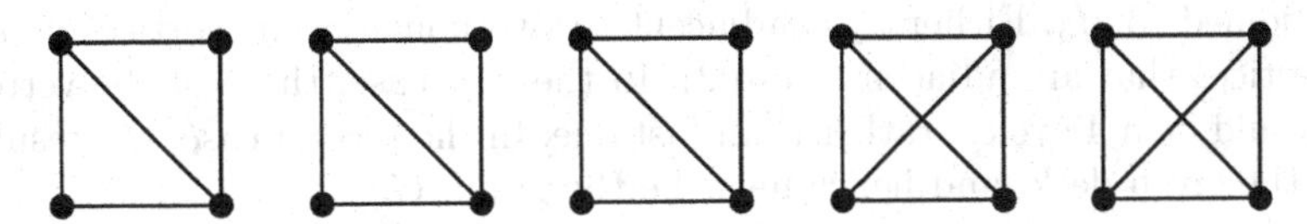

Figure 1.3.15 The deck for the graph G.

DEFINITION: The ***graph-reconstruction problem*** is to decide whether two different (i.e., *non-isomorphic*) simple graphs with three or more vertices can have the same deck.

Remark: The graph-reconstruction problem would be easy to solve if the vertex-deletion subgraphs included the vertex and edge names. However, the problem is concerned with a list of unlabeled graphs, and it is among the foremost unsolved problems in graph theory. The problem was originally formulated by P.J. Kelly and S.M. Ulam in 1941.

Example 1.3.15: In terms of Example 1.3.14, the graph-reconstruction problem asks whether the graph G in Figure 1.3.14 is the only graph (up to *isomorphism*) that has the deck shown in Figure 1.3.15.

The following results indicate that some information about the original graph can be derived from its vertex-deletion subgraph list. This information is a big help in solving small reconstruction problems.

Theorem 1.3.8: *The number of vertices and the number of edges of a simple graph G can be calculated from its vertex-deletion subgraph list.*

Proof: If the length of the vertex-deletion subgraph list is n, then clearly $|V_G| = n$. Moreover, since each edge appears only in the $n-2$ subgraphs that do not contain either of its endpoints, it follows that the sum $\sum_v |E_{G-v}|$ counts each edge $n-2$ times. Thus,

$$|E_G| = \frac{1}{n-2}\sum_v |E_{G-v}|$$

◇

Corollary 1.3.9: *The degree sequence of a graph G can be calculated from its deck.*

Proof: First calculate $|E_G|$. For each card, the degree of the missing vertex is the difference between $|E_G|$ and the number of edges on that card. ◇

The following example shows how Theorem 1.3.8 and its corollary can be used to find a graph G having a given deck. Although the uniqueness is an open question, the example shows that if only one graph can be formed when v_k is joined to H_k, for some k, then G is unique and can be retrieved.

Example 1.3.16: Suppose G is a graph with the following deck. Consider H_5. By

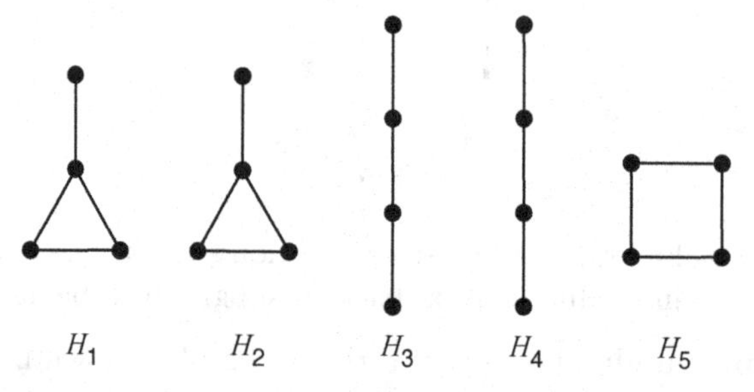

Theorem 1.3.8 and Corollary 1.3.9, $\deg(v_5) = 2$, and hence, there are two ways in which v_5 could be joined to H_5. Either v_5 is adjacent to two nonadjacent vertices of H_5 (case 1) or to two vertices that are adjacent (case 2). In the first case, three of the vertex-deletion subgraphs would be a 4-cycles, rather than just one. In the second case, the resulting graph would have the given deck, and hence must be the graph G.

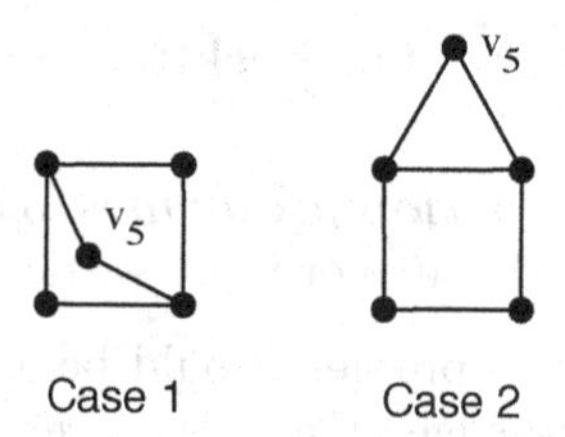

Corollary 1.3.10: *A regular graph can be reconstructed from its deck and no other graph has the same deck.*

Proof: Suppose G is a d-regular graph, for some d. By Corollary 1.3.9, we can see that G is regular and determine the value of d from the deck. On any card in the deck, there would be d vertices of degree $d-1$. Any graph reconstructed from this deck must join those vertices to the missing vertex. ◇

Remark: McKay [Mc97] and Nijenhuis [Ni77] have shown, with the aid of computers, that a counterexample to the conjecture would have to have at least 12 vertices.

EXERCISES for Section 1.3

1.3.1 Find the induced subgraphs $G(\{u, v, y\})$ and $G(\{p, r, e\})$ of the given graph.

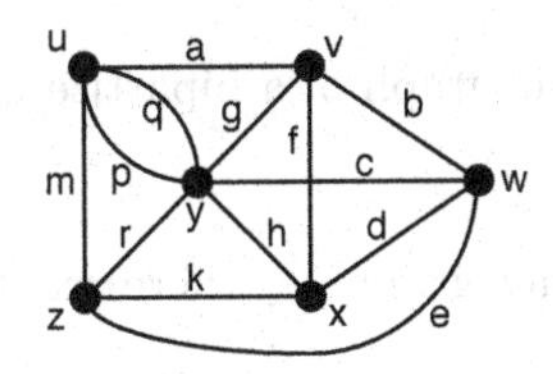

1.3.2 Find the local subgraphs $L(u)$ and $L(y)$ of the graph from Exercise 1.3.1.

1.3.3 Prove Proposition 1.3.1.

1.3.4 Find upper and lower bounds for the size of a largest independent set of vertices in an n-vertex connected graph. Then draw three 8-vertex graphs, one that achieves the lower bound, one that achieves the upper bound, and one that achieves neither.

1.3.5 Consider the following graph.

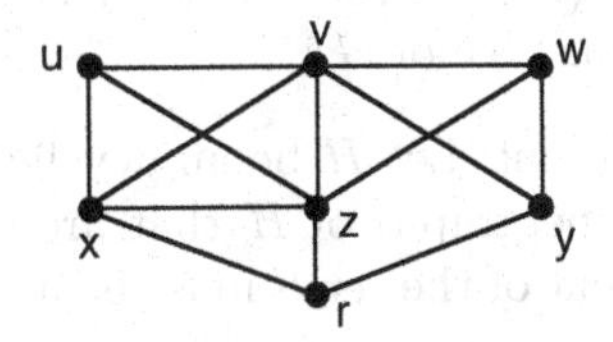

(a) Find all of the cliques and determine the clique number.

(b) Find all the maximal independent sets and determine the independence number.

(c) Find the center.

(d) Find a largest set of mutually adjacent vertices that is not a clique.

1.3.6 Consider the following graph.

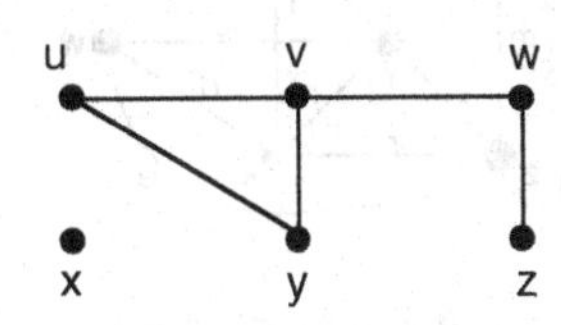

(a) Find all of the cliques and determine the clique number.

(b) Find all the maximal independent sets and determine the independence number.

(c) Find the center.

(d) Find a largest set of mutually adjacent vertices that is not a clique.

1.3.7 Find all the spanning trees of the given graph.

(a)

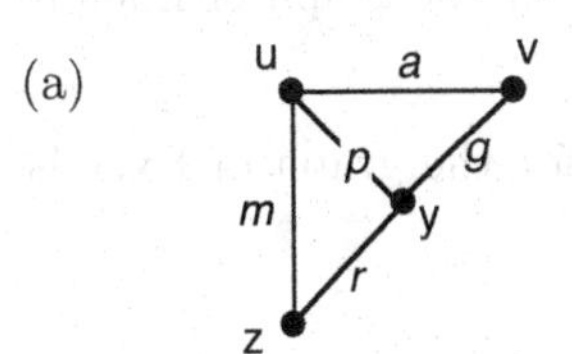

(b)

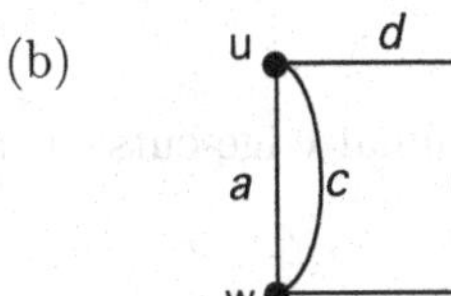

1.3.8 Draw a forest having ten vertices, seven edges, and three components.

1.3.9 Given six vertices in the plane, how many 2-component forests can be drawn having those six vertices?

1.3.10 Prove or disprove: Every subgraph of a bipartite graph is a bipartite graph.

1.3.11 Prove or disprove:

(a) For every subgraph H of any graph G, there exists a vertex subset W such that $H = G(W)$.

(b) For every subgraph H of any graph G, there exists an edge subset D such that $H = G(D)$.

1.3.12 Suppose U and W are vertex subsets and D and F are edge subsets of a graph G, under what conditions will the following be true?

(a) $G(U) \cup G(W) = G(U \cup W)$.

(b) $G(D) \cup G(F) = G(D \cup F)$.

1.3.13 Prove Proposition 1.3.2. (Hint: Prove that an edge-minimal connected spanning subgraph of G must be a spanning tree of G.)

1.3.14 Prove Proposition 1.3.3. (Hint: Let H be an acyclic subgraph of a connected graph G, and let S be the set of all supergraphs of H that are spanning subgraphs of G. Then show that an edge-minimal element of the set S must be a spanning tree of G that contains H.)

1.3.15 Prove Proposition 1.3.4.

1.3.16 Find the vertex-deletion subgraphs $G - y$ and $G - \{v, y, z\}$, and the edge-deletion subgraph $G - p$, of the graph G shown below.

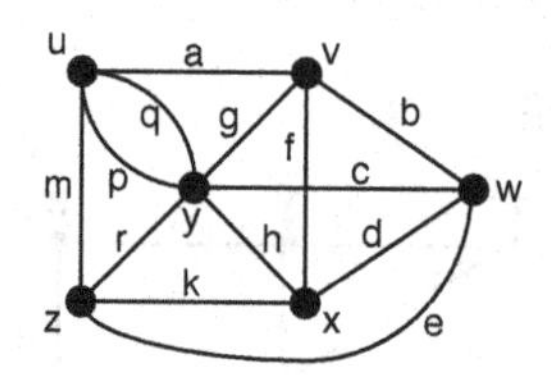

1.3.17 Find all the cut-vertices and cut-edges of the given graph.

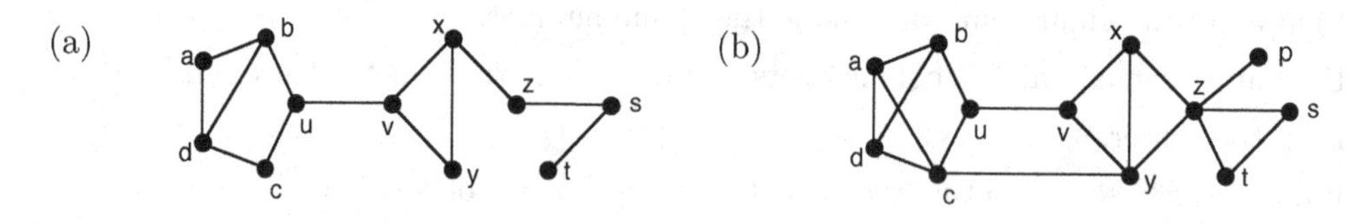

1.3.18 Find all the minimal vertex-cuts of maximum cardinality for the graph of Exercise 1.3.17(a).

1.3.19 Find all the minimal edge-cuts of maximum cardinality for the graph of Exercise 1.3.17(b).

1.3.20 Find a graph with the given vertex-deletion subgraph list.

(a)

(b)

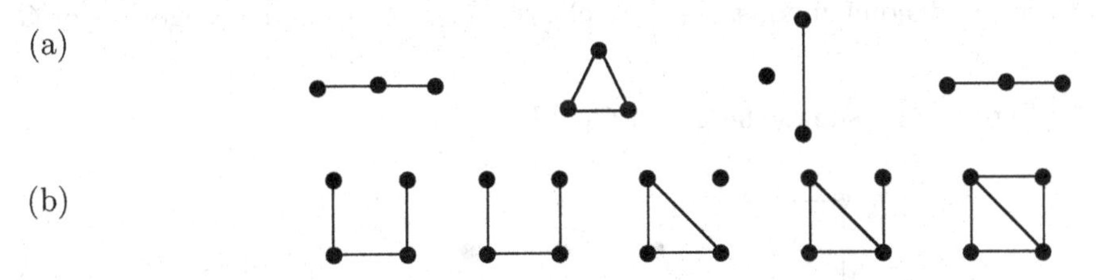

1.3.21 How would you recognize, from its deck, that a graph is bipartite? (Give a proof that your method works.)

1.3.22 How would you recognize, from its deck, that a graph on at least three vertices is connected? (Give a proof that your method works.)

1.3.23 Let v be one of the n vertices of a connected graph G. Find an upper bound for the number of components of $G - v$, and give an example that achieves the upper bound.

1.3.24 Let e be one of the edges of a connected graph G. Find an upper bound for the number of components of $G - e$, and justify your answer.

1.4 Graph Operations

Section 1.3 focused on a variety subgraphs within a graph. In this section we construct larger graphs by adding vertices or edges to a graph or by combining two graphs.

Adding Edges or Vertices

DEFINITION: ***Adding an edge*** between two vertices u and w of a graph G means creating a supergraph, denoted $G + e$, with vertex-set V_G and edge-set $E_G \cup \{e\}$, where e is a new edge with endpoints u and w.

Example 1.4.1: Figure 1.4.1 shows the effect of the operation of adding an edge.

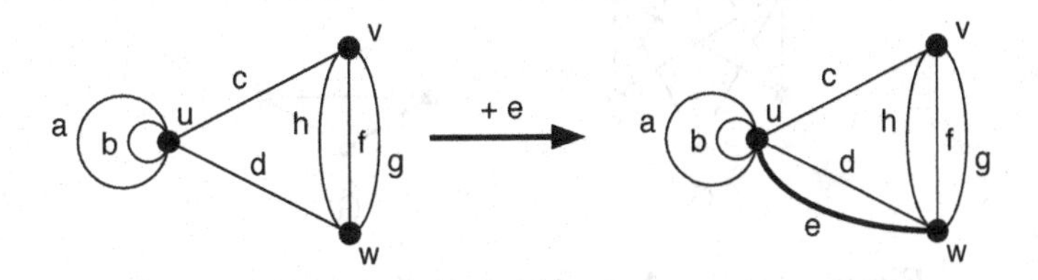

Figure 1.4.1 Adding an edge e with endpoints u and w.

DEFINITION: ***Adding a vertex*** v to a graph G, where v is a new vertex not already in V_G, means creating a supergraph, denoted $G \cup \{v\}$, with vertex-set $V_G \cup \{v\}$ and edge-set E_G.

Graph Union

The iterative application of adding vertices and edges results in the graph operation called *graph union.*

DEFINITION: The ***(graph) union*** of two graphs G and G' is the graph $G \cup G'$ whose vertex-set and edge-set are the disjoint unions, respectively, of the vertex-sets and edge-sets of G and G'.

Example 1.4.2: Figure 1.4.2 shows the graph $C_4 \cup P_3$.

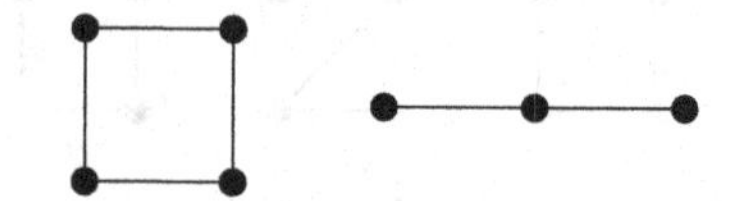

Figure 1.4.2 The graph union $C_4 \cup P_3$.

DEFINITION: The ***n-fold self-union*** of a graph G, denoted nG, is the iterated union of n copies of G.

Example 1.4.3: The graph in Figure 1.4.3 is the 4-fold self-union $K_3 \cup K_3 \cup K_3 \cup K_3$.

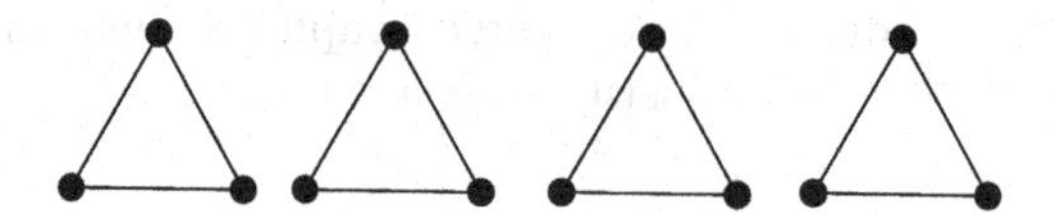

Figure 1.4.3 The 4-fold self-union $4K_3$.

Edge-Complementation

DEFINITION: Let G be a simple graph. Its ***edge-complement*** (or ***complement***) $\overline{G}$ is the graph on the same vertex-set, such that two vertices are adjacent in $\overline{G}$ if and only if they are not adjacent in G. More generally, if H is a subgraph of a graph G, the ***relative complement*** $G - H$ is the graph $G - E_H$. Thus, if G has n vertices, then $\overline{G}$ is the graph $K_n - G$.

Example 1.4.4: A graph and its edge-complement are shown in Figure 1.4.4.

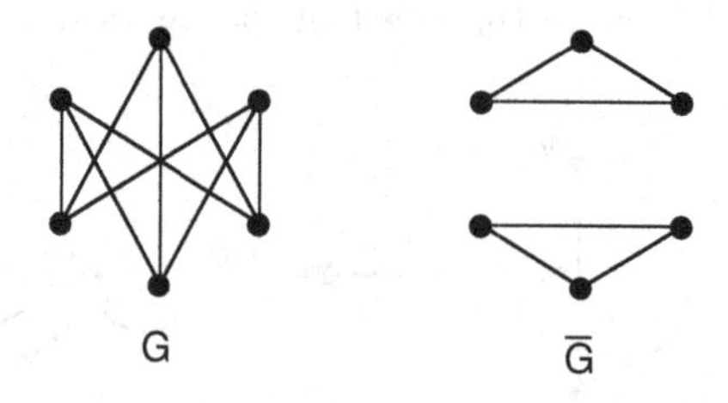

Figure 1.4.4 A graph and its complement.

Remark: Observe that the edge-complement of the edge-complement always equals the original graph, i.e., $\overline{\overline{G}} = G$.

The following theorem formulates precisely the sense in which the independence number of a graph and the clique number are complementary.

Theorem 1.4.1: *Let G be a simple graph. Then*

$$\omega(G) = \alpha(\overline{G}) \text{ and } \alpha(G) = \omega(\overline{G})$$ ◇ *(Exercises)*

Cartesian Product

DEFINITION: The ***Cartesian product*** $G \times H$ of graphs G and H is the graph whose vertex-set and edge-set are defined in terms of Cartesian products of sets:

$$V_{G\times H} = V_G \times V_H$$
$$E_{G\times H} = (V_G \times E_H) \cup (E_G \times V_H)$$

- The edge $(u, d) \in V_G \times E_H$ has endpoints $(u, y), (u, z) \in V_G \times V_H$, where y and z are the endpoints of d.
- The edge $(e, x) \in E_G \times V_H$ has endpoints $(v, x), (w, x) \in V_G \times V_H$, where v and w are the endpoints of e.

Example 1.4.5: Figure 1.4.5 illustrates the Cartesian product $P_2 \times P_3$.

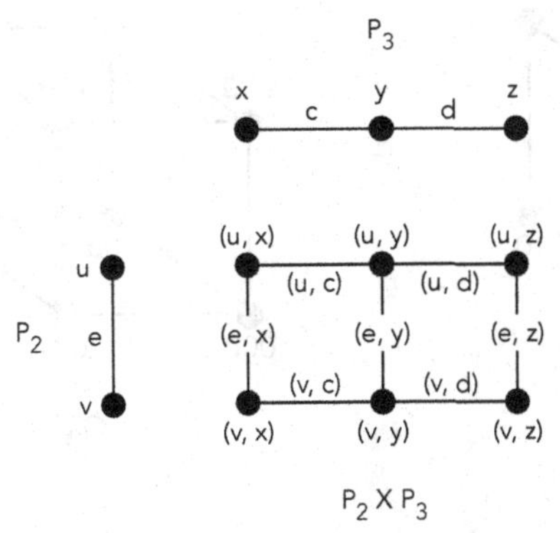

Figure 1.4.5 The Cartesian product $P_2 \times P_3$.

Example 1.4.6: Figure 1.4.6 illustrates the Cartesian product of the 4-cycle graph C_4 and the complete graph K_2.

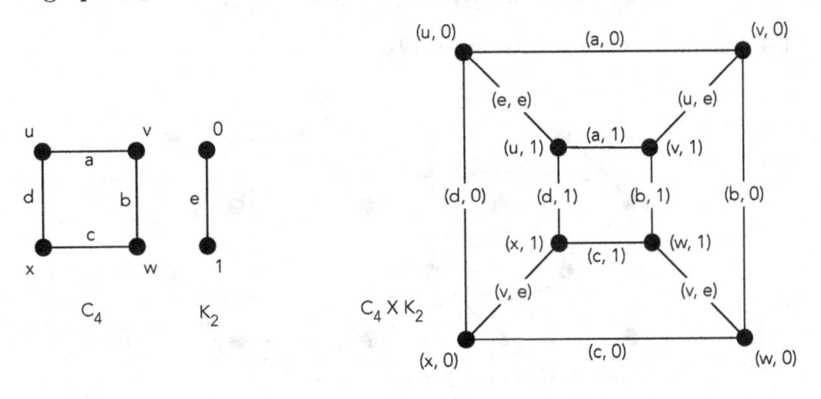

Figure 1.4.6 The labeled Cartesian product $C_4\times K_2$.

Observe that $C_4 \times K_2$ is essentially the same (i.e., *isomorphic*) as both the circular ladder graph CL_4 and the hypercube graph Q_3, introduced in §1.1. This motivates alternative definitions for CL_n and Q_n.

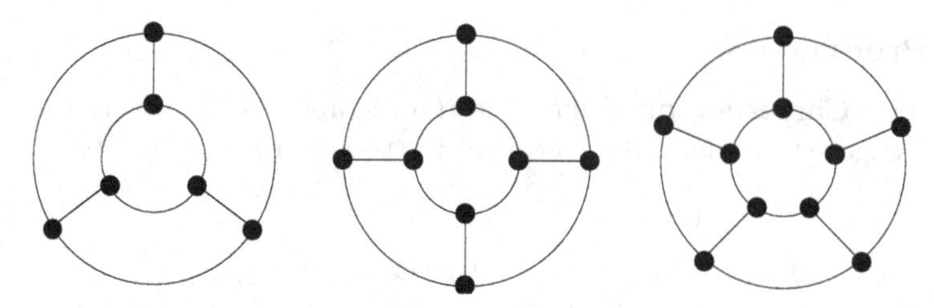

Figure 1.4.7 The circular ladders CL_3, CL_4, and CL_5.

DEFINITION: The ***circular ladder with*** n ***rungs***, denoted CL_n, is the Cartesian product $C_n \times K_2$.

In §1.1, the hypercube graph Q_n is defined as the n-regular graph whose vertex-set is the set of bitstrings of length n, such that two vertices are adjacent if and only if they differ in exactly one bit. The labeling in Figure 1.4.8 suggests how Q_n may be defined recursively.

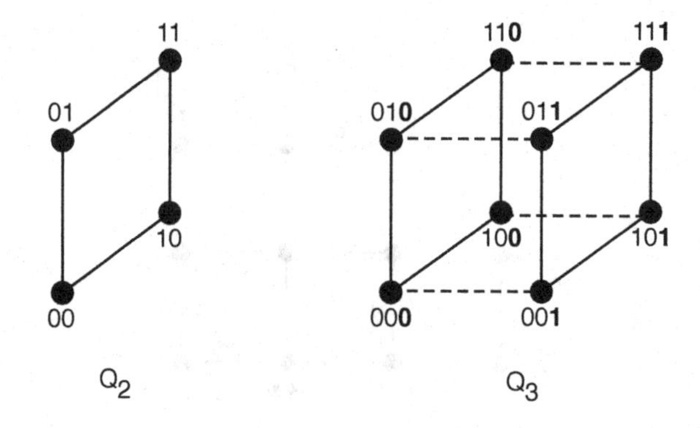

Figure 1.4.8 Bitstring labels showing $Q_3 = Q_2 \times K_2$.

RECURSIVE DEFINITION: The ***hypercube graph*** Q_n is defined recursively as follows.

$$Q_1 = K_2 \text{ and } Q_n = Q_{n-1} \times K_2, \text{ for } n \geq 2$$

Figure 1.4.9 illustrates the recursion for Q_4.

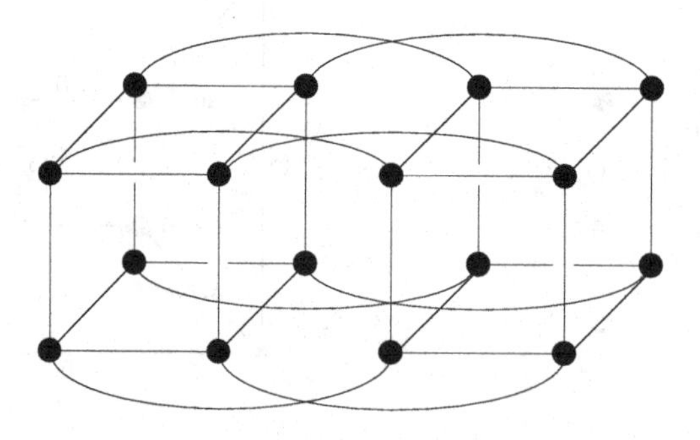

Figure 1.4.9 $Q_4 = Q_3 \times K_2$.

Observation: This recursive definition shows that Q_n is the (iterated) Cartesian product of n copies of K_2. That is,

$$Q_n = \underbrace{K_2 \times \cdots \times K_2}_{n \text{ copies}}$$

DEFINITION: The $(m_1 \times m_2 \times \cdots \times m_n)$-***mesh*** is the Cartesian product of paths, $P_{m_1} \times P_{m_2} \times \cdots \times P_{m_n}$, as shown in Figure 1.4.10.

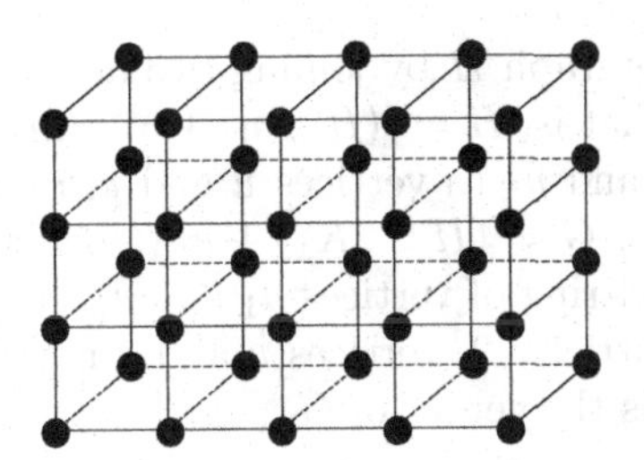

Figure 1.4.10 The $(5 \times 4 \times 2)$-mesh.

DEFINITION: The ***wraparound*** $(m_1 \times m_2 \times \cdots \times m_n)$-***mesh*** is the Cartesian product of cycles, $C_{m_1} \times C_{m_2} \times \cdots \times C_{m_n}$, as shown in Figure 1.4.11.

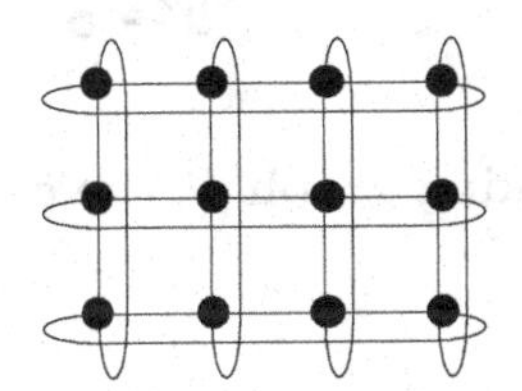

Figure 1.4.11 The wraparound (4×3)-mesh.

Joining a Vertex to a Graph

DEFINITION: If a new vertex v is joined to each of the vertices of a graph G, then the resulting graph is called the ***join of*** v ***to*** G or the ***suspension*** of G ***from*** v, and is denoted $G + v$.

DEFINITION: The n-***wheel*** W_n is the join $C_n + v$ of a single vertex to an n-cycle, as illustrated in Figure 1.4.12. (The n-cycle forms the rim of the wheel, and the additional vertex is its hub.) The graph W_n is called an ***even (odd) wheel*** if n is even (odd).

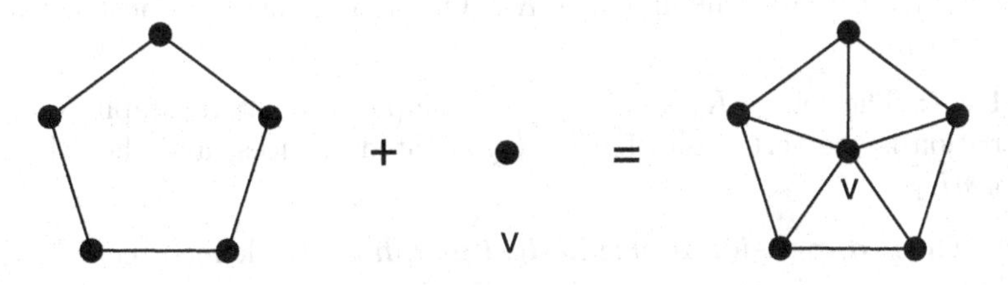

Figure 1.4.12 The 5-wheel $W_5 = C_5 + v$.

The following proposition hints at the usefulness of the join construction.

Proposition 1.4.2: *Let H be a graph. Then there is a graph G of radius 2 of which H is the center.*

Proof: First construct a supergraph $\hat{H}$ by joining two new vertices u and v to every vertex of H, but not to each other; that is, $\hat{H} = ((H + u) + v) - uv$. Then form the supergraph G by attaching new vertices u_1 and v_1 to vertices u and v, respectively, with edges uu_1 and vv_1, as in Figure 1.4.13. Thus, $G = (\hat{H} \cup 2K_1) + uu_1 + vv_1)$, where the two copies of K_1 are the two trivial graphs consisting of vertices u_1 and v_1. It is easy to verify that, in graph G, each vertex of H has eccentricity 2, vertices u and v have eccentricity 3, and u_1 and v_1 have eccentricity 4. Thus, H is the center of G. ♢

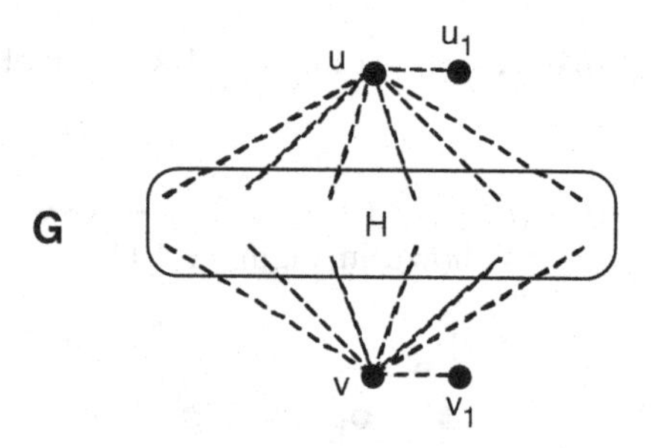

Figure 1.4.13 Making graph H the center of a graph G.

Join

DEFINITION: The ***join*** $G + H$ of the graphs G and H is obtained from the graph union $G \cup H$ by adding an edge between each vertex of G and each vertex of H.

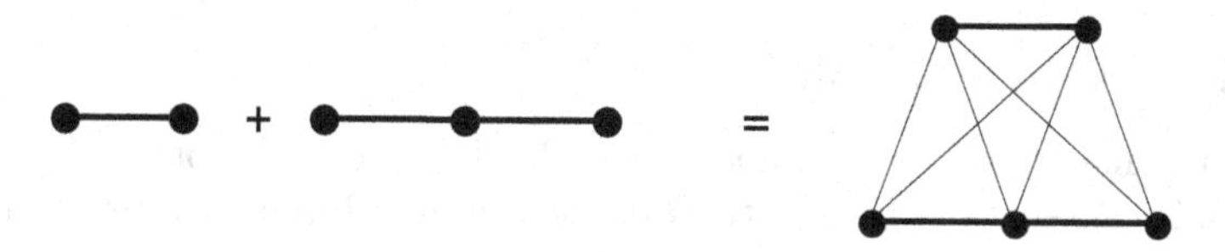

Figure 1.4.14 The join $P_2 + P_3$.

Proposition 1.4.3: *The join $K_m + K_n$ is the complete graph K_{m+n}.*

Proof: Consider any two vertices in $K_m + K_n$. If both are from K_m or if both are from K_n, then they are still adjacent in $K_m + K_n$. Otherwise, the join construction places an edge between them. ♢

Example 1.4.7: The join $mK_1 + nK_1$ is the complete bipartite graph $K_{m,n}$. One part of the bipartition is the vertex-set of mK_1 (m isolated vertices) and the other part is the vertex-set of nK_1.

DEFINITION: The ***n-dimensional octahedral graph*** $\mathcal{O}_n$, is defined recursively, using the join operation.

$$\mathcal{O}_n = \begin{cases} 2K_1 & \text{if } n = 1 \\ 2K_1 + \mathcal{O}_{n-1} & \text{if } n \geq 2 \end{cases}$$

It is also called the ***n-octahedron graph*** or, when $n = 3$, the ***octahedral graph***, because it is the 1-skeleton of the octahedron, a platonic solid.

Example 1.4.8: Figure 1.4.15 illustrates the first four octahedral graphs.

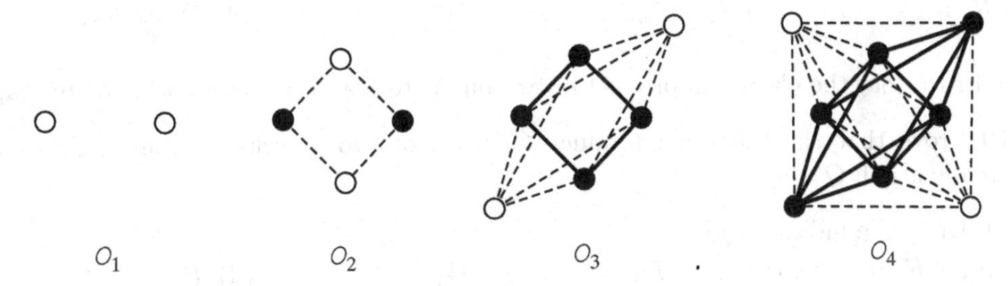

Figure 1.4.15 The octahedral graphs $\mathcal{O}_1$, $\mathcal{O}_2$, $\mathcal{O}_3$, and $\mathcal{O}_4$.

Example 1.4.9: Figure 1.4.16 illustrates that the edge-complement of the graph $3K_2$ (in K_6) is essentially the same as $\mathcal{O}_3$. Notice that the bijection that sends vertex v_i in the first graph to u_i in the second, $i = 1, \ldots, 6$, illustrates the structural similarity of the two graphs; the only missing edges are $\{v_0v_3, v_1v_4, v_2v_5\}$ in the edge-complement of $3K_2$ and $\{u_0u_3, u_1u_4, u_2u_5\}$ in $\mathcal{O}_3$. *Structural similarity* will be addressed more precisely in Chapter 2, when *graph isomorphism* is introduced. It will be shown that the complement of K_{2n} is *isomorphic* to $\mathcal{O}_n$.

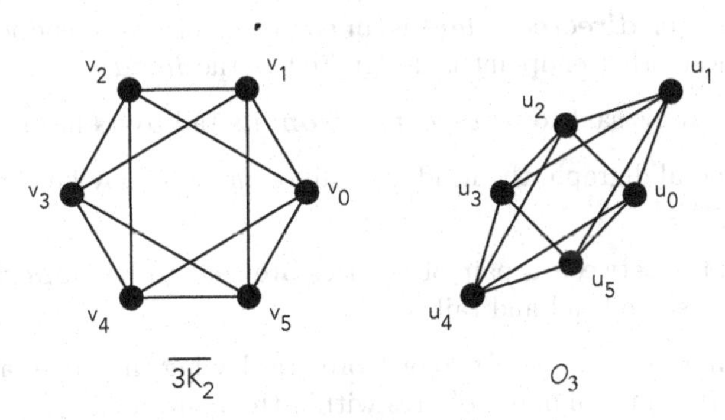

Figure 1.4.16 The graph $\overline{3K_2}$ is isomorphic to the octahedral graph $\mathcal{O}_3$.

EXERCISES for Section 1.4

1.4.1 Prove or disprove: For every graph G, the suspension $G + v$ has no cut-vertices.

1.4.2 State a necessary and sufficient condition for an n-vertex graph G to be a suspension of some subgraph H from a vertex in $V_G - V_H$.

1.4.3 Give a recursive definition for the complete graph K_n, using the join operation.

1.4.4 Explain why Theorem 1.4.1 follows immediately from the definitions.

1.4.5 Prove that for any simple graph G on 6 vertices, a copy of K_3 appears as a subgraph of either G or $\overline{G}$ (or both).

1.4.6 Prove that no vertex of a graph G can be a cut-vertex of both G and $\overline{G}$.

1.4.7 Let G be a simple graph. Prove that at least one of the graphs G and $\overline{G}$ is connected.

1.4.8 Draw the indicated Cartesian product.
(a) $K_2 \times C_5$. (b) $P_3 \times C_5$. (c) $W_5 \times P_3$. (d) $P_3 \times 2K_3$.

1.4.9 Prove that the Cartesian product of two bipartite graphs is always a bipartite graph.

1.4.10 Prove that the Cartesian product $C_4 \times C_4$ of two 4-cycles is isomorphic to the hypercube graph Q_4.

1.4.11 Draw the indicated join.
(a) $K_3 + K_3$. (b) $B_2 + K_4$. (c) $W_6 + P_2$. (d) $P_3 + K_3 + C_2$.

1.4.12 Give a necessary and sufficient condition that the join of two bipartite graphs be non-bipartite.

1.5 DIRECTED GRAPHS

Adding directions to the edges of a graph creates a *directed graph.*

DEFINITION: An ***arc*** (or ***directed edge***) is an edge, one of whose endpoints is designated as the ***tail***, and whose other endpoint is designated as the ***head***.

TERMINOLOGY: An arc is said to be ***directed from*** its tail to its head.

NOTATION: In a general digraph, the head and tail of an arc e may be denoted $head(e)$ and $tail(e)$, respectively.

DEFINITION: Two arcs between a pair of vertices are said to be ***oppositely directed*** if they do not have the same head and tail.

DEFINITION: A ***multi-arc*** is a set of two or more arcs having the same tail and same head. The ***arc multiplicity*** is the number of arcs within the multi-arc.

DEFINITION: A ***directed graph*** (or ***digraph***) is a graph, each of whose edges is directed.

DEFINITION: A digraph is ***simple*** if it has neither self-loops nor multi-arcs.

NOTATION: In a simple digraph, an arc from vertex u to vertex v may be denoted by uv or by the ordered pair (u, v).

Example 1.5.1: The digraph in Figure 1.5.1 is simple. Its arcs are uv, vu, and vw.

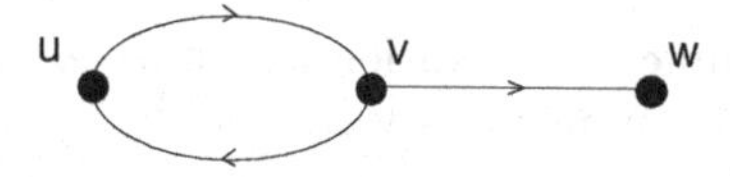

Figure 1.5.1 A simple digraph with a pair of oppositely directed arcs.

DEFINITION: A ***mixed graph*** (or ***partially directed graph***) is a graph that has both undirected and directed edges.

DEFINITION: The ***underlying graph*** of a directed or mixed graph G is the graph that results from deleting all the edge-directions.

Example 1.5.2: The digraph D in Figure 1.5.2 has the graph G as its underlying graph.

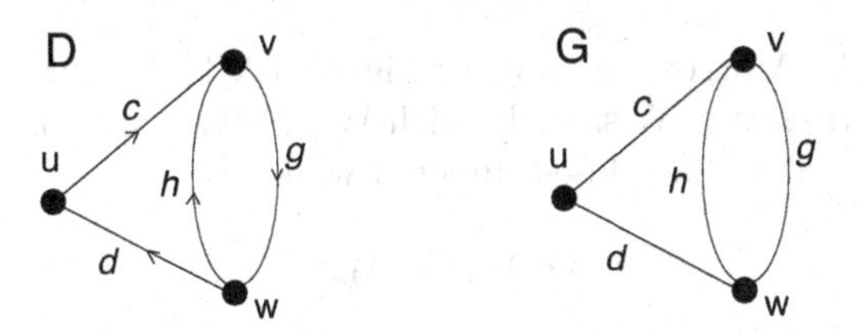

Figure 1.5.2 A digraph and its underlying graph.

Indegree and Outdegree in a Digraph

DEFINITION: The ***indegree*** of a vertex v in a digraph is the number of arcs directed to v; the ***outdegree*** of vertex v is the number of arcs directed from v. Each self-loop at v counts one toward the indegree of v and one toward the outdegree. See Figure 1.5.3.

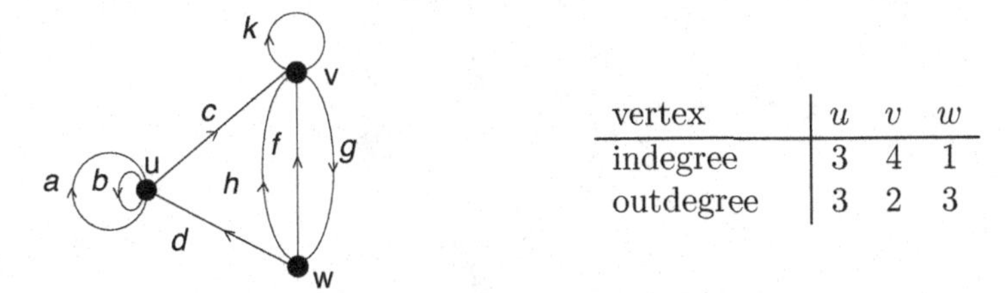

vertex	u	v	w
indegree	3	4	1
outdegree	3	2	3

Figure 1.5.3 The indegrees and outdegrees of the vertices of a digraph.

The next theorem is the digraph version of Euler's Degree-Sum Theorem 1.1.2. The proof is analogous to the proof of Theorem 1.1.2 and is left as an exercise.

Theorem 1.5.1: *In a digraph, the sum of the indegrees and the sum of the outdegrees both equal the number of directed edges.*

Application 1.5.1: *Markov Diagrams* Suppose that the inhabitants of some remote area purchase only two brands of breakfast cereal, O's and W's. The consumption patterns of the two brands are encapsulated by the *transition matrix* shown in Figure 1.5.4. For instance, if someone just bought O's, there is a 0.4 chance that the person's next purchase will be W's and a 0.6 chance it will be O's.

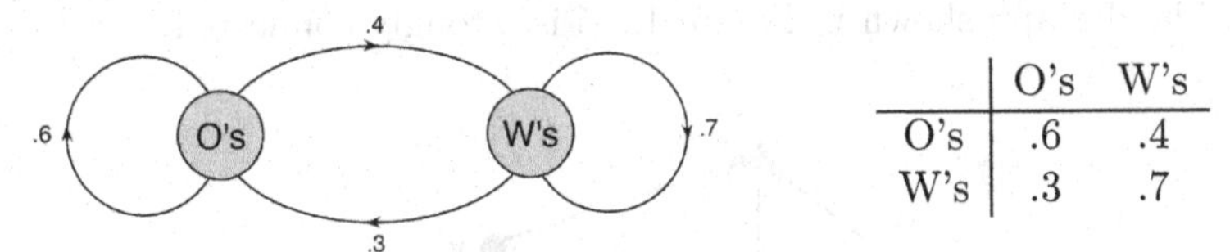

	O's	W's
O's	.6	.4
W's	.3	.7

Figure 1.5.4 Markov process from Application 1.5.1.

In a *Markov process*, the *transition probability* of going from one state to another depends only on the current state. Here, states "O" and "W" correspond to whether the most recent purchase was O's or W's, respectively. The digraph model for this Markov process, called a *Markov diagram*, is shown in Figure 1.5.4. Each arc is labeled with the transition probability of moving from the state at the tail vertex to the state at the head. Thus, the probabilities on the outgoing edges from each vertex must sum to 1. This Markov diagram is an example of a *weighted graph*.

DEFINITION: If G is a digraph, then $W = \langle v_0, e_1, v_1, e_2, \ldots, e_n, v_n \rangle$ is a ***directed walk*** if arc e_i is directed from vertex v_{i-1} to vertex v_i, $i = 1, \ldots, n$.

Example 1.5.3: In the Markov diagram in Figure 1.5.4, the choice sequence of a cereal eater who buys O's, switches to W's, sticks with W's for two more boxes, and then switches back to O's is represented by the closed directed walk

$$< \mathrm{O, W, W, W, O} >$$

The product of the transition probabilities along a walk in any Markov diagram equals the probability that the process will follow that walk during an experimental trial. Thus, the probability that this walk occurs, when starting from O's equals $.4 \times .7 \times .7 \times .3 = 0.0588$.

Example 1.5.4: In the figure below, the closed directed walk $\langle v, x, y, z, v \rangle$ is a subwalk of the open directed walk $\langle u, v, x, y, z, v, w, t \rangle$.

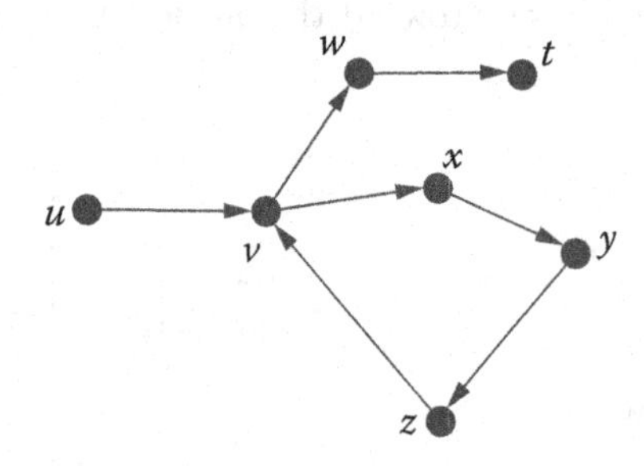

Strongly Connected Digraphs

DEFINITION: A digraph is ***connected*** if its underlying graph is connected.

TERMINOLOGY NOTE: Some authors use the term *weakly connected* digraph instead of connected digraph.

DEFINITION: A vertex v is ***reachable*** from a vertex u in a digraph D if D contains a directed u-v walk.

DEFINITION: A digraph D is ***strongly connected*** if every two of its vertices are mutually reachable.

Example 1.5.5: The digraph shown in Figure 1.5.5 is strongly connected.

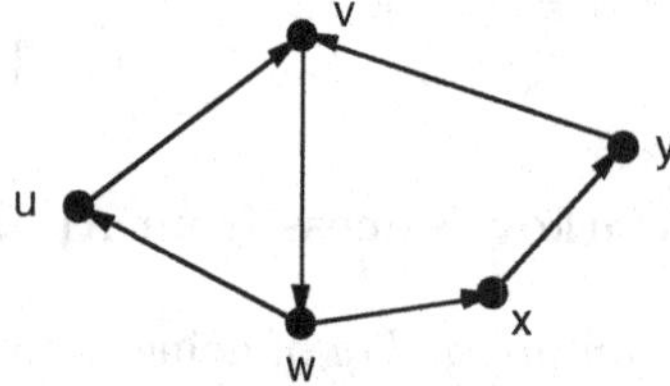

Figure 1.5.5 A strongly connected digraph.

Example 1.5.6: Suppose the graph in Figure 1.5.6 represents the network of roads in a certain national park. The park officials would like to make all of the roads one-way, but only if visitors will still be able to drive from any one intersection to any other.

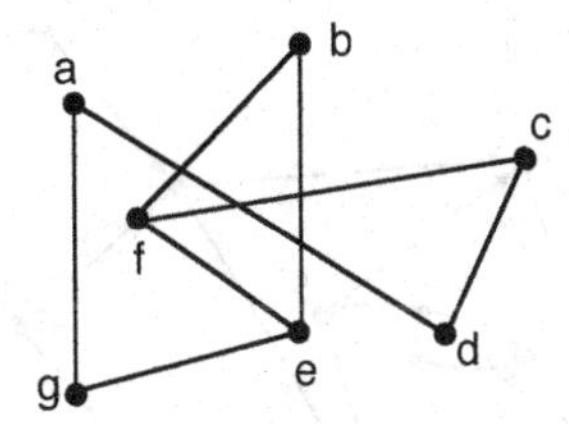

Figure 1.5.6 A national park's system of roads.

This problem can be expressed in graph-theoretic terms with the following definition.

DEFINITION: A graph is said to be ***strongly orientable*** if there is an assignment of directions to its edges so that the resulting digraph is strongly connected. Such an assignment is called a ***strong orientation***.

EXERCISES for Section 1.5

1.5.1 Draw a digraph for each of the following indegree and outdegree sequences, such that the indegree and outdegree of each vertex occupy the same position in both sequences.

(a) in: $\langle 1, 1, 1\rangle$ out: $\langle 1, 1, 1\rangle$

(b) in: $\langle 2, 1\rangle$ out: $\langle 3, 0\rangle$

DEFINITION: A pair of sequences $\langle a_1, a_2, \ldots, a_n\rangle$ and $\langle b_1, b_2, \ldots, b_n\rangle$ is called ***digraphic*** if there exists a simple digraph with vertex-set $\{v_1, v_2, \ldots, v_n\}$ such that $outdegree(v_i) = a_i$ and $indegree(v_i) = b_i$ for $i = 1, 2, \ldots, n$.

1.5.2 Determine whether the pair of sequences $\langle 3, 1, 1, 0\rangle$ and $\langle 1, 1, 1, 2\rangle$ is digraphic.

1.5.3 Establish a result analogous to Corollary 1.1.7 for a pair of sequences to be digraphic.

DEFINITION: A ***tournament*** is a digraph whose underlying graph is a complete graph.

1.5.4 Prove that every tournament has at most one vertex of indegree 0 and at most one vertex of outdegree 0.

1.5.5 Suppose that n vertices $v_1, v_2, \ldots, v_n$ are drawn in the plane. How many different n-vertex tournaments can be drawn on those vertices?

1.5.6 Prove that a digraph must have a closed directed walk if each of its vertices has nonzero outdegree.

1.5.7 Let D be a digraph. Prove that mutual reachability is an equivalence relation on V_D.

DEFINITION: A ***strong component*** of a digraph is the graph induced on an equivalence class of the *mutual reachability equivalence relation* on its vertex-set.

1.5.8 Find the strong components of the given graph.

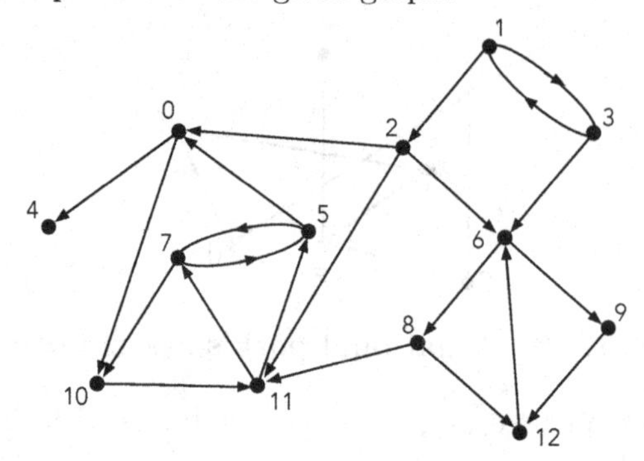

1.5.9 For each of the following graphs, either find a strong orientation or argue why no such orientation exists.

(a) (b)

1.6 FORMAL SPECIFICATIONS FOR GRAPHS AND DIGRAPHS

Except for the smallest graphs, line drawings are inadequate for describing a graph; imagine trying to draw a graph to represent a telephone network for a small city. Since many applications involve computations on graphs having hundreds, or even thousands, of vertices, another, more formal kind of specification of a graph is often needed.

The specification must include (implicitly or explicitly) a function, *endpts*, that specifies, for each edge, the subset of vertices on which that edge is incident, and for a directed graph the functions *head* and *tail*. In a simple graph, the juxtaposition notation for each edge implicitly specifies its endpoints, making formal specification simpler for simple graphs than for general graphs.

DEFINITION: The ***adjacency table for a simple graph*** has a row for each vertex, containing the list of vertices adjacent to that vertex.

Example 1.6.1: Figure 1.6.1 shows a line drawing and a formal specification for a simple graph.

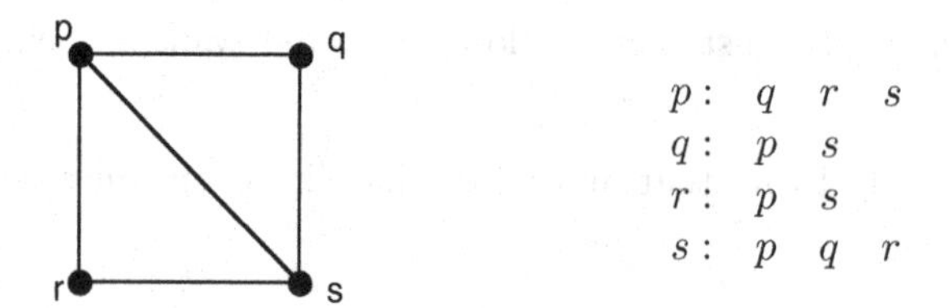

$p:$	q	r	s
$q:$	p	s	
$r:$	p	s	
$s:$	p	q	r

Figure 1.6.1 A simple graph and its adjacency table.

General Graphs

A general graph $G = (V, E, endpts)$ can be formally specified by giving a list of its vertices, a list of its edges, and an ***incidence table***, which designates the endpoints of each edge, allowing for multi-edges and self-loops. An isolated vertex will appear in the vertex list but not in any column of the table.

DEFINITION: The ***incidence table*** of a general graph has columns indexed by the edges. The entries in the column corresponding to edge e are the endpoints of e. The same endpoint appears twice if e is a self-loop.

Example 1.6.2: Figure 1.6.2 shows a line drawing and a formal specification for a general graph.

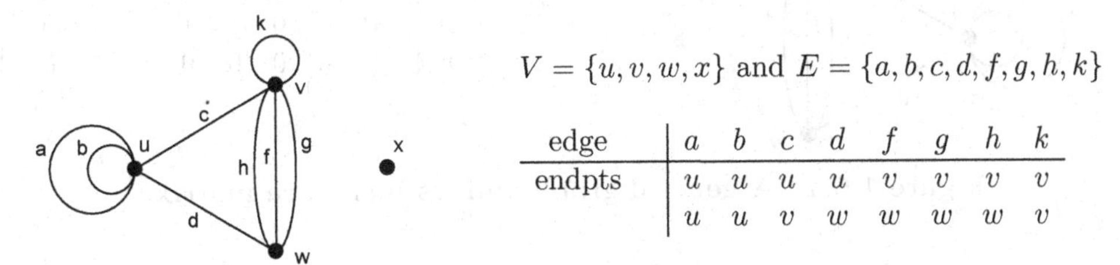

$V = \{u, v, w, x\}$ and $E = \{a, b, c, d, f, g, h, k\}$

edge	a	b	c	d	f	g	h	k
endpts	u	u	u	u	v	v	v	v
	u	u	v	w	w	w	w	v

Figure 1.6.2 A general graph and its incidence table.

DEFINITION: The ***adjacency matrix of a general graph*** G, denoted A_G, is the symmetric matrix whose rows and columns are both indexed by identical orderings of V_G, such that $A_G[u, u]$ is twice the number of self-loops on u, and for $u \neq v$ $A_G[u, v]$ is the number of edges with u and v as endpoints.

Example 1.6.3: Figure 1.6.3 shows the adjacency matrix of a general graph G, with respect to the vertex ordering u, v, w, x.

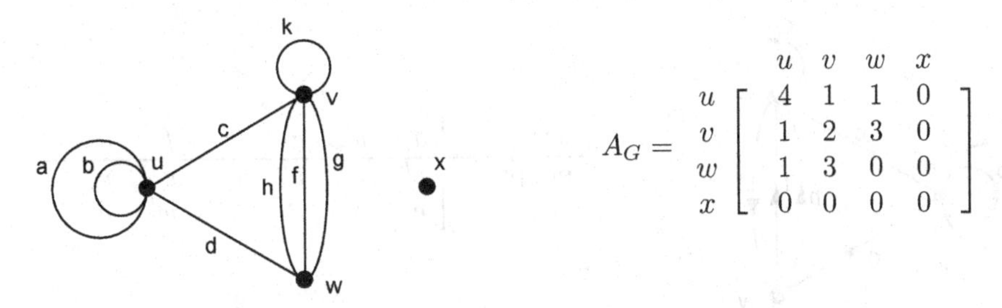

$$A_G = \begin{array}{c} \\ u \\ v \\ w \\ x \end{array} \begin{array}{c} \begin{array}{cccc} u & v & w & x \end{array} \\ \begin{bmatrix} 4 & 1 & 1 & 0 \\ 1 & 2 & 3 & 0 \\ 1 & 3 & 0 & 0 \\ 0 & 0 & 0 & 0 \end{bmatrix} \end{array}$$

Figure 1.6.3 A general graph and its adjacency matrix.

Usually the vertex order is implicit from context, in which case the adjacency matrix A_G can be written as a matrix without explicit row or column labels.

DEFINITION: The ***incidence matrix*** of a general graph G is the matrix I_G whose rows and columns are indexed by some orderings of V_G and E_G, respectively, such that†

$$I_G[v,e] = \begin{cases} 0 \text{ if } v \text{ is not an endpoint of } e \\ 1 \text{ if } v \text{ is an endpoint of } e \\ 2 \text{ if } e \text{ is self-loop at } v \end{cases}$$

Example 1.6.4: Figure 1.6.4 shows a general graph and its incidence matrix.

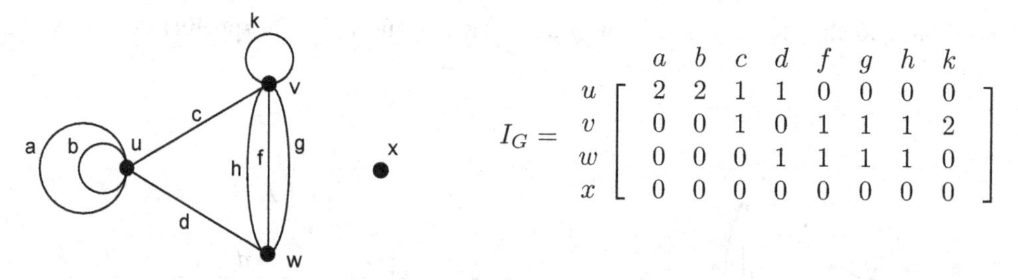

$$I_G = \begin{array}{c} \\ u \\ v \\ w \\ x \end{array} \begin{array}{c} \begin{array}{cccccccc} a & b & c & d & f & g & h & k \end{array} \\ \left[\begin{array}{cccccccc} 2 & 2 & 1 & 1 & 0 & 0 & 0 & 0 \\ 0 & 0 & 1 & 0 & 1 & 1 & 1 & 2 \\ 0 & 0 & 0 & 1 & 1 & 1 & 1 & 0 \\ 0 & 0 & 0 & 0 & 0 & 0 & 0 & 0 \end{array}\right] \end{array}$$

Figure 1.6.4 A general graph and its incidence matrix.

General Digraphs

A formal specification of a general digraph $D = (V, E, endpts, head, tail)$ is obtained from a formal specification of the underlying graph by adding the functions $head : E_G \to V_G$ and $tail : E_G \to V_G$, which designate the *head* vertex and *tail* vertex of each arc. One way to specify these designations in the incidence table is to mark in each column the endpoint that is the head of the corresponding arc.

Example 1.6.5: Figure 1.6.5 gives a formal specification for the digraph shown, including the corresponding values of the functions *head* and *tail*. A superscript "h" is used to indicate the *head* vertex.

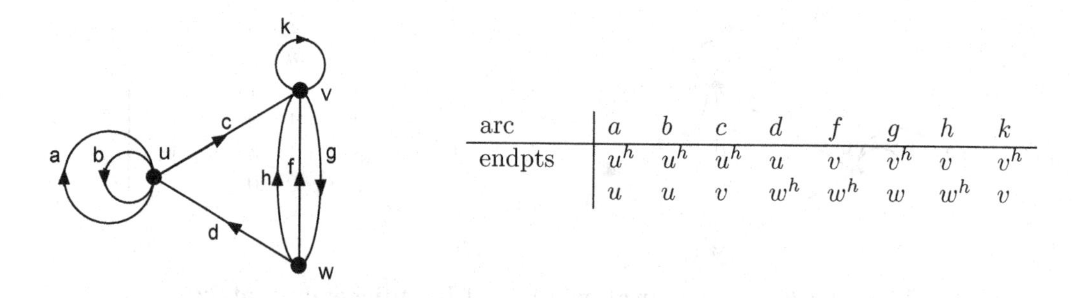

arc	a	b	c	d	f	g	h	k
endpts	u^h	u^h	u^h	u	v	v^h	v	v^h
	u	u	v	w^h	w^h	w	w^h	v

Figure 1.6.5 A general digraph and its incidence table.

†An alternative definition seen elsewhere, which uses 1 for a self-loop instead of 2, takes more effort to find the self-loops. It also has the theoretical inconsistency that row-sums are not necessarily equal to vertex-degrees and column-sums are not necessarily equal to 2.

Remark: A mixed graph is specified by simply restricting the functions *head* and *tail* to a proper subset of E_G. In this case, a column of the incidence table that has no mark means that the corresponding edge is undirected.

DEFINITION: The ***adjacency matrix of a general digraph*** D, denoted A_D, is the matrix whose rows and columns are both indexed by identical orderings of V_D, such that for distinct vertices u and v, $A_D[u, v]$ is the number of arcs from vertex u to vertex v; and $A_D[u, u]$ is twice the number of self-loops at u.

Example 1.6.6: The adjacency matrix of the digraph D in Figure 1.6.6 uses the vertex ordering u, v, w.

$$A_D = \begin{array}{c} \\ u \\ v \\ w \end{array} \begin{array}{c} \begin{array}{ccc} u & v & w \end{array} \\ \left[\begin{array}{ccc} 4 & 1 & 0 \\ 0 & 2 & 1 \\ 1 & 2 & 0 \end{array} \right] \end{array}$$

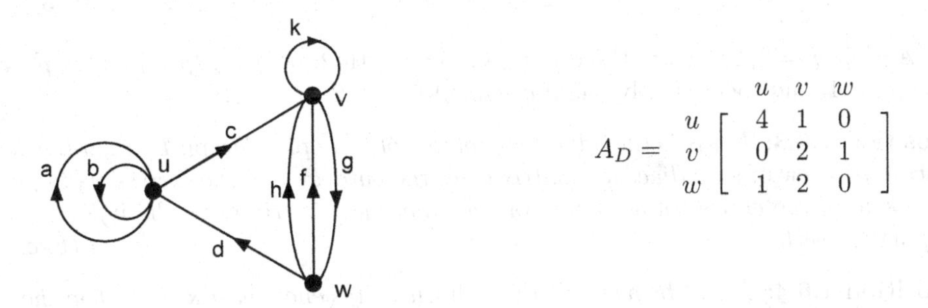

Figure 1.6.6 A digraph and its adjacency matrix.

DEFINITION: The ***incidence matrix*** of a general digraph D is the matrix whose rows and columns are indexed by some orderings of V_D and E_D, respectively, such that

$$I_D[v, e] = \begin{cases} 0 & \text{if } v \text{ is not an endpoint of } e \\ 1 & \text{if } v \text{ is the head of } e \\ -1 & \text{if } v \text{ is the tail of } e \\ 2 & \text{if } e \text{ is a self-loop at } v \end{cases}$$

Example 1.6.7: Figure 1.6.7 shows a digraph and its incidence matrix.

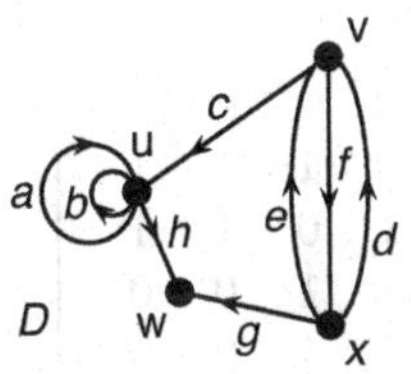

$$I_G = \begin{array}{c} \\ u \\ v \\ w \\ x \end{array} \begin{array}{c} \begin{array}{cccccccc} a & b & c & d & e & f & g & h \end{array} \\ \left[\begin{array}{cccccccc} 2 & 2 & 1 & 0 & 0 & 0 & 0 & -1 \\ 0 & 0 & -1 & 1 & 1 & -1 & 0 & 0 \\ 0 & 0 & 0 & 0 & 0 & 0 & 1 & 1 \\ 0 & 0 & 0 & -1 & -1 & 1 & -1 & 0 \end{array} \right] \end{array}$$

Figure 1.6.7 A digraph and its incidence matrix.

Properties of Graphs Using Their Specifications

Proposition 1.6.1: *If A_G is the adjacency matrix of a general graph G, then the sum of the entries of a row (or column) of A_G equals the degree of the corresponding vertex.*

$\diamondsuit$ *(Exercises)*

Proposition 1.6.2: *If A_G is the adjacency matrix of a simple graph G, then the value of element $(A_G)^r[u, v]$ of the rth power of matrix A_G equals the number of u-v walks of length r.*

Proof: The assertion holds for $r = 1$, by the definition of the adjacency matrix and the fact that walks of length 1 are the edges of the graph. The inductive step follows from the definition of matrix multiplication (see Exercises). ◇

Example 1.6.8: Squaring the adjacency matrix for the graph from Example 1.6.1 gives the number of walks of length 2.

$$A_G = \begin{array}{c} \\ p \\ q \\ r \\ s \end{array} \begin{array}{c} \begin{array}{cccc} p & q & r & s \end{array} \\ \begin{bmatrix} 0 & 1 & 1 & 1 \\ 1 & 0 & 0 & 1 \\ 1 & 0 & 0 & 1 \\ 1 & 1 & 1 & 0 \end{bmatrix} \end{array} \qquad (A_G)^2 = \begin{array}{c} \\ p \\ q \\ r \\ s \end{array} \begin{array}{c} \begin{array}{cccc} p & q & r & s \end{array} \\ \begin{bmatrix} 3 & 1 & 1 & 2 \\ 1 & 2 & 2 & 1 \\ 1 & 2 & 2 & 1 \\ 2 & 1 & 1 & 3 \end{bmatrix} \end{array}$$

Since $(A_G)^2[p,p] = 3$, there are three p-p walks of length 2: $\langle p, q, p \rangle$, $\langle p, r, p \rangle$, $\langle p, s, p \rangle$. Also, $(A_G)^2[p,q] = 1$, and there is only one p-q walk of length 2: $\langle p, s, q \rangle$.

Proposition 1.6.3: *If A_D is the adjacency matrix of a simple digraph D, then the sum of the entries of a row of the adjacency matrix gives the outdegree of the corresponding vertex and the sum of the entries of a column of the adjacency matrix gives the indegree of the corresponding vertex.* ◇ *(Exercises)*

Proposition 1.6.4: *Let D be a simple digraph with adjacency matrix A_D. Then the value of element $A_D^r[u,v]$ of the rth power of matrix A_D equals the number of directed u-v walks of length r.* ◇ *(Exercises)*

The definition of graph isomorphism implies that two graphs are isomorphic if and only if it is possible to order their respective vertex-sets so that their adjacency matrices are identical. If two graphs G and H have different degree-sequences, then they are not graph isomorphic. A brute-force algorithm to test for graph isomorphism would fix an order for V_G in non-increasing degree order and create the adjacency matrix. It would then compare this matrix to every adjacency matrix for H for which the vertices are ordered in non-increasing degree order. For all but the smallest graphs, this exhaustive method is inefficient, which is not surprising in light of the discussion of the *isomorphism problem* in §2.1.

Example 1.6.9: The graphs in Figure 1.6.8 are isomorphic, because under the orderings u, w, v, x and a, c, d, b, they have identical adjacency matrices.

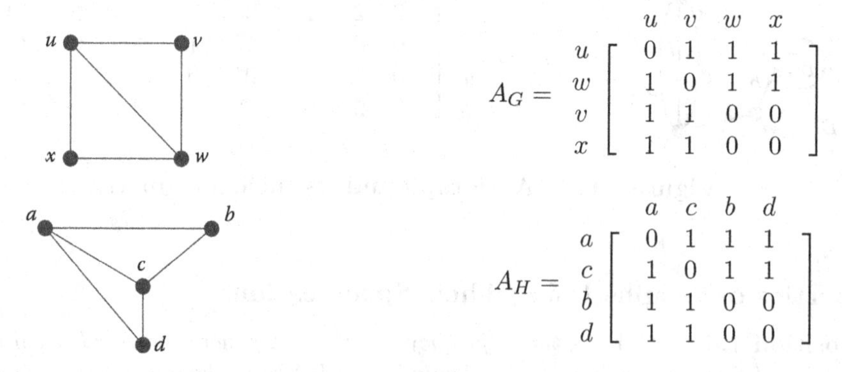

$$A_G = \begin{array}{c} \\ u \\ w \\ v \\ x \end{array} \begin{array}{c} \begin{array}{cccc} u & v & w & x \end{array} \\ \begin{bmatrix} 0 & 1 & 1 & 1 \\ 1 & 0 & 1 & 1 \\ 1 & 1 & 0 & 0 \\ 1 & 1 & 0 & 0 \end{bmatrix} \end{array}$$

$$A_H = \begin{array}{c} \\ a \\ c \\ b \\ d \end{array} \begin{array}{c} \begin{array}{cccc} a & c & b & d \end{array} \\ \begin{bmatrix} 0 & 1 & 1 & 1 \\ 1 & 0 & 1 & 1 \\ 1 & 1 & 0 & 0 \\ 1 & 1 & 0 & 0 \end{bmatrix} \end{array}$$

Figure 1.6.8 Establishing graph isomorphism using adjacency matrices.

Observe that changing the orderings of V_G and E_G permutes the rows and columns of their incidence matrix, I_G. The next two propositions follow immediately from the definition of the incidence matrix.

Proposition 1.6.5: *If I_G is the incidence matrix of a general graph G, then the sum of the entries in any row of I_G equals the degree of the corresponding vertex.* ♢ (Exercises)

Proposition 1.6.6: *If I_G is the incidence matrix of a general graph G, then the sum of the entries in any column of I_G equals 2.* ♢ (Exercises)

These two propositions lead to another proof of Euler's theorem that the sum of the degrees equals twice the number of edges. In particular, the sum of the degrees is simply the sum of the row-sums of the incidence matrix, and the sum of the column-sums equals twice the number of edges. The result follows since these two sums both equal the sum of all the entries of the matrix.

Remark: Our approach treats a digraph as an *augmented* type of graph, where each edge e of a digraph is still associated with a subset $endpts(e)$, but which now also includes a mark on one of the endpoints to specify the head of the directed edge.

This viewpoint is partly motivated by its impact on computer implementations of graphs (see the computational notes below), but it has some advantages from a mathematical perspective as well. Regarding digraphs as augmented graphs makes it easier to view certain results that tend to be established separately for graphs and for digraphs as a single result that applies to both.

EXERCISES for Section 1.6

1.6.1 Construct a line drawing for the mixed graph with vertex-set $V = \{u, v, w, x, y, z\}$, edge-set $E = \{e, f, g, h\}$, and the given incidence table.

edges	e	f	g	h
endpts	u	x^h	v	v
	w^h	u	z	u

1.6.2 For the digraph shown,

(a) Give an incidence table.

(b) Give an adjacency matrix.

(c) Give an incidence matrix.

(d) Give an adjacency matrix for the underlying graph.

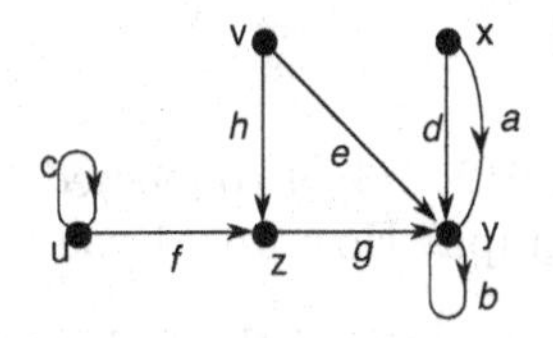

1.6.3 How many different matrices specify, as incidence matrices, general graphs on four vertices and seven edges, two of which are self-loops?

1.6.4 Given an adjacency matrix for a general graph, how would you determine the number of vertices, edges, and self-loops?

1.6.5 Which of the following incidence tables represent connected graphs?

edge	a	b	c	d
endpts	u	v	w	x
	u	w	x	v

edge	a	b	c	d
endpts	u	v	v	x
	v	w	x	v

edge	a	b	c	d
endpts	u	v	w	x
	v	v	x	x

1.6.6 Determine whether the given incidence table represents a strongly connected digraph. Assume that there are no isolated vertices.

(a)

edges	a	b	c	d
endpts	u	v	v	x
	v^h	w^h	x^h	v^h

(b)

edges	a	b	c	d
endpts	u	v^h	w	x
	v^h	w	x^h	v^h

(c)

edges	a	b	c	d	e
endpts	u	v	w^h	v	w
	v^h	u^h	x	x^h	v^h

1.7 SUPPLEMENTARY EXERCISES

1.7.1 A 20-vertex graph has 62 edges. Every vertex has degree 3 or 7. How many vertices have degree 3?

1.7.2 Draw a connected simple graph G whose line graph $L(G)$ has the degree sequence $\langle 3, 3, 2, 2, 2, 1, 1 \rangle$.

1.7.3 Draw a simple connected graph H whose line graph $L(H)$ has the degree sequence $\langle 3, 3, 2, 2, 2, 1, 1 \rangle$, such that H is *not* isomorphic to the graph G of Exercise 1.10.2.

1.7.4 What is the diameter of the Petersen graph? Explain.

1.7.5 Find an r-regular graph with diameter 2 and girth 5, or show it does not exist for the following values of r.

(a) $r = 2$ (b) $r = 3$ (c) $r = 4$

1.7.6 Is $circ(24 : 4, 9)$ connected? What is a general condition for the connectedness of a circulant graph? How would you calculate the number of components?

1.7.7 Calculate the girth of $circ(13 : 1, 5)$.

1.7.8 Prove that a graph G is bipartite if and only if it has the following property:

Every subgraph H contains an independent set of vertices whose cardinality is at least half the cardinality of V_H.

1.7.9 Prove that in a connected graph, any two longest paths must have a vertex in common.

1.7.10 Let G be a simple bipartite graph with at least 5 vertices. Prove that $\overline{G}$ is not bipartite.

1.7.11 Let G be a graph whose vertices are the bitstrings of length 4 such that two bitstrings are adjacent if they differ either in exactly one bit or in all four bits.

a. Write the vertex sequence of a minimum length cycle in the edge-complement $\overline{G}$.

b. What is the diameter of the graph $\overline{G}$? Explain.

DEFINITION: An n-vertex, m-edge simple graph G is ***graceful*** if there is a one-to-one function $g : V_G \to \{0, 1, \ldots, m\}$ such that the function $g' : E_G \to \{0, 1, \ldots, m\}$ given by the rule $uv \mapsto |g(u) - g(v)|$ is one-to-one.

1.7.12 Prove that the cycle C_5 is not graceful.

1.7.13 Prove that the cycle C_8 is graceful.

DEFINITION: A graph G is ***edge-critical*** for a property P if G has property P, but for every edge e, the graph $G - e$ does *not* have property P.

1.7.14 We consider diameter-related properties.

a. The graphs $K_{m,n}$ clearly have diameter 2 (for $m, n \geq 2$). Prove they are edge-critical graphs for this property.
b. What is the maximum increase in diameter that can be caused by deleting an edge from a 2-connected bipartite graph that is edge-critical with respect to its diameter? Explain.

1.7.15 Find the diameter of the Cartesian product $C_m \times C_n$ of two cycle graphs.

GLOSSARY

acyclic: having no cycles.

adding an edge between two vertices u and v of a graph G: creating a supergraph with vertex-set V_G and edge-set $E_G \cup \{e\}$.

adding a vertex v to a graph G: creating a supergraph, denoted $G \cup \{v\}$, with vertex-set $V_G \cup \{v\}$ and edge-set E_G.

adjacency matrix of a simple digraph D, denoted A_D: the matrix whose rows and columns are both indexed by identical orderings of V_G, such that

$$A_D[u,v] = \begin{cases} 1 & \text{if there is an edge from } u \text{ to } v \\ 0 & \text{otherwise} \end{cases}$$

adjacency matrix of a simple graph G, denoted A_G: the matrix whose rows and columns are both indexed by identical orderings of V_G, such that

$$A_G[u,v] = \begin{cases} 1 & \text{if } u \text{ to } v \text{ are adjacent} \\ 0 & \text{otherwise} \end{cases}$$

adjacency table: a table with a row for each vertex, each containing the list of neighbors of that vertex.

adjacent edges: two edges that have an endpoint in common.

adjacent vertices: two vertices joined by an edge.

arc: an edge, one of whose endpoints is designated as the ***tail***, and whose other endpoint is designated as the ***head***; a synonym for *directed edge*.

arc multiplicity: the number of arcs within a multi-arc.

bigraphical pair of sequences of nonnegative integers $\langle a_1, a_2, \ldots, a_r\rangle$ and $\langle b_1, b_2, \ldots, b_t\rangle$: there exists a bipartite graph G with bipartition subsets $U = \{u_1, u_2, \ldots, u_r\}$ and $W = \{w_1, w_2, \ldots, w_t\}$ such that $\deg(u_i) = a_i$, $i = 1, 2, \ldots, r$, and $deg(w_i) = b_i$, $i = 1, 2, \ldots, t$.

bipartite graph: a graph whose vertex-set can be partitioned into two subsets (parts) such that every edge has one endpoint in one part and one endpoint in the other part.

bipartition of a bipartite graph G: the two subsets into which V_G is partitioned.

bridge: a synonym for *cut-edge.*

card in a deck: one of the vertex-deletion subgraphs in the deck.

Cartesian product $G \times H$ of graphs G and H: the graph whose vertex-set is the Cartesian product $V_{G\times H} = V_G \times V_H$ and whose edge-set is the union $E_{G\times H} = (V_G \times E_H) \cup (E_G \times V_H)$. The endpoints of edge (u, d) are the vertices (u, x) and (u, y), where x and y are the endpoints of edge d in graph H. The endpoints of the edge (e, w) are (u, w) and (v, w), where u and v are endpoints of edge e in graph G.

center of a graph G**:** the subgraph induced on the set of *central vertices* of G (see §1.4); denoted $Z(G)$.

central vertex of a graph G: a vertex v with minimum *eccentricity*, i.e., $ecc(v) = rad(G)$.

circulant graph $circ(n : S)$ with *connections* set $S \subseteq \{1, \ldots, n-1\}$: the graph whose vertex-set is the group of integers $\mathbb{Z}_n = \{0, 1, \ldots, n-1\}$ under addition modulo n, such that two vertices i and j are adjacent if and only if there is a number $s \in S$ such that $i + s = j \bmod n$ or $j + s = i \bmod n$.

circular ladder with n rungs: the product $K_2 \times C_n$; denoted CL_n (defined informally in §1.2).

clique in a graph G: a maximal subset of mutually adjacent vertices in G.

clique number of a graph G: the number of vertices in a largest clique in G; denoted $\omega(G)$.

closed (directed) walk: a (directed) walk whose initial and final vertices are identical.

complement of a graph: shortened term for *edge-complement.*

complete bipartite graph: a simple bipartite graph such that each pair of vertices in different sides of the partition is joined by an edge.

complete graph: a simple graph such that every pair of vertices is joined by an edge.

component of a graph: a maximal connected subgraph; that is, a connected subgraph which is not contained in any larger connected subgraph.

component of a vertex v**:** a subgraph induced on the subset of all vertices reachable from v.

concatenation of walks W_1 and W_2: a walk whose traversal consists of a traversal of W_1 followed by a traversal of W_2 such that W_2 begins where W_1 ends.

connected digraph: a digraph whose underlying graph is connected.

connected graph: a graph in which every pair of distinct vertices has a walk between them.

cut-edge of a graph: an edge whose removal increases the number of components.

cutpoint: a synonym for *cut-vertex.*

cut-vertex of a graph: a vertex whose removal increases the number of components.

cycle: a closed path with at least one edge.

cycle graph: a 1-vertex *bouquet* or a simple connected graph with n vertices and n edges that can be drawn so that all of its vertices and edges lie on a single circle.

decomposition of a closed trail T: a collection $C_1, C_2, \ldots, C_m$ of edge-disjoint cycles, such that each cycle C_i is either a subwalk or a reduced walk of T, and the edge-sets of the cycles partition the edge-set of trail T.

degree of a vertex: the number of proper edges incident on that vertex plus twice the number of self-loops.

degree sequence: a list of the degrees of all the vertices in non-increasing order.

diameter of a graph G, denoted *diam*(G): given by $diam\,(G) = \max_{x,y \in V_G} \{d(x,y)\}$.

digraph: abbreviated name for *directed graph.*

digraph invariant: a property of a digraph that is preserved by isomorphisms.

digraphic pair of sequences $\langle a_1, a_2, \ldots, a_n \rangle$ and $\langle b_1, b_2, \ldots, b_n \rangle$: there exists a simple digraph with vertex-set $\{v_1, v_2, \ldots, v_n\}$ such that $indegree(v_i) = b_i$ and $outdegree(v_i) = a_i$ for $i = 1, 2, \ldots, n$.

directed distance from vertex u to v: the length of a shortest directed walk from u to v.

directed edge: a synonym for *arc.*

directed graph: a graph in which every edge is a directed edge.

directed walk from vertex u to vertex v: an alternating sequence of vertices and arcs representing a continuous traversal from u to v, where each arc is traversed from tail to head.

direction on an edge: an optional attribute that assigns the edge a one-way restriction or preference, said to be from *tail* to *head*; in a drawing, the tail is the end behind the arrow and the head is the end in front.

disjoint cycle form of a permutation: a notation for specifying a permutation; see Example 2.2.1.

distance between vertices u and v: the length of a shortest walk between u and v.

eccentricity of a vertex v in a graph G: the distance from v to a vertex farthest from v; denoted *ecc*(v).

edge: a connection between one or two vertices of a graph.

edge-complement of a simple graph G: a graph $\overline{G}$ with the same vertex-set as G, such that two vertices are adjacent in $\overline{G}$ if and only if they are *not* adjacent in G.

edge-critical graph G for a property P: G has property P, but for every edge e, the graph $G - e$ does not have property P.

edge-cut of a graph: a subset of edges whose removal increases the number of components.

edge-deletion subgraph $G - e$: the subgraph of G induced on the edge subset $E_G - \{e\}$.

edge-multiplicity: the number of edges joining a given pair of vertices or the number of self-loops at a given vertex.

edge-transitive graph: a graph G such that for every edge pair d, $e \in E_G$, there is an automorphism of G that maps d to e.

edge-weight: a number assigned to an edge (optional attribute).

endpoints of an edge: the one or two vertices that are associated with that edge.

Eulerian graph: a connected graph that has an Eulerian tour.

Eulerian tour: a closed Eulerian trail.

Eulerian trail in a graph: a trail that contains every edge of that graph.

forest: an acyclic graph.

girth of a graph G: the length of a shortest cycle in G.

graph $G = (V, E)$: a mathematical structure consisting of two sets, V and E. The elements of V are called ***vertices*** (or ***nodes***), and the elements of E are called ***edges***. Each edge has a set of one or two vertices associated to it, which are called its ***endpoints.***

graphic sequence: a non-increasing sequence $\langle d_1, d_2 \dots d_n \rangle$ that is the degree sequence of a simple graph.

Hamiltonian cycle in a graph: a cycle that uses every vertex of that graph.

head of an arc: the endpoint to which that arc is directed.

hypercube graph Q_n of dimension n: the iterated product $K_2 \times \cdots \times K_2$ of n copies of K_2. Equivalently, the graph whose vertices correspond to the 2^n bitstrings of length n and whose edges correspond to the pairs of bitstrings that differ in exactly one coordinate.

incidence: the relationship between an edge and its endpoints.

incidence matrix I_D of a digraph D: a matrix whose rows and columns are indexed by V_D and E_D, respectively, such that

$$I_D[v, e] = \begin{cases} 0 & \text{if } v \text{ is not an endpoint of } e \\ 1 & \text{if } v \text{ is the head of } e \\ -1 & \text{if } v \text{ is the tail of } e \\ 2 & \text{if } e \text{ is a self-loop at } v \end{cases}$$

incidence matrix I_G of a graph G: a matrix whose rows and columns are indexed by V_G and E_G, respectively, such that

$$I_G[v, e] = \begin{cases} 0 & \text{if } v \text{ is not an endpoint of } e \\ 1 & \text{if } v \text{ is an endpoint of } e \\ 2 & \text{if } e \text{ is a self-loop at } v \end{cases}$$

incidence table of a graph or digraph: a table that specifies the endpoints of every edge and, for a digraph, which endpoint is the head.

indegree of a vertex v: the number of arcs directed to v.

independence number of a graph G: the number of vertices in a largest independent set in G; denoted $\alpha(G)$.

independent set of vertices in a graph G: a vertex subset $W \subseteq V_G$ such that no pair of vertices in S is joined by an edge, i.e., S is a subset of mutually nonadjacent vertices of G.

induced subgraph on an edge-set D: the subgraph with edge-set D and with vertex-set consisting of the endpoints of all edges in D.

induced subgraph on a vertex-set W: the subgraph with vertex-set W and edge-set consisting of all edges whose endpoints are in W.

join of a new vertex v to a graph G: the graph that results when each of the preexisting vertices of G is joined to vertex v; denoted $G + v$.

join of two graphs G and H: the graph obtained from the graph union $G \cup H$ by adding an edge between each vertex of G and each vertex of H; denoted $G + H$.

length of a walk: the number of edge-steps in the walk sequence.

line graph $L(G)$ of a graph G: the graph that has a vertex for each edge of G, and two vertices in $L(G)$ are adjacent if and only if the corresponding edges in G have a vertex in common.

local subgraph of a vertex v: synonym of *neighborhood subgraph* of v; denoted $L(v)$.

matching in a graph G: a set of mutually nonadjacent edges in G.

mesh, ($m_1 \times m_2 \times \cdots \times m_n$-): the iterated product $P_{m_1} \times P_{m_2} \times \cdots \times P_{m_n}$ of paths.

—, **wraparound:** the iterated product $C_{m_1} \times C_{m_2} \times \cdots \times C_{m_n}$ of paths.

mixed graph: a graph that has undirected edges as well as directed edges.

Möbius ladder graph ML_n: a graph obtained from the circular ladder CL_n by deleting from the circular ladder two of its parallel curved edges and replacing them with two edges that cross; analogous to the relationship between a Möbius band and a cylindrical band.

multi-arc in a digraph: a set of two or more arcs such that each arc has the same head and the same tail.

multi-edge in a graph: a set of two or more edges such that each edge has the same set of endpoints.

mutual reachability of two vertices u and v in a digraph D: the existence of both a directed u-v walk and a directed v-u walk.

neighbor of a vertex v: any vertex adjacent to v.

(open) neighborhood subgraph of a vertex v: the subgraph induced on the neighbors of v; denoted $L(v)$. Also called *local subgraph.*

node: synonym for *vertex.*

null graph: a graph whose vertex- and edge-sets are empty.

octahedral graph, n-dimensional: the graph $\mathcal{O}_n$ defined recursively, using the join operation, as

$$\mathcal{O}_n = \begin{cases} 2K_1 & \text{if } n = 0 \\ 2K_1 + \mathcal{O}_{n-1} & \text{if } n \geq 1 \end{cases}$$

open (directed) walk: a walk whose initial and final vertex are different.

outdegree of a vertex v: the number of arcs directed from v.

partially directed graph: synonym of *mixed graph.*

path: a walk with no repeated edges and no repeated vertices (except possibly the initial and final vertex); a trail with no repeated vertices.

path graph: a connected graph that can be drawn so that all of its vertices and edges lie on a single straight line.

Petersen graph: a special 3-regular, 10-vertex graph first described by Julius Petersen.

platonic graph: a vertex-and-edge configuration of a platonic solid.

platonic solid: any one of the five regular 3-dimensional polyhedrons - tetrahedron, cube, octahedron, dodecahedron, or icosahedron.

product $G \times H$ of graphs G and H: shortened term for *Cartesian product.*

proper edge: an edge that is not a self-loop.

proper subgraph of a graph G: a subgraph of G that is neither G nor an isomorphic copy of G.

radius of a graph G, denoted $rad(G)$: the minimum of the vertex eccentricities.

reachability relation for a graph G: the equivalence relation on V_G, where two vertices are related if one is reachable from the other.

reachable from a vertex v: said of a vertex for which there is a walk from v.

deck of a graph G: the vertex-deletion subgraph list of G, with no labels on the vertices.

reconstruction problem for graphs: the unsolved problem of deciding whether two non-isomorphic graphs with three or more vertices can have the same vertex-deletion subgraph list.

reduction of a walk W by a walk W': the walk that results when all of the vertices and edges of W', except for one of the vertices common to both walks, are deleted from W.

regular graph: a graph whose vertices all have equal degree.

self-loop: an edge whose endpoints are the same vertex.

simple graph (digraph): a graph (digraph) with no multi-edges or self-loops.

spanning subgraph of a graph G: a subgraph H of G with $V_H = V_G$.

spanning tree: a spanning subgraph that is a tree.

strongly connected digraph: a digraph in which every two vertices are mutually reachable.

strong orientation of a graph: an assignment of directions to the edges of a graph so that the resulting digraph is strongly connected.

subdigraph of a digraph D: a digraph whose underlying graph is a subgraph of the underlying graph of D, and whose edge directions are inherited from D.

subgraph of a graph G: a graph H whose vertices and edges are all in G, or any graph isomorphic to such a graph.

—, **proper:** a subgraph that is neither G nor an isomorphic copy of G.

—, **spanning:** a subgraph H with $V_H = V_G$.

subwalk of a walk W: a walk consisting of a subsequence of consecutive entries of the walk sequence of W.

supergraph: the "opposite" of subgraph; that is, H is a supergraph of G if and only if G is a subgraph of H.

suspension of a graph G from a new vertex v: a synonym for *join.*

tail of an arc: the endpoint from which that arc is directed.

tournament: a digraph whose underlying graph is a complete graph.

trail: a walk with no edge that occurs more than once.

tree: a connected graph that has no cycles.

trivial walk, trail, path, or graph: a walk, trail, path, or graph consisting of one vertex and no edges.

underlying graph of a directed or mixed graph G: the graph that results from removing all the designations of *head* and *tail* from the directed edges of G (i.e., deleting all the edge directions).

(graph) union of two graphs $G = (V, E)$ and $G' = (V', E')$: the graph $G \cup G'$ whose vertex-set and edge-set are the disjoint unions, respectively, of the vertex-sets and edge-sets of G and G'.

—, ***n*-fold self-** : the iterated disjoint union $G \cup \cdots \cup G$ of n copies of the graph G; denoted nG.

valence: a synonym for *degree.*

vertex: an element of the first constituent set of a graph; a synonym for *node.*

vertex-cut of a graph: a subset of vertices whose removal increases the number of components.

vertex-deletion subgraph $G - v$**:** the subgraph of G induced on the vertex subset $V_G - \{v\}$.

vertex-deletion subgraph list: a list of the isomorphism types of the collection of vertex-deletion subgraphs.

walk from vertex u to vertex v: an alternating sequence of vertices and edges, representing a continuous traversal from vertex u to vertex v.

weighted graph: a graph in which each edge is assigned a number, called an *edge-weight.*

***n*-wheel** W_n**:** the join $K_1 + C_n$ of a single vertex and an n-cycle.

—, **even:** a wheel for n even.

—, **odd:** a wheel for n odd.

Chapter 2

ISOMORPHISMS AND SYMMETRY

INTRODUCTION

Deciding when two line drawings represent the same graph can be quite difficult for graphs containing more than a few vertices and edges. A related task is deciding when two graphs with different specifications are *structurally equivalent*, that is, whether they have the same pattern of connections. Designing a practical algorithm to make these decisions is a famous unsolved problem, called the *graph-isomorphism problem*.

2.1 GRAPH HOMOMORPHISMS AND ISOMORPHISMS

We can find relations between different graphs by defining *structure-preserving* mappings between graphs, called *graph homomorphisms*. In particular, we can use graph homomorphisms to determine if two graphs have the same structure, or if one graph is a subgraph of another.

DEFINITION: A ***graph homomorphism*** $f : G \to H$ from a graph G to a graph H is a pair of functions $(f_V : V_G \to V_H, f_E : E_G \to E_H)$ between their vertex- and edge-sets, respectively, such that for every edge $e \in E_G$, the function f_V maps the endpoints of e to the endpoints of edge $f_E(e)$. When there is a graph homomorphism $f : G \to H$, we say that G is ***homomorphic*** to H.

DEFINITION: A ***graph isomorphism*** $f : G \to H$ from a graph G to a graph H is a graph homomorphism $(f_V : V_G \to V_H, f_E : E_G \to E_H)$ for which f_V and f_E are bijections. When there is a graph isomorphism $f : G \to H$, we say that G is ***isomorphic*** to H, denoted $G \cong H$.

The relation "*is homomorphic to*" is reflexive and transitive, but not symmetric, whereas "*is isomorphic to*" is an equivalence relation (see Exercises).

Example 2.1.1: The two graphs in Figure 2.1.1 are clearly not isomorphic, since G has more edges than H, and thus there is no bijection $f_E : E_G \to E_H$. However, there is a homomorphism $f : G \to H$, where $f_E(a) = f_E(b) = f_E(c) = b'$, $f_E(d) = d'$, and $f_E(e) = e'$; and there is a homomorphism $g : H \to G$, where $g_E(b') = b$, $g_E(d') = d$, and $g_E(e') = e$.

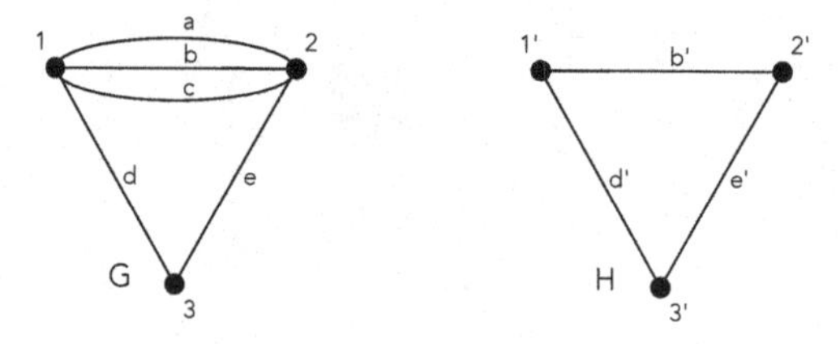

Figure 2.1.1 Two homomorphic graphs that are not isomorphic.

DEFINITION: Let G and H be two simple graphs. A vertex function $f : V_G \to V_H$ ***preserves adjacency*** if for every pair of adjacent vertices u and v in graph G, the vertices $f(u)$ and $f(v)$ are adjacent in graph H. Similarly, f ***preserves nonadjacency*** if $f(u)$ and $f(v)$ are nonadjacent whenever u and v are nonadjacent.

Theorem 2.1.1: *Let G and H be simple graphs. If (f_V, f_E) is a graph homomorphism from G to H, then f_V preserves adjacency. Conversely, if $f_V : V_G \to V_H$ preserves adjacency, then there exists a unique function $f_E : E_G \to E_H$, such that (f_V, f_E) is a graph homomorphism from G to H.*

Proof: The first statement follows from the definition of graph homomorphism. To prove the converse, let e be an arbitrary edge in E_G with endpoints v and w. Since, f_V preserves adjacency, $f_V(v)$ and $f_V(w)$ are adjacent in H, and since H is a simple graph, there is exactly one edge e' with endpoints $f_V(v)$ and $f_V(w)$. Thus, for any homomorphism (f_V, f_E) from G to H, it must be the case that $f_E(e) = e'$. ◊

Thus, when G and H are simple graphs, it is sufficient to give an adjacency-preserving function $f : V_G \to V_H$, to describe a graph homomorphism.

Theorem 2.1.2: *Let G and H be simple graphs. A bijection $f : V_G \to V_H$ gives a graph isomorphism if and only if f preserves both adjacency and nonadjacency.*

◊ *(Exercises)*

Example 2.1.2: Figure 2.1.2 specifies an isomorphism between the two simple graphs shown. Checking that the given vertex bijection is structure-preserving is left to the reader.

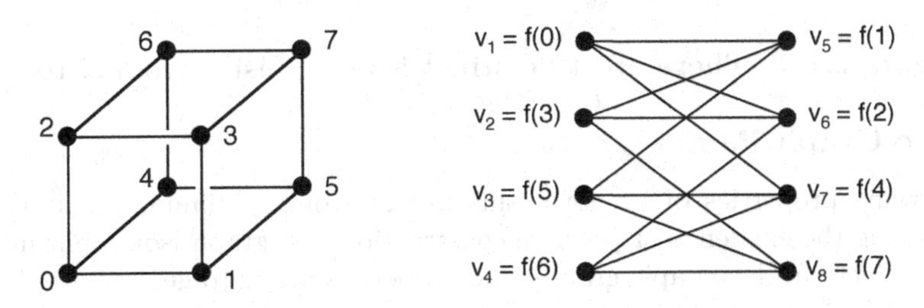

Figure 2.1.2 Specifying an isomorphism between two simple graphs.

Alternatively, one may simply relabel the vertices of the second graph with the names of the vertices in the first graph, as shown in Figure 2.1.3.

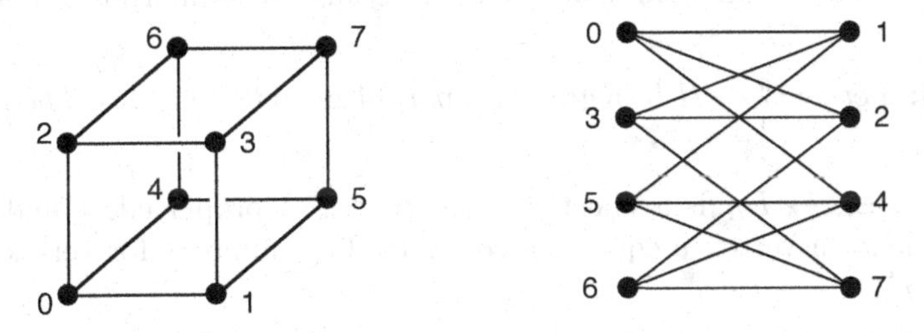

Figure 2.1.3 Another way of depicting an isomorphism.

Example 2.1.3: The vertex function $j \mapsto j + 4$ depicted in Figure 2.1.4 is bijective and adjacency-preserving, but it is not an isomorphism since it does not preserve nonadjacency. In particular, the nonadjacent pair $\{0, 2\}$ maps to the adjacent pair $\{4, 6\}$.

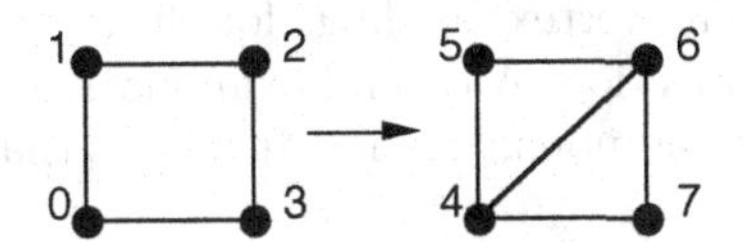

Figure 2.1.4 Bijective and adjacency-preserving, but not an isomorphism.

Example 2.1.4: The vertex function $j \mapsto j \bmod 2$ depicted in Figure 2.1.5 preserves adjacency and nonadjacency, but it is not an isomorphism, since it is not bijective.

Figure 2.1.5 Preserves adjacency and nonadjacency, but is not bijective.

Example 2.1.5: Any graph isomorphism from graph G to H shown in Figure 2.1.6, must send the edges in $\{x, y, z\}$ to edges in $\{a, b, c\}$, and thus must send vertices in $\{2, 3\}$ to vertices in $\{1, 3\}$. Thus, there are two possible vertex bijections and six possible edge bijections, giving 12 distinct graph isomorphisms from G to H.

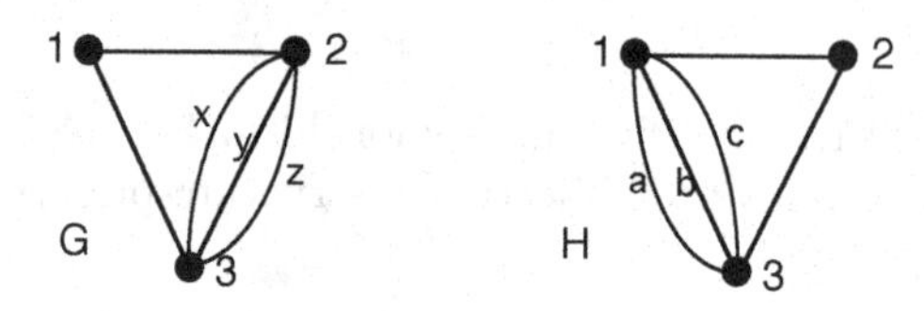

Figure 2.1.6 There are 12 distinct isomorphisms from G to H.

Isomorphic Graph Pairs

The following properties of isomorphisms provide some preliminary criteria to check when considering the existence or possible construction of a graph isomorphism. Although the examples given involve simple graphs, the properties apply to general graphs as well.

Theorem 2.1.3: *Let G and H be isomorphic graphs. Then they have the same number of vertices and the same number of edges.*

Proof: Let $f : G \to H$ be an isomorphism. The graphs G and H must have the same number of vertices and the same number of edges, because an isomorphism maps both sets bijectively. ♢

Theorem 2.1.4: *Let $f : G \to H$ be a graph isomorphism and let $v \in V_G$. Then $\deg(f(v)) = \deg(v)$.*

Proof: Since $f_E : E_G \to E_H$ is a bijection, the number of proper edges and the number of self-loops incident on vertex v equal the corresponding numbers for vertex $f(v)$. Thus, $\deg(f(v)) = \deg(v)$. ♢

Corollary 2.1.5: *Let G and H be isomorphic graphs. Then they have the same degree sequence.*

Corollary 2.1.6: *Let $f : G \to H$ be a graph isomorphism and let $e \in E_G$. Then the endpoints of edge $f(e)$ have the same degrees as the endpoints of e.*

Example 2.1.6: In Figure 2.1.7, we observe that both graphs have 8 vertices and 12 edges, and that both are 3-regular. The vertex labelings for the hypercube Q_3 and for the circular ladder CL_4 specify a vertex bijection. A careful examination reveals that this vertex bijection preserves both adjacency and nonadjacency. It follows that Q_3 and CL_4 are isomorphic graphs.

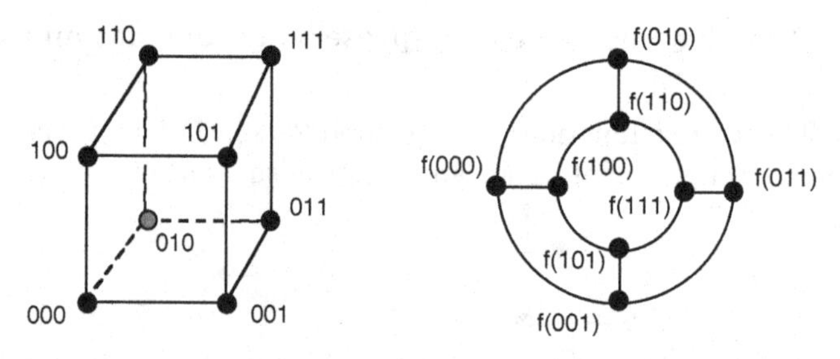

Figure 2.1.7 Hypercube graph Q_3 and circular ladder CL_4 are isomorphic.

Example 2.1.7: The complete bipartite graph $K_{3,3}$ is shown on the left in Figure 2.1.8, and the Möbius ladder ML_3 is shown on the right (obtained from the circular ladder CL_n by deleting from the circular ladder two of its parallel curved edges and replacing them with two edges that cross-match their endpoints). Both graphs have 6 vertices and 9 edges, and both are 3-regular. The vertex labelings for the two drawings specify an isomorphism.

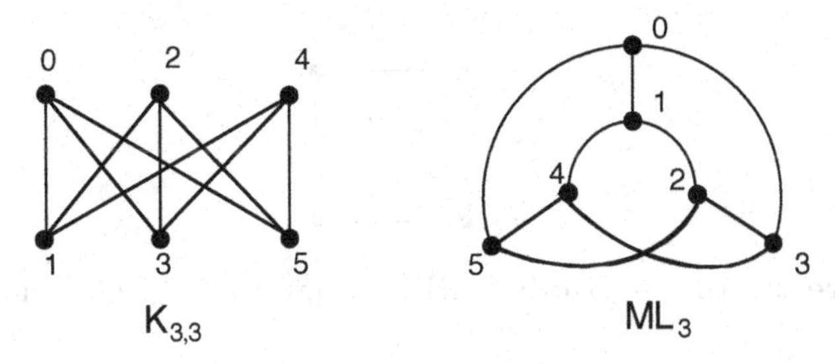

Figure 2.1.8 Bipartite graph $K_{3,3}$ and Möbius ladder ML_3 are isomorphic.

Isomorphism Type of a Graph

DEFINITION: Each equivalence class under $\cong$ is called an ***isomorphism type***.

Isomorphic graphs are graphs of the same isomorphism type. The graphs in Figure 2.1.9 represent all four possible isomorphism types for a simple graph on three vertices. Counting isomorphism types of graphs generally involves the algebra of permutation groups.

Figure 2.1.9 The four isomorphism types for a simple 3-vertex graph.

The Isomorphism Problem

It would be very useful to have a reasonably fast general algorithm that accepts as input any two n-vertex graphs and that produces as output either an isomorphism between them or a report that the graphs are not isomorphic. Checking whether a given vertex bijection is an isomorphism would require an examination of all vertex pairs, which is not in itself overwhelming. However, since there are $n!$ vertex bijections to check, this brute-force approach is feasible only for very small graphs.

DEFINITION: The ***graph-isomorphism problem*** is to devise a practical general algorithm to decide graph isomorphism, or, alternatively, to prove that no such algorithm exists.

Application 2.1.1: *Computer Chip Intellectual Property Rights* Suppose that not long after ABC Corporation develops and markets a computer chip, it happens that the DEF Corporation markets a chip with striking operational similarities. If ABC could prove that DEF's circuitry is merely a rearrangement of the ABC circuitry (i.e., that the circuitries are isomorphic), they might have the basis for a patent-infringement suit. If ABC had to check structure preservation for each of the permutations of the nodes of the DEF chip, the task would take prohibitively long. However, knowledge of the organization of the chips might enable the ABC engineers to take a shortcut.

A Broader Use of the Term "Subgraph"

The usual meaning of the phrase "H is a subgraph of G" is that H is merely isomorphic to a subgraph of G.

Example 2.1.8: The cycle graphs C_3, C_4, and C_5 are all subgraphs of the graph G of Figure 2.1.10.

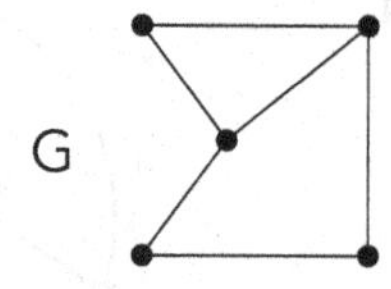

Figure 2.1.10 A graph G with subgraphs C_3, C_4, and C_5.

Observation: A graph with n vertices is Hamiltonian if and only if it contains a subgraph isomorphic to the n-cycle C_n.

Example 2.1.9: The graph G shown in Figure 2.1.11 has among its subgraphs a B_2(bouquet), a D_3 (dipole), and three copies of K_3.

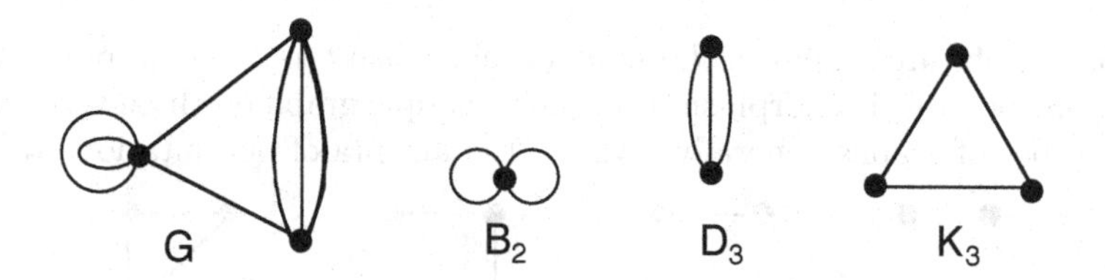

Figure 2.1.11 A graph and three of its subgraphs.

This more general meaning of subgraph gives us the following result.

Proposition 2.1.7: *A graph H is a subgraph of a graph G if and only there is a graph homomorphism $f : H \to G$, such that $f_V : V_H \to V_G$ and $f_E : E_H \to E_G$ are both one-to-one.*

Application to Parallel-Computer Computation

The following application shows how subgraph analysis might occur within a scaled-down model for porting a *distributed algorithm* from one kind of parallel architecture to another. A distributed algorithm is an algorithm that has been designed so that different steps are executed simultaneously on different processing units within a parallel computer.

Application 2.1.2: *Porting an Algorithm* Suppose that a distributed algorithm has been designed to run on an interconnection network whose architecture is an $8 \times 8 \times 32$ array. Suppose also that the only immediately available interconnection network is a 13-dimensional hypercube graph Q_{13}. If the 13-dimensional hypercube graph contains an $8 \times 8 \times 32$ mesh as a subgraph, then the algorithm could be directly ported to that hypercube.

Remark: In general, the ideal situation in which the available network (the *host*) is a supergraph of the required network (the *guest*) is not likely to occur. More realistically, one would like to associate the nodes of the guest with the nodes of the host in such a way as to minimize the decrease in performance of the algorithm.

Amalgamations

One way to construct new graphs is to paste together a few standard recognizable "graph parts", which become subgraphs of the resulting big new graph. Pasting on vertices leads directly to pasting on arbitrary subgraphs, via an isomorphism between them.

DEFINITION: Let G and H be disjoint graphs, with $u \in V_G$ and $v \in V_H$. The ***vertex amalgamation*** $(G \cup H)/\{u = v\}$ is the graph obtained from the union $G \cup H$ by merging (or amalgamating) vertex u of graph G and vertex v of graph H into a single vertex, called uv.† The vertex-set of this new graph is $(V_G - \{u\}) \cup (V_H - \{v\}) \cup \{uv\}$, and the edge-set is $E_G \cup E_H$, except that any edge that had u or v as an endpoint now has the amalgamated vertex uv as an endpoint instead.

Example 2.1.10: Figure 2.1.12 illustrates an amalgamation of a 3-cycle and a 4-cycle, in which vertex u of the 3-cycle is identified with vertex v of the 4-cycle.

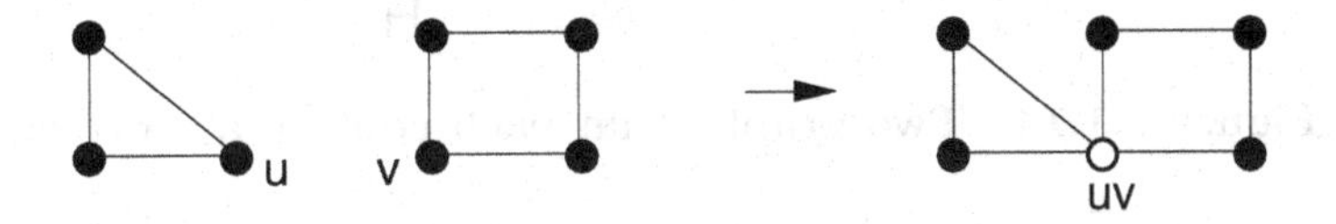

Figure 2.1.12 A vertex amalgamation of a 3-cycle and a 4-cycle.

In this example, no matter which vertices are chosen in the 3-cycle and the 4-cycle, respectively, the isomorphism type of the amalgamated graph is the same. This is due to the symmetry of the two cycles.

Example 2.1.11: Figure 2.1.13 illustrates the four different isomorphism types of graphs that can be obtained by amalgamating P_3 and P_4 at a vertex.

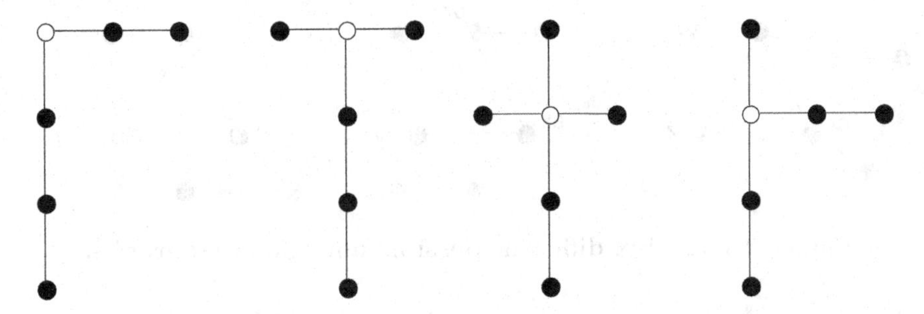

Figure 2.1.13 Four different vertex amalgamations of P_3 and P_4.

When pasting on an arbitrary pair of isomorphic subgraphs, the isomorphism type of the resulting graph may depend on exactly how the vertices and edges of the two subgraphs are matched together. The matching of subgraph to subgraph is achieved by an isomorphism.

DEFINITION: Let G and H be disjoint graphs, with X a subgraph of G and Y a subgraph of H. Let $f : X \to Y$ be an isomorphism between these subgraphs. The ***amalgamation of G and H modulo the isomorphism*** $f : X \to Y$ is the graph obtained from the union $G \cup H$ by merging each vertex u and each edge e of subgraph X with their images $f(u)$ and $f(e)$ in subgraph Y. The amalgamated vertex is generically denoted $uf(u)$, and

†In other contexts, the juxtaposition notation uv often denotes an edge with endpoints u and v. In this section, we use juxtaposition only for amalgamation.

the amalgamated edge is generically denoted $ef(e)$. The vertex-set of this new graph is $(V_G - V_X) \cup (V_H - V_Y) \cup \{uf(u)|u \in V_X\}$, and the edge-set is $(E_G - E_X) \cup (E_H - E_Y) \cup \{e\,f(e)|e \in V_X\}$, except that any edge that had $u \in V_X$ or $f(u) \in V_Y$ as an endpoint now has the amalgamated vertex $uf(u)$ as an endpoint instead. This general amalgamation is denoted $(G \cup H)/f : X \to Y$.

DEFINITION: In an amalgamated graph $(G \cup H)/f : X \to Y$, the image of the pasted subgraphs X and Y is called the ***subgraph of amalgamation***.

Example 2.1.12: Each of the six different isomorphisms from the 3-cycle in graph G to the 3-cycle in graph H in Figure 2.1.14 leads to a different amalgamated graph.

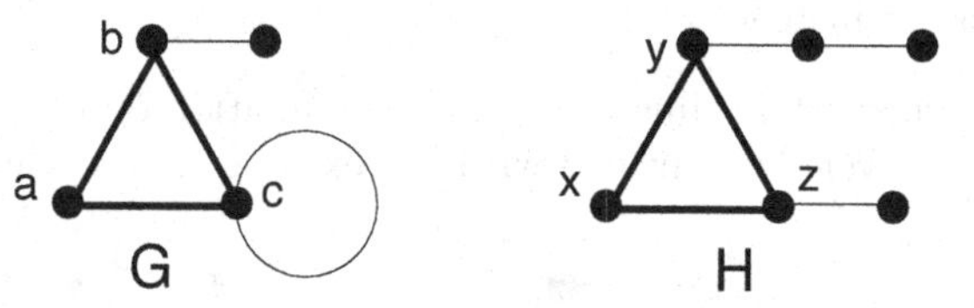

Figure 2.1.14 Two graphs that each contain a 3-cycle.

The six different possible results are illustrated in Figure 2.1.15.

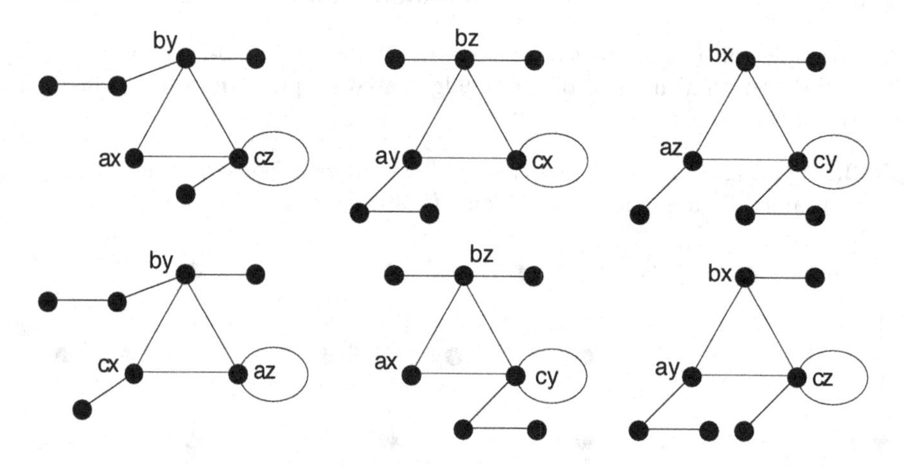

Figure 2.1.15 Six different possible amalgamated graphs.

Remark: The operation of amalgamating graphs G and H by pasting vertex $u \in V_G$ to vertex $v \in V_H$ is equivalent to amalgamating the graphs G and H modulo an isomorphism from the one-vertex subgraph u to the one-vertex subgraph v.

Homomorphism and Isomorphism of Digraphs

The definition of graph homomorphism is easily extended to digraphs.

DEFINITION: A ***digraph homomorphism*** $f : G \to H$ from a digraph G to a digraph H is a graph homomorphism from the underlying graph of G to the underlying graph of H that preserves the direction of each edge. That is, e is directed from u to v if and only if $f(e)$ is directed from $f(u)$ to $f(v)$.

DEFINITION: Two digraphs are ***isomorphic*** if there is an isomorphism f between their underlying graphs that preserves the direction of each edge.

Example 2.1.13: The four different isomorphism types of a simple digraph with three vertices and three arcs are shown in Figure 2.1.16. Notice that non-isomorphic digraphs can have underlying graphs that are isomorphic.

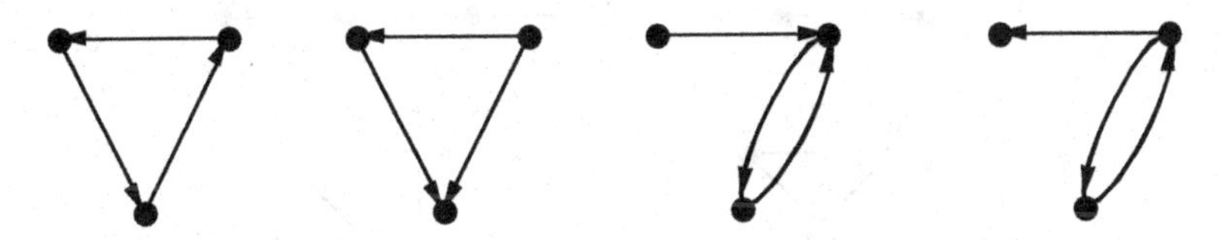

Figure 2.1.16 Four non-isomorphic digraphs.

The following results about digraphs are analogous to those for graphs, mentioned earlier in this section.

Theorem 2.1.8: *Let G and H be isomorphic digraphs. Then they have the same number of vertices and the same number of arcs.*

Theorem 2.1.9: *Let $f : G \to H$ be a digraph isomorphism and let $v \in V_G$. Then $indegree(f(v)) = indegree(v)$ and $outdegree(f(v)) = outdegree(v)$.*

Corollary 2.1.10: *Let G and H be isomorphic digraphs. Then they have the same indegree and outdegree sequences.*

Corollary 2.1.11: *Let $f : G \to H$ be a digraph isomorphism and let $d \in D_G$. Then the head of arc $f(d)$ has the same indegree and outdegree as the head of d, and the tail of arc $f(d)$ has the same indegree and outdegree as the tail of d.*

EXERCISES for Section 2.1

2.1.1 Show that C_6 is homomorphic to C_3 but that C_3 is not homomorphic to C_6.

2.1.2 Show that $K_{2,2}$ and $K_{3,3}$ are homomorphic to each other.

2.1.3 Show that *"is isomorphic to"* is an equivalence relation.

2.1.4 Show that a graph G is bipartite if and only if G is homomorphic to K_2.

2.1.5 Prove Theorem 2.1.2.

2.1.6 Find all possible isomorphism types of the given *simple* graph.

(a) A 4-vertex tree.
(b) A 4-vertex connected graph.
(c) A 5-vertex tree.
(d) A 6-vertex tree.
(e) A 5-vertex graph with exactly three edges.
(f) A 6-vertex graph with exactly four edges.

2.1.7 Find all possible isomorphism types of the given kind of *general* graph.

(a) A graph with two vertices and three edges.
(b) A graph with three vertices and two edges.
(c) A graph with three vertices and three edges.
(d) A 4-vertex graph with exactly four edges including exactly one self-loop and a multi-edge of size 2.

2.1.8 Find a vertex-bijection that specifies an isomorphism between the two graphs.

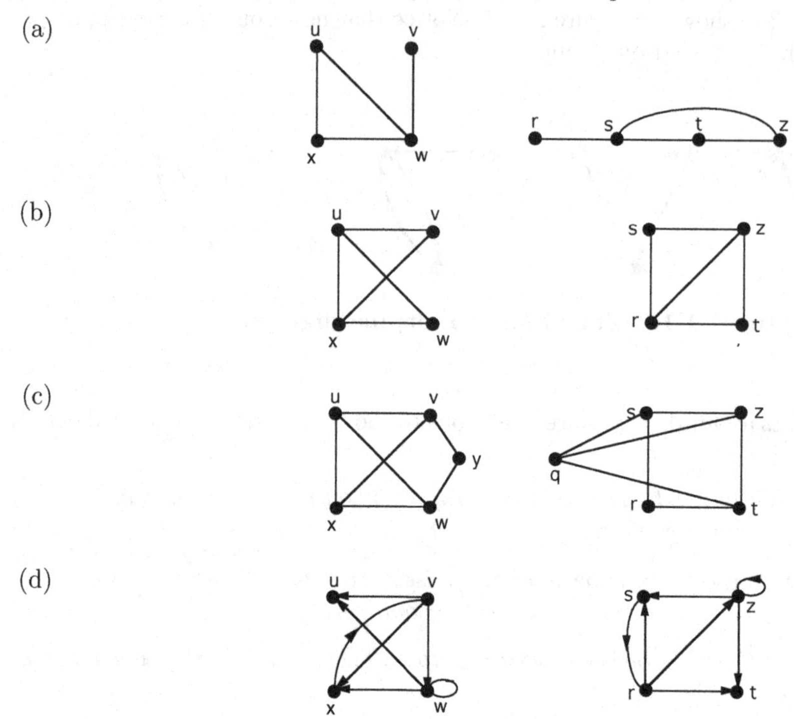

2.1.9 Give an example of two non-isomorphic 5-vertex digraphs whose underlying graphs are isomorphic.

2.1.10 Give an example of two non-isomorphic 9-vertex graphs with the same degree sequence.

2.1.11 Find all possible isomorphism types of the given kind of graph.

(a) A 6-vertex forest with exactly two components.

(b) A simple 4-vertex graph with exactly two components.

(c) A simple 5-vertex graph with exactly three components.

(d) A simple 6-vertex graph with exactly three components.

2.1.12 How many of the subgraphs of K_n are complete graphs?

2.1.13 Characterize those graphs with the given property.

(a) The local subgraph of each vertex is a complete graph.

(b) Every induced subgraph is connected.

2.2 AUTOMORPHISMS AND SYMMETRY

DEFINITION: An isomorphism from a graph G to itself is called an ***automorphism***.

Thus, an automorphism π of graph G is a structure-preserving *permutation* π_V on the vertex-set of G, along with a (consistent) permutation π_E on the edge-set of G. We may write $\pi = (\pi_V, \pi_E)$.

NOTATION: When the context is clear, the subscripts that distinguish the vertex and edge actions of an automorphism are suppressed. Thus, we may simply write π in place of π_V or of π_E.

Remark: Any structure-preserving vertex-permutation is associated with one (if simple) or more (if there are any multi-edges) automorphisms of G. Two possible measures of symmetry of a graph are (1) the number of automorphisms or (2) the proportion of vertex permutations that are structure-preserving.

Permutations and Cycle Notation

The most convenient representation of a permutation, for our present purposes, is as a *product of disjoint cycles.*

NOTATION: In specifying a permutation, we use the notation (x) to show that the object x is fixed, i.e., permuted to itself; the notation $(x\ y)$ means that objects x and y are swapped; and the notation $(x_0\ x_1\ \cdots\ x_{n-1})$ means that the objects $x_0\ x_1\ \cdots\ x_{n-1}$ are cyclically permuted, so that $x_j \mapsto x_{j+1 (\text{mod } n)}$, for $j = 0, 1, \ldots, n-1$. As explained in Appendix A4, every permutation can be written as a composition of disjoint cycles.

Example 2.2.1: The permutation

$$\pi = \begin{pmatrix} 1 & 2 & 3 & 4 & 5 & 6 & 7 & 8 & 9 \\ 7 & 4 & 1 & 8 & 5 & 2 & 9 & 6 & 3 \end{pmatrix}$$

which maps 1 to 7, 2 to 4, and so on, has the ***disjoint cycle form***

$$\pi = (1\,7\,9\,3)(2\,4\,8\,6)(5)$$

Remark: Often the disjoint cycle form excludes the cycles of length 1. Thus, the permutation π in Example 2.2.1 would typically be written $\pi = (1\,7\,9\,3)(2\,4\,8\,6)$.

Geometric Symmetry

A geometric symmetry on a graph drawing can be used to represent an automorphism on the graph.

Example 2.2.2: The graph $K_{1,3}$ has six automorphisms. Each of them is realizable by a rotation or reflection of the drawing in Figure 2.2.1.

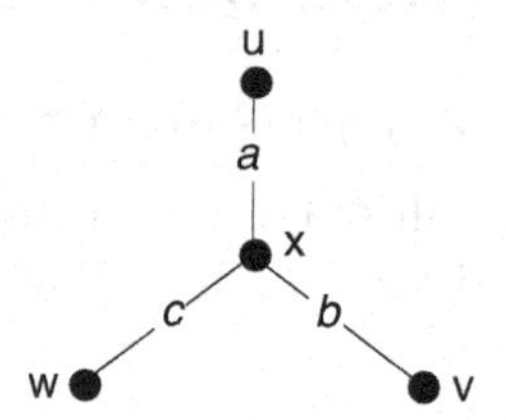

Figure 2.2.1 The graph $K_{1,3}$.

For instance, a 120° clockwise rotation of the figure corresponds to the graph automorphism with vertex-permutation $(x)(u\ v\ w)$ and edge-permutation $(a\ b\ c)$. Also, reflection through the vertical axis corresponds to the graph automorphism with vertex-permutation $(x)(u)(v\ w)$ and edge-permutation $(a)(b\ c)$. The following table lists all the automorphisms of $K_{1,3}$ and their corresponding vertex- and edge-permutations.

Symmetry	Vertex-Permutation	Edge-Permutation
identity	$(u)(v)(w)(x)$	$(a)(b)(c)$
120° rotation	$(x)(u\ v\ w)$	$(a\ b\ c)$
240° rotation	$(x)(u\ w\ v)$	$(a\ c\ b)$
refl. thru a	$(x)(u)(v\ w)$	$(a)(b\ c)$
refl. thru b	$(x)(v)(u\ w)$	$(b)(a\ c)$
refl. thru c	$(x)(w)(u\ w)$	$(c)(a\ b)$

Since the graph $K_{1,3}$ has four vertices, the total number of permutations on its vertex-set is 24. By Theorem 2.1.4, every automorphism on $K_{1,3}$ must fix the 3-valent vertex. Since there are only six permutations of four objects that fix one designated object, it follows that there can be no more than six automorphisms of $K_{1,3}$.

Remark: Except for a few exercises, the focus of this section is on simple graphs. Since each automorphism of a simple graph G is completely specified by a structure-preserving vertex-permutation, the automorphism and its corresponding vertex-permutation are often regarded as the same object.

Example 2.2.3: Figure 2.2.2 shows a graph with four vertex-permutations, each written in disjoint cyclic notation. It is easy to verify that these vertex-permutations are structure-preserving, so they are all graph automorphisms. We observe that the automorphisms λ_0, λ_1, λ_2, and λ_3 correspond, respectively, to the identity, vertical reflection, horizontal reflection, and 180° rotation. Proving that there are no other automorphisms is left as an exercise.

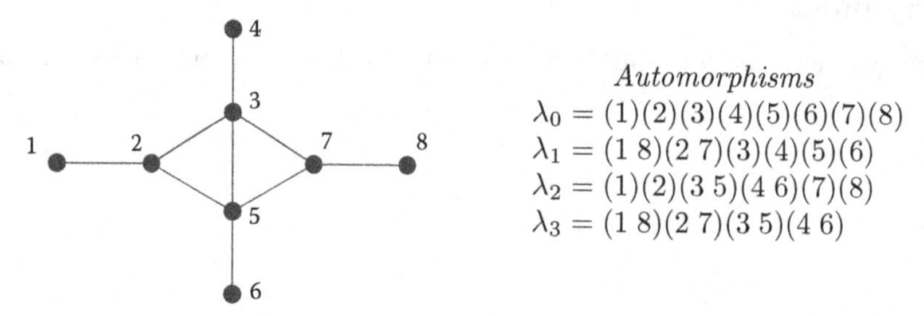

Figure 2.2.2 A graph with four automorphisms.

Limitations of Geometric Symmetry

Although looking for geometric symmetry within a drawing may help in discovering automorphisms, there may be automorphisms that are not realizable as reflections or rotations of a particular drawing.

Example 2.2.4: Figure 2.2.3 shows three different drawings of the same labeled Petersen graph (for example, vertex 0 is adjacent to vertices 1, 4, and 5 in all three). Drawing (a) has 5-fold rotational symmetry that corresponds to the automorphism (0 1 2 3 4)(5 6 7 8 9), but this automorphism does not correspond to any geometric symmetry of either of the other two drawings. The automorphism (0 5)(1 8)(4 7)(2 3)(6)(9) is realized by 2-fold reflectional symmetry in drawings (b) and (c) (about the axis through vertices 6 and 9) but is not realizable by any geometric symmetry in drawing (a). There are several other automorphisms that are realizable by geometric symmetry in at least one of the drawings but not realizable in at least one of the other ones. (See Exercises.)

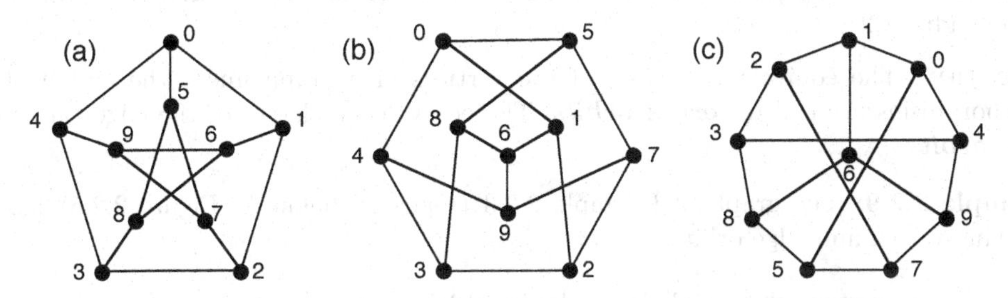

Figure 2.2.3 Three drawings of the Petersen graph.

Vertex- and Edge-Transitive Graphs

DEFINITION: A graph G is ***vertex-transitive*** if for every vertex pair $u, v \in V_G$, there is an automorphism that maps u to v.

DEFINITION: A graph G is ***edge-transitive*** if for every edge pair $d, e \in E_G$, there is an automorphism that maps d to e.

Example 2.2.5: The graph $K_{1,3}$, discussed in Example 2.2.2, is edge-transitive, but not vertex-transitive, since every automorphism must map the 3-valent vertex to itself.

Example 2.2.6: The complete graph K_n is vertex-transitive and edge-transitive for every n. (See Exercises.)

Example 2.2.7: The Petersen graph is vertex-transitive and edge-transitive. (See Exercises.)

Example 2.2.8: Every circulant graph $circ(n : S)$ is vertex-transitive. In particular, the vertex function $i \mapsto i + k \bmod n$ is an automorphism. In effect, it rotates a drawing of the circulant graph, as in Figure 2.2.4, $\frac{k}{n}$ of the way around. For instance, vertex 3 can be mapped to vertex 7 by rotating $\frac{4}{13}$ of the way around. Although $circ(13 : 1, 5)$ is edge-transitive, some circulant graphs are not. (See Exercises.)

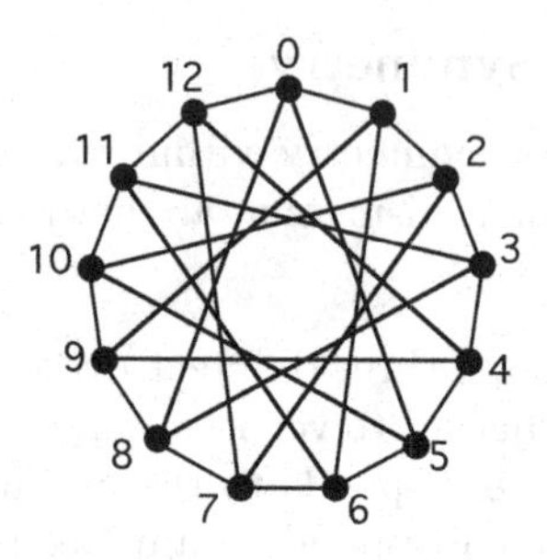

Figure 2.2.4 The circulant graph $circ(13:1,5)$**.**

Vertex Orbits and Edge Orbits

Suppose that two vertices u and v of a graph G are to be considered related if there is an automorphism that maps u to v. This is clearly an *equivalence relation* on the vertices of G. There is a similar equivalence relation on the edges. (Equivalence relations are discussed in Appendix A.2.)

DEFINITION: The equivalence classes of the vertices of a graph under the action of the automorphisms are called ***vertex orbits***. The equivalence classes of the edges are called ***edge orbits***.

Example 2.2.9: The graph of Example 2.2.3 (repeated below in Figure 2.2.5) has the following vertex and edge orbits.

vertex orbits: $\{1,8\}$, $\{4,6\}$, $\{2,7\}$, $\{3,5\}$
edge orbits: $\{12,78\}$, $\{34,56\}$, $\{23,25,37,57\}$, $\{35\}$

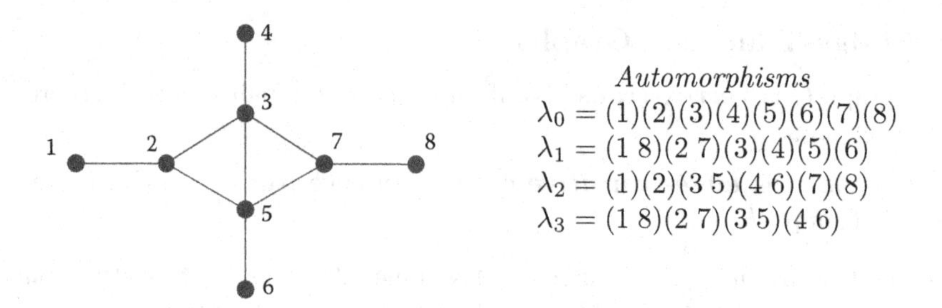

Figure 2.2.5 Graph of Example 2.2.3.

Theorem 2.2.1: *All vertices in the same orbit have the same degree.*

Proof: This follows immediately from Theorem 2.1.4. ◇

Theorem 2.2.2: *All edges in the same orbit have the same pair of degrees at their endpoints.*

Proof: This follows immediately from Corollary 2.1.6. ◇

Remark: A vertex-transitive graph is a graph with only one vertex orbit, and an edge-transitive graph has only one edge orbit.

Example 2.2.10: The complete graph K_n has one vertex orbit and one edge orbit.

Example 2.2.11: Each of the two partite sets of the complete bipartite graph $K_{m,n}$ is a vertex orbit. The graph is vertex-transitive if and only if $m = n$; otherwise it has two vertex orbits. However, $K_{m,n}$ is always edge-transitive (see Exercises).

Remark: Automorphism theory is one of several graph topics involving an interaction between graph theory and group theory.

Finding Vertex and Edge Orbits

We illustrate how to find orbits with two examples. (It is not known whether there exists a polynomial-time algorithm for finding orbits. Certainly, testing all $n!$ vertex-permutations for the adjacency preservation property is non-polynomial.) In addition to using Theorems 2.2.1 and 2.2.2, we observe that if an automorphism maps vertex u to vertex v, then it maps the neighbors of u to the neighbors of v.

Example 2.2.12: In the graph of Figure 2.2.6, vertex 0 is the only vertex of degree 2, so it is in an orbit by itself. The vertical reflection (0)(1 4)(2 3) establishes that vertices 1 and 4 are in the same orbit and that 2 and 3 are in the same orbit. Moreover, since vertices 1 and 4 have a 2-valent neighbor but vertices 2 and 3 do not, the orbit of vertices 1 and 4 must be different from that of 2 and 3. Thus, the vertex orbits are $\{0\}, \{1, 4\}$, and $\{2, 3\}$.

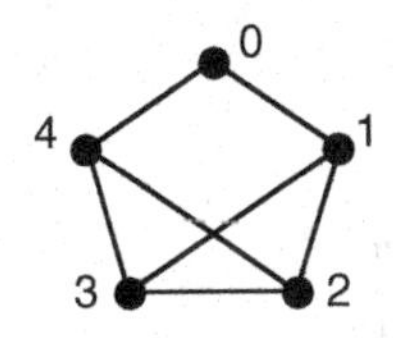

Vertex orbits: $\{0\}, \{1, 4\}$, and $\{2, 3\}$
Edge orbits: $\{23\}, \{01, 04\}$, and $\{12, 13, 24, 34\}$

Figure 2.2.6 A graph and its orbits.

Since edge 23 is the only edge, both of whose endpoints have three 3-valent neighbors, it is in an orbit by itself. The edges 01 and 04 are the only edges with a 2-valent endpoint and a 3-valent endpoint, so they could not be in the same orbit with any edges except each other. The vertical reflection establishes that the pair $\{01, 04\}$ is indeed in the same orbit, as are the pair $\{12, 34\}$ and the pair $\{13, 24\}$. The latter two pairs combine into a single edge orbit, because of the automorphism (0)(1)(4)(2 3), which swaps edges 12 and 13 as well as edges 24 and 34. Thus, the edge orbits are $\{23\}, \{01, 04\}$, and $\{12, 13, 24, 34\}$.

Example 2.2.13: Notice how the orbits in the previous example reflect the symmetry of the graph. In general, recognizing the symmetry of a graph expedites the determination of its orbits. Consider the 4-regular graph G in Figure 2.2.7.

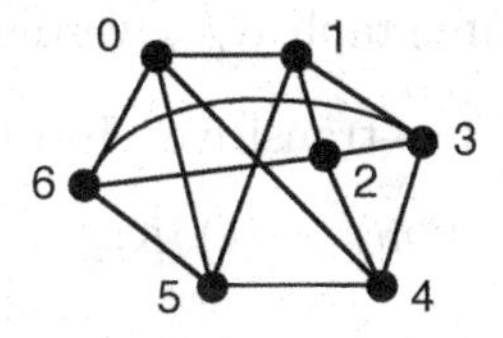

Figure 2.2.7 A 4-regular graph G whose symmetry is less apparent.

When we look at vertices 0, 2, 3, and 5, we discover that each of them has a set of 3 neighbors that are independent, while vertices 1, 4, and 6 each have two pairs of adjacent vertices. The redrawing of the graph G shown in Figure 2.2.8 reflects this symmetry.

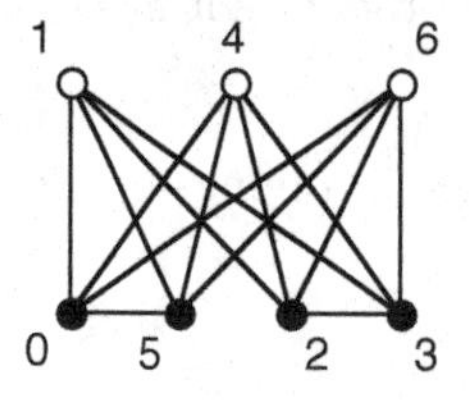

Vertex orbits: $\{0, 2, 3, 5\}$ and $\{1, 4, 6\}$
Edge orbits: $\{05, 23\}$ and $E_G - \{05, 23\}$

Figure 2.2.8 A redrawn graph and its orbits.

In that form, we see immediately that there are two vertex orbits, namely $\{0, 2, 3, 5\}$ and $\{1, 4, 6\}$. One of the two edge orbits is $\{05, 23\}$, and the other contains all the other edges.

EXERCISES for Section 2.2

2.2.1 Write down all the automorphisms of the given simple graph as vertex-permutations.

(a)
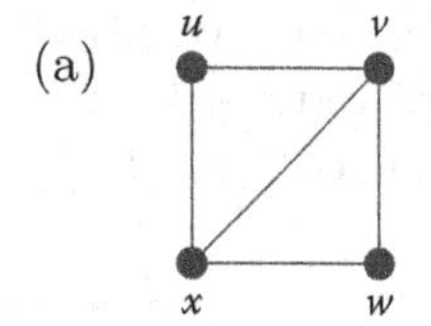

(b)
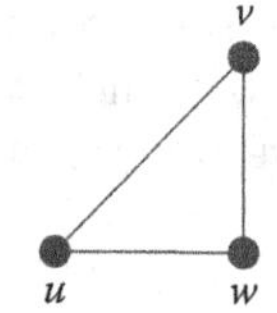

(c)
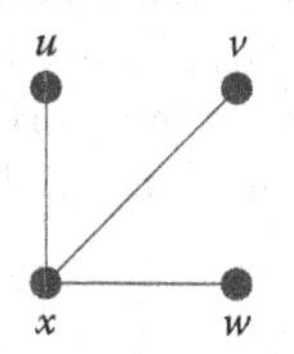

(d) u v x w

2.2.2 Specify all the automorphisms of the given graph.

(a)
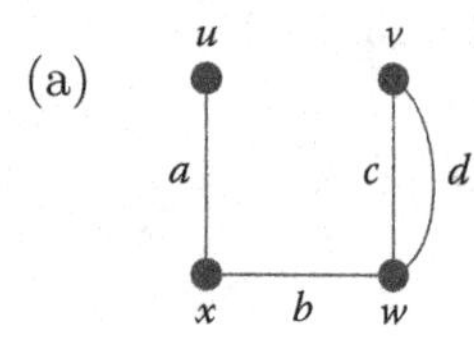

(b)
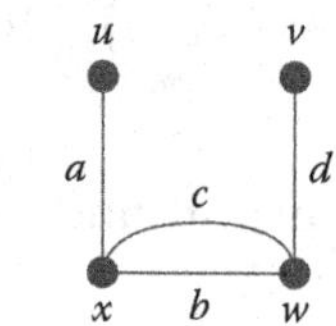

(c)
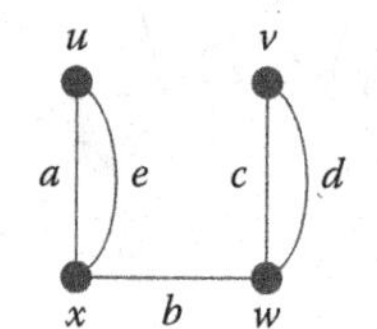

2.2.3 Determine the number of distinct automorphisms of the graph shown below.

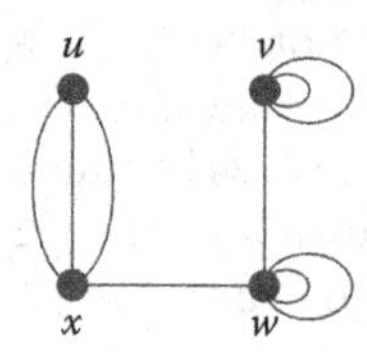

2.2.4 Prove that the specified graph is vertex-transitive.

(a) CL_n (b) $circ(n : S)$ (c) The Petersen graph

2.2.5 When is the complete bipartite graph $K_{m,n}$ vertex-transitive? Justify your answer.

2.2.6 When is the specified graph edge-transitive? Justify your answer.

(a) $circ(9 : 1, m)$ (b) $circ(13 : 1, m)$ (c) $K_{m,n}$ (d) CL_n

2.2.7 Find the vertex and edge orbits of the given graph.

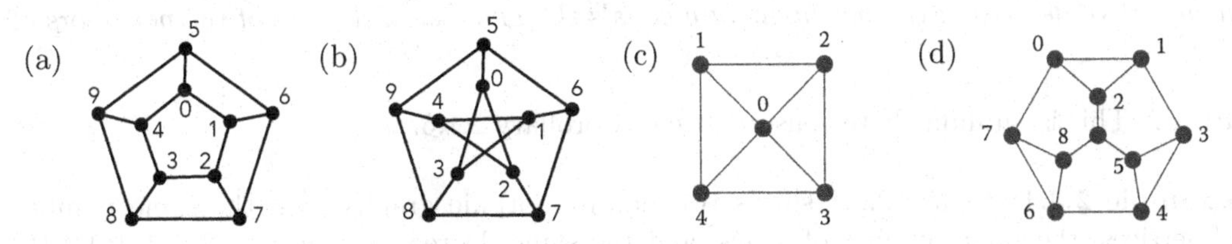

2.2.8 Prove that the list of automorphisms in Example 2.2.3 is complete.

2.2.9 Specify an automorphism of the Petersen graph that is realizable by some geometric symmetry in the middle drawing of Figure 2.2.3, but not realizable by any geometric symmetry in the other two drawings.

2.3 TESTS FOR NON-ISOMORPHISM

In §2.1 the most basic necessary conditions for two graphs to be isomorphic were derived: same number of vertices, same number of edges, and same degree sequence. We noticed also that two isomorphic graphs could differ in various artifacts of their representations, including vertex names and some kinds of differences in drawings. And we observed that a brute-force approach of considering all $n!$ possible vertex bijections for two n-vertex graphs would be too labor-intensive.

The fact is, there is no known short list of easily applied tests to decide for any possible pair of graphs whether they are isomorphic. Thus, the isomorphism problem is formidable, but not all is bleak. We will establish a collection of graph properties that are preserved under isomorphism. These will be used to show that various pairs of graphs are not isomorphic.

DEFINITION: A ***graph invariant*** (or ***digraph invariant***) is a property of graphs (digraphs) that is preserved by isomorphisms.

To show that two graphs are NOT isomorphic, the strategy is to run through a list of graph invariants until you find one that is different for the two graphs. The following invariants were established in §2.1:

1. The number of vertices. (Thm. 2.1.3)
2. The number of edges. (Thm. 2.1.3)
3. The degree sequence. (Cor. 2.1.5)

In this section, we establish several other graph invariants that are useful for isomorphism testing and in the construction of isomorphisms.

That the degree sequence is a graph invariant is an immediate consequence from Theorem 2.1.4, which establishes that the degree of a vertex equals the degree of its image under any graph isomorphism. The invariance of vertex degree under isomorphism is referred to as a "local invariant." The following theorem provides another local invariant for isomorphism, from which the fourth graph invariant in Table 2.3.1 is derived.

Theorem 2.3.1: *Let $f : G \to H$ be a graph isomorphism, and let $v \in V_G$. Then the multiset*† *of degrees of the neighbors of v equals the multiset of degrees of the neighbors of $f(v)$.*

Proof: This is an immediate consequence of Corollary 2.1.6. ◇

Example 2.3.1: Figure 2.3.1 shows two non-isomorphic graphs with the same number of vertices, the same number of edges, and the same degree sequence: $(3, 2, 2, 1, 1, 1)$. By Theorem 2.1.3, an isomorphism would have to map vertex v in graph G to vertex w in graph H since they are the only vertices of degree 3 in their respective graphs. However, the three neighbors of v have degrees 1, 1, and 2, and the structure-preservation property implies that they would have to be mapped bijectively to the three neighbors of w (i.e., *a forced match*), with degrees 1, 2, and 2. This would violate Theorem 2.3.1.

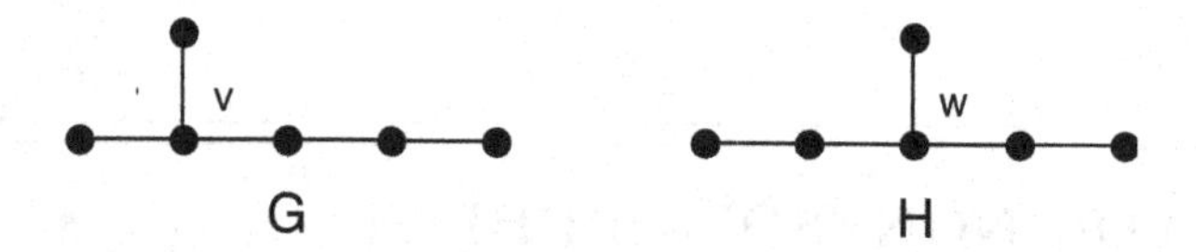

Figure 2.3.1 Non-isomorphic graphs with the same degree sequence.

The following example shows how the same strategy can be used to show that two digraphs are not isomorphic.

Example 2.3.2: Although the indegree and outdegree sequences of the digraphs in Figure 2.3.2 are identical, these digraphs are not isomorphic. To see this, first observe that vertices u, v, x, and y are the only vertices in their respective digraphs that have indegree 2. Since indegree is a local isomorphism invariant (by a digraph analogy to Theorem 2.3.1), u and v must map to y and x, respectively. But the directed path of length 3 that ends at u would have to map to a directed path of length 3 that ends at y. Since there is no such path, the digraphs are not isomorphic.

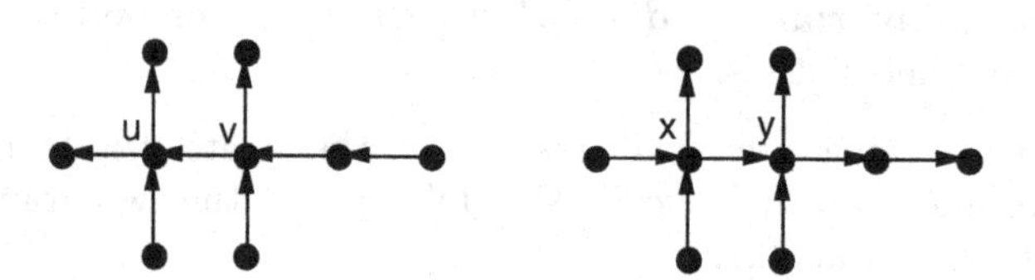

Figure 2.3.2 Non-isomorphic digraphs with identical degree sequences.

Distance Invariants

DEFINITION: Let $f : G \to H$ be a graph homomorphism. If $W = \langle v_0, e_1, v_1, \ldots, e_n, v_n \rangle$ is a walk in graph G, then the ***image of the walk*** W is the walk $f(W) = \langle f(v_0), f(e_1), f(v_1), \ldots, f(e_n), f(v_n) \rangle$ in graph H.

Theorem 2.3.2: *The homomorphic image of a graph walk is a walk of the same length.*

Proof: This follows directly from the definition of the image of a walk. ◇

†A set has no order and no repetitions, while a list has order and can have repetitions. A multiset is a hybrid, having no order, but may have repetitions.

The following two Corollaries are immediate consequences of Theorem 2.3.2.

Corollary 2.3.3: *The isomorphic image of a trail, path, or cycle is a trail, path, or cycle, respectively, of the same length.*

Corollary 2.3.4: *For each integer k, two isomorphic graphs must have the same number of trails (paths) (cycles) of length k.* ◊

Corollary 2.3.5: *The diameter, the radius, and the girth are graph invariants.*

Proof: This is an immediate consequence of Corollary 2.3.3 and the bijectivity of the isomorphism. ◊

Example 2.3.3: Figure 2.3.3 shows the circular ladder CL_4 and the Möbius ladder ML_4, which are 8-vertex graphs. Since both graphs are 3-regular, they have the same number of edges and the same degree sequence. To show they are not isomorphic, we use the diameter invariant.

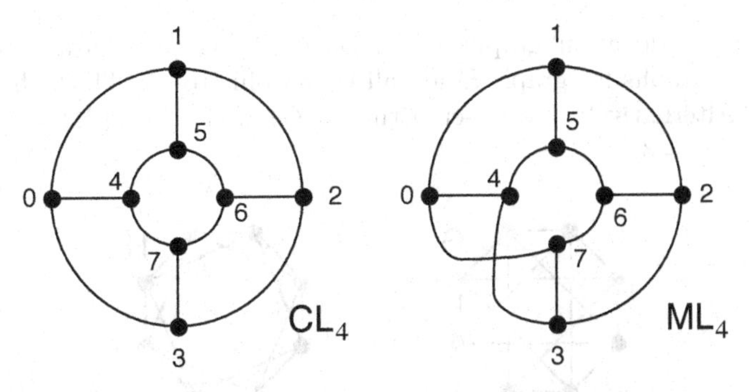

Figure 2.3.3 Non-isomorphic graphs with the same degree sequence.

To reduce the calculation of diameter from checking all $\binom{8}{2} = 28$ vertex pairs to checking the maximum distance from any one vertex, we first establish that both graphs are vertex-transitive. For the circular ladder we observe the symmetries of rotation and the isomorphism that swaps the inner cycle with the outer cycle. For the Möbius ladder, we observe that $j \mapsto j + 1 \bmod 8$ is an automorphism, whose iteration establishes vertex-transitivity.

The maximum distance from vertex 0 in CL_4 is 3, to vertex 6. The maximum distance from vertex 0 in ML_4 is 2. Thus, they are not isomorphic.

Subgraph Invariant

Theorem 2.3.6: *For each graph-isomorphism type, the number of distinct subgraphs in a graph having that isomorphism type is a graph invariant.*

Proof: Let $f : G_1 \to G_2$ be a graph isomorphism and H a subgraph of G_1. Then $f(H)$ is a subgraph of G_2, of the same isomorphism type as H. The bijectivity of f establishes the invariant. ◊

Corollary 2.3.7: *Independence number and clique number are graph invariants.*

Example 2.3.3 continued: Independence number could also have been used to show that CL_4 and ML_4 are not isomorphic, since $\alpha(CL_4) = 4$ and $\alpha(ML_4) = 3$.

Example 2.3.4: The five graphs in Figure 2.3.4 are mutually non-isomorphic, even though they have the same degree sequence. Whereas A and C have no K_3 subgraphs, B has two, D has four, and E has one. Thus, Theorem 2.3.6 implies that the only possible isomorphic pair is A and C. However, graph C has a 5-cycle, but graph A does not.

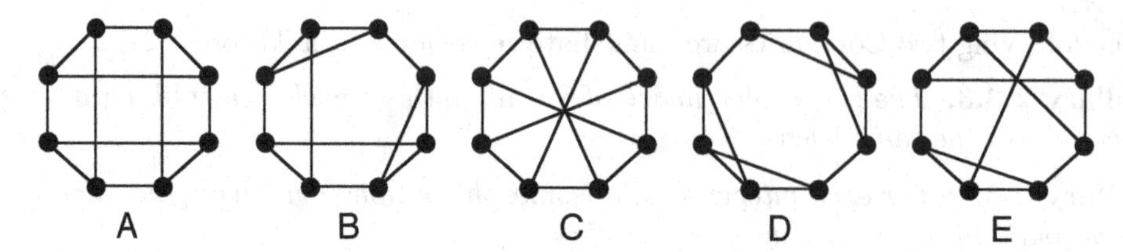

Figure 2.3.4 Five mutually non-isomorphic, 8-vertex, 3-regular graphs.

Theorem 2.3.8: *Let $f : G \to H$ be a graph isomorphism and $u \in V_G$. Then f maps the local subgraph $L(u)$ of G isomorphically to the local subgraph $L(f(u))$ of H.*

Proof: This follows since f maps the vertices of $L(u)$ bijectively to the vertices of $L(f(u))$, and since f is edge-multiplicity preserving, by Theorem 2.1.4. ◊

Example 2.3.5: The local subgraphs for graph G of Figure 2.3.5 are all isomorphic to $4K_1$. The local subgraphs for graph H are all isomorphic to P_4. Thus, the two graphs are not isomorphic. Alternatively, we observe that $\alpha(G) = 4$, but $\alpha(H) = 2$, and also that $\omega(G) = 2$, but $\omega(H) = 3$.

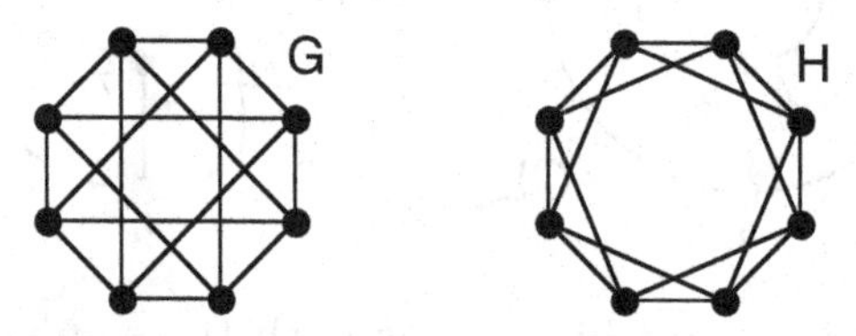

Figure 2.3.5 Two 4-regular 8-vertex graphs.

Edge-Complementation

The next invariant is particularly useful when analyzing simple graphs that are *dense* (i.e., most of the vertex pairs are adjacent).

Theorem 2.3.9: *Let G and H both be simple graphs. They are isomorphic if and only if their edge-complements are isomorphic.*

Proof: By definition, a graph isomorphism necessarily preserves nonadjacency as well as adjacency. ◊

Example 2.3.6: The two graphs in Figure 2.3.6 are relatively dense, simple graphs (both with 20 out of 28 possible edges). The edge-complement of the left graph consists of two disjoint 4-cycles, and the edge-complement of the right graph is an 8-cycle. Since these edge-complements are non-isomorphic, the original graphs must be non-isomorphic.

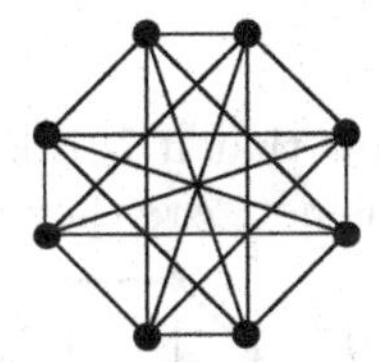
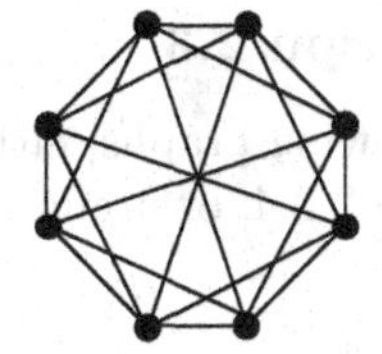

Figure 2.3.6 Two relatively dense, non-isomorphic 5-regular graphs.

Summary

Table 2.3.1 summarizes these results on graph-isomorphism invariants.

Table 2.3.1: Some graph invariants.

1.	The number of vertices
2.	The number of edges
3.	The degree sequence
4.	The multiset of degrees of the neighbors of a *forced match* (see Example 2.3.1)
5.	The multiset of local subgraphs of the neighbors of a *forced match*
6.	Diameter, radius, girth
7.	Independence number, clique number
8.	For any possible subgraph, the number of distinct copies
9.	For a simple graph, the edge-complement

Using Invariants to Construct an Isomorphism

The last example in this section illustrates how invariants can guide the construction of an isomorphism.

Example 2.3.7: The two graphs in Figure 2.3.7 both have open paths of length 9, indicated by consecutive vertex numbering $1, \ldots, 10$. By Corollary 2.3.3, any isomorphism must map the length-9 path in the left graph to some length-9 path in the right graph. The label-preserving bijection that maps the left length-9 path to the right length-9 path is a reasonable candidate for such an isomorphism. Adjacency is clearly preserved for the nine pairs of consecutively numbered vertices, $(i, i+1)$, since they represent edges along the paths. Also, it is easy to see that adjacency is preserved for the pairs of vertices that are not consecutively numbered, since each of the six remaining edges in the left graph has a corresponding edge in the right graph with matching endpoints. Thus, the label-preserving bijection is an isomorphism.

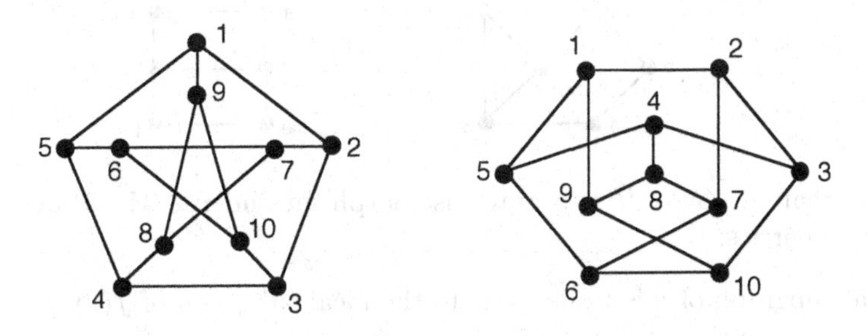

Figure 2.3.7 Two copies of the Petersen graph.

EXERCISES for Section 2.3

2.3.1 For each of the following graphs, either show that it is isomorphic to one of the five regular graphs of Example 2.3.4, or argue why it is not. Are they isomorphic to each other?

2.3.2 Determine whether the graphs in the given pair are isomorphic.

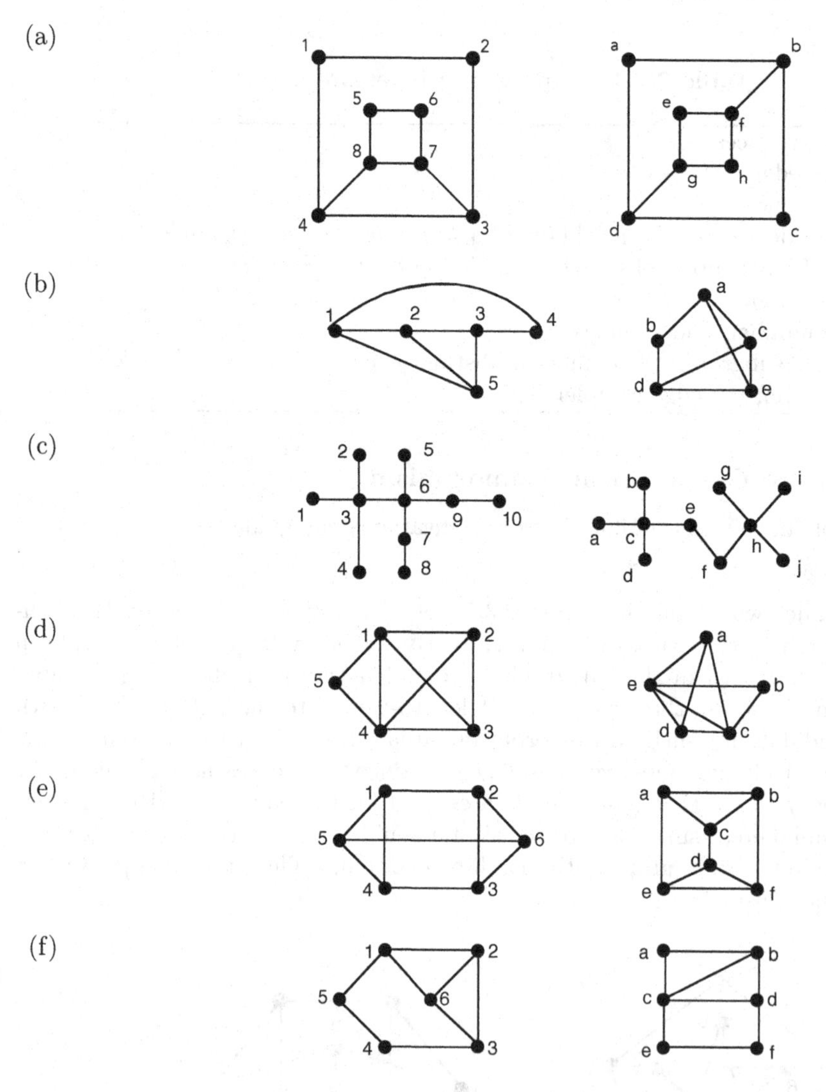

2.3.3 Suppose that $f : G \to H$ is a graph isomorphism and $v \in V_G$. Prove the following isomorphism properties.

(a) The local subgraph of v is isomorphic to the local subgraph of $f(v)$.

(b) $G - v$ is isomorphic to $H - f(v)$.

2.3.4 State and prove digraph versions of the specified assertion.

(a) Theorem 2.3.1.
(b) Theorem 2.3.2.
(c) Corollary 2.3.4.
(d) Corollary 2.3.5.
(e) Theorem 2.3.6.
(f) Theorem 2.3.9.

2.3.5 Compile a table of digraph-isomorphism invariants analogous to Table 2.3.1.

2.3.6 Determine whether the digraphs in the given pair are isomorphic.

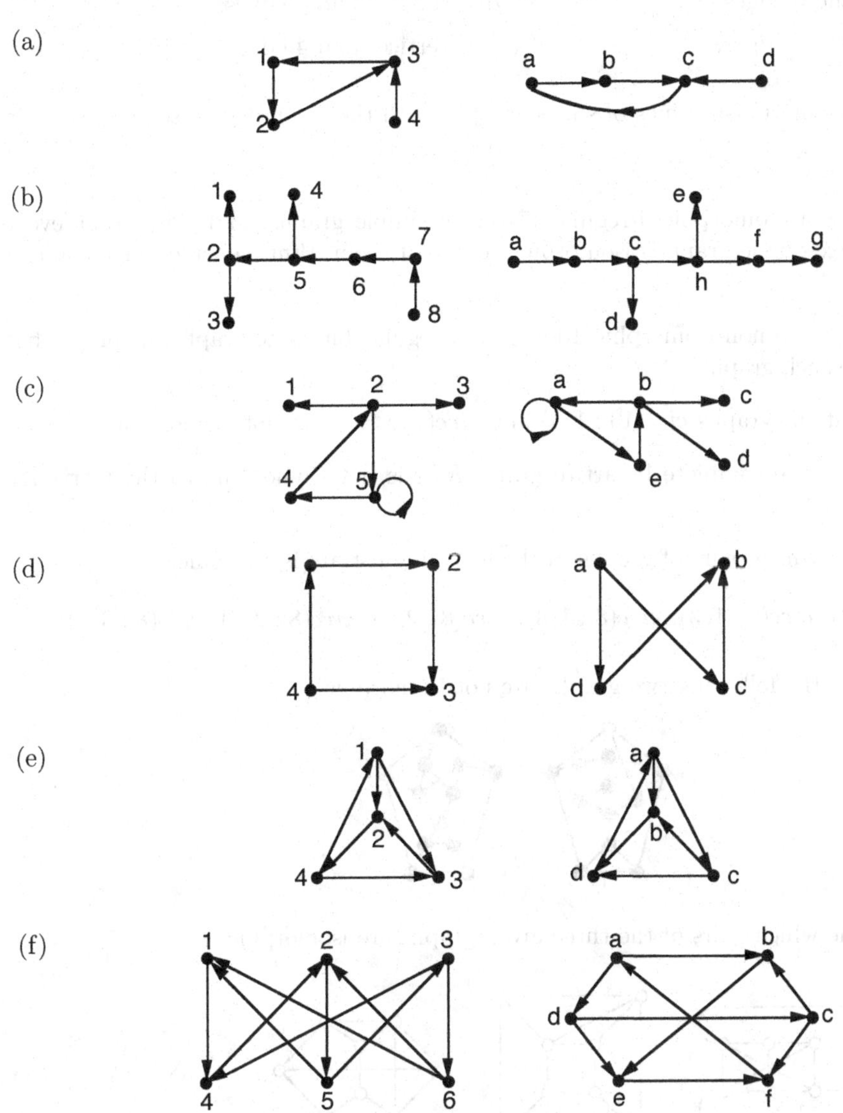

DEFINITION: A simple graph G is ***self-complementary*** if $G \cong \overline{G}$.

2.4.1 Prove that if G is an n-vertex self-complementary graph, then either $n \equiv 0 \pmod 4$ or $n \equiv 1 \pmod 4$.

2.4.2 Can a self-complementary graph be non-connected?

2.4.3 Suppose that a self-complementary graph has $4k+1$ vertices. Prove that it must have at least one vertex of degree $2k$.

2.4.4 Draw all isomorphism types of general graphs with
(a) 2 edges and no isolated vertices. (b) 2 vertices and 3 edges.

2.4.5 Draw all isomorphism types of digraphs with
(a) 2 edges and no isolated vertices. (b) 2 vertices and 3 arcs.

2.4.6 Draw all the isomorphism types of simple graphs with
(a) 6 vertices and 3 edges. (b) 6 vertices and 4 edges.
(c) 5 vertices and 5 edges. (d) 7 vertices and 4 edges.

2.4.7 Draw the isomorphism types of simple graphs with the given degree sequence.
(a) $\langle 433222 \rangle$ (b) $\langle 333331 \rangle$

2.4.8 Draw two non-isomorphic 4-regular, 7-vertex simple graphs, and prove that every 4-regular, 7-vertex simple graph is isomorphic to one of them. Hint: consider edge complements.

2.4.9 Either draw two non-isomorphic 10-vertex, 4-regular bipartite graphs, or prove that there is only one such graph.

2.4.10 Prove that the graphs $circ(13 : 1, 4)$ and $circ(13 : 1, 5)$ are not isomorphic.

2.4.11 Prove that the complete bipartite graph $K_{4,4}$ is not isomorphic to the Cartesian product graph $K_4 \times K_2$.

2.4.12 Determine which pairs of graphs in the following list are isomorphic.

$$circ(8 : 1, 2),\ circ(8 : 1, 3),\ circ(8 : 1, 4),\ circ(8 : 2, 3),\ circ(8 : 2, 4),\ circ(8 : 3, 4)$$

2.4.13 Show that the following two graphs are not isomorphic.

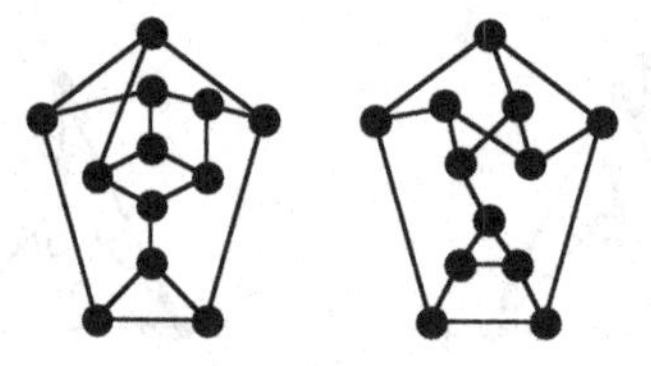

2.4.14 Determine which pairs of the three given graphs are isomorphic.

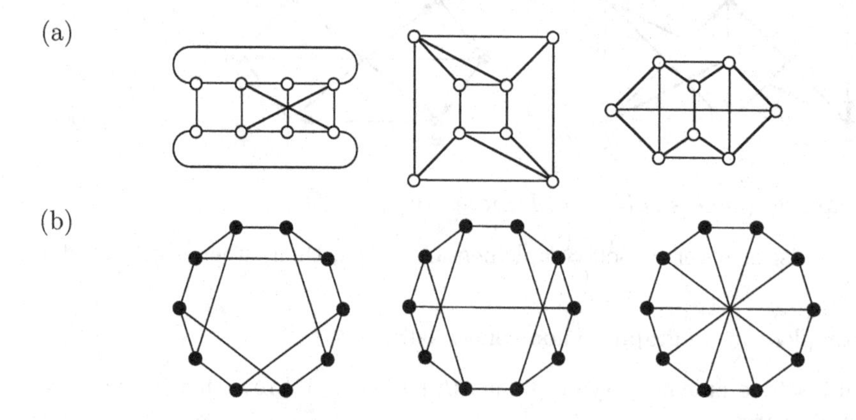

(c)

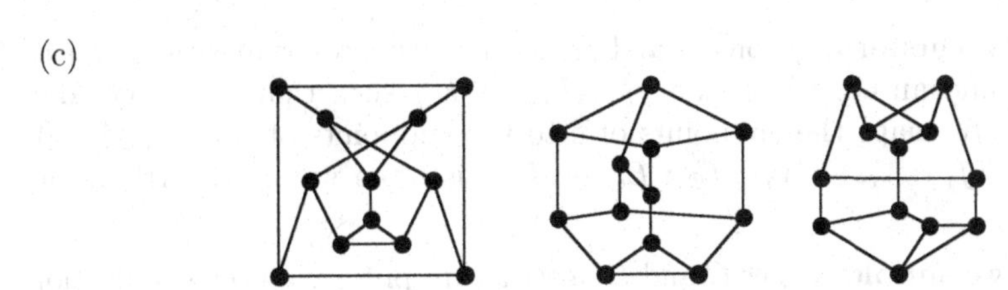

2.4.15 List the vertex orbits and the edge orbits of the given graph.

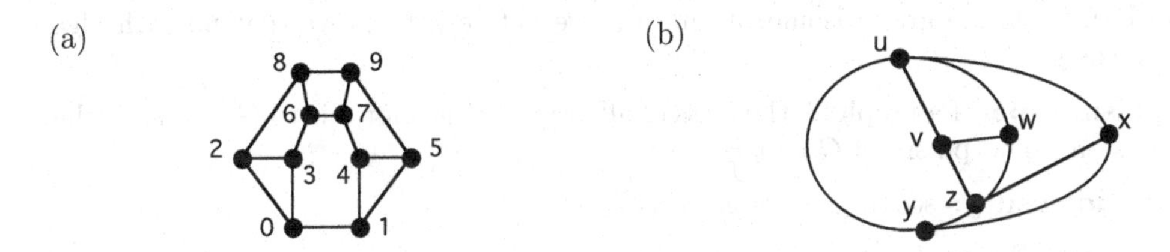

2.4.16 Draw each isomorphism type for a 4-vertex, simple graph with exactly two vertex orbits.

2.4.17 Prove that the hypercube Q_n is vertex transitive.

GLOSSARY

adjacency-preserving vertex function f $V_G \rightarrow V_H$ between two simple graphs G and H: for every pair of adjacent vertices u and v in graph G, the vertices $f(u)$ and $f(v)$ are adjacent in graph H.

—, non- : $f(u)$ and $f(v)$ are nonadjacent in H whenever u and v are nonadjacent in G.

automorphism of a graph G: an isomorphism from the graph to itself, that is, a structure-preserving permutation π_V on V_G and a consistent permutation π_E on E_G; often written as $\pi = (\pi_V, \pi_E)$.

digraph invariant: a property of a digraph that is preserved by isomorphisms.

disjoint cycle form of a permutation: a notation for specifying a permutation; see Example 2.2.1.

edge orbit of a graph G: an edge subset $F \subseteq E_G$ such that for every pair of edges d, $e \in F$, there is an automorphism of G that maps d to e. Thus, an edge orbit is an equivalence class of E_G under the action of the automorphisms of G.

edge-transitive graph: a graph G such that for every edge pair d, $e \in E_G$, there is an automorphism of G that maps d to e.

graph invariant: a property of a graph that is preserved by isomorphisms.

invariant: a shortened term for graph (or digraph) invariant.

isomorphic digraphs: two digraphs that have an isomorphism between their underlying graphs that preserves the direction of each edge.

isomorphic graphs: two graphs G and H that have a structure-preserving vertex bijection between them; denoted $G \cong H$.

isomorphism between two **general** graphs G and H: a structure-preserving vertex bijection $f_V : V_G \to V_H$ and an edge bijection $f_E : E_G \to E_H$ such that for every edge $e \in E_G$, the function f_V maps the endpoints of e to the endpoints of the edge $f_E(e)$. Such a mapping pair $(f_V : V_G \to V_H,\ f_E : E_G \to E_H)$ is often written shorthand as $f : G \to H$.

isomorphism between two **simple** graphs G and H: a structure-preserving vertex bijection $f : V_G \to V_H$

isomorphism problem for graphs: the unsolved problem of devising a practical general algorithm to decide graph isomorphism, or, alternatively, to prove that no such algorithm exists.

isomorphism type of a graph G: the class of all graphs H isomorphic to G, i.e., such that there is an isomorphism of G and H.

preserves adjacency: see *adjacency-preserving*.

preserves nonadjacency: see *adjacency-preserving, non-*.

self-complementary graph: a simple graph G such that $G \cong \overline{G}$.

structure-preserving vertex bijection between two general graphs G and H: a vertex bijection $f : V_G \to V_H$ such that

(i) the number of edges (even if 0) joining each pair of distinct vertices u and v in G equals the number of edges joining their images $f(u)$ and $f(v)$ in H, and

(ii) the number of self-loops at each vertex x in G equals the number of self-loops at the vertex $f(x)$ in H.

structure-preserving vertex bijection between two simple graphs G and H: a vertex bijection $f : V_G \to V_H$ that preserves adjacency and nonadjacency, i.e., for every pair of vertices in G,

$$u \text{ and } v \text{ are adjacent in } G \Leftrightarrow f(u) \text{ and } f(v) \text{ are adjacent in } H$$

vertex orbit of a graph G: a vertex subset $W \subseteq V_G$ such that for every pair of vertices $u, v \in W$, there is an automorphism of G that maps u to v. Thus, a vertex orbit is an equivalence class of V_G under the action of the automorphisms of G.

vertex-transitive graph: a graph G such that for every vertex pair $u, v \in V_G$, there is an automorphism of G that maps u to v.

Chapter 3

TREES AND CONNECTIVITY

INTRODUCTION

Trees are important to the structural understanding of graphs and to the algorithmics of information processing, and given their central role in the design and analysis of connected networks, they are the backbone of optimally connected networks. A main task in information management is deciding how to store data in space-efficient ways that also allow their retrieval and modification to be time-efficient, and tree-based structures are often the best way of balancing these competing goals.

Spanning trees capture the connectedness of a graph in the most efficient way, and they provide a foundation for a systematic analysis of the cycle structure of a graph. Mathematicians regard the algebraic structures underlying the collection of cycles and edge-cuts of a graph as beautiful in their own right. Establishing connections between linear algebra and graph theory (in §3.3) provides some powerful analytical tools for understanding a graph's structure.

Some connected graphs are "more connected" than others. That is, a connected graph's vulnerability to disconnection by edge- or vertex-deletion varies. Two numerical parameters, vertex-connectivity and edge-connectivity, are useful in measuring a graph's connectedness. Intuitively, a network's vulnerability should be closely related also to the number of alternative paths between each pair of nodes. There is a rich body of mathematical results concerning this relationship, many of which are variations of a classical result of Menger, and some of these extend well beyond graph theory.

3.1 CHARACTERIZATIONS AND PROPERTIES OF TREES

REVIEW FROM §1.2 : A ***tree*** is a connected graph with no cycles.

Characterizing trees in a variety of ways provides flexibility for their application. The first part of this section establishes some basic properties of trees that culminate in Theorem 3.1.8, where six different but equivalent characterizations of a tree are given.

Basic Properties of Trees

DEFINITION: A vertex of degree 1 in a tree is called a ***leaf***.

If a leaf is deleted from a tree, then the resulting graph is a tree having one vertex fewer. Thus, induction is a natural approach to proving tree properties, provided one can always find a leaf. The following proposition guarantees the existence of such a vertex.

Proposition 3.1.1: *Every tree with at least one edge has at least two leaves.*

Proof: Let $P = (v_1, v_2, \cdots, v_m)$ be a path of maximum length in a tree T. Suppose one of its endpoints, say v_1, has degree greater than 1. Then v_1 is adjacent to vertex v_2 on path P and also to some other vertex w (see Figure 3.1.1). If w is different from all of the vertices v_i, then P could be extended, contradicting its maximality. On the other hand, if w is one of the vertices v_i on the path, then the acyclic property of T would be contradicted. Thus, both endpoints of path P must be leaves in tree T. ◇

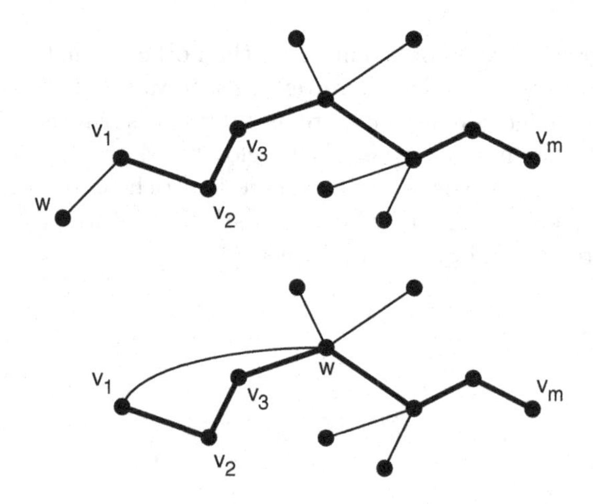

Figure 3.1.1 The two cases in the proof of Proposition 3.1.1.

Corollary 3.1.2: *If the degree of every vertex of a graph is at least 2, then that graph must contain a cycle.*

Proof: Apply Proposition 3.1.1 to any one of the components of that graph. ◇

The next proposition establishes a fundamental property of trees. Its proof is the first of several instances that demonstrate the effectiveness of the inductive approach to proving assertions about trees.

Proposition 3.1.3: *Every tree on n vertices contains exactly $n-1$ edges.*

Proof: A tree on one vertex is the trivial tree, which has no edges.

Assume for some number $k \geq 1$, as an induction hypothesis, that every tree on k vertices has exactly $k-1$ edges. Next consider any tree T on $k+1$ vertices. By Proposition 3.1.1, T contains a leaf, say v. Then the graph $T-v$ is acyclic, since deleting pieces from an acyclic graph cannot create a cycle. Moreover, $T-v$ is connected, since the vertex v has degree 1 in T. Thus, $T-v$ is a tree on k vertices, and, hence, $T-v$ has $k-1$ edges, by the induction hypothesis. But since $deg(v)=1$, it follows that $T-v$ has one edge fewer than T. Therefore, T has k edges, which completes the proof. ♢

REVIEW FROM §1.2: An acyclic graph is called a ***forest***.

REVIEW FROM §1.3: The number of components of a graph G is denoted $c(G)$.

Corollary 3.1.4: *A forest G on n vertices has $n - c(G)$ edges.*

Proof: Apply Proposition 3.1.3 to each of the components of G. ♢

Corollary 3.1.5: *Any graph G on n vertices has at least $n - c(G)$ edges.*

Proof: If G has cycle-edges, then remove them one at a time until the resulting graph $\widehat{G}$ is acyclic. Then $\widehat{G}$ has $n - c(\widehat{G})$ edges, by Corollary 3.1.4; and $c(\widehat{G}) = c(G)$, by Corollary 1.3.7. ♢

Corollary 3.1.5 provides a lower bound for the number of edges in a graph. The next two results establish an upper bound for certain simple graphs. This kind of result is typically found in *extremal graph theory.*

Proposition 3.1.6: *Let G be a simple graph with n vertices and k components. If G has the maximum number of edges among all such graphs, then*

$$|E_G| = \binom{n-k+1}{2}$$

Proof: Since the number of edges is maximum, each component of G is a complete graph. If $n=k$, then G consists of k isolated vertices, and the result is trivially true. If $n>k$, then G has at least one nontrivial component. We show that G has exactly one nontrivial component. Suppose, to the contrary, that $C_1 = K_r$ and $C_2 = K_s$, where $r \geq s \geq 2$. Then the total number of edges contained in these two components is $\binom{r}{2}+\binom{s}{2}$. However, the graph that results from replacing C_1 and C_2 by K_{r+1} and K_{s-1}, respectively, has $\binom{r+1}{2}+\binom{s-1}{2}$ edges in those two components. By expanding these formulas, it is easy to show that the second graph has $r-s+1$ more edges than the first, contradicting the maximality of the first graph. Thus, G consists of $k-1$ isolated vertices and a complete graph on $n-k+1$ vertices, which shows that $|E_G| = \binom{n-k+1}{2}$ and completes the proof. ♢

Corollary 3.1.7: *A simple n-vertex graph with more than $\binom{n-1}{2}$ edges must be connected.* ♢

Six Different Characterizations of a Tree

Trees have many possible characterizations, and each contributes to the structural understanding of graphs in a different way. The following theorem establishes some of the most useful characterizations.

Theorem 3.1.8: *Let T be a graph with n vertices. Then the following statements are equivalent.*

1. *T is a tree.*
2. *T contains no cycles and has $n-1$ edges.*
3. *T is connected and has $n-1$ edges.*
4. *T is connected, and every edge is a cut-edge.*
5. *Any two vertices of T are connected by exactly one path.*
6. *T contains no cycles, and for any new edge e, the graph $T+e$ has exactly one cycle.*

Proof: If $n=1$, then all six statements are trivially true. So assume $n \geq 2$.

($1 \Rightarrow 2$) By Proposition 3.1.3.

($2 \Rightarrow 3$) Suppose that T has k components. Then, by Corollary 3.1.4, the forest T has $n-k$ edges. Hence, $k=1$.

($3 \Rightarrow 4$) Let e be an edge of T. Since $T-e$ has $n-2$ edges, Corollary 3.1.5 implies that $n-2 \geq n-c(T-e)$. So $T-e$ has at least two components.

($4 \Rightarrow 5$) By way of contradiction, suppose that Statement 4 is true, and let x and y be two vertices that have two different paths between them, say P_1 and P_2. Let u be the first vertex from which the two paths diverge (this vertex might be x), and let v be the first vertex at which the paths meet again (see Figure 3.1.2). Then these two $u-v$ paths taken together form a cycle, and any edge on this cycle is not a cut-edge, which contradicts Statement 4.

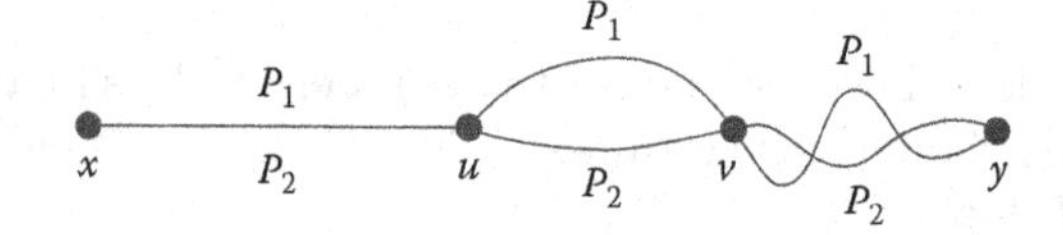

Figure 3.1.2

($5 \Rightarrow 6$) T contains no cycles since any two vertices on a cycle have two different paths between them, consisting of the opposite routes around the cycle. Furthermore, the addition of any new edge e to T will create a cycle, since the endpoints of e, say u and v, are already connected by a path in T. To show that this cycle is unique, suppose two cycles were created. They both would contain edge e, and the long way around each of these cycles would then be two different $u-v$ paths in T, contradicting Statement 5.

($6 \Rightarrow 1$) By way of contradiction, assume that T is not connected. Then the addition of an edge joining a vertex in one component to a vertex in a different component would not create a cycle, which would contradict Statement 6. ♢

The fundamental properties of a tree are summarized in Table 3.1.1 for easy reference.

Table 3.1.1: Summary of Basic Properties of a Tree T on n vertices

1.	T is connected.
2.	T contains no cycles.
3.	Given any two vertices u and v of T, there is a unique $u-v$ path.
4.	Every edge in T is a cut-edge.
5.	T contains $n-1$ edges.
6.	T contains at least two vertices of degree 1 if $n \geq 2$.
7.	Adding an edge between two vertices of T yields a graph with exactly one cycle.

Trees as Subgraphs

An arbitrary graph is likely to contain a number of different trees as subgraphs. The following theorem shows that if a simple n-vertex graph is sufficiently *dense* (i.e., it has sufficiently many edges), then it will contain every type of tree up to a certain order.

REVIEW FROM §1.1: The minimum degree of the vertices of a graph G is denoted $\delta_{\min}(G)$.

Theorem 3.1.9: *Let T be any tree on n vertices, and let G be a simple graph such that $\delta_{\min}(G) \geq n - 1$. Then T is a subgraph of G.*

Proof: The result is clearly true if $n = 1$ or $n = 2$, since K_1 and K_2 are subgraphs of every graph having at least one edge.

Assume that the result is true for some $n \geq 2$. Let T be a tree on $n + 1$ vertices, and let G be a graph with $\delta_{\min}(G) \geq n$. We show that T is a subgraph of G.

Let v be a leaf of T, and let vertex u be its only neighbor in T. The vertex-deletion subgraph $T - v$ is a tree on n vertices, so, by the induction hypothesis, $T - v$ is a subgraph of G. Since $deg_G(u) \geq n$, there is some vertex w in G but not in $T - v$ that is adjacent to u. Let e be the edge joining vertices u and w (see Figure 3.1.3). Then $T - v$ together with vertex w and edge e form a tree that is isomorphic to T and is a subgraph of G. ♢

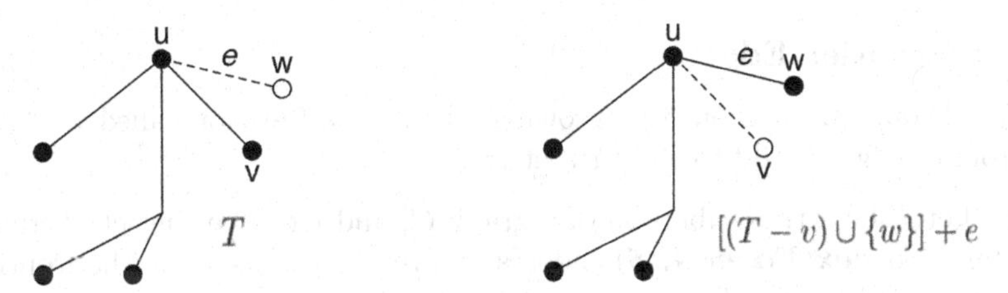

Figure 3.1.3 T **is isomorphic to** $[(T - v) \cup \{w\}] + e$.

Tree Growing

Several different problem-solving algorithms involve growing a spanning tree, one edge and one vertex at a time. All these techniques are refinements and extensions of the same basic tree-growing scheme given in this section.

TERMINOLOGY: For a given tree T in a graph G, the edges and vertices of T are called ***tree edges*** and ***tree vertices***, and the edges and vertices of G that are not in T are called ***non-tree edges*** and ***non-tree vertices***.

Frontier Edges

DEFINITION: A ***frontier edge*** for a given tree T in a graph is a non-tree edge with one endpoint in T, called its ***tree endpoint***, and one endpoint not in T, its ***non-tree endpoint***.

Example 3.1.1: For the graph in Figure 3.1.4, the tree edges of a tree T are drawn in bold. The tree vertices are black, and the non-tree vertices are white. The frontier edges for T, appearing as dashed lines, are edges a, b, c, and d. The plain edges are the non-tree edges that are not frontier edges for T.

Observe that when any one of the frontier edges in Figure 3.1.4 is added to the tree T, the resulting subgraph is still a tree. This property holds in general, as we see in the next proposition, and its iterative application forms the core of the tree-growing scheme.

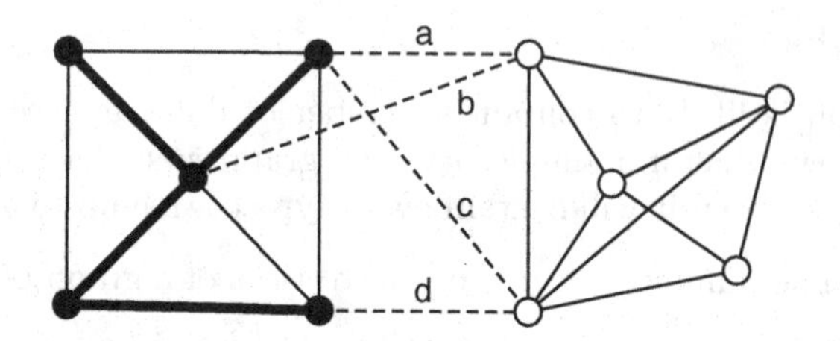

Figure 3.1.4 A tree with frontier edges $a, b, c,$ and d.

Proposition 3.1.10: *Let T be a tree in a graph G, and let e be a frontier edge for T. Then the subgraph of G formed by adding edge e to tree T is a tree.*

Proof: The addition of the frontier edge e to the tree T cannot create a cycle, since one of its endpoints is outside of T. Moreover, the vertex that was added to the tree is clearly reachable from any other vertex in the resulting tree. ◇

Remark: Formally, adding a frontier edge to a tree involves adding a new vertex to the tree, as well as the primary operation of *adding an edge*, defined in §1.4.

Choosing a Frontier Edge

An essential component of our tree-growing scheme is a function called *nextEdge*, which chooses a frontier edge to add to the current tree.

DEFINITION: Let T be a tree subgraph of a graph G, and let S be the set of frontier edges for T. The function ***nextEdge(G, S)*** chooses and returns as its value the frontier edge in S that is to be added to tree T.

Remark: Ordinarily, the function *nextEdge* is deterministic; however, it may also be randomized. In either case, the full specification of *nextEdge* must ensure that it always returns a frontier edge.

DEFINITION: After a frontier edge is added to the current tree, the procedure ***updateFrontier(G,S)*** removes from S those edges that are no longer frontier edges and adds to S those that have become frontier edges.

Example 3.1.1 continued: The current set S of frontier edges shown in the top half of Figure 3.1.5 is $S = \{a, b, c, d\}$. Suppose that the value of $nextEdge(G, S)$ is edge c. The bottom half of Figure 3.1.5 shows the result of adding edge c to tree T and applying $updateFrontier(G, S)$. Notice that $updateFrontier(G, S)$ added four new frontier edges to S and removed edges c and d, which are no longer frontier edges.

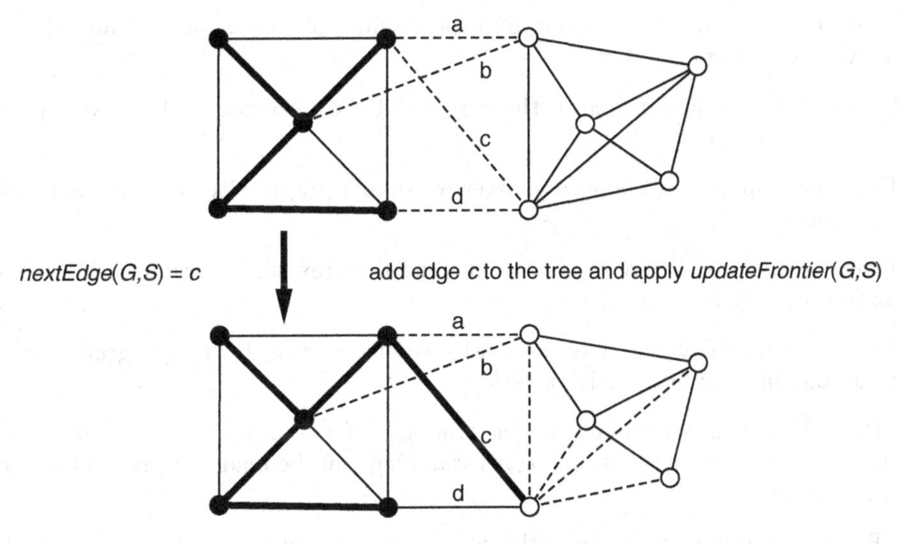

Figure 3.1.5 Result after adding edge c to the tree.

> **Algorithm 3.1.1: Tree-Growing**
> *Input*: a connected graph G, a starting vertex $v \in V_G$, and a function *nextEdge*.
> *Output*: an ordered spanning tree T of G with root v.
> Initialize tree T as vertex v.
> Initialize S as the set of proper edges incident on v.
> While $S \neq \emptyset$
> Let $e = nextEdge(G, S)$.
> Let w be the non-tree endpoint of edge e.
> Add edge e and vertex w to tree T.
> $updateFrontier(G, S)$.
> Return tree T.

EXERCISES for Section 3.1

3.1.1 Draw a 6-vertex connected graph that has exactly seven edges and exactly three cycles.

3.1.2 Draw a 12-vertex forest that has exactly 10 edges.

3.1.3 Either draw the required graph or explain why no such graph exists.

(a) A 7-vertex, 3-component, simple graph with 10 edges.

(b) A 6-vertex, 2-component, simple graph with 11 edges.

(c) An 8-vertex, 2-component, simple graph with exactly nine edges and three cycles.

(d) An 8-vertex, 2-component, simple graph with exactly 10 edges and three cycles.

(e) A 10-vertex, 2-component, forest with exactly nine edges.

(f) An 11-vertex, simple, connected graph with exactly 14 edges that contains five edge-disjoint cycles.

3.1.4 Let G be a connected simple graph on n vertices. Determine a lower bound on the average degree of a vertex, and characterize those graphs that achieve the lower bound.

3.1.5 Prove that if G is a tree having an even number of edges, then G must contain at least one vertex having even degree.

3.1.6 Suppose the average degree of the vertices of a connected graph is exactly 2. How many cycles does G have?

3.1.7 Prove or disprove: A connected n-vertex simple graph with n edges must contain exactly one cycle.

3.1.8 Prove or disprove: There exists a connected n-vertex simple graph with $n+2$ edges that contains exactly two cycles.

3.1.9 Prove or disprove: There does not exist a connected n-vertex simple graph with $n+2$ edges that contains four edge-disjoint cycles.

3.1.10 Prove that H is a maximal acyclic subgraph of a connected graph G if and only if H is a spanning tree of G. What analogous statement can be made for graphs that are not necessarily connected?

3.1.11 Prove that if a graph has exactly two vertices of odd degree, then there must be a path between them.

3.1.12 Show that any nontrivial simple graph contains at least two vertices that are not cut-vertices.

3.1.13 Let G be a simple graph on n vertices. Prove that if $\delta_{\min}(G) \geq \frac{n-1}{2}$, then G is connected.

3.1.14 Prove that if any single edge is added to a connected graph G, then at least one cycle is created.

3.1.15 Prove that a graph G is a forest if and only if every induced subgraph of G contains a vertex of degree 0 or 1.

3.1.16 Characterize those graphs with the property that every connected subgraph is an induced subgraph.

3.1.17 Indicate the set of frontier edges for the given tree.

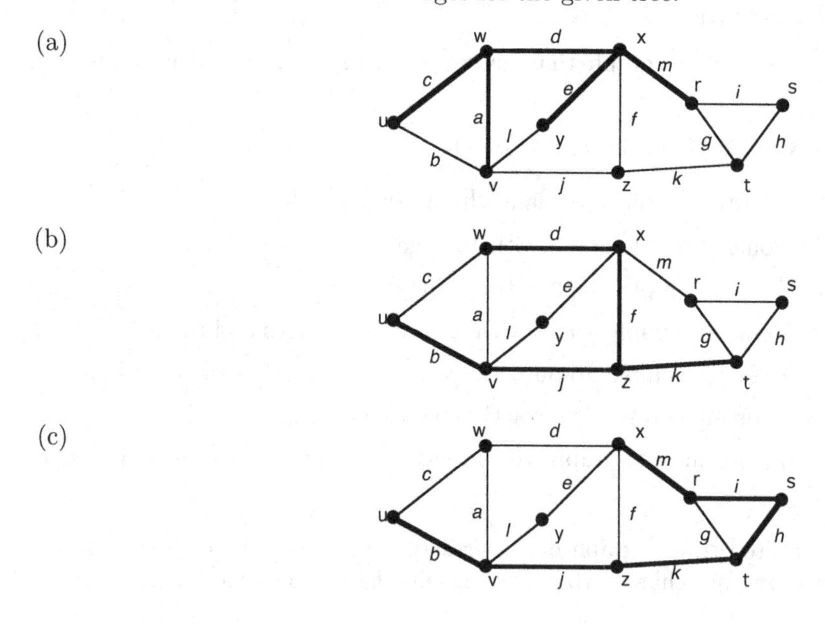

3.2 CYCLES, EDGE-CUTS, AND SPANNING TREES

This section establishes a number of properties that highlight the relationship and a certain duality between the edge-cuts and cycles of a graph.

Proposition 3.2.1: *A graph G is connected if and only if it has a spanning tree.*

Proof: If G is connected, then among the connected, spanning subgraphs of G, there is at least one, say T, with the least number of edges. If subgraph T contained a cycle, then the deletion of one of its cycle-edges would create a smaller connected, spanning subgraph, contradicting the minimality of T. Thus, T is acyclic and therefore, a spanning tree of graph G.

Conversely, if G contains a spanning tree, then every pair of vertices is connected by a path in the tree, and, hence, G is connected. ♢

Proposition 3.2.2: *A subgraph H of a connected graph G is a subgraph of some spanning tree if and only if H is acyclic.*

Proof: *Necessity* ($\Rightarrow$) A subgraph of a tree is acyclic by definition.

Sufficiency ($\Leftarrow$) Let H be an acyclic subgraph of G, and let T be any spanning tree of G. Consider the connected, spanning subgraph G_1, where $V_{G_1} = V_T \cup V_H$ and $E_{G_1} = E_T \cup E_H$. If G_1 is acyclic, then every edge of H must already be an edge of the spanning tree T, since otherwise a cycle would have been created (by the characterization theorem in §3.1). Thus, subgraph H is contained in tree T, and we are done. So suppose alternatively that G_1 has a cycle C_1. Since H is acyclic, it follows that some edge e_1 of cycle C_1 is not in H. Then the graph $G_2 = G_1 - e_1$ is still a connected, spanning subgraph of G and still contains H as a subgraph. If G_2 is acyclic, then it is the required spanning tree. Otherwise, repeat the process of removing a cycle-edge not in H until a spanning tree is obtained. ♢

Partition-Cuts and Minimal Edge-Cuts

REVIEW FROM §1.3: An ***edge-cut*** S in a graph G is a set of edges such that the edge-deletion subgraph $G - S$ has more components than G.

The following definition is closely linked to the concept of a minimal edge-cut. It is used in this and the next section.

DEFINITION: Let G be a graph, and let X_1 and X_2 form a partition of V_G. The set of all edges of G having one endpoint in X_1 and the other endpoint in X_2 is called a ***partition-cut*** of G and is denoted $\langle X_1, X_2 \rangle$.

The next two propositions make explicit the relationship between partition-cuts and minimal edge-cuts.

Proposition 3.2.3: *Let $\langle X_1, X_2 \rangle$ be a partition-cut of a connected graph G. If the subgraphs of G induced by the vertex-sets X_1 and X_2 are connected, then $\langle X_1, X_2 \rangle$ is a minimal edge-cut.*

Proof: The partition-cut $\langle X_1, X_2 \rangle$ is an edge-cut of G, since X_1 and X_2 lie in different components of $G - \langle X_1, X_2 \rangle$. Let S be a proper subset of $\langle X_1, X_2 \rangle$, and let edge $e \in \langle X_1, X_2 \rangle - S$. By definition of $\langle X_1, X_2 \rangle$, one endpoint of e is in X_1 and the other endpoint is in X_2. Thus, if the subgraphs induced by the vertex-sets X_1 and X_2 are connected, then

$G - S$ is connected. Therefore, S is not an edge-cut of G, which implies that $\langle X_1, X_2 \rangle$ is a minimal edge-cut. ◇

Proposition 3.2.4: *Let S be a minimal edge-cut of a connected graph G, and let X_1 and X_2 be the vertex-sets of the two components of $G - S$. Then $S = \langle X_1, X_2 \rangle$.*

Proof: By minimality, $S \subseteq \langle X_1, X_2 \rangle$. If $e \in \langle X_1, X_2 \rangle - S$, then its endpoints would lie in different components of $G - \langle X_1, X_2 \rangle$, contradicting the definition of S as an edge-cut. ◇

Remark: The premise of Proposition 3.2.4 assumes that the removal of a minimal edge-cut from a connected graph creates *exactly* two components. This is a generalization of Corollary 1.3.7 and can be argued similarly, using the minimality condition (see Exercises).

Proposition 3.2.5: *A partition-cut $\langle X_1, X_2 \rangle$ in a connected graph G is a minimal edge-cut of G or a union of edge-disjoint minimal edge-cuts.*

Proof: Since $\langle X_1, X_2 \rangle$ is an edge-cut of G, it must contain a minimal edge-cut, say S_1. If $\langle X_1, X_2 \rangle \neq S_1$, then let $e \in \langle X_1, X_2 \rangle - S_1$, where the endpoints v_1 and v_2 of e lie in X_1 and X_2, respectively. Since S_1 is a minimal edge-cut, the X_1-endpoints of S_1 are in one of the components of $G - S_1$, and the X_2-endpoints of S_1 are in the other component. Furthermore, v_1 and v_2 are in the same component of $G - S_1$ (since $e \in G - S_1$). Suppose, without loss of generality, that v_1 and v_2 are in the same component as the X_1-endpoints of S_1, as shown below.

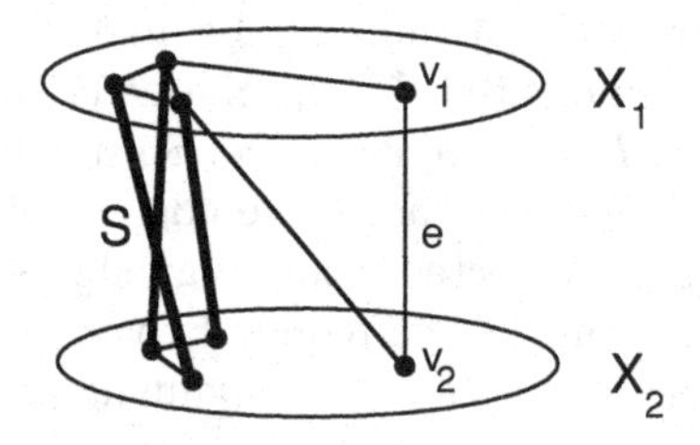

Then every path in G from v_1 to v_2 must use at least one edge of $\langle X_1, X_2 \rangle - S_1$. Thus, $\langle X_1, X_2 \rangle - S_1$ is an edge-cut of G and, hence, contains a minimal edge-cut S_2. Applying the same argument, $\langle X_1, X_2 \rangle - (S_1 \cup S_2)$ either is empty or is an edge-cut of G. Eventually, the process ends with $\langle X_1, X_2 \rangle - (S_1 \cup S_2 \cup \ldots S_r) = \emptyset$, where the S_i are edge-disjoint minimal edge-cuts of G. ◇

Fundamental Cycles and Fundamental Edge-Cuts

DEFINITION: Let G be a graph with $c(G)$ components. The ***edge-cut rank*** of G is the number of edges in a full spanning forest of G. Thus, by Corollary 3.1.4, the edge-cut rank equals $|V_G| - c(G)$.

REVIEW FROM §1.4: Let H be a fixed subgraph of a graph G. The ***relative complement*** of H (***in*** G), denoted $G - H$, is the edge-deletion subgraph $G - E(H)$.

DEFINITION: Let G be a graph with $c(G)$ components. The ***cycle rank*** (or ***Betti number***) of G, denoted $\beta(G)$, is the number of edges in the relative complement of a full spanning forest of G. Thus, the cycle rank is $\beta(G) = |E_G| - |V_G| + c(G)$.

Remark: Observe that *all* of the edges in the relative complement of a spanning forest could be removed without increasing the number of components. Thus, the cycle rank $\beta(G)$ equals the maximum number of edges that can be removed from G without increasing the number of components. Therefore, $\beta(G)$ is a measure of the *edge redundancy* with respect to the graph's connectedness.

DEFINITION: A ***full spanning forest*** of a graph G is a spanning forest consisting of a collection of trees, such that each tree is a spanning tree of a different component of G.

Example 3.2.1: The subgraph shown with bold edges in Figure 3.2.1 is a spanning forest, but is not a full spanning forest. To be a full spanning forest two more edges would need to be added.

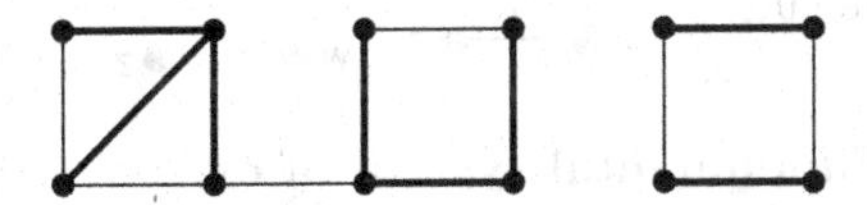

Figure 3.2.1 A spanning forest that is not a full spanning forest.

DEFINITION: Let F be a full spanning forest of a graph G, and let e be any edge in the relative complement of forest F. The cycle in the subgraph $F + e$ (existence and uniqueness guaranteed by Theorem 3.1.8 (item 6)) is called a ***fundamental cycle of*** G (***associated with the spanning forest*** F).

Remark: Each of the edges in the relative complement of a full spanning forest F gives rise to a *different* fundamental cycle.

DEFINITION: The ***fundamental system of cycles*** associated with a full spanning forest F of a graph G is the set of all fundamental cycles of G associated with F.

By the remark above, the cardinality of the fundamental system of cycles of G associated with a given full spanning forest of G is the cycle rank $\beta(G)$.

DEFINITION: Let F be a full spanning forest of a graph G, and let e be any edge of F. Let V_1 and V_2 be the vertex-sets of the two new components of the edge-deletion subgraph $F - e$. Then the partition-cut $\langle V_1, V_2 \rangle$ which is a minimal edge-cut of G by Proposition 3.2.3, is called a ***fundamental edge-cut*** (***associated with*** F).

Remark: For each edge of F, its deletion gives rise to a different fundamental edge-cut.

DEFINITION: The ***fundamental system of edge-cuts*** associated with a full spanning forest F is the set of all fundamental edge-cuts associated with F.

Thus, the cardinality of the fundamental system of edge-cuts associated with a given full spanning forest of G is the edge-cut rank of G.

Remark: If F is a full spanning forest of a graph G, then each of the components of F is a spanning tree of the corresponding component of G. Since the removal or addition of an edge in a general graph affects only one of its components, the definitions of fundamental cycle and fundamental edge-cut are sometimes given in terms of a spanning tree of a connected graph. All of the remaining assertions in this section are stated in terms of a connected graph but can easily be restated for graphs having two or more components.

Example 3.2.2: Figure 3.2.2 shows a fundamental system of cycles and a fundamental system of edge-cuts for a graph G. Both systems are associated with the spanning tree whose edges are drawn in bold. Notice that the fundamental system of edge-cuts does not contain every minimal edge-cut of graph G. For instance, the edge-cut consisting of the three edges incident on vertex v is a minimal one but is not in the fundamental system.

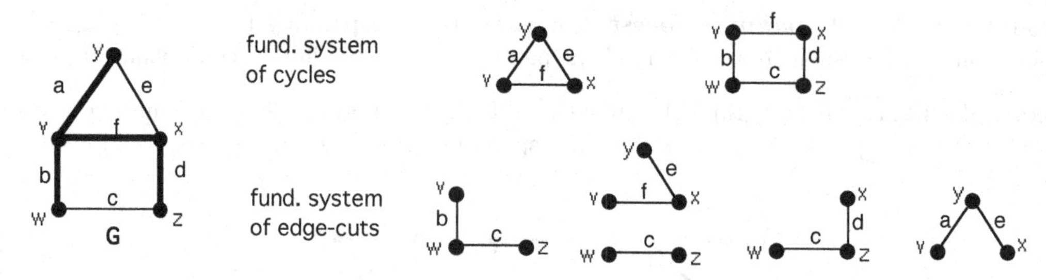

Figure 3.2.2 Fundamental system of cycles and of edge-cuts.

Relationship between Cycles and Edge-Cuts

The next series of propositions reveals a dual relationship between the cycles and minimal edge-cuts of a graph.

Proposition 3.2.6: *Let S be a set of edges in a connected graph G. Then S is an edge-cut of G if and only if every spanning tree of G has at least one edge in common with S.*

Proof: By Proposition 3.2.1, S is an edge-cut if and only if $G - S$ contains no spanning tree of G, which means that every spanning tree of G has at least one edge in common with S. ◊

Proposition 3.2.7: *Let C be a set of edges in a connected graph G. Then C contains a cycle if and only if the relative complement of every spanning tree of G has at least one edge in common with C.*

Proof: By Proposition 3.2.2, edge-set C contains a cycle if and only if C is not contained in any spanning tree of G, which means that the relative complement of every spanning tree of G has at least one edge in common with C. ◊

Proposition 3.2.8: *A cycle and a minimal edge-cut of a connected graph have an even number of edges in common.*

Proof: Let C be a cycle and S be a minimal edge-cut of a connected graph G. Let V_1 and V_2 be the vertex-sets of the two components G_1 and G_2 of $G - S$. Then each edge of S joins a vertex in V_1 to a vertex in V_2. Now consider a traversal of the edges of the cycle C. Without loss of generality, assume that the traversal begins at some vertex in V_1. Then each time the traversal uses an edge in S in moving from V_1 to V_2, it will have to return to V_2 by traversing another edge of S. This is possible only if C and S have an even number of edges in common. ◊

Example 3.2.2 continued: It is easy but tedious to check that each of the three cycles in graph G of Figure 3.2.2 has either zero or two edges in common with each minimal edge-cut of G.

Proposition 3.2.9: *Let T be a spanning tree of a connected graph, and let C be a fundamental cycle with respect to an edge e^* in the relative complement of T. Then the edge-set of cycle C consists of edge e^* and those edges of tree T whose fundamental edge-cuts contain e^*.*

Proof: Let e_1, e_2, ..., e_k be the edges of T that, with e^*, make up the cycle C, and let S_i be the fundamental edge-cut with respect to e_i, $1 \leq i \leq k$.

Edge e_i is the only edge of T common to both C and S_i (by the definitions of C and S_i). By Proposition 3.2.8, C and S_i must have an even number of edges in common, and, hence, there must be an edge in the relative complement of T that is also common to both C and S_i. But e^* is the only edge in the complement of T that is in C. Thus, the fundamental edge-cut S_i must contain e^*, $1 \leq i \leq k$.

To complete the proof, we must show that no other fundamental edge-cuts associated with T contain e^*. So let S be the fundamental edge-cut with respect to some edge b of T, different from e_1, e_2, ..., e_k. Then S does not contain any of the edges e_1, e_2, ..., e_k (by definition of S). The only other edge of cycle C is e^*; so by Proposition 3.2.8, edge-cut S cannot contain e^*. ♢

Example 3.2.2 continued: The 3-cycle in Figure 3.2.2 is the fundamental cycle obtained by adding edge e to the given spanning tree. Of the four fundamental edge-cuts associated with that spanning tree, only the second and fourth ones contain edge e. The tree edges in these two edge-cuts, namely f and a, are the other two edges of the 3-cycle.

The proof of the next proposition uses an argument similar to the one just given and is left as an exercise.

Proposition 3.2.10: *The fundamental edge-cut with respect to an edge e of a spanning tree T consists of e and exactly those edges in the relative complement of T whose fundamental cycles contain e.* ♢
(Exercises)

Eulerian Graphs

This section closes with a characterization of Eulerian graphs that dates back to as early as 1736, when Euler solved and generalized the *Königsberg Bridge Problem.*

REVIEW FROM §1.2: An ***Eulerian tour*** in a graph is a closed trail that contains every edge of that graph. An ***Eulerian graph*** is a graph that has an Eulerian tour.

Theorem 3.2.11: [***Eulerian-Graph Characterization***] *The following statements are equivalent for a connected graph G.*

1. *G is Eulerian.*
2. *The degree of every vertex in G is even.*
3. *E_G is the union of the edge-sets of a set of edge-disjoint cycles of G.*

Proof:

($1 \Rightarrow 2$) Let C be an Eulerian tour of G, and let v be the starting point of some traversal of C. The initial edge and final edge of the traversal contribute 2 toward the degree of v, and each time the traversal passes through a vertex, a contribution of 2 toward that vertex's degree also results. Thus, there is an even number of traversed edges incident with each vertex, and since each edge of G is traversed exactly once, this number is the degree of that vertex.

($2 \Rightarrow 3$) Since G is connected and every vertex has even degree, G cannot be a tree and therefore contains a cycle, say C_1. If C_1 includes all the edges of G, then statement 3 is established. Otherwise, the graph $G_1 = G - E_{C_1}$ has at least one nontrivial component. Furthermore, since the edges that were deleted from G form a cycle, the degree in G_1 of each vertex on C_1 is reduced by 2, and, hence, every vertex of G_1 still has even degree. It follows that G_1 contains a cycle C_2. If all of the edges have been exhausted, then $E_G = E_{C_1} \cup E_{C_2}$. Otherwise, consider $G_2 = G - E_{C_1} - E_{C_2}$, and continue the procedure until all the edges are exhausted. If C_n is the cycle obtained at the last step, then $E_G = E_{C_1} \cup E_{C_2} \cup \ldots \cup E_{C_n}$, which completes the proof of the implication $2 \Rightarrow 3$.

($3 \Rightarrow 1$) Assume that E_G is the union of the edge-sets of m edge-disjoint cycles of G. Start at any vertex v_1 on one of these cycles, say C_1, and consider $T_1 = C_1$ as our first closed trail. There must be some vertex of T_1, say v_2, that is also a vertex on some other cycle, say C_2. Form a closed trail T_2 by *splicing* cycle C_2 into T_1 at vertex v_2. That is, trail T_2 is formed by starting at vertex v_1, traversing the edges of T_1 until v_2 is reached, traversing all the edges of cycle C_2, and completing the closed trail T_2 by traversing the remaining edges of trail T_1. The process continues until all the cycles have been spliced in, at which point T_m is an Eulerian tour of G. ♢

Corollary 3.2.12: *A connected graph G has an open Eulerian trail if and only if it has exactly two vertices of odd degree. Furthermore, the initial and final vertices of an open Eulerian trail must be the two vertices of odd degree.*

Proof: The addition of an edge between the two odd vertices creates an Eulerian graph. The removal of that edge from any Eulerian tour, gives an open Eulerian trail on the original graph, starting at one of the odd vertices and ending at the other. ♢

Königsberg Bridges Problem

In the town of Königsberg in what was once East Prussia, the two branches of the River Pregel converge and flow through to the Baltic Sea. Parts of the town were on an island and a headland that were joined to the outer river banks and to each other by seven bridges, as shown in Figure 3.2.3. The townspeople wanted to know if it was possible to take a walk that crossed each of the bridges exactly once before returning to the starting point. The Prussian emperor, Frederick the Great, brought the problem to the attention of the famous Swiss mathematician Leonhard Euler. Euler's solution in 1736 is generally regarded as the origin of graph theory.

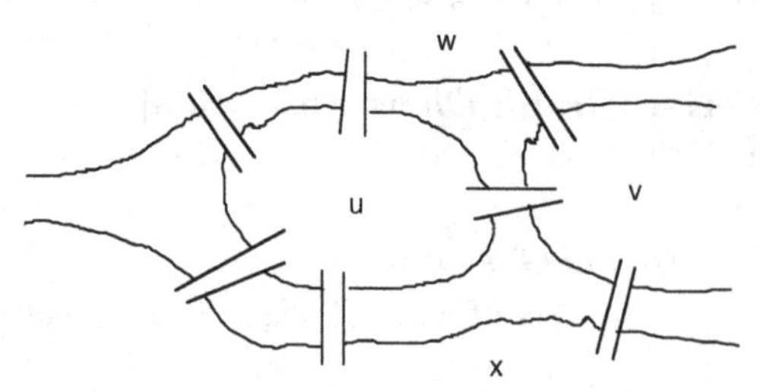

Figure 3.2.3 The seven bridges of Königsberg.

EXERCISES for Section 3.2

3.2.1 For the spanning tree of graph G induced on the given edge-set:

i. Find the associated fundamental system of cycles
ii. Find the associated fundamental system of edge-cuts
iii. Show that Proposition 3.2.6 holds for one of the fundamental edge-cuts.
iv. Show that Proposition 3.2.7 holds for one of the fundamental cycles.
v. Show that the 5-cycle in graph G has an even number of edges in common with each of the fundamental edge-cuts.
vi. Show that Proposition 3.2.9 holds for one of the fundamental cycles.
vii. Show that Proposition 3.2.10 holds for one of the fundamental edge-cuts.

(a) $\{a, e, c, d\}$
(b) $\{a, e, g, d\}$
(c) $\{e, f, c, d\}$
(d) $\{a, b, g, c\}$
(e) $\{e, f, g, d\}$

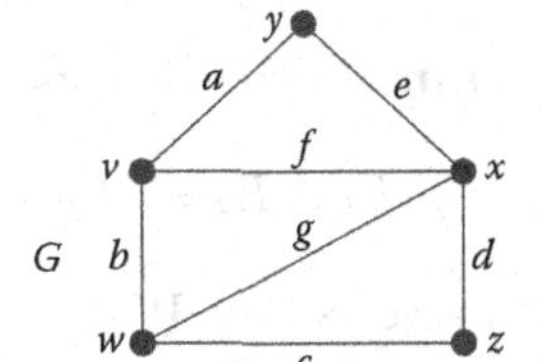

3.2.2 Prove that each edge of a connected graph G lies in a spanning tree of G.

·**3.2.3** Give an alternative proof of Proposition 3.2.2 that avoids the "repeat the process" phrase by considering a connected spanning subgraph of G that contains H, and that has the least number of edges among all such subgraphs.

3.2.4 Prove that in a tree, every vertex of degree greater than 1 is a cut-vertex.

3.2.5 Prove that every nontrivial connected graph contains a minimal edge-cut.

3.2.6 Prove that the removal of a minimal edge-cut from any graph increases the number of components by exactly 1.

3.2.7 Let T_1 and T_2 be two different spanning trees of a graph. Show that if e is any edge in tree T_1, then there exists an edge f in tree T_2 such that $T_1 - e + f$ is also a spanning tree. (Hint: Apply Proposition 3.2.6 to the fundamental edge-cut associated with T_1 and e.)

3.2.8 Prove that a subgraph of a connected graph G is a subgraph of the relative complement of some spanning tree if and only if it contains no edge-cuts of G.

3.2.9 Prove Proposition 3.2.10.

3.2.10 Give a counterexample to show that the assertion of Proposition 3.2.8 no longer holds if the minimality condition is removed.

3.2.11 What is the solution to the Königsberg Bridges Problem?

DEFINITION: The ***line graph*** $L(G)$ of a graph G has a vertex for each edge of G, and two vertices in $L(G)$ are adjacent if and only if the corresponding edges in G are adjacent.

3.2.12 Prove that if a simple graph G is Eulerian, then its line graph $L(G)$ is Eulerian.

3.3 GRAPHS AND VECTOR SPACES

The results of the previous section are used here to define a vector space associated with a graph. Two important subspaces of this vector space, the *edge-cut space* and the *cycle space*, are also identified and studied in detail. Understanding the algebraic structure underlying a graph's cycles and edge-cuts makes it possible to apply powerful and elegant results from linear algebra that lead to a deeper understanding of graphs and of certain graph algorithms. Supporting this claim is an application, appearing later this section, to a problem concerning electrical circuits.

Vector Space of Edge Subsets

NOTATION: For a graph G, let $W_E(G)$ denote the set of all subsets of E_G.

DEFINITION: The ***ring sum*** of two elements of $W_E(G)$, say E_1 and E_2, is defined by

$$E_1 \oplus E_2 = (E_1 - E_2) \cup (E_2 - E_1)$$

The next proposition asserts that $W_E(G)$ under the ring-sum operation forms a vector space over the field of scalars $GF(2)$, where the scalar multiplication $*$ is defined by $1 * S$ = S and $0 * S = \emptyset$ for any S in $W_E(G)$. Its proof is a straightforward verification of each of the defining properties of a vector space and is left as an exercise. (The basic definitions and properties of a vector space and of the finite field $GF(2)$ appear in Appendix A.4.)

Proposition 3.3.1: *$W_E(G)$ is a vector space over $GF(2)$.* ♢ *(Exercises)*

TERMINOLOGY: The vector space $W_E(G)$ is called the **edge space** of G.

Proposition 3.3.2: *Let $E_G = \{e_1, e_2, \ldots, e_m\}$ be the edge-set of a graph G. Then the subsets $\{e_1\}, \{e_2\}, \ldots, \{e_m\}$ form a basis for the edge space $W_E(G)$. Thus, $W_E(G)$ is an m-dimensional vector space over $GF(2)$.*

Proof: If $H = \{e_{i_1}, e_{i_2}, \ldots, e_{i_r}\}$ is any vector in $W_E(G)$, then $H = \{e_{i_1}\} \oplus \{e_{i_2}\} \oplus \cdots \oplus \{e_{i_r}\}$. Clearly, the elements $\{e_1\}, \{e_2\}, \ldots, \{e_m\}$ are also linearly independent. ♢

DEFINITION: Let $s_1, s_2 \ldots, s_n$ be any sequence of objects, and let A be a subset of $S = \{s_1, s_2 \ldots, s_n\}$. The ***characteristic vector*** of the subset A is the n-tuple whose jth component is 1 if $s_j \in A$ and 0 otherwise.

A general result from linear algebra states that every m-dimensional vector space over a given field F is isomorphic to the vector space of m-tuples over F. For the vector space $W_E(G)$, this result may be realized in the following way. If $E_G = \{e_1, e_2, \ldots, e_m\}$, then the mapping *charvec* that assigns to each subset of E_G its *characteristic vector* is an isomorphism from $W_E(G)$ to the vector space of m-tuples over GF(2). Proving this assertion first requires showing that the mapping *preserves the vector-space operations.* If E_1 and E_2 are two subsets of E_G, then the definitions of the ring-sum operator $\oplus$ and mod 2 component-wise addition $+_2$ imply

$$charvec\,(E_1 \oplus E_2) = charvec\,(E_1) +_2 charvec\,(E_2)$$

The remaining details are left to the reader.

Remark: Each subset E_i of E_G uniquely determines a subgraph of G, namely, the edge-induced subgraph $G_i = G(E_i)$. Thus, the vectors of $W_E(G)$ may be viewed as the edge-induced subgraphs G_i instead of as the edge subsets E_i. Accordingly, $1 * G_i = G_i$ and $0 * G_i = \emptyset$, where G_i is any edge-induced subgraph of G and where $\emptyset$ now refers to the *null graph* (with no vertices and no edges). This vector space will still be denoted $W_E(G)$.

Furthermore, the only subgraphs of a graph that are *not* edge-induced subgraphs are those that contain isolated vertices. Since isolated vertices play no role in the discussions in this section and in the section that follows, the adjective "edge-induced" will no longer be used when referring to the elements of the edge space $W_E(G)$.

The Cycle Space of a Graph

DEFINITION: The ***cycle space*** of a graph G, denoted $W_C(G)$, is the subset of the edge space $W_E(G)$ consisting of the null set (graph) $\emptyset$, all cycles in G, and all unions of edge-disjoint cycles of G.

Example 3.3.1: Figure 3.3.1 shows a graph G and the seven non-null elements of the cycle space $W_C(G)$.

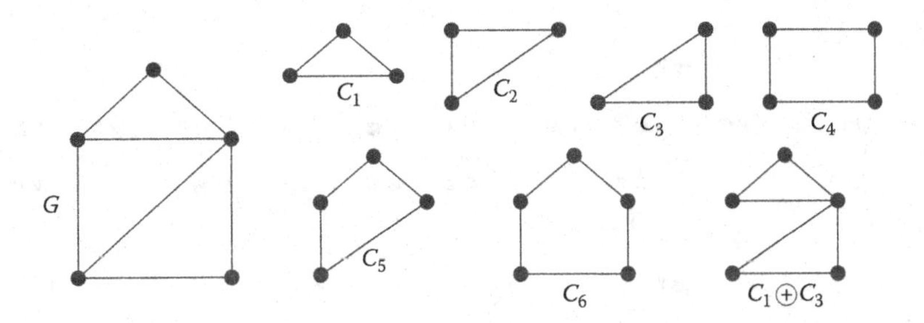

Figure 3.3.1 A graph G and the non-null elements of cycle space $W_C(G)$.

Notice that each vector of $W_C(G)$ is a subgraph having no vertices of odd degree, and that the sum of any two of the vectors of $W_C(G)$ is again a vector of $W_C(G)$. The first of these observations follows directly from the characterization of Eulerian graphs from the previous section (Theorem 3.2.11). The next result shows that the second property of $W_C(G)$ is also true in general.

Proposition 3.3.3: *Given a graph G, the cycle space $W_C(G)$ is a subspace of the edge space $W_E(G)$.*

Proof: It suffices to show that the elements of $W_C(G)$ are closed under $\oplus$. So consider any two distinct members C_1 and C_2 of $W_C(G)$ and let $C_3 = C_1 \oplus C_2$. By Theorem 3.2.11, it must be shown that $deg_{C_3}(v)$ is even for each vertex v in C_3.

Consider any vertex v in C_3, and let X_i denote the set of edges incident with v in C_i for $i = 1, 2, 3$. Since $|X_i|$ is the degree of v in C_i, $|X_1|$ and $|X_2|$ are both even, and $|X_3|$ is nonzero. Since $C_3 = C_1 \oplus C_2$, $X_3 = X_1 \oplus X_2$. But this implies that $|X_3| = |X_1| + |X_2| - 2\,|X_1 \cap X_2|$, which shows that $|X_3|$ must be even. ♢

The Edge-Cut Subspace of a Graph

DEFINITION: The ***edge-cut space*** of a graph G, denoted $W_S(G)$, is the subset of the edge space $W_E(G)$ consisting of the null graph $\emptyset$, all minimal edge-cuts in G, and all unions of edge-disjoint minimal edge-cuts of G.

Example 3.3.2: Figure 3.3.2 shows a graph G and three of its edge-cuts. Since R and T are edge-disjoint minimal ones, all three edge-cuts are in $W_S(G)$.

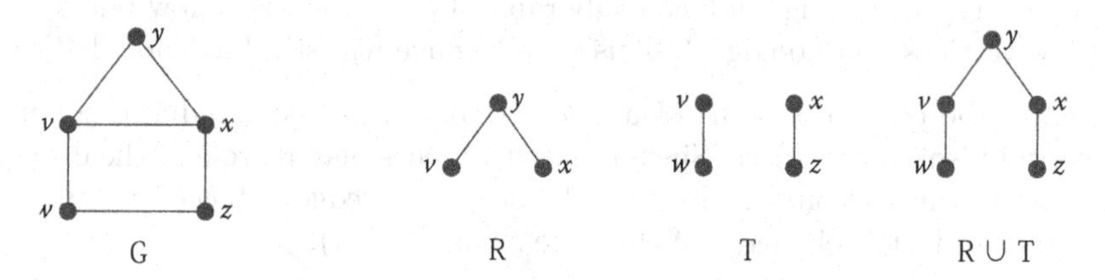

Figure 3.3.2 A graph G and three edge-cuts in $W_S(G)$.

Example 3.3.2 continued: Figure 3.3.3 shows graph G and the 15 non-null elements of $W_S(G)$. The non-minimal edge-cuts appear in the bottom row.

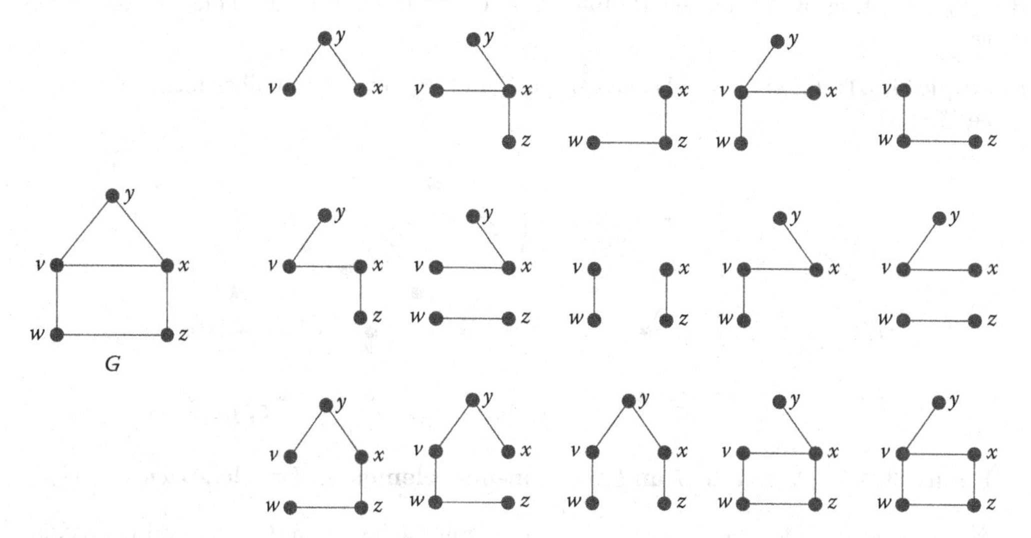

Figure 3.3.3 Graph G and the 15 non-null elements of $W_S(G)$.

Proposition 3.3.4: *Given a graph G, the edge-cut space $W_S(G)$ is closed under the ring-sum operation $\oplus$ and, hence, is a subspace of the edge space $W_E(G)$.*

Proof: By Proposition 3.2.3, it suffices to show that the ring sum of any two partition-cuts in a graph G is also a partition-cut in G. So let $S_1 = \langle X_1, X_2\rangle$ and $S_2 = \langle X_3, X_4\rangle$ be any two partition-cuts in G, and let $V_{ij} = X_i \cap X_j$, for $i = 1, 2$ and $j = 3, 4$.
Then the V_{ij} are mutually disjoint, with

$$S_1 = \langle V_{13} \cup V_{14}, V_{23} \cup V_{24}\rangle = \langle V_{13}, V_{23}\rangle \cup \langle V_{13}, V_{24}\rangle \cup \langle V_{14}, V_{23}\rangle \cup \langle V_{14}, V_{24}\rangle \quad \text{and}$$
$$S_2 = \langle V_{13} \cup V_{23}, V_{14} \cup V_{24}\rangle = \langle V_{13}, V_{14}\rangle \cup \langle V_{13}, V_{24}\rangle \cup \langle V_{23}, V_{14}\rangle \cup \langle V_{23}, V_{24}\rangle$$

Hence,

$$S_1 \oplus S_2 = \langle V_{13}, V_{23}\rangle \cup \langle V_{14}, V_{24}\rangle \cup \langle V_{13}, V_{14}\rangle \cup \langle V_{23}, V_{24}\rangle$$

But

$$\langle V_{13}, V_{23}\rangle \cup \langle V_{14}, V_{24}\rangle \cup \langle V_{13}, V_{14}\rangle \cup \langle V_{23}, V_{24}\rangle = \langle V_{13} \cup V_{24}, V_{14} \cup V_{23}\rangle$$

which is a partition-cut in G, and the proof is complete. ◇

Example 3.3.3: To illustrate Proposition 3.3.4, it is easy to check that the elements of $W_S(G)$ in Figure 3.3.3 are closed under ring sum $\oplus$ and that each element is a partition-cut. For instance, if S_1 denotes the last element in the second row of five elements, and S_2 and S_3 are the first and last elements of the third row, then $S_1 = \langle\{v, w\}, \{x, y, z\}\rangle$; $S_2 = \langle\{v, x, y\}, \{w, z\}\rangle$; and $S_3 = S_1 \oplus S_2 = \langle\{x, y, w\}, \{v, z\}\rangle$.

Bases for the Cycle and Edge-Cut Spaces

Theorem 3.3.5: *Let T be a spanning tree of a connected graph G. Then the fundamental system of cycles associated with T is a basis for the cycle space $W_C(G)$.*

Proof: By the construction, each fundamental cycle associated with T contains exactly one non-tree edge that is not part of any other fundamental cycle associated with T. Thus, no fundamental cycle is a ring sum of some or all of the other fundamental cycles. Hence, the fundamental system of cycles is a linearly independent set.

To show the fundamental system of cycles spans $W_C(G)$, suppose H is any element of $W_C(G)$. Now let $e_1, e_2, \ldots, e_r$ be the non-tree edges of H, and let C_i be the fundamental cycle in $T + e_i$, $i = 1, \ldots, r$. The completion of the proof requires showing that $H = C_1 \oplus C_2 \oplus \cdots \oplus C_r$, or equivalently, that $B = H \oplus C_1 \oplus C_2 \oplus \cdots \oplus C_r$ is the null graph.

Since each e_i appears only in C_i, and e_i is the only non-tree edge in C_i, B contains no non-tree edges. Thus, B is a subgraph of T and is therefore acyclic. But B is an element of $W_C(G)$ (since $W_C(G)$ is closed under ring sum), and the only element of $W_C(G)$ that is acyclic is the null graph. ♢

Example 3.3.1 continued: Figure 3.3.4 shows a spanning tree and the associated fundamental system of cycles $\{C_3, C_4, C_6\}$ for the graph of Figure 3.3.1. Each of the other four non-null elements of cycle space $W_C(G)$ can be expressed as the ring sum of some or all of the fundamental cycles. In particular,

$$\begin{aligned} C_1 &= C_4 \oplus C_6 \\ C_2 &= C_3 \oplus C_4 \\ C_5 &= C_3 \oplus C_6 \\ C_1 \oplus C_3 &= C_3 \oplus C_4 \oplus C_6 \end{aligned}$$

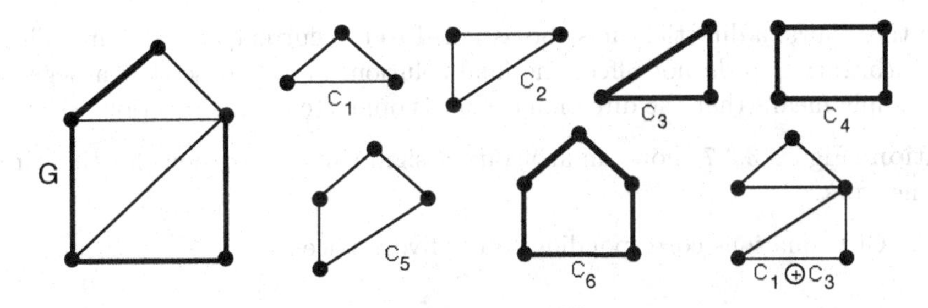

Figure 3.3.4 Fundamental system $\{C_3, C_4, C_6\}$ is a basis for $W_C(G)$.

The next theorem and its proof are analogous to Theorem 3.3.5 and its proof.

Theorem 3.3.6: *Let T be a spanning tree of a connected graph G. Then the fundamental system of edge-cuts associated with T is a basis for the edge-cut space $W_S(G)$.*

♢ (*Exercises*)

Example 3.3.2 continued: Figure 3.3.5 shows a spanning tree and the associated fundamental system of edge-cuts $\{S_1, S_2, S_3, S_4\}$ for the graph of Figure 3.3.3. It is easy to verify that each of the 15 non-null elements of the edge-cut space $W_S(G)$ can be expressed as the ring sum of some or all of the fundamental edge-cuts S_1, S_2, S_3, S_4 (see Exercises). For instance, the edge-cut appearing on the lower right in Figure 3.3.3 is equal to $S_1 \oplus S_2 \oplus S_3 \oplus S_4$.

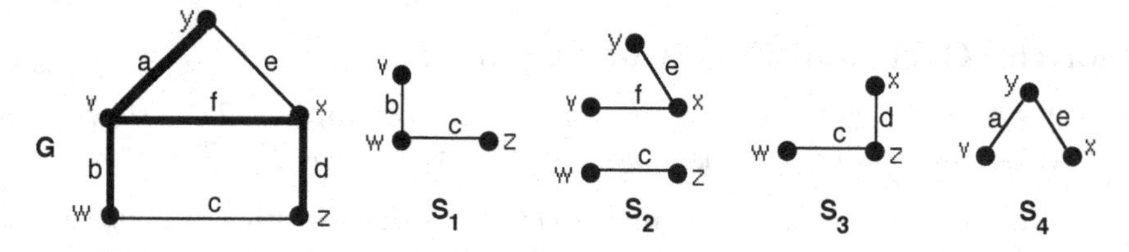

Figure 3.3.5 Fundamental system $\{S_1, S_2, S_3, S_4\}$ is a basis for $W_S(G)$.

Application 3.3.1: *Applying Ohm's and Kirchhoff's Laws* Suppose that the graph shown in Figure 3.3.6 represents an electrical network with a given voltage E on wire e_1, oriented as shown. Also let R_j be the resistance on e_j.

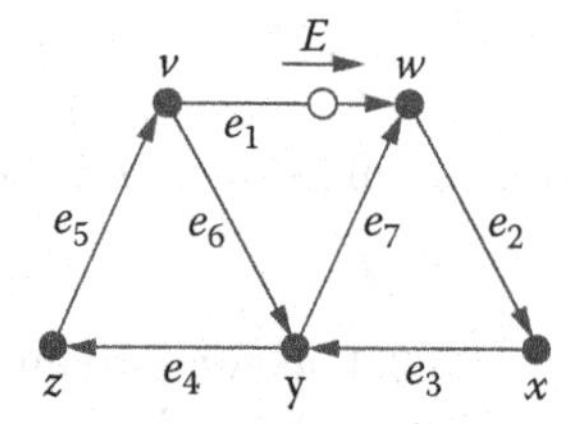

Figure 3.3.6 An electrical network.

The problem is to determine the electric current i_j for wire e_j, using Ohm's Law, ***Kirchhoff's current law (KCL)***, and ***Kirchhoff's voltage law (KVL)***, given by:

- **Ohm's Law:** For a current i flowing through a resistance r, the voltage drop v across the resistance satisfies $v = ir$.
- **KCL:** The algebraic sum of the currents at each vertex is zero.
- **KVL:** The algebraic sum of voltage drops around any cycle is zero.

To apply these laws, a direction must be assigned to the current in each wire. These directions are arbitrary and do not affect the final solution, in the sense that a negative value for an i_j simply means that the direction of flow is opposite to the direction assigned for e_j.

Illustration: Figure 3.3.7 shows an arbitrary assignment of directions for the wires in the example network.

The five (KCL)-equations corresponding to the five vertices are

$$\begin{aligned} -i_1 + i_5 - i_6 &= 0 \\ i_2 - i_3 &= 0 \\ i_4 - i_5 &= 0 \\ i_3 - i_4 + i_6 - i_7 &= 0 \\ i_1 - i_2 + i_7 &= 0 \end{aligned}$$

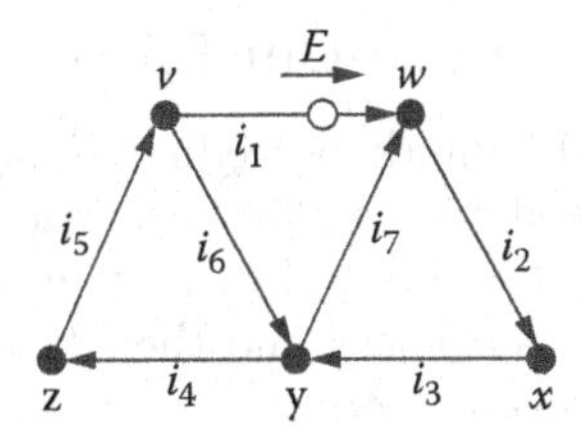

Figure 3.3.7 An electrical network.

Notice that the sum of these equations is the equation $0 = 0$, which indicates that one of them is redundant. Furthermore, if one circuit is the sum of other circuits, then its (KVL)-equation is redundant. For instance, the (KVL)-equations for the circuits $\langle v, w, y, v\rangle$, $\langle w, x, y, w\rangle$, and $\langle v, w, x, y, v\rangle$ are, respectively,

$$\begin{aligned} i_1R_1 - i_7R_7 - i_6R_6 - E &= 0 \\ i_2R_2 + i_3R_3 + i_7R_7 &= 0 \\ i_1R_1 + i_2R_2 + i_3R_3 - i_6R_6 - E &= 0 \end{aligned}$$

The third equation is the sum of the first two, since the third circuit is the sum of the first two circuits.

A large network is likely to have a huge number of these redundancies. The objective for an efficient solution strategy is to find a minimal set of circuits whose corresponding equations, together with the equations from Kirchhoff's circuit law, are just enough to solve for the i_j's. Since a fundamental system of cycles is a basis for the cycle space, their corresponding equations will constitute a full set of linearly independent equations and will meet the objective.

Illustration: One of the spanning trees of the example network is shown below.

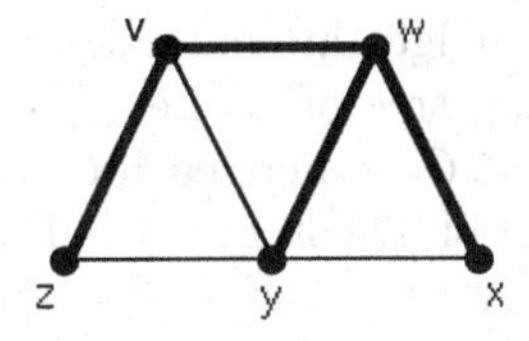

The fundamental system of cycles and their corresponding equations for this spanning tree are as follows:

$$\begin{aligned} \langle z, v, w, y, z\rangle &: \quad i_5R_5 + i_1R_1 - i_7R_7 + i_4R_4 - E = 0 \\ \langle v, w, y, v\rangle &: \quad i_1R_1 - i_7R_7 - i_6R_6 - E = 0 \\ \langle w, x, y, w\rangle &: \quad i_2R_2 + i_3R_3 + i_7R_7 = 0 \end{aligned}$$

These three equations, together with any four of the five (KCL)-equations will determine the i_j's. If, for example, the voltage $E = 28$ and all the R_j's are taken to be 1, then it is easy to verify that the solution for the currents is

$$i_1 = 12; \quad i_2 = i_3 = 4; \quad i_4 = i_5 = 4; \quad i_6 = i_7 = -8$$

Dimension of the Cycle and Edge-Cut Spaces

In light of the discussion and remark preceding Example 3.2.2, Theorems 3.3.5 and 3.3.6 extend easily to non-connected graphs. For such graphs, the fundamental systems are associated with a full spanning forest instead of a spanning tree. The resulting corollary establishes the dimensions of the cycle space and the edge-cut space, thereby justifying the use of the terms *cycle rank* and *edge-cut rank*.

Corollary 3.3.7: *Let G be a graph with $c(G)$ components. Then the dimension of the cycle space $W_C(G)$ is the cycle rank $\beta(G) = |E_G| - |V_G| + c(G)$, and the dimension of the edge-cut space $W_S(G)$ is the edge-cut rank $|V_G| - c(G)$.*

Proof: This follows directly from Theorems 3.3.5 and 3.3.6 and the definitions of cycle rank and edge-cut rank. ♢

Relationship between the Cycle and Edge-Cut Spaces

The next two propositions characterize the elements of the cycle space and the edge-cut space of a graph in terms of each other. The proof of the second characterization is similar to that of the first and is left as an exercise.

Proposition 3.3.8: *A subgraph H of a graph G is in the cycle space $W_C(G)$ if and only if it has an even number of edges in common with every subgraph in the edge-cut space $W_S(G)$.*

Proof: *Necessity* ($\Rightarrow$) Each subgraph in the cycle space is a union of edge-disjoint cycles, and each subgraph in the edge-cut space is a union of edge-disjoint edge-cuts. Thus, necessity follows from Proposition 3.2.8.
Sufficiency ($\Leftarrow$) It may be assumed without loss of generality that G is connected, since the argument that follows may be applied separately to each of the components of G if G is not connected.

Suppose H has an even number of edges in common with each subgraph in the edge-cut space of G, and let T be a spanning tree of G. Let $e_1, e_2, \ldots, e_r$ be the non-tree edges of H, and consider $C = C_1 \oplus C_2 \oplus \ldots \oplus C_r$, where each C_i is the fundamental cycle in $T + e_i$, $i = 1, \ldots, r$. Arguing as in the proof of Theorem 3.3.5, $H \oplus C$ has no non-tree edges. Thus, the only possible edges in $H \oplus C$ are edges of T. So suppose b is an edge in both T and $H \oplus C$. Now if S is the fundamental edge-cut associated with b, then b is the only edge in $H \oplus C \oplus S$. But, since $C \in W_C(G), C$ has an even number of edges in common with each subgraph in the edge-cut space of G, as does H. This implies that $H \oplus C$ has an even number of edges in common with each subgraph in the edge-cut space. In particular, $H \oplus C \oplus S$ must have an even number of edges. This contradiction shows that $H \oplus C$ must be the null graph, that is, $H = C$. Hence, H is a subgraph in the cycle space of G. ♢

Proposition 3.3.9: *A subgraph H of a graph G is an element of the edge-cut space of G if and only if it has an even number of edges in common with every subgraph in the cycle space of G.* ♢ *(Exercises)*

Orthogonality of the Cycle and Edge-Cut Spaces

DEFINITION: Two vectors in the edge space $W_E(G)$ are said to be ***orthogonal*** if the dot product of their characteristic vectors equals 0 in $GF(2)$. Thus, two subsets of edges are orthogonal if they have an even number of edges in common.

Theorem 3.3.10: *Given a graph G, the cycle space $W_C(G)$ and the edge-cut space $W_S(G)$ are orthogonal subspaces of $W_E(G)$.*

Proof: Let H be a subgraph in the cycle space, and let K be a subgraph in the edge-cut space. By either Proposition 3.3.8 or Proposition 4.6.9, H and K have an even number of edges in common and, hence, are orthogonal. ♢

Theorem 3.3.11: *Given a graph G, the cycle space $W_C(G)$ and the edge-cut space $W_S(G)$ are orthogonal complements in $W_E(G)$ if and only if $W_C(G) \cap W_S(G) = \{\emptyset\}$.*

Proof: If $W_C(G) \oplus W_S(G)$ denotes the direct sum of the subspaces $W_C(G)$ and $W_S(G)$, then

$$dim\,(W_C(G) \oplus W_S(G)) = dim\,(W_C(G)) + dim\,(W_S(G)) - dim\,(W_C(G) \cap W_S(G))$$

By Corollary 3.3.7, $dim\,(W_C(G)) + dim\,(W_S(G)) = |E_G|$. Thus, $dim\,(W_C(G) \oplus W_S(G)) = |E_G| = dim\,(W_E(G))$ if and only if $dim\,(W_C(G) \cap W_S(G)) = 0$. ♢

Vector Space of Vertex Subsets

This section concludes by showing that a spanning tree of a simple graph G corresponds to a basis of the column space of the incidence matrix of G.

FROM LINEAR ALGEBRA: The column space of a matrix M is the set of column vectors that are linear combinations of the columns of M. When the entries of M come from $GF(2)$, linear combinations are simply mod 2 sums. (Also see Appendix A.4.)

Let G be a simple graph with n vertices and m edges, and let $\langle e_1, e_2, \ldots, e_m \rangle$ and $\langle v_1, v_2, \ldots, v_n \rangle$ be fixed orderings of E_G and V_G, respectively.

REVIEW FROM §1.6: The incidence matrix I_G of G is the matrix whose (i,j)th entry is given by

$$I_G[i,j] = \begin{cases} 0, & \text{if } v_i \text{ is not an endpoint of } e_j \\ 1, & \text{otherwise} \end{cases}$$

Analogous to the edge space $W_E(G)$, the collection of vertex subsets of V_G under ring sum forms a vector space over $GF(2)$, which is called the ***vertex space of*** G and is denoted $W_V(G)$. Each element of the vertex space $W_V(G)$ may be viewed as an n-tuple over $GF(2)$.

In this setting, I_G represents a linear transformation from edge space $W_E(G)$ to vertex space $W_V(G)$, mapping the characteristic vectors of edge subsets to characteristic vectors of vertex subsets.

Example 3.3.4: Consider the graph G and its corresponding incidence matrix I_G shown in Figure 3.3.8.

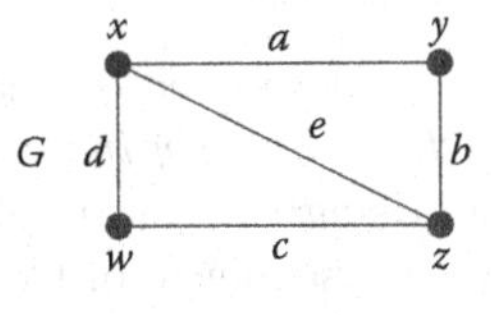

$$I_G = \begin{array}{c} \\ x \\ y \\ z \\ w \end{array} \begin{array}{c} \begin{matrix} a & b & c & d & e \end{matrix} \\ \begin{pmatrix} 1 & 0 & 0 & 1 & 1 \\ 1 & 1 & 0 & 0 & 0 \\ 0 & 1 & 1 & 0 & 1 \\ 0 & 0 & 1 & 1 & 0 \end{pmatrix} \end{array}$$

Figure 3.3.8 A graph G and its incidence matrix I_G.

The characteristic vector of the image of the subset $E_1 = \{a, c, e\}$ under the mapping is obtained by multiplying the characteristic vector of E_1 by I_G (mod 2), as follows:

$$\begin{pmatrix} 1 & 0 & 0 & 1 & 1 \\ 1 & 1 & 0 & 0 & 0 \\ 0 & 1 & 1 & 0 & 1 \\ 0 & 0 & 1 & 1 & 0 \end{pmatrix} \begin{pmatrix} 1 \\ 0 \\ 1 \\ 0 \\ 1 \end{pmatrix} = \begin{pmatrix} 0 \\ 1 \\ 0 \\ 1 \end{pmatrix}$$

Thus, E_1 is mapped to the vertex subset $V_1 = \{y, w\}$. Notice that V_1 consists of the endpoints of the path formed by the edges of E_1. It is not hard to show that every open path in G is mapped to the vertex subset consisting of the path's initial and terminal vertices. For this reason, I_G is sometimes called a *boundary operator*.

TERMINOLOGY NOTE: Algebraists and topologists sometimes refer to the edge subsets of $W_E(G)$ and the vertex subsets of $W_V(G)$ as *1-chains* and *0-chains*, respectively, and denote these vector spaces $C_1(G)$ and $C_0(G)$. In this context, the chains are typically represented as sums of edges or vertices. For example, the expression $e_1 + e_2 + e_3$ would represent the edge subset $\{e_1, e_2, e_3\}$.

In general, the image under I_G of any edge subset is obtained by simply computing the mod 2 sum of the corresponding set of columns of I_G. For instance, in Example 3.3.4, I_G maps the edge subset $E_2 = \{c, d, e\}$ to $\emptyset$ (which is the zero element of the vertex space $W_V(G)$), since the mod 2 sum of the third, fourth, and fifth columns of I_G is the column vector of all zeros. It is not a coincidence that E_2 comprises the edges of a cycle, as the next proposition confirms.

NOTATION: For any edge subset D of a simple graph G, let C_D denote the corresponding set of columns of I_G.

Proposition 3.3.12: *Let D be the edge-set of a cycle in a simple graph G. Then the mod 2 sum of the columns in C_D is the zero vector.*

Proof: For a suitable ordering of the edges and vertices of G, a cycle of length k corresponds to the following submatrix of I_G. ♢

$$\begin{pmatrix} 1 & 0 & \cdots & 0 & 1 \\ 1 & 1 & \ddots & \vdots & 0 \\ 0 & 1 & \ddots & 0 & \vdots \\ \vdots & \ddots & \ddots & 1 & 0 \\ 0 & \cdots & 0 & 1 & 1 \end{pmatrix}$$

Corollary 3.3.13: *A set D of edges of a simple graph G forms a cycle or a union of edge-disjoint cycles if and only if the mod 2 sum of the columns in C_D is the zero vector.* ♢ (*Exercises*)

Each column of I_G may be regarded as an element of the vector space of all n-tuples over $GF(2)$. Therefore, a subset S of the columns of I_G is *linearly independent* over $GF(2)$ if no non-empty subset of S sums to the zero vector. Using this terminology, the following characterization of acyclic subgraphs is an immediate consequence of Corollary 3.3.13.

Proposition 3.3.14: *An edge subset D of a simple graph G forms an acyclic subgraph of G if and only if the column set C_D is linearly independent over $GF(2)$.*

The next proposition is needed to complete the characterization of spanning trees promised earlier.

Proposition 3.3.15: *Let D be a subset of edges of a connected, simple graph G. Then D forms (induces) a connected spanning subgraph of G if and only if column set C_D spans the column space of I_G.*

Proof: The column set C_D spans the column space of I_G if and only if each column of I_G can be expressed as the sum of the columns in a subset of C_D. But each column of I_G corresponds to some edge $e = xy$ of G, and that column is a sum of the columns in some subset of C_D if and only if there is a path from x to y whose edges are in D. ◇

Proposition 3.3.16: *Let G be a connected, simple graph. Then an edge subset D induces a spanning tree of G if and only if the columns corresponding to the edges of D form a basis for the column space of I_G over $GF(2)$.*

Proof: This follows directly from Propositions 3.3.14 and 3.3.15. ◇

EXERCISES for Section 3.3

3.3.1 Consider the given graph and the spanning tree T defined by the specified edge subset of the graph G.

i. Find the non-null elements of the cycle space $W_C(G)$ for the given graph G.

ii. Find the non-null elements of the edge-cut space $W_S(G)$.

iii. Show that the fundamental system of cycles associated with T is a basis for the cycle-space $W_C(G)$.

v. Show that the fundamental system of edge-cuts associated with T is a basis for the edge-cut space $W_S(G)$.

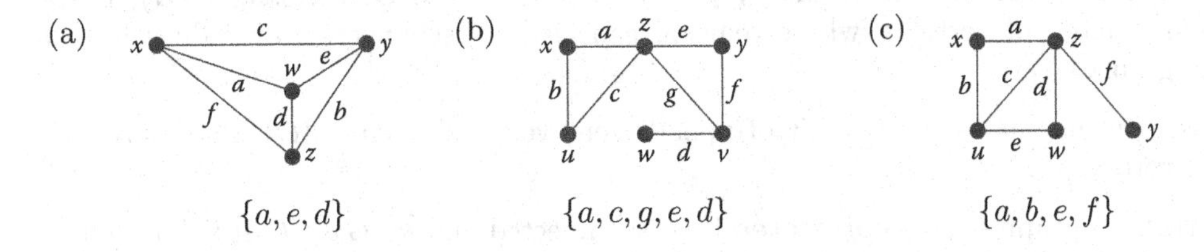

3.3.2 For the given graph G:

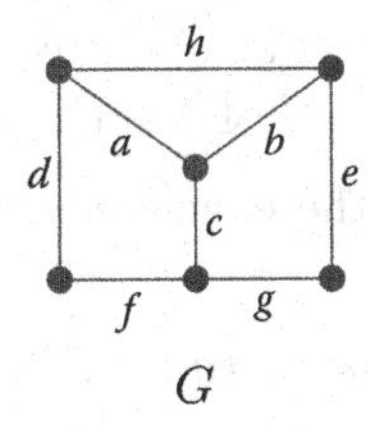

(a) Show that the collection $\{\{a, c, d, f\}, \{b, c, e, g\}, \{a, b, h\}\}$ of edge subsets of E_G forms a basis for the cycle space $W_C(G)$ of G.

(b) Find a different basis by choosing some spanning tree and using the associated fundamental system of cycles.

3.3.3 Express each of the 15 non-null elements of $W_S(G)$ in Figure 3.3.3 as the ring sum of some or all of the four fundamental edge-cuts in Figure 3.3.5.

3.3.4 For Application 3.3.1, reverse the directions assigned to i_1, i_2, and i_3, and show that the solution for the actual currents does not change.

3.3.5 Verify that Corollary 3.3.13 and Proposition 3.3.16 hold for the given edge subset of the following graph.

(a) $\{a, b, c, d\}$.

(b) $\{a, b, e, f\}$.

(c) $\{a, b, e, d\}$.

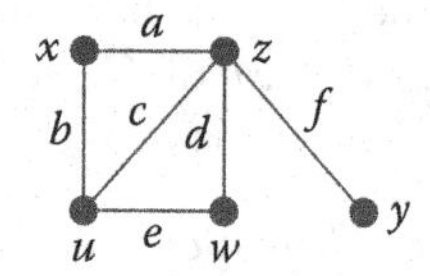

3.3.6 Prove Proposition 3.3.1.

3.3.7 Prove Theorem 3.3.6.

3.3.8 Prove Proposition 3.3.9.

3.3.9 Prove Corollary 3.3.13.

3.3.10 Test whether the cycle and edge-cut spaces for the graph in Figure 3.3.3 are orthogonal complements of the edge space $W_E(G)$.

3.3.11 Express the graph of Figure 3.3.3 as the ring sum of a subgraph in the cycle space with a subgraph in the edge-cut space of G.

3.4 VERTEX- AND EDGE-CONNECTIVITY

DEFINITION: The ***vertex-connectivity*** of a connected graph G, denoted $\kappa_v(G)$, is the minimum number of vertices, whose removal can either disconnect G or reduce it to a 1-vertex graph.

Thus, if G has at least one pair of nonadjacent vertices, then $\kappa_v(G)$ is the size of a smallest vertex-cut.

DEFINITION: A graph G is ***k-connected*** if G is connected and $\kappa_v(G) \geq k$. If G has nonadjacent vertices, then G is k-connected if every vertex-cut has at least k vertices.

DEFINITION: The ***edge-connectivity*** of a connected graph G, denoted $\kappa_e(G)$, is the minimum number of edges whose removal can disconnect G.

Thus, if G is a connected graph, the edge-connectivity $\kappa_e(G)$ is the size of a smallest edge-cut.

DEFINITION: A graph G is ***k-edge-connected*** if G is connected and every edge-cut has at least k edges (i.e., $\kappa_e(G) \geq k$].

Example 3.4.1: In the graph G shown in Figure 3.4.1, the vertex-set $\{x, y\}$ is one of three different 2-element vertex-cuts, and it is easy to see that there is no cut-vertex. Thus, $\kappa_v(G) = 2$. The edge-set $\{a, b, c\}$ is the unique 3-element edge-cut of graph G, and there is no edge-cut with fewer than three edges. Therefore, $\kappa_e(G) = 3$.

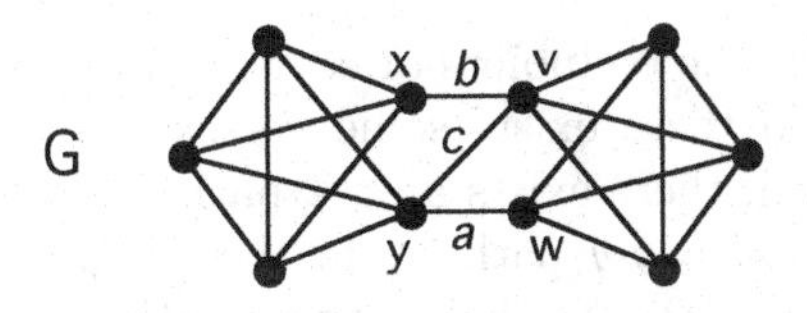

Figure 3.4.1 A graph G with $\kappa_v(G) = 2$ and $\kappa_e(G) = 3$.

Application 3.4.1: *Network Survivability* The connectivity measures κ_v and κ_e are used in a quantified model of *network survivability*, which is the capacity of a network to retain connections among its nodes after some edges or nodes are removed.

Remark: Since neither the vertex-connectivity nor the edge-connectivity of a graph is affected by the existence or absence of self-loops, we will assume that all graphs under consideration throughout this chapter are loopless, unless otherwise specified.

Proposition 3.4.1: *Let G be a graph. Then the edge-connectivity $\kappa_e(G)$ is less than or equal to the minimum degree $\delta_{\min}(G)$.*

Proof: Let v be a vertex of graph G, with degree $k = \delta_{\min}(G)$. Then the deletion of the k edges that are incident on vertex v separates v from the other vertices of G. ◇

REVIEW FROM §3.2: A ***partition-cut*** $\langle X_1, X_2 \rangle$ is an edge-cut each of whose edges has one endpoint in each of the vertex bipartition sets X_1 and X_2.

The following proposition characterizes the edge-connectivity of a graph in terms of the size of its partition-cuts.

Proposition 3.4.2: *A graph G is k-edge-connected if and only if every partition-cut contains at least k edges.*

Proof: ($\Rightarrow$) Suppose that graph G is k-edge-connected. Then every partition-cut of G has at least k edges, since a partition-cut is an edge-cut.

($\Leftarrow$) Suppose that every partition-cut contains at least k edges. By Proposition 3.2.4, every minimal edge-cut is a partition-cut. Thus, every edge-cut contains at least k edges.

Relationship Between Vertex- and Edge-Connectivity

The next few results concern the relationship between vertex-connectivity and edge-connectivity. They lead to a characterization of 2-connected graphs, first proved by Hassler Whitney in 1932.

Proposition 3.4.3: *Let e be any edge of a k-connected graph G, for $k \geq 3$. Then the edge-deletion subgraph $G - e$ is $(k-1)$-connected.*

Proof: Let $W = \{w_1, w_2, \ldots, w_{k-2}\}$ be any set of $k-2$ vertices in $G - e$, and let x and y be any two different vertices in $(G - e) - W$. It suffices to show the existence of an x-y walk in $(G - e) - W$.

First, suppose that at least one of the endpoints of edge e is contained in set W. Since the vertex-deletion subgraph $G - W$ is connected (in fact, 2-connected), there is an x-y path in $G - W$. This path cannot contain edge e and, hence, it is an x-y path in the subgraph $(G - e) - W$. Next, suppose that neither endpoint of edge e is in set W. Then there are two cases to consider.

Case 1: Vertices x and y are the endpoints of edge e. Graph G has at least $k+1$ vertices (since G is k-connected). So there exists some vertex $z \in G - \{w_1, w_2, \ldots, w_{k-2}, x, y\}$. Since graph G is k-connected, there exists an x-z path P_1 in the vertex-deletion subgraph $G - \{w_1, w_2, \ldots, w_{k-2}, y\}$ and a z-y path P_2 in the subgraph $G - \{w_1, w_2, \ldots, w_{k-2}, x\}$ (shown on the left in Figure 3.4.2). Neither of these paths contains edge e, and, therefore, their concatenation is an x-y walk in the subgraph $G - \{w_1, w_2, \ldots, w_{k-2}\}$.

Case 2: At least one of the vertices x and y, say x, is not an endpoint of edge e. Let u be an endpoint of edge e that is different from vertex x. Since graph G is k-connected, the subgraph $G - \{w_1, w_2, \ldots, w_{k-2}, u\}$ is connected. Hence, there is an x-y path P in $G - \{w_1, w_2, \ldots, w_{k-2}, u\}$ (shown on the right in Figure 3.4.2). It follows that P is an x-y path in $G - \{w_1, w_2, \ldots, w_{k-2}\}$ that does not contain vertex u and, hence, excludes edge e (even if P contains the other endpoint of e, which it could). Therefore, P is an x-y path in $(G - e) - \{w_1, w_2, \ldots, w_{k-2}\}$. ◇

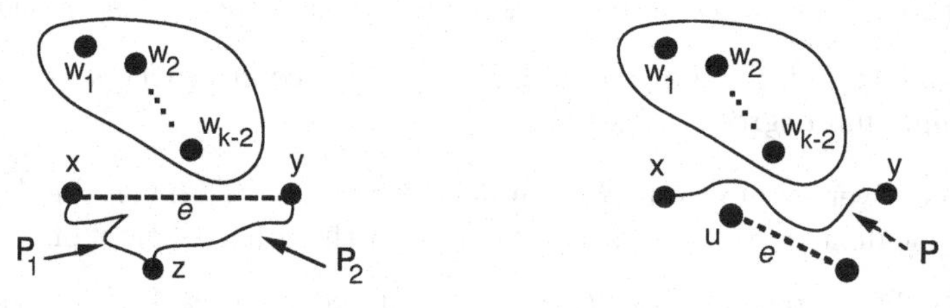

Figure 3.4.2 The existence of an x-y walk in $(G - e) - \{w_1, w_2, \ldots, w_{k-2}\}$.

Corollary 3.4.4: *Let G be a k-connected graph, and let D be any set of m edges of G, for $m \leq k-1$. Then the edge-deletion subgraph $G - D$ is $(k-m)$-connected.*

Proof: The result follows by the iterative application of Proposition 3.4.3. ◇

Corollary 3.4.5: *Let G be a connected graph. Then $\kappa_e(G) \geq \kappa_v(G)$.*

Proof: Let $k = \kappa_v(G)$, and let S be any set of $k-1$ edges in graph G. Since G is k-connected, the graph $G - S$ is 1-connected, by Corollary 3.4.4. Thus, edge subset S is not an edge-cut of graph G, which implies that $\kappa_e(G) \geq k$. ◇

Corollary 3.4.6: *Let G be a connected graph. Then $\kappa_v(G) \leq \kappa_e(G) \leq \delta_{\min}(G)$.*

Proof: The assertion simply combines Proposition 3.4.1 and Corollary 3.4.5. ◇

Internally Disjoint Paths and Vertex-Connectivity: Whitney's Theorem

A communications network is said to be *fault-tolerant* if it has at least two alternative paths between each pair of vertices. This notion actually characterizes 2-connected graphs, as the next theorem demonstrates. The theorem was proved by Hassler Whitney in 1932 and is a prelude to his more general result for k-connected graphs, which appears in §3.5.

TERMINOLOGY: A vertex of a path P is an ***internal vertex*** of P if it is neither the initial nor the final vertex of that path.

DEFINITION: Let u and v be two vertices in a graph G. A collection of u-v paths in G is said to be ***internally disjoint*** if no two paths in the collection have an internal vertex in common.

Theorem 3.4.7: [***Whitney's 2-Connected Characterization***] *Let G be a connected graph with three or more vertices. Then G is 2-connected if and only if for each pair of vertices in G, there are two internally disjoint paths between them.*

Proof: ($\Leftarrow$) Arguing by contrapositive, suppose that graph G is not 2-connected. Then let v be a cut-vertex of G. Since $G - v$ is not connected, there must be two vertices x and y such that there is no x-y path in $G - v$. It follows that v is an internal vertex of every x-y path in G.

($\Rightarrow$) Suppose that graph G is 2-connected, and let x and y be any two vertices in G. We use induction on the distance $d(x, y)$ to prove that there are at least two vertex-disjoint x-y paths in G. If there is an edge e joining vertices x and y, (i.e., $d(x, y) = 1$), then the edge-deletion subgraph $G - e$ is connected, by Corollary 3.4.4. Thus, there is an x-y path P in $G - e$. It follows that path P and edge e are two internally disjoint x-y paths in G.

Next, assume for some $k \geq 2$ that the assertion holds for every pair of vertices whose distance apart is less than k. Suppose $d(x, y) = k$, and consider an x-y path of length k. Let w be the vertex that immediately precedes vertex y on this path, and let e be the edge between vertices w and y. Since $d(x, w) < k$, the induction hypothesis implies that there are two internally disjoint x-w paths in G, say P and Q. Also, since G is 2-connected, there exists an x-y path R in G that avoids vertex w. This is illustrated in the figure below for the two possibilities for path Q: either it contains vertex y (as shown on the right), or it does not (as on the left).

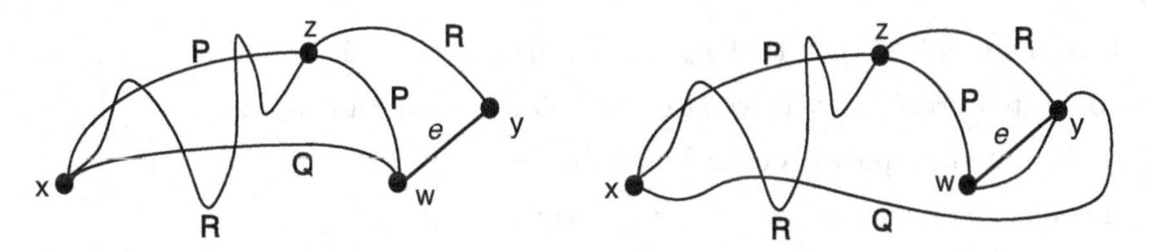

Let z be the last vertex on path R that precedes vertex y and is also on one of the paths P or Q (z might be vertex x). Assume, without loss of generality, that z is on path P. Then G has two internally disjoint x-y paths. One of these paths is the concatenation of the subpath of P from x to z with the subpath of R from z to y. If vertex y is not on path Q, then a second x-y path, internally disjoint from the first one, is the concatenation of path Q with the edge joining vertex w to vertex y. If y is on path Q, then the subpath of Q from x to y can be used as the second path. $\diamondsuit$

Corollary 3.4.8: *Let G be a graph with at least three vertices. Then G is 2-connected if and only if any two vertices of G lie on a common cycle.*

Proof: This follows from Theorem 3.4.7, since two vertices x and y lie on a common cycle if and only if there are two internally disjoint x-y paths. $\diamondsuit$

Remark: Theorem 3.4.7 is a prelude to Whitney's more general result for k-connected graphs, which appears in §3.5. Corollary 3.4.8 is used in Chapter 4 in the proof of Kuratowski's characterization of graph planarity.

The following theorem extends the list of characterizations of 2-connected graphs. Its proof uses reasoning similar to that used in the proof of the last two results (see Exercises).

Theorem 3.4.9: [***Characterization of 2-Connected Graphs***] *Let G be a connected graph with at least three vertices. Then the following statements are equivalent.*

1. *Graph G is 2-connected.*
2. *For any two vertices of G, there is a cycle containing both.*
3. *For any vertex and any edge of G, there is a cycle containing both.*
4. *For any two edges of G, there is a cycle containing both.*
5. *For any two vertices and one edge of G, there is a path from one of the vertices to the other, that contains the edge.*
6. *For any sequence of three distinct vertices of G, there is a path from the first to the third that contains the second.*
7. *For any three distinct vertices of G, there is a path containing any two of them which does not contain the third.*

EXERCISES for Section 3.4

3.4.1 Find the other two 2-element vertex-cuts in the graph of Example 3.4.1.

3.4.2 Either draw a graph meeting the given specifications or explain why no such graph exists.

(a) A 6-vertex graph G such that $\kappa_v(G) = 2$ and $\kappa_e(G) = 2$.

(b) A connected graph with 11 vertices and 10 edges and no cut-vertices.

(c) A 3-connected graph with exactly one bridge.

(d) A 2-connected 8-vertex graph with exactly two bridges.

3.4.3 Determine the vertex- and edge-connectivity of the given graph.

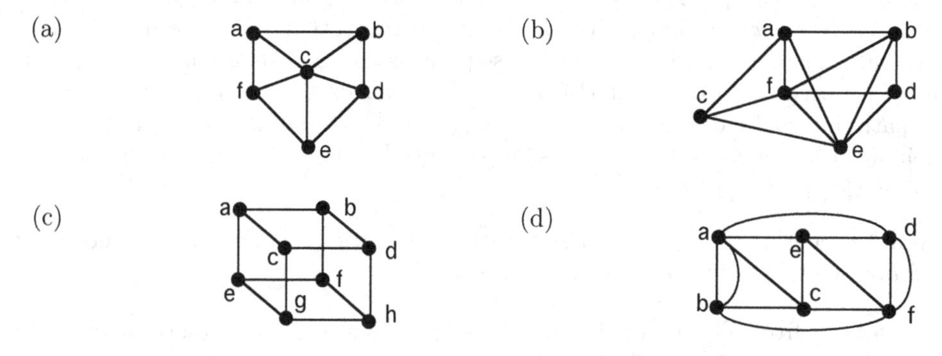

(e) Complete bipartite graph $K_{4,7}$.

(f) Hypercube graph Q_4.

3.4.4 Determine the vertex- and edge-connectivity of the complete bipartite graph $K_{m,n}$.

3.4.5 Determine the vertex-connectivity and edge-connectivity of the Petersen graph (§1.1), and justify your answer. (Hint: Use the graph's symmetry to reduce the number of cases to consider.)

3.4.6 Give an example of a graph G satisfying the given conditions.

(a) $\kappa_v(G) = \kappa_e(G) = \delta_{\min}(G)$ (b) $\kappa_v(G) = \kappa_e(G) < \delta_{\min}(G)$

(c) $\kappa_v(G) < \kappa_e(G) = \delta_{\min}(G)$ (d) $\kappa_v(G) < \kappa_e(G) < \delta_{\min}(G)$

3.4.7 Let $v_1, v_2, \ldots, v_k$ be k distinct vertices of a k-connected graph G, and let G^W be the graph formed from G by joining a new vertex w to each of the v_i's. Show that $\kappa_v(G^w) = k$.

3.4.8 Let G be a k-connected graph, and let v be a vertex not in G. Prove that the suspension $H = G + v$ (§1.4) is $(k+1)$-connected.

3.4.9 Prove that there exists no 3-connected simple graph with exactly seven edges.

3.4.10 Let a, b, and c be positive integers with $a \leq b \leq c$. Show that there exists a graph G with $\kappa_v(G) = a$, $\kappa_e(G) = b$, $\delta_{\min}(G) = c$.

DEFINITION: A ***unicyclic*** graph is a connected graph with exactly one cycle.

3.4.11 Show that the edge-connectivity of a unicyclic graph is no greater than 2.

3.4.12 Characterize those unicyclic graphs whose vertex-connectivity equals 2.

3.4.13 Prove the characterization of 2-connected graphs given by Theorem 3.4.9.

3.4.14 Prove that if G is a connected graph, then $\kappa_v(G) = 1 + \min_{v \in V}\{\kappa_v(G - v)\}$.

3.4.15 Find a lower bound on the number of vertices in a k-connected graph with diameter d (§1.2) and a graph that achieves that lower bound (thereby showing that the lower bound is *sharp*).

3.5 MAX-MIN DUALITY AND MENGER'S THEOREMS

Borrowing from operations research terminology, we consider certain *primal-dual pairs* of optimization problems that are intimately related. Usually, one of these problems involves the maximization of some objective function, while the other is a minimization problem. A feasible solution to one of the problems provides a bound for the optimal value of the other problem (this is sometimes referred to as *weak duality*), and the optimal value of one problem is equal to the optimal value of the other (*strong duality*). Menger's Theorems and their many variations epitomize this primal-dual relationship. The following terminology is used throughout this section.

DEFINITION: Let u and v be distinct vertices in a connected graph G. A vertex subset (or edge subset) S is u-v ***separating*** (or ***separates*** u and v), if the vertices u and v lie in different components of the deletion subgraph $G - S$.

Thus, a u-v separating vertex-set is a vertex-cut, and a u-v separating edge-set is an edge-cut. When the context is clear, the term u-v ***separating set*** will refer either to a u-v separating vertex-set or to a u-v separating edge-set.

Example 3.5.1: For the graph in Figure 3.5.1, the vertex-cut $\{x, w, z\}$ is a u-v separating set of vertices of minimum size, and the edge-cut $\{a, b, c, d, e\}$ is a u-v separating set of edges of minimum size. Notice that a minimize-size u-v separating set of edges (vertices) need not be a minimum-size edge-cut (vertex-cut). For instance the set $\{a, b, c, d, e\}$ is not a minimum-size edge-cut, because the set of edges incident on the 3-valent vertex y is an edge-cut of size 3.

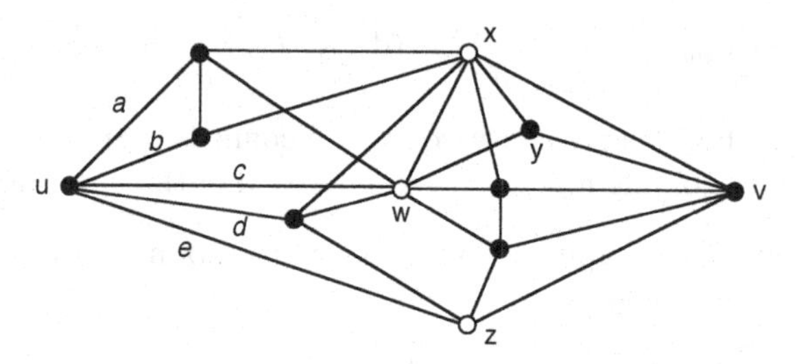

Figure 3.5.1 Vertex- and edge-cuts that are u-v separating sets.

A Primal-Dual Pair of Optimization Problems

The discussion in §3.4 suggests two different interpretations of a graph's connectivity. One interpretation is the number of vertices or edges it takes to disconnect the graph, and the other is the number of alternative paths joining any two given vertices of the graph.

Corresponding to these two perspectives are the following two optimization problems for two nonadjacent vertices u and v of a connected graph G.

Maximization Problem: Determine the maximum number of internally disjoint u-v paths in graph G.

Minimization Problem: Determine the minimum number of vertices of graph G needed to separate the vertices u and v.

Proposition 3.5.1: *(Weak Duality) Let u and v be any two nonadjacent vertices of a connected graph G. Let P_{uv} be a collection of internally disjoint u-v paths in G, and let S_{uv} be a u-v separating set of vertices in G. Then $|P_{uv}| \leq |S_{uv}|$.*

Proof: Since S_{uv} is a u-v separating set, each u-v path in P_{uv} must include at least one vertex of S_{uv}. Since the paths in P_{uv} are internally disjoint, no two of them can include the same vertex. Thus, the number of internally disjoint u-v paths in G is at most $|S_{uv}|$ (by the pigeonhole principle). ◊

Corollary 3.5.2: *Let u and v be any two nonadjacent vertices of a connected graph G. Then the maximum number of internally disjoint u-v paths in G is less than or equal to the minimum size of a u-v separating set of vertices of G.*

The main result of this section is Menger's Theorem, which states that these two quantities are in fact equal. But even the weak duality result by itself provides *certificates of optimality*, as the following corollary shows. It follows directly from Proposition 3.5.1.

Corollary 3.5.3: [***Certificate of Optimality***] *Let u and v be any two nonadjacent vertices of a connected graph G. Suppose that P_{uv} is a collection of internally disjoint u-v paths in G, and that S_{uv} is a u-v separating set of vertices in G, such that $|P_{uv}| = |S_{uv}|$. Then P_{uv} is a maximum-size collection of internally disjoint u-v paths, and S_{uv} is a minimum-size u-v separating set.* ◊ *(Exercises)*

Example 3.5.2: Consider the graph G shown below. The vertex sequences $\langle u, x, y, t, v \rangle$, $\langle u, z, v \rangle$, and $\langle u, r, s, v \rangle$ represent a collection P of three internally disjoint u-v paths in G, and the set $S = \{y,$ s, $z\}$ is a u-v separating set of size 3. Therefore, by Corollary 3.5.3, P is a maximum-size collection of internally disjoint u-v paths, and S is a minimum-size u-v separating set.

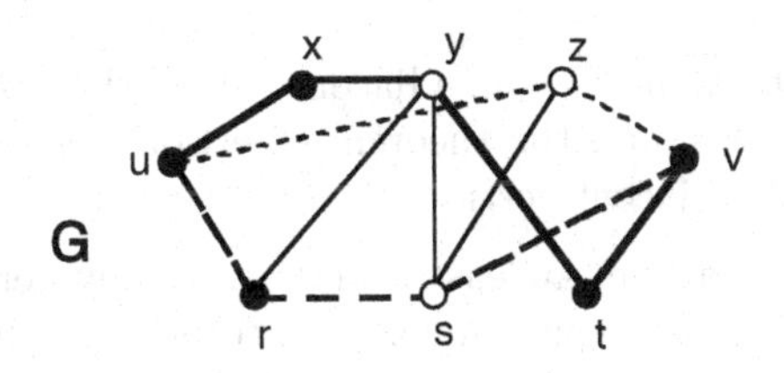

Menger's Theorem

The next theorem, proved by K. Menger in 1927, establishes a *strong duality* between the two optimization problems introduced earlier. The theorem and its variations are closely related (and in many cases, equivalent) to several other *max-min* duality results for graphs and directed networks and to several results outside graph theory as well. Some of these are stated in this section.

The proof, which involves several steps, appears at the end of this section so that we may first present a number of consequences and related results.

Theorem 3.5.4: [**Menger**] *Let u and v be distinct, nonadjacent vertices in a connected graph G. Then the maximum number of internally disjoint u-v paths in G equals the minimum number of vertices needed to separate u and v.*

Variations and Consequences of Menger's Theorem

The vertex-connectivity of a graph can be expressed in terms of the *local connectivity* between a given pair of vertices, and this relationship is used in the proof of Whitney's Theorem given below.

DEFINITION: Let s and t be nonadjacent vertices of a connected graph G. Then the ***local vertex-connectivity*** between s and t, denoted $\kappa_v(s, t)$, is the size of a smallest s-t separating vertex-set in G.

Lemma 3.5.5: *Let G be a connected graph containing at least one pair of nonadjacent vertices. Then the vertex-connectivity $\kappa_v(G)$ is the minimum of the local vertex-connectivity $\kappa_v(s, t)$, taken over all pairs of nonadjacent vertices s and t.*

Proof: Since each s-t separating vertex-set of the graph G is a vertex-cut, it follows that $\kappa_v(G) \leq \kappa_v(s, t)$ for all pairs of nonadjacent vertices s and t. Thus, $\kappa_v(G)$ is less than or equal to the minimum of $\kappa_v(s, t)$ over all nonadjacent s and t. On the other hand, if S is a vertex-cut of size $\kappa_v(G)$, then there are at least two vertices, say s and t, that lie in different components of the vertex-deletion subgraph $G - S$. But then $\kappa_v(s, t) \leq \kappa_v(G)$, which implies that the minimum of $\kappa_v(s, t)$ over all nonadjacent s and t is less than or equal to $\kappa_v(G)$. ◇

The following variation of Menger's Theorem was published by Whitney in 1932. It generalizes the characterization of 2-connected graphs given by Theorem 3.4.7.

Theorem 3.5.6: [***Whitney's k-Connected Characterization***] *A nontrivial graph G is k-connected if and only if for each pair u, v of vertices, there are at least k internally disjoint u-v paths in G.*

Proof: If every two vertices in G are adjacent, then the special case of the vertex-connectivity definition applies, and the theorem assertion is immediately true. So assume that G has at least two nonadjacent vertices.

If G is k-connected, then there are at least k vertices in any vertex-cut of G. Thus, there are at least k vertices in any u-v separating set. Theorem 3.5.4 implies that the maximum number of internally disjoint u-v paths is at least k. Hence, there are at least k internally disjoint u-v paths.

Conversely, if for each pair of vertices u and v, there are at least k internally disjoint, u-v paths, then Proposition 3.5.1 implies that $\kappa_v(u, v) \geq k$, for each pair u, v of nonadjacent vertices. Therefore, $\kappa_v(G) \geq k$, by Lemma 3.5.5. ◇

Corollary 3.5.7: *Let G be a k-connected graph and let $u, v_1, v_2, \ldots, v_k$ be any $k+1$ distinct vertices of G. Then there are paths P_i from u to v_i, for $i = 1, \ldots, k$, such that the collection $\{P_i\}$ is internally disjoint.*

Proof: Construct a new graph G^w from graph G by adding a new vertex w to G together with an edge joining w to v_i, for $i = 1, \ldots, k$, as in Figure 3.5.2.

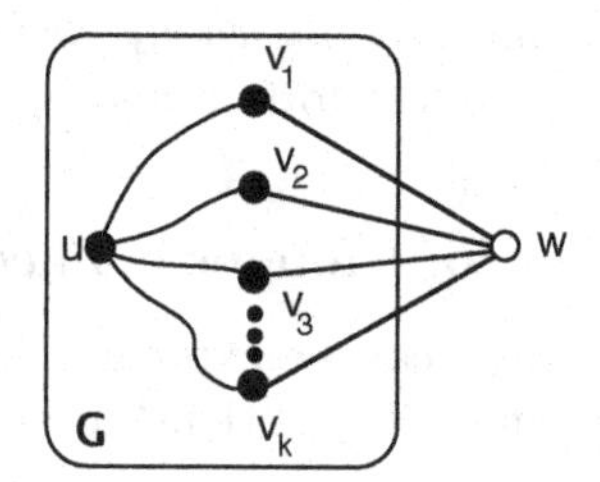

Figure 3.5.2 The graph G^w is constructed from G.

Since graph G is k-connected, it follows that graph G^w is also k-connected (by Exercise 5.1.18). By Theorem 3.5.6, there are k internally disjoint u-w paths in G^w. The u-v_i portions of these paths are k internally disjoint paths in G. ◇

The following theorem, proved by Dirac in 1960, generalizes one-half of the characterization of 2-connected graphs given in Corollary 3.4.8.

Theorem 3.5.8: [***Dirac Cycle Theorem***] *Let G be a k-connected graph with at least $k+1$ vertices, for $k \geq 3$, and let U be any set of k vertices in G. Then there is a cycle in G containing all the vertices in U.*

Proof: Let C be a cycle in G that contains the maximum possible number of vertices of set U, and suppose that $\{v_1, \ldots, v_m\}$ is the subset of vertices of U that lie on C. By Corollary 3.4.8, $m \geq 2$. If there were a vertex $u \in U$ not on cycle C, then by Corollary 3.5.7, there would exist a set of internally disjoint paths P_i from u to v_i, one for each $i = 1, \ldots, m$. But then cycle C could be extended to include vertex u, by replacing the cycle edge between v_1 and v_2 by the paths P_1 and P_2 (see Figure 3.5.3), and this extended cycle would contradict the maximality of cycle C. ◇

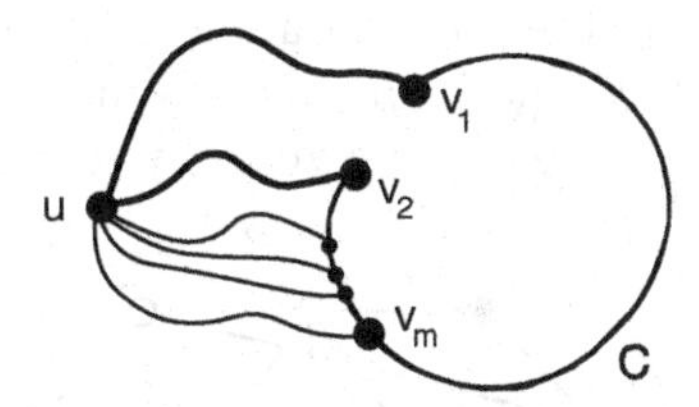

Figure 3.5.3 Extending cycle C to include vertex u.

Analogues of Menger's Theorem

Some of the results given at the beginning of this section have the following edge analogues. Each can be proved by mimicking the proof of the corresponding vertex version. The proofs are left as exercises.

Proposition 3.5.9: [***Edge Form of Certificate of Optimality***] *Let u and v be any two vertices of a connected graph G. Suppose P_{uv} is a collection of edge-disjoint u-v paths in G, and S_{uv} is a u-v separating set of edges in G, such that $|P_{uv}| = |S_{uv}|$. Then P_{uv} is the largest possible collection of edge-disjoint u-v paths, and S_{uv} is the smallest possible u-v separating set. In other words, each is an optimal solution to its respective problem.* ♢ *(Exercises)*

Theorem 3.5.10: [***Edge Form of Menger's Theorem***] *Let u and v be any two distinct vertices in a graph G. Then the minimum number of edges of G needed to separate u and v equals the maximum size of a set of edge-disjoint u-v paths in G.* ♢ (Exercises)

Theorem 3.5.11: [***Whitney's k-Edge-Connected Characterization***] *A nontrivial graph G is k-edge-connected if and only if for every two distinct vertices u and v of G, there are at least k edge-disjoint u-v paths in G.* ♢ *(Exercises)*

Example 3.5.3: For the graph shown below, it is easy to find four edge-disjoint u-v paths and a u-v separating edge-set of size 4. Thus, the maximum number of edge-disjoint u-v paths and the minimum size of a u-v separating set are both 4.

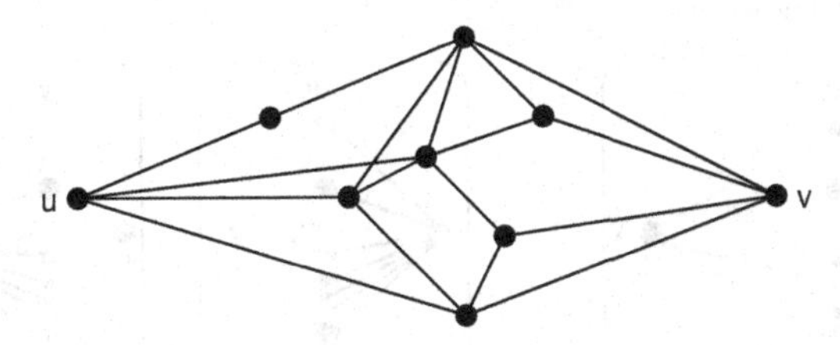

Proof of Menger's Theorem

DEFINITION: Let W be a set of vertices in a graph G and x another vertex not in W. A ***strict*** x-W ***path*** is a path joining vertex x to a vertex in W and containing no other vertex of W. A ***strict*** W-x ***path*** is the reverse of a strict x-W path (i.e., its sequence of vertices and edges is in reverse order).

Example 3.5.4: Corresponding to the u-v separating set $W = \{y, s, z\}$ in the graph shown below, the vertex sequences $\langle u, x, y\rangle$, $\langle u, r, y\rangle$, $\langle u, r, s\rangle$, and $\langle u, z\rangle$ represent the four strict u-W paths, and the three strict W-v paths are given by $\langle z, v\rangle$, $\langle y, t, v\rangle$, and $\langle s, v\rangle$.

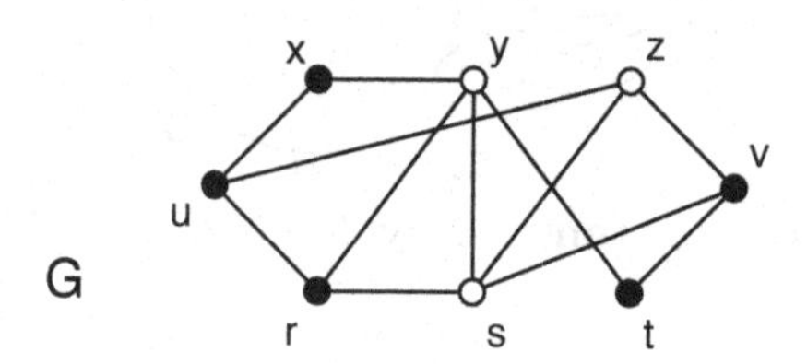

Proof: The proof uses induction on the number of edges. The smallest graph that satisfies the premises of the theorem is the path graph from u to v of length 2, and the theorem is trivially true for this graph. Assume that the theorem is true for all connected graphs having fewer than m edges, for some $m \geq 3$.

Now suppose that G is a connected graph with m edges, and let k be the minimum number of vertices needed to separate the vertices u and v. By Corollary 5.3.2, it suffices to show that there exist k internally disjoint u-v paths in G. Since this is clearly true if $k = 1$ (since G is connected), we may assume that $k \geq 2$.

Assertion 5.3.4a: If G contains a u-v path of length 2, then G contains k internally disjoint u-v paths.

Proof of 5.3.4a: Suppose that $P = \langle u, e_1, x, e_2, v\rangle$ is a path in G of length 2. Let W be a smallest u-v separating set for the vertex-deletion subgraph $G - x$. Since $W \cup \{x\}$ is a u-v separating set for G, the minimality of k implies that $|W| \geq k - 1$.

By the induction hypothesis, there are at least $k-1$ internally disjoint $u - v$ paths in $G - x$. Path P is internally disjoint from any of these, and, hence, there are k internally disjoint u-v paths in G. ◇ (Assertion 5.3.4a)

If there is a u-v separating set that contains a vertex adjacent to *both* vertices u and v, then Assertion 5.3.4a guarantees the existence of k internally disjoint u-v paths in G. The argument for *distance* $(u, v) \geq 3$ is broken into two cases, according to the kinds of u-v separating sets that exist in G.

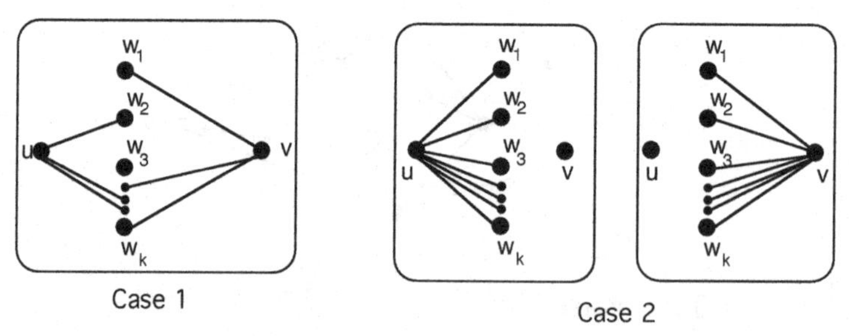

Figure 3.5.4 The two cases remaining in the proof of Menger's Theorem.

In Case 1, there exists a u-v separating set W, as depicted on the left in Figure 3.5.4, where neither u nor v is adjacent to every vertex of W. In Case 2, no such separating set exists. Thus, in every u-v separating set for Case 2, either every vertex is adjacent to u or every vertex is adjacent to v, as shown on the right in the figure.

Case 1: There exists a u-v separating set $W = \{w_1, w_2, \ldots, w_k\}$ of vertices in G of minimum size k, such that neither u nor v is adjacent to every vertex in W.

Let G_u be the subgraph induced on the union of the edge-sets of all strict u-W paths in G, and let G_v be the subgraph induced on the union of edge-sets of all strict W-v paths (see Figure 3.5.5).

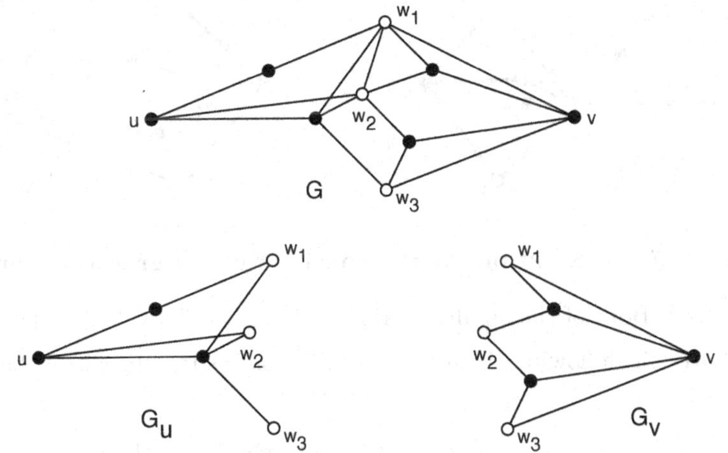

Figure 3.5.5 An example illustrating the subgraphs G_u and G_v.

Assertion 5.3.4b: Both of the subgraphs G_u and G_v have more than k edges.

Proof of 5.3.4b: For each $w_i \in W$, there is a u-v path P_{w_i} in G on which w_i is the only vertex of W (otherwise, $W - \{w_i\}$ would still be a u-v separating set, contradicting the minimality of W). The u-w_i subpath of P_{w_i} is a strict u-W path that ends at w_i. Thus, the final edge of this strict u-W path is different for each w_i. Hence, G_u has at least k edges.

The only way G_u could have exactly k edges would be if each of these strict u-W paths consisted of a single edge joining u and w_i, $i = 1, \ldots, k$. But this is ruled out by the condition for Case 1. Therefore, G_u has more than k edges. A similar argument shows that G_v also has more than k edges. (Assertion 5.3.4b)

Assertion 5.3.4c: The subgraphs G_u and G_v have no edges in common.

Proof of 5.3.4c: By way of contradiction, suppose that the subgraphs G_u and G_v have an edge in common. By the definitions of G_u and G_v, edge e is an edge of both a strict u-W path and a strict W-v path. Hence, at least one of its endpoints, say x, is not a vertex in the u-v separating set W (see Figure 3.5.6). But this implies the existence of a u-v path in $G - W$, which contradicts the definition of W. ♢ (Assertion 5.3.4c)

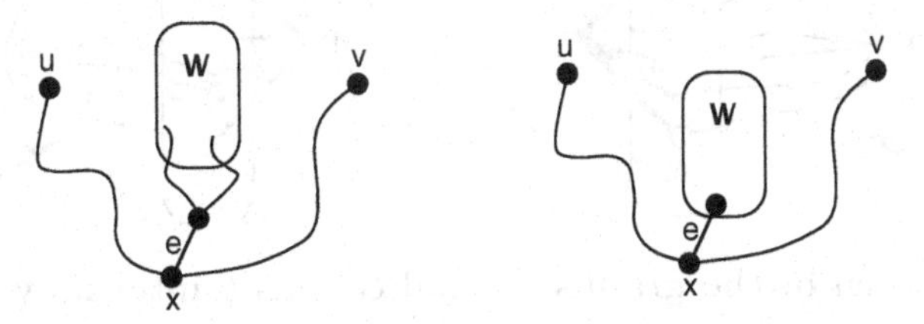

Figure 3.5.6 At least one of the endpoints of edge e lies outside W.

We now define two auxiliary graphs G_u^* and G_v^* : G_u^x is obtained from G by replacing the subgraph G_v with a new vertex v^* and drawing an edge from each vertex in W to v^* and G_v^* is obtained by replacing G_u with a new vertex u^* and drawing an edge from u^* to each vertex in W (see Figure 3.5.7).

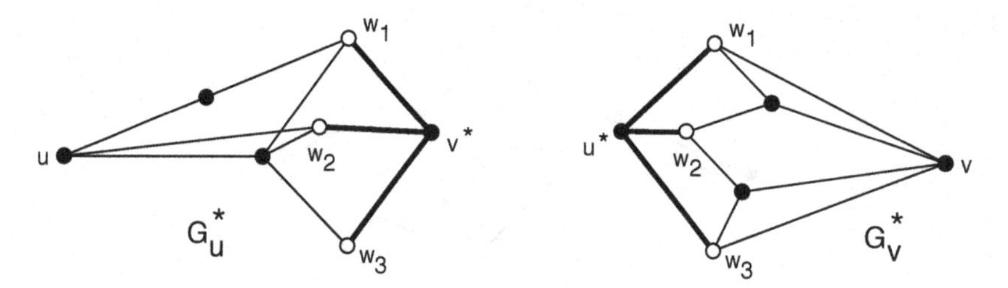

Figure 3.5.7 Illustration for the construction of graphs G_u^* and G_v^*.

Assertion 5.3.4d: Both of the auxiliary graphs G_u^* and G_v^* have fewer edges than G.

Proof of 5.3.4d: The following chain of inequalities shows that graph G_u^* has fewer edges than G.

$$\begin{aligned} |E_G| &\geq |E_{G_u \cup G_u}| \quad \text{(since } G_u \cup G_v \text{ is a subgraph of } G\text{)} \\ &= |E_{G_u}| + |E_{G_v}| \quad \text{(by Assertion 5.3.4c)} \\ &> |E_{G_u}| + k \quad \text{(by Assertion 5.3.4b)} \\ &= |E_{G_u^*}| \quad \text{(by the construction of } G_u^*\text{)} \end{aligned}$$

A similar argument shows that G_v^* also has fewer edges than G. ♢ (Assertion 5.3.4d)

By the construction of graphs G_u^* and G_v^*, every u-v^* separating set in graph G_u^* and every u^*-v separating set in graph G_v^* is a u-v separating set in graph G. Hence, the set W is a smallest u-v^* separating set in G_u^* and a smallest u^*-v separating set in G_v^*. Since G_u^* and G_v^* have fewer edges than G, the induction hypothesis implies the existence of two collections, P_u^* and P_v^*, of k internally disjoint u-v^* paths in G_v^* and k internally disjoint u^*-v paths in G_v^*, respectively (see Figure 3.5.8 below). For each w_i, one of the paths in P_u^* consists of a u-w_i path P_i' in G plus the new edge from w_i to v^*, and one of the paths in P_v^* consists of the new edge from u^* to w_i followed by a w_i-v path P_i'' in G.

Let P_i be the concatenation of paths P_i' and P_i'', for $i = 1, \ldots, k$. Then the set $\{P_i\}$ is a collection of k internally disjoint u-v paths in G. ♢ (Case 1)

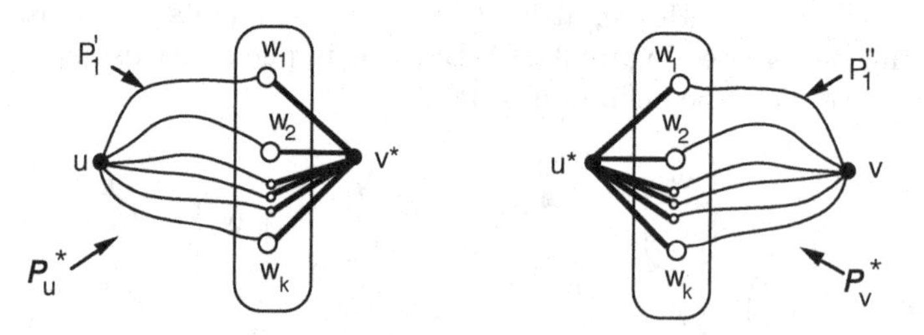

Figure 3.5.8 Each of the graphs G_u^* and G_v^* has k internally disjoint paths.

Case 2: Suppose that for each u-v separating set of size k, one of the vertices u or v is adjacent to all the vertices in that separating set.

Let $P = \langle u, e_1, x_1, e_2, x_2, \ldots, v\rangle$ be a shortest u-v path in G. By Assertion 5.3.4a, we can assume that P has length at least 3 and that vertex x_1 is not adjacent to vertex v. By Proposition 3.4.3, the edge-deletion subgraph $G - e_2$ is connected. Let S be a smallest u-v separating set in subgraph $G - e_2$ (see Figure 3.5.9). Then S is a u-v separating set in the vertex-deletion subgraph $G - x_1$ (since $G - x_1$ is a subgraph of $G - e_2$). Thus, $S \cup \{x_1\}$ is a u-v separating set in G, which implies that $|S| \geq k - 1$, by the minimality of k. On the other hand, the minimality of $|S|$ in $G - e_2$ implies that $|S| \leq k$, since every u-v separating set in G is also a u-v separating set in $G - e_2$.

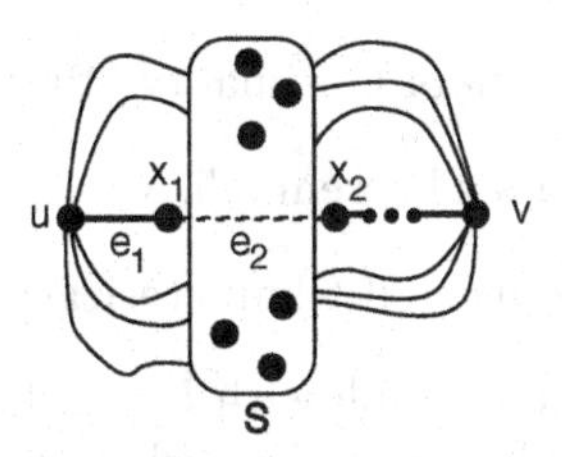

Figure 3.5.9 Completing Case 2 of Menger's Theorem.

If $|S| = k$, then, by the induction hypothesis, there are k internally disjoint u-v paths in $G - e_2$ and, hence, in G. If $|S| = k - 1$, then $x_i \notin S$, $i = 1, 2$ (otherwise $S - \{x_1\}$ would be a u-v separating set in G, contradicting the minimality of k). Thus, the sets $S \cup \{x_1\}$ and $S \cup \{x_2\}$ are both of size k and both u-v separating sets of G. The condition for Case 2 and the fact that vertex x_1 is not adjacent to v imply that every vertex in S is adjacent to vertex u. Hence, no vertex in S is adjacent to v (lest there be a u-v path of length 2). But then condition for Case 2 applied to $S \cup \{x_2\}$ implies that vertex x_2 is adjacent to vertex u, which contradicts the minimality of path P and completes the proof. ◊

EXERCISES for Section 3.5

3.5.1 For the given graph:

i. Find the maximum number of internally disjoint u-v paths for the given graph, and use Certificate of Optimality (Corollary 3.5.3) to justify your answer.

ii. Find the maximum number of edge-disjoint u-v paths for the specified graph, and use Edge Form of Certificate of Optimality (Proposition 3.5.9) to justify your answer.

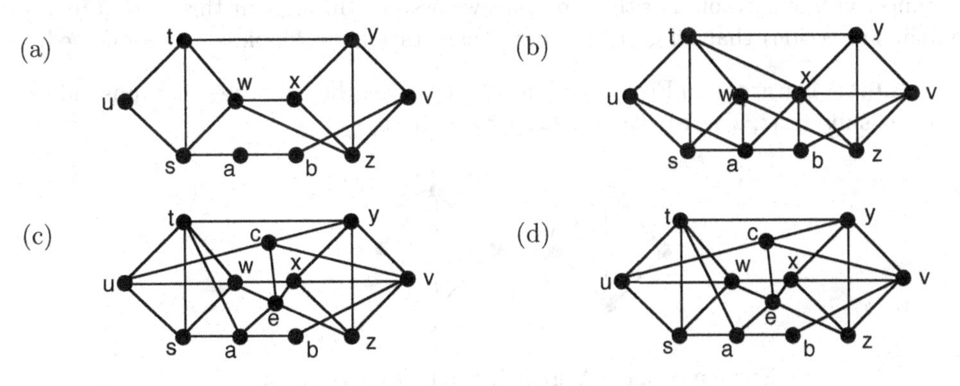

3.5.2 Find the maximum number of arc-disjoint directed u-v paths for the given graph, and use the digraph version of Proposition 3.5.9 to justify your answer. That is, find k arc-disjoint u-v paths and a set of k arcs that separate vertices u and v, for some integer k.

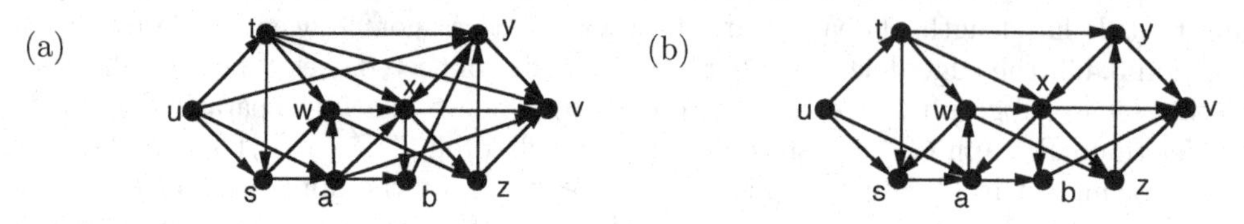

3.5.3 Prove Certificate of Optimality (Corollary 3.5.3).

3.5.4 Prove Edge Form of Certificate of Optimality (Proposition 3.5.9).

3.5.5 Prove Edge Form of Menger's Theorem 3.5.10.

3.5.6 Prove Whitney's k-Edge-Connected Characterization Theorem 3.5.11.

3.5.7 Let G be a simple k-connected graph with $k \geq 2$. Let S be a set of two edges and W a set of $k-2$ vertices. Prove that there exists a cycle in G containing the elements of S and W.

3.5.8 Let G be a simple k-connected graph. Let $W = \{w_1, w_2, \ldots, w_k\}$ be a set of k vertices and $v \in V_G - W$. Prove that there exists a v-w_i path P_i, $i = 1, \ldots, k$, such that the collection $\{P_i\}$ of k paths is internally disjoint.

3.6 BLOCK DECOMPOSITIONS

DEFINITION: A ***block*** of a loopless graph is a maximal connected subgraph H such that no vertex of H is a cut-vertex of H.

Thus, if a block has at least three vertices, then it is a maximal 2-connected subgraph. The only other types of blocks (in a loopless graph) are isolated vertices or dipoles (2-vertex graphs with a single edge or a multi-edge).

Remark: The blocks of a graph G are the blocks of the components of G and can therefore be identified one component of G at a time. Also, self-loops (or their absence) have no effect on the connectivity of a graph. For these reasons we assume throughout this section (except for the final subsection) that all graphs under consideration are loopless and connected.

Example 3.6.1: The graph in Figure 3.6.1 has four blocks; they are the subgraphs induced on the vertex subsets $\{t, u, w, v\}$, $\{w, x\}$, $\{x, y, z\}$, and $\{y, s\}$.

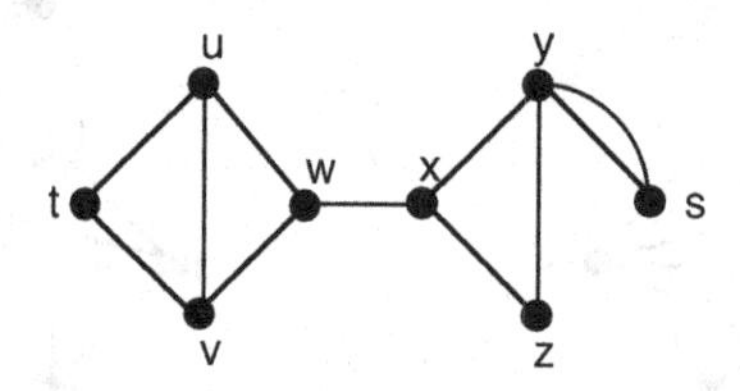

Figure 3.6.1 A graph with four blocks.

Remark: By definition, a block H of a graph G has no cut-vertices (of H), but H may contain vertices that are cut-vertices of G. For instance, in the above figure, the vertices w, x, and y are cut vertices of G.

The complete graphs K_n have no cut-vertices. The next result concerns the other extreme.

Proposition 3.6.1: *Every nontrivial connected graph G contains two or more vertices that are not cut-vertices.*

Proof: Choose two 1-valent vertices of a spanning tree of graph G. ♢

Proposition 3.6.2: *Two different blocks of a graph can have at most one vertex in common.*

Proof: Let B_1 and B_2 be two different blocks of a graph G, and suppose that x and y are vertices in $B_1 \cap B_2$. Since the vertex-deletion subgraph $B_1 - x$ is a connected subgraph of B_1, there is a path in $B_1 - x$ between any given vertex w_1 in $B_1 - x$ and vertex y. Similarly, there is a path in $B_2 - x$ from vertex y to any given vertex w_2 in $B_2 - x$ (see Figure 3.6.2). The concatenation of these two paths is a w_1-w_2 walk in the vertex-deletion subgraph $(B_1 \cup B_2) - x$, which shows that x is not a cut-vertex of the subgraph $B_1 \cup B_2$. The same argument shows that no other vertex in $B_1 \cap B_2$ is a cut-vertex of $B_1 \cup B_2$. Moreover, none of the vertices that are in exactly one of the B_i's is a cut-vertex of $B_1 \cup B_2$, since such a vertex would be a cut-vertex of that block B_i. Thus, the subgraph $B_1 \cup B_2$ has no cut-vertices, which contradicts the maximality of blocks B_1 and B_2. ♢

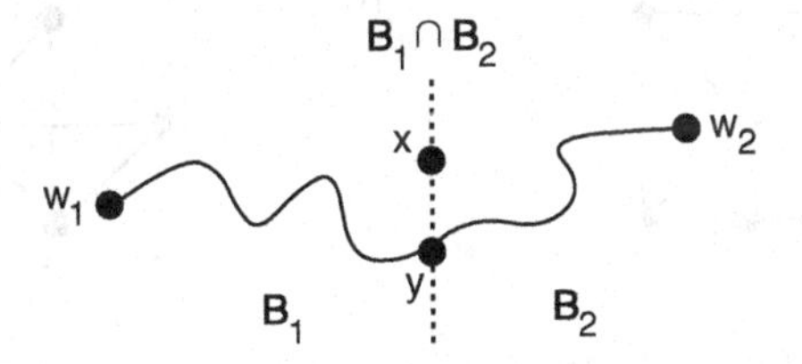

Figure 3.6.2 Two blocks cannot have more than one vertex in common.

The following assertions are immediate consequences of Proposition 3.6.2.

Corollary 3.6.3: *The edge-sets of the blocks of a graph G partition E_G.* ♢ *(Exercises)*

Corollary 3.6.4: *Let x be a vertex in a graph G. Then x is a cut-vertex of G if and only if x is in two different blocks of G.* ♢ *(Exercises)*

Corollary 3.6.5: *Let B_1 and B_2 be distinct blocks of a connected graph G. Let y_1 and y_2 be vertices in B_1 and B_2, respectively, such that neither is a cut-vertex of G. Then vertex y_1 is not adjacent to vertex y_2.* ♢ *(Exercises)*

DEFINITION: The ***block graph*** of a graph G, denoted $BL(G)$, is the graph whose vertices correspond to the blocks of G, such that two vertices of $BL(G)$ are joined by a single edge whenever the corresponding blocks have a vertex in common.

Example 3.6.2: Figure 3.6.3 shows a graph G and its block graph $BL(G)$.

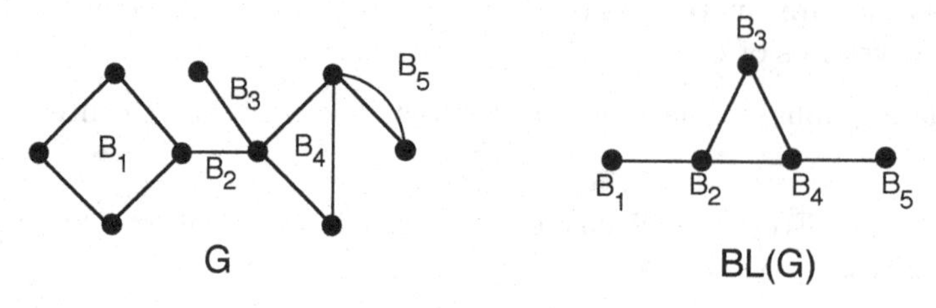

Figure 3.6.3 A graph and its block graph.

DEFINITION: A ***leaf block*** of a graph G is a block that contains exactly one cut-vertex of G.

The following result is used in §6.1 to prove *Brooks's Theorem* concerning the *chromatic number* of graph.

Proposition 3.6.6: *Let G be a connected graph with at least one cut-vertex. Then G has at least two leaf blocks.* ♢ *(Exercises)*

EXERCISES for Section 3.6

3.6.1 Identify the blocks in the given graph and draw the block graph.

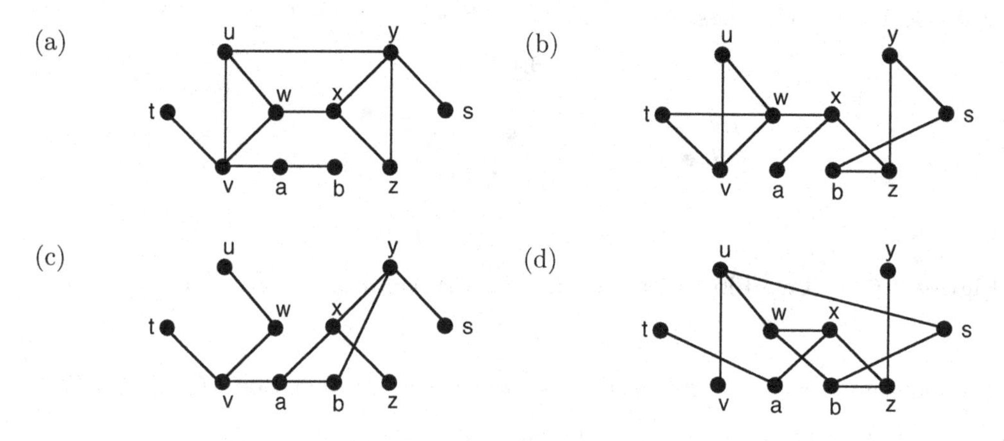

3.6.2 Draw a graph whose block graph is the complete graph K_3.

3.6.3 Find two non-isomorphic connected graphs with six vertices, six edges, and three blocks.

3.6.4 Find a graph whose block graph is the n-cycle graph C_n.

3.6.5 How many non-isomorphic simple connected graphs are there that have seven vertices, seven edges, and three blocks?

3.6.6 Prove or disprove: Every simple graph is the block graph of some graph.

3.6.7 Prove or disprove: Two graphs are isomorphic if and only if their block graphs are isomorphic.

3.6.8 Prove Corollary 3.6.3.

3.6.9 Prove Corollary 3.6.4.

3.6.10 Prove Corollary 3.6.5.

DEFINITION: Let G be a simple connected graph with at least two blocks. The ***block-cutpoint graph*** $bc(G)$ of G is the bipartite graph with vertex bipartition $\langle V_b, V_c \rangle$, where the vertices in V_b bijectively correspond to the blocks of G and the vertices in V_c bijectively correspond to the cut-vertices of G, and where vertex V_b is adjacent to vertex v_c if cut-vertex c is a vertex of block b.

3.6.11 Draw the block-cutpoint graph $bc(G)$ of the specified graph G, and identify the vertices in $bc(G)$ that correspond to leaf blocks of G.

(a) The graph of Exercise 3.6.1(a). (b) The graph of Exercise 3.6.1(b).
(c) The graph of Exercise 3.6.1(c). (d) The graph of Exercise 3.6.1(d).

3.6.12 Let G be a simple connected graph with at least two blocks. Prove that the block-cutpoint graph bc(G) is a tree.

3.6.13 Prove Proposition 3.6.6. (**Hint**: See Exercise 3.6.12.)

3.7 SUPPLEMENTARY EXERCISES

3.7.1 Draw every tree T such that the edge-complement $\bar{T}$ is a tree.

3.7.2 What are the minimum and maximum independence numbers of an n-vertex tree?

3.7.3 What are the minimum and maximum number of vertex orbits in an n-vertex tree?

3.7.4 Calculate the vertex-connectivity of $K_{4,7}$.

3.7.5 Prove that the complete bipartite graph $K_{m,m}$ is m-connected.

3.7.6 Prove that the vertex-connectivity of the hypercube graph Q_n is n.

3.7.7 In the n-dimensional hypercube graph Q_n, find n edge-disjoint paths from $(0, 0, \ldots, 0)$ to $(1, 1, \ldots, 1)$.

3.7.8 Prove that deleting two vertices from the graph below is insufficient to separate vertex s from vertex t.

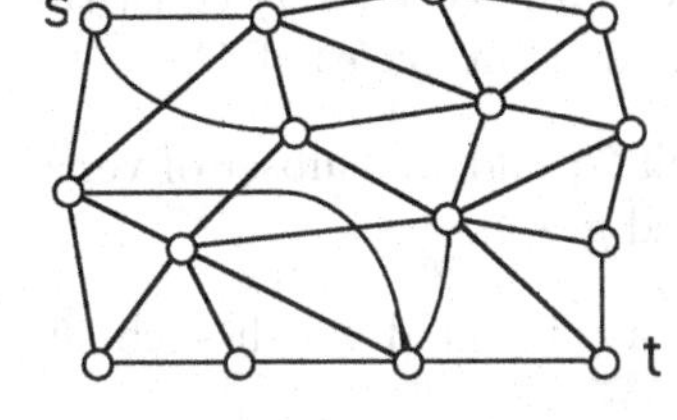

3.7.9 How many vertices must be removed from the graph below to separate vertex s from vertex t?

s

t

Glossary

Betti number of a graph: synonym for *cycle rank.*

block of a loopless graph G: a maximal connected subgraph H of G such that no vertex of H is a cut-vertex of H; in a graph with self-loops, each self-loop and its endpoint are regarded as a distinct block, isomorphic to the bouquet B_1.

—, leaf: a block that contains exactly one cut-vertex.

block-cutpoint graph $bc(G)$ of a simple connected graph G with at least two blocks: the bipartite graph with vertex bipartition $\langle V_b, V_c \rangle$, where the vertices in V_b bijectively correspond to the blocks of G and the vertices in V_c bijectively correspond to the cut-vertices of G, and where vertex v_b is adjacent to vertex v_c if cut-vertex c is a vertex of block b.

block graph $BL(G)$ of a graph G: the graph whose vertices correspond to the blocks of G, such that two vertices of $BL\,(G)$ are joined by a single edge whenever the corresponding blocks have a vertex in common.

bridge: synonym for *cut-edge.*

component of a vertex v in a graph G, denoted $C_G(v)$: the subgraph of G induced on the set of vertices that are reachable from v.

k**-connected graph** G**:** a connected graph with $\kappa_v(G) \geq k$.

cut-edge of a graph: an edge whose removal increases the number of components.

cutpoint: a synonym for *cut-vertex.*

cut-vertex of a graph: a vertex whose removal increases the number of components.

cycle addition to a graph G: the addition of a cycle that has exactly one vertex in common with G.

cycle rank of a graph G: the number of edges in the complement of a full spanning forest of G; the quantity $|E_G| - |V_G|$ plus the number of components of G.

cycle space of a graph G: the subspace $W_C(G)$ consisting of the null graph $\emptyset$, all cycles in G, and all unions of edge-disjoint cycles of G.

disconnects a connected graph G: said of a subset of vertices or edges whose deletion from G results in a non-connected graph.

k**-edge-connected graph** G**:** a connected graph such that every edge-cut has at least k edges (i.e., $\kappa_e(G) \geq k$.

edge-connectivity of a connected graph G: the minimum number of edges whose removal can disconnect G; denoted $\kappa_e(G)$.

edge-cut in a graph G: a subset D of edges such that $G - D$ has more components than G.

edge-cut rank of a graph G: the quantity $|V_G|$ minus the number of components of G; the number of edges in a spanning forest of G.

edge-cut space of a graph G: the subspace $W_S(G)$ consisting of the null graph $\emptyset$, all minimal edge-cuts of G, and all unions of edge-disjoint minimal edge-cuts of G.

edge space $W_E(G)$ of a graph G: the vector space over $GF(2)$ consisting of the collection of edge subsets of E_G under ring sum.

Eulerian graph: a graph that has an Eulerian tour.

Eulerian tour in a graph: a closed trail that contains every edge of that graph.

forest: an acyclic graph.

full spanning forest of a graph G: a spanning forest consisting of a collection of trees, such that each tree is a spanning tree of a different component of G.

fundamental cycle of a graph G associated with a full spanning forest F and an edge e not in F: the unique cycle that is created when the edge e is added to the forest F.

fundamental edge-cut of a graph G associated with a full spanning forest F and an edge e in F: the unique partition-cut $\langle V_1, V_2 \rangle$ of G, where V_1 and V_2 are the vertex-sets of the two components of the subgraph $F - e$.

fundamental system of cycles of a graph G associated with a full spanning forest F: the collection of fundamental cycles of the graph G that result from each addition of an edge to the spanning forest F.

fundamental system of edge-cuts of a graph G associated with a full spanning forest F: the collection of fundamental edge-cuts of the graph G that result from each removal of an edge from the spanning forest F.

internal vertex in a rooted tree: a vertex that is not a leaf.

internally disjoint paths: paths that have no internal vertex in common.

leaf block: see *block, leaf.*

leaf in an undirected tree: a vertex of degree 1.

local vertex-connectivity between nonadjacent vertices s and t: the size of a smallest s-t separating vertex-set; denoted $\kappa_v(s,t)$.

partition-cut $\langle X_1, X_2 \rangle$ of a graph G: the set of all edges of G having one endpoint in X_1 and the other endpoint in X_2, where X_1 and X_2 form a partition of V_G.

relative complement of a subgraph H in a graph G, denoted $G - H$: the edge-deletion subgraph $G - E(H)$.

ring sum of two sets A and B: the set $(A - B) \cup (B - A)$, denoted $A \oplus B$.

u-v **separating set** S of vertices or of edges: a set S such that the vertices u and v lie in different components of the deletion subgraph $G - S$.

spanning forest of a graph G: an acyclic spanning subgraph of G.

—, full: a spanning forest consisting of a collection of trees, such that each tree is a spanning tree of a different component of G.

spanning tree of a (connected) graph: a spanning subgraph that is a tree.

strict x-W path: a path joining vertex x to a vertex in vertex-set W and containing no other vertex of W. A strict W-x path is the reverse of a strict x-W path.

unicyclic graph: a connected graph with exactly one cycle.

vertex-connectivity of a connected graph G: the minimum number of vertices whose removal can disconnect G or reduce it to a 1-vertex graph; denoted $\kappa_v(G)$.

—, local between nonadjacent vertices s and t: the size of a smallest s-t separating vertex-set; denoted $\kappa_v(s,t)$.

vertex-cut in a graph G: a subset U of vertices such that $G - U$ has more components than G.

vertex space $\mathcal{W}_V(G)$ of a graph G: the vector space over $GF(2)$ consisting of the collection of vertex subsets of V_G under ring sum.

Chapter 4

PLANARITY AND KURATOWSKI'S THEOREM

INTRODUCTION

The central theme of this chapter is the topological problem of deciding whether a given graph can be drawn in the plane or sphere with no edge-crossings. Some planarity tests for graphs are in the form of algebraic formulas (see §4.5) based on the numbers of vertices and edges. These are the easiest tests to apply, yet they are one-way tests, and there are difficult cases in which they are inconclusive.

A celebrated result of the Polish mathematician Kasimir Kuratowski transforms the planarity decision problem into the combinatorial problem of calculating whether the given graph contains a subgraph *homeomorphic* (defined in §4.2) to the complete graph K_5 or to the complete bipartite graph $K_{3,3}$. Directly searching for K_5 and $K_{3,3}$ would be quite inefficient, but this chapter includes a simple, practical algorithm (see §4.6) to test for planarity.

The relationship of planarity to topological graph theory is something like the relationship of plane geometry to what geometers call geometry, where mathematicians long ago began to develop concepts and methods to progress far beyond the plane. Whereas planarity consists mostly of relatively accessible ideas that were well understood several decades ago, topological graph theorists use newer methods to progress to all the other surfaces.

4.1 PLANAR DRAWINGS AND SOME BASIC SURFACES

Our approach to drawings of graphs on surfaces begins with our intuitive notions of what we mean by a drawing and a surface. This chapter provides some precise examples of a surface.

Planar Drawings

Consistent with our temporary informality, we introduce the main topic of this chapter with the following definition.

DEFINITION: A ***planar drawing*** of a graph is a drawing of the graph in the plane without edge-crossings.

DEFINITION: A graph is said to be ***planar*** if there exists a planar drawing of it.

Example 4.1.1: Two drawings of the complete graph K_4 are shown in Figure 4.1.1. The planar drawing on the right shows that K_4 is a planar graph.

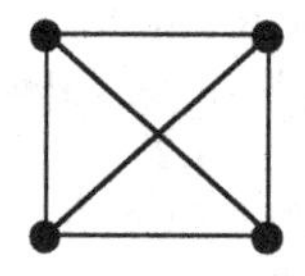

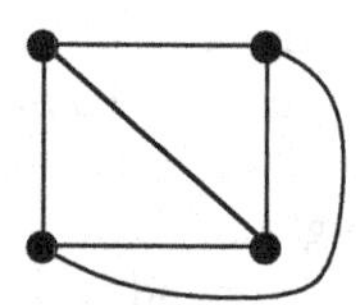

Figure 4.1.1 A non-planar drawing and a planar drawing of K_4.

Example 4.1.2: An instance of the problem of determining whether a given graph is planar occurs in the form of a well-known puzzle, called the ***utilities problem***, in which three houses are on one side of a street and three utilities (electricity, gas, and water) are on the other. The objective of the puzzle is to join each of the three houses to each of the three utilities without having any crossings of the utility lines. Later in this section, we show that this is impossible, by proving that $K_{3,3}$ is non-planar.

Remark: A graph G and a given drawing of G are categorically different objects. That is, a graph is combinatorial and a drawing is topological. In particular, the *vertices* and *edges* in a drawing of a graph are actually *images* of the vertices and edges in that graph. Yet, to avoid excessive formal phrasing, these distinctions are relaxed when it is discernible from context what is intended.

TERMINOLOGY: Intuitively, we see that in a planar drawing of a graph, there is exactly one *exterior* (or *infinite*) region whose area is infinite.

Example 4.1.3: The exterior region, R_e, and the three *finite* regions of a planar drawing of a graph are shown in Figure 4.1.2.

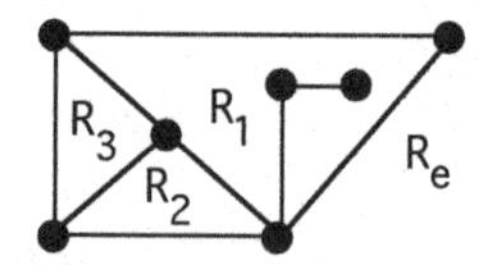

Figure 4.1.2 The four regions of a planar drawing of a graph.

Remark: If we consider a planar drawing of a graph on a piece of paper, then the intuitive notion of *region* corresponds to the pieces of paper that result from cutting the paper along the length of every edge in the drawing. If one adds new edges to a graph without crossing an existing edge, then a region may be subdivided into more regions.

TERMINOLOGY NOTE: We restrict the use of the word "regions" to the case of crossing-free drawings, since many assertions that are true in that case may be untrue when there are edge-crossings.

Three Basic Surfaces

All of the surfaces under consideration in this chapter are subsets of Euclidean 3-space. Although we assume that the reader is familiar with the following surfaces, it is helpful to think about their mathematical models.

DEFINITION: A ***plane*** in Euclidean 3-space $\mathbb{R}^3$ is a set of points (x, y, z) such that there are numbers a, b, c, and d with $ax + by + cz = d$.

DEFINITION: A ***sphere*** is a set of points in $\mathbb{R}^3$ equidistant from a fixed point, the *center*.

DEFINITION: The ***standard torus*** is the surface of revolution obtained by revolving a circle of radius 1 centered at $(2, 0)$ in the xy-plane around the y-axis in 3-space, as depicted in Figure 4.1.3. The solid inside is called the ***standard donut***.

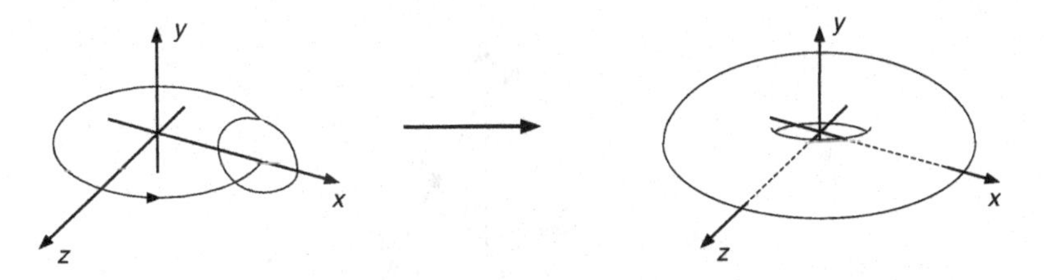

Figure 4.1.3 Creating a torus.

DEFINITION: The circle of intersection, as in Figure 4.1.4(a), of the standard torus with the half-plane $\{(x, y, z) \mid x = 0,\ z \geq 0\}$ is called the ***standard meridian***. We observe that the standard meridian bounds a disk inside the standard donut.

DEFINITION: The circle of tangent intersection of the standard torus with the plane $y = 1$ is called the ***standard longitude***. Figure 4.1.4(b) illustrates the standard longitude. We observe that the standard longitude bounds a disk in the plane $y = 1$ that lies outside the standard donut.

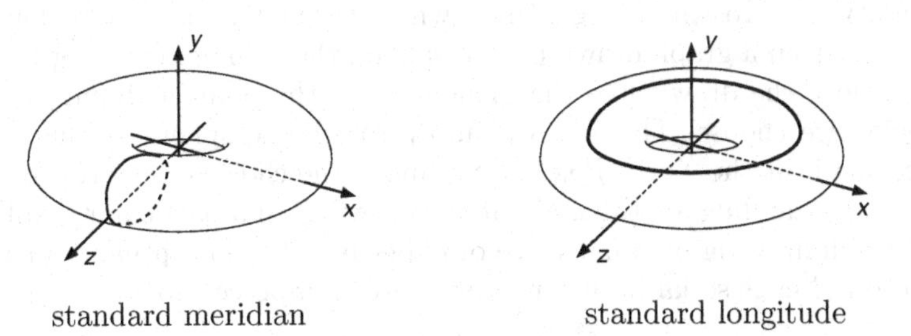

Figure 4.1.4

TERMINOLOGY: Any closed curve that circles the torus once in the meridian direction (the "short" direction) without circling in the longitude direction (the "long" direction) is called a ***meridian***. Any closed curve that circles the torus once in the longitude direction without circling in the meridian direction is called a ***longitude***.

Remark: Surfaces generally fall into two infinite sequences. The three surfaces described above are all relatively uncomplicated and lie in the sequence of orientable surfaces. Surfaces such as the *Möbius band* and the *Klein bottle* are examples of unorientable surfaces.

Riemann Stereographic Projection

Riemann observed that deleting a single point from a sphere yields a surface that is equivalent to the plane for many purposes, including that of drawing graphs.

DEFINITION: The ***Riemann stereographic projection*** is the function ρ that maps each point w of the unit-diameter sphere (tangent at the origin $(0,0,0)$ to the xz-plane in Euclidean 3-space) to the point $\rho(w)$ where the ray from the North Pole $(0,1,0)$ through point w intersects the xz-plane.

Under the Riemann projection, the "southern hemisphere" of the sphere is mapped continuously onto the unit disk. The "northern hemisphere" (minus the North Pole) is mapped continuously onto the rest of the plane. The points nearest to the North Pole are mapped to the points farthest from the origin. Figure 4.1.5 illustrates the construction.

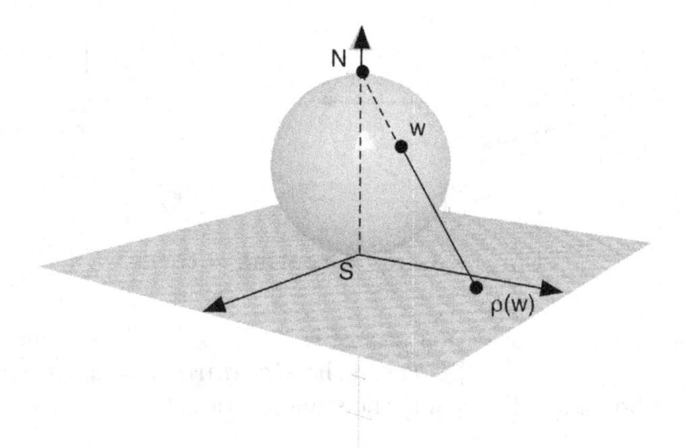

Figure 4.1.5 The Riemann stereographic projection.

Remark: Whereas a planar drawing of a graph has one exterior region (containing the "point at infinity"), a crossing-free graph drawn on the sphere has one region that contains the North Pole. Given a graph drawn on the sphere, the Riemann stereographic projection enables us to move the drawing to the plane so that the point at infinity is deleted from whatever region we choose. That is, we simply rotate the sphere so that a point in the designated region is at the North Pole. Accordingly, the choice of exterior region is usually irrelevant to understanding anything about the drawings of a particular graph in the plane. Moreover, a graph drawing on a flat piece of paper may be conceptualized as a drawing on a sphere whose radius is so large that its curvature is imperceptible.

Proposition 4.1.1: *A graph is planar if and only if it can be drawn without edge-crossings on the sphere.*

Proof: This is an immediate consequence of the Riemann stereographic projection. ◇

Jordan Separation Property

The objective of the definitions in this subsection is to lend precision to the intuitive notion of using a closed curve to separate a surface. Most of these definitions can be generalized.

DEFINITION: By a ***Euclidean set***, we mean a subset of any Euclidean space $\mathbb{R}^n$.

DEFINITION: An ***open path*** from s to t in a Euclidean set X is the image of a continuous bijection f from the unit interval $[0, 1]$ to a subset of X such that $f(0) = s$ and $f(1) = t$. (One may visualize a path as the trace of a particle traveling through space for a fixed length of time.)

DEFINITION: A ***closed path*** or ***closed curve*** in a Euclidean set is the image of a continuous function 1 from the unit interval $[0, 1]$ to a subset of that space such that $f(0) = f(1)$, but which is otherwise a bijection. (For instance, this would include a "knotted circle" in space.)

Example 4.1.4: Figure 4.1.6 shows an open path and a closed curve in the plane.

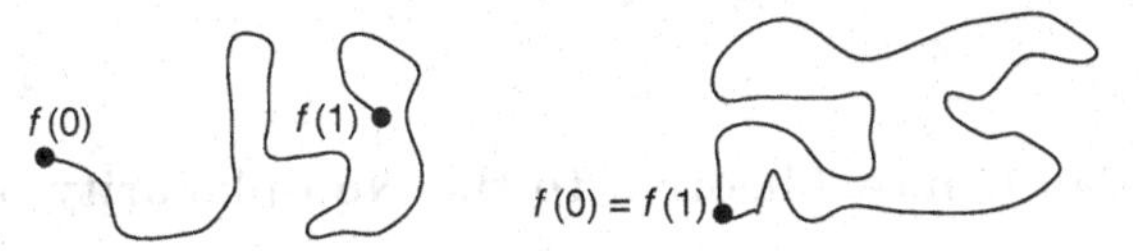

Figure 4.1.6 Open path and closed curve (= closed path).

Example 4.1.5: In a crossing-free drawing of a graph on a surface, a cycle of the graph is a closed curve.

DEFINITION: A Euclidean set X is ***connected*** † if for every pair of points $s, t \in X$, there exists a path within X from s to t.

DEFINITION: The Euclidean set X ***separates*** the connected Euclidean set Y if there exist a pair of points s and t in $Y - X$, such that every path in Y from s to t intersects the set X.

Example 4.1.6: Figure 4.1.7 shows a meandering closed curve in the plane. It separates the white part from the shaded part.

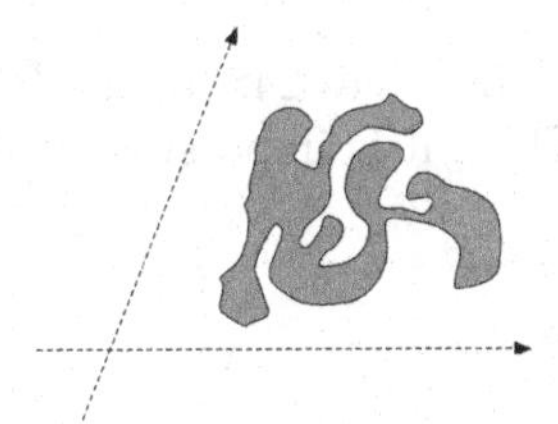

Figure 4.1.7 A closed curve separating the plane.

† A topologist would say *path-connected.*

What makes the plane and the sphere the simplest surfaces for drawing graphs is the *Jordan separation property.* By invoking this property, we can prove that K_5 and $K_{3,3}$ cannot be drawn without edge-crossings in the plane or the sphere.

DEFINITION: A Euclidean set X has the ***Jordan separation property*** if every closed curve in X separates X.

Theorem 4.1.2: [***Jordan Curve Theorem***] *Every closed curve in the sphere (plane) has the Jordan separation property, that is, it separates the sphere (plane) into two regions, one of which contains the North Pole (contains "infinity").*† ◇ *(proof omitted)*

Corollary 4.1.3: *A path from one point on the boundary of a disk through the interior to another point on the boundary separates the disk.* ◇ *(proof omitted)*

We again emphasize that the Jordan separation property distinguishes the plane and sphere from other surfaces. For instance, although it is possible to draw a closed curve on a torus that separates the surface into two parts (e.g., just draw a little circle around a point), a meridian does not, nor does a longitude, as one sees clearly in Figure 4.1.4.

The Jordan Curve Theorem is quite difficult to prove in full generality; in fact, Jordan himself did it incorrectly in 1887. The first correct proof was by Veblen in 1905. A proof for the greatly simplified case in which the closed curve consists entirely of straight-line segments, so that it is a closed polygon, is given by Courant and Robbins in *What Is Mathematics?*.

Applying the Jordan Curve Theorem to the Non-planarity of K_5 and $K_{3,3}$

It is possible to prove that a particular graph is non-planar directly from the Jordan Curve Theorem. In what follows, we continue to rely temporarily on an intuitive notion of the *regions* of a crossing-free drawing of a graph on a sphere or plane. A more formal definition of *region* is given at the beginning of §4.3.

Theorem 4.1.4: *Every drawing of the complete graph K_5 in the sphere (or plane) contains at least one edge-crossing.*

Proof: Label the vertices $0, \ldots, 4$. By the Jordan Curve Theorem, any drawing of the cycle $(1, 2, 3, 4, 1)$ separates the sphere into two regions. Consider the region with vertex 0 in its interior as the "inside" of the cycle. By the Jordan Curve Theorem, the edges joining vertex 0 to each of the vertices 1, 2, 3, and 4 must also lie entirely inside the cycle, as illustrated in Figure 4.1.8. Moreover, each of the 3-cycles $(0, 1, 2, 0)$, $(0, 2, 3, 0)$, $(0, 3, 4, 0)$, and $(0, 4, 1, 0)$ also separates the sphere, and, hence, edge 24 must lie to the exterior of the cycle $(1, 2, 3, 4, 1)$, as shown.

It follows that the cycle formed by edges 24, 40, and 02 separates vertices 1 and 3, again by the Jordan Curve Theorem. Thus, it is impossible to draw edge 13 without crossing an edge of that cycle. ◇

† By the Schönfliess theorem, both regions on a sphere and the interior region in a plane are topologically equivalent to open disks. For a proof, see [Ne54].

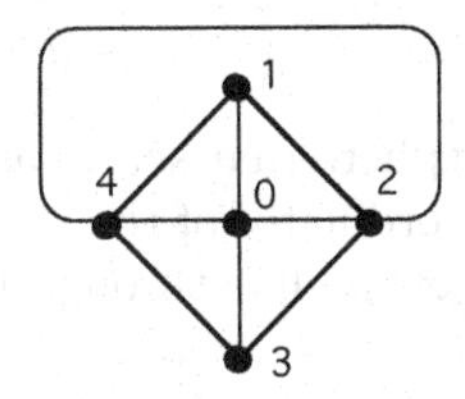

Figure 4.1.8 Drawing most of K_5 in the sphere.

Theorem 4.1.5: *Every drawing of the complete bipartite graph $K_{3,3}$ in the sphere (or plane) contains at least one edge-crossing.*

Proof: Label the vertices of one partite set 0, 2, 4, and of the other 1, 3, 5. By the Jordan Curve Theorem, cycle $(2, 3, 4, 5, 2)$ separates the sphere into two regions, and, as in the previous proof, we regard the region containing vertex 0 as the "inside" of the cycle. By the Jordan Curve Theorem, the edges joining vertex 0 to each of the vertices 3 and 5 lie entirely inside that cycle, and each of the cycles $(0, 3, 2, 5, 0)$ and $(0, 3, 4, 5, 0)$ separates the sphere, as illustrated in Figure 4.1.9.

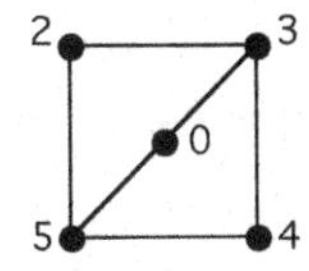

Figure 4.1.9 Drawing most of $K_{3,3}$ in the sphere.

Thus, there are three regions: the exterior of cycle $(2, 3, 4, 5, 2)$, and the inside of each of the other two cycles. It follows that no matter which region contains vertex 1, there must be some even-numbered vertex that is not in that region, and, hence, the edge from vertex 1 to that even-numbered vertex would have to cross some cycle edge. ◊

Corollary 4.1.6: *If either K_5 or $K_{3,3}$ is a subgraph of a graph G, then every drawing of G in the sphere (or plane) contains at least one edge-crossing.*

DEFINITION: The complete graph K_5 and the complete bipartite graph $K_{3,3}$ are called the ***Kuratowski graphs***, as shown in Figure 4.1.10.

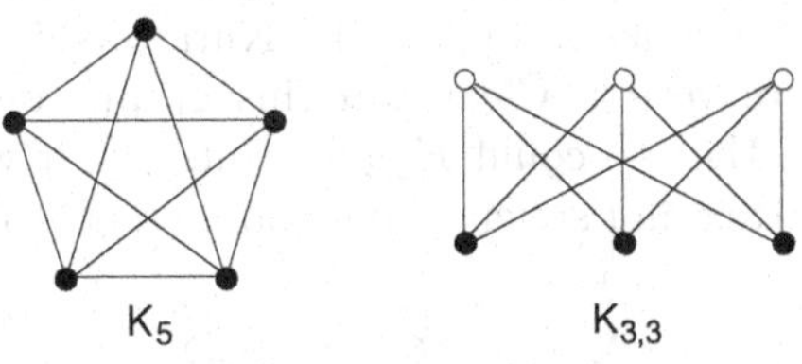

Figure 4.1.10 The Kuratowski graphs.

Example 4.1.7: *Figure 4.1.11 illustrates that K_5 can be drawn without edge-crossings on the torus, even though this is impossible on the sphere. It can be shown that for every graph there is a surface on which a crossing-free drawing is possible.*

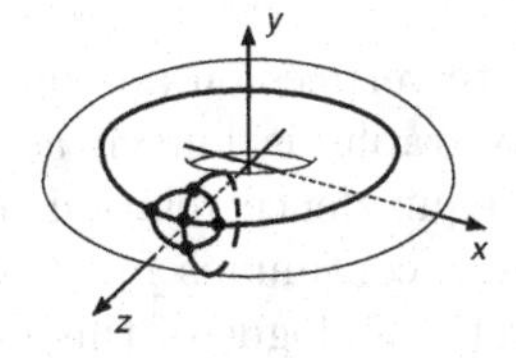

Figure 4.1.11 A crossing-free drawing of K_5 on the torus.

EXERCISES for Section 4.1

4.1.1 For the given point w on the sphere that serves as the domain of the Riemann stereographic projection, calculate the coordinates of the point $\rho(w)$ in the xz-plane to which it is projected. (Hint: Write the locus of the line through the North Pole and w.)

(a) $(\frac{1}{2}, \frac{1}{2}, 0)$ (b) $\left(\frac{1}{4}, \frac{3}{4}, \frac{\sqrt{2}}{4}\right)$

(c) $\left(\frac{1}{5}, \frac{9}{10}, \frac{\sqrt{5}}{10}\right)$ (d) $\left(\frac{1}{3}, \frac{3}{4}, \frac{\sqrt{11}}{12}\right)$

4.1.2 For the given point w in the xz-plane, calculate the coordinates of the point $\rho^{-1}(w)$ in the unit sphere under the inverse Riemann stereographic projection. (Hint: Use analytic geometry, as in the previous four exercises.)

(a) $\left(\frac{\sqrt{3}}{3}, 0, 0\right)$ (b) $(3, 0, 4)$

(c) $(1, 0, 1)$ (d) $(\frac{1}{2}, 0, 0)$

4.1.3 Explain why every complete graph K_n, $n \geq 5$, and every complete bipartite graph $K_{m,n}$, $m, n \geq 3$, is not planar.

4.1.4 Determine if the hypercube Q_3 is planar.

4.1.5 Determine if the complete tripartite graph $K_{2,2,2} = 2K_1 + K_{2,2}$ is planar.

4.1.6 Determine if the complete tripartite graph $K_{2,2,3} = 2K_1 + K_{2,3}$ is planar.

4.1.7 Determine if the complete 4-partite graph $K_{2,2,2,2} = K_{2,2} + K_{2,2}$ is planar.

4.2 SUBDIVISION AND HOMEOMORPHISM

The graph in Figure 4.2.1 looks a lot like the Kuratowski graph $K_{3,3}$, but one of the "edges" has an intermediate vertex. Clearly, if this graph could be drawn on a surface without any edge-crossings, then so could $K_{3,3}$,. Intuitively, it would be an easy matter of erasing the intermediate vertex and splicing the "loose ends" together.

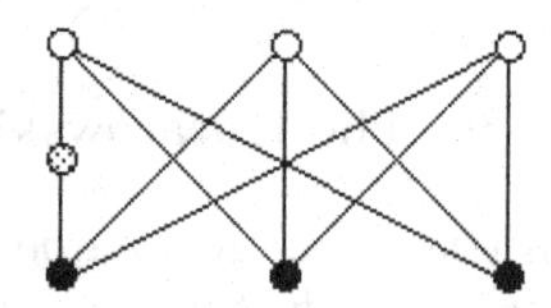

Figure 4.2.1 Subdividing an edge.

Placing one or more intermediate vertices on an edge is formally known as *subdivision*. The relationship between the two graphs is known as *homeomorphism*. The concept of homeomorphism originated as an equivalence relation of topological spaces. It is defined in topology to be a one-to-one, onto, continuous function with a continuous inverse. This section introduces the graph-theoretic analogue of this topological notion.

Graph Subdivision

DEFINITION: Let e be an edge in a graph G. ***Subdividing the edge*** e means that a new vertex w is added to V_G, and that edge e is replaced by two edges. If e is a proper edge with endpoints u and v, then it is replaced by e' with endpoints u and w and e'' with endpoints w and v. If e is a self-loop incident on vertex u, then e' and e'' form a multi-edge between w and u.

Figure 4.2.2 Subdividing an edge.

DEFINITION: Let w be a 2-valent vertex in a graph G, such that two proper edges e' and e'' are incident on w. ***Smoothing away*** (or ***smoothing out***) vertex w means replacing w and edges e' and e'' by a new edge e. If e' and e'' have only the endpoint w in common, then e is a proper edge joining the other endpoints of e' and e''. If e' and e'' have endpoints w and u in common (i.e., they form a multi-edge), then e is a self-loop incident on u.

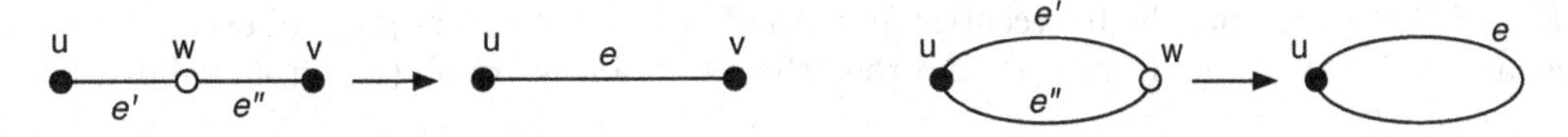

Figure 4.2.3 Smoothing away a vertex.

Example 4.2.1: The $(n+1)$-cycle graph can be obtained from the n-cycle graph by subdividing any edge, as illustrated in Figure 4.2.4. Inversely, smoothing away any vertex on the $(n+1)$-cycle, for $n \geq 1$, yields the n-cycle. It is not permitted to smooth away the only vertex of the 1-cycle C_1 (= the bouquet B_1).

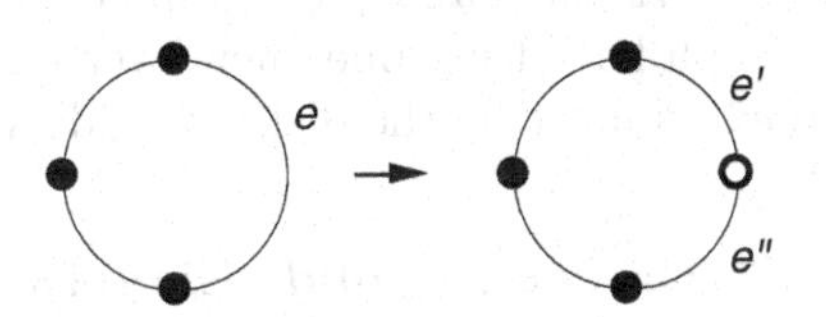

Figure 4.2.4 Subdividing an edge of the 3-cycle yields the 4-cycle.

DEFINITION: ***Subdividing a graph*** G means performing a sequence of edge-subdivision operations. The resulting graph is called a ***subdivision*** of the graph G.

Example 4.2.2: Performing any k subdivisions on the n-cycle graph C_n yields the $(n+k)$-cycle graph C_{n+k}, and C_n can be obtained by any k smoothing operations on C_{n+k}. Thus, C_{n+k} is a subdivision of C_k, for all $n \geq k \geq 1$.

Proposition 4.2.1: *A subdivision of a graph can be drawn without edge-crossings on a surface if and only if the graph itself can be drawn without edge-crossings on that surface.*

Proof: When the operations of subdivision and smoothing are performed on a copy of the graph already drawn on the surface, they neither introduce nor remove edge-crossings. ◊

Barycentric Subdivision

The operation of subdivision can be used to convert a general graph into a simple graph. This justifies a brief digression on our path toward a proof of Kuratowski's theorem.

DEFINITION: The ***(first) barycentric subdivision*** of a graph is the subdivision in which one new vertex is inserted in the interior of each edge.

Example 4.2.3: The concept of barycentric subdivision and the next three propositions are illustrated by Figure 4.2.5.

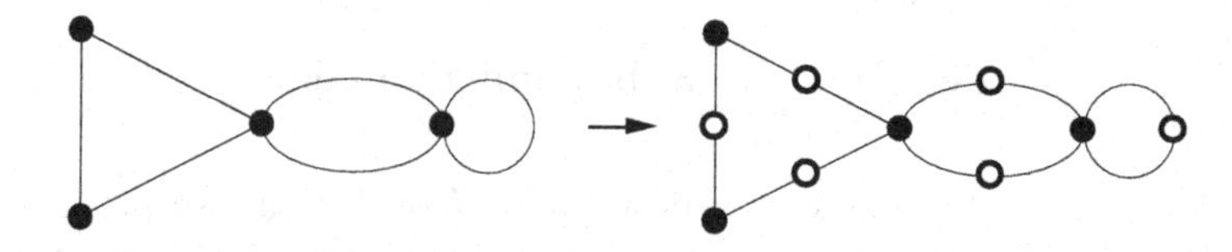

Figure 4.2.5 A graph and its barycentric subdivision.

Proposition 4.2.2: *The barycentric subdivision of any graph is a bipartite graph.*

Proof: Let G' denote the barycentric division of graph G. One endpoint of each edge in G' is an "old" vertex (i.e., from V_G), and the other endpoint is "new" (i.e., from subdividing). ♢

Proposition 4.2.3: *Barycentric subdivision of any graph yields a loopless graph.*

Proof: By Proposition 4.2.2, a barycentric subdivision of a graph is bipartite, and a bipartite graph has no self-loops. ♢

Proposition 4.2.4: *Barycentric subdivision of any loopless graph yields a simple graph.*

Proof: Clearly, barycentric subdivision of a loopless graph cannot create loops. If two edges of a barycentric subdivision graph have the same "new" vertex w as an endpoint, then the other endpoints of these two edges must be the distinct "old" vertices that were endpoints of the old edge subdivided by w. ♢

DEFINITION: For $n \geq 2$, the nth **barycentric subdivision** of a graph is the first barycentric subdivision of the $(n-1)^{st}$ barycentric subdivision.

Proposition 4.2.5: *The second barycentric subdivision of any graph is a simple graph.*

Proof: By Proposition 4.2.3, the first barycentric subdivision is loopless, and thus, by Proposition 4.2.4, the second barycentric subdivision is simple. See Figure 4.2.6. ♢

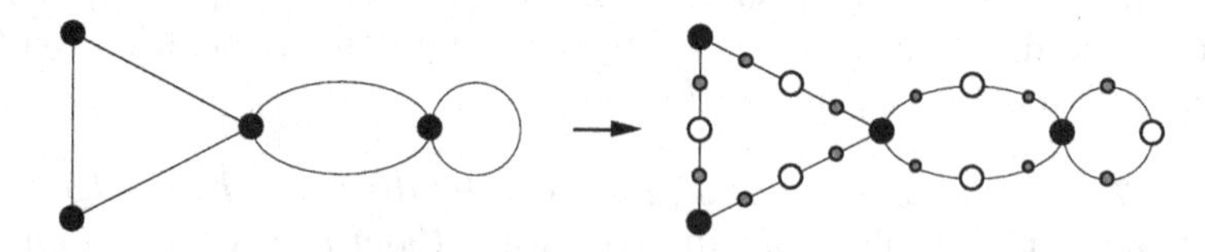

Figure 4.2.6 A graph and its second barycentric subdivision.

Graph Homeomorphism

DEFINITION: The graphs G and H are ***homeomorphic graphs*** if there is an isomorphism from a subdivision of G to a subdivision of H.

Example 4.2.4: Graphs G and H in Figure 4.2.7 cannot be isomorphic, since graph G is bipartite and graph H contains a 3-cycle. However, they are homeomorphic, because if edge d in graph G and edge e in graph H are both subdivided, then the resulting graphs are isomorphic.

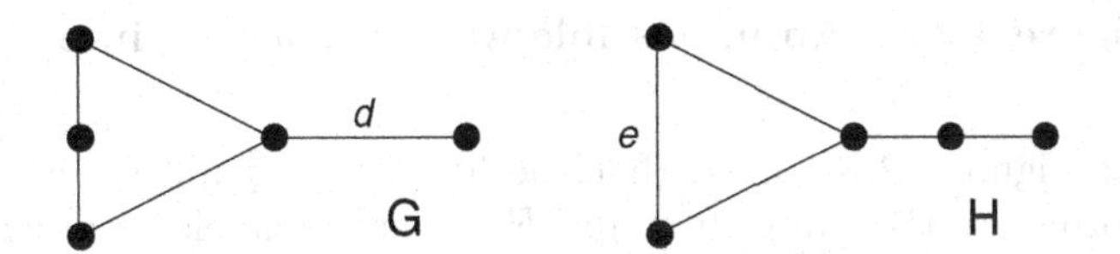

Figure 4.2.7 Two non-isomorphic graphs that are homeomorphic.

Remark: Notice in Example 4.2.4 that no subdivision of graph G is isomorphic to graph H, and that no subdivision of graph H is isomorphic to graph G.

Proposition 4.2.6: *Let G and H be homeomorphic graphs. Then G can be drawn without edge-crossings on a surface S if and only if H can be drawn on S without edge-crossings.*

Proof: This follows from iterated application of Proposition 4.2.1. ◇

Proposition 4.2.7: *Every graph is homeomorphic to a bipartite graph.*

Proof: By Proposition 4.2.2, the barycentric subdivision of a graph is bipartite. Of course, a graph is homeomorphic to a subdivision of itself. ◇

Subgraph Homeomorphism Problem

Deciding whether a graph G contains a subgraph that is homeomorphic to a target graph H is a common problem in graph theory.

Example 4.2.5: Figure 4.2.8 shows that the complete bipartite graph $K_{3,3}$ contains a subgraph that is homeomorphic to the complete graph K_4. Four of the edges of K_4 are represented by the cycle $0 - 1 - 2 - 3 - 0$, and the other two are represented by the two paths of length 2 indicated by broken lines.

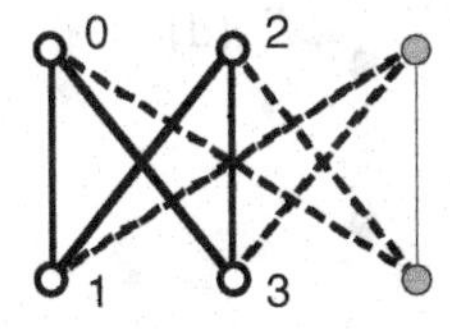

Figure 4.2.8 A homeomorphic copy of K_4 in $K_{3,3}$.

Notice in Figure 4.2.8 that each side of the bipartition of $K_{3,3}$ contains two of the images of vertices of K_4. A homeomorphic copy of K_4 in $K_{3,3}$ cannot have three vertex images on one side of the bipartition. To see this, suppose (without loss of generality) that vertex images 0, 1, and 2 are on one side, and that vertex 3 is on the other side, as illustrated in Figure 4.2.9 below. A homeomorphic copy of K_4 would require three internally disjoint paths joining the three possible pairs of vertices 0, 1, and 2. But this is impossible, since there are only two remaining vertices on the other side that can be used as internal vertices of these paths.

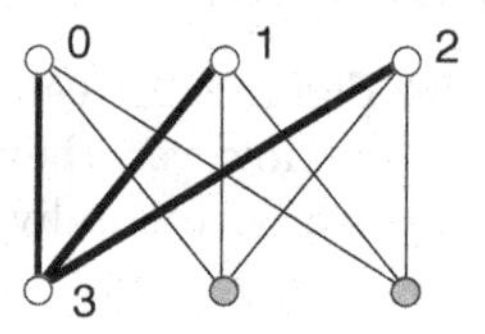

Figure 4.2.9 An impossible way to place K_4 into $K_{3,3}$.

Example 4.2.6: Figure 4.2.10 shows that the hypercube graph Q_4 contains a subgraph that is homeomorphic to the complete graph K_5. The five labeled white vertices represent vertices of K_5. The 10 internally disjoint paths joining the 10 different pairs of these vertices are given bold edges. Gray vertices are internal vertices along such paths.

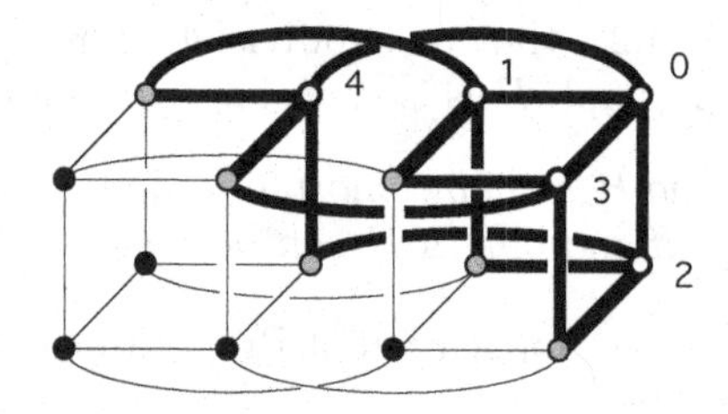

Figure 4.2.10 A homeomorphic copy of K_5 in Q_4.

EXERCISES for Section 4.2

4.2.1 For the given pair of graphs, either prove that they are homeomorphic or prove that they are not.

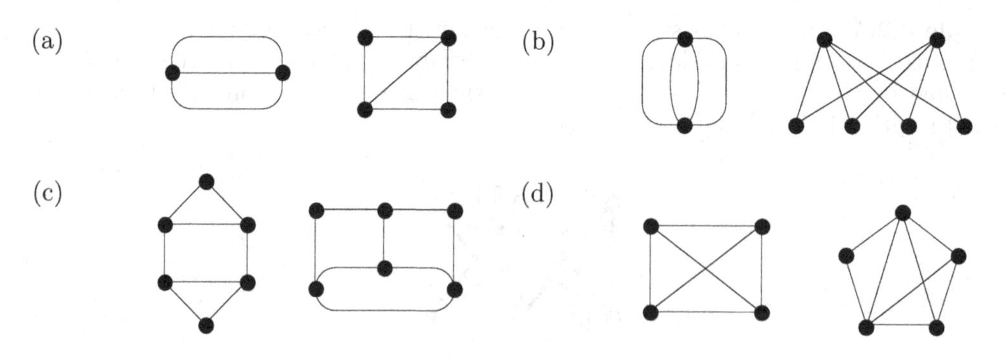

4.2.2 What is the minimum number of subdivision vertices required to make the bouquet B_n simple? (Hint: The second barycentric subdivision is not the minimum such subdivision.)

4.2.3 Show that two graphs are homeomorphic if and only if they are isomorphic to subdivisions of the same graph.

4.2.4 Prove that a homeomorphic copy of the first of the given graphs is contained in the second graph.

(a) K_4 in $K_{3,4}$.

(b) $K_{3,3}$ in $K_{4,5}$.

(c) W_4 in $K_{4,5}$.

(d) K_5 in $K_{4,5}$.

4.2.5 Prove that the second given graph does NOT contain a homeomorphic copy of the first graph.

(a) K_4 in $K_{2,3}$.

(b) $K_{3,3}$ in $K_{4,4} - 4K_2$.

(c) K_5 in $K_{4,4}$.

(d) Q_3 in $K_{3,5}$.

4.3 EXTENDING PLANAR DRAWINGS

We continue to regard a *drawing* of a graph on a surface as an intuitive notion. A more precise definition appears in the next chapter. The following definitions become mathematically precise as soon as the notion of a *drawing* is precise.

DEFINITION: An ***imbedding of a graph*** G on a surface S is a drawing without any edge-crossings. We denote the imbedding $\iota : G \to S$.

NOTATION: When the surface of the imbedding is a plane or sphere, we use S_0 to denote the surface.

Thus, a planar drawing of a graph G is an imbedding, $\iota : G \to S_0$, of G on the plane S_0.

DEFINITION: A ***region*** of a graph imbedding $\iota : G \to S$ is a component of the Euclidean set that results from deleting the image of G from the surface S.

DEFINITION: The ***boundary of a region*** of a graph imbedding $\iota : G \to S$ is the subgraph of G that comprises all vertices that abut that region and all edges whose interiors abut the region.

Remark: When a 2-connected graph is imbedded on the sphere, the boundary of every region is a cycle on the perimeter of the region. However, it is a dangerous misconception to imagine that this is the general case.

DEFINITION: A ***face*** of a graph imbedding $\iota : G \to S$ is the union of a region and its boundary.

Remark: A drawing does *not* have faces unless it is an imbedding.

Planar Extensions of a Planar Subgraph

A standard way to construct a planar drawing of a graph is to draw a subgraph in the plane, and then to extend the drawing by adding the remaining parts of the graph. This section gives the terminology and some basic results about the planar extensions of a subgraph. These results are helpful in proving Kuratowski's theorem and in specifying a planarity algorithm in the final two sections of this chapter.

Proposition 4.3.1: *A planar graph G remains planar if a multiple edge or self-loop is added to it.*

Proof: Draw graph G in the plane. Alongside any existing edge e of the drawing, another edge can be drawn between the endpoints of e, sufficiently close to e that it does not intersect any other edge. Moreover, at any vertex v, inside any region with v on its boundary, it is possible to draw a self-loop with endpoint v, sufficiently small that it does not intersect another edge. (See Figure 4.3.1.) ◊

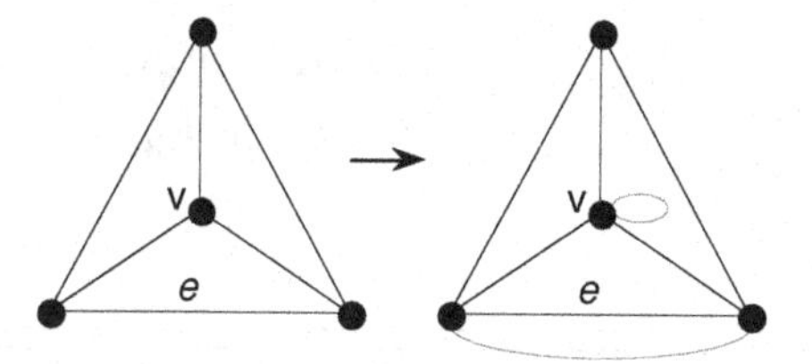

Figure 4.3.1 Adding a self-loop or a multi-edge preserves planarity.

Corollary 4.3.2: *A non-planar graph remains non-planar if a self-loop is deleted or if one edge of a multi-edge is deleted.*

Proposition 4.3.3: *A planar graph G remains planar if any edge is subdivided or if a new edge e is attached to a vertex $v \in V_G$ (with the other endpoint of e added as a new vertex).*

Proof: Clearly, placing a dot in the middle of an edge in a planar drawing of G to indicate subdivision does not create edge-crossings. Moreover, it is easy enough to insert edge e into any region that is incident on vertex v. (See Figure 4.3.2.) ◊

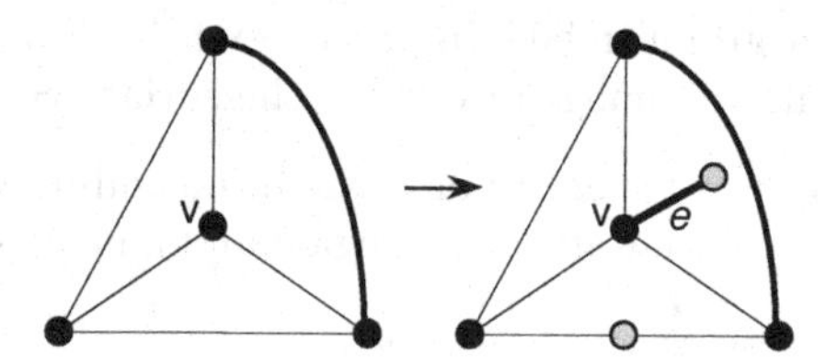

Figure 4.3.2 Subdividing an edge or adding a spike preserves planarity.

The following proposition applies to imbeddings in all surfaces, not just to imbeddings in the sphere.

Proposition 4.3.4: *Let $\iota : G \to S$ be a graph imbedding on a sphere or on any other surface. Let d be an edge of G with endpoints u and v. Then the imbedding of the graph $G - d$ obtained by deleting edge d from the imbedding $\iota : G \to S$ has a face whose boundary contains both of the vertices u and v.*

Proof: In the imbedding $\iota : G \to S$, let f be a face whose boundary contains edge d. When edge d is deleted, face f is merged with whatever face lies on the other side of edge d, as illustrated in Figure 4.3.3. On surface S, the boundary of the merged face is the union of the boundaries of the faces containing edge d, minus the interior of edge d. (This is true even when the same face f lies on both sides of edge d.) Vertices u and v both lie on the boundary of that merged face, exactly as illustrated. ◊

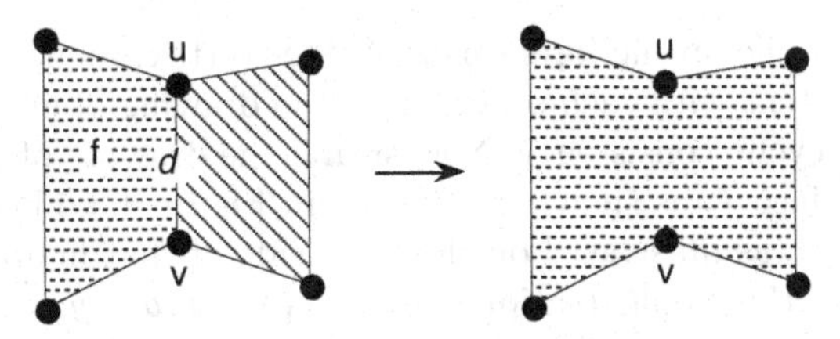

Figure 4.3.3 Merging two regions by deleting an edge.

Amalgamating Planar Graphs

We recall from §2.1 that amalgamating two graphs means pasting a subgraph in one graph to an isomorphic subgraph in the other.

Proposition 4.3.5: *Let f be a face of a planar drawing of a connected graph G. Then there is a planar drawing of G in which the boundary walk of face f bounds a disk in the plane that contains the entire graph G. That is, in the new drawing, face f is the "outer" face.*

Proof: Copy the planar drawing of G onto the sphere so that the North Pole lies in the interior of face f. Then apply the Riemann stereographic projection. ◊

Proposition 4.3.6: *Let f be a face of a planar drawing of a graph H, and let $u_1, \ldots, u_n$ be a subsequence of vertices in the boundary walk of f. Let f' be a face of a planar drawing of a graph J, and let $w_1, \ldots, w_n$ be a subsequence of vertices in the boundary walk of f'. Then the amalgamated graph $(H \cup J)/\{u_1 = w_1 \cdots, u_n = w_n\}$ is planar.*

Proof: Planar drawings of graphs H and J with $n = 3$ are illustrated in Figure 4.3.4.

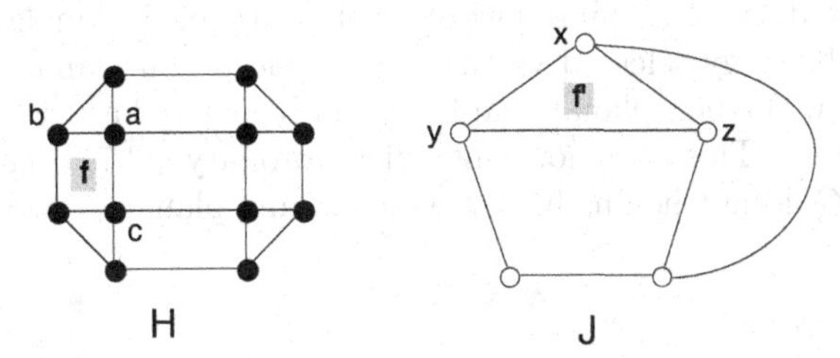

Figure 4.3.4 Two planar graph drawings.

First redraw the plane imbedding of H so that the unit disk lies wholly inside face f. Next redraw graph J so that the boundary walk of face f' surrounds the rest of graph J, which is possible according to Proposition 4.3.5. Figure 4.3.5 shows these redrawing of graphs H and J.

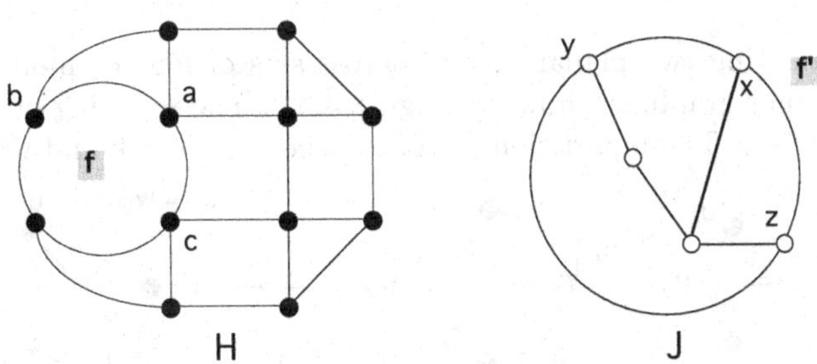

Figure 4.3.5 Redrawings of the two planar graph imbeddings.

We may assume that the cyclic orderings of the vertex sequences $\{u_1, \ldots, u_n\}$ and $\{w_1, \ldots, w_n\}$ are consistent with each other, since the drawing of graph J can be reflected, if necessary, to obtain cyclic consistency. Now shrink the drawing of graph J so that it fits inside the unit disk in face f, as shown on the left in Figure 4.3.6 below. Then stretch the small copy of J outward, as illustrated on the right side of that figure, thereby obtaining a crossing-free drawing of the amalgamation $(H \cup J)/\{a = x, b = y, c = z\}$. ◇

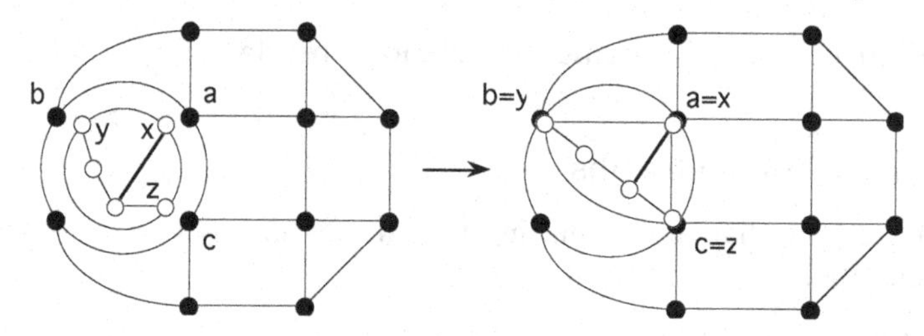

Figure 4.3.6 Position the two graphs and then amalgamate by stretching.

Corollary 4.3.7: *Let H and J be planar graphs. Let U be a set of one, two, or three vertices in the boundary of a face f of the drawing of H, and let W be a set of the same number of vertices in the boundary of a face f' of the drawing of J. Then the amalgamated graph $(H \cup J)/\{U = W\}$ is planar.*

Proof: Whenever there are at most three vertices in the vertex subsequences to be amalgamated, there are only two possible cyclic orderings. Since reflection of either of the drawings is possible, the vertices of sets U and W in the boundaries of faces f and f', respectively, can be aligned to correspond to any bijection $U \to W$. ◇

Remark: The requirement that vertices of amalgamation be selected in their respective graphs from the boundary of a single face cannot be relaxed. Amalgamating two planar graphs across two arbitrarily selected vertices may yield a non-planar graph. For instance, Figure 4.3.7 shows how the non-planar graph K_5 can be derived as the amalgamation of the planar graphs G and K_2. This does not contradict Corollary 4.3.7, because the two vertices of amalgamation in G do not lie on the same face of any planar drawing of G.

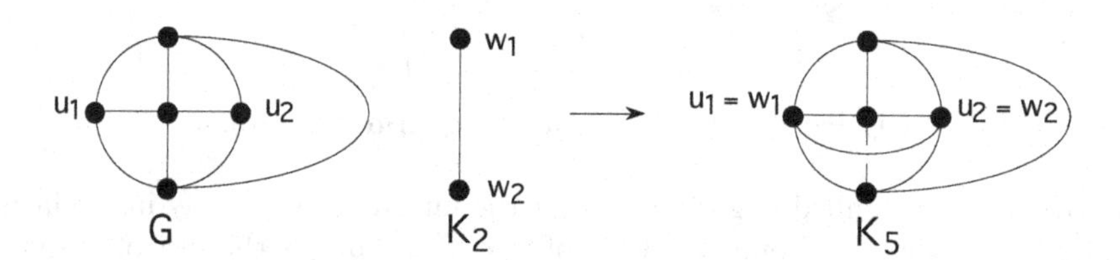

Figure 4.3.7 A 2-vertex amalgamation of two planar graphs into K_5.

Remark: Amalgamating two planar graphs across sets of four or more vertices per face may yield a non-planar graph, as shown in Figure 4.3.8. The resulting graph shown is $K_{3,3}$ with two doubled edges. The bipartition of $K_{3,3}$ is shown in black and white.

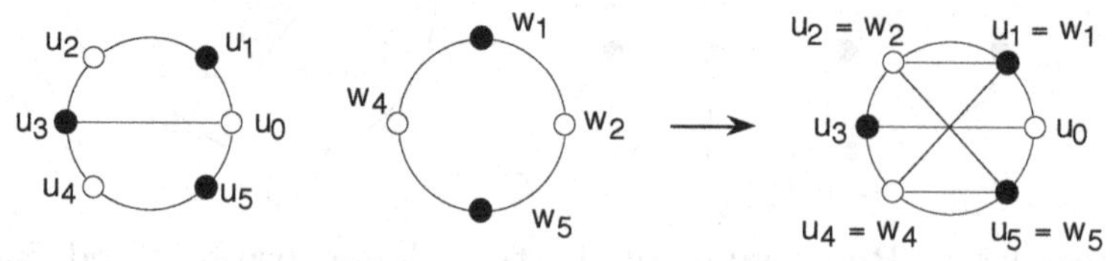

Figure 4.3.8 A 4-vertex amalgamation of two planar graphs into $K_{3,3}$.

Appendages to a Subgraph

In an intuitive sense, subgraph H of a graph G *separates* two edges if it is impossible to get from one edge to the other without going through H. The following definition make this precise.

DEFINITION: Let H be a subgraph of a connected graph G. Two edges e_1 and e_2 of $E_G - E_H$ are ***unseparated by subgraph*** H if there exists a walk in G that contains both e_1 and e_2, but whose internal vertices are not in H.

Remark: The relation *unseparated by subgraph* H is an equivalence relation on $E_G - E_H$; that is, it is reflexive, symmetric, and transitive.

DEFINITION: Let H be a subgraph of a graph G. Then an ***appendage to subgraph*** H is the induced subgraph on an equivalence class of edges of $E_G - E_H$ under the relation *unseparated by H*.

DEFINITION: Let H be a subgraph of a graph. An appendage to H is called a ***chord*** if it contains only one edge. Thus, a chord joins two vertices of H but does not lie in the subgraph H itself.

DEFINITION: Let H be a subgraph of a graph, and let B be an appendage to H. Then a ***contact point*** of B is a vertex of both B and H.

Example 4.3.1: In the graph G of Figure 4.3.9, subgraph H is the bold cycle. Appendage B_1 is the broken-edge subgraph with contact points a_1, a_2, and a_3. Appendage B_2 is the exterior subgraph with contact points a_3 and a_4. Appendage B_3 is the interior subgraph with contact points a_1, a_5, and a_6. Appendage B_4 is an exterior chord. Notice that within each of the three non-chord appendages, it is possible to get from any edge to any other edge by a walk in which none of the internal vertices is a contact point.

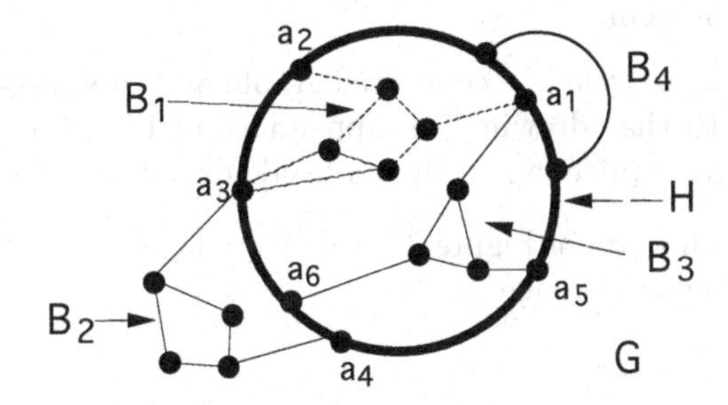

Figure 4.3.9 A subgraph H and its appendages B_1, B_2, B_3, and B_4.

Also notice that each non-chord appendage contains a single component of the deletion subgraph $G - V_H$. In addition to a component of $G - V_H$, a non-chord appendage also contains every edge extending from that component to a contact point, and the contact point as well.

Remark: Every subgraph of a graph (not just cycles) has appendages. Even a subgraph comprising a set of vertices and no edges would have appendages.

Overlapping Appendages

The construction of a planar drawing of a connected graph G commonly begins with the selection of a "large" cycle to be the subgraph whose appendages constitute the rest of G.

In this case, some special terminology describes the relationship between two appendages, in terms of their contact points.

DEFINITION: Let C be a cycle in a graph. The appendages B_1 and B_2 of C ***overlap*** if either of these conditions holds:

i. Two contact points of B_1 alternate with two contact points of B_2 on cycle C.
ii. B_1 and B_2 have three contact points in common.

Example 4.3.2: Both possibilities (i) and (ii) for overlapping appendages are illustrated in Figure 4.3.10.

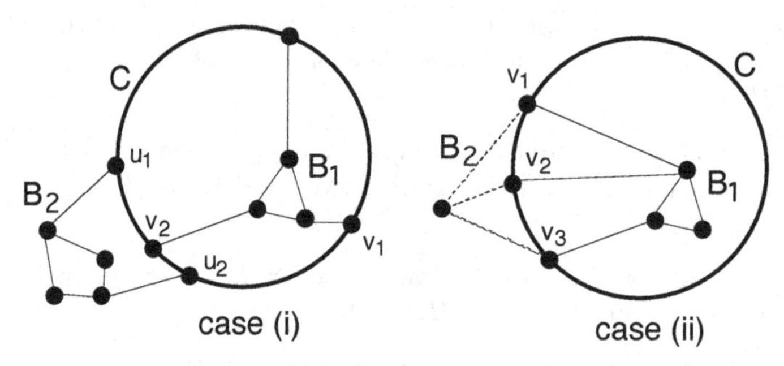

Figure 4.3.10 The two ways that appendages can overlap.

Proposition 4.3.8: *Let C be a cycle in a planar drawing of a graph, and let B_1 and B_2 be overlapping appendages of C. Then one appendage lies inside cycle C and the other outside.*

Proof: Overlapping appendages on the same side of cycle C would cross, by Corollary 4.1.3 to the Jordan Curve Theorem. ◊

TERMINOLOGY: Let C be a cycle of a connected graph, and suppose that C has been drawn in the plane. Relative to that drawing, an appendage of C is said to be *inner* or *outer*, according to whether that appendage is drawn inside or outside of C.

Example 4.3.3: In both parts of Figure 4.3.10, appendage B_1 is an inner appendage, and appendage B_2 is an outer appendage.

EXERCISES for Section 4.3

4.3.1 The given graphs are to be amalgamated so that each vertex labeled u_i is matched to the vertex v_i with the same subscript. Draw a planar imbedding of the resulting graph.

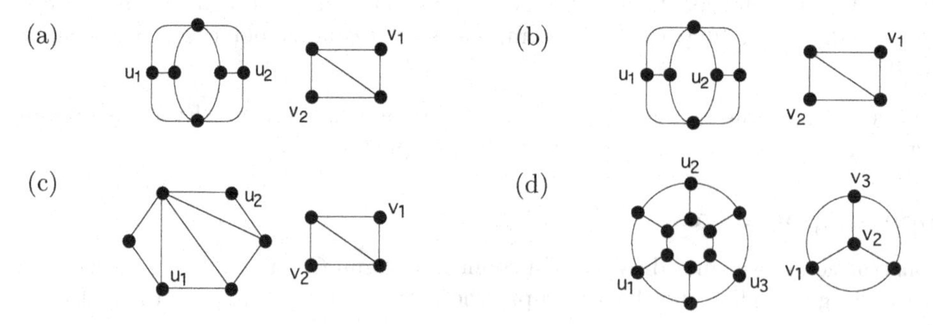

4.3.2 The given graphs are to be amalgamated so that each vertex labeled u_i is matched to the vertex v_i with the same subscript. Either draw a planar imbedding of the resulting graph or prove that it is non-planar.

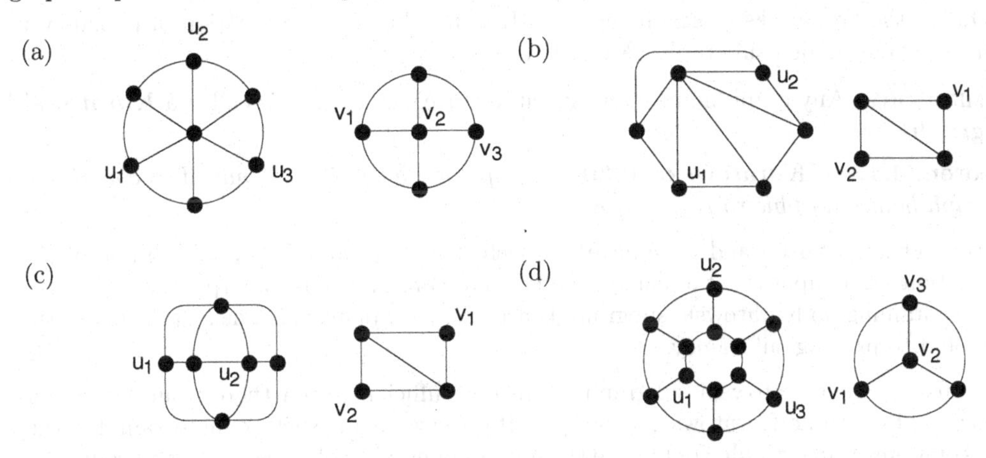

4.3.3 Relative to the subgraph shown in bold,

i. specify each appendage by listing its vertex-set and its edge-set;
ii. list the contact-point sets of each appendage;
iii. determine which appendages are overlapping.

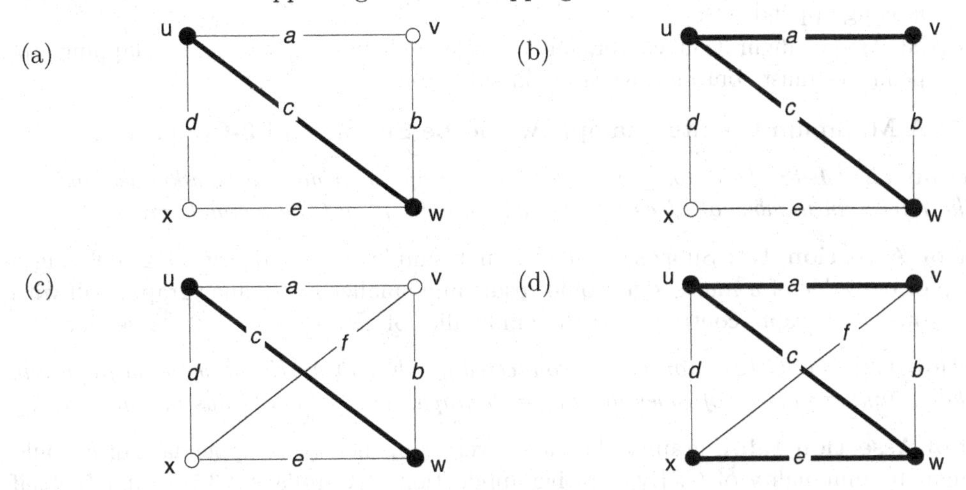

4.3.4 Given two graph imbeddings $\iota : G \to S_0$ and $\iota' : H \to S_0$ in the sphere, suppose that two vertices u_1 and u_2 are selected from G that do not lie on the same face boundary of the imbedding of G, and two vertices v_1 and v_2 are selected from H that do not lie on the same face boundary of the imbedding of H. Decide whether the amalgamation of G and H across these two pairs of vertices must be non-planar. Explain your answer.

4.4 KURATOWSKI'S THEOREM

One of the landmarks of graph theory is Kuratowski's characterization of planarity in terms of two forbidden subgraphs, K_5 and $K_{3,3}$.

TERMINOLOGY: Any graph homeomorphic either to K_5 or to $K_{3,3}$ is called a ***Kuratowski subgraph***.

Theorem 4.4.1: **[*Kuratowski*, 1930]** *A graph is planar if and only if it contains no subgraph homeomorphic to K_5 or to $K_{3,3}$.*

Proof: Theorems 4.1.4 and 4.1.5 have established that K_5 and $K_{3,3}$ are both non-planar. Proposition 4.2.6 implies that a planar graph cannot contain a homeomorphic copy of either. Thus, containing no Kuratowski subgraph is necessary for planarity. The rest of this section is devoted to proving sufficiency.

If the absence of Kuratowski subgraphs were not sufficient, then there would exist non-planar graphs with no Kuratowski subgraphs. If there were any such counterexamples, then some counterexample graph would have the minimum number of edges among all counterexamples. The strategy is to derive some properties that this minimum counterexample would have to have, which ultimately establish that it could not exist. The main steps within this strategy are proofs of three statements:

Step 1: The minimum counterexample would be simple and 3-connected.

Step 2: The minimum counterexample would contain a cycle with three mutually overlapping appendages.

Step 3: Any configuration comprising a cycle and three mutually overlapping appendages must contain a Kuratowski subgraph.

Step 1: A Minimum Counterexample would be Simple and 3-Connected

Assertion 1.1: *Let G be a non-planar connected graph with no Kuratowski subgraph and with the minimum number of edges for any such graph. Then G is a simple graph.*

Proof of Assertion 1.1: Suppose that G is not simple. By Corollary 4.3.2, deleting a self-loop or one edge of a multi-edge would result in a smaller non-planar graph, still with no Kuratowski subgraph, contradicting the minimality of G. ♢ (Assertion 1.1)

Assertion 1.2: *Let G be a non-planar connected graph with no Kuratowski subgraph and with the minimum number of edges for any such graph. Then graph G has no cut-vertex.*

Proof of Assertion 1.2: If graph G had a cut-vertex v, then every appendage of v would be planar, by minimality of G. By iterative application of Corollary 4.3.7, graph G itself would be planar, a contradiction. ♢ (Assertion 1.2)

Assertion 1.3: *Let G be a non-planar connected graph containing no Kuratowski subgraph, with the minimum number of edges for any such graph. Let $\{u, v\}$ be a vertex-cut in G, and let L be a non-chord appendage of $\{u, v\}$. Then there is a planar drawing of L with a face whose boundary contains both vertices u and v.*

Proof of Assertion 1.3: Since $\{u, v\}$ is a vertex-cut of graph G, the graph $G - \{u, v\}$ has at least two components. Thus, the vertex-set $\{u, v\}$ must have at least one more non-chord appendage besides L. Since, by Assertion 1.2, $\{u, v\}$ is a minimal cut, it follows that both u and v must be contact points of this other non-chord appendage. Since this other non-chord appendage is connected and has no edge joining u and v, it must contain a u-v path P of length at least 2, as illustrated in Figure 4.4.1.

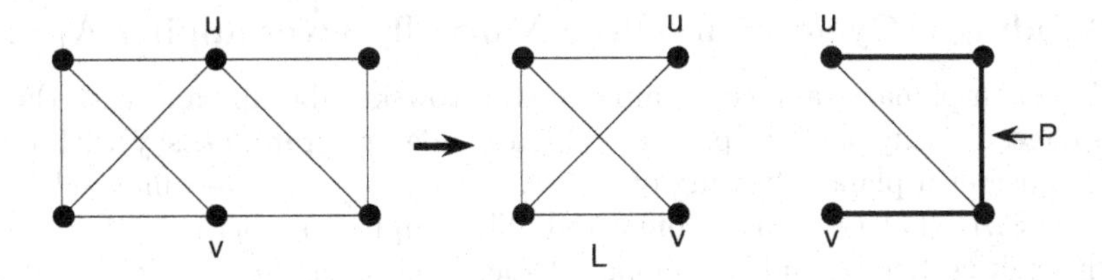

Figure 4.4.1 A u-v path P in an appendage other than L.

The subgraph $H = (V_L \cup V_P, E_L \cup E_P)$ (obtained by adding path P to subgraph L) has no Kuratowski subgraph, because it is contained in graph G, which by premise contains no Kuratowski subgraphs. Let d be the edge obtained from path P by smoothing away all the internal vertices. Then the graph $L + d$ contains no Kuratowski subgraphs (because $L+d$ is homeomorphic to subgraph H). Moreover, the graph $L+d$ has fewer edges than the minimal counterexample G (because P has at least one internal vertex). Thus, the graph $L + d$ is planar.

In every planar drawing of the graph $L + d$, vertices u and v both lie on the boundary of each face containing edge d. Discarding edge d from any such planar drawing of $L + d$ yields a planar drawing of appendage L such that vertices u and v lie on the same face (by Proposition 4.3.4). ♢ (Assertion 1.3)

Assertion 1.4: *Let G be a non-planar connected graph containing no Kuratowski subgraph, with the minimum number of edges for any such graph. Then graph G has no vertex-cut with exactly two vertices.*

Proof of Assertion 1.4: Suppose that graph G has a minimal vertex-cut $\{u, v\}$. Clearly, a chord appendage of $\{u, v\}$ would have a planar drawing with only one face, whose boundary contains both the vertices u and v. By Assertion 1.3, every non-chord appendage of $\{u, v\}$ would also have a planar drawing with vertices u and v on the same face boundary. Figure 4.4.2 illustrates the possible decomposition of graph G into appendages.

By Proposition 4.3.6, when graph G is reassembled by iteratively amalgamating these planar appendages at vertices u and v, the result is a planar graph. ♢ (Assertion 1.4)

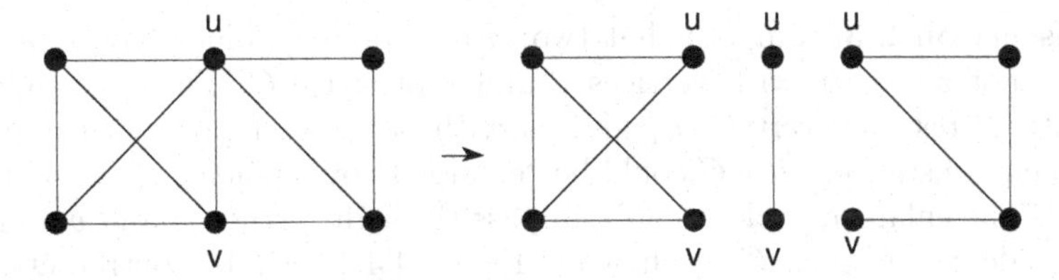

Figure 4.4.2 Decomposition of a graph at vertices u and v.

Completion of Step 1: *Let G be a non-planar connected graph containing no Kuratowski subgraph and with the minimum number of edges for any such graph. Then graph G is simple and 3-connected.*

Proof: This summarizes the four preceding assertions. The connectedness premise serves only to avoid isolated vertices. ♢ (Step 1)

Step 2: Finding a Cycle with Three Mutually Overlapping Appendages

Let G be a non-planar graph containing no Kuratowski subgraph and with the minimum number of edges for any such graph. Let e be any edge of graph G, say with endpoints u and v, and consider a planar drawing of $G - e$. Since graph G is 3-connected (by Step 1), it follows (see §5.1) that $G - e$ is 2-connected. This implies (see §5.1) that there is a cycle in $G - e$ through vertices u and v. Among all such cycles, choose cycle C, as illustrated in Figure 4.4.3, so that the number of edges "inside" C is as large as possible. The next few assertions establish that cycle C must have two overlapping appendages in $G - e$ that both overlap edge e.

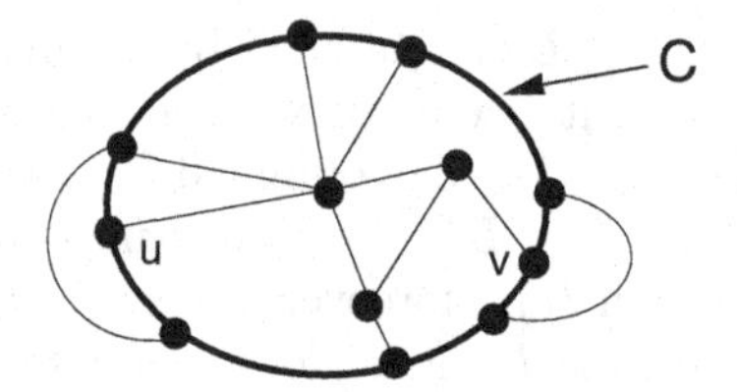

Figure 4.4.3 Cycle C has the maximum number of edges inside it.

Assertion 2.1: *Cycle C has at least one outer appendage.*

Proof of Assertion 2.1: Otherwise, edge e could be drawn in the outer region, thereby completing a planar drawing of G. ♢ (Assertion 2.1)

Assertion 2.2: *Let B be an appendage of C that has only two contact points, neither of which is u or v. Then appendage B is a chord.*

Proof of Assertion 2.2: If B were a non-chord appendage, then those two contact points of B would separate vertices u and v in G from the other vertices of B, which would contradict the 3-connectivity of G. ♢ (Assertion 2.2)

Assertion 2.3: *Let d be an outer appendage of cycle C. Then d is a chord, and its endpoints alternate on C with u and v, so that edges e and d are overlapping chords of cycle C.*

Proof of Assertion 2.3: Suppose that two of the contact points, say a_1 and a_2, of appendage d do not alternate with vertices u and v on cycle C. Then appendage d would contain a path P between vertices a_1 and a_2 with no contact point in the interior of P. Under such a circumstance, cycle C could be "enlarged" by replacing its arc between a_1 and a_2 by path P. The enlarged cycle would still pass through vertices u and v and would have more edges inside it than cycle C, as shown in Figure 4.4.4, thereby contradicting the choice of cycle C. Moreover, if outer appendage d had three or more contact points, then there would be at least one pair of them that does not alternate with u and v. Thus, appendage d has only two contact points, and they alternate with u and v. By Assertion 2.2, such an appendage is a chord. ♢ (Assertion 2.3)

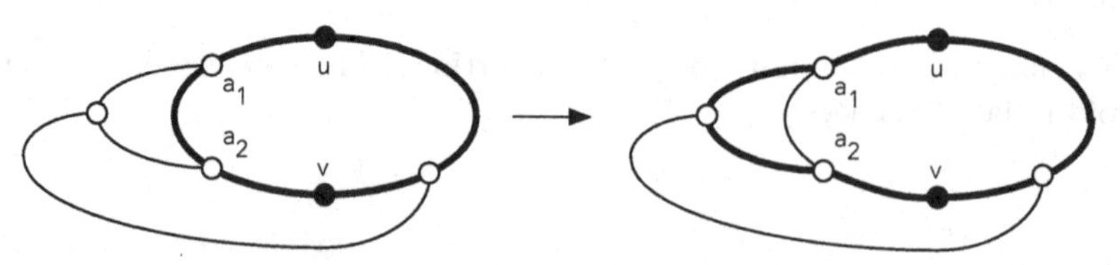

Figure 4.4.4 Using a hypothetical outer appendage to extend cycle C.

Assertion 2.4: *There is an inner appendage of cycle C that overlaps edge e and also overlaps some outer chord.*

Proof of Assertion 2.4: One or more inner appendages must overlap chord e, because otherwise edge e could be drawn inside cycle C without crossing any inner appendages, thereby completing a planar drawing of graph G. Moreover, at least one inner appendage that overlaps e also overlaps some outer appendage d. Otherwise, every such inner appendage could be redrawn outside cycle C, which would permit edge e to be drawn inside, thereby completing the drawing of G. By Assertion 2.3, outer appendage d is necessarily a chord. ◇ (Assertion 2.4)

Completion of Step 2: *Let G be a non-planar graph containing no Kuratowski subgraph and with the minimum number of edges for any such graph. Then graph G contains a cycle that has three mutually overlapping appendages, two of which are chords.*

Proof: The cycle C selected at the start of Step 2 meets the requirements of this concluding assertion. In particular, one of the appendages of cycle C is the chord e designated at the start of Step 2. Assertion 2.4 guarantees the existence of a second appendage B that not only overlaps chord e, but also overlaps some outer chord d. By Assertion 2.3, chord d also overlaps chord e. Thus, the three appendages e, B, and d are mutually overlapping.

◇ (Step 2)

Step 3: Analyzing the Cycle-and-Appendages Configuration

The concluding step in the proof of Kuratowski's theorem is to show that the cycle-and-appendages configuration whose existence is guaranteed by Step 2 must contain a Kuratowski subgraph.

Step 3: *Let C be a cycle in a connected graph G. Let edge e be an inner chord, edge d an outer chord, and B an inner appendage, such that e, d, and B are mutually overlapping. Then graph G has a Kuratowski subgraph.*

Proof: Let u and v be the contact points of inner chord e, and let x and y be the contact points of outer chord d. These pairs of contact points alternate on cycle C, as shown in Figure 4.4.5. Observe that the union of cycle C, chord e, and chord d forms the complete graph K_4. There are two cases to consider, according to the location of the contact points of appendage B.

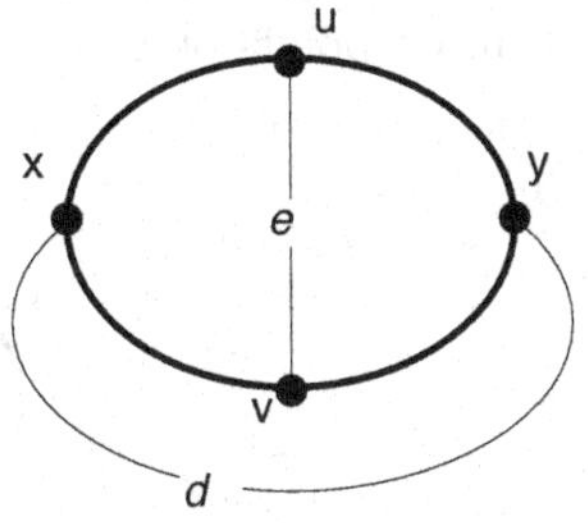

Figure 4.4.5 **Forming K_4 with cycle C and chords e and d.**

Case 1. Suppose that appendage B has at least one contact point s that differs from u, v, x, and y.

By symmetry of the C-e-d configuration, it suffices to assume that contact point s lies between vertices u and x on cycle C. In order to overlap chord e, appendage B must have a contact point t on the other side of u and v from vertex s. In order to overlap chord d, appendage B must have a contact point t' on the other side of x and y from vertex s.

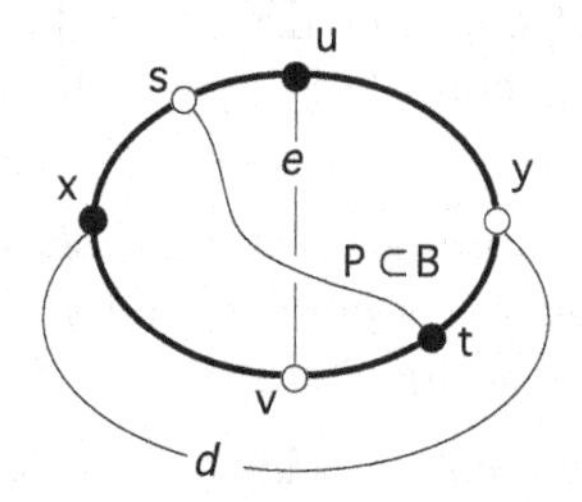

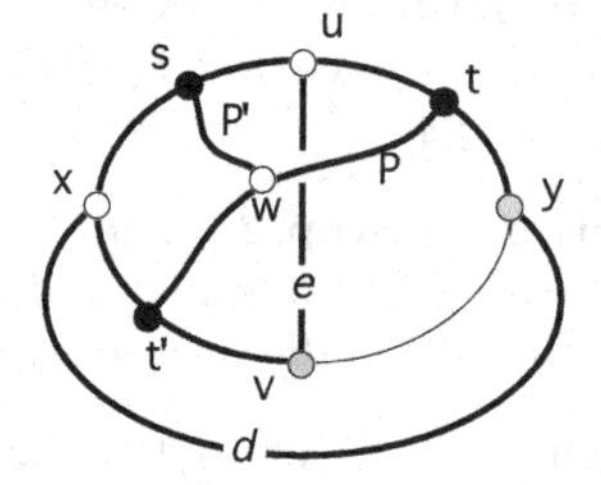

Figure 4.4.6 Subcases 1a and 1b.

Subcase 1a. Contact point t lies in the interior of the arc between v and y on cycle C, in which case it may be considered that $t' = t$, as in Figure 4.4.6. In this subcase, let P be a path in appendage B between contact points s and t. Then the union of cycle C, path P, and chords e and d forms a homeomorphic copy of $K_{3,3}$. ♢ (Subcase 1a)

Subcase 1b. Contact point t lies on arc uy, with $t \neq u$ (so that appendage B overlaps chord e), and contact point t' lies on arc xv with $t' \neq x$ (so that B overlaps chord d), as in Figure 4.4.6. In this subcase, let P be a path in appendage B between contact points t' and t. Let w be an internal vertex on path P such that there is a w-s path P' in B with no internal vertices in P. (Such a path P' exists, because appendage B is connected.) Then this configuration contains a homeomorph of $K_{3,3}$ in which each of the vertices w, x, and u is joined to each of the vertices s, t, and t'. As is apparent in Figure 4.4.6, it does not matter if $t = y$ or if $t' = v$. ♢ (Subcase 1b)

Case 2. Appendage B has no contact points other than u, v, x, and y.

Vertices x and y must be contact points of B so that it overlaps chord e. Vertices u and v must be contact points of B, so that it overlaps chord d. Thus, all four vertices u, v, x, and y must be contact points of appendage B. Appendage B has a u-v path P and an x-y path Q whose internal vertices are not contact points of B.

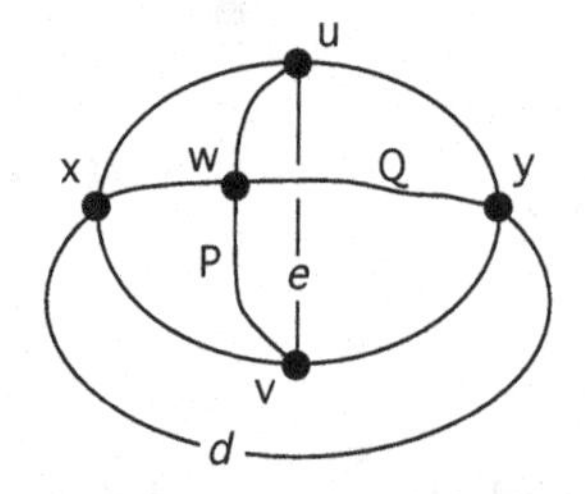

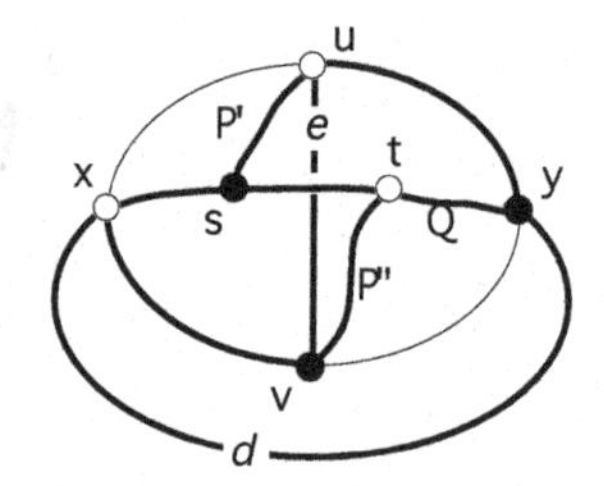

Figure 4.4.7 Subcases 2a and 2b.

Subcase 2a. Paths P and Q intersect in a single vertex w, as shown in Figure 4.4.7. Then the union of the C-e-d configuration with paths P and Q yields a configuration homeomorphic to K_5, as illustrated, in which the five mutually linked vertices are u, v, y, x, and w. ♢ (Subcase 2a)

Subcase 2b. Paths P and Q intersect in more than one vertex.

Let s be the intersection nearest on path P to contact point u, and t the intersection nearest on P to contact point v, as in Figure 4.4.7. Also, let P' be the us-subpath of P and P'' the tv-subpath of P. Then the union of chords e and d with the paths Q, P', and P'' and with arcs vx and uy on cycle C form a subgraph homeomorphic to $K_{3,3}$ with bipartition $(\{u, x, t\}, (\{v, y, s\})$. ♢ (Subcase 2b) ♢ (Theorem 4.4.1)

The following Corollary sometimes expedites planarity testing.

Corollary 4.4.2: *Let G be a non-planar graph formed by the amalgamation of subgraphs H and J at vertices u and v, such that J is planar. Then the graph obtained from graph H by joining vertices u and v is non-planar.*

Proof: By Theorem 4.4.1, graph G must contain a Kuratowski subgraph K. All the vertices of the underlying Kuratowski graph would have to lie on the same side of the cut $\{u, v\}$, because K is 3-connected. At most, a subdivided edge of the Kuratowski subgraph crosses through u and back through v. The conclusion follows. ♢

Finding $K_{3,3}$ or K_5 in Small Non-planar Graphs

A simple way to find a homeomorphic copy of $K_{3,3}$ in a small graph is to identify two cycles C and C' (necessarily of length at least 4) that meet on a path P, such that there is a pair of paths between $C - P$ and $C' - P$ that "cross", thereby forming a subdivided Möbius ladder ML_3, as illustrated in Figure 4.4.8(a).

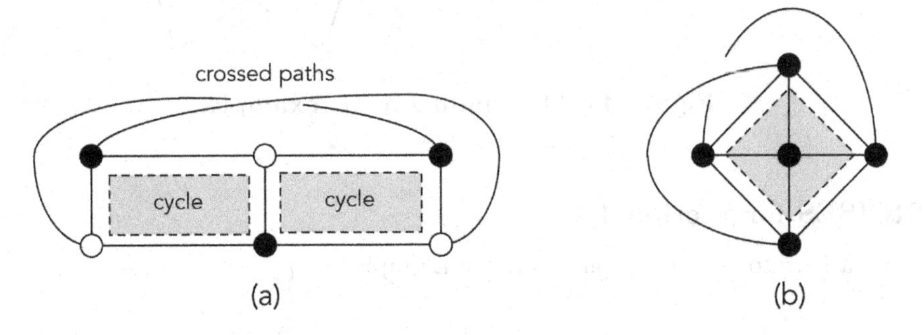

Figure 4.4.8 Finding a $K_{3,3}$ (a) or a K_5 (b) by visual inspection.

To find a homeomorphic copy of K_5, look for a 4-wheel with a pair of disjoint paths joining two pairs of vertices that alternate on the rim, as illustrated in Figure 4.4.8(b).

Example 4.4.1: The bold edges show the crossed paths joining a white vertex on one cycle with a black vertex on the other in Figure 4.4.9, thereby forming a subdivided $K_{3,3}$.

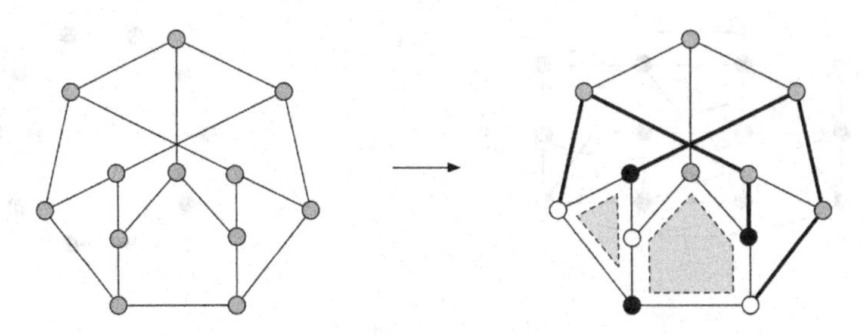

Figure 4.4.9 Finding a $K_{3,3}$, example.

Example 4.4.2: The bold edges show the crossed paths joining a white vertex on one cycle with a black vertex on the other in Figure 4.4.10, thereby forming a subdivided $K_{3,3}$.

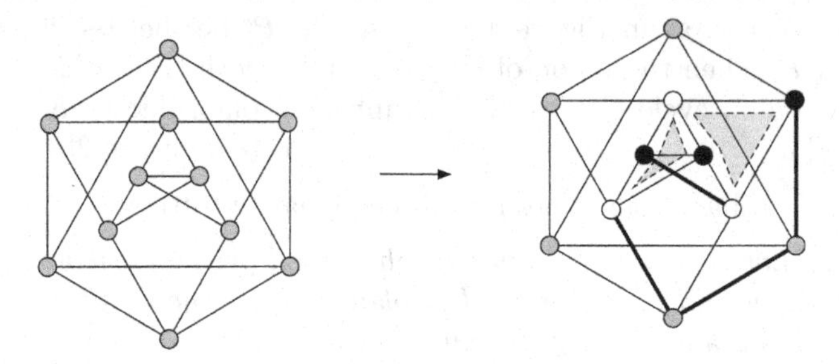

Figure 4.4.10 Finding a $K_{3,3}$, example.

Example 4.4.3: The bold edges show the crossed paths joining vertices that alternate on the rim of the wheel in Figure 4.4.11, thereby forming a subdivided K_5.

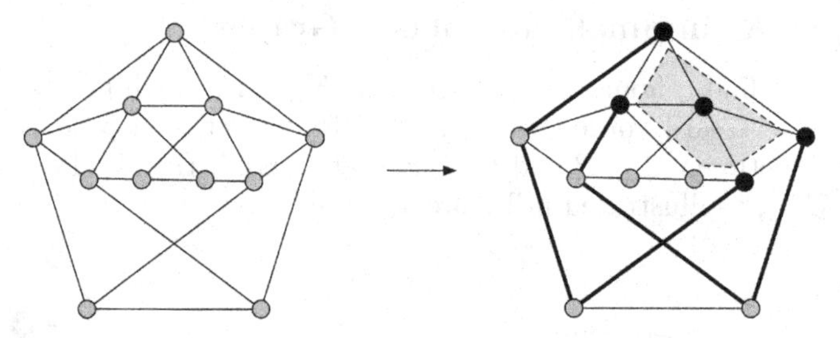

Figure 4.4.11 Finding a K_5, example.

EXERCISES for Section 4.4

4.4.1 Find a Kuratowski subgraph in the given graph.

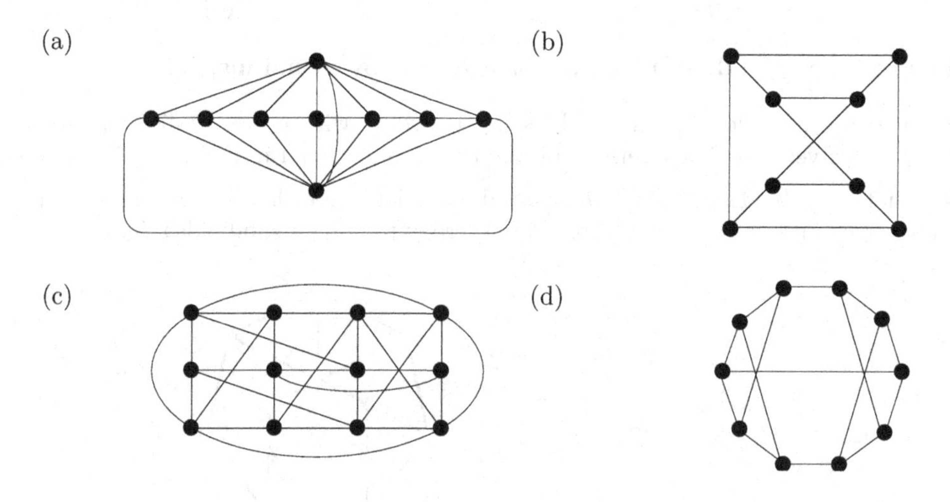

(e)

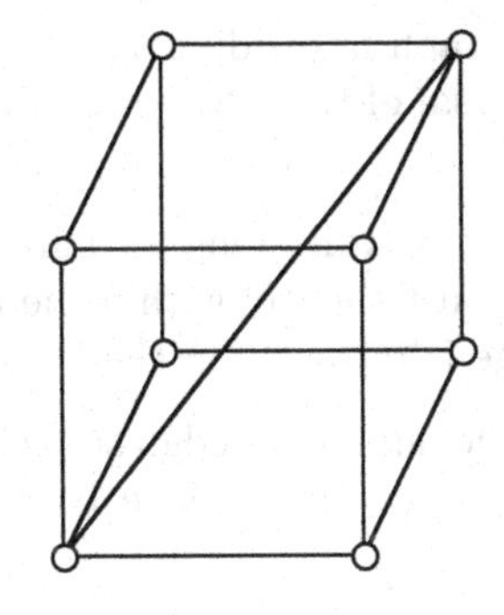

(f)

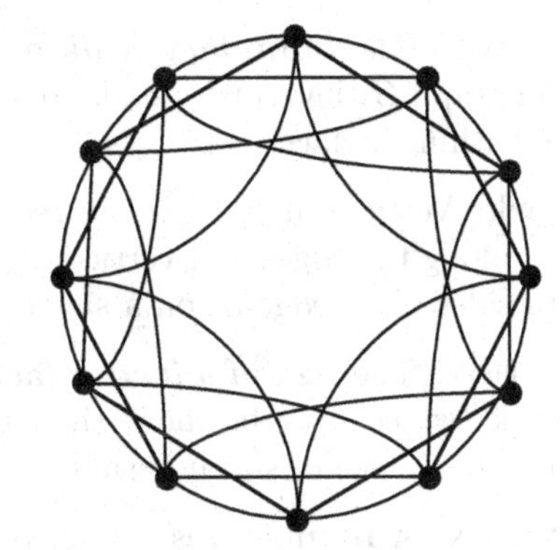

(g) $K_9 - K_{4,5}$.

(h) $(K_4 - K_2) \times P_3$.

(i) $C_4 + K_2$.

(j) $(K_5 - K_3) \times P_2$.

(k) $K_2 \times K_4$.

(l) $C_5 \times C_5$.

(m) $Q_3 + K_1$.

(n) $K_7 - C_7$.

4.5 ALGEBRAIC TESTS FOR PLANARITY

The most basic type of graph-imbedding problem asks whether a particular graph can be imbedded in a particular surface. In this chapter, the surface of particular concern is the sphere (or plane). Transforming such problems into algebraic problems is the inspired approach of algebraic topology. This section derives the fundamental relations used in algebraic analysis of graph-imbedding problems.

The algebraic methods now introduced apply to all surfaces. Although our present version of the Euler Polyhedral Equation is only for the sphere, it can be generalized to all surfaces.

About Faces

DEFINITION: The ***face-set of a graph imbedding*** $\iota : G \to S$ is the set of all faces of the imbedding, formally denoted F_ι. Informally, when there is no ambiguity regarding the imbedding, the face-set is often denoted F_G or F.

Example 4.5.1: The graph imbedding in Figure 4.5.1 has four faces, labeled f_1, f_2, f_3, and f_4. Observe that the boundary of face f_4 is not a cycle.

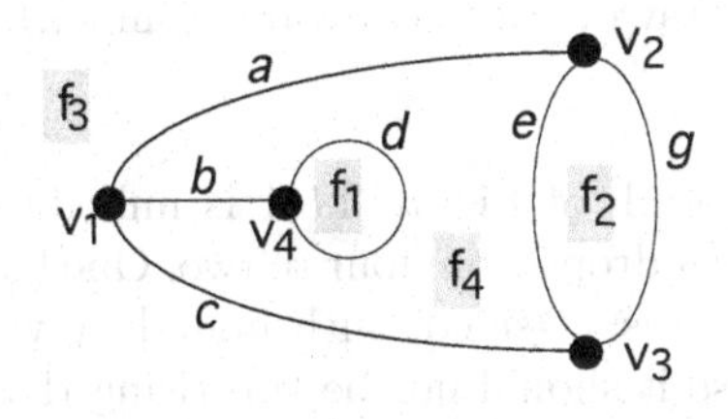

Figure 4.5.1 A graph imbedding in the sphere.

DEFINITION: The ***boundary walk of a face*** f of a graph imbedding $\iota : G \to S$ is a closed walk in graph G that corresponds to a complete traversal of the perimeter of the polygonal region within the face.

Remark: Vertices and edges can reoccur in a boundary walk. One visualizes one's hand tracing along the edges and vertices in the walk, from just slightly within the region. Thus, if both sides of an edge lie on a single region, the edge is retraced on the boundary walk.

DEFINITION: The ***size of a face*** or ***face-size*** means the number of edge-steps in the boundary walk (which may be more than the number of edges on the face boundary, due to retracings). A face of size n is said to be ***n-sided***.

DEFINITION: A ***monogon*** is a 1-sided region.

DEFINITION: A ***digon*** is a 2-sided region.

Example 4.5.2: In Figure 4.5.1, face f_1 is 1-sided (a monogon) with boundary walk $\langle d \rangle$, face f_2 is 2-sided (a digon) with boundary walk $\langle e, g \rangle$, and face f_3 is 3-sided with boundary walk $\langle a, g, c \rangle$. Even though only five edges lie on the boundary of face f_4, face f_4 is 6-sided with boundary walk $\langle a, b, d, b, c, e \rangle$. Two sides of face f_4 are pasted together across edge b, and that edge is a cut-edge of the graph.

Proposition 4.5.1: *Let $\iota : G \to S$ be a planar graph imbedding. Let e be an edge of G that occurs twice on the boundary walk of some face f. Then e is a cut-edge of G.*

Proof: Choose any point x in the interior of edge e. Let C be a curve in face f from the instance of point x on one occurrence of edge e to the instance of x on the other occurrence of edge e, as illustrated in Figure 4.5.2.

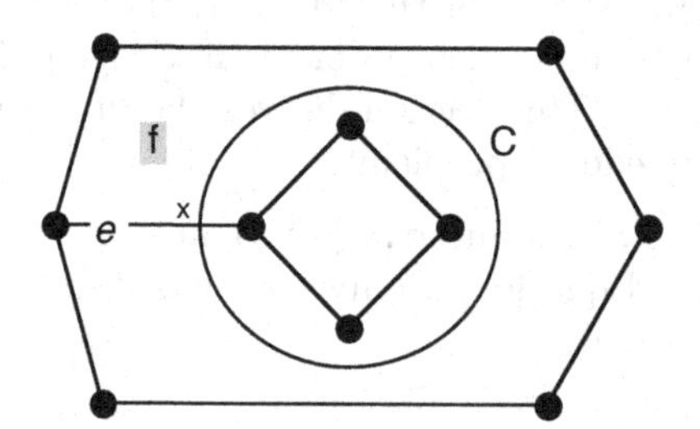

Figure 4.5.2 A curve C from one instance of point x to the other.

Since curve C starts and stops at the same point x, it is a closed curve, so it separates the plane (by the Jordan Curve Theorem). Thus, curve C separates whatever subgraph of G lies on one side of curve C from whatever subgraph lies on the other side. Since curve C intersects graph G only in edge e, it follows that deleting edge e would separate graph G. In other words, edge e is a cut-edge. ♢

The number of faces of a drawing of G can vary from surface to surface, as we observe in the following example.

Example 4.5.3: When the graph of Figure 4.5.1 is imbedded in the torus, as shown in Figure 4.5.3, the number of faces drops from four to two. One face is the monogon $\langle d \rangle$ shaded with wavy lines. The other face is 11-sided, with boundary walk $\langle a, b, d, b, c, e, g, c, a, e, g \rangle$. The surface is not the plane, so it should not be surprising that even though edge a occurs twice on the boundary of the 11-sided face, edge a is not a cut-edge.

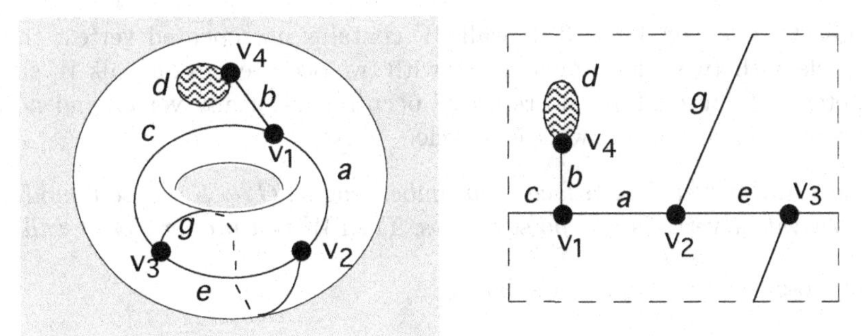

Figure 4.5.3 Two views of the same graph imbedding in the torus.

Face-Size Equation

Theorem 4.5.2: [***Face-Size Equation***] *Let $\iota : G \to S$ be an imbedding of graph G into surface S. Then*

$$2\,|E_G| = \sum_{f \in F} size(f)$$

Proof: Each edge either occurs once in each of two different face boundary walks or occurs twice in the same boundary walk. Thus, by definition of face-size, each edge contributes two sides to the sum. ◇

Example 4.5.4: The graph in Figure 4.5.4 has six edges, so the value of the left side of the equation is $2\,|E| = 12$. The value of the right side of the equation is

$$\sum_{f \in F} size(f) = size(f_1) + size(f_3) + size(f_4)$$
$$= 1 + 2 + 3 + 6 = 12$$

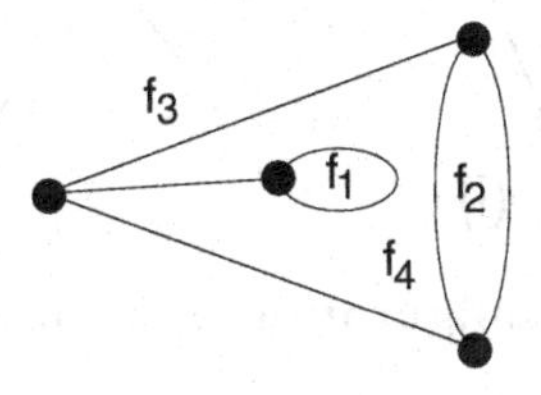

Figure 4.5.4 A graph imbedding in the sphere.

Edge-Face Inequality

DEFINITION: A ***proper walk*** in a graph is a walk in which no edge occurs immediately after itself.

Proposition 4.5.3: *Every proper closed walk W in a graph contains a subwalk (possibly W itself) that is a cycle.*

Proof: If $length(W) = 1$, then walk W itself is a cycle. If $length(W) = 2$, then walk W has the form

$$W = \langle v, d, v', e, v \rangle$$

where $d \neq e$, since walk W is proper. If $v' = v$, then the subwalk v, d, v of length 1 is a cycle. Otherwise, walk W itself is a 2-cycle.

Now suppose that $length(W) \geq 3$. If walk W contains no repeated vertex, then walk W itself is a cycle. Otherwise, let v be a vertex with two occurrences on walk W such that the subwalk between them contains no repeated occurrences of any vertex and no additional occurrences of v. Then that subwalk is a cycle. ♢

Proposition 4.5.4: *Let f be a face of an imbedding $\iota : G \to S$ whose boundary walk W contains no 1-valent vertices and no self-loops. Then W is a proper closed walk.*

Proof: Suppose that the improper subwalk

$$\langle \ldots, u, e, v, e, w, \ldots \rangle$$

occurs in boundary walk W. This means that face f meets itself on edge e, and that the two occurrences of e are consecutive sides of f when f is regarded as a polygon. Clearly, the direction of the second traversal of edge e is opposite to that of the immediately preceding traversal, since e is not a self-loop. Moreover, since edge e has only one endpoint other than v, it follows that $u = w$. This situation is illustrated by Figure 4.5.5(i) below.

Consider a closed curve C in surface S that begins at some point x on edge e near vertex v, and runs down the end of e around vertex v and back out the other side of that end of e to that same point x on the other side of edge e. Since all of curve C except point x lies in the interior of face f, it follows that no edge other than e intersects curve C. Since curve C separates surface S (one side is a topological neighborhood of an end of edge e, and the other is the rest of surface S), it follows that no edge except e is incident on v, as shown in Figure 4.5.5(ii). Thus, vertex v is 1-valent, which contradicts the premise. ♢

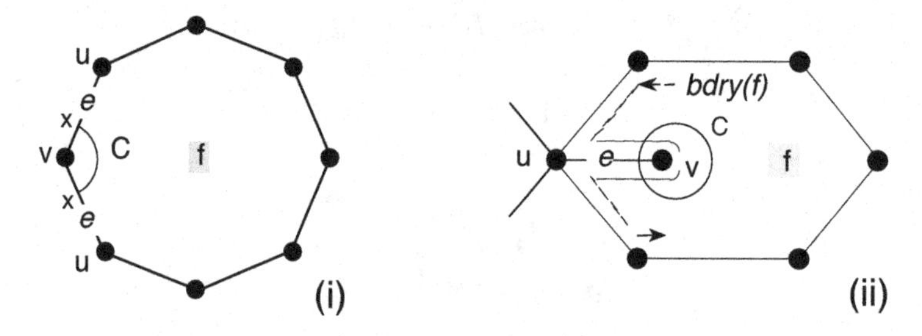

Figure 4.5.5 Surface configuration of an improper boundary walk.

We recall that the ***girth*** of a non-acyclic graph G is defined to be the length of a smallest cycle in G.

Proposition 4.5.5: *The number of sides of each face f of an imbedding $\iota : G \to S$ of a connected graph G that is not a tree is at least as large as girth(G).*

Proof: (Case 1) If the boundary walk of face f is a proper walk, then its length is at least as large as the girth, by Proposition 4.5.3.

(Case 2) If the boundary walk of face f contains a self-loop, then

$$girth(G) = 1 \leq size(f)$$

since 1 is the minimum possible size of any face.

(Case 3) By Proposition 4.5.4, the remaining case to consider is when the boundary walk contains one or more 1-valent vertices and no self-loop. In this case, consider the imbedding $\iota' : G' \to S'$ that results from iteratively contracting every edge with a 1-valent endpoint, until no such edges remain, with f' denoting the face that results from deleting all these edges that "spike" into face f. Then

$$\begin{aligned} size(f) &\geq size(f') \\ &\geq girth(G') && \text{by case 1} \\ &= girth(G) && \text{since contracting a cut-edge preserves the girth} \end{aligned}$$

◇

Theorem 4.5.6: **[*Edge-Face Inequality*]** *Let G be a connected graph that is not a tree, and let $\iota : G \to S$ be an imbedding. Then*

$$2\,|E| \geq girth(G) \cdot |F|$$

Proof: In the Face-Size Equation

$$2\,|E| = \sum_{f \in F} size(f)$$

each term in the sum on the right side is at least $girth(G)$, and there are exactly $|F|$ terms. ◇

Euler Polyhedral Equation

We observe that a tetrahedron has 4 vertices, 6 edges, and 4 faces, and we calculate that $4 - 6 + 4 = 2$. We also observe that a cube has 8 vertices, 12 edges, and 6 faces, and that $8 - 12 + 6 = 2$.

Remark: Robin Wilson [Wi04] offers an historical account of the generalization of this observation. In 1750, Euler communicated the formula $|V| - |E| + |F| = 2$ by letter to Goldbach. The first proof, in 1794, was by Legendre, who used metrical properties. Lhuilier gave a topological proof in 1811, and he also derived the formula $|V| - |E| + |F| = 0$ for graphs on a torus. Poincaré generalized the formula to other surfaces in a series of papers of 1895–1904.

Theorem 4.5.7: **[*Euler Polyhedral Equation*]** *Let $\iota : G \to S$ be an imbedding of a connected graph G into a sphere. Then*

$$|V| - |E| + |F| = 2$$

Proof: If the cycle rank $\beta(G)$ is zero, then graph G is a tree, because $\beta(G)$ equals the number of non-tree edges for any spanning tree in G. Of course, a tree contains no cycles. Thus, in a drawing of graph G on the sphere, the original single region is not subdivided. It follows that $|F| = 1$. Thus,

$$\begin{aligned} |V| - |E| + |F| &= |V| - (|V| - 1) + |F| && \text{by Prop. 3.1.3} \\ &= |V| - (|V| - 1) + 1 = 2 \end{aligned}$$

This serves as the base case for an induction on cycle rank. Next suppose that the Euler Polyhedral Equation holds for $\beta(G) = n$, for some $n \geq 0$. Consider an imbedding in which graph G has cycle rank $\beta(G) = n + 1$. Let T be a spanning tree in G, and let $e \in E_G - E_T$, so that $\beta(G - e) = n$. Clearly,

(1) $|V_G| = |V_{G-e}|$ and

(2) $|E_G| = |E_{G-e}| + 1$

Figure 4.5.6 illustrates the relationship between the face sets.

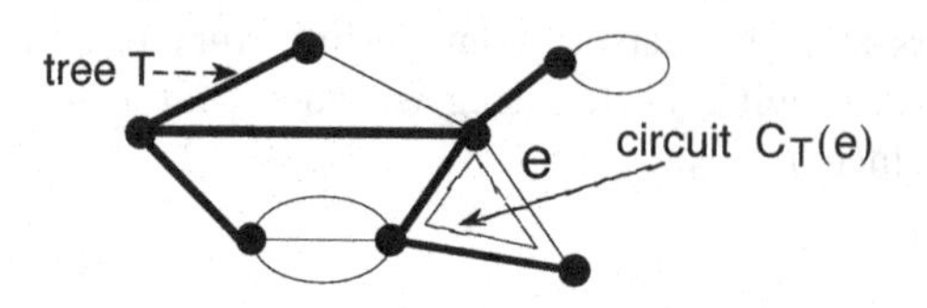

Figure 4.5.6 Erasing a non-tree edge in a sphere.

By Theorem 3.1.8, the subgraph $T+e$ contains a unique cycle, $C_T(e)$. By the Jordan Curve Theorem, the cycle $C_T(e)$ separates the sphere, which implies that two different faces meet at edge e. Thus, deleting edge e from the imbedding merges those two faces. This yields an imbedding of the connected graph $G-e$ in the sphere. Therefore,

(3) $|F_G| = |F_{G-e}| + 1$

Accordingly,

$$\begin{aligned}
|V_G| - |E_G| + |F_G| &= |V_{G-e}| - |E_G| + |F_G| && \text{by (1)}\\
&= |V_{G-e}| - (|E_{G-e}| + 1)\,|F_G| && \text{by (2)}\\
&= |V_{G-e}| - (|E_{G-e}| + 1) + (|F_{G-e}| + 1) && \text{by (3)}\\
&= |V_{G-e}| - |E_{G-e}| + |F_{G-e}|\\
&= 2 && \text{(by the induction hypothesis)}
\end{aligned}$$ ♢

Theorem 4.5.7 specifies the Euler equation only for an imbedding of a connected graph in the sphere, not for other kinds of drawings. Figure 4.5.7 illustrates two drawings for which the equation does not apply.

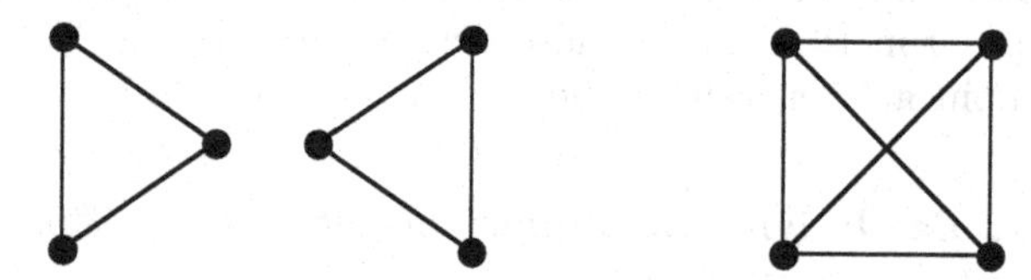

Figure 4.5.7 (a) Non-connected graph; (b) drawing with edge-crossings.

Example 4.5.5: The non-connected graph $2K_3$ in Figure 4.5.7(a) has 6 vertices and 6 edges. When it is imbedded in the plane, there are 3 faces, so the Euler equation does not hold. We observe that there is a face (the exterior face) with two boundary components, which cannot happen with a connected graph in the plane. We also observe that the equation does not hold for the drawing in Figure 4.5.7(b), because it has edge-crossings and, thus, is not an imbedding.

Poincaré Duality

The French mathematician Henri Poincaré [1900] invented a method by which, for every (*cellular*) imbedding $\iota : G \to S$ of any graph G in any (*closed*) surface S, one can construct a new imbedding $\iota^* : G^* \to S$ of a new graph C^* in that same surface S. The formal definition of ***cellular imbedding*** and ***closed surface*** is beyond the scope of this chapter. The sphere is closed, and an imbedding of a connected graph on the sphere is cellular.

DEFINITION: Whatever graph imbedding $\iota : G \to S$ is to be supplied as input to the duality process (whose definition follows below) is called the ***primal graph imbedding***. Moreover, the graph G is called the ***primal graph***, the vertices of G are called ***primal vertices***, the edges of G are called ***primal edges***, and the faces of the imbedding of G are called ***primal faces***.

DEFINITION: Given a primal graph imbedding $\iota : G \to S$, the ***(Poincaré) duality construction*** of a new graph imbedding is a two-step process, one for the dual vertices and one for the dual edges:

dual vertices: Into the interior of each primal face f, insert a new vertex f^* (as in Figure 4.5.8), to be regarded as dual to face f. The set $\{f^* | f \in F_\iota\}$ is denoted V^*.

Figure 4.5.8 Inserting dual vertices into a primal imbedding.

dual edges: Through each primal edge e, draw a new edge e^* (as in Figure 4.5.9), joining the dual vertex in the primal face on one side of that edge to the dual vertex in the primal face on the other side, to be regarded as dual to edge e. If the same primal face f contains both sides of primal edge e, then dual edge e^* is a self-loop through primal edge e from the dual vertex f^* to itself. The set $\{e^* | e \in E_G\}$ is denoted E^*.

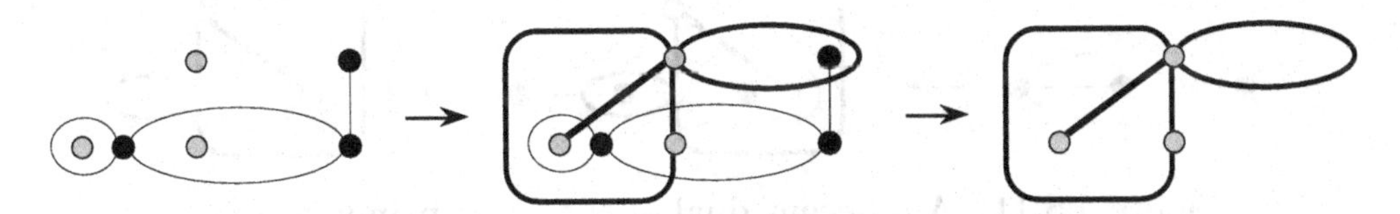

Figure 4.5.9 Inserting dual edges completes the dual imbedding.

DEFINITION: In Poincaré duality, the ***graph*** is the graph G^* with vertex-set V^* and edge-set E^*. (The vertices and edges of the primal graph are not part of the dual graph.)

DEFINITION: In Poincaré duality, the ***dual imbedding*** is the imbedding $\iota^* : G^* \to S$ of the dual graph G^* that is constructed in the dualization process.

DEFINITION: In the Poincaré duality construction, the ***dual faces*** are the faces of the dual imbedding. The dual face containing the vertex v is denoted v^*. (For each vertex $v \in V_G$, there is only one such face.) The set $\{v^* | v \in V_G\}$ of all dual faces is denoted F^*.

NOTATION: Occasionally, a dual graph (or its vertex-set, edge-set, or face-set) is subscripted to distinguish it from the dual graph of other imbeddings of the same graph.

Proposition 4.5.8: *(Numerics of Duality) Let $\iota : G \to S$ be a cellular graph imbedding. Then the following numerical relations hold.*

(i) $|V^*| = |F|$

(ii) $|E^*| = |E|$

(iii) $|F^*| = |V|$

Proof: Parts (i) and (ii) are immediate consequences of the definition of the Poincaré duality construction. Part (iii) holds because the dual edges that cross the cyclic sequence of edges emanating from a primal vertex v of graph G form a closed walk that serves as boundary of the dual face v^* that surrounds primal vertex v, as illustrated in Figure 4.5.10. ♢

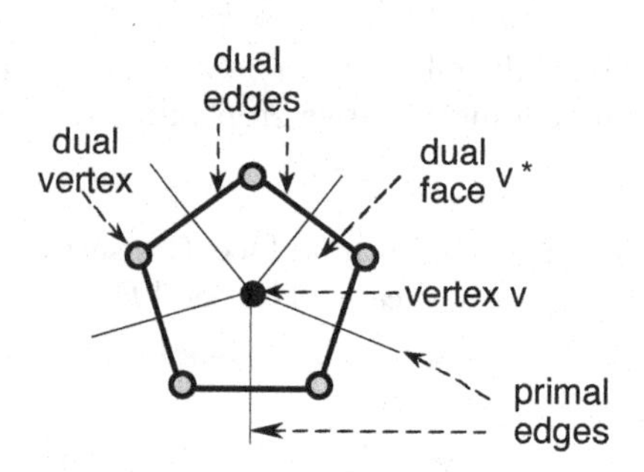

Figure 4.5.10 A dual face surrounding its primal vertex.

Example 4.5.6: The same graph may have many different cellular imbeddings in a given surface and many more over the full range of possible surfaces. Figure 4.5.11 shows a different imbedding of the primal graph from the imbedding of Figure 4.5.9. The dual graph of the previous imbedding has degree sequence $\langle 1, 2, 5\rangle$, and the dual graph of Figure 4.5.11 has degree sequence $\langle 1, 3, 4\rangle$, so the dual graphs are not isomorphic.

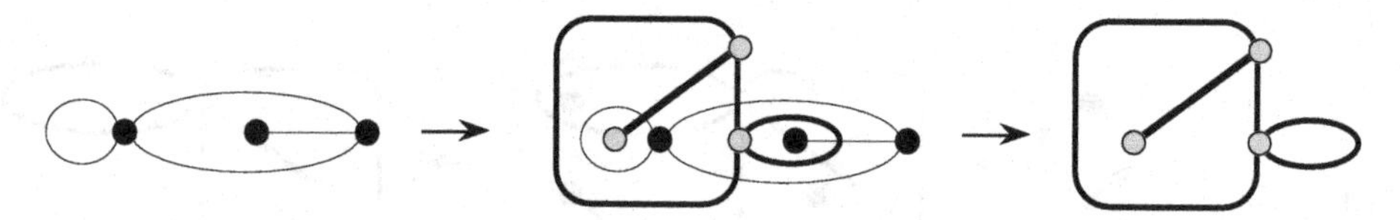

Figure 4.5.11 A different dual of the same primal graph.

Remark: Applying the Poincaré duality construction to the dual graph imbedding simply reconstructs the primal graph and the primal imbedding. This restoration of the original object through redualization is an essential feature of any "duality" construction.

Remark: The Edge-Face Inequality (Theorem 4.5.6) is dual to Euler's theorem on degree-sum (Theorem 1.1.2).

Algebraic Proofs of Non-planarity

The Euler Polyhedral Equation $|V| - |E| + |F| = 2$ amounts to an algebraization of the Jordan property. Upon the Euler equation, we can construct inequalities that serve as quick tests for planarity, which we can apply to graphs far too large for ad hoc application of the Jordan property. These are one-way tests, and there are difficult cases in which they are inconclusive.

This algebraic approach to such non-imbeddability proofs illustrates one of the dominant themes of 20th-century mathematics — the transformation of various kinds of other problems, especially topological and geometric problems, into algebraic problems. The formula derived in the following theorem illustrates this theme, by providing a necessary algebraic condition for graph planarity.

Theorem 4.5.9: *Let G be any connected simple graph with at least three vertices and an imbedding $\iota : G \to S_0$ in the sphere or plane. Then*

$$|E_G| \leq 3\,|V_G| - 6$$

Proof: The Euler Polyhedral Equation is the starting point.

(1)	$\|V_G\| - \|E_G\| + \|F_\iota\| = 2$	Euler Polyhedral Equation
(2)	$girth(G) \cdot \|F_\iota\| \leq 2\,\|E_G\|$	Edge-Face Inequality
(3)	$3\,\|F_\iota\| \leq 2\,\|E_G\|$	G simple bipartite $\Rightarrow 3 \leq girth(G)$
(4)	$\|F_\iota\| \leq \dfrac{2\,\|E_G\|}{3}$	
(5)	$\|V_G\| - \|E_G\| + \dfrac{2\,\|E_G\|}{3} \geq 2$	substitute (4) into (1)
(6)	$\|V_G\| - \dfrac{\|E_G\|}{3} \geq 2$	
(7)	$\dfrac{\|E_G\|}{3} \leq \|V_G\| - 2$	
(8)	$\|E_G\| \leq 3\,\|V_G\| - 6$	

◇

Theorem 4.5.10: *The complete graph K_5 is non-planar.*

Proof: Since $3\,|V| - 6 = 9$ and $|E| = 10$, the inequality $|E| > 3\,|V| - 6$ is satisfied. According to Theorem 4.5.9, the graph K_5 must be non-planar. ◇

Remark: The Jordan Curve Theorem was used to derive the Euler Polyhedral Equation, which was used, in turn, to derive the planarity criterion $|E_G| \leq 3\,|V_G| - 6$. It remains the ultimate reason why K_5 is not imbeddable in the sphere.

Whereas an ad hoc proof of the non-planarity of K_5 establishes that fact alone, the inequality of Theorem 4.5.9, derived by the algebraic approach, can be used to prove the non-planarity of infinitely many different graphs. The next proposition is another application of that inequality.

Proposition 4.5.11: *Let G be any 8-vertex simple graph with $|E_G| \geq 19$. Then graph G has no imbeddings in the sphere or plane.*

Proof: Together, the calculation $3\,|V_G| - 6 = 3 \cdot 8 - 6 = 18$ and the premise $|E_G| \geq 19$ imply that $|E_G| > 3\,|V_G| - 6$. By Theorem 4.5.9, graph G is non-planar. ◇

Remark: According to Table A1 of [Ha69], there are 838 connected simple 8-vertex graphs with 19 or more edges. Proposition 4.5.11 establishes their non-planarity collectively with one short proof. An ad hoc approach could not so quickly prove the non-planarity of even one such graph, much less all of them at once.

A More Powerful Non-planarity Condition for Bipartite Graphs

For a bipartite graph, the same algebraic methods used in Theorem 4.5.9 yield an inequality that can sometimes establish non-planarity when the more general inequality is not strong enough.

Example 4.5.7: The graph $K_{3,3}$ has 6 vertices and 9 edges. Thus, $|E| \leq 3\,|V| - 6$, even though $K_{3,3}$ is not planar.

Theorem 4.5.12: *Let G be any connected bipartite simple graph with at least three vertices and an imbedding $\iota : G \to S_0$ in the sphere or plane. Then*

$$|E_G| \leq 2\,|V_G| - 4$$

Proof: This proof is exactly like that of the previous theorem.

(1)	$|V_G| - |E_G| + |F_\iota| = 2$	Euler Polyhedral Equation
(2)	$girth(G) \cdot |F_\iota| \leq 2\,|E_G|$	Edge-Face Inequality
(3)	$4\,|F_\iota| \leq 2\,|E_G|$	G simple bipartite $\Rightarrow 4 \leq girth(G)$
(4)	$|F_\iota| \leq \dfrac{2\,|E_G|}{4}$	
(5)	$|V_G| - |E_G| + \dfrac{2\,|E_G|}{4} \geq 2$	substitute (4) into (1)
(6)	$|V_G| - \dfrac{|E_G|}{2} \geq 2$	
(7)	$\dfrac{|E_G|}{2} \leq |V_G| - 2$	
(8)	$|E_G| \leq 2\,|V_G| - 4$	

◇

Theorem 4.5.13: *The complete bipartite graph $K_{3,3}$ is non-planar.*

Proof: Since $2\,|V| - 4 = 8$ and $|E| = 9$, the inequality $|E| > 2\,|V| - 4$ is satisfied. According to Theorem 4.5.12, the graph $K_{3,3}$ must be non-planar. ◇

REVIEW FROM §7.1: The complete graph K_5 and the complete bipartite graph $K_{3,3}$ are called the ***Kuratowski graphs***.

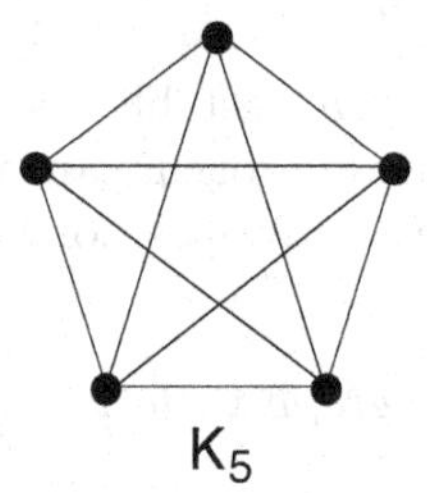

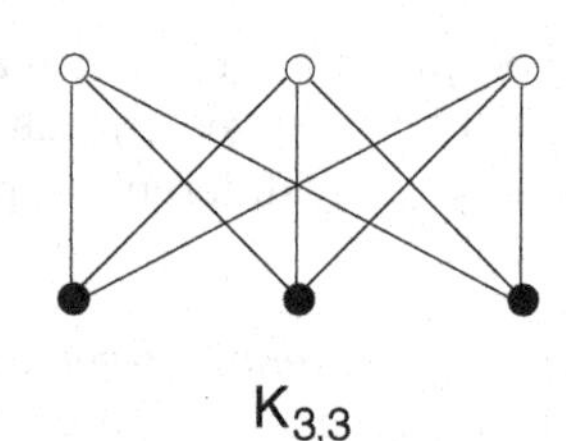

Figure 4.5.12 The Kuratowski graphs.

Non-planar Subgraphs

Another way to prove a graph is non-planar, besides applying Theorem 4.5.9 or Theorem 4.5.12, is to establish that it contains a non-planar subgraph.

Theorem 4.5.14: *If a graph contains a non-planar subgraph, then that graph is non-planar.*

Proof: Every planar drawing of a graph remains free of edge-crossings after the deletion (e.g., by erasure) of any set of edges and vertices. This establishes the contrapositive, that every subgraph of a planar graph G is planar. ◇

Example 4.5.8: The graph $G = K_7 - E(K_4)$ is obtained by deleting the six edges joining any four designated vertices in K_7. Since $|V_G| = 7$ and $|E_G| = 21 - 6 = 15$, it follows that

$$|E_G| \leq 3\,|V_G| - 6$$

Thus, the planarity formula does not rule out the planarity of $K_7 - E(K_4)$. However, $K_7 - E(K_4)$ contains the Kuratowski graph $K_{3,3}$ as indicated by Figure 4.5.13. Hence, $K_7 - E(K_4)$ is non-planar, by Theorem 4.5.14.

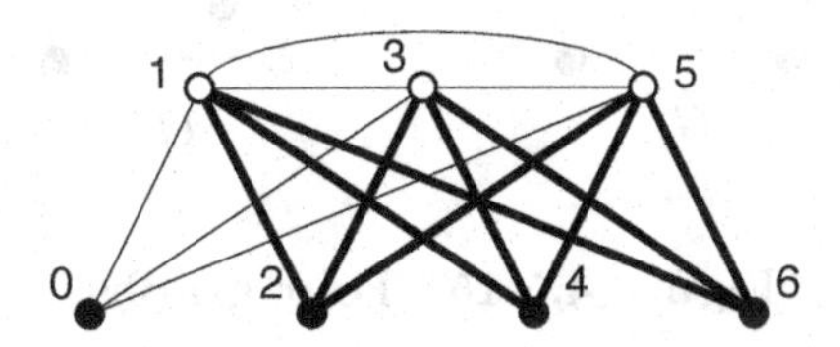

Figure 4.5.13 $K_7 - E(K_4)$ **contains** $K_{3,3}$**.**

Proposition 4.5.15: *Planarity is invariant under subdivision.*

Proof: Suppose that G is a planar graph and that graph H is homeomorphic to G. Then, in a planar drawing of G, smooth whatever vertices are needed and subdivide whatever edges are needed to transform G into H. These operations simultaneously transform the planar drawing of G into a planar drawing of H. ♢

Proposition 4.5.16: *Suppose that a graph G contains a subgraph H that is homeomorphic to a non-planar graph. Then G is non-planar.*

Proof: This follows immediately from the preceding theorem and proposition. ♢

As important as the non-planarity inequalities of Theorem 4.5.9 and Theorem 4.5.12 may be, they provide only one-sided tests. That is, if a graph has too many edges according to either theorem, then it is non-planar; however, not having too many edges does not imply that the graph is planar. The usual ways to prove planarity are to exhibit a planar drawing or to apply a planarity algorithm, as in §7.6.

Heuristic 4.5.17: *To construct a non-planar, non-bipartite graph with a specified number of vertices and edges, start with a Kuratowski graph and add vertices, add edges, and subdivide edges.*

Example 4.5.9: To construct an 8-vertex non-planar, non-bipartite graph with at most 13 edges, simply subdivide K_5 or $K_{3,3}$, as shown in Figure 4.5.14. Observe that the two graphs resulting from these subdivisions have too few edges to make a conclusive application of the numerical non-planarity formulas.

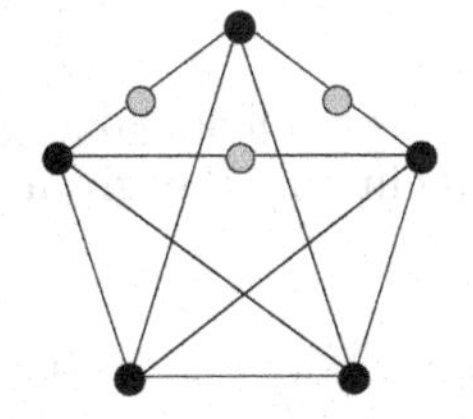
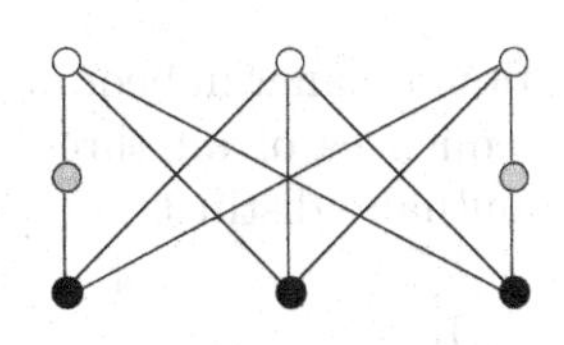

Figure 4.5.14 Two graphs with too few edges for the non-planarity formula.

EXERCISES for Section 4.5

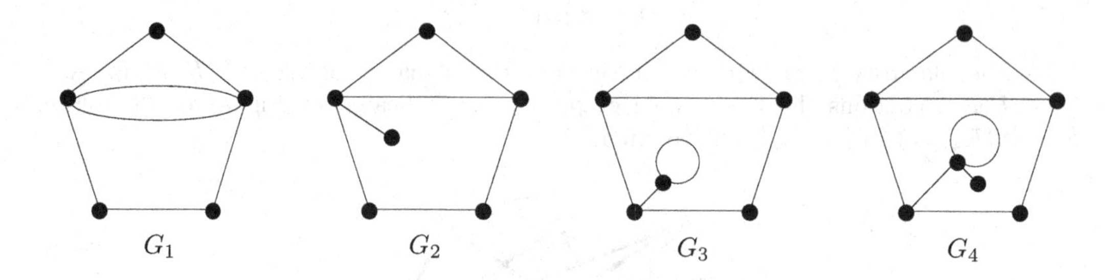

Figure 4.5.15 Four graphs.

4.5.1 Do the following for the given graph in Figure 4.5.15:

i. redraw the given graph;

ii. write the face-size into each region;

iii. show that the sum of the face sizes equals twice the number of edges.

(a) G_1 (b) G_2 (c) G_3 (d) G_4

4.5.2 Do the following for the given graph in Figure 4.5.15:

i. calculate the girth of the graph;

ii. show that the product of the girth and the number of faces is less than or equal to twice the number of edges.

(a) G_1 (b) G_2 (c) G_3 (d) G_4

4.5.3 Do the following for the given imbedded graph in Figure 4.5.15:

i. calculate $|V|$, $|E|$, and $|F|$;

ii. show that the Euler Polyhedral Equation $|V| - |E| + |F| = 2$ holds.

(a) G_1 (b) G_2 (c) G_3 (d) G_4

4.5.4 Draw the Poincaré dual of the given planar graph imbedding in Figure 4.5.15:.

(a) G_1 (b) G_2 (c) G_3 (d) G_4

4.5.5 Draw two additional planar imbeddings of the given imbedded graph in Figure 4.5.15, so that the degree sequences of all three duals (including the given imbedding) of the indicated graph are mutually distinct.

(a) G_1 (b) G_2 (c) G_3 (d) G_4

DEFINITION: A graph G is ***self-dual*** in the sphere if there is an imbedding of G such that the Poincaré dual graph G^* for that imbedding is isomorphic to G.

4.5.**6** Either draw an imbedding establishing self-duality of the given graph, or prove that no such imbedding is possible.

(a) K_4 (b) $K_4 - K_2$ (c) D_2
(d) $K_5 - K_2$ (e) $B_1 + K_1$ (f) $K_6 - 3K_2$

4.5.**7** Prove that the given graph is non-planar.

(a) $K_8 - 4K_2$ (b) $K_{3,6} - 3K_2$ (c) $K_8 - 2C_4$
(d) $K_8 - K_{3,3}$ (e) $K_2 + Q_3$ (f) $K_{6,6} - C_{12}$

4.5.**8** Either draw a plane graph that meets the given description or prove that no such graph exists.

(a) A simple graph with 6 vertices and 13 edges.
(b) A non-simple graph with 6 vertices and 13 edges.
(c) A bipartite simple graph with 7 vertices and 11 edges.
(d) A bipartite non-simple graph with 7 vertices and 11 edges.
(e) A simple planar graph with 144 vertices and 613 edges.
(f) A simple planar graph with 1728 vertices and 5702 edges.
(g) A bipartite simple graph with 36 vertices and 68 edges.
(h) A planar simple graph with 36 vertices and 102 edges.

4.5.**9** Draw a bipartite graph G with 15 vertices and 18 edges that is not planar, even though it satisfies the formula $|E_G| \leq 2|V_G| - 4$.

4.5.**10** Show that the given graph satisfies the planarity inequality $|E_G| \leq 3|V_G| - 6$, and that it is nonetheless non-planar.

(a) $K_{3,3} + K_1$. (b) $K_{3,3} \times K_2$.
(c) $K_8 - (K_5 \cup K_3)$. (d) $W_5 \times C_4$.

4.6 PLANARITY ALGORITHM

Several good planarity algorithms have been developed. The planarity algorithm described here (from [DeMaPe64]) has been chosen for simplicity of concept and of implementability. It starts by drawing a cycle, and adds to it until the drawing is completed or until further additions would force an edge-crossing in the drawing. This section first gives a prose description of this algorithm, then pseudo-code for the algorithm.

Blocked and Forced Appendages of a Subgraph

The body of the main loop of the planarity algorithm attempts to extend a subgraph already drawn in the plane by choosing a plausible appendage of that subgraph to be used to augment the drawing.

DEFINITION: Let H be a subgraph of a graph. In a planar drawing of H, an appendage of H is ***undrawable in region*** R if the boundary of region R does not contain all the contact points of that appendage.

DEFINITION: Let H be a subgraph of a graph. In a planar drawing of H, an appendage of H is ***blocked*** if that appendage is undrawable in every region.

Example 4.6.1: In Figure 4.6.1, a subgraph isomorphic to $K_{2,3}$ is shown with black vertices and solid edges, and its only appendage is shown with a white vertex and broken edges. The appendage is blocked, because no region boundary contains all three contact points.

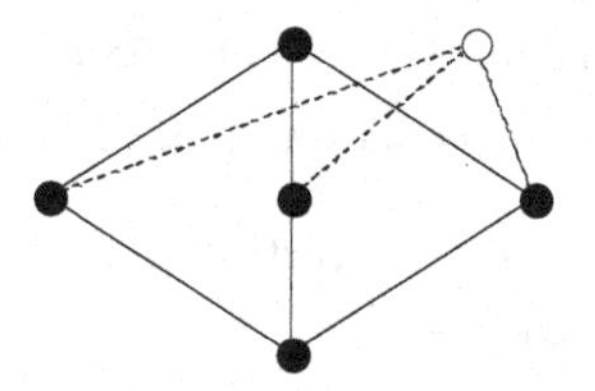

Figure 4.6.1 A subgraph with a blocked appendage.

Proposition 4.6.1: *Let $\iota : H \to S_0$ be a planar drawing of a subgraph H of a graph G, such that H has a blocked appendage B. Then it is impossible to extend $\iota : H \to S_0$ to a planar drawing of G.*

Proof: Let u be a contact point in B, and let e be an edge in B incident on u. In any extension of $\iota : H \to S_0$ to a planar drawing of G, e must be drawn in some region R whose boundary contains u. Since B is a blocked appendage, it contains a contact point v that does not lie on the boundary of R and an edge e' incident on v. It follows that e' must be drawn in a different region than R. Also, there is a walk W in B containing edges e and e', such that no internal vertex of W is in H. Since e and e' lie in different regions, that walk must cross some edge on the boundary of R, by the Jordan Curve Theorem.

◊

DEFINITION: Let H be a subgraph of a graph. In a drawing of H on any surface, an appendage of H is ***forced into region*** R if R is the only region whose boundary contains all the contact points of that appendage.

Example 4.6.2: In Figure 4.6.2, a subgraph isomorphic to $K_{2,3}$ is shown with black vertices and solid edges, and its only appendage is shown with a white vertex and broken edges. The appendage is forced into region R, because it is the only region whose boundary contains all three contact points.

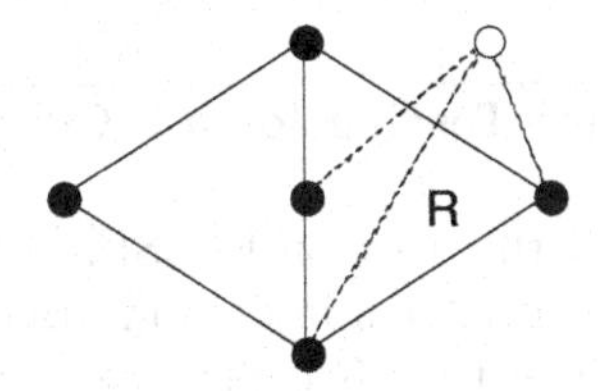

Figure 4.6.2 A subgraph with a forced appendage.

Algorithmic Preliminaries

By iterative application of Corollary 4.3.7, a graph is planar if and only if all its blocks are planar. Performing a prior decomposition into blocks (see §5.4) frees the main part of the algorithm from a few cluttersome details and permits a focus on the most interesting case, in which the graph to be tested for planarity is 2-connected.

In testing the planarity of a 2-connected graph G, the basis for a Whitney synthesis (see §5.2) is to find a cycle and to draw it in the plane. (If there is no cycle, then the graph is a tree, and it is planar.) The cycle is designated as the basis subgraph G_0.

Selecting the Next Path Addition

Before each iteration of the body of the main loop, there is a two-part exit test. One exit condition is that the entire graph is already drawn in the plane, in which case it is decided, of course, that the graph is planar. The other exit condition is that some appendage of subgraph G_j is blocked. If so, then graph G is declared non-planar and the algorithm terminates. If neither exit is taken, then there are two possible cases of appendage selection.

In the main loop, graph G_{j+1} is obtained from graph G_j by choosing an appendage of G_j and adding a path joining two contact points of that appendage to the planar drawing of G_j. The main technical concern is that of choosing an appropriate appendage at each iteration.

In the first case, it is determined that some appendage is forced. If so, then a path between two of its contact points is drawn into the one region whose boundary contains all contact points of that forced appendage, thereby extending the drawing of G_j to a drawing of G_{j+1}.

In the second case, there are no forced appendages. Under this circumstance, an arbitrary appendage is chosen, and a path between two of its contact points is drawn into any region whose boundary contains all contact points of that appendage, thereby extending the drawing of G_j to a drawing of G_{j+1}.

After extension to a drawing of G_{j+1}, the loop returns to the two-fold exit test and possibly continues with the next attempt at extending the drawing.

The Algorithm

Naively searching for a homeomorphic copy of K_5 or of $K_{3,3}$ in an n-vertex graph would require exponentially many steps. The planarity algorithm given here requires $O(n^2)$ execution steps.

Algorithm 4.6.1: Planarity-Testing for a 2-Connected Graph
Input: a 2-connected graph G
Output: a planar drawing of G, or the decision FALSE.
{*Initialize*} Find an arbitrary cycle G_0 in G, and draw it in the plane.
While $G_j \neq G$ {this exit implies that G is planar}
  If any appendage is blocked
    return FALSE
  Else
    If some appendage is forced
      $B :=$ that appendage
    Else
      $B :=$ any appendage whatsoever
      $R :=$ any region whose boundary contains all contact points of B
      Select any path between two contact points of B
      Draw that path into region R to obtain G_{j+1}
      Continue with next iteration of while-loop
  {End of while-loop body}
Return (planar drawing of G)

Outline of Correctness of Planarity Algorithm

To establish the correctness of this algorithm, there is a two-step proof that when no appendage is forced, the particular choice of an appendage to extend the subgraph G_j and a region in which to draw it does not affect the eventual decision of the algorithm. The first step is to identify a configuration under which there is an appendage A with no forced choice between two regions R and R', as illustrated in Figure 4.6.3.

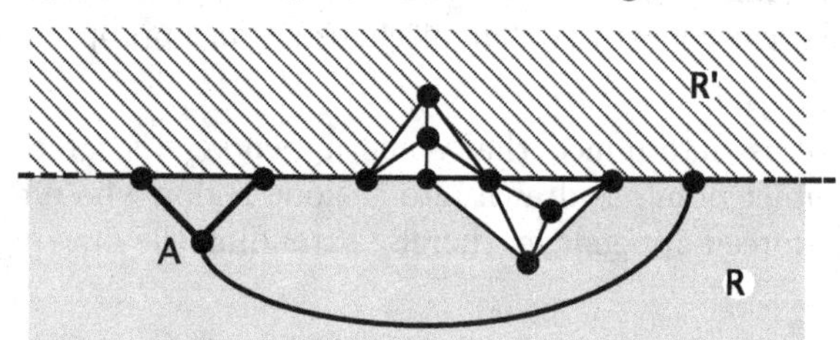

Figure 4.6.3 Appendage A has option of region R or region R'.

The second step is to demonstrate that regardless of which region is chosen, the eventual algorithmic decision on planarity is the same. For further details, see §9.8 of [BoMu76] or §6.3 of [We0l].

Hopcroft and Tarjan [HoTa74] have published a linear-time planarity algorithm. Topological arguments of Gross and Rosen [GrRo8l] enabled them to achieve a linear time reduction of planarity-testing for *2-complexes* [GrR079] to a planarity-test for graphs.

EXERCISES for Section 4.6

4.6.1 Apply Algorithm 4.6.1 to the designated graph.

(a) The graph of Exercise 4.4.1
(b) The graph of Exercise 4.4.2
(c) The graph of Exercise 4.4.3
(d) The graph of Exercise 4.4.4
(e) The graph of Exercise 4.4.5
(f) The graph of Exercise 4.4.6
(g) $K_{3,3}$
(h) K_5
(i) $K_3 + K_1$
(j) $(K_4 - K_2) \times P_3$
(k) $K_9 - K_{4,5}$
(l) $(K_5 - K_3) \times P_2$

4.6.2 [Computer Project] Implement Algorithm 4.6.1 on a computer.

4.7 CROSSING NUMBERS AND THICKNESS

Two quantifications of the non-planarity of a graph involve only the plane, without going to higher-order surfaces. One is to determine the minimum number of edge-crossings needed in a drawing of that graph in the plane. the other partitions the edge-set of a graph into "layers" of planar subgraphs and establishes the minimum number of layers needed.

Crossing Numbers

Although it is easy to count the edge-crossings in a graph drawing, it is usually very difficult to prove that the number of crossings in a particular drawing is the smallest possible. The context of most such calculations is that the drawing has been *normalized* so that no edge crosses another more than once and that at most two edges cross at any point.

DEFINITION: Let G be a simple graph. The ***crossing number*** $cr(G)$ is the minimum number of edge-crossings that can occur in a normal drawing of G in the plane.

The usual way to calculate the crossing number of a graph G has two parts. A lower bound $cr(G) \geq b$ is established by a theoretical proof. An upper bound $cr(G) \leq b$ is established by exhibiting a drawing of G in the plane with b crossings. If one is able to draw the given graph with only one edge-crossing, then calculating the crossing number reduces to a planarity test.

Proposition 4.7.1: $cr(K_5) = 1$.

Proof: By Theorem 4.1.4, every drawing of K_5 in the plane must have at least one crossing. Thus, $cr(K_5) \geq 1$. Since the drawing of K_5 in Figure 4.7.1 has only one crossing, it follows that $cr(K_5) \leq 1$. ♢

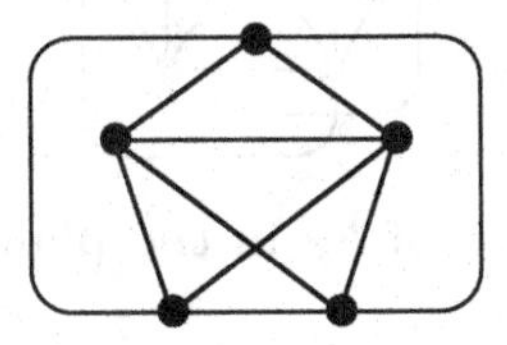

Figure 4.7.1 A drawing of K_5 in the plane with only one crossing.

Proposition 4.7.2: $cr(K_{3,3}) = 1$.

Proof: By Theorem 4.1.5, it follows that $cr(K_{3,3}) \geq 1$. The drawing of $K_{3,3}$ in Figure 4.7.2 with only one crossing implies that $cr(K_{3,3}) \geq 1$. ♢

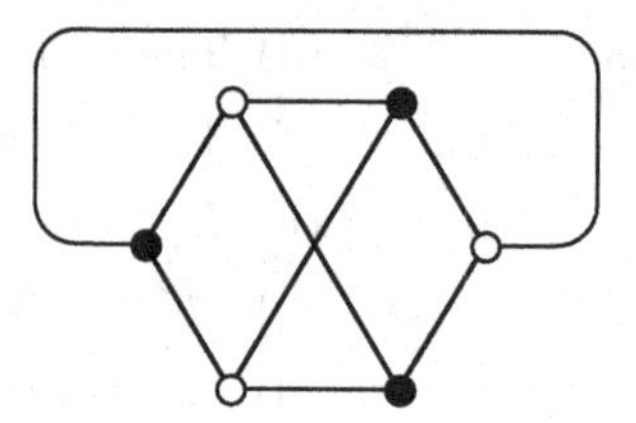

Figure 4.7.2 A drawing of $K_{3,3}$ in the plane with only one crossing.

Lower Bounds for Crossing Numbers

The main interest in crossing numbers of graphs is for cases in which the minimum number is more than 1.

Theorem 4.7.3: *Let G be a connected simple graph. Then*

$$cr(G) \geq |E_G| - 3\,|V_G| + 6$$

Proof: Let H be a planar spanning subgraph of G with the maximum number of edges. Theorem 4.5.9 implies that $|E_H| \leq 3\,|V_H| - 6$, from which it follows that $|E_H| \leq 3\,|V_G| - 6$, since $V_H = V_G$. Since H is maximum planar, we infer that no matter how each of the remaining $|E_G| - |E_H|$ edges is added to a planar drawing of H, it crosses at least one edge of H. Thus, $cr(G) \geq |E_G| - |E_H| \geq |E_G| - 3\,|V_G| + 6$. ♢

Proposition 4.7.4: $cr(K_6) = 3$.

Proof: Since $|E(K_6)| = 15$ and $|V(K_6)| = 6$, Theorem 4.7.3 implies that

$$cr(K_6) \geq 15 - 3 \cdot 6 + 6 = 3$$

The drawing of K_6 in Figure 4.7.3 with three crossings establishes that $cr(K_6) \leq 3$. ♢

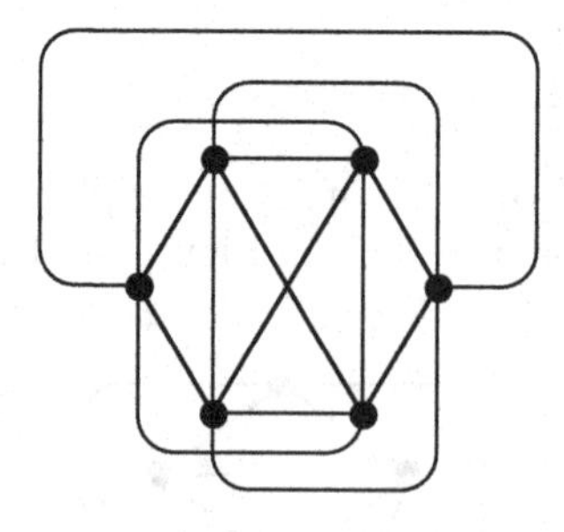

Figure 4.7.3 A drawing of K_6 in the plane with three crossings.

Theorem 4.7.5: *Let G be a connected simple bipartite graph. Then*

$$cr(G) \geq |E_G| - 2|V_G| + 4$$

Proof: Let H be a planar spanning subgraph of G with the maximum number of edges. Theorem 4.5.12 implies that $|E_H| \leq 2|V_H| - 4$, from which it follows that $|E_H| \leq 2|V_G| - 4$, since $V_H = V_G$. Since H is maximum planar, it follows that no matter how each of the remaining $|E_G| - |E_H|$ edges is added to a planar drawing of H, it crosses at least one edge of H. Thus, $cr(G) \geq |E_G| - |E_H| \geq |E_G| - 2|V_G| + 4$. ♢

Proposition 4.7.6: $cr(K_{3,4}) = 2$.

Proof: Since $|E(K_{3,4})| = 12$ and $|V(K_{3,4})| = 7$, Theorem 4.7.5 implies that

$$cr(K_{3,4}) \geq 12 - 2 \cdot 7 + 4 = 2$$

The drawing of $K_{3,4}$ in Figure 4.7.4 with two crossings implies that $cr(K_{3,4}) \leq 2$. ♢

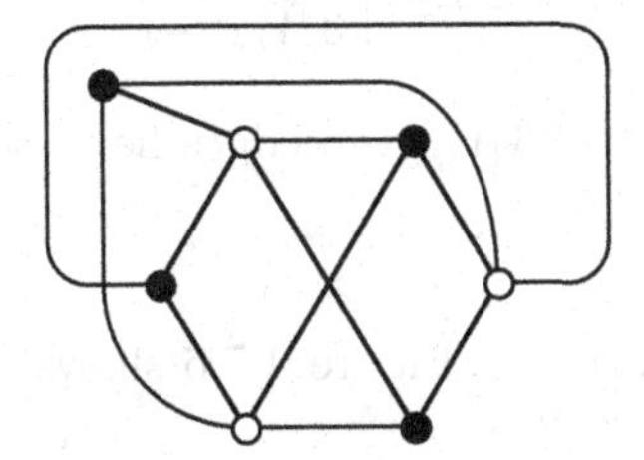

Figure 4.7.4 A drawing of $K_{3,4}$ in the plane with two crossings.

Example 4.7.1: The requirement that no more than two edges cross at the same intersection is necessary for these results to hold. Figure 4.7.5 shows that if a triple intersection is permitted, then $K_{3,4}$ can be drawn in the plane with only one crossing.

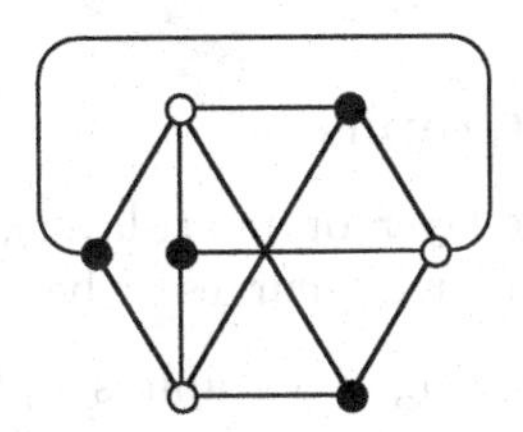

Figure 4.7.5 A drawing of $K_{3,4}$ in the plane with one triple-crossing.

Thickness

Application 4.7.1: *Partitioning a Network into Planar Layers* One technique involved in miniaturizing a non-planar electronic network is to sandwich layers of insulation between planar layers of uninsulated wires, which connect nodes that pierce through all the layers. Minimizing the number of layers tends to reduce the size of the chip. Moreover, in applying the technique to mass production, minimizing the number of layers tends to reduce the cost of fabrication.

DEFINITION: The ***thickness*** of a simple graph G, denoted $\theta(G)$, is the smallest cardinality of a set of planar spanning subgraphs of G whose edge-sets partition E_G.

Example 4.7.2: The two planar graphs in Figure 4.7.6 below both span K_8. The union of their edge-sets is E_{K_8}. Thus, a chip implementing K_8 would need only two planar layers.

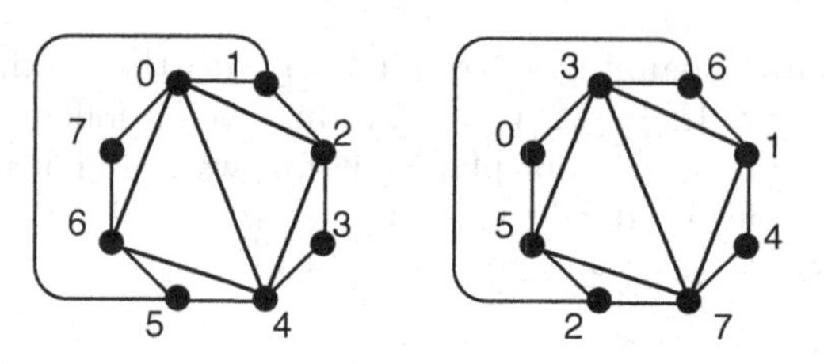

Figure 4.7.6 Two planar spanning subgraphs of K_8.

Theorem 4.7.7: *Let G be a connected simple graph. Then*

$$\theta(G) \geq \left\lceil \frac{|E_G|}{3\,|V_G| - 6} \right\rceil$$

Proof: By Theorem 4.5.9, at most $3\,|V_G| - 6$ edges lie in a planar subgraph of G. ♢

Proposition 4.7.8: $\theta(K_8) = 2$.

Proof: By Theorem 4.7.7, $\theta(K_8) \geq 2$. Figure 4.7.6 shows two planar graphs whose union is K_8, which implies that $\theta(K_8) \leq 2$. ♢

Theorem 4.7.9: *Let G be a connected simple bipartite graph. Then*

$$\theta(G) \geq \left\lceil \frac{E_G}{2\,|V_G| - 4} \right\rceil$$

Proof: By Theorem 4.5.12, at most $2\,|V_G| - 4$ edges lie in a planar subgraph of G. ♢

Straight-Line Drawings of Graphs

In circuit design (either single-layer or multi-layer), it is convenient for wires to be straight-line segments, rather than having curves or bends.

DEFINITION: A ***straight-line drawing*** of a graph is a planar drawing in which every edge is represented by a straight-line segment.

Wagner [Wa36] and Fary [Fa48] proved the following result, which is known as Fary's theorem. Thomassen [Th80, Th81] developed a proof of Kuratowski's theorem that simultaneously yields Fary's theorem.

Theorem 4.7.10: *Let G be a simple planar graph. Then G has a straight-line drawing without crossings.* ♢

EXERCISES for Section 4.7

4.7.1 Derive this lower bound for the crossing number of the complete graph K_n.

$$cr(K_n) \geq \left\lceil \frac{(n-3)(n-4)}{2} \right\rceil$$

4.7.2 Derive this lower bound for the crossing number of the complete bipartite graph $K_{m,n}$.

$$cr(K_{m,n}) \geq (m-2)(n-2)$$

4.7.3 Draw the Petersen graph in the plane with two (normal) crossings.

4.7.4 Draw K_7 in the plane with nine (normal) crossings.

4.7.5 Calculate the crossing number of a given graph.

(a) $K_{4,4}$ (b) $K_{4,5}$

(c) $Q_3 + K_1$ (d) $C_4 + K_2$

4.7.6 Prove that $\theta(K_{m,n}) \geq \left\lceil \frac{mn}{2(m+n-2)} \right\rceil$.

4.7.7 Prove that $\theta(K_{6,6}) = 2$.

4.7.8 Prove that $\theta(K_9) = 3$.

Application 4.7.1, continued: A straightforward approach to multi-layer circuit design is to use "simultaneous straight-line drawings" of the spanning subgraphs that induce the edge-set partition. This means that the nodes in each layer are in a fixed location in the plane, so that each layer is given a straight-line drawing. Figure 4.7.7 shows a two-layer partition of the complete graph K_5 by simultaneous straight-line drawings.

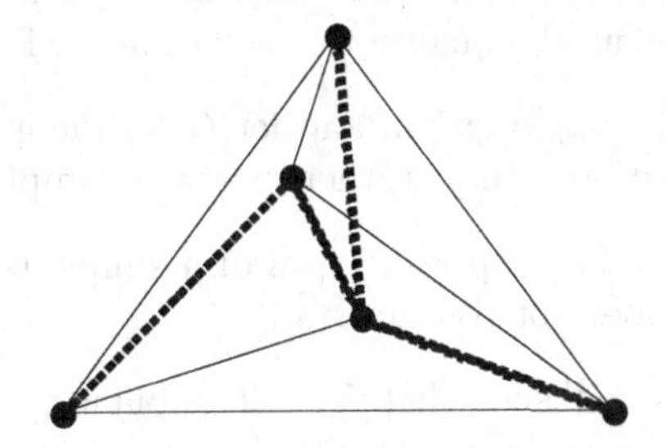

Figure 4.7.7 Two-layer straight-line partition of K_5.

4.7.9 Partition K_6 into two simultaneous straight-line layers.

4.7.10 Partition $K_{4,4}$ into two simultaneous straight-line layers.

4.7.11 Partition $K_8 - K_4$ into three simultaneous straight-line layers.

4.7.12 Partition K_8 into four simultaneous straight-line layers.

4.8 SUPPLEMENTARY EXERCISES

4.8.1 Consider a plane drawing of an n-vertex simple graph G with fewer than $3n-6$ edges.
a) Use the Euler Polyhedral Equation and the Edge-Face Inequality to prove that there exists a face f with more than 3 sides.
b) Use the Jordan Curve Theorem to prove that there are two non-consecutive vertices on the boundary of face f that are not adjacent in graph G.

4.8.2 Draw a self-dual imbedding of a planar simple graph with degree sequence $\langle 4,4,4,3,3,3,3\rangle$.

4.8.3 Draw a non-self-dual imbedding of a planar simple graph with degree sequence $\langle 4,4,4,3,3,3,3\rangle$.

4.8.4 A connected graph G with p vertices and q edges is imbedded in the sphere. Its dual is drawn. The dual vertex in each primal face is joined to every vertex on the face boundary, and every crossing of a primal and dual edge is converted into a new vertex, so that the result is a graph $G^{\#}$ imbedded in the sphere.
a) How many vertices and edges does $G^{\#}$ have?
b) For $G = C_3$, draw the edge-complement of $G^{\#}$.

4.8.5 Draw a 4-regular simple 9-vertex planar graph.

4.8.6 Suppose that a simple graph has degree sequence $\langle 5,5,5,5,5,5,4,4\rangle$. Prove that it cannot be planar.

4.8.7 Prove that every simple planar graph with at least four vertices has at least four vertices of degree less than 6.

4.8.8 Can a 3-connected simple graph of girth 3 be drawn in the plane so that every face has at least four sides? Either give an example or prove it is impossible.

4.8.9 Prove that the Petersen graph has no 3-cycles or 4-cycles. Then use the Euler Polyhedral Equation and the Edge-Face Inequality to show that the Petersen graph is non-planar.

4.8.10 Let H be a 4-regular simple graph, and let G be the graph obtained by joining H to K_1. Decide whether G can be planar. Either give an example or prove impossibility.

4.8.11 Draw each of the four isomorphism types of a simple 6-vertex graph that contains K_5 homeomorphically, but does not contain $K_{3,3}$.

4.8.12 a. Draw a 9-vertex tree T such that $K_2 + T$ is planar.
b. Prove that there is only one such tree.

4.8.13 Decide whether the following graphs contain a homeomorphic copy of $K_{3,3}$, of K_5, of both, or of neither.

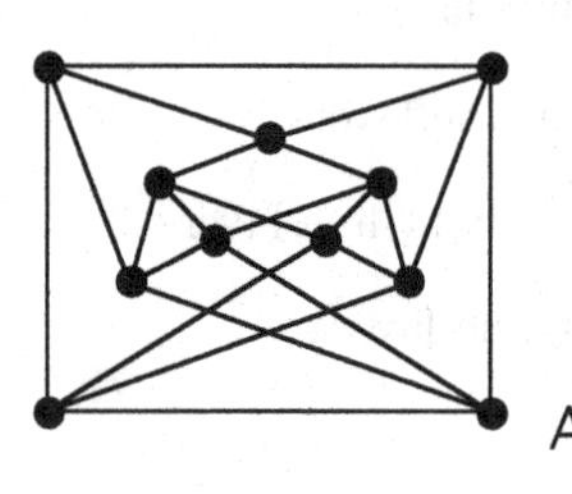

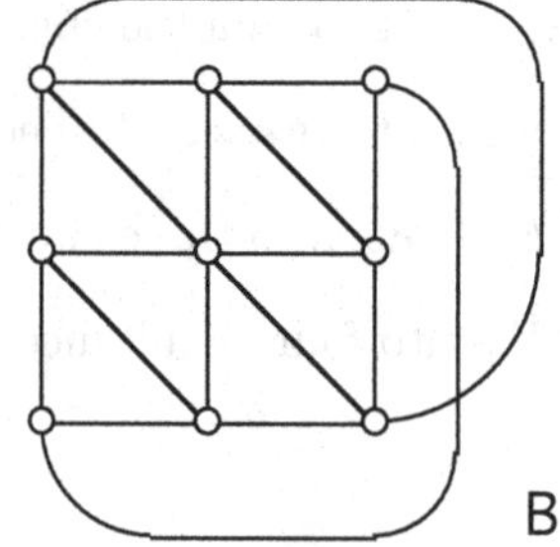

4.8.14 Decide which of the graphs below are planar.

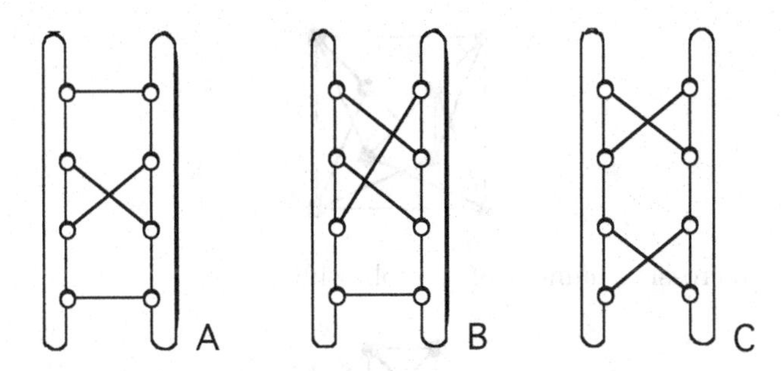

4.8.15 Decide which of the graphs below are planar.

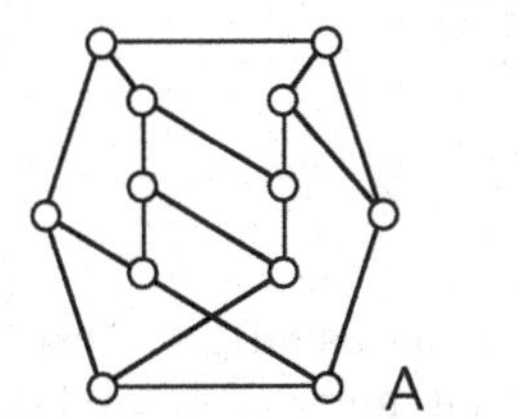

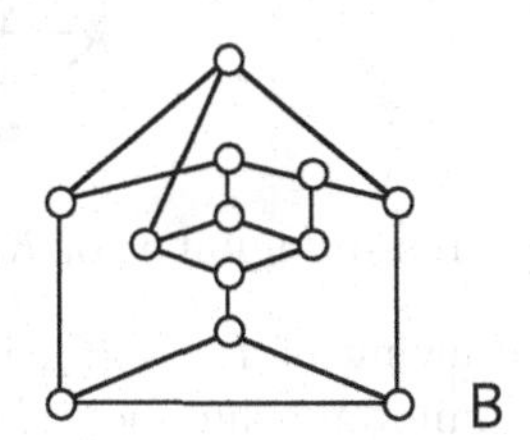

4.8.16 a. Show how to add one edge to graph A so that the resulting graph is non-planar (and prove non-planarity). b. Show how to delete one edge from graph B so that the resulting graph is planar.

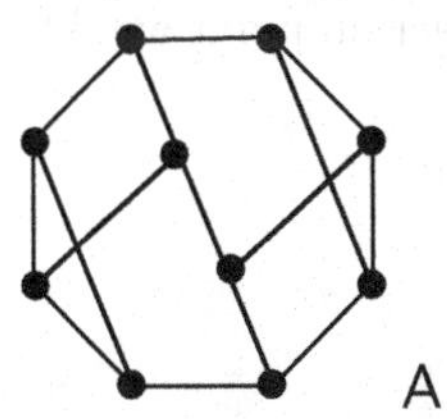

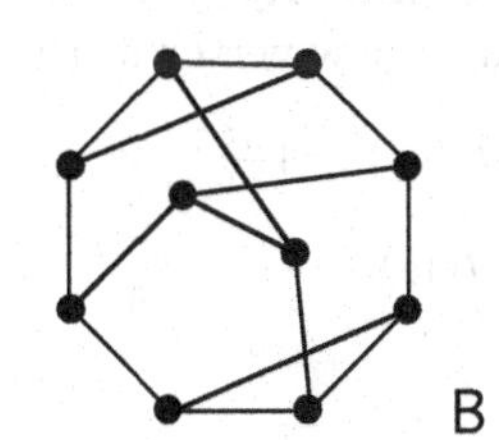

4.8.17 Decide which of the graphs below are planar.

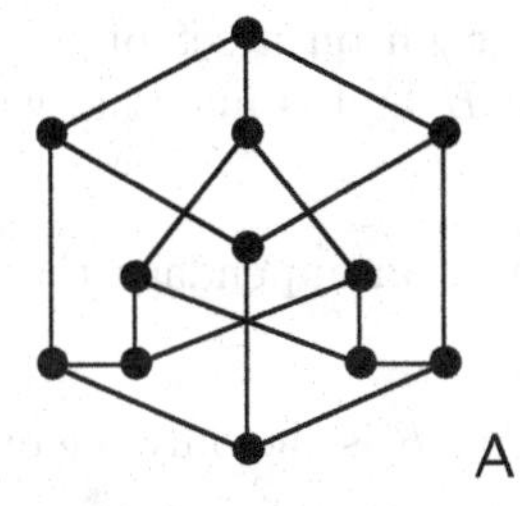

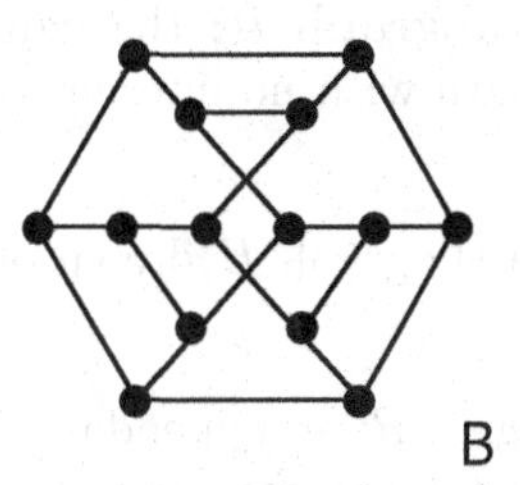

4.8.18 Use the planarity algorithm to show that the circulant graph $circ(7 : 1, 2)$ is non-planar. Find a Kuratowski subgraph.

4.8.19 Use the planarity algorithm to show that the circulant graph $circ(9 : 1, 3)$ is non-planar. Find a Kuratowski subgraph.

4.8.20 Redraw the following graph with only two crossings.

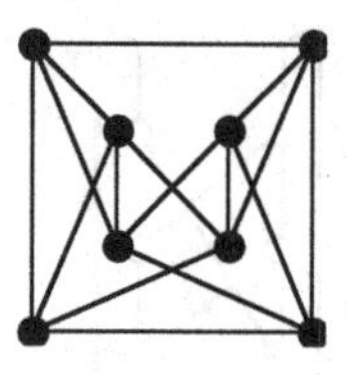

4.8.21 Calculate the crossing number of the following graph.

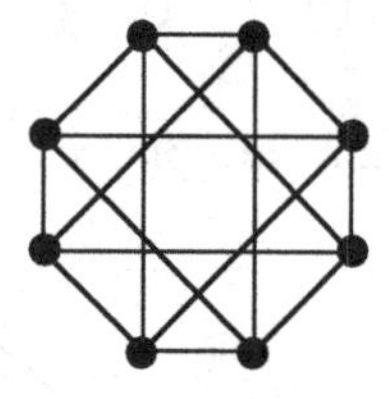

4.8.22 Calculate the crossing number of $K_6 - K_2$.

4.8.23 In a planar drawing of $K_6 - K_2$, in which (as usual) at most two edges cross at a point, what is the minimum number of edges that must cross at least one other edge? (Hint: be careful; for $K_{3,3}$, the answer is two – not one, since two edges cross at the single crossing.)

4.8.24 Suppose that a triple edge-crossing were permitted. Prove that even then, it would still be impossible to draw $K_6 - K_2$ in the plane with only one crossing. (Hint: the graph obtained by inserting a new vertex at the triple-intersection point would have to be planar.)

4.8.25 Prove that $\theta(K_n) \geq \lfloor \frac{n+7}{6} \rfloor$.

4.8.26 Calculate $\theta(K_{12})$.

Glossary

appendage to a subgraph H**:** the induced subgraph on a set of edges such that each pair lies on a path with no internal vertices in H and is maximal with respect to this property.

—, **blocked:** for a subgraph H drawn on a surface, an appendage that is undrawable in every region.

—, **forced** into region R: an appendage B such that R is the only region whose boundary contains all the contact points of B.

—, **overlapping pair for a cycle** C**:** appendages B_1 and B_2 such that either of these conditions holds:

(i) Two contact points of B_1 alternate with two contact points of B_2 on cycle C.

(ii) B_1 and B_2 have three contact points in common.

—, **undrawable** for region R: for a subgraph H drawn on a surface, an appendage B such that the boundary of region R does not contain all the contact points of B.

barycentric subdivision, (first) of a graph G: the subdivision of G such that every edge of G is subdivided exactly once.

—, n^{th}: the first barycentric subdivision of the $(n-1)^{st}$ barycentric subdivision.

boundary of a region of an imbedding of a graph G: the subgraph of G that comprises all vertices that abut the region and all edges whose interiors abut the region.

boundary walk of a face of an imbedding of a graph G: a dosed walk in graph G that corresponds to a complete traversal of the perimeter of the polygonal region within the face.

cellular imbedding $\iota : G \to S$: a graph imbedding such that every region is topologically equivalent to an open disk.

chord of a subgraph H: an appendage of H that consists of a single edge.

connected Euclidean set: see *Euclidean set.*

contact point of an appendage B to a subgraph H: a vertex in both B and H.

crossing number $cr(G)$ of a simple graph G: the minimum number of edge-crossings that can occur in a normal planar drawing of G.

curve, closed or open in a Euclidean set: see *path.*

digon: a 2-sided region.

dual edges: the edges of the dual graph.

dual faces: the faces of the dual imbedding.

dual graph and dual imbedding: the new graph and its imbedding derived, starting with a cellular imbedding of a graph, by inserting a new vertex into each existing face, and by then drawing through each existing edge a new edge that joins the new vertex in the region on one side to the new vertex in the region on the other side.

dual vertices: the vertices of the dual graph.

edges separated in a graph G by the subgraph H: two edges e_1 and e_2 of $E_G - E_H$ such that every walk in G from an endpoint of e_1 to an endpoint of e_2 contains a vertex of H.

edges unseparated in a graph G by the subgraph H: two edges e_1 and e_2 of $E_G - E_H$ that are not separated.

Euclidean set X: a subset of any Euclidean space $\mathbb{R}^n$.

—, connected: a Euclidean set X such that for every pair of points $s, t \in X$, there exists a path within X from s to t.

—, separation of X by a subset $W \subset X$: the condition that there exists a pair of points s and t in $X - W$, such that every path in X from s to t intersects the set W.

face of a graph imbedding $\iota : G \to S$: the union of a region and its boundary. (A drawing does *not* have faces unless it is an imbedding.)

face-set of a graph imbedding $\iota : G \to S$: the set of all faces of the imbedding, denoted F_ι, F_G, or F, depending on the context.

face-size: the number of edge-steps in the closed walk around its boundary (in which some edges may count twice). A face of size n is said to be *n-sided.*

girth of a non-acyclic graph: the length of the smallest closed cycle (undefined for an ayclic graph).

homeomorphic graphs: a pair of graphs G and H such that there is an isomorphism from a *subdivision* of G to a subdivision of H.

imbedding of a graph G on a surface S: a drawing without any edge-crossings.

Jordan separation property: for a Euclidean set X, the property that every closed curve in X separates X.

Kuratowski graphs: the complete graph K_5 and the complete bipartite graph $K_{3,3}$.

Kuratowski subgraph of a graph: a subgraph homeomorphic either to K_5 or $K_{3,3}$.

longitude on the standard torus: a closed curve that bounds a disk in the space exterior to the solid donut (and goes around in the "long" direction).

meridian on the standard torus: a closed curve that bounds a disk inside the solid donut it surrounds (and goes around the donut in the "short" direction).

monogon: a 1-sided region.

path from s to t in a Euclidean set: the image of a continuous function f from the unit interval $[0, 1]$ to a subset of that space such that such that $f(0) = s$ and $f(1) = t$, that is a bijection on the interior of $[0, 1]$. (One may visualize a path as the trace of a particle traveling through space for a fixed length of time.)

—, **closed:** a path such that $f(0) = f(1)$, i.e., in which $s = t$. (For instance, this would include a "knotted circle" in space.)

—, **open:** a path such that $f(0) \neq f(1)$, i.e., in which $s \neq t$.

planar drawing of a graph: a drawing of the graph in the plane without edge-crossings.

planar graph: a graph such that there exists a planar drawing of it.

plane in Euclidean 3-space $\mathbb{R}^3$: a set of points (x, y, z) such that there are numbers a, b, c, and d with $ax + by + cz = d$.

Poincaré dual: *see* dual graph imbedding.

primal graph and primal imbedding: whatever graph and imbedding are supplied as input to the Poincaré duality process.

primal vertices, primal edges, and primal faces: the vertices, edges, and faces of the primal graph and primal imbedding.

proper walk: a walk in which no edge occurs immediately after itself.

region of a graph imbedding $\iota : G \to S$: a component of the Euclidean set $S \to \iota(G)$ which does *not* contain its boundary.

Riemann stereographic projection: a function ρ that maps each point w of the unit-diameter sphere (tangent at the origin $(0,0,0)$ to the xz-plane in Euclidean 3-space) to the point $\rho(x)$ where the ray from the North Pole $(0,1,0)$ through the point w intersects the xz-plane.

self-dual graph: a graph G with an imbedding in the sphere (or sometimes another surface) such that the Poincaré dual graph G^* for that imbedding is isomorphic to G.

separated Euclidean set: see *Euclidean set.*

size of a face: *see* face-size.

smoothing out a 2-valent vertex v**:** replacing two different edges that meet at v by a new edge that joins their other endpoints.

sphere: a set of points in $\mathbb{R}^3$ equidistant from a fixed point.

standard donut: the surface of revolution obtained by revolving a disk of radius 1 centered at (2,0) in the xy-plane around the y-axis in 3-space.

standard torus: the surface of revolution obtained by revolving a circle of radius 1 centered at $(2,0)$ in the xy-plane around the y-axis in 3-space. It is the surface of the standard donut.

straight-line drawing of a graph: a planar drawing in which every edge is represented by a straight-line segment.

subdividing an edge e**:** with endpoints u and v:] replacing that edge e by a path u, e', w, e'', v, where edges e' and e'' and vertex w are new to the graph.

subdividing a graph: performing a sequence of edge-subdivision operations.

thickness $\theta(G)$ of a simple graph G: the smallest cardinality of a set of planar spanning subgraphs of G whose edge-sets partition E_G.

Chapter 5

DRAWING GRAPHS AND MAPS

INTRODUCTION

We are ready to say with more precision what it means to *draw* a graph on a surface. For that purpose, it is necessary to have spatial models for surfaces and graphs, in which the edges of a graph have length, just as they do in drawings.

In practical applications, most graph drawings are on the plane or the sphere. Both these surfaces have the simplifying feature described in §4.1, called the *Jordan separation property*, that every closed curve separates it into two parts. As described there, applying the Jordan separation property provides an elementary method for solving the problem of deciding which graphs can be drawn on the plane or sphere.

However, some elementary methods of *algebraic topology* for solving this drawability problem, once mastered, yield correct results with far less effort, and they also apply to the more complicated surfaces. Chapter 4 presented the most basic principles for such endeavors, in particular, the Euler Polyhedral Equation, the face-size equation, the edge-face inequality, and Poincaré duality. These principles are applied in §5.4 to the problem of drawing highly symmetric maps on the sphere. They are subsequently generalized to the higher-order surfaces in §5.5, and then applied to the problem of drawing graphs on higher-order surfaces, where the Jordan Curve Theorem is of no avail. This chapter may be regarded as the starting point of topological graph theory.

5.1 TOPOLOGY OF LOW DIMENSIONS

The formal definition of a drawing as a continuous function depends on a spatial model for a graph (the domain) and a spatial model for a surface (the codomain), in which the edges of a graph have length, just as they do in drawings. Surfaces are defined intrinsically, starting here and continuing into §5.2, with the aid of *topological equivalence*, and they need not be attached to some solid that they bound.

Some Subsets of Euclidean 2-Space and 3-Space

DEFINITION: ***Euclidean n-space*** $\mathbb{R}^n$ is the set of n-tuples $(x_1, \ldots, x_n)$ with the usual Euclidean distance metric

$$\delta((x_1, \ldots, x_n), (y_1, \ldots, y_n)) = \sqrt{(x_1 - y_1)^2 + \cdots + (x_n - y_n)^2}$$

DEFINITION: A ***Euclidean set*** is a subset of a Euclidean space.

DEFINITION: The ***plane*** is another name for Euclidean 2-space $\mathbb{R}^2$. (This is consistent with the definition in §4.1 of a plane in $\mathbb{R}^3$.)

DEFINITION: The ***open unit disk*** is the plane Euclidean set containing the points $\{(x_1, x_2) \mid x_1^2 + x_2^2 < 1\}$, i.e., the points inside the unit circle.

DEFINITION: The ***closed unit disk*** is the plane Euclidean set containing the points, $\{(x_1, x_2) \mid x_1^2 + x_2^2 \leq 1\}$ i.e., the points inside and on the unit circle.

DEFINITION: The ***standard half-disk*** is the plane Euclidean set containing the points $\{(x_1, x_2) \mid x_1 \geq 0 \text{ and } x_1^2 + x_2^2 < 1\}$.

Example 5.1.1: As illustrated in Figure 5.1.1, the open unit disk and the half-disk are subsets of the closed unit disk. The half-disk contains the segment of its frontier that is an open interval on the x_2-axis, but it does not contain the semi-circle that is its other frontier segment in $\mathbb{R}^2$.

Figure 5.1.1 An open disk, closed disk, and standard half-disk.

DEFINITION: The ***unit sphere*** is the 3-dimensional Euclidean set containing the points $\{(x_1, x_2, x_3) \mid x_1^2 + x_2^2 + x_3^2 = 1\}$.

DEFINITION: The ***unit cylinder*** is the 3-dimensional Euclidean set containing the points $\{(x_1, x_2, x_3) \mid 0 \leq x_1 \leq 1 \text{ and } x_2^2 + x_3^2 = 1\}$. See Figure 5.1.2.

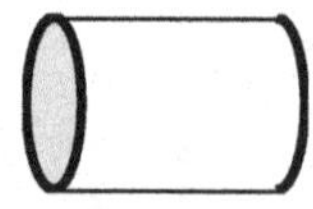

Figure 5.1.2 The unit cylinder.

Topological Equivalences of Euclidean Sets

Intuitively, a function from one Euclidean set to another is continuous if it always maps "nearby" points of the domain to "nearby" points of the codomain. In theoretical calculus and point-set topology, this can be expressed precisely with epsilons and deltas. It is possible to gain an elementary grasp of graph placements on surfaces without a digression into the technicalities of continuous functions.

DEFINITION: A continuous function $f : X \to Y$ between two Euclidean sets is called a ***topological equivalence*** if it is one-to-one and onto, and if the inverse function $f^{-1} : Y \to X$ is continuous.

Remark: Intuitively, a topological equivalence is a function that deforms its domain into its codomain, without tearing the domain and without compressing any set of two or more points down to a single point.

Example 5.1.2: The Riemann stereographic projection (see §4.1) is a topological equivalence between a sphere minus its North Pole and the plane.

Example 5.1.3: A topological equivalence between the open unit disk and the entire plane, which illustrates how topological equivalence permits spaces to be stretched, is given by the function

$$f((x,y)) = \left(\frac{x}{1-x^2-y^2}, \frac{y}{1-x^2-y^2}\right)$$

Example 5.1.4: The *interior* region R in the plane enclosed by any polygon but not including the polygon itself is topologically equivalent to the open unit disk. Writing out the continuous bijection with precise formulas is tedious. Instead, we describe the equivalence informally.

Case 1: a convex polygon P. Place the geometric center (or any other interior point) of the polygon at the origin. Next observe that the ray at angle α from the origin to infinity intersects the open disk in a half-open segment I_D^α and intersects region R in a half-open segment I_R^α, as illustrated in Figure 5.1.3. Let the function f map I_D^α to I_R^α by linear stretching or compression. Then f is a topological equivalence from the unit open disk to the polygonal region R.

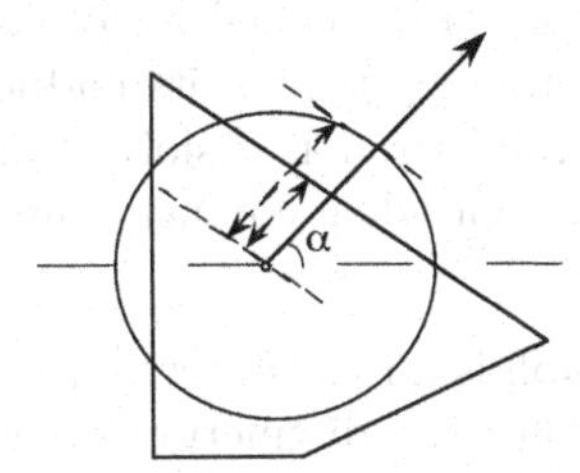

Figure 5.1.3 Equivalence of a convex polygonal region to a unit disk.

Case 2: a non-convex polygon P. First fill polygon P with "mathematical helium" that "inflates" it into a convex polygon with the same cycle of side-lengths, as illustrated in Figure 5.1.4. This inflation process is a topological equivalence. Then apply the method of the convex case.

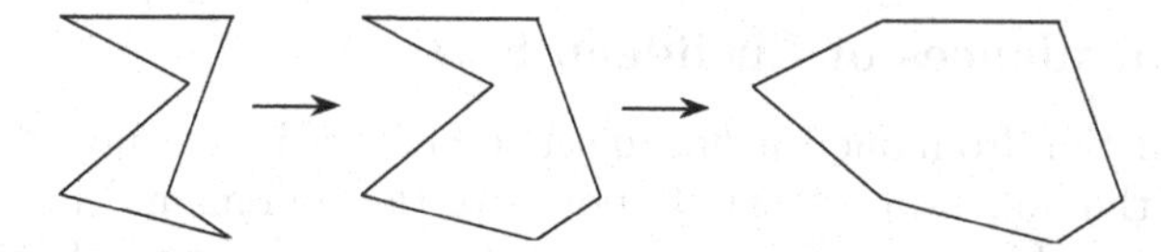

Figure 5.1.4 Mapping a non-convex region to a convex region.

DEFINITION: A ***sphere*** is any Euclidean set that is topologically equivalent to the unit sphere.

Example 5.1.5: The surface of any convex 3-dimensional solid is a sphere. This follows by a construction analogous to the convex-polygon example above. For instance, Figure 5.1.5 illustrates a ball, ovoid, and cube, each a 3-dimensional convex solid. Thus, the surface of each of these solids is a sphere.

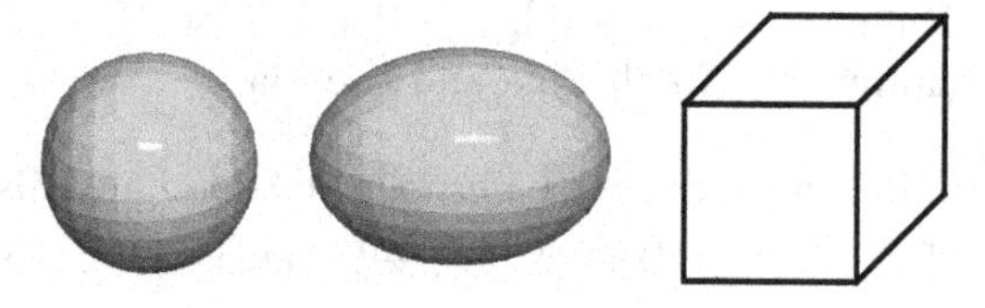

Figure 5.1.5 The surfaces of these convex solids are spheres.

Topological Model of a Graph

To draw graphs on surfaces, except for simple graphs in the plane, we need a model of an edge.

DEFINITION: A ***space curve*** between two points x and y in 3-space is the image of a continuous function $f : [0,1] \to \mathbb{R}^3$ from the unit interval into 3-space such that $f(0) = x$ and $f(1) = y$, which is one-to-one, except that possibly $x = y$. That is, the space curve is the set $\{f(t) \mid t \in [0,1]\}$. The ***interior of the space curve*** is the subset $\{f(t) \mid t \in (0,1)\}$.

DEFINITION: In a ***topological model*** (or ***carrier***) of a graph $G = (V, E)$, each vertex $v \in V$ is represented by a point p_v in $\mathbb{R}^3$ called a ***space vertex***, and each proper edge $e \in E$ is represented by a space curve q_e joining its endpoints, called a ***space edge***. The interior of each space edge q_e of the carrier is disjoint from all the other space edges and also from all the space vertices. Except when confusion might result, the topological model of a graph G is also denoted G.

The topological model of a graph is a rich source of topological and geometric intuition. For instance, it immediately clarifies the distinction between the ends of an edge, which means intuitively the two minute parts of the edge "near" its endpoints, and the endpoints themselves, which are vertices.

DEFINITION: The ***0-end of a space edge*** of a graph is the part near the image of the 0-end of the unit interval $[0,1]$ in the space curve.

DEFINITION: The ***1-end of a space edge*** of a graph is the part near the image of the 1-end of the unit interval $[0,1]$ in the space curve.

Remark: In this topological sense, every edge has two distinct *ends*, even if it is a self-loop with only one *endpoint*. If a large segment $\{f(t) \mid t \in (\epsilon, 1-\epsilon)\}$ of the interior of a space edge is discarded, what remains is the two whisker-like ends, as illustrated in Figure 5.1.6.

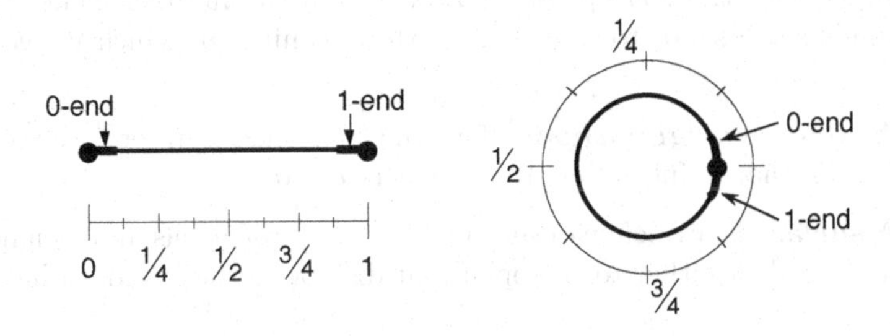

Figure 5.1.6 The 0-end and the 1-end of a proper edge and of a self-loop.

Remark: The *direction* of a space edge may coincide with parametric increase or with parametric decrease. In the incidence table, a marker on the upper copy of the endpoint of a self-loop designates direction of parametric increase from the 0-end to the 1-end, and a marker on the lower copy of the endpoint designates direction of parametric decrease from the 1-end to the 0-end. Thus, even though a self-loop may have only one possible direction in a standard combinatorial model of a graph, it has two directions in a topological model.

EXERCISES for Section 5.1

5.1.1 Specify a function from the open unit interval $(0,1)$ to the entire real line that is a topological equivalence.

5.1.2 Specify a function from the annulus $\{(x,y) \mid 1 \leq x^2+y^2 \leq 2\}$ to the unit cylinder that is a topological equivalence.

5.1.3 How many topologically inequivalent Euclidean sets can be formed by identifying a single point of an open interval to a point in an open disk?

5.1.4 Four different Euclidean sets (i.e., topologically inequivalent) can be formed by identifying a single point of an closed real interval to a point of a closed disk. Describe them.

5.2 HIGHER-ORDER SURFACES

Topology generalizes the concept of a surface, from the most elementary meaning, in which it surrounds some solid, to a more powerful meaning, in which it need not enclose anything.

DEFINITION: An ***open ϵ-neighborhood*** of a point in a Euclidean set is the set of all points whose distance from that point is less than ϵ, where $\epsilon > 0$.

DEFINITION: A ***surface*** is a Euclidean set in which every point has an ϵ-neighborhood that is topologically equivalent either to the open unit disk or to the standard half-disk.

Torus and Möbius Band

One possible kind of complication in a surface is apparent in the *torus*, and another kind is apparent in the *Möbius band*. Interestingly, these two kinds of complication are the only kinds of complication.

REVIEW FROM §4.1: The ***standard torus*** is the surface of revolution obtained by revolving a circle of radius 1 centered at $(2, 0)$ in the xy-plane around the y-axis in 3-space.

DEFINITION: A ***torus*** is any Euclidean set that is topologically equivalent to the standard torus.

Example 5.2.1: The surface of a knotted solid donut is a torus (as in Figure 5.2.1).

Figure 5.2.1 A knotted donut in 3-space.

Akin to the way that the Riemann stereographic projection enables us to represent drawings on a sphere by drawings on a flat piece of paper, there is a way to represent toroidal drawings on a flat piece of paper.

The rectangle in Figure 5.2.2 represents a torus. The top edge and the bottom edge of the rectangle are both marked with the letter "a". This indicates that these two edges are to be identified with each other, that is, pasted together. This pasting of edges creates a cylinder.

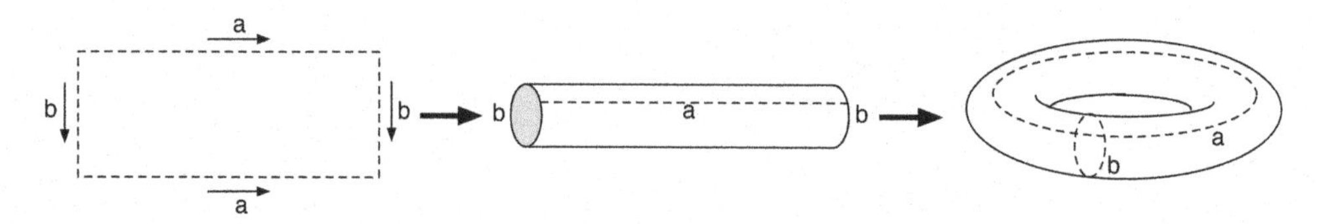

Figure 5.2.2 Folding a rectangle into a torus.

Once the top and bottom edges are pasted together, the left and right edges become closed curves, both marked with the letter "b". When the left end of the cylinder is pasted to the right end, the resulting surface is a torus.

Remark: Pasting the left edge to the right edge converts line "a" into a closed curve. Indeed, on the resulting torus, what was once edge "a" has become a longitude, and what was once edge "b" has become a meridian.

DEFINITION: A ***Möbius band*** is a space formed from a rectangular strip with a half-twist, as illustrated in Figure 5.2.3, by identifying the left edge with the right edge, to form a continuous band, as shown in Figure 5.2.4.

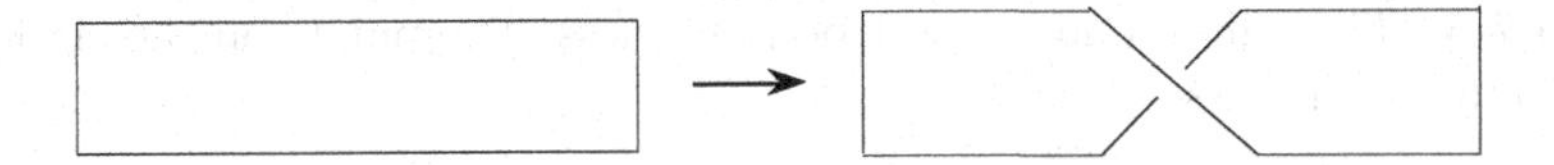

Figure 5.2.3 A rectangular strip with a half-twist.

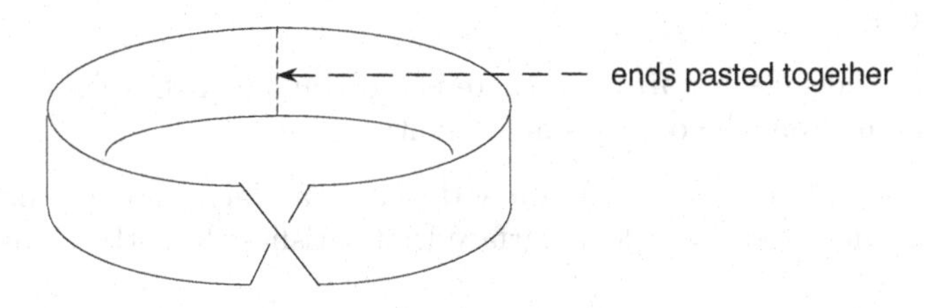

Figure 5.2.4 A Möbius band.

DEFINITION: The ***central circuit of a Möbius band*** (Figure 5.2.5) is the result of matching one end of the line halfway between the top and bottom of the rectangular strip to the other end, as the right and left ends are identified.

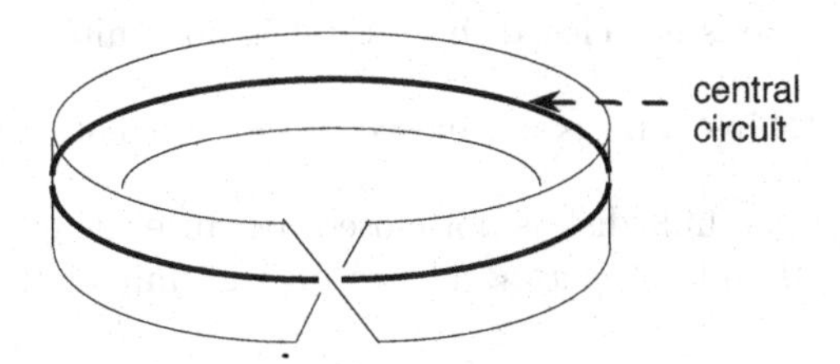

Figure 5.2.5 A Möbius band.

Theorem 5.2.1: *The Möbius band does not have the Jordan separation property.*

Proof: The central circuit does not separate a Möbius band into two parts. ◇

Remark: To illustrate that the Jordan Curve Theorem does not hold for the Möbius band, make a paper model of a Möbius band. Cut the Möbius band open along the central circuit. Notice that the result is a single connected surface, not two surfaces.

Bounded and Boundaryless Surfaces

DEFINITION: An ***interior point of a surface*** is a point that has an ϵ-neighborhood topologically equivalent to an open disk.

DEFINITION: A ***boundary point of a surface*** is a point that is not an interior point.

DEFINITION: A surface is ***boundaryless*** if every point is an interior point.

Example 5.2.2: The Euclidean plane is a boundaryless surface.

Example 5.2.3: A sphere is a boundaryless surface. Although a sphere may bound a solid ball, the sphere itself is boundaryless.

Example 5.2.4: The closed unit disk is not boundaryless. The points on the unit circle are its boundary points.

Example 5.2.5: The unit cylinder is not boundaryless. The points on the circles at either end are its boundary points.

Example 5.2.6: The Möbius band is not a boundaryless surface.

Closed Surfaces

DEFINITION: A Euclidean set is ***finite*** if there is a real number M such that the maximum distance of any point from the origin is at most M.

DEFINITION: A connected surface S is ***closed*** if it is a Euclidean set that satisfies or is topologically equivalent to a Euclidean surface that satisfies these three conditions:

(i) The surface is finite.

(ii) The surface is boundaryless.

(iii) The endpoints of every open arc in S are in surface S itself.

Example 5.2.7: The sphere and the torus are closed surfaces.

Example 5.2.8: The plane is not closed, because it is not finite.

Example 5.2.9: The Möbius band is not closed, because it has boundary points.

Example 5.2.10: The open unit disk is not closed, because the endpoints of the open line segment $\{(x_1, x_2)| x_1 = 0 \text{ and } -1 < x_2 < 1\}$ are on the unit circle, not in the open disk itself.

TERMINOLOGY NOTE: As readers familiar with topology may recognize, the present usage of "closed" in the geometric-topology sense of a closed surface differs from the usage of "closed" in point-set topology.

DEFINITION: A surface is ***non-orientable*** if it contains a subspace that is topologically equivalent to a Möbius band, and ***orientable*** otherwise.

Remark: Suppose that a small square is placed with its bottom edge on the central circuit of a Möbius band in 3-space. Suppose that the square is slid forward around the band exactly *once* (not twice). Then the top of the square will be positioned below the bottom. This reversability of top and bottom is the intuitive idea of "non-orientability".

Classification of Surfaces

DEFINITION: By ***attaching a handle to a surface***, we mean punching two separate holes in the surface and connecting one to the other with a tube.

DEFINITION: The ***sequence of orientable surfaces*** S_0, S_1, $S_2, \ldots$ is defined recursively. S_0 is the sphere, and S_1 is the torus. Surface S_{n+1} is obtained by attaching a handle to surface S_n. This sequence is illustrated in Figure 5.2.6.

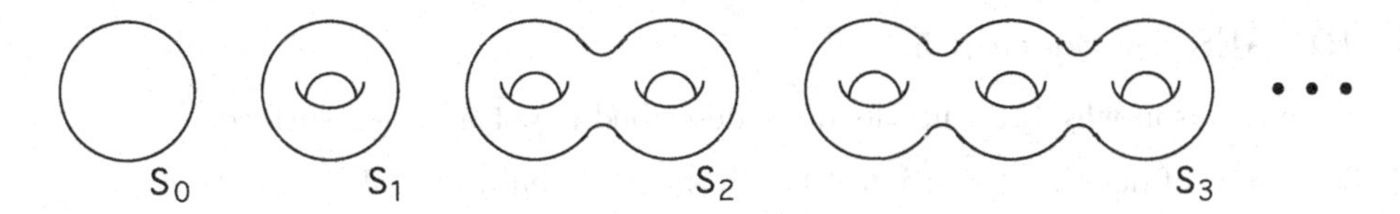

Figure 5.2.6 The sequence of closed orientable surfaces.

Theorem 5.2.2: [***Classification of Closed Orientable Surfaces***] *Every closed orientable surface is topologically equivalent to exactly one of the surfaces in the infinite sequence* $S_0, S_1, S_2, \ldots$ ◇ *(proof omitted here; see [GrTh87])*

DEFINITION: The ***genus of a closed orientable surface*** is the subscript of the surface S_g to which it is topologically equivalent. That is, the genus is the "number of handles" on the surface.

DEFINITION: By ***adding a crosscap to a surface***, we mean punching a hole in the surface and reclosing it by attaching a Möbius band. That is, the boundary of the Möbius band is identified with the boundary of the hole by a topological equivalence, which specifies exactly how to make the attachment.

Remark: The topological operation of replacing a disk on a sphere or other surface by a Möbius band can be readily performed in 4-space, but not in 3-space. Fortunately, there is a way to conceptualize the operation without thinking 4-dimensionally. The *result* of the operation is simply that whenever we cross through from the "main part" of the surface through the boundary of a removed disk, we cross into the Möbius band that replaced the disk.

Remark: Trying to draw a surface from 4-space in the plane would be a drop of two dimensions. (In other words, it would be a loss of detail comparable to representing a 3-dimensional figure by a subset of a single line.) The circle with the $\times$ is the graphic device by which the result of the crosscap operation is represented.

DEFINITION: The ***sequence of non-orientable surfaces*** N_1, N_2, $N_3, \ldots$ is defined recursively, starting from the sphere, which in this context is denoted N_0, even though it is orientable. Surface N_{n+1} is obtained by adding a crosscap to surface N_n. Surface N_1 is called the ***projective plane*** and surface N_2 the ***Klein bottle***. See Figure 5.2.7.

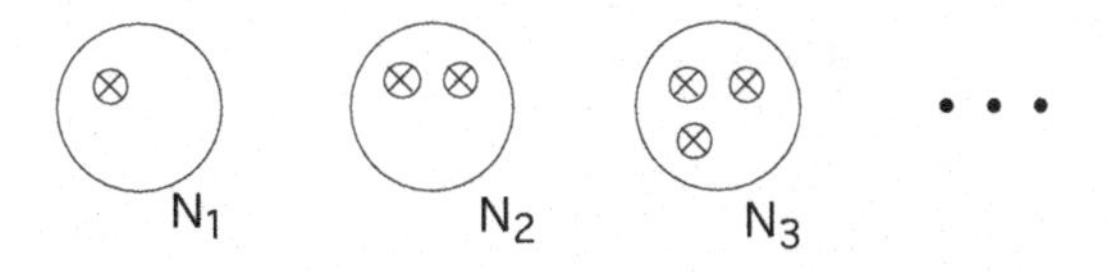

Figure 5.2.7 The sequence of closed non-orientable surfaces.

Theorem 5.2.3: [***Classification of Closed Non-Orientable Surfaces***] *Every closed non-orientable surface is topologically equivalent to exactly one of the surfaces in the infinite sequence N_1, N_2, $N_3, \ldots$* ◇ *(proof omitted here; see [GrTh87])*

DEFINITION: The ***crosscap number*** (or ***non-orientable genus***) ***of a closed non-orientable surface*** is the subscript of the surface N_k to which it is topologically equivalent. That is, it is the number of (disjoint) Möbius bands on the surface.

EXERCISES for Section 5.2

5.2.**1** Give a reason why the Euclidean set described is not a closed surface.

(a) The result of deleting the origin $(0, 0)$ from the open unit disk.

(b) The result of deleting the origin $(0, 0)$ from the closed unit disk.

(c) The result of deleting the point $(1, 0)$ from the closed unit disk.

(d) The result of deleting a single point from a torus in 3-space.

(e) The result of deleting a meridian from the standard torus in 3-space.

5.2.**2** Determine whether the Euclidean set described is a boundaryless surface.

(a) The result of deleting the origin $(0, 0)$ from the open unit disk.

(b) The result of deleting the origin $(0, 0)$ from the closed unit disk.

(c) The result of deleting the point $(1, 0)$ from the closed unit disk.

(d) The result of deleting a single point from a torus in 3-space.

(e) The result of deleting a meridian from the standard torus in 3-space.

5.2.**3** Cut a paper model of a Möbius band open along the central circuit, and see that the Jordan Curve Theorem does not hold for the Möbius band, thereby confirming Theorem 5.2.1.

5.2.**4** Draw a closed circuit on a flat polygon representation of the torus that traverses the meridian direction once and the longitudinal direction once. Does this circuit separate the torus?

5.2.**5** Draw a closed circuit on the torus that traverses the torus once in the meridian direction and once in the longitudinal direction.

5.2.**6** Draw a closed circuit on the torus that traverses the torus three times in the meridian direction and twice in the longitudinal direction.

5.3 MATHEMATICAL MODEL FOR DRAWING GRAPHS

REVIEW FROM ELEMENTARY SET THEORY: The ***image*** of a point or subset of the domain of a function is the point or subset in the codomain onto which it is mapped.

REVIEW FROM §4.1:

- A ***path from*** s ***to*** t in a Euclidean set X is the image of a continuous function f from the unit interval $[0,1]$ into X such that $f(0) = s$, $f(1) = t$, and either
  - f is one-to-one, or
  - the restriction of f to $[0,1)$ is one-to-one and $s = t$.

  In the first case, the path is an ***open path***, and in the second, it is a ***closed path***.
- An ***open path from*** s ***to*** t in a Euclidean set X is the image of a continuous one-to-one function f from the unit interval $[0,1]$ into X such that $f(0) = s$ and $f(1) = t$.
- A ***closed path from*** s ***to*** s in a Euclidean set X is the image of a continuous function f from the unit interval $[0,1]$ to a subset of X such that the restriction of f to $[0,1)$ is one-to-one and $f(0) = f(1) = s$.

One may visualize a path as the trace of a particle traveling through space for a fixed length of time.

TERMINOLOGY NOTE: A ***singularity*** of a continuous function is a point of the image where the function is not one-to-one. We use the term "singularity" for a drawing of a graph on a surface when different "points" on a graph get identified with the same point on the surface.

Designing a mathematical model of a drawing of a graph on a surface starts by representing each vertex as a point on the surface and representing each edge as an open path in the surface between the images of its endpoints.

DEFINITION: A ***drawing of a graph*** $G = (V, E)$ on a surface S is the union of a set of points of S, one for each vertex of G; a set of open paths, one for each proper edge of G; and a set of closed paths, one for each self-loop of G. Each path corresponding to an edge joining the images of the endpoints of that edge. That is, there is a one-to-one function $p : V \to S$; and for each edge $e \in E$, a path q_e connecting the endpoints of e (open if e is proper and closed if e is a self-loop) with no interior points in $p(V)$. The drawing of G on S is the union $\bigcup_{e \in E} q_e$.

Three types of ***forbidden singularities*** that occasionally occur in representations of graphs on surfaces are illustrated in Figure 5.3.1 but are otherwise absent from this book.

(i) Two vertex images $p(u)$ and $p(v)$ are equal.

(ii) An edge image q_e has a self-intersection or other singularities.

(iii) The interior of an edge image intersects a vertex image.

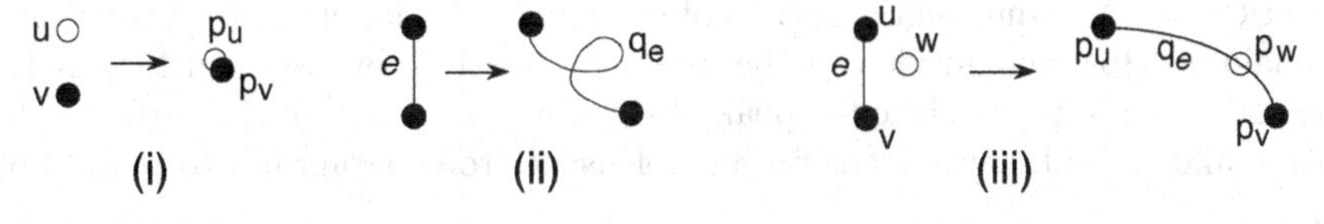

Figure 5.3.1 Three forbidden singularity types for graph drawings.

DEFINITION: An ***edge-crossing*** in a graph drawing is a singularity such that the images of two different edges meet at interior points of the edges.

Normalized Drawings and Imbeddings

DEFINITION: An ***abnormality in a drawing of a graph*** on a surface is any of these three types of singularities:

(i) The interiors of the images of two edges meet at more than one point of the surface.
(ii) The interiors of the images of two different edges meet but fail to cross each other.
(ii) The interiors of the images of three or more edges meet at the same point of the surface.

Figure 5.3.2 illustrates the remedies for the three kinds of abnormalities: (i) an edge that crosses another several times can stay on the same side until it gets to the final crossing; (ii) if one edge just touches another, then it can be pulled away from the other edge; (iii) if an edge crosses a place where two other edges already cross, then it can be rerouted to go around that preexisting crossing.

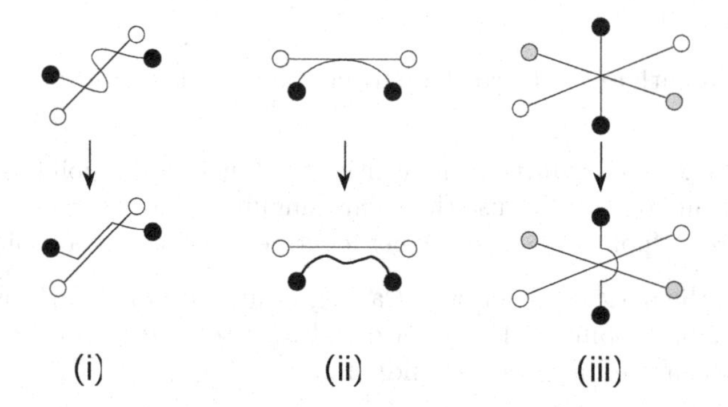

Figure 5.3.2 Remedies for the three types of abnormalities in drawings.

DEFINITION: A ***normal drawing of a graph*** on a surface is a drawing that is free of all three types of abnormalities.

Unless explicitly specified otherwise in context, it is assumed throughout this book that a drawing is normalized. (We occasionally relax rule (iii).)

DEFINITION: An ***imbedding of a graph*** G on a surface S is a drawing with no edge-crossings at all; that is, it is the image of a topological equivalence $\iota : G \to S$ from the topological model of G to a subset of S.

TERMINOLOGY NOTE: More generally, an *imbedding* is a one-to-one function from one topological space to another, for instance, between arbitrary Euclidean sets. In this sense, the *standard torus* is imbedded in $\mathbb{R}^3$.

TERMINOLOGY NOTE: One usually speaks of the image of a function $f : X \to Y$ as lying *in* the codomain Y. However, in view of the notion that one draws *on* a surface, it is natural when the codomain Y is a surface to speak of the image as lying *on* the surface. This is not a rigid rule, and indeed, some inconsistency of usage from paragraph to paragraph is not unusual.

Eliminating Edge-Crossings

Frequently, it is desirable to keep the number of edge-crossings to a minimum, which may involve some redrawing.

Example 5.3.1: Sometimes a drawing of a graph can be changed into an imbedding by rerouting the image of one or more edges to different positions. For instance, in Figure 5.3.3, a drawing of the complete graph K_4 with one edge-crossing is changed into an imbedding by moving the image of edge e.

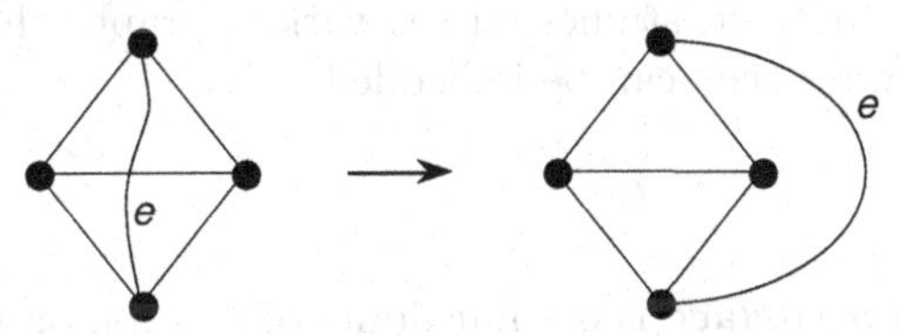

Figure 5.3.3 Eliminating an edge-crossing by moving the image of an edge.

Example 5.3.2: Adding a handle to a surface is another way to eliminate an edge-crossing from a drawing of a given graph, as illustrated in Figure 5.3.4. The handle is added from one side of edge e to the other, and then the image of edge d is rerouted so that it lies on the new handle, instead of crossing edge e. Of course, the handle-adding operation changes the surface on which the graph is drawn.

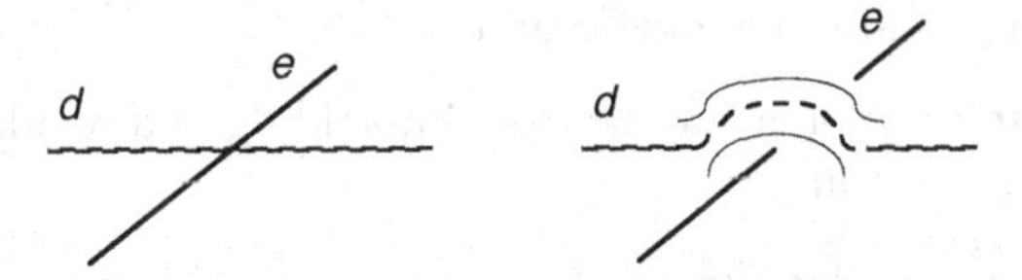

Figure 5.3.4 Eliminating an edge-crossing by adding a handle.

Remark: Every graph can be imbedded in 3-space. First draw the graph in the plane, and then eliminate the crossings by adding handles, as in Example 5.3.2. Different handles can be at different heights off the plane, so they do not intersect.

Application 5.3.1: *Printed Circuit Boards* A planar electronic circuit can be printed onto a board. Since the wires of a planar circuit need not cross, no insulation is needed. When a circuit is non-planar, one practical approach, described in §4.7, is to partition its edges into a few layers.

EXERCISES for Section 5.3

5.3.1 Show how to remove the abnormality from the drawing.

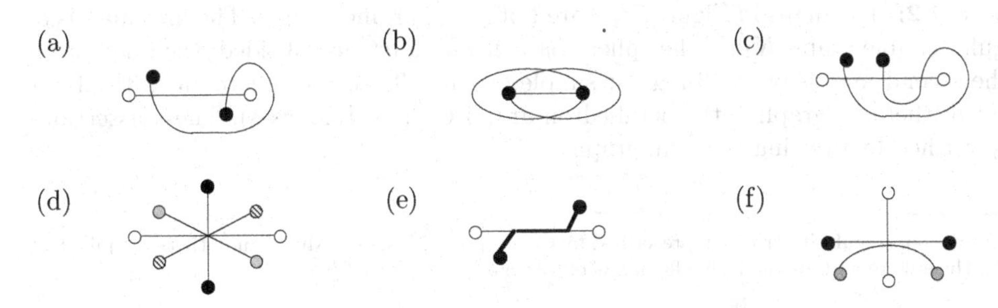

5.4 REGULAR MAPS ON A SPHERE

The difference between *imbedding* and *map* is largely nuance. The notation $G \to S$ is the same. When one discusses the imbeddings of a graph G, the graph is fixed and the surface S may vary. When one discusses the maps on a surface S, the graph G may vary and the surface is fixed. Whereas the *imbedding theory* of a given graph studies all the various ways that graph can be imbedded in its range of possible closed surfaces, the *map theory* of a given surface, by way of contrast, studies all the various graphs that can be imbedded on that surface and all the ways they can be imbedded.

Map Theory

DEFINITION: A ***map on a surface*** is an imbedding of a graph on that surface.

TERMINOLOGY NOTE: Whereas the term *mapping* is a generic synonym for function, the term *map* refers to a function from a graph to a surface.

REVIEW FROM §4.5: The (***Poincaré***) ***dual graph*** and the ***dual imbedding*** are derived, starting with a cellular imbedding of a graph, by inserting a new vertex into each existing face, and by then drawing through each existing edge a new edge that joins the new vertex in the region on one side to the new vertex in the region on the other side.

DEFINITION: A ***simple map*** on a surface is an imbedding of a simple graph on that surface, such that the dual graph is also a simple graph.

DEFINITION: A ***regular map*** on a surface is an imbedding of a regular graph such that the dual graph is also a regular graph.†

Example 5.4.1: The graphs for the *sphere maps* (i.e., maps on a sphere) in Figure 5.4.1 are both simple and regular. However, the dual graph of map (i) is the dipole D_3, which is regular but not simple. Furthermore, the dual graph of map (ii) is the graph $K_5 - e$, which is simple but not regular. These assertions are easily verified by drawing the dual graphs.

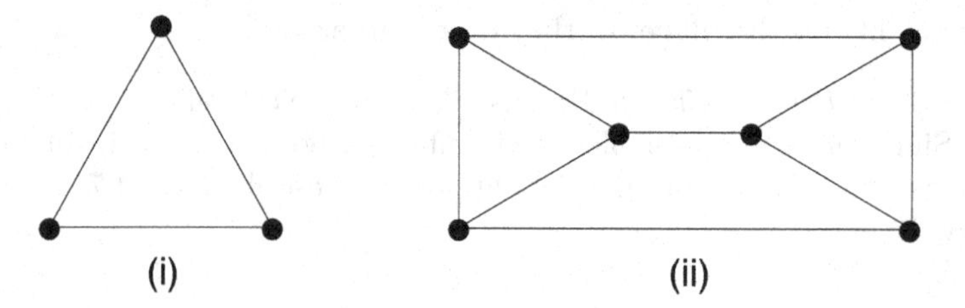

Figure 5.4.1 Two sphere maps: (i) one non-simple, (ii) the other non-regular.

Example 5.4.2: The maps in Figure 5.4.2 are both regular and simple. The first map is of the 3-regular simple graph K_4 on the sphere S_0, with all four faces 3-sided; the dual graph is K_4. The second map is of the 3-regular simple graph CL_4 on the sphere S_0, with all six faces 4-sided; the dual graph is the octahedron graph $\mathcal{O}_3$ (see Exercises). These assertions are easily verified by drawing the dual graphs.

† The algebraic map-regularity criterion prescribed by Coxeter and Moser [CoM072] and others is equivalent for maps on the sphere and less inclusive for maps on other surfaces.

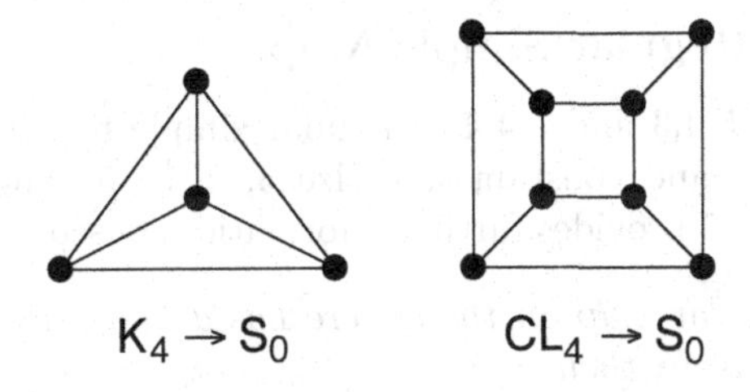

Figure 5.4.2 Two regular simple maps on the sphere.

Remark: These definitions of regular and simple also apply to maps on surfaces other than the sphere.

Proposition 5.4.1: *A map $\iota : G \to S$ is regular if and only if graph G is regular, and every face has the same number of sides.*

Proof: ($\Rightarrow$) Assume that the map $\iota : G \to S$ is regular. Then graph G is regular. Moreover, since the dual graph G^* is regular, each dual vertex has the same degree. Since the degree of a dual vertex equals the number of sides of a primal face, each primal face has the same number of sides.
($\Leftarrow$) Assume that graph G is regular and that every face has the same number of sides. The latter condition implies that the dual graph G^* is regular. Thus, the map $\iota : G \to S$ is regular. ♢

DEFINITION: A ***self-dual map*** on a surface is a map $\iota : G \to S$ such that there is a topological equivalence of the surface S to itself that takes graph G onto the dual graph G^*.

Example 5.4.3: The map $K_4 \to S_0$ in Figure 5.4.2 is self-dual, but the map $CL_4 \to S_0$ is not.

Degrees and Face-Sizes of Regular Maps

The following straightforward application of the numerical relations in the previous section yields an upper bound on the average degree $\delta_{avg}(G)$ of a graph G that can be imbedded on the sphere.

Theorem 5.4.2: *Let $\iota : G \to S_0$ be an imbedding of a simple graph on the sphere. Then $\delta_{avg}(G) < 6$.*

Proof: Since $G = (V, E)$ is simple, $\frac{2|E|}{|V|} \leq 6 - \frac{12}{|V|}$, by Theorem 4.5.9. Therefore, by Theorem 1.1.2,

$$\delta_{avg}(G) = \frac{1}{|V|} \sum_{v \in V} \deg(v) = \frac{2|E|}{|V|} \leq 6 - \frac{12}{|V|} < 6$$ ♢

Corollary 5.4.3: *The only possible degrees of a graph with a regular simple map on the sphere are 3, 4, and 5.*

Proof: Theorem 5.4.2 precludes the possibility of any degree greater than 5. ♢

Corollary 5.4.4: *The only possible face-sizes of a regular simple map on the sphere are 3, 4, and 5.*

Proof: Apply Corollary 5.4.3 to the dual map. ♢

Construction of the Regular Simple Maps

In view of Corollaries 5.4.3 and 5.4.4, a regular simple map on the sphere must have constant degree 3, 4, or 5 and constant face-size 3, 4, or 5. That seems to permit nine combinations. Theorem 5.4.5 provides further information, based on the numerical relations.

Theorem 5.4.5: *If a regular map on the sphere has d and r for its constant degree and constant face-size, respectively, then*

$$|V| = \frac{4r}{2(d+r)-dr} \qquad |E| = \frac{2dr}{2(d+r)-dr} \qquad |F| = \frac{4d}{2(d+r)-dr}$$

Proof: Solving the following three linear equations for the "unknowns" $|V|$, $|E|$, and $|F|$ in terms of d and r yields the result.

$|V| - |E| + |F| = 2$ Euler polyhedral equation, (Theorem 4.5.7)
$2|E| = r|F|$ face-size equation, (Theorem 4.5.2)
$2|E| = d|V|$ degree-sum equation, (Theorem 1.1.2)

◇

The following table gives the numbers of vertices, edges, and faces that would correspond to each of the nine possible combinations of degree d and face-size r in a regular simple map on the sphere. The common denominator in the conclusion of Theorem 5.4.5 is denoted Y. In the rightmost column, the table names the polyhedron whose 1-skeleton and surface realize the map if it exists.

d	r	Y	$\frac{4r}{Y}$	$\frac{2dr}{Y}$	$\frac{4d}{Y}$	**Name of polyhedron**
3	3	3	4	6	4	**Tetrahedron**
3	4	2	8	10	6	**Cube**
3	5	1	20	30	12	**Dodecahedron**
4	3	2	6	12	8	**Octahedron**
4	4	0	undef	undef	undef	no solution
4	5	−2	−10	−20	−8	no solution
5	3	1	12	30	20	**Icosahedron**
5	4	−2	−8	−20	−10	no solution
5	5	−5	−4	−10	−4	no solution

DEFINITION: A ***platonic solid*** is a *geometrically regular* 3-dimensional polyhedron. Geometrically regular means here that its 1-skeleton is a regular graph and also that each of the faces is geometrically a regular polygon.

DEFINITION: A ***platonic graph*** is the 1-skeleton of a platonic solid.

DEFINITION: A ***platonic map*** is the imbedding of the 1-skeleton of a platonic solid into its surface.

Example 5.4.4: The five platonic solids, graphs, and maps are illustrated in Figure 5.4.3.

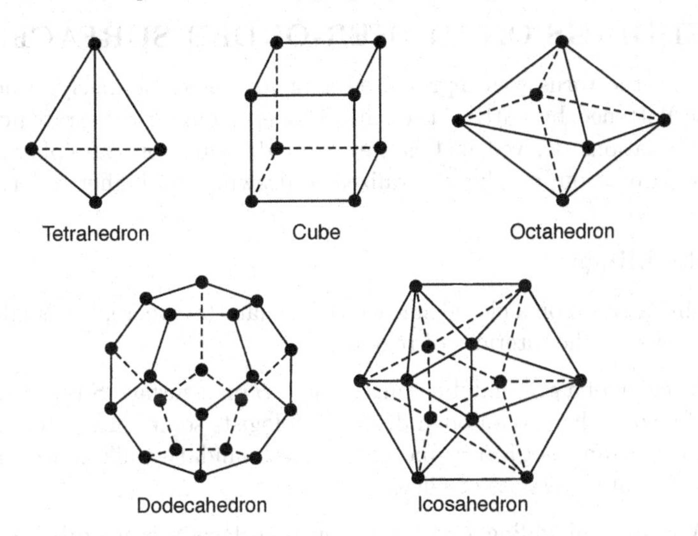

Figure 5.4.3 The five platonic graphs.

Remark: The five maps in Figure 5.4.3 are the only regular simple maps on the sphere. The illustration proves their existence.

Remark: Since the three numerical equations used in the proof of Theorem 5.4.5 are true for non-simple graphs, the resulting formulas can be used to determine possibilities for non-simple regular maps on the sphere.

EXERCISES for Section 5.4

5.4.1 Draw a regular map of a possibly non-simple graph on the sphere that conforms to the given description.

(a) Regular of degree 1.

(b) Regular of degree 2 with face-size 4.

(c) Regular of degree 4 with face-size 2.

(d) Regular of degree 2 with face-size 3.

5.4.2 Find all non-simple regular maps on the sphere with degree at least 3.

5.4.3 Draw the dual of the map $CL_4 \to S_0$ (see Figure 5.4.2), and prove that the dual graph is isomorphic to the octahedron graph O_3.

5.4.4 Prove that for degree 2, there is a regular map on the sphere with any positive integer specified as the face-size.

5.4.5 Prove that for face-size 2, there is a regular map on the sphere with any positive integer specified as the degree.

5.5 IMBEDDINGS ON HIGHER-ORDER SURFACES

In Section 5.2, the torus was represented as a rectangle. Similarly, other higher-order surfaces can be flattened by cutting the surface open along a few strategic closed curves. Some algebraic relations derived in Chapter 4 for drawings on the sphere, including the Euler Polyhedral Equation, can be generalized to drawings on higher-order surfaces.

Cellular Imbeddings

When drawing a graph on a closed surface other than the sphere, we usually avoid having handles or crosscaps in the interiors of regions.

DEFINITION: A region of a graph imbedding $\iota : G \to S$ on a surface S is a ***cellular region*** if it is topologically equivalent to an open disk. (Topologists sometimes call a disk a "2-cell".) It is a ***strongly cellular region*** if, moreover, the boundary walk is a cycle in the graph (that is, has no repeated vertices or edges).

DEFINITION: A graph imbedding $\iota : G \to S$ on a surface S is a ***cellular imbedding*** if every region is cellular. It is a ***strongly cellular imbedding*** if every region is strongly cellular.

Example 5.5.1: In Figure 5.5.1, the imbeddings of the bouquets B_1 and B_2 are both non-cellular. In Figure 5.5.1(a), the "big" region is not cellular, because it does not have the Jordan separation property. In Figure 5.5.1(b), the big region is topologically equivalent to the unit cylinder minus its two boundary components.

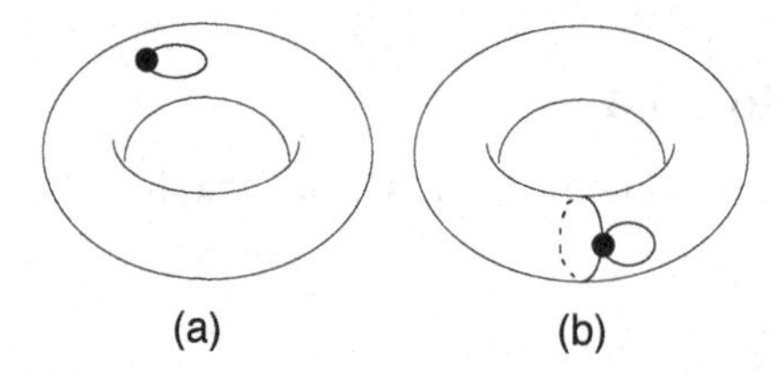

Figure 5.5.1 Two non-cellular imbeddings on the torus.

Any non-cellular imbedding of a graph on a closed surface can be obtained by adding handles and crosscaps to a cellular imbedding. Largely for this reason, we concentrate on cellular imbeddings of graphs. For instance, the non-cellular imbedding in Figure 5.5.1(a) could be obtained by adding a handle to one region of a cellular imbedding of the bouquet B_1 in the sphere as shown in Figure 5.5.2(a). Moreover, the non-cellular imbedding in Figure 5.5.1(b) could be obtained from a cellular imbedding of the bouquet B_2 on the sphere by adding a handle joining the digon to one of the monogons as shown in 5.5.2(b).

Proposition 5.5.1: *The boundary of a face f of a connected graph G imbedded on the sphere is connected.*

Proof: A closed loop drawn just inside one of the boundary components of face f would separate the sphere, by the Jordan Curve Theorem. Since graph G is connected, all of it lies on the same side of that closed loop as that boundary component. Thus, there is no other boundary component. ♢

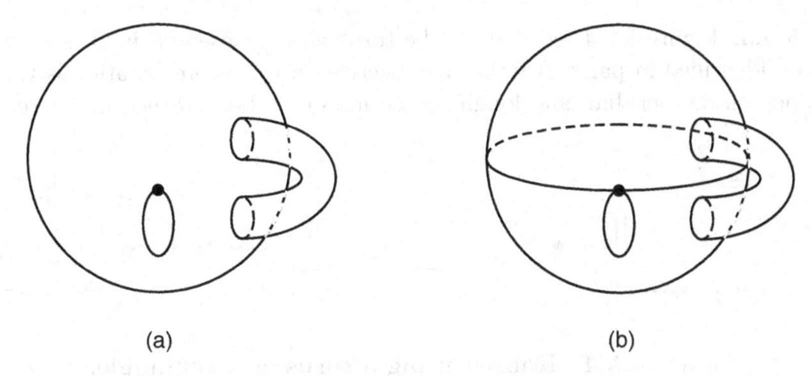

Figure 5.5.2 Two non-cellular imbeddings on the torus.

Proposition 5.5.2: *Every imbedding of a connected graph on the sphere is a cellular imbedding.*

Proof: This follows from Proposition 5.5.1 and the Schoenfliess theorem, that every closed curve on a sphere bounds a disk (see the footnote for Theorem 4.1.2). ♢

Example 5.5.2: In the planar imbedding of K_4 in Figure 5.5.3(a), there are four regions, all strongly cellular. In the toroidal imbedding of K_4 in Figure 5.5.3(b), there are two regions. Centered in drawing (b) is a 4-sided strongly cellular region. The other (8-sided) region is cellular, but its boundary walk includes the vertical and horizontal edges (which wrap around the sides and top/bottom of the picture) twice and each vertex twice. Accordingly, that region is not strongly cellular.

Observe in drawing (b) that the partial edge incident on the topmost vertex meets the partial edge incident on the bottommost vertex. These two parts of the same edge are joined when the top broken edge of the rectangle is identified with the bottom broken edge. Similarly, the partial edge incident on the leftmost vertex is part of the same edge as the partial edge incident on the rightmost vertex.

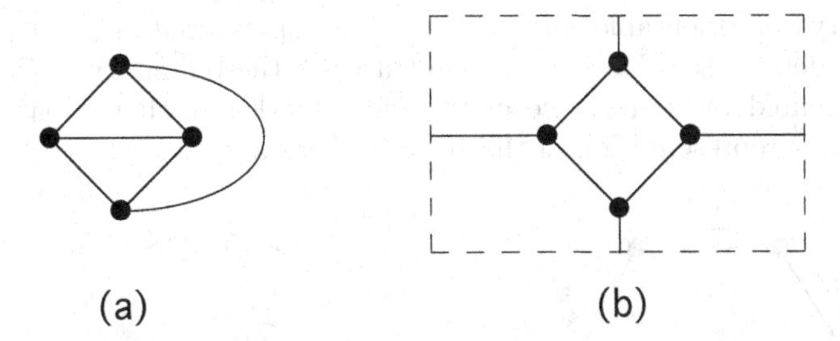

Figure 5.5.3 Two imbeddings of K_4: (a) on the plane; (b) on the torus.

Flat Polygon Drawings

DEFINITION: A ***flat polygon representation*** of a surface S is a drawing of a polygon with markings to match its sides in pairs, such that when the sides are pasted together as the markings indicate, the resulting surface obtained is topologically equivalent to S.

Example 5.5.3: Figure 5.5.4 recalls how the torus was represented in §5.2 as a rectangle with its sides identified in pairs. Another perspective on this representation is that cutting the torus open on its meridian and longitude permits it to be flattened into a rectangle.

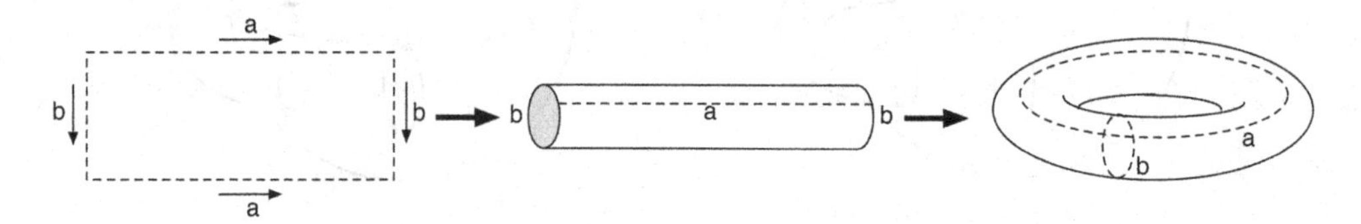

Figure 5.5.4 Representing a torus as a rectangle.

Every orientable surface S_g can be represented as a flat polygon with $4g$ sides (for a detailed proof, see [Ma67]). As one traverses the boundary of the polygon, each handle is represented by a sequence of four consecutive sides, which are marked with the pasting pattern $sts^{-1}t^{-1}$.

Example 5.5.4: Figure 5.5.5 shows a representation of the double torus S_2 as an octagon, from which one handle is formed by the four sides marked $aba^{-1}b^{-1}$ and the other by the four sides marked $cdc^{-1}d^{-1}$.

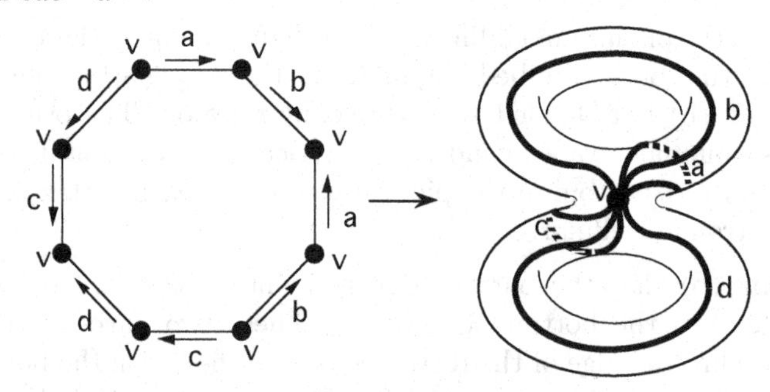

Figure 5.5.5 Representing a double torus as an octagon.

Similarly, every non-orientable surface N_k can be represented as a flat polygon with $2k$ sides (for further details, see [Ma67]). As one traverses the boundary of the polygon, each crosscap is represented by a sequence of two sides marked with the pasting pattern ss. Figure 5.5.6 shows a representation of the surface N_3 as a hexagon.

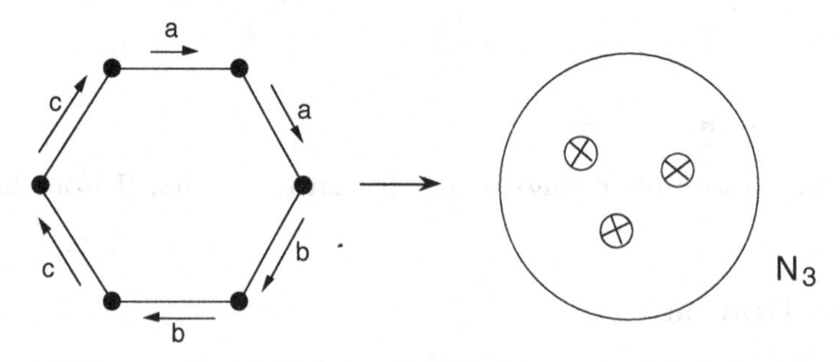

Figure 5.5.6 Representing the surface N_3 as a hexagon.

In a drawing of a graph on the flat polygon representation of a surface, an edge may cross through one copy of a matched side to the other copy. For instance, Figure 5.5.7 shows an imbedding of $K_{3,3}$ on the torus with three hexagonal regions. One edge crosses through the meridian cut a, two edges cross through the longitudinal cut b, and one edge cuts through the point where the meridian and longitude intersect.

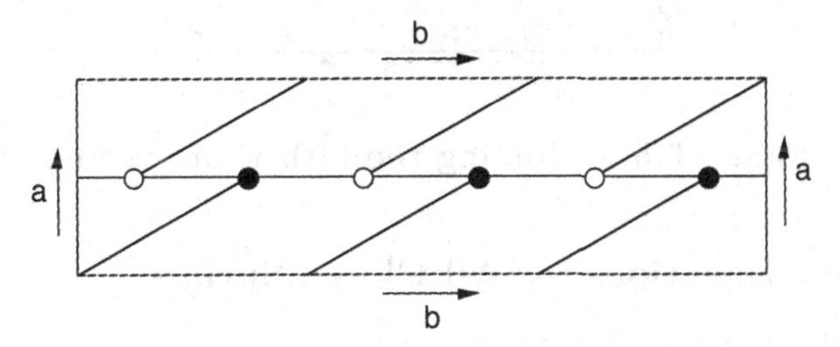

Figure 5.5.7 Toroidal imbedding of $K_{3,3}$.

Surgery on Imbeddings

Trying to draw graph imbeddings on flat polygon representations of surfaces more complicated than a torus (S_1) or a Klein bottle (N_2) is often quite frustrating. It is frequently easier to draw most of the graph on a simpler surface and to complete the imbedding by attaching extra handles and drawing edges across the extra handles. Informally, such modifications are called surgery.

Example 5.5.5: Drawing $K_{4,5}$ on a flat polygon representation of S_2 is no easy task. (Try it and see!) A surgical approach to this imbedding problem is to start by drawing $K_{4,4}$ on S_1, as shown in Figure 5.5.8.

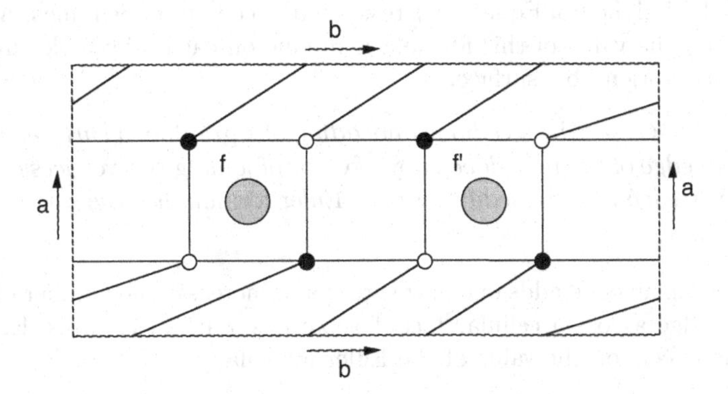

Figure 5.5.8 Toroidal imbedding of $K_{4,4}$, with holes punched in two regions.

Two regions f and f' of this imbedding are selected so that the union of their face-boundary walks contains all the black vertices. Next, a hole is punched in regions f and f', shown by shaded disks in Figure 5.5.8. Then a tube is attached from one hole to the other, thereby reclosing the surface, so that a single new *non-cellular* region is formed from the two regions with the holes plus the tube. The resulting surface is S_2, and the new region has two boundary components, at opposite ends of the tube. Finally, a fifth white vertex is drawn, as shown in Figure 5.5.9, on the tube (depicted as shaded) and joined to both black vertices on the boundary components. That final result of this surgery is a cellular imbedding of $K_{4,5}$ on S_2.

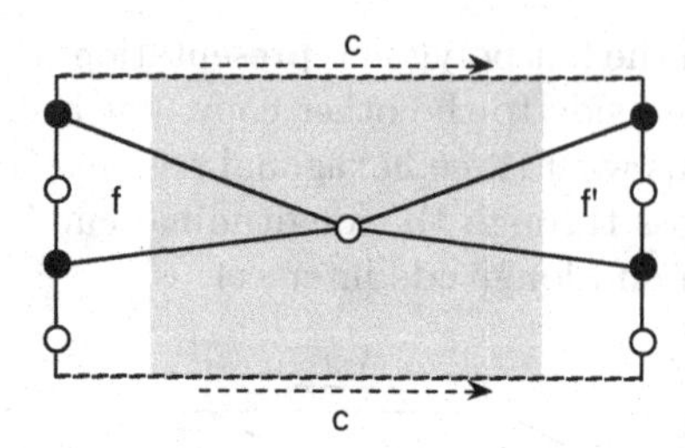

Figure 5.5.9 Joining the fifth white vertex.

Euler Polyhedral Equations for All Closed Surfaces

We observe that the

Face-Size Equation: $$2\,|E_G| = \sum_{f \in F} size(f)$$

holds for every graph imbedding $\iota : G \to S$, cellular or not, in every surface. Similarly, the

Edge-Face Inequality: $$2|E| \geq girth(G) \cdot |F|$$

holds whenever the graph is connected but not a tree (for which girth is undefined), in every surface, even if the imbedding is non-cellular. Moreover, for any cellular graph imbedding $\iota : G \to S$, the following Poincaré duality relations hold.

(i) $|V^*| = |F|$
(ii) $|E^*| = |F|$
(iii) $|F^*| = |V|$

However, the Euler Polyhedral Equation is restricted to cellular imbeddings. As the following theorems indicate, the value of this formula is unique on each orientable surface, and it is unique on each non-orientable surface.

Lemma 5.5.3: *Let $G \to S$ be a cellular imbedding of a graph in a surface. Then the result of subdividing an edge of the imbedded copy of G, or of joining two vertices on the boundary of a face by a new edge is an imbedding whose Euler formula has the same value as for the imbedding $G \to S$.*

Proof: Subdividing an edge adds one new vertex and increases the number of edges by one. Drawing a new edge across a cellular face f separates f into two faces. Either operation causes offsetting effects on the value of the Euler formula. $\diamond$

Theorem 5.5.4: *Let a graph G be cellularly imbedded in the orientable surface S_g. Then $|V| - |E| + |F| = 2 - 2g$.*

Proof: For imbeddings on the sphere S_0, we proved in §4.5 that the value of the Euler formula $|V| - |E| + |F|$ must be equal to 2. This serves as the base case for an induction. As an inductive hypothesis, assume that the equation is correct for every cellular imbedding in the surface S_g. We consider a cellular imbedding of a graph G in the surface S_{g+1}.

Draw a meridian on the rightmost handle of S_{g+1} so that it meets graph G in finitely many points, each in the interior of some edge of G. If necessary, we subdivide edges of G so that no edge crosses the meridian more than once. By Lemma 5.5.3, the value of the Euler formula in the resulting graph imbedding is unchanged from the original.

Next we thicken the meridian to an annulus A, as shown at the left of Figure 5.5.10, and we subdivide each edge of G that crosses the annulus at both of its intersection points with a boundary component of the annulus, which, by Lemma 5.5.3, also preserves the value of the Euler formula. Then we augment the edge-set of the graph by adding to it each segment of annulus boundary that joins two vertices on it. By Lemma 5.5.3, the resulting graph imbedding has the same value of the formula $|V| - |E| + |F|$ as the original imbedding.

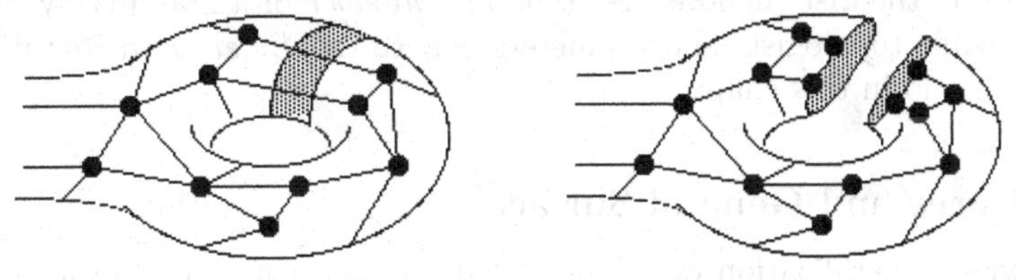

Figure 5.5.10 Thickening a meridian and subsequent surgery.

In the interior of annulus A, edges and faces alternate, so there are equally many. Thus, excising the interior of the annulus preserves the value of the formula $|V| - |E| + |F|$. Finally, by capping the two resulting holes in the surface with disks, as shown at the right of Figure 5.5.10, we obtain a surface that is topologically equivalent to S_g and a cellular imbedding with vertex-set V', edge-set E', and face-set F', such that $|V'| - |E'| + |F'| = |V| - |E| + |F| + 2$. By the induction hypothesis, $|V'| - |E'| + |F'| = 2 - 2g$. It follows that $|V| - |E| + |F| = -2g = 2 - 2(g + 1)$. ◊

Theorem 5.5.5: *Let a graph G be cellularly imbedded in the non-orientable surface N_k. Then $|V| - |E| + |F| = 2 - k$.*

Proof: This proof follows the same pattern as that of Theorem 5.5.4, except that the role of the meridian is filled by the central cycle on a Möbius band. ◊

Example 5.5.6: The imbedding $K_{3,3} \to S_1$ in Figure 5.5.7 above is cellular, and the value of the Euler polyhedral formula $|V| - |E| + |F|$ is $6 - 9 + 3 = 0 = 2 - 2g$.

Example 5.5.5, continued: The imbedding $K_{4,5} \to S_2$ in Figures 5.5.8 and 5.5.9 above is cellular, and the value of the Euler polyhedral formula $|V| - |E| + |F|$ is $9 - 20 + 9 = -2 = 2 - 2g$.

Example 5.5.7: The imbedding of K_6 into N_2 in Figure 5.5.11 has nine regions, each marked with a distinguishing numeral. Thus, $|V| - |E| + |F| = 6 - 15 + 9 = 0 = 2 - k$.

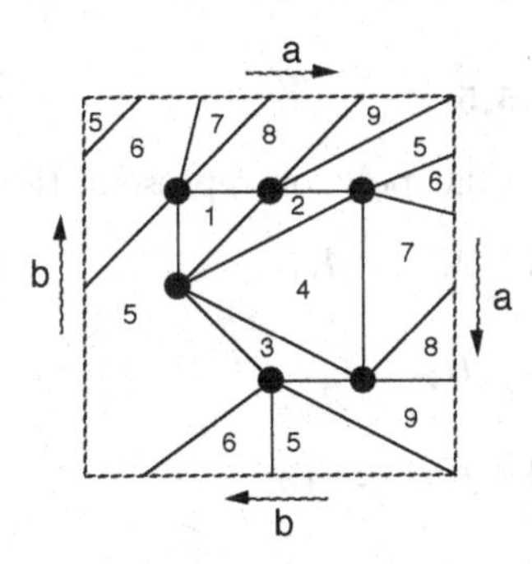

Figure 5.5.11 An imbedding $K_6 \to N_2$ with nine regions.

DEFINITION: The ***Euler characteristic*** for a surface equals the value of the Euler polyhedral formula $|V| - |E| + |F|$ for any cellular imbedding on that surface. It is denoted $\chi(S)$. That is,

$$\chi(S) = \begin{cases} 2 - 2g & \text{for the orientable surface } S_g \\ 2 - k & \text{for the non-orientable surface} N_k \end{cases}$$

NOTATION: Graph theorists denote the *chromatic number* of a graph G by $\chi(G)$, as we do in Chapter 6, while topologists and geometers denote the *Euler characteristic* of a surface S by $\chi(S)$, as we do in this chapter.

Average Degree and General Surfaces

The following generalization of Theorem 5.4.2 is often useful in proving that a given graph cannot be imbedded in some given surface.

Theorem 5.5.6: *Let $\iota : G \to S$ be an imbedding of a simple graph G on a surface S of Euler characteristic $\chi(S)$. Then $\delta_{\text{avg}}(G) \le 6 - \frac{6\cdot\chi(S)}{|V|}$.*

Proof: Since the girth of a simple graph is at least 3,

$$\chi(S) = |V| - |E| + |F| \le |V| - |E| + \frac{2|E|}{F} = |V| - \frac{|E|}{3}$$

by the edge-face inequality. That is,

$$\frac{2|E|}{|V|} \le 6 - \frac{6 \cdot \chi(G)}{|V|}$$

Therefore, by Theorem 1.1.2,

$$\delta_{\text{avg}}(G) = \frac{1}{|V|}\sum_{v\in V} \deg(v) = \frac{2|E|}{|V|} \le 6 - \frac{6\cdot\chi(G)}{|V|}$$

◇

Example 5.5.8: Since $\chi(N_1) = 1$, the average degree of an imbedding in the non-orientable surface N_1 is less than 6. Thus, K_7 cannot be imbedded in N_1.

Example 5.5.9: The average degree of $C_4 + C_5$ is $6\frac{4}{9}$, and $\chi(N_2) = 0$. Theorem 5.5.6 implies that $C_4 + C_5$ cannot be imbedded in N_2.

EXERCISES for Section 5.5

5.5.1 Draw the given graph on a flat polygon representation of the torus.

(a) $Q_3 + K_1$ (b) $ML_3 + K_1$ (c) K_6

(d) $K_{4,4}$ (e) K_7 (f) $K_8 - 4K_2$

(g) $K_8 - (K_3 \cup K_2)$ (h) $K_8 - 2P_3$ (i) $K_9 - C_9$

(j) $K_8 - C_5$

5.5.2 Draw the given graph on a flat polygon representation of the surface N_1.

(a) K_6 (b) $K_{3,4}$ (c) ML_4 (d) ML_5

5.5.3 Draw the given graph on a flat polygon representation of the surface N_2.

(a) $K_7 - K_2$ (b) $K_{4,4}$ (c) $K_9 - K_6$ (d) $K_8 - C_5$

5.5.4 Draw a regular map of the designated graph on the torus.

5.5.5 B_2 **5.5.6** B_3 **5.5.7** $C_3 \times C_3$ **5.5.8** $K_{3,3}$

5.5.9 K_5 **5.5.10** K_7 **5.5.11** $K_6 - 3K_2$ **5.5.12** Q_4

5.5.13 Prove that $C_4 + C_4$ cannot be imbedded in N_1.

5.5.14 Prove that $K_9 - 4K_2$ cannot be imbedded in S_1 or in N_2.

5.5.15 Prove that K_{19} cannot be imbedded in S_{19}.

5.6 GEOMETRIC DRAWINGS OF GRAPHS

Geometric drawing of graphs is a topic studied in computational geometry. Unlike topological graph drawings, geometric graph drawings are concerned with exact coordinates of the images of the vertices and edges of a graph, with lengths and areas, and with the angles formed in the drawings. Computer drawings are geometric drawings. Interest in geometric graphs is motivated in part by their many applications, which include computer graphics, pattern recognition, and communication networks. The intent, of this brief, optional section is to introduce some types of geometric drawings, some properties of graph drawings, and some categories of problems involving geometric graph drawings.[†]

Geometric graph drawings avoid all three types of *forbidden singularities.* However, they need not be *normal drawings.* (See §5.3.)

Some Types of Geometric Graph Drawings

Polyline drawings are piecewise-linear approximations of graph drawings with curved edges. The choice of polyline or straight-line drawings depends on the application.

DEFINITION: In a ***straight-line drawing*** of a graph, each edge is a single line segment.

DEFINITION: In a ***polyline drawing of a graph*** in a plane, each edge is a chain of line segments (see Figure 5.6.1(a)).

DEFINITION: A ***bend in a polyline drawing*** is a point where two segments belonging to the same edge meet.

DEFINITION: A ***grid drawing*** is a polyline drawing such that the vertices, crossings, and bends all have integer coordinates.

[†] No subsequent sections depend on this section.

DEFINITION: In an ***orthogonal drawing*** of a graph in the plane, each edge is a chain of horizontal and vertical segments (see Figure 5.6.1(b) below).

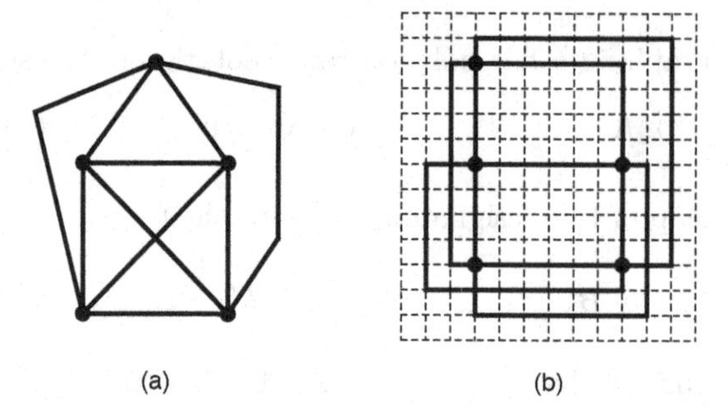

Figure 5.6.1 Two drawings of K_5: (a) general polyline; (b) orthogonal grid.

Some Properties of Graph Drawings

It is desirable to optimize various geometric properties of a graph, for instance, for the sake of miniaturization on a chip or of appearance on a video screen. One typically seeks to maximize the *angular resolution* and to minimize the other measures.

DEFINITION: The ***area of a geometric drawing*** is usually the area either of its convex hull or of the smallest rectangle with vertical and horizontal sides that covers the drawing.

DEFINITION: A ***resolution rule for a drawing*** is a constraint that prevents it from being arbitrarily scaled down, for instance, a minimum unit distance between vertices.

DEFINITION: The ***aspect ratio of a drawing*** is the ratio of the length of the longer side to the shorter side of the smallest rectangle covering the drawing.

DEFINITION: The ***angular resolution*** of a polyline drawing is the smallest angle formed by any two segments of an edge at a bend or between any two edges incident at the same vertex. It does not usually take into account the angle at which two edges cross.

DEFINITION: The ***total edge-length of a drawing*** is the sum of the lengths of its edges.

Minimizing Total Edge Length

DEFINITION: A ***(geometric) spanning tree for a set S of points in $\mathbb{R}^n$*** is a drawing of a tree that has S as its vertex-set.

DEFINITION: A ***(geometric) minimum spanning tree for a set of points in $\mathbb{R}^n$*** is a geometric spanning tree that minimizes the total edge length. (This is provably a straight-line drawing.)

DEFINITION: The ***Euclidean MST problem*** is to produce a geometric minimum spanning tree for a set of points in $\mathbb{R}^n$ (as in Figure 5.6.2), that is, to determine the pairs of points that are endpoints of the edges of the minimum spanning tree.

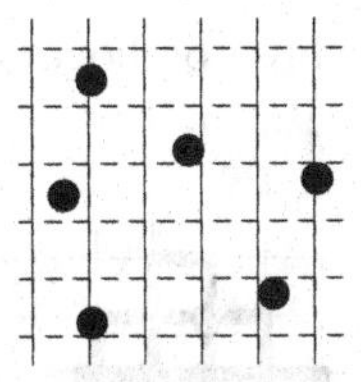

Figure 5.6.2 Find the Euclidean minimum spanning tree.

Remark: For a set of points in $\mathbb{R}^2$, knowing the locations of the points permits a solution in $O(n \log n)$ time, i.e., faster than Prim's algorithm.

DEFINITION: A ***convex set*** S in $\mathbb{R}^n$ is a set such that for every pair of points $x, y \in S$, every point of the straight-line segment joining x and y is also in S.

DEFINITION: The ***convex hull*** of a set of points in $\mathbb{R}^n$ is the smallest convex set that contains all the points.

DEFINITION: A ***triangulation*** of a set S of points in $\mathbb{R}^2$ is a straight-line graph drawing whose vertices are the points in S, with the property that the graph subdivides the convex hull of S into triangular faces, as in Figure 5.6.3.

DEFINITION: A triangulation of a set of points in $\mathbb{R}^n$ is a ***minimum weight triangulation*** if it minimizes the total edge length.

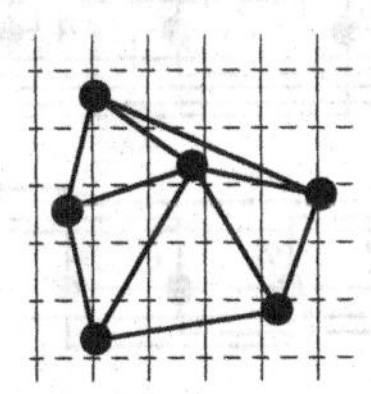

Figure 5.6.3 Triangulation of a geometric set.

Remark: It is not presently known ([LiTa04]) whether the problem of finding minimum weight triangulations is *NP*-hard.

Minimizing Area

There are algorithms to shrink various kinds of polyline drawings. Doing it ad hoc is something of an art.

Example 5.6.1: Figure 5.6.4 shows three grid drawings of K_5, progressively reduced in area.

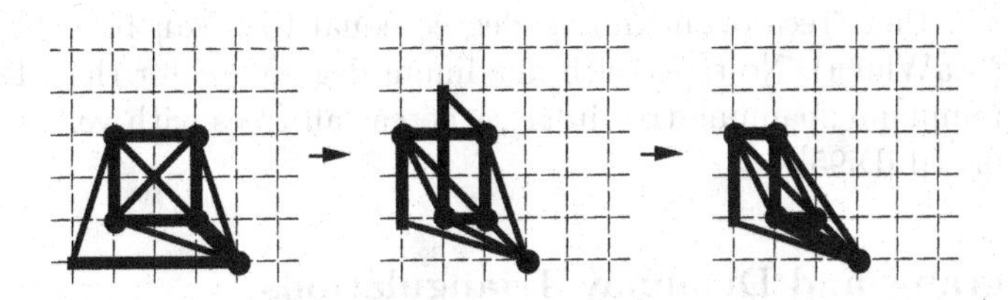

Figure 5.6.4 Shrinking a grid drawing.

Example 5.6.2: Figure 5.6.5 shows three orthogonal grid drawings of K_5, progressively reduced in area.

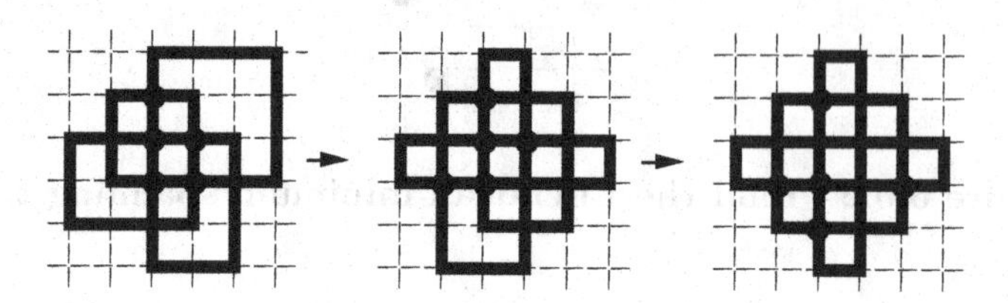

Figure 5.6.5 Shrinking an orthogonal grid drawing.

Representability

We seek combinatorial characterizations of graphs that are amenable to drawings of various types and with various geometric properties.

BY ORTHOGONAL DRAWING. Obviously, a graph with a vertex of degree more than 4 has no orthogonal drawing. To visualize how to construct an orthogonal drawing for an arbitrary graph with maximum degree at most 4, consider the orthogonal drawings of $circ(7:2,3)$ in Figure 5.6.6.

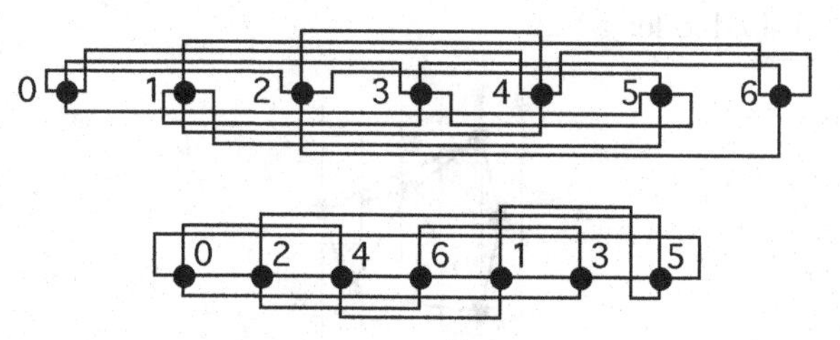

Figure 5.6.6 Two orthogonal drawings of $circ(7:2,3)$.

The lower drawing indicates savings in total edge-length and area if the vertices are placed judiciously.

BY MINIMUM SPANNING TREE. When drawing a tree T in the plane, quite possibly the geometric minimum spanning tree for the assigned locations of the vertices of T is not the tree T itself.

DEFINITION: A tree T is ***minimum-weight drawable*** if there exists a set of points whose geometric minimum spanning tree is isomorphic to T.

Remark: Each tree with maximum vertex degree at most 5 can be drawn as a minimum spanning tree of some set of vertices, by a linear-time algorithm, but no tree with maximum degree greater than 6 can be drawn as a minimum spanning tree ([MoSu92]). It is *NP*-hard to decide whether trees of maximum degree equal to 6 can be drawn as minimum spanning trees ([EaWh96]). No trees with maximum degree greater than 12 can be drawn as a Euclidean minimum spanning tree in $\mathbb{R}^3$, whereas all trees with vertex degree at most 9 are $\mathbb{R}^3$-drawable ([LiDi95]).

Voronoi Diagrams and Delaunay Triangulations

Voronoi diagrams and Delaunay triangulations can be used to construct geometric minimum spanning trees.

DEFINITION: An n-dimensional ***convex polytope*** is the intersection of a finite number of half-spaces

$$\{(x_1, \ldots, x_n) \mid a_1 x_1 + \cdots + a_n x_n \geq b\}$$

It may be finite or infinite.

DEFINITION: The ***Voronoi diagram*** for a set $S = \{p_1, \ldots, p_n\}$ of points in $\mathbb{R}^n$ is a partition of $\mathbb{R}^n$ into n convex polytopes $V(p_1), \ldots, V(p_n)$ such that the interior of the region $V(p_j)$ contains all points that are closer to p_j than to any other point in S.

DEFINITION: The ***Delaunay graph*** for a set $S = \{p_1, \ldots, p_n\}$ of points in the plane has the set S as its vertices. Two vertices p_i and p_j are joined by a straight line (representing an edge) if and only if the Voronoi regions $V(p_i)$ and $V(p_j)$ share an edge.

It follows from the definition that the Delaunay graph of a set of points is isomorphic to the topological dual of the Voronoi diagram for that set. Figure 5.6.7 shows the Voronoi diagram for a set of points in the plane and the corresponding Delaunay graph. Observe that their straight-line geometric specification may cause a Delaunay edge to traverse a region containing neither of its endpoints.

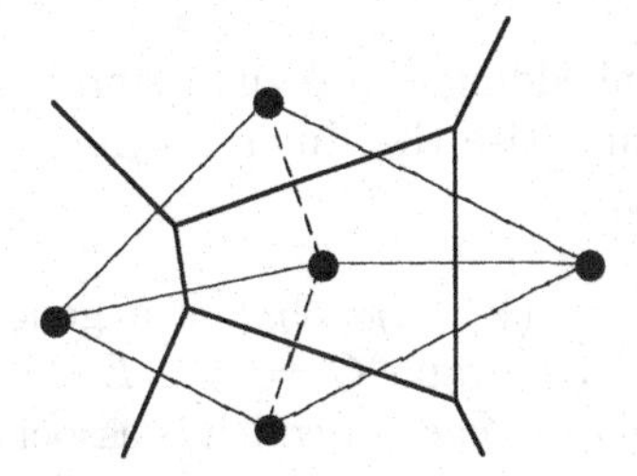

Figure 5.6.7 Voronoi diagram and Delaunay graph.

DEFINITION: A ***Delaunay triangulation*** of a point set P in $\mathbb{R}^2$ is a straight-line triangulation with all internal faces triangles, such that three points are the vertices of a face if and only if their convex hull does not contain any other point of P.

DEFINITION: A graph is ***Delaunay drawable*** if it admits a drawing that is a Delaunay triangulation.

Remark: The Voronoi diagram of a point set can be constructed in $O(n \log n)$-time and used in an $O(n \log n)$-time algorithm to construct the geometric minimum spanning tree of the set. (See [Ch00].)

Remark: Finding a complete combinatorial characterization of Delaunay drawable graphs is an open problem. (See [LiTa04].)

Remark: For a set of points in $\mathbb{R}^2$, knowing the locations of the endpoints permits a solution in $\Omega(n \log n)$ time, i.e., faster than Prim's algorithm. (See [AgWe88].)

5.7 SUPPLEMENTARY EXERCISES

5.7.1 Draw an imbedding of the Möbius ladder ML_4 in the torus.

5.7.2 Consider the following specifications for a graph imbedding: three 3-sided faces and four 4-sided faces. Either draw such an imbedding in the sphere or prove it is impossible.

5.7.3 Draw two cellular imbeddings of the following graph on a torus, such that the dual graphs are not isomorphic.

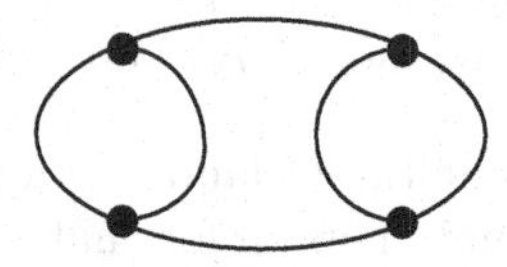

5.7.4 Draw the two dual graphs of Exercise 5.7.3 in the plane.

5.7.5 Draw a self-dual map of $C_3 \times C_3$ on the torus.

5.7.6 Either draw a self-dual imbedding of a regular simple graph in S_2 or prove that such an imbedding cannot exist. (Hint: Use the Euler Polyhedral Equation and self-duality to determine the number of edges.)

DEFINITION: A ***2-complex*** $K = (V_K, E_K, R_K)$ is a generalization of a graph to a 2-dimensional object. It consists of a graph $G = (V_K, E_K)$, called the ***1-skeleton of the complex***, and a set R_K of *regions*, to each of which is associated a closed walk in the graph that is the boundary walk of the polygon.

5.7.7 Draw a 2-complex whose 1-skeleton is planar, but which cannot be drawn in the sphere with no overlap of edges or regions.

5.7.8 What is the minimum number of vertices in a 2-complex that cannot be drawn in the sphere with no overlap of edges or regions.

Glossary

abnormality in a drawing of a graph on a surface: any of these three types of singularities:

(i) The images of two edges meet at more than one point of the surface.

(ii) The images of two different edges meet, but fail to actually cross each other (i.e., a tangency).

(iii) Images of three or more edges meet at the same point of the surface.

angular resolution of a polyline drawing: the smallest angle formed by any two segments of an edge at a bend or between any two edges incident at the same vertex. It does not usually take into account the angle at which two edges cross.

area of a geometric drawing: usually the area either of its convex hull or of the smallest rectangle with vertical and horizontal sides that covers the drawing.

aspect ratio of a drawing: the ratio of the length of the longer side to the shorter side of the smallest rectangle covering the drawing.

bend in a polyline drawing: a point where two segments belonging to the same edge meet.

boundary point of a surface: a point that is not an interior point.

boundary walk of a face f: a closed walk that corresponds to a complete traversal of the perimeter of the face.

boundaryless surface: a surface such that every point is an interior point.

carrier of a graph: synonym for *topological model.*

cellular imbedding $\iota : G \to S$: a graph imbedding such that every region is topologically equivalent to an open disk.

—, **strongly:** a cellular imbedding whose regions are strongly cellular.

cellular region of a graph imbedding: a region that is topologically equivalent to an open disk.

—, **strongly:** a cellular region whose boundary walk is a cycle in the graph.

central circuit of a Möbius band: the circuit formed from the line halfway between the top and bottom of the rectangular strip when the right and left edges are identified.

closed surface: a connected surface that is finite and boundaryless, such that the endpoints of every open arc are in the surface itself.

closed unit disk: the plane set $\{(x_1,\ x_2) \mid x_1^2 + x_2^2 \leq 1\}$.

2-complex $K = (V_K, E_K, R_K)$: a generalization of a graph to a 2-dimensional object, comprising a graph $G = (V_K, E_K)$, called the *1-skeleton of the complex*, and a set R_K of *regions*, to each of which is associated a closed walk in the graph that is the boundary walk of the polygon.

convex hull of a set of points in $\mathbb{R}^n$: the smallest convex set that contains all the points.

convex polytope: the intersection of a finite number of half-spaces

$$\{(x_1, \ldots, x_n) \mid a_1x_1 + \cdots + a_nx_n \geq b\}$$

It may be of finite size or of infinite measure.

convex set in $\mathbb{R}^n$: a set S such that for every pair of points $x, y \in S$, every point of the straight-line segment joining x and y is also in S.

crosscap on a surface: a Möbius band that occurs as a subspace of that surface.

—, **adding a:** the operation on a surface of replacing a subspace that is homeomorphic to a closed disk by a Möbius band.

crosscap number of a closed non-orientable surface: the subscript of the surface N_k to which it is topologically equivalent, i.e., the number of Möbius bands on the surface.

Delaunay drawable graph: a graph that admits a drawing that is a *Delaunay triangulation.*

Delaunay graph for a set $S = \{p_1, \ldots, p_n\}$ of points in the plane: a graph whose vertex-set is the set S, and such that two vertices p_i and p_j are joined by a straight line (representing an edge) if and only if the *Voronoi regions* $V(p_i)$ and $V(p_j)$ share an edge.

Delaunay triangulation of a point set P in $\mathbb{R}^2$: a *straight-line triangulation* with all internal faces triangles, such that three points are the vertices of a face if and only if their convex hull does not contain any other point of P.

donut, standard: the solid obtained by revolving the xy-plane disk of radius 1 centered at $(0, 2)$ around the z-axis.

drawing of a graph $G = (V, E)$**:** the continuous image on a surface of a topological model of G, that is one-to-one everywhere except possibly the interior of one edge crossing the interior of another.

—, **grid drawing:** a polyline drawing such that the vertices, crossings, and bends all have integer coordinates.

—, **polyline:** a plane drawing in which each edge is a chain of line segments.

—, **orthogonal:** a drawing in which each edge is a chain of horizontal and vertical segments.

—, **straight-line:** a drawing such that each edge is a single line segment.

dual graph and dual imbedding: the new graph and its imbedding derived, starting with a cellular imbedding of a graph, by inserting a new vertex into each existing face, and by then drawing through each existing edge a new edge that joins the new vertex in the region on one side to the new vertex in the region on the other side.

edge-crossing in a graph drawing: a point on the surface where the images of two different edges meet.

ends of a space edge: the parts of the space edge nearest to the image of the endpoint(s).

—, **0-end:** the part of the space edge near the image of the 0-end of the unit interval $[0, 1]$.

—, **1-end:** the part of the space edge near the image of the 1-end of the unit interval $[0, 1]$.

Euclidean metric: the usual distance metric on real n-dimensional space

$$\delta\left((x_1, \ldots, x_n), (y_1, \ldots, y_n)\right) = \sqrt{(x_1 - y_1)^2 + \cdots + (x_n - y_n)^2}$$

Euclidean MST problem: to produce a geometric minimum spanning tree for a set S of points in $\mathbb{R}^n$, that is, to determine the pairs of points that are endpoints of the edges of the minimum spanning tree.

Euclidean set: a subset X of a Euclidean space.

—, **connected:** a Euclidean set X such that for every pair of points $s, t \in X$, there exists a path in X from s to t.

—, **finite:** a Euclidean set in which the maximum distance of any point from the origin is at most M, for some real number M.

Euclidean space: the usual real vector space $\mathbb{R}^n$, for any positive integer n, with the Euclidean metric.

Euler characteristic of a surface: the value of the Euler polyhedral formula $|V|-|E|+|F|$ for any cellular imbedding on that surface; it is usually denoted by $\chi(S)$. Thus,

$$\chi(S) = \begin{cases} 2-2g & \text{for the orientable surface } S_g \\ 2-k & \text{for the non-orientable surface } N_k \end{cases}$$

flat polygon representation of a surface S: a drawing of a polygon with markings to match its sides in pairs, such that when the sides are pasted together as the markings indicate, the resulting surface obtained is topologically equivalent to S.

forbidden types of singularities in a graph drawing: illustrated in §5.3 but are otherwise absent from this book.

(i) Two vertex images $p(u)$ and $p(v)$ are equal.
(ii) An edge image q_e has a self-intersection or other singularities.
(iii) The interior of an edge image intersects a vertex image.

genus of a closed orientable surface: the subscript of the surface S_g to which it is topologically equivalent, i.e., the "number of handles" on the surface.

geometric minimum spanning tree for a set S of points in $\mathbb{R}^n$: a geometric spanning tree that minimizes the total edge length.

geometric spanning tree for a set S of points in $\mathbb{R}^n$: a drawing of a tree that has S as its vertex-set.

half-disk, standard: the plane set $\{(x_1, x_2) \mid x_1 \geq 0 \text{ and } x_1^2 + x_2^2 < 1\}$.

handle on a surface: a subspace topologically equivalent to the result of removing the interior of a closed disk from a torus.

—, **adding a:** the operation on a surface of replacing a subspace that is homeomorphic to a closed disk by a handle, or equivalently, by punching two separate holes in the surface S_n and then connecting them with a "tube".

imbedding of a graph G**:** a drawing with no edge-crossings at all.

interior of a space curve: all of the space curve except for its endpoints.

interior point of a surface: a point that has an ϵ-neighborhood topologically equivalent to an open disk.

Klein bottle: the non-orientable surface N_2 with two crosscaps.

map on a surface: an imbedding of a graph on that surface.

—, **regular:** an imbedding of a regular graph such that the dual graph is also a regular graph.

—, **self-dual:** an imbedding such that there is a topological equivalence from the imbedding surface to itself that takes the primal graph to the dual graph.

—, **simple:** an imbedding of a simple graph, such that the dual graph is also a simple graph.

mapping: a generic synonym for function.

minimum-weight drawable tree: a tree such that there exists a set S of points in $\mathbb{R}^2$ whose geometric minimum spanning tree is isomorphic to T.

Möbius band: the space formed from a rectangular strip with a half-twist, by identifying the left edge to the right edge, to form a continuous band.

neighborhood of a point y **in a Euclidean set** X**:** a subset N of X that contains the set $\{z \in X : |z - y| < epsilon\}$, for some positive number ϵ.

—, open ϵ**-:** the set of all points whose distance from y is less than ϵ, where $\epsilon > 0$.

non-orientable genus: a synonym for *crosscap number*.

non-orientable surface: a surface that contains a Möbius band.

non-orientable surfaces, classification sequence $N_1, N_2, N_3, \ldots$**:** a recursively defined sequence starting from the sphere, which in this context is denoted N_0, even though it is orientable. The surface N_{k+1} is obtained from the surface N_k by *adding a crosscap.*

normal drawing of a graph G **in a surface** S**:** a drawing such that (i) two different edges cross at most once; (ii) tangent edge touchings are not permitted; and (iii) images of three distinct edges never meet at the same point of the surface. It is usually assumed in graph theory that a drawing is normalized.

open unit disk: the plane set $\{(x_1,\ x_2) \mid x_1^2 + x_2^2 < 1\}$.

orientable surface: a surface that does not contain a Möbius band.

orientable surfaces, classification sequence $S_0, S_1, S_2, \ldots$**:** a recursively defined sequence such that S_0 is the sphere, S_1 is the torus, and surface S_{n+1} is obtained from S_n by *adding a handle.*

path from s **to** t **in a Euclidean set:** the image of a continuous function f from the unit interval $[0, 1]$ into X such that $f(0) = s$, $f(1) = t$, and either

- f is a one-to-one, or
- the restriction of f to $[0, 1)$ is one-to-one and $s = t$.

(One may visualize a path as the trace of a particle traveling through space for a fixed length of time.)

—, closed: a path such that $f(0) = f(1)$, i.e., in which $s = t$. (For instance, this would include a "knotted circle" in space.)

—, open: a path such that $f(0) \neq f(1)$, i.e., in which $s \neq t$.

planar graph: a graph that has an imbedding in the plane.

plane: Euclidean 2-space.

platonic graph: the 1-skeleton of a platonic solid.

platonic map: the imbedding of the 1-skeleton of a platonic solid on its surface.

platonic solid: any of the five *geometrically regular* 3-dimensional polyhedra: tetrahedron, octahedron, cube, dodecahedron, and icosahedron.

projective plane: the non-orientable surface N_1 with one crosscap.

resolution rule for a drawing: a constraint that prevents the drawing from being arbitrarily scaled down, for instance, a minimum unit distance between vertices.

singularity in a graph drawing: a place where the drawing is not one-to-one.

1-skeleton of a 2-complex $K = (V_K, E_K, R_K)$: the graph $G = (V_K, E_K)$.

space curve between two points x and y in 3-space: the image of a continuous 1-to-1 function from the unit interval [0, 1] into 3-space that maps the endpoints 0 and 1 of $[0, 1]$ to x and y.

space edge: *see* topological model of a graph.

space vertex: *see* topological model of a graph.

sphere: any Euclidean set that is topologically equivalent to the unit sphere.

subspace of a Euclidean space: any subset of the space, plus structure implicit in the Euclidean metric.

surface: a Euclidean set in which every point has a neighborhood that is topologically equivalent either to the open unit disk or to the standard half-disk.

—, **boundaryless:** a surface such that every point is an interior point.

topological equivalence between two Euclidean sets X and Y: a continuous function $f : X \to Y$ that is one-to-one and onto, and whose inverse function $f^{-1} : Y \to X$ is continuous.

topological model of a graph $G = (V, E)$: a model of G in Euclidean 3-space $\mathbb{R}^3$, where each vertex $v \in V$ is represented by a point p_v in 3-space, called a space vertex, and each proper edge $e \in E$ is represented by a space curve q_e joining its endpoints, called a *space edge*. The topological model of a graph G is usually also denoted G.

torus: any Euclidean set that is topologically equivalent to the standard torus.

total edge-length of a drawing: the sum of the lengths of its edges.

triangulation of a set S of points in $\mathbb{R}^2$: a straight-line graph drawing whose vertices are the points in S, with the property that the graph subdivides the convex hull of S into triangular faces.

—, **minimum weight:** a triangulation that minimizes the total edge length.

unit cylinder: the Euclidean set $\{(x_1, x_2, x_3) \mid 0 \leq x_1 \leq 1 \text{ and } x_2^2 + x_3^2 = 1\}$.

unit sphere: the Euclidean set $\{(x_1, x_2, x_3) \mid x_1^2 + x_2^2 + x_3^2 = 1\}$.

Voronoi diagram for a set $S = \{p_1, \ldots, p_n\}$ of points in $\mathbb{R}^n$:] a partition of $\mathbb{R}^n$ into n convex polytopes $V(p_1), \ldots, V(p_n)$ such that the interior of the region $V(p_j)$ contains all points that are closer to p_j than to any other point in S.

Chapter 6

GRAPH COLORINGS

INTRODUCTION

The first known mention of coloring problems was in 1852, when Augustus De Morgan, Professor of Mathematics at University College, London, wrote Sir William Rowan Hamilton in Dublin about a problem posed to him by a former student, named Francis Guthrie. Guthrie noticed that it was possible to color the counties of England using four colors so that no two adjacent counties were assigned the same color. The question raised thereby was whether four colors would be sufficient for all possible decompositions of the plane into regions.

The Poincaré duality construction transforms this question into the problem of deciding whether it is possible to color the vertices of every planar graph with four colors so that no two adjacent vertices are assigned the same color. Wolfgang Haken and Kenneth Appel provided an affirmative solution in 1976.

Numerous practical applications involving graphs and some form of coloring are described in this chapter. Factorization is included here, because, like edge-coloring, it involves a partitioning of the edge-set of a graph.

6.1 VERTEX-COLORINGS

In the most common kind of graph coloring, colors are assigned to the vertices. From a standard mathematical perspective, the subset comprising all the vertices of a given color would be regarded as a cell of a partition of the vertex-set. Drawing the graph with colors on the vertices is simply an intuitive way to represent such a partition.

NOTATION: The maximum degree of a vertex in a graph G is denoted $\delta_{\max}(G)$, or simply by $\delta_{\max}$ when the graph of reference is evident from context.

Minimization Problem for Vertex-Colorings

Most applications involving *vertex-colorings* are concerned with determining the minimum number of colors required under the condition that the endpoints of an edge cannot have the same color.

DEFINITION: A ***vertex k-coloring*** is an assignment $f : V_G \to C$ from its vertex-set onto a k-element set C whose elements are called *colors* (typically, $C = \{1, 2, \ldots k\}$). For any k, such an assignment is called a **vertex-coloring**.

TERMINOLOGY: Since vertex-colorings arise more frequently than edge-colorings or map-colorings, one often says *coloring*, instead of *vertex-coloring*, when the context is clear.

DEFINITION: Let c be a vertex-coloring of a graph G. The ***color class for color i***, denoted c_i, is the subset of V_G containing all the vertices assigned color i. That is, $c_i = \{x \in V_G : c(x) = i\}$.

DEFINITION: A ***proper vertex-coloring*** of a graph is a vertex-coloring such that the endpoints of each edge are assigned two different colors.

Proposition 6.1.1: *A vertex k-coloring of a graph is a proper vertex-coloring if and only if each color class is an independent set of vertices.*

Proof: This follows directly from the definitions of *proper* and *independent*. ♢

DEFINITION: A graph is said to be ***vertex k-colorable*** if it has a proper vertex k-coloring.

Example 6.1.1: The vertex-coloring shown in Figure 6.1.1 demonstrates that the graph is vertex 4-colorable.

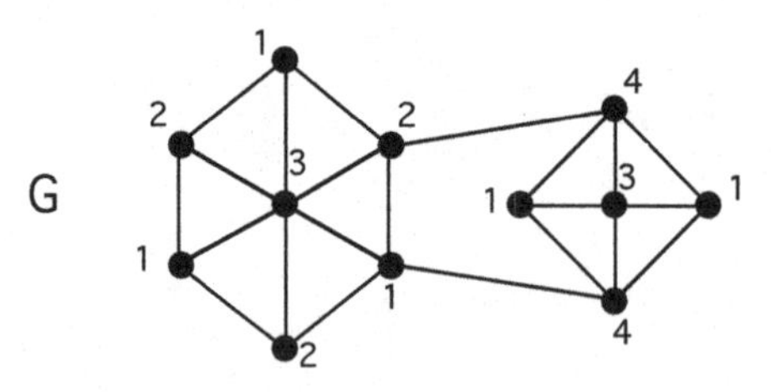

Figure 6.1.1 A proper vertex 4-coloring of a graph.

DEFINITION: The ***(vertex) chromatic number*** of a graph G, denoted $\chi(G)$, is the minimum number of different colors required for a proper vertex-coloring of G. A graph G is ***(vertex) k-chromatic*** if $\chi(G) = k$. A ***chromatic coloring*** of G is a proper coloring of G that uses $\chi(G)$ colors.

Thus, $\chi(G) = k$, if graph G is k-colorable but not $(k-1)$-colorable.

Example 6.1.1, continued: The 3-coloring in Figure 6.1.2 shows that the graph G in Figure 6.1.1 is 3-colorable, which means that $\chi(G) \leq 3$. However, graph G contains three mutually adjacent vertices and hence is not 2-colorable. Thus, G is 3-chromatic.

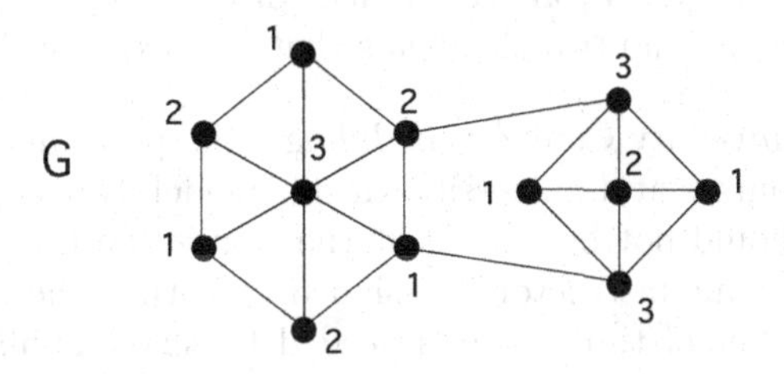

Figure 6.1.2 A proper 3-coloring of graph G.

Example 6.1.2: The 4-coloring of the graph G shown in Figure 6.1.3 establishes that $\chi(G) \leq 4$, and the K_4-subgraph (drawn in bold) shows that $\chi(G) \geq 4$. Hence, $\chi(G) = 4$.

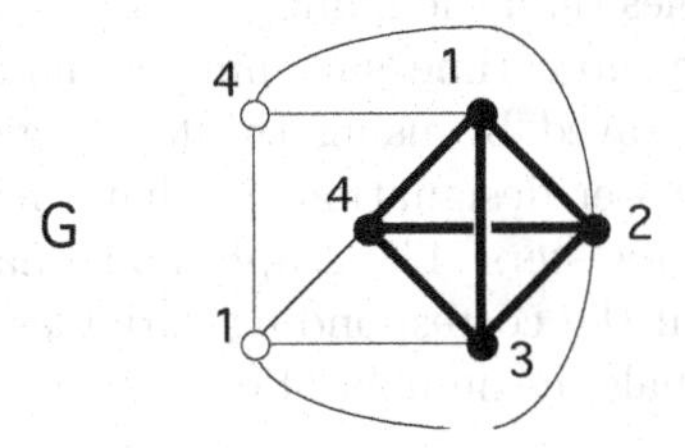

Figure 6.1.3 Graph G has chromatic number 4.

Remark: The study of vertex-colorings of graphs is customarily restricted to simple, connected graphs for the following reasons:

i. A graph with a self-loop is uncolorable, since the endpoint of a self-loop is adjacent to itself.
ii. A multi-edge requires two different colors for its endpoints in the same way that a single edge does.
iii. The chromatic number of a graph is simply the maximum of the chromatic numbers of its components.

Modeling Applications as Vertex-Coloring Problems

When an application is modeled as a vertex-coloring problem, the vertices in each color class typically represent individuals or items that do not compete or conflict with each other.

Application 6.1.1: *Assignment of Radio Frequencies* Suppose that the vertices of a graph G represent transmitters for radio stations. In this model, two stations are considered adjacent when their broadcast areas overlap, which would result in interference if they broadcast at the same frequency. Then two "adjacent" stations should be assigned different transmission frequencies. Regarding the frequencies as colors transforms the situation into a graph-coloring problem, in which each color class contains vertices representing stations with no overlap. In this model, $\chi(G)$ equals the minimum number of transmission frequencies required to avoid broadcast interference.

Application 6.1.2: *Separating Combustible Chemical Combinations* Suppose that the vertices of a graph represent different kinds of chemicals needed in some manufacturing process. For each pair of chemicals that might explode if combined, there is an edge between the corresponding vertices. The chromatic number of this graph is the number of different storage areas required so that no two chemicals that mix explosively are stored together.

Application 6.1.3: *University Course Scheduling* Suppose that the vertices of a simple graph G represent the courses at a university. In this model, two vertices are adjacent if the classes they represent should not be offered at the same time (i.e., some students need to take both classes or the same instructor is assigned to both). The chromatic number $\chi(G)$ gives the minimum number of time periods needed to schedule all the classes so that no time conflicts occur.

Application 6.1.4: *Fast-Register Allocation for Computer Programming* In some computers there are a limited number of special "registers" that permit faster execution of arithmetic operations than ordinary memory locations. The program variables that are used most often can be declared to have "register" storage class. Unfortunately, if the programmer declares more "register" variables than the number of hardware registers available, then the program execution may waste more time swapping variables between ordinary memory and the fast registers than is saved by using the fast registers. One solution is for the programmer to control the register designation, so that variables that are simultaneously active are assigned to different registers. The graph model has one vertex for each variable, and two vertices are adjacent if the corresponding variables can be simultaneously active. Then the chromatic number equals the number of registers needed to avoid the overswapping phenomenon.

Sequential Vertex-Coloring Algorithm

There is a naive (brute-force) algorithm to decide, for some fixed k, whether a given n-vertex graph is k-colorable. Just check to see if any of the k^n possible k-colorings is proper and return YES if so. By iterating this decision procedure, starting with $k = 1$, until a YES is returned, one obtains an algorithm for calculating the exact value of the chromatic. However, its running time is exponential in the number of vertices.

Alternatively, Algorithm 6.1.1 is a *sequential* algorithm that quickly produces a proper coloring of any graph; yet the coloring it produces is unlikely to be a minimum one. Moreover, it seems unlikely that *any* polynomial-time algorithm can accomplish this, since the problem of calculating the chromatic number of a graph is known to be NP-hard [GaJo79]. In fact, deciding whether a graph has a 3-coloring is an NP-complete problem.

Algorithm 6.1.1: Sequential Vertex-Coloring
Input: a graph G with vertex list $v_1, v_2, \ldots, v_p$.
Output: a proper vertex-coloring $f : V_G \to \{1, 2, \ldots\}$.
For $i = 1, \ldots, p$
Let $f(v_i) :=$ the smallest color number not used on any of the smaller-subscripted neighbors of v_i.
Return vertex-coloring f.

Example 6.1.3: When the sequential vertex-coloring algorithm is applied to the graph of Figure 6.1.4, the result is a 4-coloring.

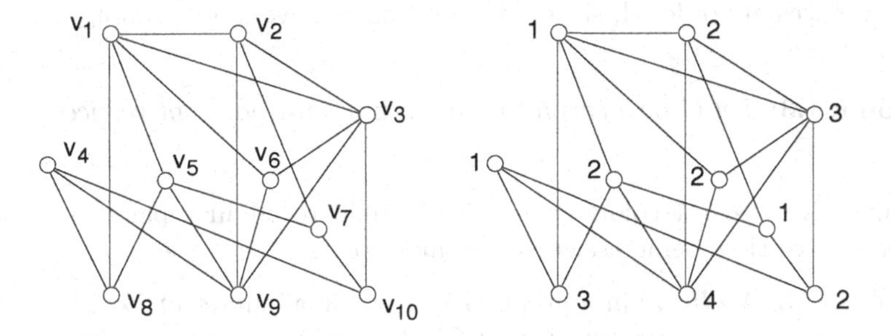

Figure 6.1.4 A graph and a sequential 4-coloring.

However, if the coloring assignment relaxes its obsessiveness with using the smallest possible color number at vertex v_4 and at vertex v_7, then a 3-coloring can be obtained for the same graph, as shown in Figure 6.1.5.

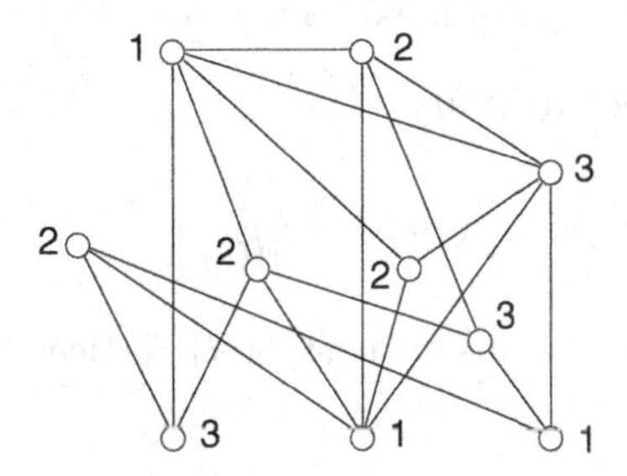

Figure 6.1.5 A non-sequential 3-coloring for the same graph.

Remark: Notice that if the vertices of the graph in Figure 6.1.5 are ordered so that v_1, v_2, v_3 are the vertices with color 1, vertices v_4, v_5, v_6, v_7 are the ones with color 2, and v_8, v_9, v_{10} are the vertices with color 3, then the sequential coloring algorithm would have produced that minimum coloring. More generally, for a given minimum coloring for a graph G (obtained by whatever means), if the vertices of G are linearly ordered so that all the vertices with color i precede all the vertices with color j whenever $i < j$, then, when G along with this vertex ordering is supplied as input to the sequential-coloring algorithm, the result is a minimum coloring. Thus, for any n-vertex graph, there is at least one vertex ordering (of the $n!$ orderings) for which Algorithm 6.1.1 will produce a minimum coloring.

Basic Principles for Calculating Chromatic Numbers

A few basic principles recur in many chromatic-number calculations. They combine to provide a direct approach involving two steps, as already seen for Examples 6.1.2 and 6.1.3.

- *Upper Bound*: Show $\chi(G) \leq k$, most often by exhibiting a proper k-coloring of G.
- *Lower Bound*: Show $\chi(G) \geq k$, most especially, by finding a subgraph that requires k colors.

The following result is an easy upper bound for $\chi(G)$; it is complemented by the easy lower bound of clique number $\omega(G)$. Brooks's Theorem, appearing later in this section, sharpens this upper bound for a large class of graphs.

Proposition 6.1.2: *Let G be a simple graph. Then $\chi(G) \leq \delta_{\max}(G) + 1$.*

Proof: The sequential coloring algorithm never uses more than $\delta_{\max}(G)+1$ colors, no matter how the vertices are ordered, since a vertex cannot have more than $\delta_{\max}(G)$ neighbors. ♢

Proposition 6.1.3: *Let G be a graph that has k mutually adjacent vertices. Then $\chi(G) \geq k$.*

Proof: Using fewer than k colors on graph G would result in a pair from the mutually adjacent set of k vertices being assigned the same color. ♢

REVIEW FROM §1.3: A ***clique*** in a graph G is a maximal subset of V_G whose vertices are mutually adjacent. The ***clique number*** $\omega(G)$ of a graph G is the number of vertices in a largest clique in G.

Corollary 6.1.4: *Let G be a graph. Then $\chi(G) \geq \omega(G)$.*

REVIEW FROM §1.3: The ***independence number*** $\alpha(G)$ of a graph G is the number of vertices in an independent set in G of maximum cardinality.

Proposition 6.1.5: *Let G be any graph. Then*

$$\chi(G) \geq \left\lceil \frac{|V_G|}{\alpha(G)} \right\rceil$$

Proof: Since each color class contains at most $\alpha(G)$ vertices, the number of different color classes must be at least $\left\lceil \frac{|V_G|}{\alpha(G)} \right\rceil$. ♢

Example 6.1.4: Consider the graph $G = circ(7 : 1, 2)$, shown in Figure 6.1.6 (§1.1). If G had an independent set of size 3 (i.e., if $\alpha(G) \geq 3$), then its edge-complement $\overline{G}$ would contain a 3-cycle. However, $\overline{G} = circ(7 : 3)$ is a 7-cycle. Thus, $\alpha(G) = 2$, and, consequently, by Proposition 6.1.5,

$$\chi(G) \geq \left\lceil \frac{|V_G|}{\alpha(G)} \right\rceil = \left\lceil \frac{7}{2} \right\rceil = 4$$

There is a proper 4-coloring with color classes $\{0, 3\}$, $\{1, 4\}$, $\{2, 5\}$, and $\{6\}$.

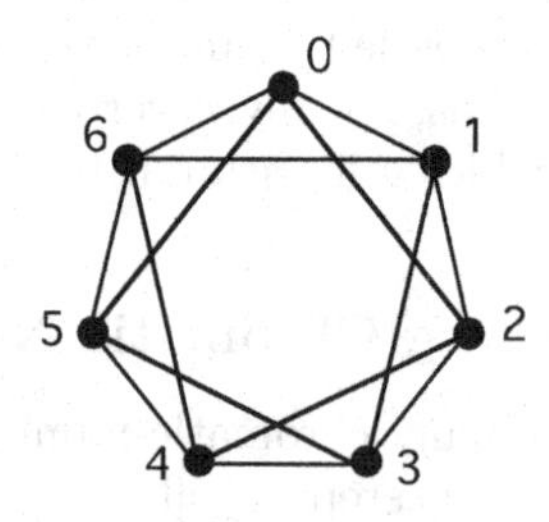

Figure 6.1.6 $circ(7 : 1, 2)$.

Proposition 6.1.6: *Let H be a subgraph of graph G. Then $\chi(G) \geq \chi(H)$.*

Proof: Whatever colors are used on the vertices of subgraph H in a minimum coloring of graph G can also be used in a coloring of H by itself. ◊

REVIEW FROM §1.4: The ***join*** $G + H$ of the graphs G and H is obtained from the graph union $G \cup H$ by adding an edge between each vertex of G and each vertex of H.

Lemma 6.1.7: *For any proper coloring of the join of two graphs, $G + H$, each color class lies entirely in G or in H.*

Proof: If a color class had vertices in both G and H, the coloring would not be proper, since every vertex in G is adjacent to every vertex in H. ◊

Proposition 6.1.8: *The join of graphs G and H has chromatic number*

$$\chi(G + H) = \chi(G) + \chi(H)$$

Proof: *Lower Bound*: For any proper coloring of the join $G + H$, no color used on the subgraph G can be the same as a color used on the subgraph H, by Lemma 6.1.7. Since $\chi(G)$ colors are required for subgraph G and $\chi(H)$ colors are required for subgraph H, it follows that $\chi(G + H) \geq \chi(G) + \chi(H)$.

Upper Bound: Using $\chi(G)$ colors to color the graph G, and $\chi(H)$ different colors to color the graph H creates a proper $(\chi(G) + \chi(H))$-coloring of $G + H$, establishing the upper bound. ◊

Chromatic Numbers for Common Graph Families

By using the basic principles given above, it is straightforward to establish the chromatic numbers of graphs in some of the most common graph families, which are summarized in Table 6.1.1.

Table 6.1.1: Chromatic numbers for common graph families

Graph G	$\chi(G)$
trivial graph	1
bipartite graph	2
nontrivial path graph P_n	2
nontrivial tree T	2
cube graph Q_n	2
even cycle graph C_{2n}	2
odd cycle graph C_{2n+1}	3
even wheel W_{2n}	3
odd wheel W_{2n+1}	4
complete graph K_n	n

Proposition 6.1.9: *A graph G has $\chi(G) = 1$ if and only if G has no edges.*

Proof: The endpoints of an edge must be colored differently. ♢

Proposition 6.1.10: *A bipartite graph G has $\chi(G) = 2$, unless G is edgeless.*

Proof: A 2-coloring is obtained by assigning one color to every vertex in one of the bipartition parts and another color to every vertex in the other part. If G is not edgeless, then $\chi(G) \geq 2$, by Proposition 6.1.3. ♢

Corollary 6.1.11: *Path Graphs: $\chi(P_n) = 2$, for $n \geq 2$.*

Proof: P_n is bipartite. Apply Proposition 6.1.10. ♢

Remark: Just as a bipartite graph is a graph whose vertices can be partitioned into two color classes, we define a ***multipartite graph*** for any number of color classes.

DEFINITION: A ***k-partite graph*** is a loopless graph whose vertices can be partitioned into k independent sets, which are sometimes called the ***partite sets*** of the partition.

Corollary 6.1.12: *Trees: $\chi(T) = 2$, for any nontrivial tree T.*

Proof: Trees are bipartite. ♢

Corollary 6.1.13: *Cube Graphs: $\chi(Q_n) = 2$.*

Proof: The cube graph Q_n is bipartite. ♢

Corollary 6.1.14: *Even Cycles: $\chi(C_{2n}) = 2$.*

Proof: An even cycle is bipartite. ♢

Proposition 6.1.15: *Odd Cycles: $\chi(C_{2n+1}) = 3$.*

Proof: Clearly, $\alpha(C_{2n+1}) = n$. Thus, by Proposition 6.1.5,

$$\chi(C_{2n+1}) \geq \left\lceil \frac{|V(C_{2n+1})|}{\alpha(V(C_{2n+1}))} \right\rceil = \left\lceil \frac{2n+1}{n} \right\rceil = 3$$

♢

REVIEW FROM §1.4: The wheel graph $W_n = K_1 + C_n$ is called an ***odd wheel*** if n is odd, and an ***even wheel*** if n is even.

Proposition 6.1.16: *Even Wheels: $\chi(W_{2m}) = 3$.*

Proof: Using the fact that $W_{2m} = C_{2m} + K_1$, Proposition 6.1.8 and Corollary 6.1.14 imply that $\chi(W_{2m}) = \chi(C_{2m}) + \chi(K_1) = 2 + 1 = 3$. ♢

Proposition 6.1.17: *Odd Wheels: $\chi(W_{2m+1}) = 4$, for all $m \geq 1$.*

Proof: Using the fact that the wheel graph W_{2m+1} is the join $C_{2m+1} + K_1$, Propositions 6.1.8 and 6.1.15 imply that $\chi(W_{2m+1}) = \chi(C_{2m+1}) + \chi(K_1) = 3 + 1 = 4$, for all $m \geq 1$. ♢

Proposition 6.1.18: *Complete Graphs: $\chi(K_n) = n$.*

Proof: By Proposition 6.1.2, $\chi(K_n) \leq n$ and by Proposition 6.1.3, $\chi(K_n) \geq n$. ♢

Chromatically Critical Graphs

Proposition 6.1.19: *Let G be a graph.*

(i) For any vertex v, $\chi(G) - 1 \leq \chi(G - v) \leq \chi(G)$.

(ii) For any edge e, $\chi(G) - 1 \leq \chi(G - e) \leq \chi(G)$.

Proof: The second inequality in each statement follows from Proposition 6.1.6. The first inequalities are left as an exercise. ♢

DEFINITION: An edge e in a graph G is ***(chromatically) critical*** if $\chi(G - e) = \chi(G) - 1$. A vertex v in a graph G is ***(chromatically) critical*** if $\chi(G - v) = \chi(G) - 1$.

DEFINITION: A graph G is ***k-critical*** if G is connected, $\chi(G) = k$, and every edge is critical.

The next two results will be used in Section 6.2.

Example 6.1.5:

1. The complete graph K_n is n-critical.
2. An odd cycle is 3-critical.

Proposition 6.1.20: *If e is a critical edge of a graph G, and c is a chromatic coloring of $G - e$, then c must assign the same color to both endpoints of e.*

Proof: If c assigned different colors to the endpoints of edge e, then c would be a chromatic coloring of G, contradicting the criticality of e. ♢

Example 6.1.6: The odd wheel W_5 is 4-chromatic, by Proposition 6.1.17, and Figure 6.1.7 illustrates how removing either a spoke-edge or a rim-edge reduces the chromatic number to 3. Thus, W_5 is 4-critical. The argument that all odd wheels W_{2m+1} with $m \geq 1$ are 4-critical is essentially the same (see Exercises).

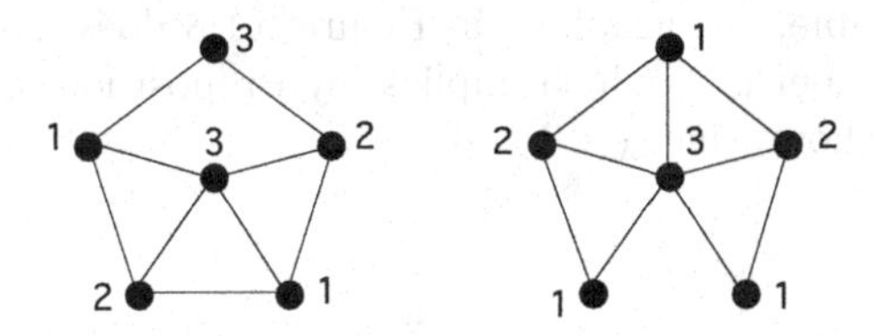

Figure 6.1.7 The graph $W_5 - e$ is 3-colorable, for any edge e.

Proposition 6.1.21: *If G is a k-critical graph, then each vertex is critical.*

♢ *(Exercises)*

Theorem 6.1.22: *If G is a k-critical graph, then every vertex has degree greater than or equal to $k - 1$.*

Proof: By way of contradiction, suppose a vertex v had degree less than $k - 1$. Consider a $(k - 1)$-coloring of the vertex-deletion subgraph $G - v$, which exists by Proposition 6.1.21. Since vertex v has fewer than $k - 1$ neighbors, there is a color that is not assigned to any neighbor of v. Assigning v that color gives a $(k - 1)$-coloring of G, which is a contradiction. ♢

Obstructions to k-Chromaticity

For a relatively small graph G, it is usually easier to calculate a plausible upper bound for $\chi(G)$ than a lower bound. Finding a configuration, called an *obstruction*, that forces the number of colors to exceed some value is the general approach to establishing a lower bound. The presence of either a subgraph or a graph property can serve as an obstruction.

DEFINITION: An ***obstruction to k-chromaticity*** (or **k-obstruction**) is a subgraph that forces every graph that contains it to have chromatic number greater than k.

Example 6.1.7: The complete graph K_{k+1} is an obstruction to k-chromaticity. (This is simply a restatement of Proposition 6.1.3.)

Proposition 6.1.23: *Every $(k+1)$-critical graph is an edge-minimal obstruction to k-chromaticity.*

Proof: If any edge is deleted from a $(k+1)$-critical graph, then, by definition, the resulting graph is not an obstruction to k-chromaticity. ♢

DEFINITION: A set $\{G_j\}$ of chromatically $(k+1)$-critical graphs is a ***complete set of obstructions*** if every $(k+1)$-chromatic graph contains at least one member of $\{G_j\}$ as a subgraph.

Example 6.1.8: The singleton set $\{K_2\}$ is a complete set of obstructions to 1-chromaticity.

REVIEW FROM §1.2: [Characterization of bipartite graphs] A graph is bipartite if and only if it contains no cycles of odd length.

Example 6.1.9: The bipartite-graph characterization above implies that the family $\{C_{2j+1} \mid j = 1, 2, \dots\}$ of all odd cycles is a complete set of 2-obstructions.

Example 6.1.10: Although the odd-wheel graphs W_{2m+1} with $m \geq 1$ are 4-critical, they do not form a complete set of 3-obstructions, since there are 4-chromatic graphs that contain no such wheel. For example, the graph G in Figure 6.1.8 does not contain an odd wheel, but its independence number is 2, which implies, by Proposition 6.1.5, that $\chi(G) \geq 4$. The 4-coloring shown below shows that $\chi(G) = 4$.

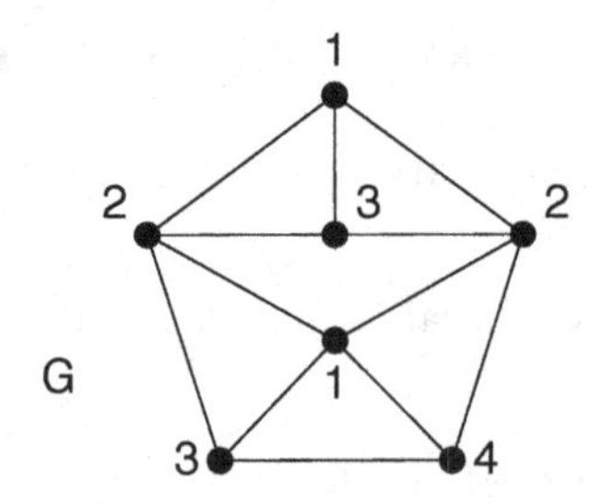

Figure 6.1.8 A 4-chromatic graph that contains no odd wheel.

Brooks's Theorem

REVIEW FROM §3.4: The ***vertex-connectivity*** of a connected graph G, denoted $\kappa_v(G)$, is the minimum number of vertices, whose removal can either disconnect G or reduce it to a 1-vertex graph.

REVIEW FROM §3.6:

- A ***block*** of a loopless graph is a maximal connected subgraph H such that no vertex of H is a cut-vertex of H.
- A ***leaf block*** of a graph G is a block that contains exactly one cut-vertex of G.

Lemma 6.1.24: *Let G be a non-complete, k-regular 2-connected graph with $k \geq 3$. Then G has a vertex x with two nonadjacent neighbors y and z such that $G - \{y, z\}$ is a connected graph.*

Proof: Let w be any vertex in graph G. First suppose that subgraph $G - w$ is 2-connected. Since the graph G is regular and non-complete, there is a vertex z at distance 2 from vertex w. If x is a vertex adjacent to both w and z, then vertices x, w, and z satisfy the conditions of the assertion (with w playing the role of y).

If $G - w$ is not 2-connected, then it has at least one cut-vertex. By Proposition 3.6.6, there exist two leaf blocks B_1 and B_2 of $G - w$. Since graph G has no cut-vertices, it follows that vertex w is adjacent to some vertex y_1 in B_1 that is not a cut-vertex of $G - w$ and, likewise, adjacent to some vertex y_2 in B_2 that is not a cut-vertex of $G - w$. By Corollary 3.6.5, vertex y_1 is not adjacent to vertex y_2. Then w, y_1, and y_2 satisfy the conditions of the assertion (with w playing the role of x). ♢

Theorem 6.1.25: [***Brooks*, 1941**] *Let G be a non-complete, simple connected graph with maximum vertex degree $\delta_{\max}(G) \geq 3$. Then $\chi(G) \leq \delta_{\max}(G)$.*

Proof: In this proof of [Lo75], the sequential coloring algorithm enables us to make some simplifying reductions. Suppose that $|V_G| = n$.

Case 1. G is not regular.
First choose vertex v_n to be any vertex of degree less than $\delta_{\max}(G)$. Next grow a spanning tree (see §3.1) from v_n, assigning indices in decreasing order. In this ordering of V_G, every vertex except v_n has a higher-indexed neighbor along the tree path to v_n. Hence, each vertex has at most $\delta_{\max}(G) - 1$ lower-index neighbors. It follows that the sequential coloring algorithm uses at most $\delta_{\max}(G)$ colors.

Case 2. G is regular, and G has a cut-vertex x.
Let $C_1, \ldots, C_m$ be the components of the vertex-deletion subgraph $G - x$, and let G_i be the subgraph of G induced on the vertex-set $V_{C_i} \cup \{x\}$, $i = 1, \ldots, m$. Then the degree of vertex x in each subgraph G_i is clearly less than $\delta_{\max}(G)$. By Case 1, each subgraph G_i has a proper vertex-coloring with $\delta_{\max}(G)$ colors. By permuting the names of the colors in each such subgraph so that vertex x is always assigned color 1, one can construct a proper coloring of G with $\delta_{\max}(G)$ colors.

Case 3. G is regular and 2-connected.
By Lemma 6.1.24, graph G has a vertex, which we call v_n, with two nonadjacent neighbors, which we call v_1 and v_2, such that $G - \{v_1, v_2\}$ is connected. Grow a spanning tree from v_n in $G - \{v_1, v_2\}$, assigning indices in decreasing order. As in Case 1, each vertex except v_n has at most $\delta_{\max}(G) - 1$ lower-indexed neighbors. It follows that the sequential coloring algorithm uses at most $\delta_{\max}(G)$ colors on all vertices except v_n. Moreover, the sequential coloring algorithm assigns the same colors to v_1 and v_2 and at most $k - 2$ colors on the other $k - 2$ neighbors of v_n. Thus, one of the $\delta_{\max}(G)$ colors is available for v_n. ♢

Heuristics for Vertex-Coloring

Many vertex-coloring heuristics for graphs are based on the intuition that a vertex of large degree will be more difficult to color later than one of smaller degree [ma81, Ca86, Br79], and that if two vertices have equal degree, then the one having the denser *neighborhood subgraph* will be harder to color later. These three references discuss variations and combinations of these ideas.

The following algorithm is one such combination. Assume that the colors are named $1, 2, \ldots$. As the algorithm proceeds, the *colored degree* of a vertex v is the number of different colors that have thus far been assigned to vertices adjacent to v.

Algorithm 6.1.2: Vertex-Coloring: Largest Degree First
Input: an n-vertex graph G.
Output: a vertex-coloring f of graph G.
While there are uncolored vertices of G
  Among the uncolored vertices with maximum degree,
   choose vertex v with maximum colored degree.
  Assign smallest possible color k to vertex v : $f(v) := k$.
Return graph G with vertex-coloring f.

Example 6.1.11: Descending degree order for the graph of Figure 6.1.9 (left) is

$$v_1,\ v_3,\ v_9,\ v_2,\ v_5,\ v_4,\ v_6,\ v_7,\ v_8,\ v_{10}$$

Applying the largest-degree-first algorithm yields a 3-coloring, as shown in 6.1.9 (right).

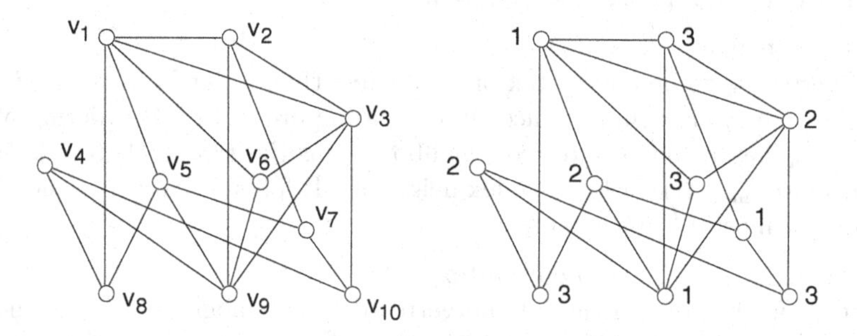

Figure 6.1.9 A largest-degree-first sequential coloring.

Coloring the Vertices of an Edge-Weighted Graph

Application 6.1.5: *Timetabling with Unavoidable Conflicts* [KiYe92] Suppose that the evening courses at a certain school must be scheduled in k timeslots, $t_1, t_2, \ldots, t_k$. The school would like to schedule the courses so that, whenever at least three students are preregistered for both of a particular pair of courses, the two courses are assigned different timeslots. The graph model has a vertex-set corresponding to the set of courses and has an edge between a pair of vertices if the corresponding pair of courses should be scheduled in different timeslots.

Typically, the chromatic number of such a graph is greater than k, which means that it is impossible to avoid *conflicts* (i.e., assigning the same timeslot to certain pairs of courses that should have been scheduled in different timeslots). A more realistic objective under such circumstances is to schedule the courses so that the total number of conflicts is minimized. Moreover, since some conflicts have greater impact than others, the number of students that preregistered for a given pair of courses is assigned as a weight to the corresponding edge. A further refinement of the model may adjust these edge-weights according to other factors not related to numbers of students. For instance, an edge corresponding to a pair of courses that are both required for graduation by some students or are to be taught by the same instructor should be assigned a large enough positive integer so as to preclude their being assigned to the same timeslot.

Suppose that $w(e)$ denotes the positive integer indicating the deleterious impact of scheduling the two courses corresponding to the endpoints of edge e in the same timeslot. For a given k-coloring f (not necessarily *proper*), let z_f be the total edge-weight of those edges whose endpoints are assigned the same color. Then the objective is to find a k-coloring f for which z_f is as small as possible.

[KiYe92] develops a number of heuristics based on coloring the "most difficult" vertices as early as possible, where "most difficult" means having the greatest impact on the remaining vertices to be colored.

[WeYe14] extends the graph model to 2-component edge-weights in order to include a secondary objective – creating compact schedules.

EXERCISES for Section 6.1

6.1.1 The chromatic number of an acquaintance network tells the minimum number of groups into which the persons in that network must be partitioned so that no two persons in a group have prior acquaintance. Calculate the chromatic number of the following acquaintance network.

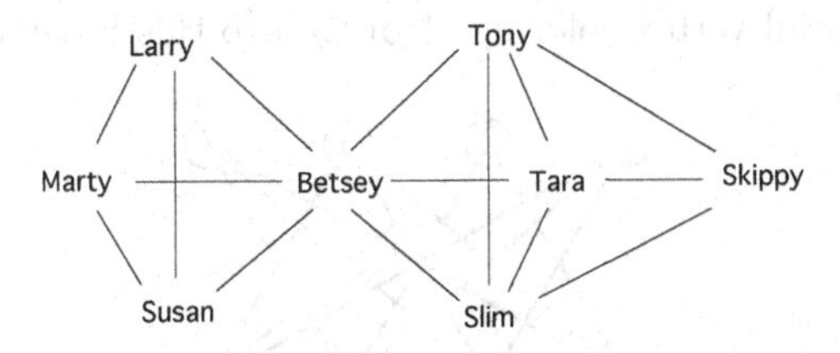

6.1.2 Calculate the number of different radio frequencies needed to avoid interference among the stations in the following configuration in which two stations interfere if they are within 100 miles of each other.

	B	C	D	E	F	G
A	55	110	108	60	150	88
B		87	142	133	98	139
C			77	91	85	93
D				75	114	82
E					107	41
F						123

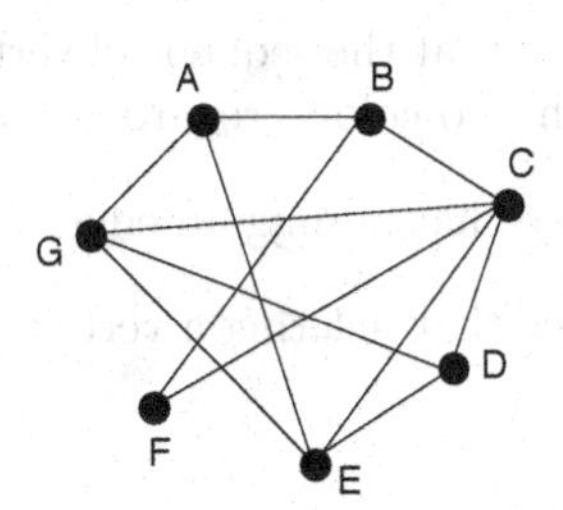

6.1.**3** To the given graph:

i. Assign a minimum vertex-coloring, and prove that it is a minimum coloring.

ii. Apply the largest-degree-first heuristic algorithm.

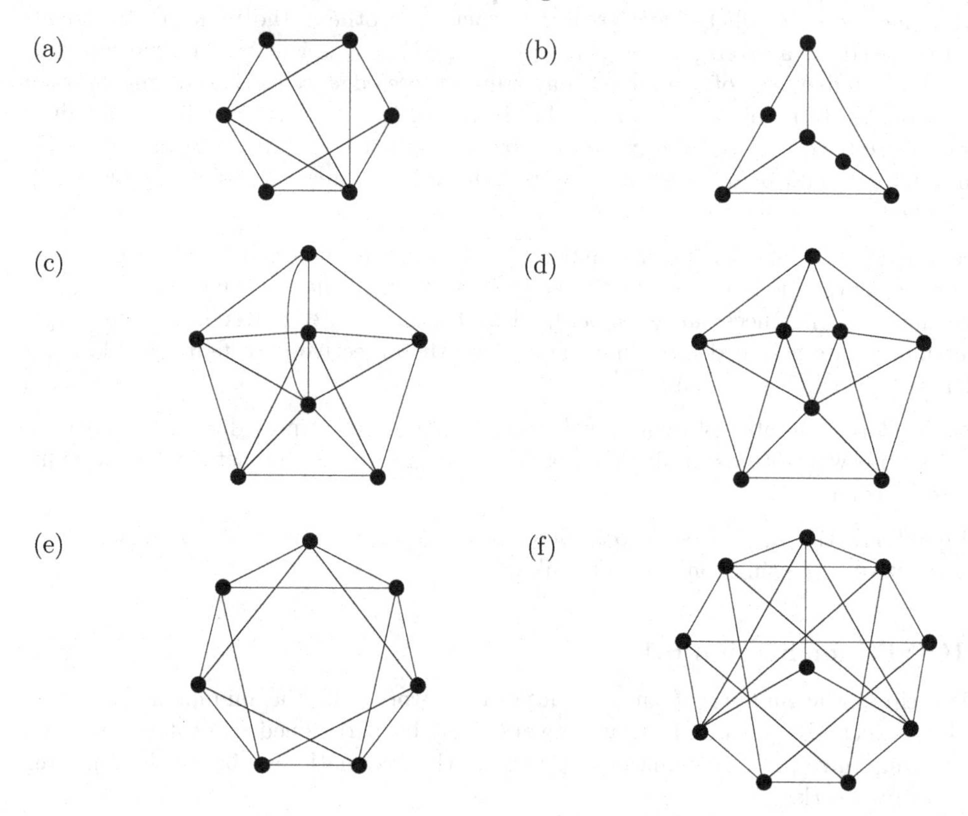

6.1.**4** Apply the sequential vertex-coloring algorithm to this bipartite graph.

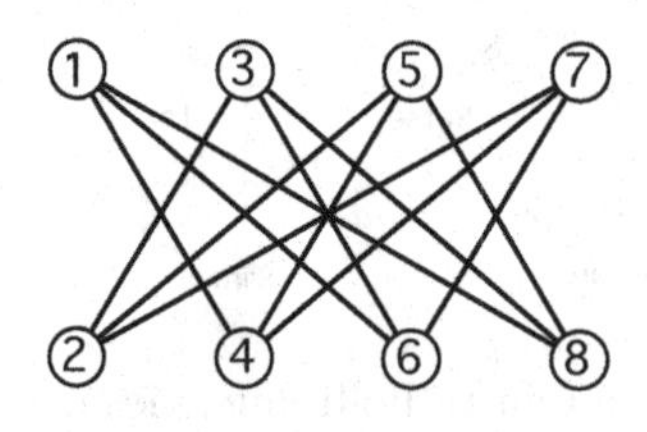

6.1.**5** Give orderings to the vertices of the graphs of Figure 6.1.7 so that the sequential vertex-coloring algorithm will yield 3-colorings.

6.1.**6** Prove that the sequential vertex-coloring algorithm always colors a complete bipartite graph with two colors, regardless of the order of its vertices in the input list.

6.1.**7** Prove that adding an edge to a graph increases its chromatic number by at most one.

6.1.**8** Prove that deleting a vertex from a graph decreases the chromatic number by at most one.

6.1.9 Apply the sequential vertex-coloring algorithm to this graph, with vertices in order of ascending index.

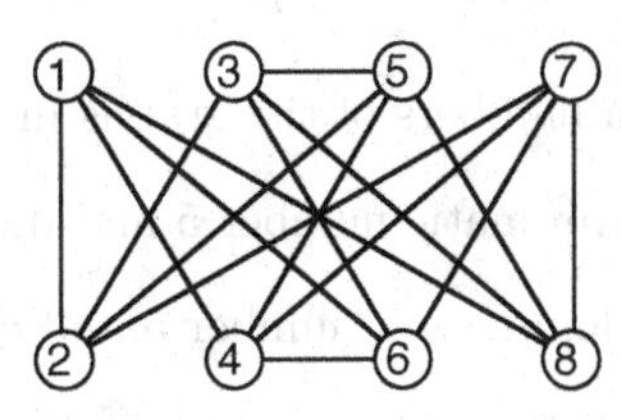

6.1.10 Apply the largest-degree-first heuristic to the graph of Exercise 6.1.9.

6.1.11 Prove Proposition 6.1.21. (Hint: Use Proposition 6.1.6.)

6.1.12 (a) Label the vertices of the following 3-chromatic graph so that the sequential vertex-coloring algorithm uses 4 colors.

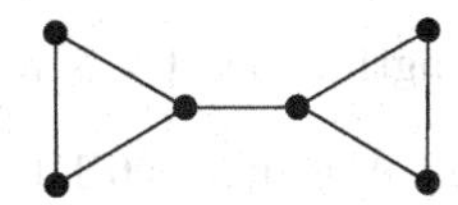

(b) Label the vertices of the following 3-chromatic graph so that the sequential vertex-coloring algorithm uses 5 colors.

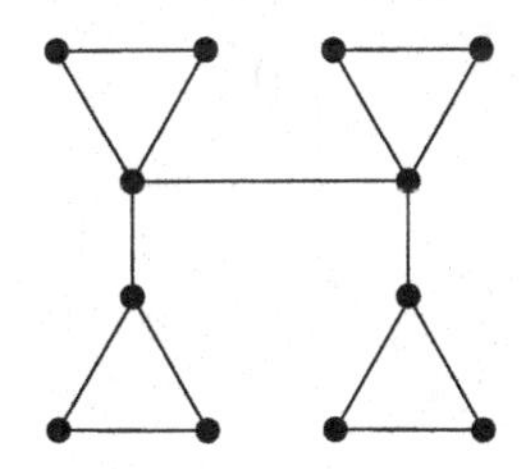

6.1.13 Give an inductive argument, based on the strategy used in Exercise 6.1.26, to show that the number of colors used by the sequential vertex-coloring algorithm can be arbitrarily larger than the chromatic number of a graph.

6.1.14 Prove that the wheel W_4 is not chromatically criticial.

6.1.15 Prove that all odd wheels are chromatically 4-critical.

6.1.16 Prove that for any graph G, G is chromatic k-critical if and only if $G + K_1$ is chromatic $(k+1)$-critical.

6.1.17 Prove that $K_7 - C_7$ is not 4-critical.

6.1.18 Prove that the join of two chromatically critical graphs is a chromatically critical graph.

6.1.19 Calculate the independence numbers of the graphs in Exercise 6.1.3.

6.1.20 Describe how to construct a connected graph G with independence number $\alpha(G) = a$ and chromatic number $\chi(G) = c$, for arbitrary values $a \geq 1$ and $c \geq 2$.

DEFINITION: The ***domination number*** of a graph G, denoted $\gamma(G)$ is the cardinality of a minimum set S of vertices such that every vertex of G is either in S or a neighbor of a vertex in S.

6.1.**21** Calculate the domination numbers of the graphs in Exercise 6.1.3.

6.1.**22** Construct a graph with chromatic number 5 and domination number 2.

6.1.**23** Construct a graph with domination number 5 and chromatic number 2.

6.1.**24** Describe how to construct, for arbitrary values $c \geq 2$ and $m \geq 1$, a connected graph G with chromatic number $\chi(G) = c$ and domination number $\gamma(G) = m$.

DEFINITION: A graph G is ***perfect*** if for every *induced subgraph* (§1.3) H in G, the chromatic number $\chi(H)$ equals the clique number $\omega(H)$.

6.1.**25** Prove that every bipartite graph is perfect.

6.1.**26** Prove that an odd cycle of length at least 5 is not perfect.

6.1.**27** [*Computer Project*] Implement Algorithms 6.1.1 and 6.1.2 and compare their results on the graphs of Exercise 6.1.3.

6.1.**28** Find a vertex 2-coloring of the given weighted graph such that the total weight of the edges whose endpoints are assigned the same color is minimized.

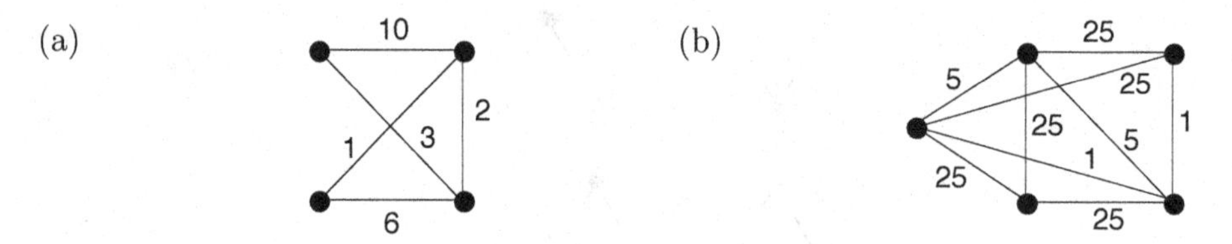

6.1.**29** [*Computer Project*] Use the ideas discussed in the last subsection of this section to design and write a computer program to construct a vertex k-coloring of a weighted graph that tries to minimize the total weight of the edges whose endpoints receive the same color. Test your program on various edge-weighted graphs, including the ones in Exercise 6.1.28, and compare its vertex-colorings to the ones you obtained by hand.

6.1.**30** **Application 6.1.6:** *Examination Scheduling* Suppose that four final exams are to be scheduled in three timeslots. It would be ideal if no two of the exams were assigned the same timeslot, but that would require four timeslots. The table shown below assigns a number to each exam pair, indicating the penalty for scheduling those two exams in the same timeslot. Find an examination schedule that minimizes the total penalty.

$$\begin{array}{c} \\ e_1 \\ e_2 \\ e_3 \\ e_4 \end{array} \begin{array}{c} \begin{array}{cccc} e_1 & e_2 & e_3 & e_4 \end{array} \\ \begin{pmatrix} - & 4 & 16 & 4 \\ 4 & - & 4 & 16 \\ 1 & 16 & - & 4 \\ 4 & 1 & 4 & - \end{pmatrix} \end{array}$$

6.2 LOCAL RECOLORINGS

This section expands on Application 6.1.3 *University Course Scheduling*, where vertices represent courses and the endpoints of each edge represent a pair of courses that cannot be scheduled in the same time slot [KiYe92, WeYe14]. If each timeslot is represented by a different color, then a conflict-free schedule corresponds to a proper coloring of the graph. As remarked early in Section 6.1, the study of vertex-colorings of graphs is customarily restricted to simple, connected graphs. This is reasonable in the context of this section, since there can be no time conflict between a course and itself, and the scheduling of different components of the graph can be done independently.

The problem introduced in this section arises from a naturally occurring situation in the construction of a course schedule. Suppose a schedule has been completed and announced, and it turns out the instructor of one of the courses is unable to teach at the timeslot assigned to that course. In terms of the associated proper coloring, the color assigned to a particular vertex v must change. Ideally, one would prefer to change that color assignment only, but if that is not possible, one would like to make the change with minimum perturbation. In particular, is there a new proper coloring of the graph such that the color assignment of vertex v changes, and the only other allowable changes are to color assignments of vertices that are adjacent to v?

DEFINITION: Let c be a proper k-coloring of a graph G and let v be a vertex of G. A proper k-coloring c' of G is a ***local recoloring of*** V ***with respect to*** c if

i. $c'(v) \neq c(v)$ and

ii. for all $x \in V_G - N[v]$, $c'(x) = c(x)$.

The vertex v is said to be ***locally recolorable with respect to c***.

Example 6.2.1: Consider the proper 3-coloring shown in Figure 6.2.1. Each of the vertices u, v, y, and z is locally recolorable. For example, to show vertex u is locally recolorable, let $c'(u) = 1$ and $c'(w) = c'(x) = 2$. Vertex v can be locally recolored by simply letting $c'(v) = 1$. However, neither w nor x is locally recolorable. For example, since v and x are not adjacent to w, their colors do not change in a local recoloring of w, and hence, the color of u cannot change either. Therefore, the color of w cannot be changed to 2. Similarly, the color of z cannot change as it is adjacent to vertices not adjacent to w with both colors 1 and 2. Therefore, the color of w cannot be changed to 3. Thus, w is not locally recolorable.

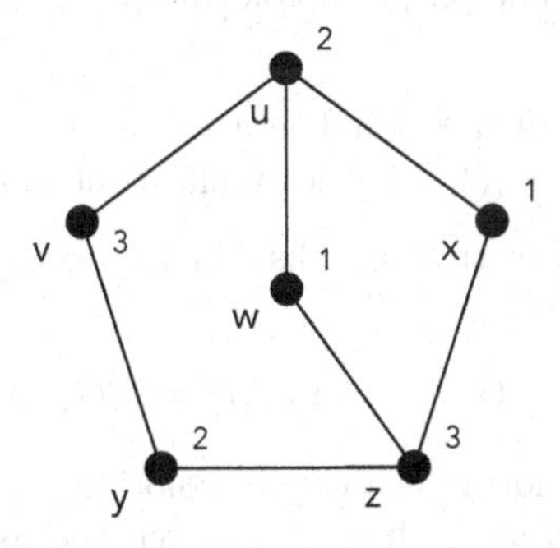

Figure 6.2.1 Neither vertex w nor x is locally recolorable.

Example 6.2.2: Each of the six vertices is locally recolorable with respect to the proper 3-coloring shown in Figure 6.2.2. Vertex w can be locally recolored by letting $c'(w) = 3$ and $c'(z) = 2$. The local recolorability of each of the other five vertices is left as an exercise.

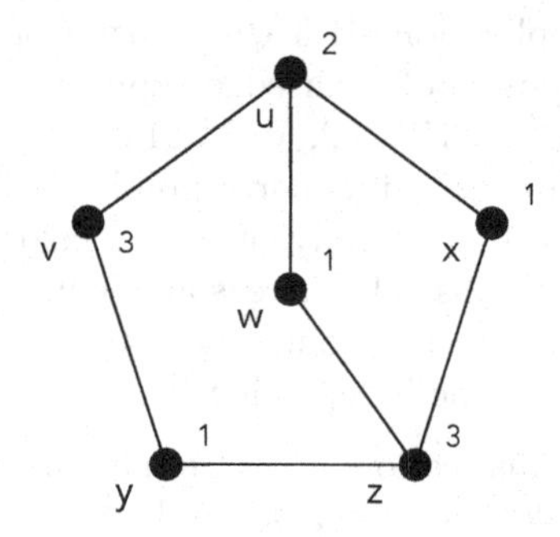

Figure 6.2.2 Each of the six vertices is locally recolorable.

Proposition 6.2.1: *Suppose c is a proper k-coloring of a graph G, where $k \geq 2$. If a vertex v is the only vertex in its color class, then v is locally recolorable with respect to c.*

Proof: Without loss of generality, assume $c(v) = 1$. Define the k-coloring c' by

$$c'(x) = \begin{cases} 2, & \text{if } x = v \\ 1, & \text{if } c(x) = 2 \text{ and } x \in N(v) \\ c(x), & \text{otherwise} \end{cases}$$

Since every neighbor of v with color 2 was changed to color 1, v is not adjacent to any other vertex assigned color 2 by c'. Also, since v is the only vertex that was assigned color 1 by c, all vertices assigned color 1 by c' are nonadjacent. Therefore, c' is a proper k-coloring. It is a local recoloring of v, since the color assigned to v has changed and only vertices in its closed neighborhood have changed. ◊

Robust Colorings

DEFINITION: A proper k-coloring c of a graph G is **robust** if every vertex of G is locally recolorable with respect to c.

Example 6.2.2, continued: The proper 3-coloring in Figure 6.2.1 is not robust while the one in Figure 6.2.2 is.

DEFINITION: A graph G is **k-robust** if it has a robust k-coloring, and the **robust chromatic number** of G, denoted $\chi_R(G)$, is the smallest integer k such that G is k-robust.

The following theorem shows that graphs fall into one of two categories in terms of robust coloring.

Theorem 6.2.2: *For any graph G, either $\chi_R(G) = \chi(G)$ or $\chi_R(G) = \chi(G) + 1$.*

Proof: Since every robust k-coloring is a proper coloring, $\chi_R(G) \geq \chi(G)$, for any graph G. Also, a proper $\chi(G)$-coloring of a graph G may be viewed as a proper $(\chi(G) + 1)$-coloring having a color that is not assigned to any vertex. That unused color can then be used to locally recolor any vertex. Therefore, $\chi_R(G) \leq \chi(G) + 1$. ◊

DEFINITION: A graph G is **χ-robust** if $\chi_R(G) = \chi(G)$.

The next two propositions give examples of graphs in each category.

Proposition 6.2.3: *The following graphs are χ-robust.*

1. *Every complete graph K_n, for $n \geq 2$.*
2. *Every odd cycle.*

Proof:

1. For a proper n-coloring of K_n, any vertex v can be locally recolored by interchanging the colors assigned to v and any one of its neighbors. Therefore, $\chi_R(K_n) = n = \chi(K_n)$.

2. Let C_n be an odd cycle with vertices $v_1, v_2, v_3, \ldots, v_n$, where v_i is adjacent to v_{i+1}, for $i = 1, 2, \ldots, n-1$ and v_n is adjacent to v_1. Without loss of generality, we show that v_1 is locally recolorable. If $c(v_n) = c(v_2)$, then v_1 can be locally recolored by assigning it the color not in $\{c(v_1), c(v_2)\}$. If $c(v_n) \neq c(v_2)$, then v_1 can be locally recolored by assigning it the color $c(v_2)$ and assigning v_2 the color not in $\{c(v_2), c(v_3)\}$. Therefore, $\chi_R(C_n) = 3 = \chi(C_n)$.

Remark: Note that K_1 is not χ-robust, since its chromatic number is 1, but a second color is needed to locally recolor the vertex.

Lemma 6.2.4: *Let e be a critical edge of a graph G. If c is a chromatic coloring of $G - e$, then neither endpoint of e is locally recolorable with respect to c.*

Proof: Let x and y be the endpoints of edge e. Suppose c' is a local recoloring of x with respect to c. Since c and c' are chromatic colorings of $G - e$, $c(x) = c(y)$ and $c'(x) = c'(y)$ by Proposition 6.1.20. However, any local recoloring of x with respect to c would change the color of x but not y, since x and y are not adjacent. Thus, $c'(x) \neq c'(y)$, a contradiction. ◇

Proposition 6.2.5: *The following graphs are not χ-robust.*

1. *Every complete graph minus an edge, $K_n - e$, for $n \geq 2$.*
2. *Every bipartite graph with a vertex of degree greater than 1.*

Proof:

1. This follows from Lemma 6.2.4, since K_n is n-critical.

2. Suppose v is a vertex in a bipartite graph G with neighbors u and w, and let c be a proper 2-coloring of G. Since u and w are both adjacent to v, $c(u) = c(w)$. If c' were a local recoloring of u with respect to c, then $c'(u) \neq c(u)$, and $c'(w) = c(w) = c(u)$, since w is not adjacent to u. Therefore, $c'(v)$ must be equal to either $c'(u)$ or $c'(w)$, and so c' is not a proper 2-coloring of G.

Proposition 6.2.6: *Let c be a proper k-coloring of a graph $G \neq K_1$. If v is a vertex in G such that every vertex in the open neighborhood $N(v)$ has degree less than k, then v is locally recolorable with respect to c.*

Proof: Assume, without loss of generality, that $c(v) \neq 1$. Let N be the set of neighbors of v with color 1. We create a local recoloring of v with respect to c by changing the color of v to 1 and then for each vertex w in N, changing its color from 1 to a color not assigned to any of its neighbors, which is possible since $\deg(w) < k$. ◇

The following corollary is a generalization of Proposition 6.2.3.

Corollary 6.2.7: *If c is a proper k-coloring of a graph $G \neq K_1$ with $\delta_{max}(G) < k$, then c is robust.*

Theorem 6.2.8: *For any simple connected graph $G \neq K_1$, $\chi_R(G) \leq \delta_{max}(G) + 1$.*

Proof: By Theorem 6.2.2, $\chi_R(G) \leq \chi(G) + 1$.

- If $G = K_n$, with $n \geq 2$, then by Proposition 6.2.3,
$$\chi_R(G) = \chi(G) = n = \delta_{max}(G) + 1$$
- If $\delta_{max}(G) = 2$, then G is either an odd cycle or is bipartite. By Theorem 6.2.2 and Propositions 6.2.3 and 6.2.5,
$$\chi_R(G) = 3 = \delta_{max}(G) + 1$$
- If G is non-complete and $\delta_{\max}(G) \geq 3$, then
$$\begin{aligned} \chi_R(G) &\leq \chi(G) + 1, \text{ by Theorem 6.2.2} \\ &\leq \delta_{max}(G) + 1, \text{ by Brooks Theorem (Theorem 6.1.25)} \end{aligned}$$
◇

Robust Critical

In Section 6.1, we examined the impact of deleting an edge or a vertex on the chromatic number of a graph. Here we examine the impact on the robust chromatic number.

By Proposition 6.1.21, a k-critical connected graph is one for which every vertex and every edge is k-critical. Although the removal of a vertex from a graph never increases the robust chromatic number, the removal of an edge can, as shown later in this section.

NOTATION: If c is a k-coloring of a graph G and H is a subgraph of G, then $c|_H$ denotes the restriction of c to H.

Lemma 6.2.9: *For any graph G and any vertex v of G, if c is a robust k-coloring of G, then $c|_{G-v}$ is a robust k-coloring of $G - v$.*

Proof: Let w be any vertex in $G - v$. If c' is a local recoloring of w with respect to c, then $c'|_{G-v}$ is a local recoloring of w with respect to $c|_{G-v}$. ◇

Theorem 6.2.10: *For any graph G and any vertex v of G, $\chi_R(G-v) = \chi_R(G)$ or $\chi_R(G-v) = \chi_R(G) - 1$.*

Proof: By Lemma 6.2.9, $\chi_R(G - v) \leq \chi_R(G)$. To show $\chi_R(G - v) \geq \chi_R(G) - 1$, we show that G has a robust $(k + 1)$-coloring, where $k = \chi_R(G - v)$.

Let c be a robust k-coloring of $G - v$. This can be extended to the following proper $(k+1)$-coloring d of G.
$$d(x) = \begin{cases} k+1 & \text{, if } x = v \\ c(x) & \text{, otherwise} \end{cases}$$
To show d is robust, let w be any vertex in G. There are two cases to consider.

Case 1: If w is in $G - v$, then there exists a local recoloring c' of w with respect to c. This can be extended to the following local recoloring d' of w with respect to d.
$$d'(x) = \begin{cases} k+1 & \text{, if } x = v \\ c'(x) & \text{, otherwise} \end{cases}$$

Case 2: If $w = v$, then the following $(k+1)$-coloring of G is a local recoloring of v with respect to d.

$$d'(x) = \begin{cases} 1 & ; \text{ if } x = v \\ k+1 & , \text{ if } d(x) = 1 \text{ and } x \in N(v) \\ d(x) & , \text{ otherwise} \end{cases}$$

$\diamondsuit$

Corollary 6.2.11: *If H is an induced subgraph of G, then $\chi_R(H) \leq \chi_R(G)$.*

The following example shows that deleting an edge can increase the robust chromatic number.

Example 6.2.3: Consider the graph $\Gamma \cong K_2 \times K_3$, shown in Figure 6.2.3.

- Up to symmetry, there is only one proper 3-coloring of Γ. To show Γ is 3-robust, it suffices, due to the symmetry of Γ and its proper 3-coloring, to show that any one vertex is locally recolorable. Figure 6.2.3 shows a local recoloring of the white vertex, and therefore, $\chi_R(\Gamma) = 3$.
- The subgraph $\Gamma - e$, shown in Figure 6.2.4, has two proper 3-colorings, up to symmetry. For each of those colorings, the vertex shown in white cannot be locally recolored with respect to the coloring, and therefore, $\chi_R(\Gamma - e) = 4$.

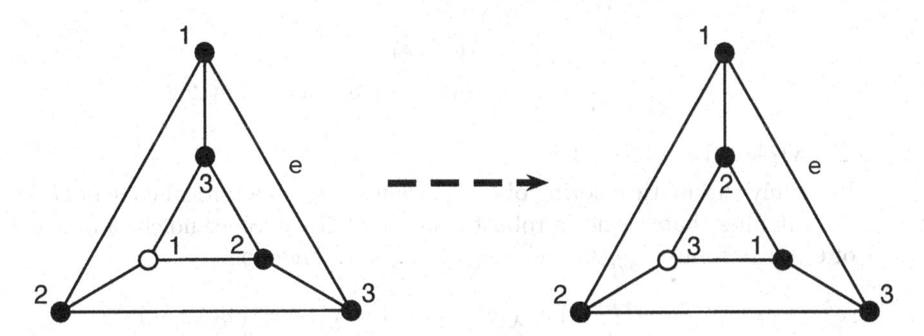

Figure 6.2.3 Every proper 3-coloring of $K_2 \times K_3$ is robust.

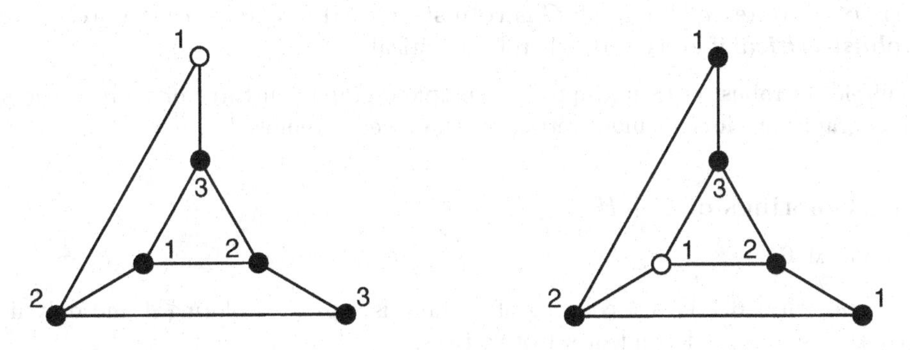

Figure 6.2.4 The two proper 3-colorings of $(K_2 \times K_3) - e$.

Theorem 6.2.12: *For any graph G and any edge e in G,*

$$\chi_R(G) - 1 \le \chi_R(G - e) \le \chi_R(G) + 1.$$

Proof: The following establishes the right-hand inequality.

$$\begin{aligned}\chi_R(G - e) &\le \chi(G - e) + 1 \qquad \text{(by Theorem 6.2.2)}\\ &\le \chi(G) + 1 \qquad \text{(by Proposition 6.1.6)}\\ &\le \chi_R(G) + 1 \qquad \text{(by Theorem 6.2.2)}\end{aligned}$$

To establish the left-hand inequality, there are two cases to consider, by Theorem 6.2.2.

Case 1: $\chi_R(G) = \chi(G)$.

$$\begin{aligned}\chi_R(G) - 1 &= \chi(G) - 1\\ &\le \chi(G - e), \text{by Proposition 6.1.6}\\ &\le \chi_R(G - e), \text{by Theorem 6.2.2}\end{aligned}$$

Case 2: $\chi_R(G) = \chi(G) + 1$. There are two subcases to consider by Proposition 6.1.6.

Case 2a: $\chi(G - e) = \chi(G)$

$$\begin{aligned}\chi_R(G) - 1 &= \chi(G)\\ &= \chi(G - e)\\ &\le \chi_R(G - e), \text{by Theorem 6.2.2}\end{aligned}$$

Case 2b: $\chi(G - e) = \chi(G) - 1$.

If c is any chromatic coloring of $G - e$, then since e is a critical edge of G, Lemma 6.2.4 implies that c is not a robust coloring of $G - e$. Since no chromatic coloring of $G - e$ is robust, $\chi_R(G - e) = \chi(G - e) + 1$. Therefore,

$$\chi_R(G - e) = \chi(G - e) + 1 = \chi(G) = \chi_R(G) - 1 \qquad \diamondsuit$$

Since the removal of a vertex never increases the robust chromatic number, but the removal of an edge may, we do not define *robust critical* for edges.

DEFINITION: A vertex v of a graph G is **robust-critical** if $\chi_R(G - v) < \chi_R(G)$. A graph G is ***robust-critical*** if every vertex is robust-critical.

Examples of robust-critical graphs are complete graphs on three or more vertices and complete graphs on three or more vertices with one edge removed.

Robust Colorings of $G + H$

REVIEW FROM 6.1

- **Proposition 6.1.1**. A k-coloring of a graph is a proper coloring if and only if each color class is an independent set of vertices.
- **Lemma 6.1.7**. For any proper coloring of the join of two graphs, $G + H$, each color class lies entirely in G or in H.
- **Proposition 6.1.8**. $\chi(G + H) = \chi(G) + \chi(H)$.

Lemma 6.1.7 is the key idea in the proof of Proposition 6.1.8 and is also the central idea for writing a similar formula for the robust chromatic number of the join of two graphs. If c is a proper coloring of $G + H$ and c' is a local recoloring of a vertex v in G with respect to c, then the lemma applies to c', since c' is a proper-coloring of $G + H$.

Lemma 6.2.13: *Let c be a proper coloring of $G + H$, and let c' be a local recoloring of a vertex $v \in V_G$ with respect to c. If $c_j \subseteq V_H$ and $c'_j \subseteq V_G$, then $c'_j \subseteq N[v]$, the closed neighborhood of v.*

Proof: Suppose $x \in c'_j$, i.e., $c'(x) = j$. Since $x \in V_G$ and $c_j \subseteq V_H$, $c(x) \neq j$. Thus, the local recoloring c' changed the color of x, and hence, $x \in N[v]$. ◇

Proposition 6.2.14: *Suppose c is a chromatic coloring of the join $G + H$ of graphs G and H, with $G \neq K_1$. A vertex $v \in V_G$ is locally recolorable with respect to c if and only if it is locally recolorable with respect to $c|_G$, the restriction of c to G.*

Proof: A local recoloring of v with respect to $c|_G$ extends to a local recoloring with respect to c, by keeping the color assignments of vertices in H fixed.

Conversely, suppose v is locally recolorable with respect to c, and assume that c assigns the colors $1, 2, \ldots, k$ to vertices in G and the colors $k + 1, k + 2, \ldots, k + m$ to vertices in H, where $k = \chi(G)$ and $m = \chi(H)$.

If v is the only vertex in its color class with respect to $c|_G$, then by Proposition 6.2.1, it is locally recolorable with respect to $c|_G$.

Assume v is not the only vertex in its color class with respect to $c|_G$. Among all local recolorings of v with respect to c, let c' be one with a minimum number of vertices in H assigned colors from $\{1, 2, \ldots, k\}$. If this minimum number is 0, then c' assigns the colors $1, 2, \ldots, k$ to vertices in G and the other m colors to vertices in H, and therefore, $c'|_G$ is a local recoloring of v with respect to $c|_G$.

To show that this minimum number is, in fact, 0, suppose, by way of contradiction, a color i, $1 \leq i \leq k$, is assigned by c' to at least one vertex in H. By Lemma 6.1.7, $c'_i \subseteq V_H$, and so no vertex in G is assigned the color i by c'. Since $\chi(G) = k$, c' must assign a color j, $k + 1 \leq j \leq k + m$ to at least one vertex in G. Consider the following $(k + m)$-coloring c'' of $G + H$, which changes the vertices in c'_i to color j and those in c'_j to color i.

$$c''(x) = \begin{cases} j, & \text{if } c'(x) = i \\ i, & \text{if } c'(x) = j \\ c'(x), & \text{otherwise} \end{cases}$$

Showing that c'' is a local recoloring of v with respect to c will complete the proof, since c'' assigns fewer vertices in H colors from $\{1, 2, \ldots, k\}$, thereby contradicting the minimality of c'.

By Proposition 6.1.1, c'' is a proper coloring, since switching the colors of two color classes does not affect their independence. It remains to show: (i) $c''(v) \neq c(v)$; and (ii) for all $x \in V_G - N[v]$, $c''(x) = c(x)$.

(i) By the definition of c'', there are three cases to consider: $c'(v) = i$, $c'(v) = j$, or neither.

- The first case cannot occur since $c'_i \subseteq V_H$ and $v \in V_G$.
- If $c'(v) = j$, then $c''(v) = i$. Since v is not the only vertex in its color class with respect to $c|_G$, there is a vertex $w \neq v$ with $c(w) = c(v)$. Therefore, for this case

$$\begin{aligned} c''(v) &= i \\ &\neq c'(w) \qquad \text{(since } w \in V_G \text{ and } c'_i \subseteq V_H\text{)} \\ &= c(w) \qquad \text{(since } v \text{ and } w \text{ are not adjacent)} \\ &= c(v) \end{aligned}$$

- If $c''(v) = c'(v)$, then $c''(v) \neq c(v)$, since $c'(v) \neq c(v)$.

(ii) Let $x \in V_G - N[v]$. Since c' is a local recoloring of v with respect to c and $x \notin N[v]$, we have $c'(x) = c(x)$. Thus, to show $c''(x) = c(x)$, it suffices to show $c''(x) = c'(x)$. By definition of c'', this is equivalent to showing (a) $c'(x) \neq i$ and (b) $c'(x) \neq j$.

(a) Since $c'_i \subseteq V_H$ and $x \in V_G$, $x \notin c'_i$, i.e., $c'(x) \neq i$.

(b) Since $c_j \subseteq V_H$ and $c'_j \subseteq V_G$, we have $c'_j \subseteq N[v]$ by Lemma 6.2.13. But $x \notin N[v]$, and hence, $x \notin c'_j$, , i.e., $c'(x) \neq j$, which completes the proof.

$\diamondsuit$

Theorem 6.2.15: *For any graphs $G, H \neq K_1$, $G + H$ is χ-robust if and only if G and H are both χ-robust.*

Proof: Let c be a chromatic $(k+m)$-coloring of $G + H$, where $k = \chi(G)$ and $m = \chi(H)$.

The coloring c is robust

$\iff$ every vertex in $G + H$ is locally recolorable with respect to c

$\iff$ every vertex in G is locally recolorable with respect to $c|_G$ and every vertex in H is locally recolorable with respect to $c|_H$ (by Prop. 6.2.14)

$\iff$ $c|_G$ and $c|_H$ are both robust colorings.

But since $c|_G$ is a chromatic k-coloring of G and $c|_H$ is a chromatic m-coloring of H, it follows that there exists a robust chromatic coloring of $G + H$ if and only if there exist robust chromatic colorings of both G and H. $\diamondsuit$

Corollary 6.2.16: *For any graphs $G, H \neq K_1$,*

$$\chi_R(G+H) = \begin{cases} \chi_R(G) + \chi_R(H) - 1, & \textit{if neither } G \textit{ nor } H \textit{ is } \chi\textit{-robust} \\ \chi_R(G) + \chi_R(H), & \textit{otherwise} \end{cases}$$

Proof: This is an immediate consequence of Theorems 6.2.15 and 6.2.2. $\diamondsuit$

Theorem 6.2.17: *For any graph $G \neq K_1$, $G + K_1$ is χ-robust if and only if G is χ-robust.*

Proof: Let c be a chromatic $(k+1)$-coloring of $G + K_1$.

c is robust $\iff$ every vertex in $G + K_1$ is locally recolorable with respect to c

$\iff$ every vertex in G is locally recolorable with respect to c (by Prop. 6.2.1)

$\iff$ every vertex in G is locally recolorable with respect to $c|_G$ (by Prop. 6.2.14)

$\iff$ $c|_G$ is robust

$\iff$ G is χ-robust (since $c|_G$ is a chromatic coloring of G)

Corollary 6.2.18: *For any graph $G \neq K_1$, $\chi_R(G + K_1) = \chi_R(G) + 1$.*

Proof: If G is χ-robust, then $G + K_1$ is χ-robust and

$$\chi_R(G) + 1 = \chi(G) + 1 = \chi(G + K_1) = \chi_R(G + K_1)$$

If G is not χ-robust, then $G + K_1$ is not χ-robust and

$$\chi_R(G) + 1 = (\chi(G) + 1) + 1 = \chi(G + K_1) + 1 = \chi_R(G + K_1)$$

$\diamond$

Theorem 6.2.19: *Suppose $H = K_1$ or H is χ-robust. For all $G \neq K_1$ or K_2, if $G + H$ is robust-critical, then G is robust-critical.*

Proof: Suppose $G + H$ is robust-critical and w is a vertex in G.

Case 1: $H = K_1$

$$\begin{aligned}\chi_R(G - w) &= \chi_R((G - w) + K_1) - 1, \text{ by Corollary 6.2.18}\\ &= \chi_R((G + H) - w) - 1\\ &= \chi_R(G + H) - 1 - 1, \text{ since } G + H \text{ is robust-critical}\\ &= \chi_R(G) - 1, \text{ by Corollary 6.2.18}\end{aligned}$$

Case 2: H is χ-robust.

$$\begin{aligned}\chi_R(G - w) &= \chi_R((G - w) + H) - \chi_R(H), \text{ by Corollary 6.2.16}\\ &= \chi_R((G + H) - w) - \chi_R(H)\\ &= \chi_R(G + H) - 1 - \chi_R(H), \text{ since } G + H \text{ is robust-critical}\\ &= \chi_R(G) - 1, \text{ by Corollary 6.2.16}\end{aligned}$$

Therefore, G is robust-critical. $\diamond$

Lemma 6.2.20: *If a graph G is χ-robust and robust-critical, then for any vertex in G, $G - v$ is χ-robust.*

Proof: Let v be a vertex in G. By definition, $\chi_R(G - v) \geq \chi(G - v)$. For the reverse inequality,

$$\begin{aligned}\chi_R(G - v) &= \chi_R(G) - 1 && (G \text{ is robust-critical})\\ &= \chi(G) - 1 && (G \text{ is } \chi\text{-robust})\\ &\leq \chi(G - v) && (\text{Proposition 6.1.19})\end{aligned}$$

Therefore, $\chi_R(G - v) = \chi(G - v)$. $\diamond$

Theorem 6.2.21: *Suppose $H = K_1$, $H = K_2$, or H is robust-critical. For all $G \neq K_1$ or K_2, if G is robust-critical, then $G + H$ is robust-critical.*

Proof: Suppose G is robust-critical.

Case 1: $H = K_1$.

If v is the vertex of H, then

$$\begin{aligned}\chi_R((G + H) - v) &= \chi_R(G)\\ &= \chi_R(G + H) - 1 && (\text{by Corollary 6.2.18})\end{aligned}$$

If v is a vertex in G, then

$$\begin{aligned}\chi_R((G+H)-v) &= \chi_R((G-v)+H) && \text{(by definition of join)}\\ &= \chi_R(G-v)+1 && \text{(by Corollary 6.2.18)}\\ &= \chi_R(G)-1+1 && \text{(since } G \text{ is robust-critical)}\\ &= \chi_R(G+H)-1 && \text{(by Corollary 6.2.18)}\end{aligned}$$

Case 2: $H = K_2$

If v is a vertex in H, then

$$\begin{aligned}\chi_R((G+H)-v) &= \chi_R(G+(H-v))\\ &= \chi_R(G+K_1)\\ &= \chi_R(G)+1 && \text{(by Corollary 6.2.18)}\\ &= \chi_R(G)+\chi_R(H)-1 && \text{(by Proposition 6.2.3)}\\ &= \chi_R(G+H)-1 && \text{(by Corollary 6.2.16)}\end{aligned}$$

If v is a vertex in G, then

$$\begin{aligned}\chi_R((G+H)-v) &= \chi_R((G-v)+H)\\ &= \chi_R(G-v)+2 && \text{(by Corollary 6.2.16)}\\ &= \chi_R(G)-1+2 && \text{(since } G \text{ is robust-critical)}\\ &= \chi_R(G+H)-1 && \text{(by Corollary 6.2.16)}\end{aligned}$$

Case 3: H is robust-critical.

Without loss of generality, we can assume v is in G, since both G and H are robust-critical. There are three cases to consider.

Case 3a: Neither G nor H is χ-robust.

$$\begin{aligned}\chi_R((G+H)-v) &= \chi_R((G-v)+H)\\ &= \chi_R(G-v)+\chi_R(H)-1 && \text{(by Corollary 6.2.16)}\\ &= \chi_R(G)+\chi_R(H)-2 && \text{(since } G \text{ is robust-critical)}\\ &= \chi_R(G+H)-1 && \text{(by Corollary 6.2.16)}\end{aligned}$$

Case 3b: Either G or H is χ-robust.

By Lemma 6.2.20, $G-v$ is also χ-robust. Therefore, either $G-v$ or H is χ-robust. Therefore,

$$\begin{aligned}\chi_R((G+H)-v) &= \chi_R((G-v)+H)\\ &= \chi_R(G-v)+\chi_R(H) && \text{(by Corollary 6.2.16)}\\ &= \chi_R(G)-1+\chi_R(H) && \text{(since } G \text{ is robust-critical)}\\ &= \chi_R(G+H)-1 && \text{(by Corollary 6.2.16)}\end{aligned}$$

◇

EXERCISES for Section 6.2

6.2.1 Show that every vertex in the graph of Example 6.2.2 is locally recolorable with respect to the coloring c.

6.2.2 Determine if the given graph is χ-robust. Justify your answer by either giving a robust 3-coloring, or showing that no such coloring exists.

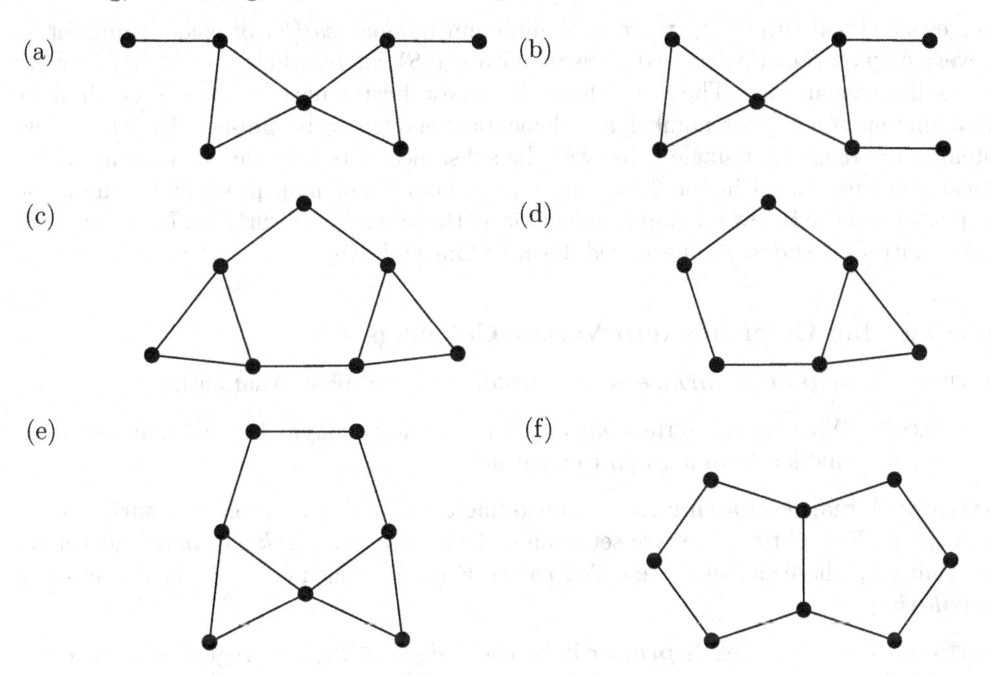

6.2.3 For what values of n is the wheel W_n χ-robust?

6.2.4 Consider the graph $K_7 - C_6$.

(a) What is the robust chromatic number?

(b) Is it possible to remove an edge and decrease the robust chromatic number?

(c) Is it possible to remove an edge and leave the robust chromatic number the same?

(d) Is it possible to remove an edge and increase the robust chromatic number?

6.2.5 By Theorem 6.1.22, the minimum degree of a k-critical graph is at least $k - 1$? Is there an analogous result for k-robust critical graphs? The graphs below may be helpful.

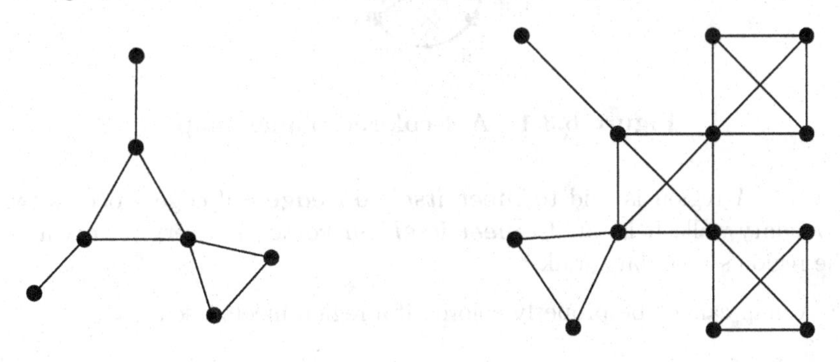

6.3 MAP-COLORINGS

Francis Guthrie, a South African mathematician, found that exactly four colors were needed to color a map of the English counties, so that no two counties that shared a border received the same color. In 1852, he proposed that four colors sufficed for *any* map.

For every closed surface S, there is a minimum number $chr(S)$ of colors sufficient so that every map on S can be colored *properly* with $chr(S)$ colors, which means that no color meets itself across an edge. The methods needed to establish a narrow range of possibilities for that minimum sufficient number are elementary enough to be presented in this book. Tightening the range to a single value was the substance of two of the outstanding mathematical problems solved in the 20th century, the Four-Color map problem for the plane (and sphere) and the Heawood map problem for all the other closed surfaces. In this section, the possibilities for $chr(S)$ are narrowed down to four and five.

Dualizing Map-Colorings into Vertex-Colorings

DEFINITION: A ***map on a surface*** is an imbedding of a graph on that surface.

TERMINOLOGY: Whereas the term *mapping* is a generic synonym for function, the term map refers to a function from a graph to a surface.

DEFINITION: A ***map k-coloring*** for an imbedding $\iota : G \to S$ of a graph on a surface is an assignment $f : F \to C$ from the face-set F onto the set $C = \{1, \ldots, k\}$, or onto another set of cardinality k, whose elements are called *colors*. For any k, such an assignment is called a ***map-coloring***.

DEFINITION: A map-coloring is ***proper*** if for each edge $e \in E_G$, the regions that meet on edge e are colored differently.

DEFINITION: The ***chromatic number of a map*** $\iota : G \to S$ is the minimum number $chr(\iota)$ of colors needed for a proper coloring.

Example 6.3.1: Figure 6.3.1 below shows a proper 4-coloring of a planar map. Observe that in a proper coloring of this map, no two regions can have the same color, since every pair of regions meets at an edge. Thus, this map requires four colors.

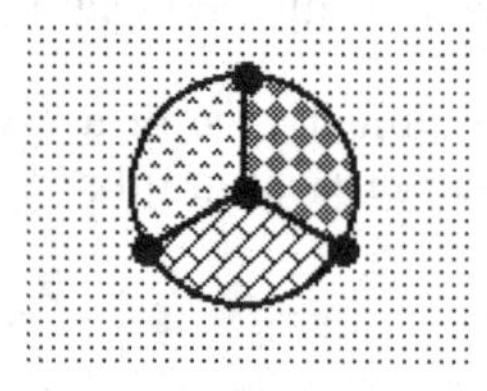

Figure 6.3.1 A 4-colored planar map.

TERMINOLOGY: A region is said to ***meet itself on edge*** e if edge e occurs twice in the region's boundary walk. It is said to ***meet itself on vertex*** v if vertex v occurs more than once in the region's boundary walk.

Remark: A map cannot be properly colored if a region meets itself.

Example 6.3.2: The map in Figure 6.3.2 has no proper coloring, since region f_2 meets itself on an edge (which cannot occur in a geographic map). Observe also that region f_1 meets region f_4 in two distinct edges.

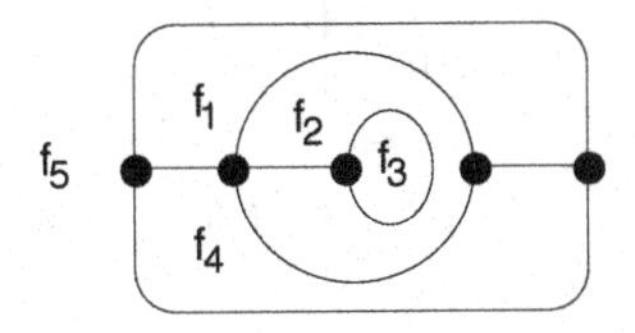

Figure 6.3.2 A map with no proper coloring.

Long ago, it was realized that a coloring problem for the regions of a map on any closed surface can be converted by Poincaré duality (see §4.5) into a vertex-coloring problem for the dual graph. The dual of the map in Figure 6.3.2 is the graph in Figure 6.3.3, with dual vertex v_j corresponding to primal face f_j, for $j = 1, \ldots, 5$. Thus, the self-adjacent region f_2 dualizes to the self-adjacent vertex v_2.

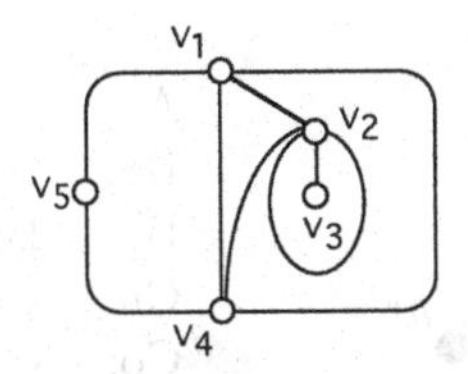

Figure 6.3.3 The dual graph for the map of Figure 9.2.2.

Proposition 6.3.1: *The chromatic number of a map equals the chromatic number of its dual graph.* ◊

Geographic Maps

When the surface is the plane, sometimes a collection of contiguous regions are colored, and the other regions are ignored.

Application 6.3.1: *Political Cartography* In the cartography of political maps, various interesting configurations arise. For instance, France and Spain meet on two distinct borders, one from Andorra to the Atlantic Ocean, the other from Andorra to the Mediterranean Sea, as represented in Figure 6.3.4. Similar configurations occur where Switzerland meets Austria twice around Liechtenstein. Moreover, India and China have a triple adjacency around Nepal and Bhutan. Multiple adjacency of regions does not affect the rules for coloring a map.

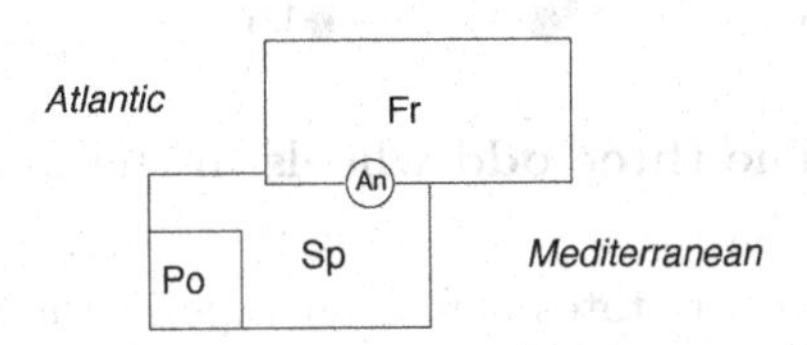

Figure 6.3.4 Double adjacency of France and Spain around Andorra.

Remark: Two faces that meet at a vertex but not along an edge may have the same color in a proper map coloring. Thus, a checkerboard configuration such as the Four Corners, USA, represention in Figure 6.3.5, may be properly colored with only two colors.

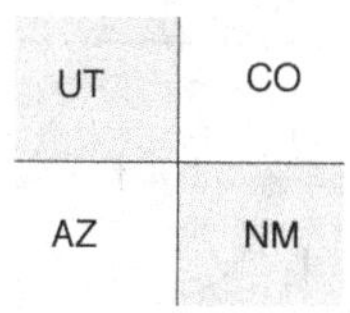

Figure 6.3.5 A proper 2-coloring at Four Corners, USA.

Example 6.3.3: The chromatic number of the map of the countries of South America is equal to 4. In the dual graph, which appears in Figure 6.3.6, there is a 5-wheel with Bolivia as the hub and a 3-wheel with Paraguay as hub. Thus, by Proposition 6.1.17, South America requires at least four colors. We leave it as an exercise to give a 4-coloring of South America (see Exercises).

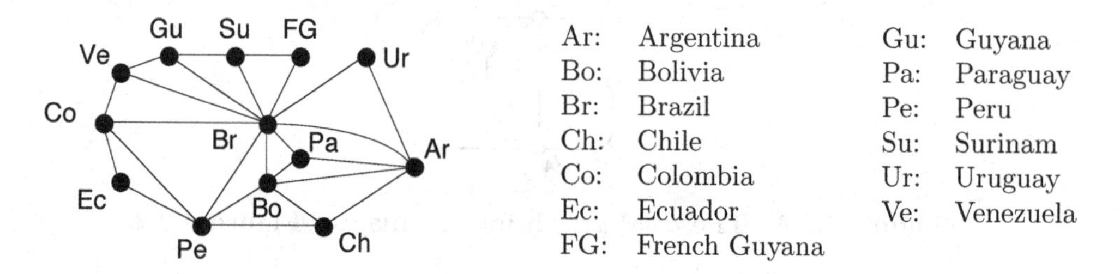

Figure 6.3.6 The dual graph of the map of South America.

Example 6.3.4: The chromatic number of the map of the United States of America is four. The graph of the USA contains three odd wheels, as illustrated in Figure 6.3.7 below. West Virginia and Nevada are each encircled by five neighbors, and Kentucky is encircled by seven neighbors.

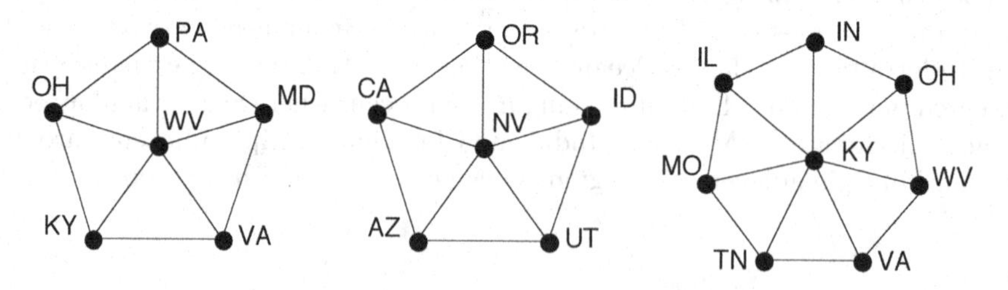

Figure 6.3.7 The three odd wheels in the map of the USA.

Remark: Utah meets five other states across an *edge* in the map, but these five do not quite encircle Utah, since Arizona and Colorado do not meet at an *edge*, even though they do meet at Four Corners. Thus, the Utah configuration is the join $P_5 + K_1$, and not the wheel W_5.

Five-Color Theorem for Planar Graphs and Maps

An early investigation of the Four Color Problem by Kempe [1879] introduced a concept that enabled Heawood [1890] to prove without much difficulty that five colors are sufficient. Heawood's proof is the main concern of this section.

DEFINITION: The ***$\{i, j\}$-subgraph*** of a graph G with a vertex-coloring that has i and j in its color set is the subgraph of G induced on the subset of all vertices that are colored either i or j.

DEFINITION: A ***Kempe i-j chain*** for a vertex-coloring of a graph is a component of the $\{i, j\}$-subgraph.

Example 6.3.5: Figure 6.3.8 illustrates two Kempe 1-3 chains in a graph coloring. The edges in the Kempe chains are dashed.

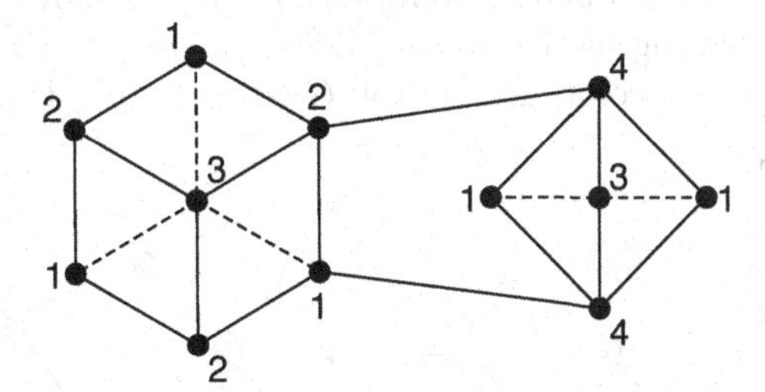

Figure 6.3.8 A graph coloring with two Kempe 1-3 chains.

The next theorem gives an upper bound on the average degree $\delta_{avg}(G)$ of a graph G imbedded in the sphere.

Theorem 6.3.2: *For any connected simple planar graph G, with at least three vertices,* $\delta_{avg}(G) < 6$.

Proof: By Theorem 4.5.9,

$$\frac{2\,|E_G|}{|V_G|} \leq 6 - \frac{12}{|V_G|}$$

Therefore,

$$\begin{aligned}
\delta_{avg}(G) &= \frac{\sum_{v \in V_G} deg(v)}{|V_G|} && \text{definition of average} \\
&= \frac{2\,|E_G|}{|V_G|} && \text{by Theorem 1.1.2} \\
&\leq 6 - \frac{12}{|V_G|} && \text{by Theorem 4.5.9} \\
&< 6
\end{aligned}$$

◇

Theorem 6.3.3: **[Heawood, 1890]** *The chromatic number of a planar simple graph is at most 5.*

Proof: Starting with an arbitrary planar graph G, edges and vertices can be removed until a chromatically critical subgraph is obtained having the same chromatic number as G. Therefore, we may assume, without loss of generality that G is chromatically critical. We may also assume G is connected by the remark in Section §6.1. It suffices to prove that G is 5-colorable.

By Theorem 6.3.2, there is a vertex $w \in V_G$ of degree at most 5 and therefore, by Theorem 6.1.22, $\chi(G) \leq 6$. Since G is chromatically critical, the vertex-deletion subgraph $G - w$ is 5-colorable, by Proposition 6.1.21.

Next, consider any 5-coloring of subgraph $G - w$. If not all five colors were used on the neighbors of vertex w, then the 5-coloring of $G - w$ could be extended to graph G by assigning to w a color not used on the neighbors of w. Thus, we can assume that all five colors are assigned to the neighbors of vertex w. Moreover, there is no loss of generality in assuming that these colors are consecutive in counterclockwise order, as shown on the left in Figure 6.3.9. Consider the $\{2, 4\}$-subgraph shown on the left in Figure 6.3.9 with dashed edges, and let K be the Kempe 2-4 chain that contains the *2-neighbor* of vertex w (i.e., the neighbor that was assigned color 2).

Case 1. Suppose that Kempe chain K does not also contain the 4-neighbor of vertex w. Then colors 2 and 4 can be swapped in Kempe chain K, as shown on the right in Figure 6.3.9. The result is a 5-coloring of $G - w$ that does not use color 2 on any neighbor of w. This 5-coloring extends to a 5-coloring of graph G when color 2 is assigned to vertex w.

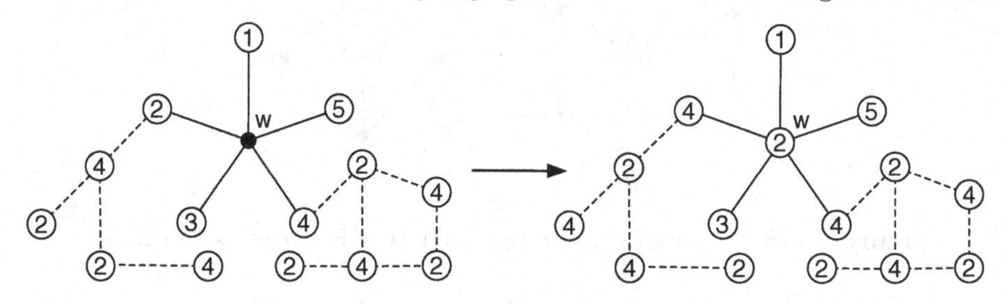

Figure 6.3.9 Swapping colors in a Kempe 2-4 chain.

Case 2. Suppose that Kempe chain K contains both the 4-neighbor and the 2-neighbor of vertex w. Then there is a path in Kempe chain K from the 2-neighbor to the 4-neighbor, as illustrated with a bold broken path on the left in Figure 6.3.10. Appending the edges between vertex w and both these neighbors extends that path to a cycle, as depicted on the right in Figure 6.3.10. By the Jordan Curve Theorem (§4.1), this cycle separates the plane.

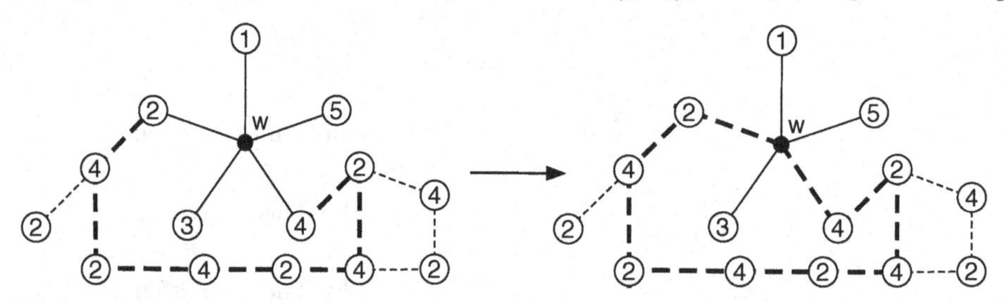

Figure 6.3.10 Extending a path in a Kempe 2-4 chain to a cycle.

Since the 3-neighbor and 5-neighbor of w are on different sides of the separation, it follows that the Kempe 3-5 chain L containing the 5-neighbor cannot also contain the 3-neighbor. Thus, it is possible to swap colors in Kempe chain L and assign color 5 to vertex w, thereby completing a 5-coloring of G. $\diamondsuit$

Theorem 6.3.4: [***Appel and Haken*, 1976**] *Every planar graph is 4-colorable.* ◊

Remark: The proof by Appel and Haken [ApHa76] of the Four Color Theorem is highly specialized, intricate, and long. Following an approach initiated by Heesch, Appel and Haken first reduced the seemingly infinite problem of considering every planar graph to checking a finite, unavoidable set of (over 1900) reducible configurations. Over 1200 hours of computer time were used. Eventually, a more concise proof was derived by Robertson, Sanders, Seymour, and Thomas [RoSaSeTh97].

EXERCISES for Section 6.3

6.3.1 Draw a minimum proper coloring of the following map, excluding the exterior region, and prove it is a minimum coloring.

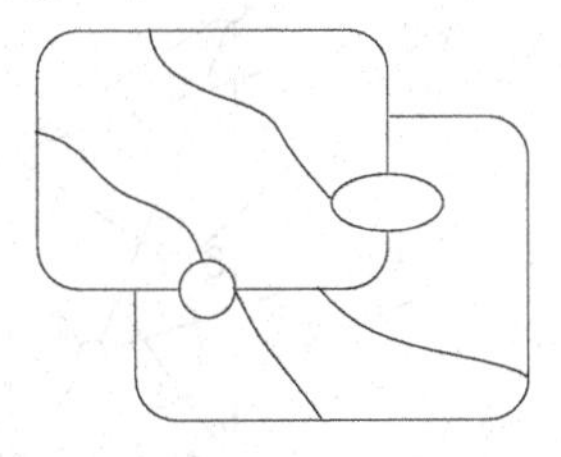

6.3.2 Draw a minimum proper coloring of the map of Exercise 6.2.1, including the exterior region, and prove it is a minimum coloring.

6.3.3 Draw a minimum proper coloring of the following map, excluding the exterior region, and prove it is a minimum coloring.

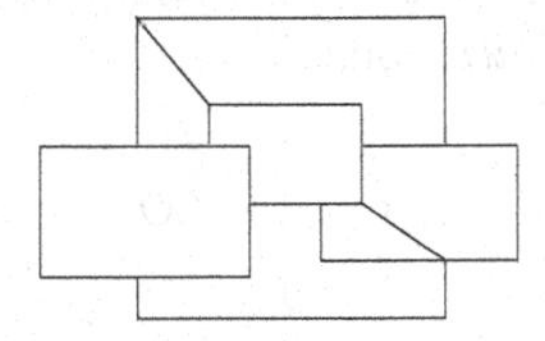

6.3.4 Draw a minimum proper coloring of the map of Exercise 6.2.3, including the exterior region, and prove it is a minimum coloring.

6.3.5 Is it possible for a map in the plane to be 4-chromatic when the exterior region is included, but 3-chromatic when it is excluded? Explain your answer.

6.3.6 Draw the dual graph of the map of Exercise 6.2.3, and give it a minimum vertex-coloring.

6.3.7 Draw a minimum proper coloring of the following map, excluding the exterior region, and prove it is a minimum coloring.

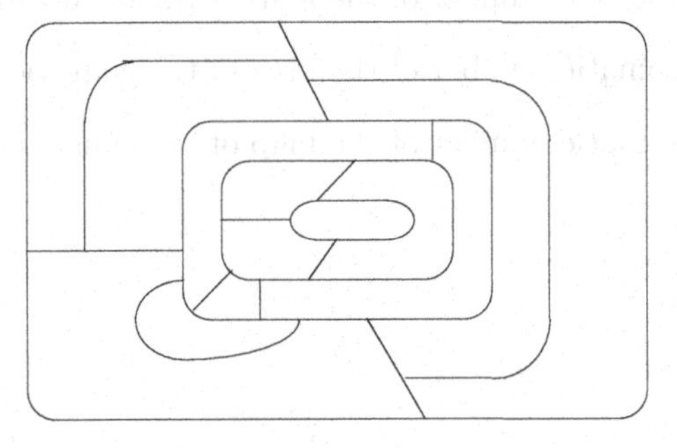

6.3.8 Find the Kempe 1-4 chains in the following graph.

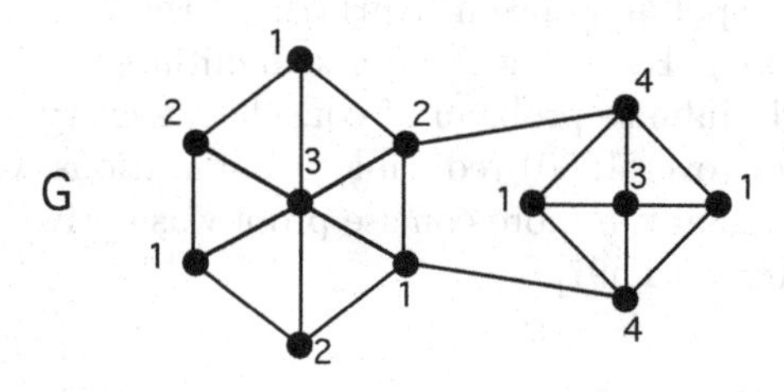

6.3.9 Find the Kempe 1-3 chains in the following graph.

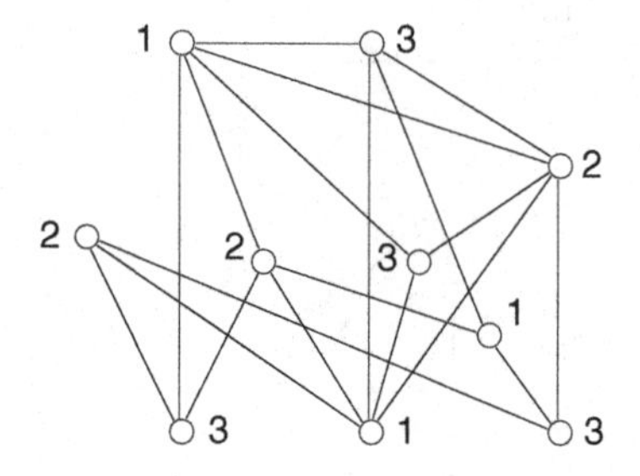

REVIEW FROM §4.1: The *Riemann stereographic projection* provides a correspondence between the plane and the sphere.

6.3.10 Convert the Four Corners map to a sphere map, by redrawing it so that the "infinite lines" all meet at the restored *infinity* point.

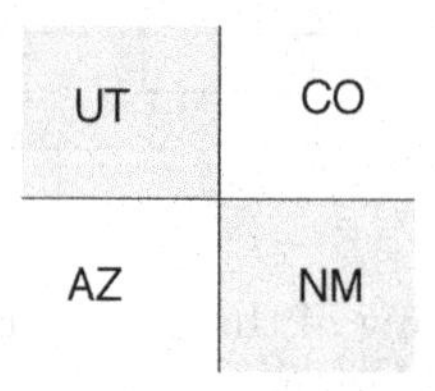

6.3.11 Give a 4-coloring of the map (see Figure 6.3.6) of the countries of South America, or of the dual graph.

6.3.12 Consult a map of Europe, if necessary, and draw a map representing the following countries: France, Belgium, Netherlands, Luxembourg, Germany, and Switzerland. What is the chromatic number of this map?

6.3.13 Calculate the chromatic number of the map of the countries of North America.

6.3.14 Calculate the chromatic number of the map of the countries of Africa.

6.3.15 Calculate the chromatic number of the map of the countries of Asia.

6.3.**16** Calculate the chromatic number of the following map on the torus.

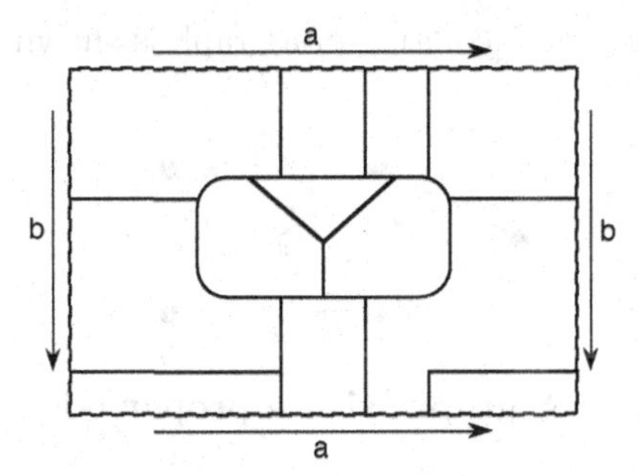

6.3.**17** Calculate the chromatic number of the following map on the torus.

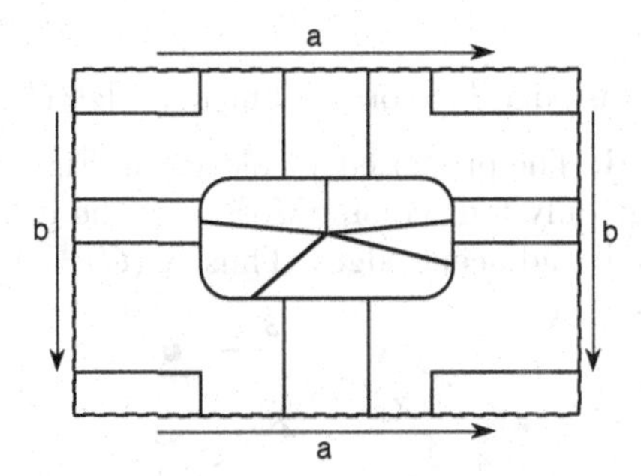

6.4 EDGE-COLORINGS

For certain problems, the most natural graph model for a problem might involve edge-colorings instead of vertex-colorings. Analogous to vertex-colorings, an edge-coloring partitions the edge-set of a graph into color classes. Although the *line-graph* transformation converts an edge-coloring problem into a vertex-coloring problem, the theory of edge-colorings has some special aspects.

The Minimization Problem for Edge-Colorings

REVIEW FROM §1: Two different edges are ***adjacent*** if they have at least one endpoint in common.

DEFINITION: An ***edge k-coloring*** of a graph G is an assignment $f : E_G \to C$ from its edge-set onto a k-element set C whose elements are called *colors* (typically, $C = \{1, 2, \ldots k\}$). For any k, such an assignment is called an ***edge-coloring***.

DEFINITION: An ***edge color class*** in an edge-coloring of a graph G is a subset of E_G containing all the edges of a given color.

DEFINITION: A ***proper edge-coloring*** of a graph is an edge-coloring such that adjacent edges are assigned different colors.

Remark: Whereas multi-edges have no bearing on the proper vertex-colorings of a graph, they have an obvious effect on the proper edge-colorings and cannot be ignored. Graphs with self-loops are excluded from the present discussion.

DEFINITION: A graph is said to be ***edge k-colorable*** if it has a proper edge k-coloring.

Example 6.4.1: A proper edge 5-coloring of a graph is shown in Figure 6.4.1.

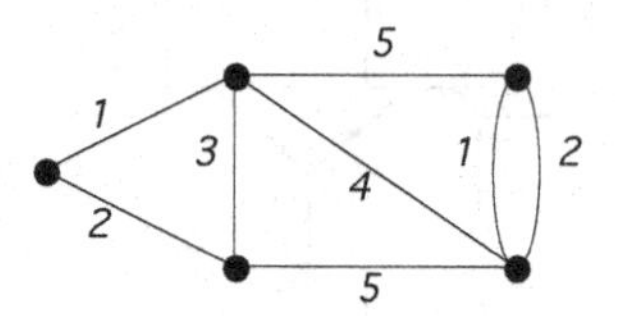

Figure 6.4.1 A graph with a proper edge 5-coloring.

DEFINITION: The ***edge-chromatic number*** of a graph G, denoted $\chi'(G)$, is the minimum number of different colors required for a proper edge-coloring of G. A graph G is edge ***k-chromatic*** if $\chi'(G) = k$.

Thus, $\chi'(G) = k$ if graph G is edge k-colorable but not edge $(k-1)$-colorable.

Example 9.3.1, continued: The proper edge-coloring in Figure 6.4.2 improves on the one in Figure 9.3.1, since it uses only four colors. Moreover, the graph is not edge 3-colorable, since it contains four mutually adjacent edges. Thus, $\chi'(G) = 4$.

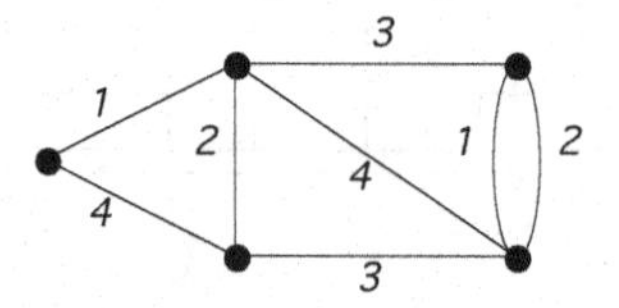

Figure 6.4.2 A proper edge 4-coloring of the graph from Figure 6.4.1.

Modeling Applications as Edge-Coloring Problems

Application 6.4.1: *Circuit Boards* Some electronic devices $x_1, x_2, \ldots, x_n$ are on a board. The connecting wires emerging from each device must be colored differently, so that they can be distinguished. The least number of colors required is the edge-chromatic number of the associated network.

Application 6.4.2: *Scheduling Class Times* A high school has teachers $t_1, \ldots, t_m$ to teach courses $s_1, \ldots, s_n$. In particular, teacher t_j must teach $s_{j,k}$ sections of course s_k. PROBLEM: Calculate the minimum number of time periods required to schedule all the courses so that no two sections of the same course are taught at the same time. SOLUTION: Form a bipartite graph on the two sets $\{t_1, \ldots, t_m\}$ and $\{s_1, \ldots, s_n\}$ so that there are $s_{j,k}$ edges joining t_j and s_k, for all j and k, as shown in Figure 6.4.3.

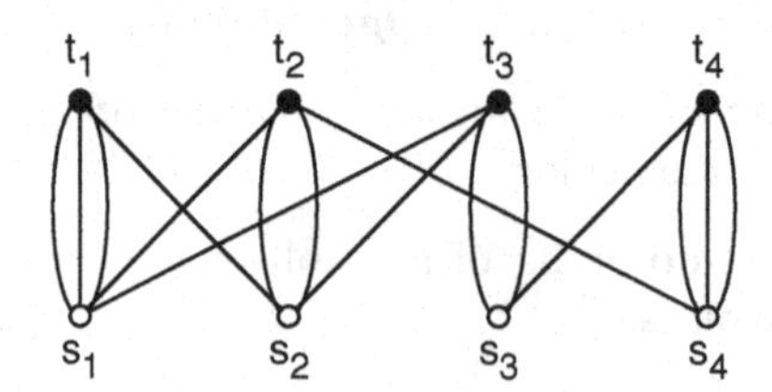

Figure 6.4.3 Representing a scheduling problem by a bipartite graph.

A matching of teachers to courses can be realized in a time period. If each edge-color represents a timeslot in the schedule, then an edge-coloring of the bipartite graph represents a feasible timetable for sections of courses. A minimum edge-coloring, as shown in Figure 6.4.4, uses the smallest number of time periods.

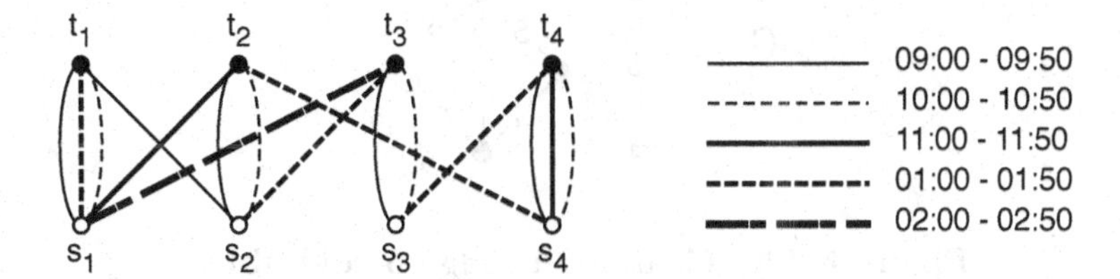

Figure 6.4.4 A minimum proper edge-coloring for the graph of Figure 6.4.3.

Sequential Edge-Coloring Algorithm

There is a sequential edge-coloring algorithm analogous to the sequential vertex-coloring algorithm of §6.1.

DEFINITION: A ***neighbor of an edge*** e is another edge that shares one or both of its endpoints with e.

Algorithm 6.4.1: Sequential Edge-Coloring
Input: a graph with edge list $e_1, e_2, \ldots, e_p$.
Output: a proper edge-coloring f, with positive integers as colors
  For $i = 1, \ldots, p$
    Let $f(e_i) :=$ the smallest color number not used on any of the smaller-subscripted neighbors of e_i.
  Return edge-coloring f.

Basic Principles for Calculating Edge-Chromatic Numbers

Edge-chromatic-number calculations are largely based on a few simple principles, mostly analogous to those used in the two-step vertex-chromatic-number calculations.

- *Upper Bound*: Show $\chi'(G) \leq k$ by exhibiting a proper edge k-coloring of G.
- *Lower Bound*: Show $\chi'(G) \geq k$ by using properties of graph G.

The next three results help in establishing a lower bound for the edge-chromatic number. They are immediate consequences of the definitions.

Proposition 6.4.1: *Let G be a graph that has k mutually adjacent edges. Then* $\chi'(G) \geq k$. ♢

Corollary 6.4.2: *For any graph G,* $\chi'(G) \geq \delta_{\max}(G)$. ♢

Proposition 6.4.3: *Let H be a subgraph of graph G. Then* $\chi'(G) \geq \chi'(H)$. ♢

Example 6.4.2: *The edge 5-coloring of the graph G in Figure 6.4.5 establishes that $\chi'(G) \leq 5$, and the existence of a 5-valent vertex shows that $\chi'(G) \geq 5$, by Corollary 6.4.2. Hence, $\chi'(G) = 5$.*

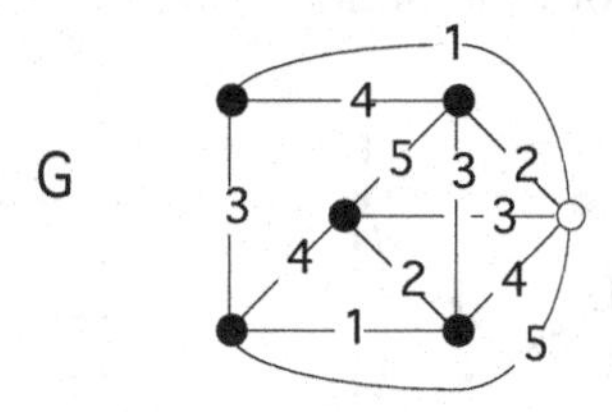

Figure 6.4.5 Graph G is edge 5-colorable.

Matchings

A *matching* is the edge analogue of an independent vertex-set. It can be used to obtain another lower bound for the edge-chromatic number.

DEFINITION: A ***matching*** (or ***independent set of edges***) of a graph G is a subset of edges of G that are mutually nonadjacent.

DEFINITION: A ***maximum matching*** in a graph is a matching with the maximum number of edges.

NOTATION: The cardinality of a maximum matching in a graph G is denoted $\alpha'(G)$, analogous to the independence number $\alpha(G)$ for the vertices.

Remark: It follows immediately from the definition that each color class of a proper edge-coloring of a graph G is a matching of G. The following proposition provides a lower bound on the edge-chromatic number that is based on the size of a maximum matching.

Proposition 6.4.4: *For any graph G, $\chi'(G) \geq \left\lceil \frac{|E_G|}{\alpha'(G)} \right\rceil$.* ◇ *(Exercises)*

The following algorithm constructs a proper edge-coloring by iteratively finding maximum matchings.

Algorithm 6.4.2: Edge-Coloring by Maximum Matching
Input: a graph G.
Output: a proper edge k-coloring f.
  Initialize color number $k := 0$.
  While $E_G \neq \emptyset$
    $k := k + 1$
    Find a maximum matching M of graph G.
    For each edge $e \in M$
      $f(e) := k$
    $G := G - M$ (edge-deletion subgraph)
  Return edge-coloring f.

Edge-Chromatic Numbers for Common Graph Families

It is now possible to derive the edge-chromatic numbers for the same graph families, summarized below in Table 6.4.1, for which the vertex-chromatic numbers were derived in §6.1. The first of the following six results are analogous to their companion results for vertex-chromatic number and are left as exercises.

Proposition 6.4.5: *A graph G has $\chi'(G) = 1$ if and only if $\delta_{\max}(G) = 1$.* ♢ *(Exercises)*

Proposition 6.4.6: *Path Graphs: $\chi'(P_n) = 2$, for $n \geq 3$.* ♢ *(Exercises)*

Proposition 6.4.7: *Even Cycle Graphs: $\chi'(C_{2n}) = 2$.* ♢ *(Exercises)*

Proposition 6.4.8: *Odd Cycle Graphs: $\chi'(C_{2n+1}) = 3$.* ♢ *(Exercises)*

Proposition 6.4.9: *Trees: $\chi'(T) = \delta_{\max}(T)$, for any tree T.* ♢ *(Exercises)*

Proposition 6.4.10: *Hypercube Graphs: $\chi'(Q_n) = n$.* ♢ *(Exercises)*

Proposition 6.4.11: *Wheel Graphs: $\chi'(W_n) = n$, for $n \geq 3$.* ♢ *(Exercises)*

Example 6.4.3: Figure 6.4.6 illustrates an edge 5-coloring of the wheel W_5.

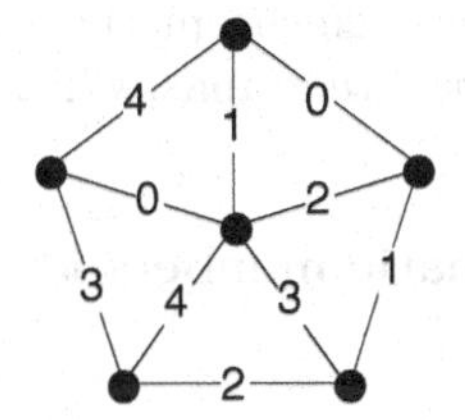

Figure 6.4.6 A proper edge 5-coloring of the wheel W_5.

Proposition 6.4.12: *Odd Complete Graphs: $\chi'(K_n) = n$ for all odd $n \geq 3$.*

Proof: *Upper bound*: Draw the complete graph K_n so that its vertices are the vertices of a regular n-gon, labeled $0, 1, 2, \ldots, n-1$ clockwise around the n-gon (illustrated in Figure 6.4.7 for $n = 7$). Observe that the edge joining vertices 0 and 1 along with all the other edges whose endpoints sum to 1 (mod n) (depicted as bold edges in Figure 6.4.7) form a matching and so they can all be assigned the color 1.

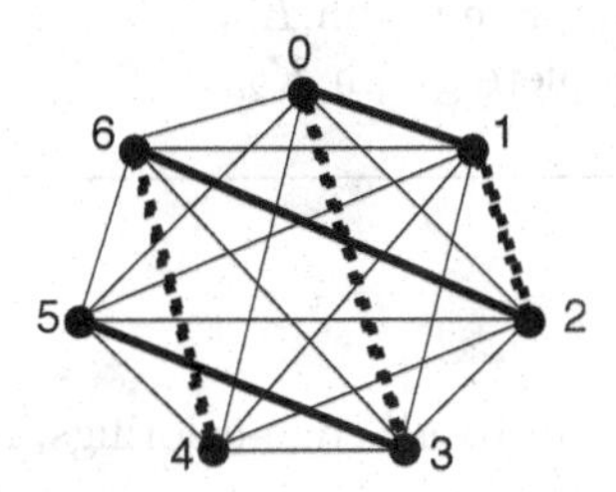

Figure 6.4.7 Two matchings in K_7.

Similarly, the edges whose endpoints sum to 3 (mod n) form a matching (dashed edges in Figure 6.4.7) and can be assigned the color 3. In all, there are n sets $S_1, S_2, \ldots, S_n$, where S_k is the matching consisting of those edges whose endpoints sum to k (mod n). Thus, if each of the edges in set S_k is assigned color k, then a proper edge n-coloring of K_n is obtained.

Lower bound: The size of a maximum matching in K_n with n odd is $\frac{n-1}{2}$. Since K_n contains $\binom{n}{2} = \frac{n(n-1)}{2}$ edges, Proposition 6.4.4 implies that $\chi'(K_n) \geq n$. ♢

Corollary 6.4.13: *Even Complete Graphs:* $\chi'(K_n) = n - 1$ *for all even* n.

Proof: The even complete graph K_n is the join of the odd complete graph K_{n-1} with a single vertex x. The proof of Proposition 6.4.12 constructs a proper edge n-coloring of K_{n-1} in which each edge-color is missing at exactly one vertex. Thus, the edge-coloring of K_{n-1} can be extended to an edge-coloring of K_n by assigning the missing color at each vertex v in K_{n-1} to the edge joining vertex v to vertex x (see Figure 6.4.8 below).

◇

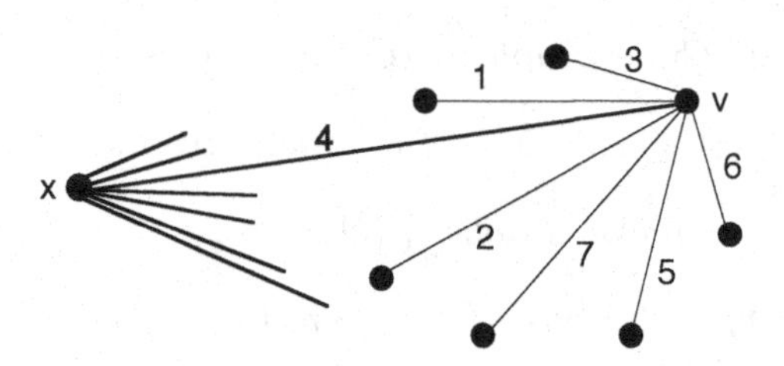

Figure 6.4.8 Extending the edge n-coloring from K_{n-1} to K_n.

Table 6.4.1 below summarizes the edge-chromatic numbers of the common graph families considered here, and also of the bipartite graphs, which we discuss later in this section.

Table 6.4.1: Edge-chromatic numbers for common graph families

Graph G	$\chi'(G)$
graph G with $\delta_{\max}(G) = 1$	1
path graph P_n, $n \geq 3$	2
even cycle graph C_{2n}	2
odd cycle graph C_{2n+1}	3
bipartite graph G	$\delta_{\max}(G)$
tree T	$\delta_{\max}(T)$
cube graph Q_n	n
wheel W_n, $n \geq 3$	n
even complete graph K_{2n}	$n - 1$
odd complete graph K_{2n+1}	n

Chromatic Incidence

The next few definitions pertain to *all* edge-colorings, not just to proper ones.

DEFINITION: For a given edge-coloring of a graph, color i is an ***incident edge-color*** on vertex v if some edge incident on v has been assigned color i. Otherwise, color i is an ***absent edge-color*** at vertex v.

DEFINITION: The ***chromatic incidence*** at v of a given edge-coloring f is the number of different edge-colors incident on vertex v. It is denoted $ecr_v(f)$.

DEFINITION: The ***total chromatic incidence*** for an edge-coloring f of a graph G, denoted $ecr(f)$, is the sum of the chromatic incidences of all the vertices. That is,

$$ecr(f) = \sum_{v \in V_G} ecr_v(f)$$

Example 6.4.4: For the edge-colorings shown in Figure 6.4.9 below, the three different edge-colors are represented by dashed, regular, and bold edges, instead of color numbers. This is to avoid confusion with the chromatic incidence numbers on the vertices. For the edge-coloring f on the left, the total chromatic incidence is 13, and for the edge-coloring g on the right, the total chromatic incidence is 15.

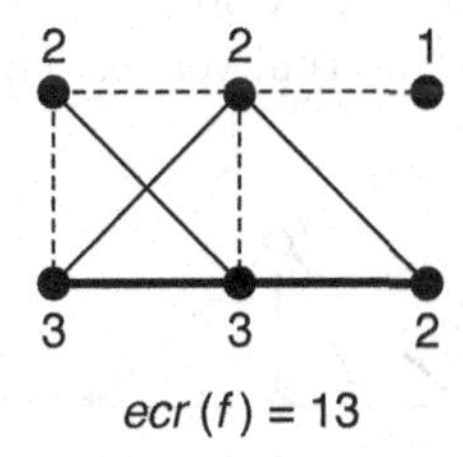

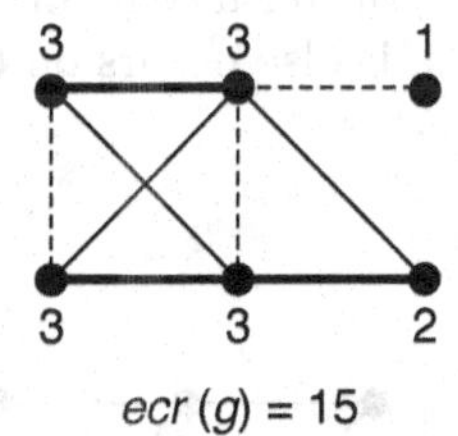

Figure 6.4.9 Total-chromatic-incidence calculations.

The following four assertions are immediate consequences of the definitions.

Proposition 6.4.14: *Let f be any edge-coloring of a graph G. Then for every $v \in V_G$,*

$$ecr_v(f) \leq deg(v)$$ ◊

Corollary 6.4.15: *Let f be any edge-coloring of a graph G. Then*

$$\sum_{v \in V_G} ecr_v(f) \leq \sum_{v \in V_G} deg(v)$$ ◊

Proposition 6.4.16: *An edge-coloring f of a graph G is proper if and only if for every vertex $v \in V_G$,*

$$ecr_v(f) = deg(v)$$ ◊

Corollary 6.4.17: *An edge-coloring f of a graph G is proper if and only if*

$$\sum_{v \in V_G} ecr_v(f) \leq \sum_{v \in V_G} deg(v)$$ ◊

Edge-Coloring of Bipartite Graphs

Deriving a formula for the edge-chromatic number of a bipartite graph G is not quite as easy as for the vertex-chromatic number. Nonetheless, the eventual formula is uncomplicated: $\chi'(G) = \delta_{\max}(G)$. The characterization of bipartite graphs as the graphs without odd cycles is crucial to the derivation.

The following two lemmas establish facts about the chromatic degree that are used in the derivation. They involve edge-colorings that are *not* assumed to be proper. The first lemma makes use of the properties of an *Eulerian graph.*

REVIEW FROM §1.2: An ***Eulerian tour*** in a graph is a closed trail that contains every edge of that graph. An ***Eulerian graph*** is a graph that has an Eulerian tour.

Lemma 6.4.18: *Let G be a connected graph with at least two edges. If G is not an odd-cycle graph, then G has an edge 2-coloring such that both colors are incident on every vertex of degree at least 2.*

Proof: *Case* 1: G is an even cycle. The edge 2-coloring obtained by assigning two edge-colors that alternate around the cycle meets the requirement.

Case 2: G is Eulerian but not a cycle. By Theorem 3.2.11, every vertex in G has even degree and since G is not a cycle, it has a vertex v with degree at least 4. Consider an Eulerian tour that starts (and ends) at v. Assign color 1 to the edges that occur as odd terms in the edge sequence of the tour, and assign color 2 to the even-term edges. Then the two colors are incident at least once on each internal vertex of the tour, since each such vertex is an endpoint of both an odd-term edge and an even-term edge. Moreover, since the start vertex has degree at least 4, it also occurs on the tour as an internal vertex. Thus, both colors are incident on every vertex.

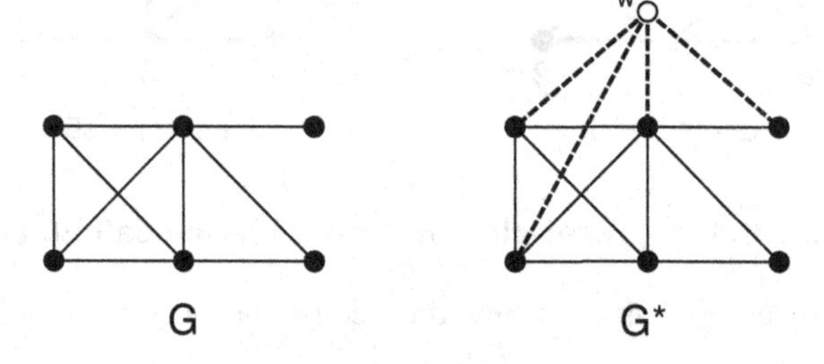

Figure 6.4.10 Constructing the auxiliary graph for Case 3.

Case 3: G is not Eulerian. Construct an auxiliary graph G^* by joining a new vertex w to every odd-degree vertex of G, thereby making each such vertex have even-degree in G^* (see Figure 6.4.10 above). By Corollary 1.1.3, every graph has an even number of odd-degree vertices, so vertex w has even degree. Thus, the auxiliary graph G^* is Eulerian, by Theorem 3.2.11. Now let f be an edge 2-coloring of graph G^*, as specified in Case 2. Then it is easy to verify that the edge-coloring $f|_G$ of f restricted to the edges of graph G is an edge-coloring such that both colors are incident on each vertex of G of degree at least 2. ◇

DEFINITION: In a graph G with a (possibly improper) edge-coloring, a ***Kempe i-j edge-chain*** is a component of the subgraph of G induced on all the i-colored and j-colored edges.

Example 6.4.5: An edge 4-coloring is shown in Figure 6.4.11, and the two Kempe 1-2 edge-chains are shown as dashed edges.

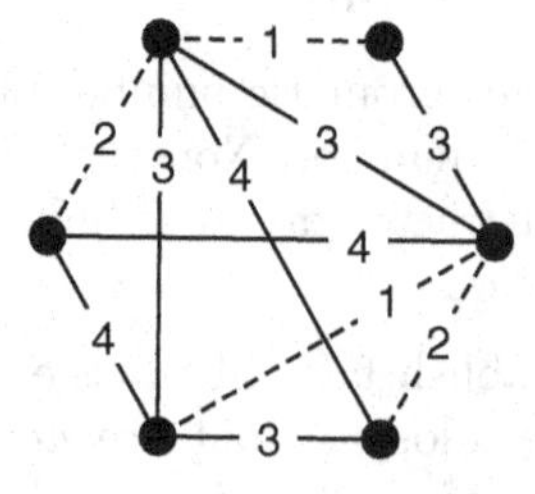

Figure 6.4.11 An edge 4-coloring with two Kempe 1-2 edge-chains.

Lemma 6.4.19: *Let f be an edge k-coloring of a graph G with the largest possible total chromatic incidence. Let v be a vertex on which some color i is incident at least twice and on which some color j is not incident at all. Then the Kempe i-j edge-chain K containing vertex v is an odd cycle.*

Proof: By Lemma 6.4.18, if the Kempe i-j edge-chain K incident on vertex v were not an odd cycle, then we could rearrange edge colors i and j within K so that the chromatic incidence of the coloring of K would be 2 at every vertex. The edge-coloring for G thereby obtained would have higher chromatic incidence at vertex v and at-least-equal chromatic incidence at every other vertex of G. This would contradict the premise that edge-coloring f has the maximum possible total chromatic incidence. ◇

Theorem 6.4.20: [***König*, 1916**] *Let G be a bipartite graph. Then $\chi'(G) = \delta_{\max}(G)$.*

Proof: Let $\Delta = \delta_{\max}(G)$, and, by way of contradiction, suppose that $\chi'(G) \neq \Delta$. Then, by Corollary 6.4.2, $\Delta < \chi'(G)$. Next, let f be an edge Δ-coloring of graph G for which the total chromatic incidence $ecr_G(f)$ is maximum. Since f is not a proper edge-coloring, there is a vertex v such that $ecr_v(f) < deg(v)$ (by Proposition 6.4.16). Thus, some color occurs on at least two edges incident on v. But there are $\Delta - 1$ other colors and at most $\Delta - 2$ other edges incident on v, which means that some other color is not incident on vertex v. It follows by Lemma 6.4.19, that graph G contains an odd cycle, which contradicts the fact that G is bipartite. ◇

Vizing's Theorem

Complementing the lower bound $\chi'(G) \geq \delta_{\max}(G)$ for a simple graph, provided by Corollary 6.4.2, Vizing's theorem provides a sharp upper bound that narrows the range for $\chi'(G)$ to two possible values.

DEFINITION: Let G be a graph, and let f be a proper edge k-coloring of a subset S of the edges of E_G. Then f is a ***blocked partial edge k-coloring*** if for each uncolored edge e, every color has already been assigned to the edges that are adjacent to e. Thus, f cannot be extended to any edge outside subset S.

Example 6.4.6: In Figure 6.4.12, the edge 5-coloring of all but one of the edges of K_5 is blocked, since all five colors have been assigned to the neighbors of the uncolored edge.

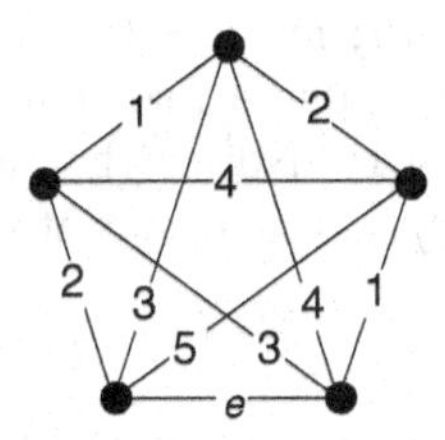

Figure 6.4.12 Attempted edge 5-coloring of K_5 that is blocked at edge e.

Lemma 6.4.21: *Let i and j be two of the colors used in a proper edge-coloring of a graph. Then every Kempe i-j edge-chain K is a path (open or closed).*

Proof: Every vertex of Kempe chain K has degree at most 2 (since the edge-coloring is proper), and, by definition, K is a connected subgraph. ◇

Theorem 6.4.22: **[*Vizing*, 1964, 1965] [*Gupta*, 1966]** *Let G be a simple graph. Then there exists a proper edge-coloring of G that uses at most $\delta_{\max}(G)+1$ colors.*

Proof: To construct a $(\delta_{\max}(G)+1)$-edge-coloring, start by successively coloring edges, using any method (e.g., Algorithm 6.4.1) until the coloring is blocked or complete. If the set of uncolored edges is empty, then the construction is complete. Otherwise, there is some edge e, with endpoints u and v, that remains uncolored. It will be shown that by recoloring some edges, the blocked coloring can be transformed into one that can be extended to edge e. The process can then be repeated until all edges have been colored.

Since the number of colors exceeds $\delta_{\max}(G)$, it follows that at each vertex, at least one of the colors is absent. Let c_0 be a color absent at vertex u, and c_1 a color absent at vertex v. Color c_1 cannot also be absent at vertex u, since if it were, edge e would not have remained uncolored. (For the same reason, color c_0 must occur at vertex v.) So let e_1 be the c_1-edge incident on vertex u, and let v_1 be its other endpoint. Next, let c_2 be a color absent at v_1. If c_2 is also absent at vertex u, then the color of edge e_1 can be changed from c_1 to c_2, thereby permitting the assignment of color c_1 to edge e, as illustrated in Figure 6.4.13. A missing color c at a vertex is indicated by placing c alongside that vertex. Several missing colors may be grouped in braces.

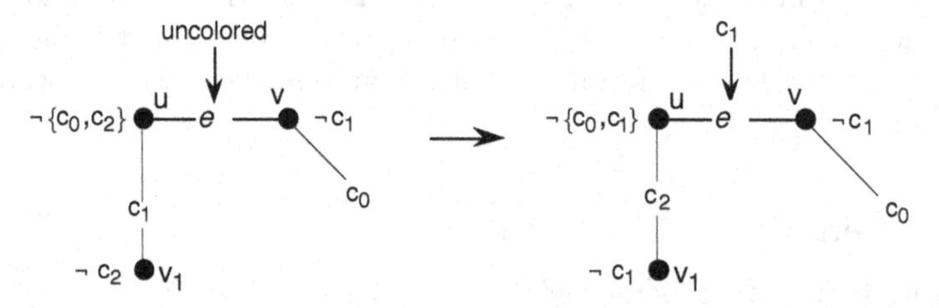

Figure 6.4.13 Extending an edge-coloring to edge e by recoloring edge e_1.

If color c_2 does occur at vertex u, then let e_2 be the c_2-edge incident on vertex u, let v_2 be its other endpoint, and let c_3 be a color absent at vertex v_2. Continue iteratively in this way, so that at the jth iteration, e_j is the c_j-edge incident on vertex u, v_j is its other endpoint, and c_{j+1} is the color absent at vertex v_j. Let ℓ be the smallest j such that vertex v_l has a missing color c_{l+1} and that c_{l+1} is also absent at vertex u or is one of the colors in the list $c_1, \ldots, c_l$ (such an l exists, since the set of colors is finite).

Case 1: Color c_{l+1} is absent at both vertex v_l and vertex u. *Color Shift.*
Then perform the following *color shift*: for $j = 1, \ldots, l$, change the color of edge e_j from c_j to c_{j+1}. This releases color c_1 from edge e_1, so that it can be reassigned to edge e. The color shift is illustrated in Figure 6.4.14. Notice that it maintains a proper edge-coloring, because, by the construction, color c_{j+1} was absent at both endpoints of edge e_j before the shift.

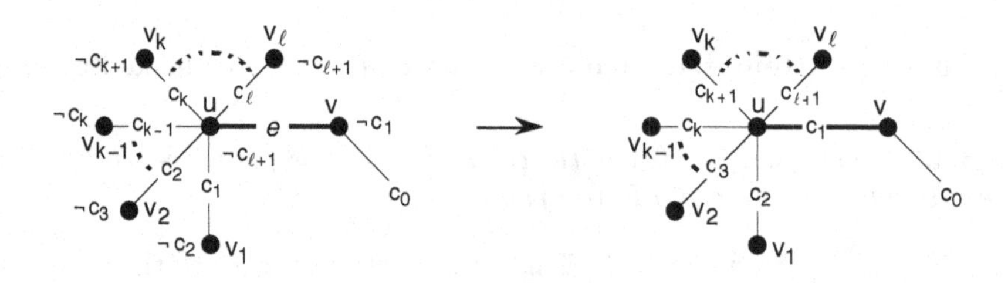

Figure 6.4.14 Case 1: *color shift* to free color c_1 for edge e.

Case 2: Color $c_{l+1} = c_k$, where $1 \leq k \leq l$. *Swap and Shift.*
Let K be the Kempe c_0-c_k edge-chain incident on vertex v_l. By definition, K includes the c_0-edge incident on v_l, but there is no c_k-edge incident on vertex v_l (by definition of l). By Lemma 6.4.21, Kempe chain K is a path, and one end of this path is vertex v_l. There are three subcases to consider, according to where the other end of the path is. In each of the three subcases, the two colors are swapped so that a Case 1 color shift can then be performed.

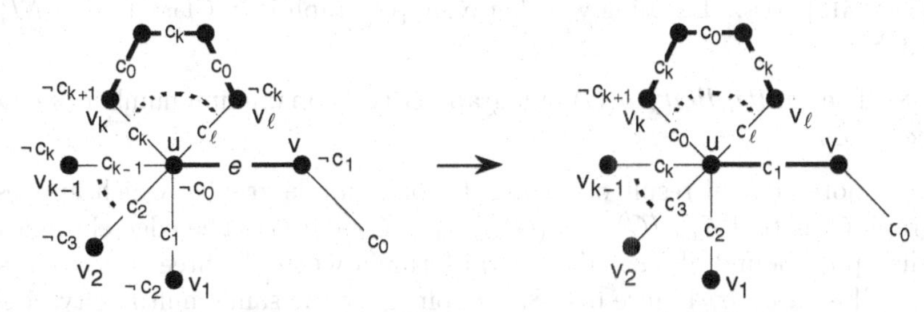

Figure 6.4.15 Case 2a: swap and shift.

Case 2*a*. Path K reaches vertex v_k.
Then swap colors c_0 and c_k along path K. As a result of the swap, color c_k no longer occurs at vertex u. This configuration permits a Case 1 color shift that releases color c_1 for edge e. The swap and shift are illustrated in Figure 6.4.15 above.

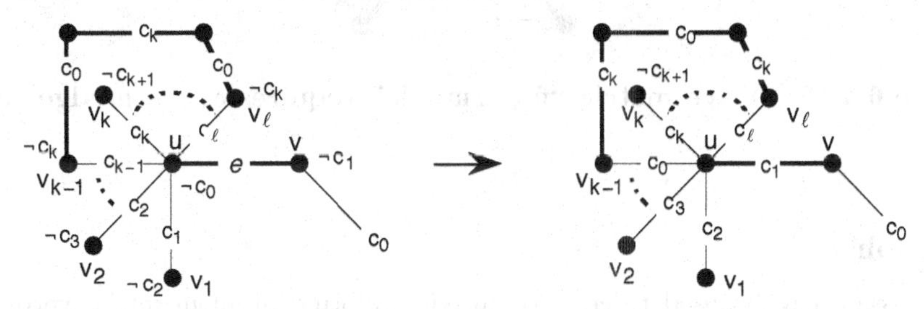

Figure 6.4.16 Case 2b: swap, recolor edge e_1, and then shift.

Case 2*b*. Path K reaches vertex v_{k-1}.
Then swap colors c_0 and c_k along path K. As a result of the swap, color c_0 no longer occurs at vertex v_{k-1}. Thus, edge e_{k-1} can be recolored c_0, as in Figure 6.4.16 above. A color shift can now be performed to release color c_1 for edge e.

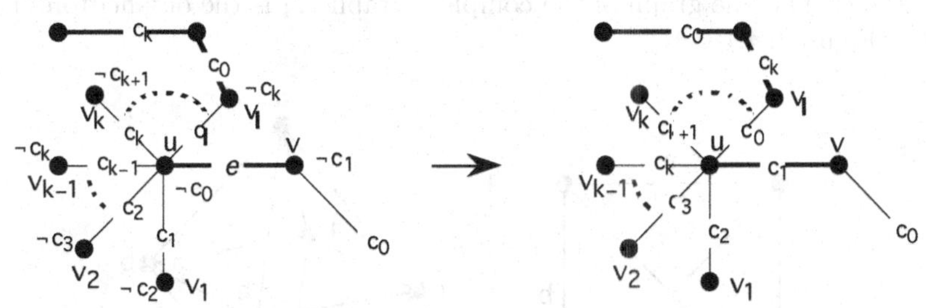

Figure 6.4.17 Case 2c: swap and shift.

Case 2c. Path K never reaches vertex v_{k-1} or vertex v_k.
Since color c_0 does not occur at vertex u, and since color c_k occurs at u only on the edge from v_k, it follows that path K does not reach vertex u. Then swap colors c_0 and c_k along path K, so that color c_0 no longer occurs at vertex v_l. Now perform a Case 1 color shift that releases color c_1 for edge e, as in Figure 6.4.17 above. $\diamond$

Corollary 6.4.23: *Let G be a simple graph. Then either $\chi'(G) = \delta_{\max}(G)$ or $\chi'(G) = \delta_{\max}(G) + 1$.*

Proof: This follows immediately from Vizing's theorem and Corollary 6.4.2. ♢

DEFINITION: ***Class 1*** is the set of non-empty simple graphs G such that $\chi'(G) = \delta_{\max}(G)$. ***Class 2*** is the set of simple graphs G such that $\chi'(G) = \delta_{\max}(G) + 1$.

COMPUTATIONAL NOTE: Deciding whether a simple graph is in Class 1 is an *NP*-complete problem [Ho81].

DEFINITION: The ***multiplicity*** $\mu(G)$ of a graph G is the maximum number of edges joining two vertices of G.

Remark: A more general result of Vizing, beyond simple graphs, which applies to every loopless graph G, is that $\delta_{\max}(G) \leq \chi'(G) \leq \delta_{\max}(G) + \mu(G)$. The edge-chromatic number achieves the upper bound of Vizing's general formula when all three vertex pairs of a "fat triangle", as illustrated by Figure 6.4.18, are joined by the same multiplicity of edges.

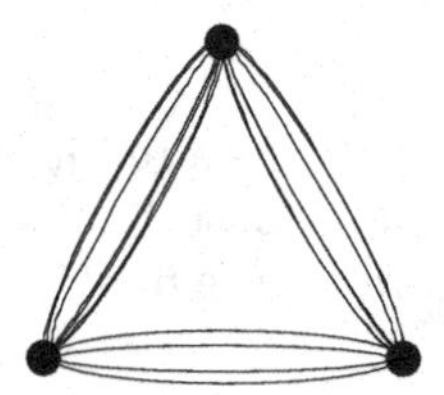

Figure 6.4.18 A symmetric "fat triangle" requires $\delta_{\max} + \mu$ edge colors.

Line Graphs

A line graph can be used to convert an edge-coloring problem into a vertex-coloring problem.

REVIEW FROM §1.1: The ***line graph*** of a graph G is the graph $L(G)$ whose vertices correspond bijectively to the edges of G, and such that two of these vertices are adjacent if and only if their corresponding edges in G have a vertex in common.

Example 6.4.7: The line graph of the complete graph K_4 is the octahedron graph $\mathcal{O}_3$, as illustrated in Figure 6.4.19.

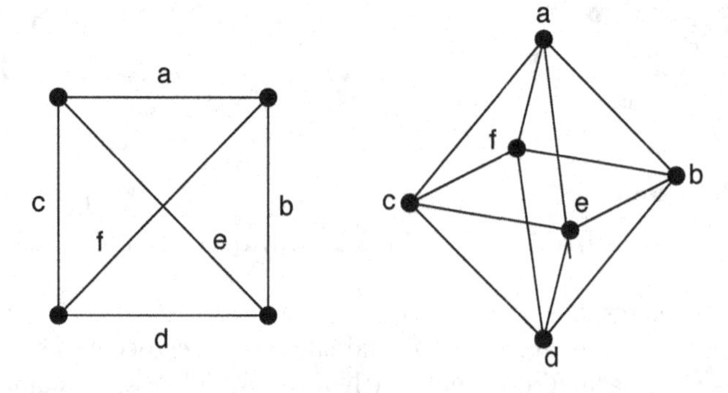

Figure 6.4.19 The complete graph K_4 and its line graph $\mathcal{O}_3$.

Proposition 6.4.24: *The edge-chromatic number of a graph G equals the vertex-chromatic number of its line graph $L(G)$.*

Proof: This follows immediately from the definitions. ◊

Remark: Beineke [Be68] proved that a simple graph G is a line graph of some simple graph if and only if G does not contain any of the graphs in Figure 6.4.20 as an induced subgraph. Since it is an *NP*-complete problem to decide this subgraph problem, much of the theory of edge-colorings has prospered separately from the theory of vertex-colorings.

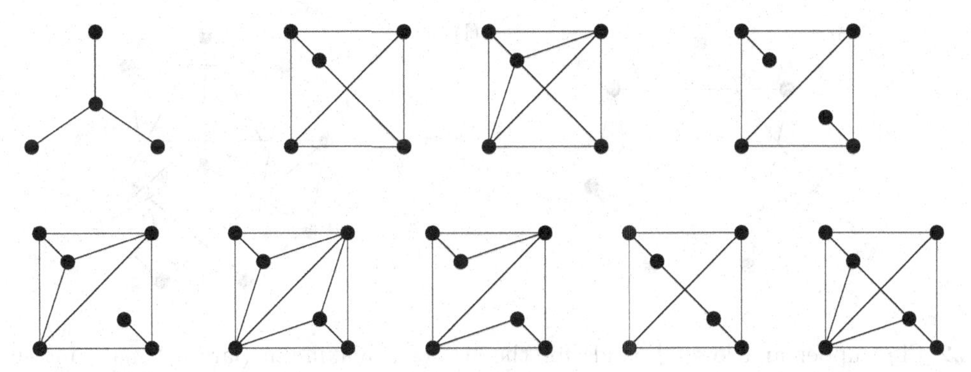

Figure 6.4.20 The nine forbidden induced subgraphs of line graphs.

EXERCISES for Section 6.4

6.4.1 For the given graph, assign a minimum edge-coloring and prove that it is a minimum coloring.

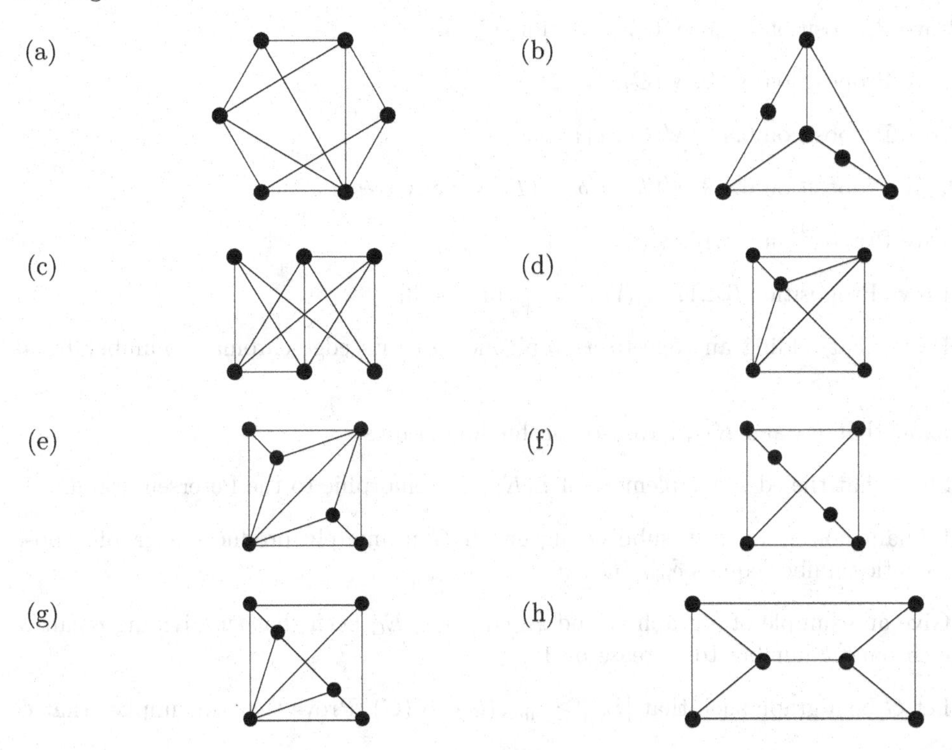

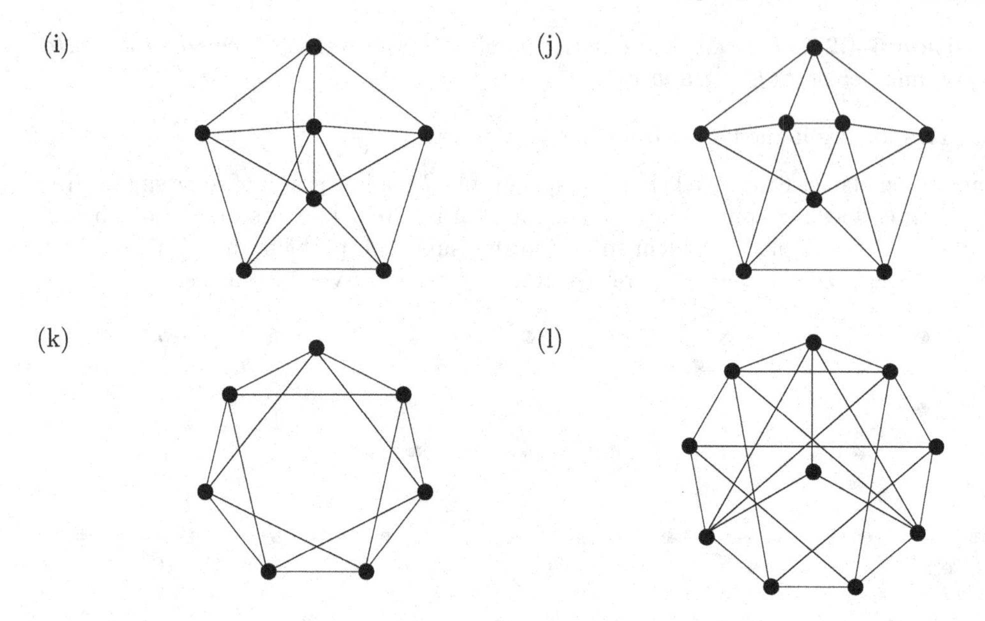

6.4.**2** Find upper and lower bounds for the size of a maximum (largest) matching in an n-vertex connected graph. Then draw three 8-vertex graphs, one that achieves the lower bound, one that achieves the upper bound, and one that achieves neither.

6.4.**3** Prove Proposition 6.4.4. Let $\alpha'(G)$ be the size of a maximum matching in a graph G. Then $\chi'(G) \geq \left\lceil \frac{|E_G|}{\alpha'(G)} \right\rceil$.

6.4.**4** Prove Proposition 6.4.5. A graph G has $\chi'(G) = 1$ if and only if $\delta_{\max}(G) = 1$.

6.4.**5** Prove Proposition 6.4.6. $\chi'(P_n) = 2$, for $n \geq 3$.

6.4.**6** Prove Proposition 6.4.7. $\chi'(C_{2n}) = 2$.

6.4.**7** Prove Proposition 6.4.8. $\chi'(C_{2n+1}) = 3$.

6.4.**8** Prove Proposition 6.4.9. $\chi'(T) = \delta_{\max}(T)$, for any tree T.

6.4.**9** Prove Proposition 6.4.10. $\chi'(Q_n) = n$.

6.4.**10** Prove Proposition 6.4.11. $\chi'(W_n) = n$, for $n \geq 3$.

6.4.**11** Prove that adding an edge to a graph increases its edge-chromatic number by at most 1.

6.4.**12** Show that K_3 and $K_{1,3}$ have isomorphic line graphs.

6.4.**13** Show that the edge-complement of $L(K_5)$ is isomorphic to the Petersen graph.

6.4.**14** Explain how iteratively subdividing graph G ultimately produces a graph whose edge-chromatic number equals $\delta_{max}(G)$.

6.4.**15** Give an example of a graph G and an edge $e \in E_G$ such that subdividing e causes the edge-chromatic number to increase by 1.

6.4.**16** Let G be a graph such that $|E_G| \geq \delta_{\max}(G) \cdot \alpha'(G)$. Prove that this implies that G is in Class 2.

6.4.17 Let G be a regular graph with an odd number of vertices. Prove that G is in Class 2.

6.4.18 Let G be a 3-regular graph with a cut-edge. Prove that G is in Class 2.

6.4.19 Give an example of a graph for which Algorithm 6.4.1 does not produce a minimum edge-coloring.

6.4.20 [*Computer Project*] Implement Algorithms 6.4.1 and 6.4.2 and compare their results on the graphs in Exercises 6.4.1 through 6.4.12.

6.5 FACTORIZATION

We observe that an edge-coloring of a graph is a partitioning of the edge-set into cells of mutually nonadjacent edges, and we now finish this chapter by considering an additional topic concerned with partitioning an edge-set, called *factorization.* In a factorization, each cell of the partition of the edge-set induces a spanning subgraph. We are able to prove a classical result of W. Tutte.

Factors

DEFINITION: A ***factor*** of a graph is a spanning subgraph.

DEFINITION: A ***factorization*** of a graph G is a set of factors whose edge-sets form a partition of the edge-set E_G.

Example 6.5.1: Figure 6.5.1 shows a factorization of the complete graph K_6 into three spanning paths.

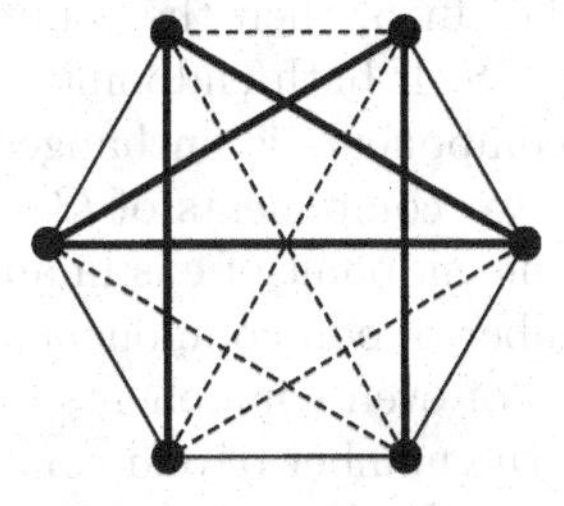

Figure 6.5.1 Factorization of K_6 into three paths.

DEFINITION: A ***k-factor*** of a graph G is a k-regular factor of G.

DEFINITION: A ***k-factorization*** of a graph G is a factorization into k-factors.

Example 6.5.2: Figure 6.5.2 shows two 2-factorizations of the complete graph K_7. The factors of factorization (a) are all spanning cycles. In factorization (b), two factors are spanning cycles, but the factor represented by bold dashes is the union of a 3-cycle and a 4-cycle.

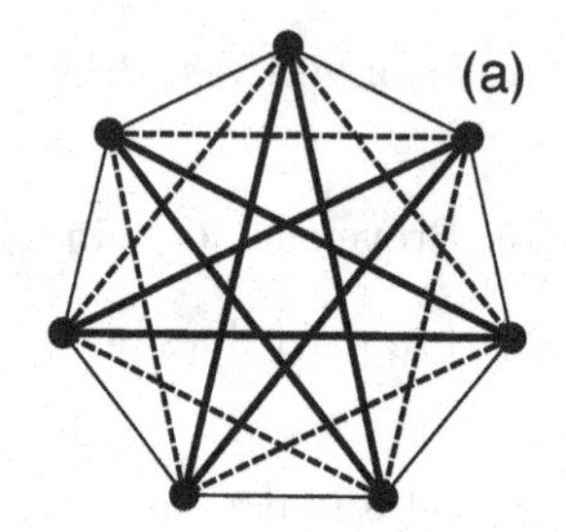

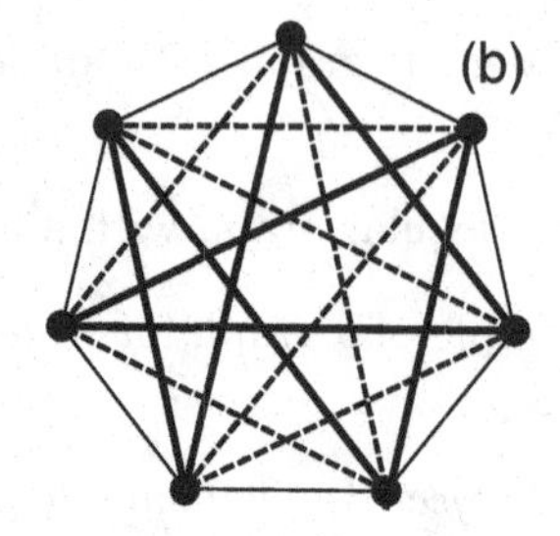

Figure 6.5.2 Two factorizations of K_7 into 2-factors.

TERMINOLOGY: A 1-factor of a graph G is also called a ***perfect matching***. Two vertices are said to be ***matched*** with respect to a 1-factor if they are the endpoints of an edge in the 1-factor.

Tutte's 1-Factor Theorem

Plummer [Pl04] characterizes Tutte's 1-Factor Theorem as the most influential theorem in the study of 1-factors. The definitions and lemmas that precede Tutte's theorem help to simplify the proof.

DEFINITION: An ***odd component of a graph*** is a component with an odd number of vertices.

DEFINITION: ***Tutte's condition*** on a graph G is that for every subset $S \subset V_G$, the number of odd components of $G - S$ does not exceed $|S|$.

Lemma 6.5.1: *Tutte's condition is preserved under edge addition.*

Proof: Let G be a graph that satisfies Tutte's condition, and let e be an edge added to G between two nonadjacent vertices. To show that $G + e$ satisfies Tutte's condition, let $S \subset V_G$. If either endpoint of e lies in S, then the components of the graph $(G + e) - S$ are exactly the components of $G - S$. If both endpoints of e lie in the same component of $G - S$, then the number of odd components is unchanged. This reduces our consideration to the circumstance that e joins two components of $G - S$. If both endpoints of e are in even components of $G - S$, or if one endpoint of e is in an odd component and the other in an even component, then the number of odd components stays the same, that is, less than or equal to $|S|$ (while the number of even components decreases by 1). If both endpoints of e are in odd components, then the number of odd components decreases by 2 (while the number of even components increases by 1). ◇

Lemma 6.5.2: *Let G be a connected graph with evenly many vertices and evenly many edges, with one vertex v of degree 3, one vertex u of degree 1, and all other vertices of degree 2. Then G has a 1-factor.*

Proof: By Corollary 3.2.12, G has an Eulerian trail from v to u. If we color the edges alternatingly red and blue, starting with red at vertex v, then the trail terminates with a blue edge at u (since there are evenly many edges). We observe that at every vertex except u, both colors are present because by whatever color edge the trail enters a vertex, it leaves by an edge of the other color. (Thus, there are two red edges at v and one blue edge.) It follows that the blue edges form a 1-factor. ◇

Example 6.5.3: A graph that satisfies the premises of Lemma 6.5.2 looks something like a *polygon kite*, as illustrated in Figure 6.5.3. In order for the number of vertices in the graph to be even, the number of edges in the tail and the number of edges in the polygon must have the same parity. The dark edges indicate the 1-factor promised by Lemma 6.5.2.

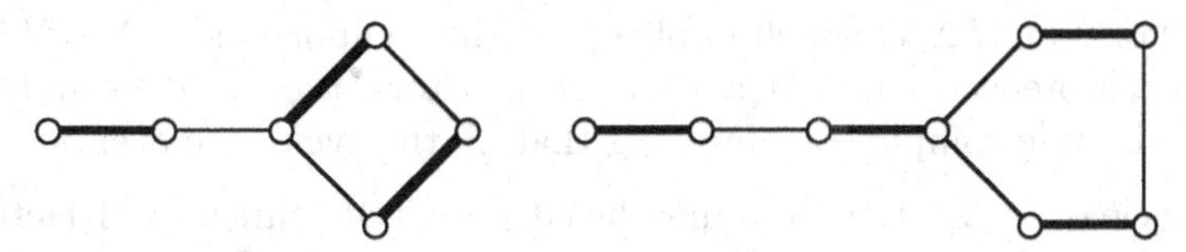

Figure 6.5.3 Two polygon kites with their 1-factors.

DEFINITION: Let M and N be spanning subgraphs of the same graph G. The ***symmetric difference*** $M \Delta N$ is the spanning subgraph of G whose edge-set is $(E_M \cup E_N) - (E_M \cap E_N)$.

Tutte's 1-Factor Theorem characterizes graphs having a 1-factor as those satisfying *Tutte's condition*. The following proof is due to Lovász [L075].

Theorem 6.5.3: [***Tutte's 1-Factor Theorem***] *A nontrivial graph G has a 1-factor if and only if for every subset $S \subset V_G$, the number of odd components of $G - S$ does not exceed $|S|$.*

Proof: ($\Rightarrow$) Suppose that G has a 1-factor. Then in each odd component of $G - S$, there is at least one vertex that is not matched to another vertex within that component, and hence, such a vertex is matched to a vertex of S. It follows that $|S|$ is at least as large as the number of components of $G - S$.

($\Leftarrow$) By way of contradiction, suppose that there exists a graph that satisfies Tutte's condition but has no 1-factor. By adding edges, one at a time, until it is impossible to do so without creating a 1-factor, we obtain a graph H that still satisfies Tutte's condition (by Lemma 6.5.1) and is edge maximal with respect to having no 1-factor. We now pursue the contradiction that H does contain a 1-factor.

Let W be the vertex subset given by

$$W = \{\, w \in V_H \,|\, w \text{ is adjacent to every other vertex of } H\}$$

Case 1. Suppose that every component of $H - W$ is a complete graph. Then we may match the vertices of $H - W$ in pairs, except for one leftover vertex in each odd component of $H - W$. Every vertex in W is adjacent to every one of these leftover vertices (by the definition of set W). Moreover, Tutte's condition implies that $|W|$ is at least as large as the number of odd components of $H - W$. Hence, we may pair each of the leftover vertices from the odd components with a vertex of W. A 1-factor H will exist if the remaining (unpaired) vertices of W can be matched into pairs. These remaining vertices are mutually adjacent (again, by the definition of W) and hence, can be matched if there are evenly many of them. Since we have previously matched evenly many vertices of H, it suffices to show that $|V_H|$ is even. Tutte's condition, with $S = \emptyset$, implies that H has no odd components, which implies that H has evenly many vertices.

Case 2. Suppose that some component of $H - W$ is not a complete graph. In this case, there is a pair of nonadjacent vertices u and v in that component with a common neighbor $y \notin W$. Moreover, since $y \notin W$, it follows that some vertex z of H is not adjacent to y. By the definition of graph H, adding an edge creates a 1-factor. In particular, let M and N be

1-factors in the graphs $H + uv$ and $H + yz$, respectively. We shall show that the symmetric difference $M\Delta N$ has a 1-factor that contains neither uv nor yz, and is, thus, also a 1-factor of H.

Since every vertex of H has degree 1 in M and degree 1 in N, it follows that every vertex of H has degree 0 or 2 in $M\Delta N$, which implies that the components of $M\Delta N$ are cycles and isolated vertices. Moreover, since M and N are 1-factors, the M-edges and N-edges must alternate on each cycle component, implying that all the cycles are even.

Let C be the cycle of $M\Delta N$ that contains the edge uv. (Of course, the 1-factor M contains uv, because we have specified that H has no 1-factor.) If cycle C does not also contain the edge yz, then the union of the set of the N-edges in cycle C with the set of M-edges not in cycle C forms a 1-factor of H.

If cycle C does contain the edge yz, then, using the fact that we chose vertices u and v to have the common neighbor y, we consider the subgraph $J = (C + uy + vy - uv - yz)$ of the graph H. See Figure 6.5.4.

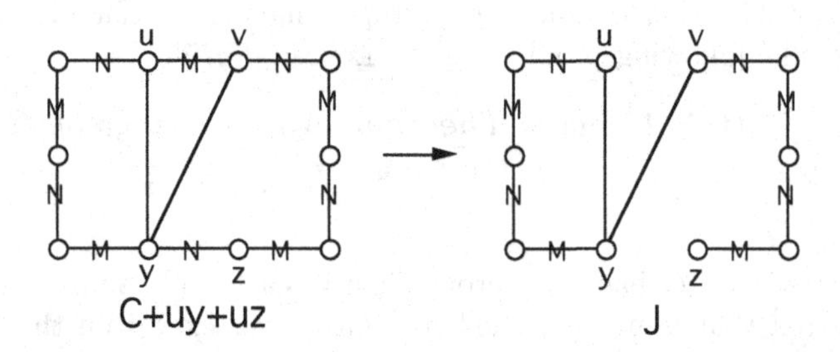

Figure 6.5.4 Constructing the subgraph J.

Since C has evenly many vertices and edges, so does J. Moreover, in subgraph J, vertex y has degree 3, vertex z has degree 1, and all other vertices have degree 2. It follows from Lemma 6.5.2 that the subgraph J has a 1-factor. By combining the 1-factor in subgraph J with the edges of M that are not part of cycle C, we obtain a 1-factor of the graph H.

Petersen's 1-Factor Theorem

NOTATION: The number of odd components of a graph G is denoted $oc(G)$.

Theorem 6.5.4: [*Petersen's 1-Factor Theorem*] *Every 2-edge-connected 3-regular graph G has a 1-factor.*

Proof: By Tutte's 1-Factor Theorem, it suffices to show that G satisfies Tutte's condition. Let S be an arbitrary subset of vertices of G, and let k be the number of edges between S and the odd components of $G - S$. We first observe that

$$k \leq 3\,|S|$$

because each vertex of S has degree 3, since G is 3-regular. For any odd component H (odd H) of $G - S$, let k_H be the number of edges joining H with S. Summing over all the odd components,

$$\sum_{H \text{ odd}} k_H = k$$

The sum of the vertex degrees in H is $3\,|V_H| - k_H$. By Euler's Degree-Sum Theorem (Theorem 1.1.2), applied to the graph H, this sum is even. Since $|V_H|$ is odd, so is $3\,|V_H|$, which implies that k_H is odd. Since G has no cut-edge, we have $k_H > 1$, from which it now follows that $k_H \geq 3$, and in turn, that

$$3oc(G - S) \leq \sum_{H \text{ odd}} k_H = k$$

Thus,

$$3oc(G - S) \leq 3\,|S|$$

which implies Tutte's condition $oc(G - S) \leq |S|$. ♢

Corollary 6.5.5: *Every 2-edge-connected 3-regular graph G has a 2-factor.*

Proof: By Petersen's 1-Factor Theorem, the graph G has a 1-factor. Since G is 3-regular, the edge-complement of that 1-factor is a 2-factor. ♢

Remark: Bäbler ([Bä38]) proved that every $(r-1)$-edge-connected r-regular graph has a 1-factor, and also that every 2-edge-connected $(2r+1)$-regular graph without self-loops has a 1-factor.

Remark: The following two results will be proved in Chapter 10. Their proofs use *Hall's Theorem for Bipartite Graphs*, which is also proved in Chapter 10 with the aid of network flows.

- ***König's 1-Factorization Theorem*** [Kö16] Every r-regular bipartite graph G with $r > 0$ is 1-factorable.
- ***Petersen's 2-Factorization Theorem*** [Pe189I]. Every regular graph G of even degree is 2-factorable.

EXERCISES for Section 6.5

6.5.1 Draw a 3-regular simple graph that has no 1-factor or 2-factor

6.5.2 Draw a 5-regular simple graph that has no 1-factor or 2-factor.

6.5.3 Prove that if two graphs G and H each have a k-factor, then their join has a k-factor.

6.5.4 Draw a connected simple graph that is decomposable into a 2-factor and a 1-factor, but is non-Hamiltonian.

6.5.5 Prove that if a graph has a k-factor, then its Cartesian product with any other graph has a k-factor.

6.5.6 Prove that the complete graph K_{2r} is 1-factorable. Hint: Use induction on r.

6.5.7 Prove that an r-regular bipartite graph G can be decomposed into k-factors if and only if k divides r.

6.5.8 Describe a 2-factorable graph whose edge-complement contains no 2-factor.

6.5.9 Suppose that two graphs with a 1-factor are amalgamated across an edge. Does the resulting graph necessarily have a 1-factor?

6.5.10 Draw two graphs that have no 1-factor, but whose Cartesian product does have a 1-factor.

6.5.11 Draw two graphs that have no 2-factor, but whose Cartesian product does have a 2-factor.

6.6 SUPPLEMENTARY EXERCISES

6.6.1 Let S be the set of connected graphs with 8 vertices and 17 edges.

(a) Prove that no graph in S is 2-chromatic.

(b) Draw a 3-chromatic graph from S, and prove it is 3-chromatic.

(c) Draw a 4-chromatic graph from S, and prove it is 4-chromatic.

(d) Draw a 5-chromatic graph from S, and prove it is 5-chromatic.

(e) Draw a 6-chromatic graph from S, and prove it is 6-chromatic.

(f) Prove that no graph in S is 7-chromatic.

6.6.2 Prove that for any two graphs G and H, $\chi(G+H) = \chi(G) + \chi(H)$.

6.6.3 Draw a simple 6-vertex graph G such that $\chi(G) + \chi(\overline{G}) = 5$, where $\overline{G}$ is the edge-complement of G.

6.6.4 Draw a simple 6-vertex graph G such that $\chi(G) + \chi(\overline{G}) = 7$.

6.6.5 Construct a non-complete graph of chromatic number 6 that is chromatically critical.

6.6.6 Draw a minimum vertex-coloring and a minimum edge-coloring of the given graph and prove their minimality.

(a)

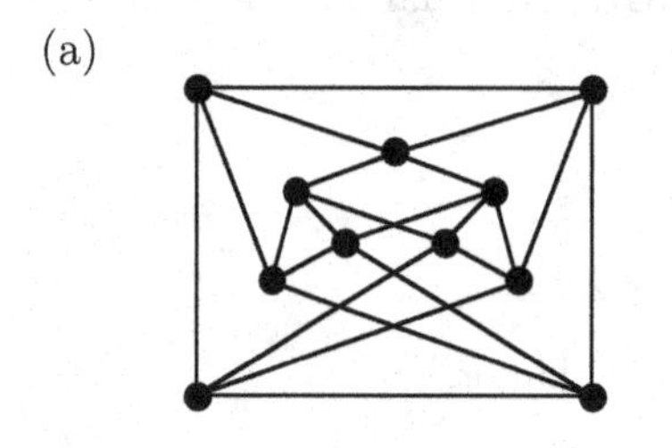

(b)

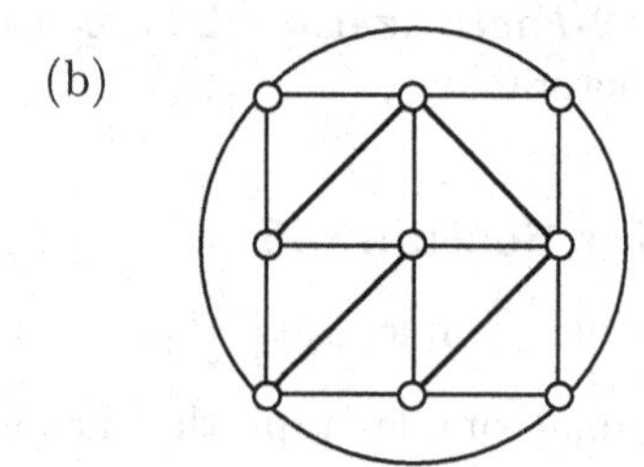

6.6.7 Calculate the vertex chromatic number and edge chromatic number of the graph below. Is every edge critical with respect to the vertex chromatic number? Is every edge critical with respect to the edge chromatic number?

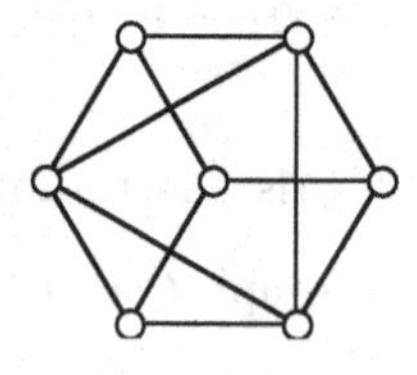

6.6.8 Prove that the minimum chromatic number among all 4-regular 9-vertex graphs is three. (Hint: First prove that chromatic number two is impossible, and then draw a 3-chromatic 4-regular 9-vertex graph.)

6.6.9 Calculate the maximum possible number of edges of a simple 3-colorable planar graph on 12 vertices. Be sure to prove that your number is achievable.

6.6.10 Calculate the chromatic number of the circulant graph $circ(13 : 1, 5)$.

6.6.11 (a) Prove that the graph below has chromatic number 5.

(b) Mark two edges whose removal would make the graph 3-chromatic.

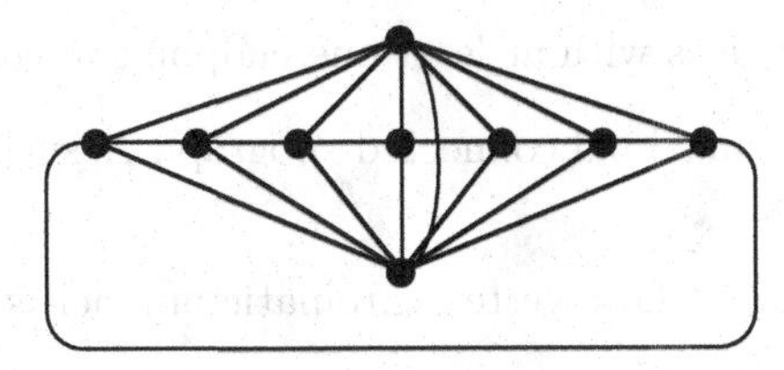

6.6.12 Calculate the chromatic number of the graph $C_3 \times C_3 \times C_3 \times C_3 \times C_3$.

6.6.13 Give a proper 4-coloring of the ***Grötzsch graph*** (a Mycielski graph), shown below. Why must at least three different colors be used on the outer cycle in a proper coloring? Why must at least three different colors be used on the five hollow vertices? (This implies that a fourth color is needed for the central vertex.)

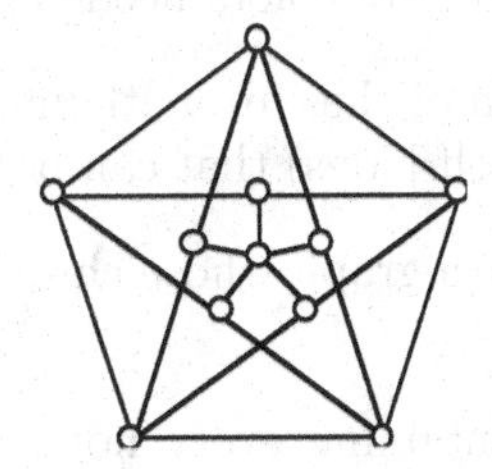

6.6.14 Calculate the chromatic number of the product graph $K_4 \times K_4$.

6.6.15 Among all graphs with 8 vertices and independence number 4, determine the largest possible number of edges and draw such a graph.

6.6.16 Among all graphs with 8 vertices and clique number 4, draw a graph with the largest possible number of edges. Write the number.

6.6.17 Calculate the independence number of $C_5 \times C_5$.

6.6.18 Calculate the clique number of the circulant graph $circ(9 : 1, 3, 4)$.

6.6.19 Calculate the independence number and chromatic number of the given graph.

(a)

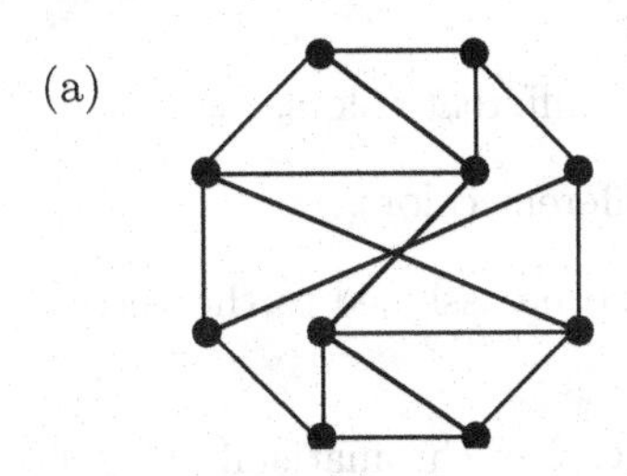

(b)

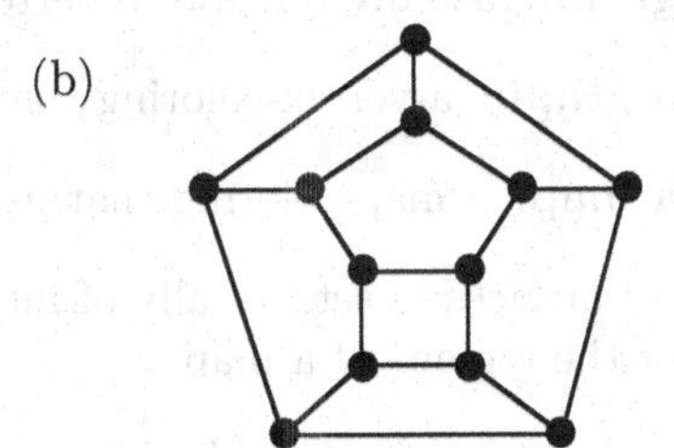

6.6.20 Prove that every Hamiltonian 3-regular graph is 3-edge-colorable.

6.6.21 Draw as many copies of the cube Q_3 as needed, each with a different 1-factor, to give a complete repetition-free list of all the possible 1-factors.

Glossary

adjacent edges: different edges with at least one endpoint in common.

block of a loopless graph: a maximal connected subgraph H such that no vertex of H is a cut-vertex of H.

k-chromatic graph: a graph whose vertex chromatic number is k.

chromatic incidence of an edge-coloring f at a vertex v: the number of different edge-colors present at v, denoted by $ecr_v(f)$.

chromatic number:

—, of a graph G: the minimum number of different colors required for a *proper vertex-coloring* of G, usually denoted by $\chi(G)$.

—, of a map: the minimum number of colors needed for a *proper map-coloring.*

—, of a surface S: the maximum of the chromatic numbers of the maps on S, or equivalently, of the graphs (without self-loops) that can be imbedded in S, denoted by $chr(S)$.

chromatically k-critical graph: a graph whose chromatic number would decrease if any edge were deleted.

class one: the class of graphs containing every non-empty graph whose edge-chromatic number equals its maximum degree.

class two: the class of graphs containing every graph whose edge-chromatic number is one more than its maximum degree.

clique in a graph: a maximal subset of mutually adjacent vertices.

clique number of a graph: the number of vertices in the largest clique.

color class in a vertex-coloring of a graph G: a subset of V_G containing all the vertices of some color.

k-colorable graph: a graph that has a proper vertex k-coloring.

coloring of a graph: usually refers to a vertex-coloring.

k-coloring of a graph: a vertex-coloring that uses exactly k different colors.

k-coloring of a map: a map-coloring that uses exactly k different colors.

colors of vertices or faces: a set, usually of integers $1, 2, \ldots$, to be assigned to the vertices of a graph or the regions of a map.

complete set of obstructions to k-chromaticity: a set $\{G_j\}$ of chromatically $(k+1)$-critical graphs such that every $(k+1)$-chromatic graph contains at least one graph G_j as a subgraph.

domination number $\gamma(G)$ of a graph G: the cardinality of a minimum set S of vertices such that every vertex of G is either in S or a neighbor of a vertex in S.

edge k-chromatic graph G: a graph with $\chi'(G) = k$.

edge-chromatic number $\chi'(G)$ of a graph: the minimum number of different colors required for a proper edge-coloring of a graph G.

edge k-colorable graph: a graph that has a proper edge k-coloring.

edge-coloring of a graph: an assignment to its edges of "colors" from any set.

edge k-coloring of a graph: an edge-coloring that uses exactly k different colors.

—, blocked partial: the circumstance in which a proper subset of edges has a proper edge k-coloring, and every uncolored edge is adjacent to at least one edge of each of the k colors.

edge-independence number $ind_E(G)$ of a graph G: the maximum cardinality of an independent set of edges.

Eulerian graph: a graph that has an Eulerian tour.

Eulerian tour in a graph: a closed trail that contains every edge of that graph.

even wheel: a wheel graph $W_n = K_1 + C_n$ such that n is even.

factor of a graph: a spanning subgraph.

k-factor of a graph: a spanning subgraph of that is regular of degree k.

factorization of a graph G: a set of factors whose edge-sets form a partition of the edge-set E_G.

k-factorization of a graph G: a factorization of G into k-factors.

independence number of a graph G: the maximum cardinality of an independent set of vertices, denoted by $\alpha(G)$.

independent set of edges: a set of mutually nonadjacent edges.

independent set of vertices: a set of mutually nonadjacent vertices.

join of two graphs G and H: the graph $G + H$ obtained from the graph union $G \cup H$ by adding an edge between each vertex of G and each vertex of H.

Kempe i-j chain for a vertex-coloring of a graph: a component of the subgraph induced on the set of all vertices colored either i or j.

Kempe i-j edge-chain in an edge-colored graph: a component of the subgraph induced on all the i-colored and j-colored edges.

leaf block of a graph G: a block that contains exactly one cut-vertex of G.

line graph of a graph G: the graph L(G) whose vertices are the edges of G, such that edges with a common endpoint in G are adjacent in $L(G)$.

local recoloring of vertex v with respect to a proper coloring c of a graph G: a proper coloring c' such that

i. $c'(v) \neq c(v)$ and

ii. for all $x \in V_G - N[v]$, $c'(x) = c(x)$.

locally recolorable vertex with respect to a proper coloring c: a vertex that has a local recoloring with respect to c.

map on a surface: an imbedding of a graph on that surface.

map-coloring for an imbedding of a graph: a function from the set of faces to a set whose elements are regarded as *colors*.

matching in a graph G: a subset of edges of G that are mutually nonadjacent.

—, maximum: a matching with the maximum number of edges.

multipartite graph: a loopless graph whose vertices can be partitioned into k independent sets, which are sometimes called the ***partite sets***, is said to be k-partite.

multiplicity $\mu(G)$ of a graph: the maximum number of edges joining two vertices.

neighbor of an edge e**:** another edge that shares one or both of its endpoints with edge e.

obstruction to k**-chromaticity:** a graph whose presence as a subgraph forces the chromatic number to exceed k.

odd component of a graph: a component with an odd number of vertices.

odd wheel: a wheel graph $W_n = K_1 + C_n$ such that n is odd.

partite sets: see *multipartite graph.*

perfect graph: a graph G such that every induced subgraph H has its chromatic number $\chi(H)$ equal to its clique number $\omega(H)$.

proper edge-coloring of a graph: an edge-coloring such that if two edges have a common endpoint, then they are assigned two different colors.

proper map-coloring: a coloring such that if two regions meet at an edge, then they are colored differently.

proper vertex-coloring: a coloring such that the endpoints of each edge are assigned two different colors.

robust k**-coloring** of a graph: a k-coloring for which every vertex is locally recolorable.

k**-robust graph:** a graph that has a robust k-coloring.

χ**-robust graph:** a graph G for which $\chi_R(G) = \chi(G)$.

robust chromatic number of a graph G: the smallest integer k such that G is k-robust, denoted $\chi_R(G)$.

robust-critical graph: a graph in which every vertex is robust-critical.

robust-critical vertex: a vertex v in a graph G for which $\chi_R(G - v) < \chi_R(G)$.

total chromatic incidence of an edge-coloring on a graph: the sum of the chromatic incidences at the vertices.

Tutte's condition on a graph G: the condition that for every subset $S \subset V_G$, the number of odd components of $G - S$ does not exceed $|S|$.

vertex k-colorable graph: a graph that has a proper vertex k-coloring.

vertex-coloring: a function from the vertex-set of a graph to a set whose members are called *colors.*

vertex-k-coloring: a vertex-coloring that uses exactly k different colors.

(vertex) k-chromatic graph: a graph G with $\chi(G) = k$.

(vertex) chromatic number of a graph G, denoted $\chi(G)$: the minimum number of different colors required for a proper vertex-coloring of G.

Chapter 7

MEASUREMENT AND MAPPINGS

INTRODUCTION

This chapter provides the fundamental concepts and some basic results in graph-theoretic topics motivated by measurement considerations foreshadowed in earlier chapters. Distance-related invariants, including radius, diameter, and girth are treated systematically in the first section, domination in the second, and *bandwidth*, from the perspective of vertex labelings, in the third. The fourth section focuses on intersection graphs.

The final two sections discuss graph mappings that are generalizations of isomorphism. In §7.6, we see how these mappings provide a model for a software process called *emulation*, under which distributed algorithms for one parallel architecture are ported to another architecture. Emulation involves additional forms of measurement called *load*, *congestion*, and *dilation*.

7.1 DISTANCE IN GRAPHS

REVIEW FROM §1.2: The ***distance*** $d(s,t)$ from a vertex s to a vertex t in a graph is the length of a shortest s-t path if one exists; otherwise, $d(s,t) = \infty$.

NOTATION: When there is more than one graph under consideration, we sometimes use a subscript (as in $d_G(s,t)$) to indicate the graph in which the distance is taken.

Proposition 7.1.1: *Let G be a connected graph. The distance function d is a* ***metric*** *on V_G. That is, it satisfies the following four conditions:*

- $d(x,y) \geq 0$ *for all* $x, y \in V_G$.
- $d(x,y) = 0$ *if and only if* $x = y$.
- $d(x,y) = d(y,x)$ *for all* $x, y \in V_G$. ***(symmetry property)***
- $d(x,y) + d(y,z) \geq d(x,z)$ *for all* $x, y, z \in V_G$. ***(triangle inequality)***

Proof: The first three conditions follow immediately from the definition of distance. The triangle inequality holds because the concatenation of a shortest x-y path with a shortest y-z path forms an x-z walk whose length is certainly no smaller than a shortest x-z walk. ♢

Eccentricity, Diameter, and Radius

REVIEW FROM §1.2, §1.3, §1.4, AND §3.6:

- The ***eccentricity*** of a vertex v in a graph G, denoted $ecc(v)$, is the distance from v to a vertex farthest from v. That is, $ecc(v) = \max_{x \in V_G} d(v,x)$.
- The ***diameter*** of a graph G, denoted $diam(G)$, is the maximum of the vertex eccentricities in G or, equivalently, the maximum distance between two vertices in G. That is, $diam(G) = \max_{x \in V_G} ecc(x) = \max_{x,y \in V_G} d(x,y)$.
- The ***radius*** of a graph G, denoted $rad(G)$, is the minimum of the vertex eccentricities. That is, $rad(G) = \max_{x \in V_G} ecc(x)$.
- A ***central vertex*** v of a graph G is a vertex with minimum eccentricity, i.e., such that $ecc(v) = rad(G)$.
- The ***center of a graph*** G, denoted $Z(G)$, is the subgraph induced on the set of central vertices of G.
- ***Proposition 1.4.2:*** For every graph H, there is a connected graph G such that $H = Z(G)$.
- A ***block*** in a graph G is a maximal connected subgraph that contains no cut-vertices of G.

Example 7.1.1: The graph G in Figure 7.1.1 has diameter $diam(G) = 4$ and radius $rad(G) = 2$. Its center $Z(G)$ is isomorphic to K_2 (induced on the vertex subset $\{x, y\}$).

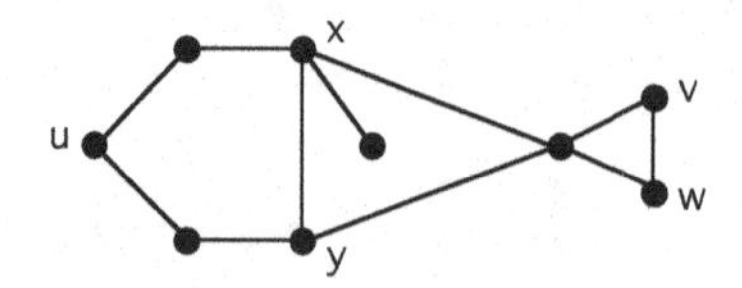

Figure 7.1.1 A graph G with $diam(G) = 4$, $rad(G) = 2$, and $Z(G) \cong K_2$.

Proposition 7.1.2: *Let G be a connected graph. Then*

$$rad(G) \leq diam(G) \leq 2 \cdot rad(G)$$

Proof: The first inequality is an immediate consequence of the definitions of radius and diameter. The second follows from the triangle inequality. ◊

Remark: Both inequalities in Proposition 7.1.2 are *tight* (or *best possible*). That is, there are graphs for which the inequality is strict, and there are graphs for which equality holds. (See Exercises.)

Proposition 7.1.3: [HaNo53] *The center of every connected graph G lies within a single block.*

Proof: Suppose, to the contrary, that the center $Z(G)$ has a pair of vertices u and w that do not lie in the same block of G. Then G has a cut-vertex v, distinct from u and w, such that u and w lie in different components of $G - v$, say C_u and C_w, respectively. Let x be a vertex of G that is farthest from cut-vertex v (i.e., $d(x,v) = ecc(v)$). If x does not lie in component C_w, then v is on every x-w path. It follows that $ecc(w) \geq d(x,w) > d(x,v) = ecc(v)$, which contradicts the fact that w is a central vertex. A similar contradiction arises if x does not lie in component C_u. ◊

Periphery

The *periphery* of a graph is the "opposite" of the center.

DEFINITION: A ***peripheral vertex*** v of a graph G is a vertex with maximum eccentricity, i.e., $ecc(v) = diam(G)$.

DEFINITION: The ***periphery of a graph*** G, denoted $per(G)$, is the subgraph induced on the set of peripheral vertices of G.

Example 7.1.2: For the graph G in Figure 7.1.1, the peripheral vertices are u, v, and w, and $per(G) \cong K_1 \cup K_2$.

In §1.4, we established that every graph is the center of some graph. The following result characterizes when a graph is the periphery of some graph.

Proposition 7.1.4: [BySy83] *A nontrivial graph G is the periphery of some connected graph if and only if G is complete or no vertex of G has eccentricity 1.*

Proof: ($\Leftarrow$) If G is complete, then $per(G) = G$, i.e., G is the periphery of itself. Alternatively, if no vertex of G has eccentricity 1, then let $G^* = G + v$ be the join (see §1.4) of G with a new vertex v. Since $ecc_G(x) > 1$ for every $x \in V_G$, it follows that $ecc_{G^*}(x) = 2$. Moreover, $ecc_{G^*}(v) = 1$, and hence, $per(G^*) = G$.

($\Rightarrow$) Suppose that $G = per(G^*)$, for some connected graph G^*. If $diam(G^*) = 1$, then every vertex in G has eccentricity 1, and hence, G is complete. Suppose that $diam(G^*) \geq 2$, and let x be any vertex in G. Since x is in $per(G^*)$, there exists y in G^* such that $d_{G^*}(x,y) = diam(G^*) \geq 2$. Then y is also a vertex in $per(G^*) = G$, and, hence, $d_G(x,y) \geq d_{G^*}(x,y) \geq 2$. Thus, $ecc_G(x) \geq 2$. ◊

Girth and Circumference

Girth and *circumference* are opposite measures.

REVIEW FROM §1.2: The ***girth*** of a graph G with at least one cycle is the length of a shortest cycle in G and is denoted $girth(G)$. The girth of G is undefined if G is acyclic.

DEFINITION: The ***circumference*** of a graph G with at least one cycle is the length of a longest cycle in G and is denoted $circum(G)$. The circumference of G is undefined if G is acyclic.

Example 7.1.3: The graph in Figure 7.1.2 has girth 3 and circumference 7.

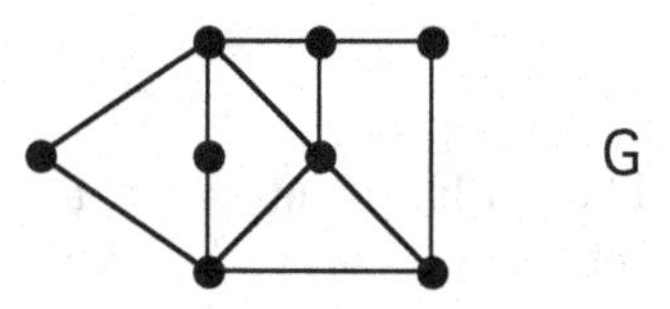

Figure 7.1.2 A graph G with $girth(G) = 3$ and $circum(G) = 7$.

Convexity

DEFINITION: A ***geodesic*** between vertices u and v is a u-v path of minimum length.

DEFINITION: For any two vertices u, v of a connected graph G, the ***closed interval*** $I[u, v]$ is the union of the sets of vertices of G that lie on u-v geodesics.

DEFINITION: For any vertex subset S of a connected graph, the ***closed interval*** $I[S]$ is the union

$$\bigcup_{u,v \in S} I[u, v]$$

Example 7.1.4: For the graph G shown in Figure 7.1.3, $I[u, v] = \{u, v, w, z\}$ and $I[\{v, x, y\}] = \{v, x, y, w, z\}$.

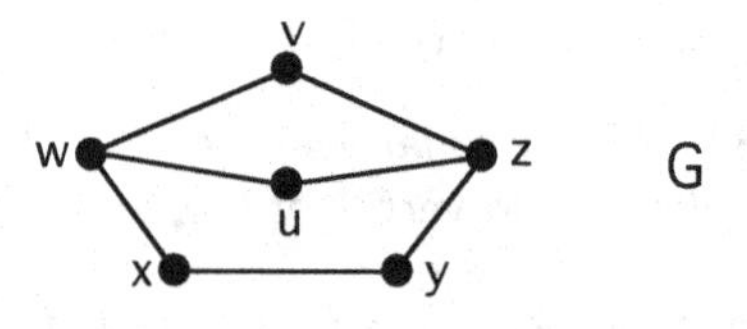

Figure 7.1.3

DEFINITION: A subset S of vertices of a graph is ***convex*** if $I[S] = S$.

DEFINITION: The ***convex hull*** of a subset S of vertices of a graph is the smallest convex set that contains S.

Example 7.1.4, continued: For the graph G in Figure 7.1.3, the convex hull of $\{u, v\}$ is {u,v,w,z}, and the convex hull of $\{v, x, y\}$ is V_G (since $\{v, x, y, w, z\}$ is not convex).

Steiner Distance

The distance between two vertices in a graph is the number of edges in a shortest path between them. *Steiner distance* generalizes this concept to subsets of two or more vertices in a graph.

DEFINITION: Let U be a subset of vertices in a connected graph G.

- A ***Steiner tree*** for U is a smallest tree subgraph of G that contains all the vertices of U.
- The ***Steiner distance*** of U, denoted $sd(U)$, is the number of edges in a Steiner tree for U.

Example 7.1.5: In the graph of Figure 7.1.4, a Steiner tree for the vertex subset $U = \{x, y, z\}$ is shown with edges in bold. Thus, $sd(U) = 5$.

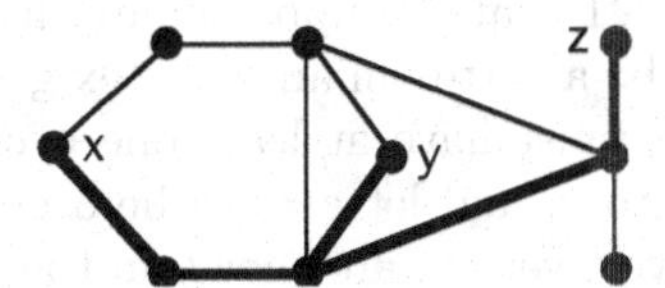

Figure 7.1.4 A Steiner tree for $\{x, y, z\}$.

Total Distance and Medians

DEFINITION: The ***total distance*** $td(v)$ of a vertex v in a connected graph G is given by

$$td(v) = \sum_{w \in V_G} d(v, w)$$

Example 7.1.6: In Figure 7.1.5, each vertex in the graph is labeled with its distance from the vertex z. Summing these distances yields $td(z) = 9$.

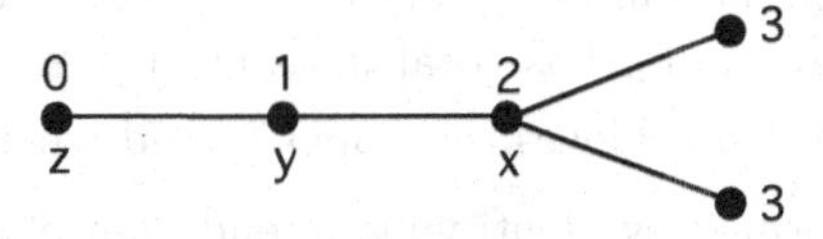

Figure 7.1.5 Total distance $td(z) = 9$.

Proposition 7.1.5: *Let v be any vertex of a connected graph with n vertices and m edges. Then the inequality*

$$n - 1 \leq td(v) \leq \frac{(n-1)(n+2)}{2} - m$$

provides tight upper and lower bounds on total distance.

Proof: Since the distance between distinct vertices is always greater than or equal to 1, the lower bound holds. The lower bound is achieved in any graph having a vertex that is adjacent to every other vertex.

To establish the upper bound, we use induction on the number of edges, m. Observe that $m \geq n-1$, since G is connected. If $m = n-1$, then G is a tree. For any vertex v, let d_i be the number of vertices at distance i from v. Then

$$td(v) = \sum_i id_i \quad \text{and} \quad \sum_i d_i = n-1$$

Clearly, if $d_i = 0$, then $d_{i+1} = 0$. It follows that the sum $\sum_i id_i$ is maximum when $d_i = 1$ for all nonzero d_i (i.e., for $1 \leq i \leq n-1$). Thus,

$$td(v) = \sum_{i=1}^{n-1} id_i = \sum_{i=1}^{n-1} i = \frac{n(n-1)}{2} = \frac{(n-1)(n+2)}{2} - (n-1)$$

which establishes the base of the induction.

Next, assume for some $m \geq n-1$ that the upper bound holds for any n-vertex connected graph with m edges, and let v be a vertex in an n-vertex graph G with $m+1$ edges. Since G contains at least n edges, it must have at least one cycle. Among all cycle vertices in G, let x be one that is closest to v, and let $e = xy$ be a cycle edge. The two possibilities, depending on whether v is a cycle vertex, are shown in the figure below.

In either case, the choice of x implies that $d_{G-e}(v,y) \geq d_G(v,y) + 1$, and, hence, $td_G(v) \leq td_{G-e}(v) - 1$. Then, applying the induction hypothesis to the graph $G - e$, we have

$$td_G(v) \leq td_{G-e}(v) - 1 \leq \frac{(n-1)(n+2)}{2} - m - 1 = \frac{(n-1)(n+2)}{2} - (m+1)$$

which establishes the upper bound.

To show that the upper bound can be achieved, first observe that the vertex at either end of the n-vertex path graph P_n has total distance $1 + 2 + \cdots + (n-1) = \frac{(n-1)n}{2} = \frac{(n-1)(n+2)}{2} - (n-1)$, and, thus, achieves the upper bound when $m = n-1$.

Next, consider a graph G formed by identifying an endpoint of a path of length $(n-c)$ and one of the vertices of the complete graph K_c. Then G has n vertices and $m = \binom{c}{2} + n - c = \frac{c(c-3)+2n}{2}$ edges, and it is straightforward to verify that its univalent vertex v has total distance

$$td(v) = 1 + 2 + \cdots + (n-c) + (c-1)(n-c+1) = \frac{(n-1)(n+2)}{2} - m$$

Thus, the upper bound is achieved by such a graph whenever there is positive integer c that satisfies $m = \frac{c(c-3)+2n}{2}$ (that is, if $c = \frac{3+\sqrt{8(m-n)+9}}{2}$ is an integer). When there is no such c, the upper bound is achieved by a graph formed by amalgamating a path of length $(n-c)$ and and K_c, where $c = \lfloor \frac{3+\sqrt{8(m-n)+9}}{2} \rfloor$ by identifying the vertex x of degree one in the path with a vertex y in the complete graph and adding edges between neighbors of y in the complete graph to the neighbor of x in the path. For the details, see, for example, [BuHa90]. ◇

Example 7.1.7: Both graphs in Figure 7.1.6 have 11 vertices. Graph G_1 has 13 edges and $13 = \frac{4(4-3)+2(11)}{2}$. However, graph G_2 has 15 edges and $\frac{4(4-3)+2(11)}{2} < 15 < \frac{5(5-3)+2(11)}{2}$. The total distance of v in G_1 is $td(v) = 1+2+\cdots+7+3(8) = 52 = \frac{(11-1)(11+2)}{2} - 13$. The total distance of v in G_2 is $td(v) = 1+2+\cdots+6+3(7)+1(8) = 50 = \frac{(11-1)(11+2)}{2} - 15$. You can see how each added edge reduced the total distance by one.

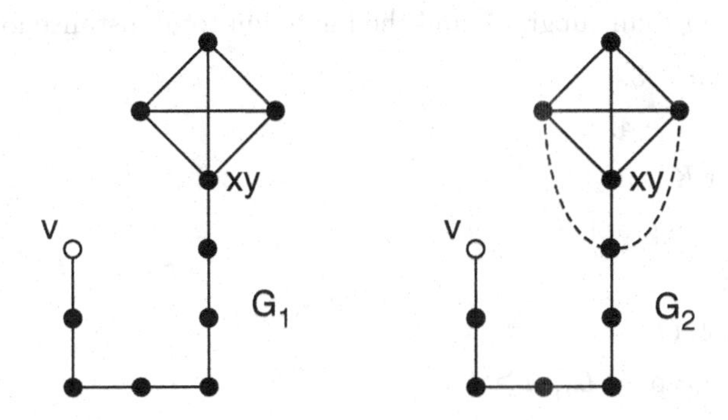

Figure 7.1.6 Two graphs with maximum total distance.

DEFINITION: A ***median vertex*** in a connected graph is a vertex whose total distance is minimum.

DEFINITION: The ***median subgraph*** of a connected graph G is the subgraph $M(G)$ induced on the median vertices of G.

Example 7.1.6, continued: Vertex x of the graph in Figure 7.1.5 has total distance 5 and is the only median vertex. Observe that y is a central vertex but is not a median vertex, since $td(y) = 6$.

Proposition 7.1.6: *Every graph is the median subgraph of some graph.*

Proof: Let G be a graph with $V_G = \{v_1, v_2, \ldots, v_n\}$. Form a supergraph H as follows: for each $i = 1, 2 \ldots, n$, join a new vertex w_i to v_i and to each vertex of G not adjacent to v_i. It is straightforward to show that $M(H) = G$ (see Exercises). ◇

EXERCISES for Section 7.1

7.1.1 Determine the circumference of the given graph.

(a) Complete bipartite graph $K_{3,7}$.

(b) Complete bipartite graph $K_{m,n}$, $m \geq n \geq 3$.

(c) Complete graph K_n.

(d) Hypercube graph Q_3. Can you generalize to Q_n?

(e) Circular ladder graph CL_6. Can you generalize to CL_n?

(f) The Petersen graph.

(g) Circulant graph $circ(n : 2)$.

(h) Circulant graph $circ(n : m)$.

7.1.2 Find the periphery of the given graph.

(a) The graph in Figure 7.1.2.

(b) The Petersen graph.

7.1.3 Prove or disprove: If a graph G is vertex-transitive, then $per(G) = G$.

7.1.4 Determine the median subgraph and the minimum total distance for the given graph.

(a) Path graph P_n, $n \geq 3$.

(b) Cycle graph C_n, $n \geq 4$.

(c) Complete graph K_n, $n \geq 3$.

(d) Complete bipartite graph $K_{n,n}$, $n \geq 3$.

(e) Petersen graph.

(f) Hypercube graph Q_3.

(g) Circular ladder graph CL_n, $n \geq 4$.

7.1.5 Both inequalities in Proposition 7.1.2 are tight. Each of the following parts confirms a different aspect of this assertion. Find a family of graphs for which:

(a) The first inequality is always strict.

(b) The first inequality is always an equality.

(c) The second inequality is always strict.

(d) The second inequality is always an equality.

7.1.6 Write out a complete proof of Proposition 7.1.2.

7.1.7 Construct a graph whose diameter, girth, and circumference are all equal.

7.1.8 Let G be a graph with at least two components. Prove that the edge-complement graph $\overline{G}$ has diameter $diam\left(\overline{G}\right) \leq 2$.

7.1.9 Let x and y be any two adjacent vertices in a connected graph. Prove that their eccentricities differ by at most 1.

7.1.10 Let G be a connected graph with $rad(G) \geq 3$. Prove $rad\left(\overline{G}\right) \leq 2$.

7.1.11 Let r and d be positive integers such that $r \leq d \leq 2r$. Construct a graph with radius r and diameter d. (Hint: Try a graph with exactly one cycle.)

7.1.12 Let vertices u and v be two different neighbors of a vertex w in a tree. Prove that $2ecc(w) \leq ecc(u) + ecc(v)$.

7.1.13 Complete the proof of Proposition 7.1.6, by showing that G is the median subgraph of H.

7.2 DOMINATION IN GRAPHS

TERMINOLOGY: A vertex is said to ***dominate*** itself and each of its neighbors.

The applicability of *domination theory* to such fields as network design and analysis, linear algebra, and optimization has attracted researchers from a broad range of disciplines. The origins are ascribed to Berge [Be62] and Ore [Or62]. For a quick survey of domination, see [HaHe04]. For a comprehensive treatment, see [HaHeS98].

DEFINITION: Let G be a graph and let $D \subseteq V_G$. A vertex subset D ***dominates*** a graph G (or is a ***dominating set***) if every vertex of G is in D or is adjacent to at least one vertex in D.

DEFINITION: A ***minimal dominating set*** of a graph G is a dominating set such that every proper subset is non-dominating.

DEFINITION: The ***domination number*** of a graph G, denoted $dom(G)$ (elsewhere, often $\gamma(G)$), is the cardinality of a minimum dominating set of G.

Example 7.2.1: Figure 7.2.1 shows a graph with minimal dominating sets (solid vertices) of two different cardinalities. It is straightforward to show that the dominating set on the left is a minimum one (i.e., there are no 2-vertex dominating sets). Thus, the dominating number of the graph is 3.

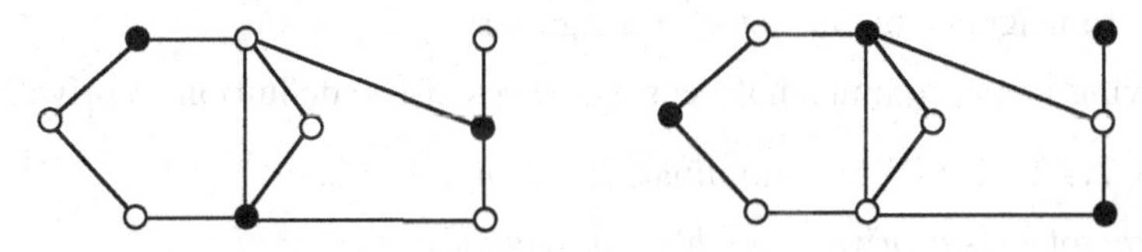

Figure 7.2.1 Two minimal dominating sets for a graph.

Example 7.2.2: For the 5-vertex path graph P_5, we have $dom(P_5) = 2$. The second and fourth vertices on the path form the unique minimum dominating set.

Three Applications of Domination

Here are three applications in which the problem of finding a minimum dominating set occurs naturally. It also has independent mathematical interest for many researchers.

Application 7.2.1: [Be73] *Surveillance* The minimum number of radar stations needed to keep a set of strategic locations under surveillance is the domination number of the associated graph.

Application 7.2.2: *Emergency First Aid Stations* Suppose that a natural disaster has struck some region consisting of many small villages. The vertices of a graph represent the villages in the region. An edge joining two vertices indicates that an emergency first-aid station set up in one of the corresponding villages can also serve the other one. Then a minimum dominating set of the graph would prescribe a way of serving the entire region with a mininum number of first aid stations.

Application 7.2.3: *Chess* Consider the problem of placing queens on a chessboard so that every square is either occupied by a queen or can be reached in one move by a queen. Determining the minimum number of queens is equivalent to finding the domination number of a 64-vertex graph, where two vertices are adjacent if and only if their corresponding squares lie on the same diagonal, same row, or same column. (See Exercises.)

Private Neighbors

Both dominating sets in Figure 7.2.1 are minimal, because each vertex in each set dominates at least one vertex that no other vertex in that set dominates.

REVIEW FROM §1.1:

- The ***open neighborhood of a vertex*** v in a graph, denoted $N(v)$, is the set of all the neighbors of v.
- The ***closed neighborhood of a vertex*** v, denoted $N[v]$, is given by $N[v] = N(v) \cup \{v\}$.

DEFINITION: Let D be a dominating set of a graph, and let $v \in D$. A vertex w is a ***private neighbor*** of v *relative to* D if v is the only vertex in D that dominates w. That is, $N[w] \cap D = \{v\}$.

CAUTION REGARDING TERMINOLOGY: Because a vertex dominates itself, it is possible for a vertex to be a private neighbor of itself, even if that vertex is not adjacent to itself. In such a case, a private neighbor might not be a neighbor!

The following facts are immediate consequences of the definition of private neighbor.

Proposition 7.2.1: *Let D be a dominating set of a graph.*

(i) No vertex of D is a private neighbor of any other vertex of D.

(ii) Vertex $v \in D$ is a private neighbor of itself if and only if v is not adjacent to any other vertex in D.

Two of the earliest results on domination, both due to Ore [Or62], establish basic properties of a minimal dominating set.

Proposition 7.2.2: [Or62] *Let D be a dominating set of a graph G. Then D is a minimal dominating set of G if all only if every vertex in D has at least one private neighbor.*

Proof: ($\Rightarrow$) By way of contrapositive, suppose that some vertex $v \in D$ does not have a private neighbor. Then every vertex dominated by v is also dominated by at least one other vertex in D. Thus, $D - \{v\}$ dominates G, which implies that D is not a minimal dominating set.

($\Leftarrow$) Suppose that every vertex in D has a private neighbor. Let v be any vertex in D, and let w be a private neighbor of v. Then $D - \{v\}$ does not dominate w. Therefore, D is a minimal dominating set. ◇

Proposition 7.2.3: [Or62] *Let G be a graph with no isolated vertices, and let D be a minimal dominating set of G. Then the complementary vertex-set $V_G - D$ is also a dominating set of G.*

Proof: We show that $V_G - D$ dominates an arbitrary vertex $v \in V_G$. If $v \in V_G - D$, then we are done. Alternatively, if $v \in D$, then it follows from Proposition 7.2.2 that vertex v has a private neighbor w. By Proposition 7.2.1(i), either $w = v$ or $w \in V_G - D$. If $w \in V_G - D$, then we are done. If $w = v$, then v is not adjacent to any vertex in D (by Proposition 7.2.1(ii)), and since v is not an isolated vertex, it must be adjacent to some vertex in $V_G - D$. ♢

Corollary 7.2.4: *Let G be an n-vertex graph with no isolated vertices. Then we have the upper bound $dom(G) \leq n/2$.* ♢ *(Exercises)*

Bounds on the Domination Number

Theorem 7.2.5: [WaAcSa79] *Let G be an n-vertex graph. Then*

$$\left\lceil \frac{n}{1+\delta_{\max}(G)} \right\rceil \leq dom(G)$$

♢ *(Exercises)*

Theorem 7.2.6: [Be73] *Let G be an n-vertex graph. Then*

$$dom(G) \leq n - \delta_{\max}(G)$$

♢ *(Exercises)*

Corollary 7.2.7: [WaAcSa79] *Let G be an n-vertex graph. Then*

$$dom(G) \leq n - \kappa_v(G)$$

where $\kappa_v(G)$ is the vertex-connectivity of G (§3.4).

Proof: By Corollary 3.4.6, $\kappa_v(G) \leq \delta_{\max}(G)$. ♢

Independent Domination

REVIEW FROM §1.3:

- A subset S of V_G is said to be an ***independent set*** if no two vertices in S are adjacent.
- The ***independence number*** of a graph G, denoted $\alpha(G)$, is the cardinality of a largest independent set of G.

DEFINITION: An ***independent dominating set*** of a graph G is an independent set of vertices that is also a dominating set of G.

DEFINITION: The ***independent domination number*** of a graph G, denoted $i\text{-}dom(G)$, is the cardinality of a minimum independent dominating set of G.

Example 7.2.3: The solid vertices of the graph G in Figure 7.2.2 form a minimum independent dominating set. Thus, $i\text{-}dom(G) = 3$.

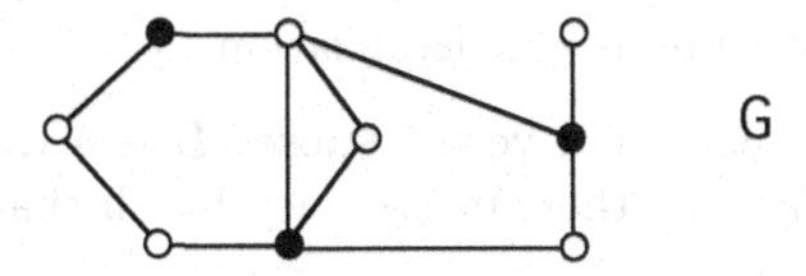

Figure 7.2.2 A minimum independent dominating set.

Proposition 7.2.8: *For every graph G, $dom(G) \leq i\text{-}dom(G)$.*

Proof: This follows from the definition of independent domination number. ♢

Proposition 7.2.9: *Every independent dominating set is a minimal dominating set.*

Proof: Let D be an independent dominating set of a graph. Then each vertex in D is its own private neighbor, and, hence, D is minimal by Proposition 7.2.2. ♢

Proposition 7.2.10: *A set of vertices in a graph is a maximal independent set if and only if it is an independent dominating set.* ♢ *(Exercises)*

Corollary 7.2.11: *For every graph G, $i\text{-}dom(G) \leq \alpha(G)$.*

Proof: Let S be a maximum independent set of G (i.e., $|S| = \alpha(G)$). Then S is an independent dominating set, by Proposition 7.2.10, and, hence, $i\text{-}dom(G) \leq |S|$. ♢

Corollary 7.2.12: *For every graph G, $dom(G) \leq \alpha(G)$.*

Corollary 7.2.13: *Every maximal independent set is a minimal dominating set.*

Proof: This is an immediate consequence of Propositions 7.2.9 and 7.2.10. ♢

Connected Domination

DEFINITION: A ***connected dominating set*** of a connected graph G is a dominating set D such that the subgraph induced on D is connected.

DEFINITION: The ***connected domination number*** of a connected graph G, denoted $c\text{-}dom(G)$, is the cardinality of a minimum connected dominating set of G.

Example 7.2.4: The solid vertices in the graph G of Figure 7.2.3 form a minimum connected dominating set. Thus, $c\text{-}dom(G) = 4$.

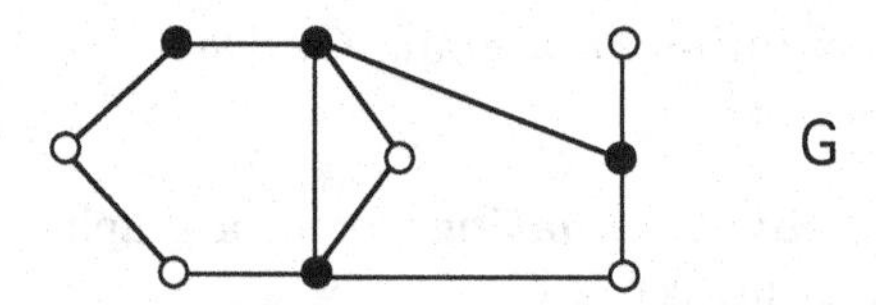

Figure 7.2.3 A minimum connected dominating set.

Distance-k Domination

Whereas independent domination and connected domination are special cases of ordinary domination, *distance-k domination* is a generalization.

DEFINITION: Given any integer $k \geq 1$, vertex subset D is a ***distance-k dominating set*** of a graph G if for all $v \in V_G - D$, there exists $x \in D$ such that $d(v, x) \leq k$.

DEFINITION: The ***distance-k domination number*** of a graph G, denoted $d_k\text{-}dom(G)$, is the cardinality of a minimum distance-k dominating set of G.

Observe that a d_1-dominating set is an ordinary dominating set, and observe also that $d_1\text{-}dom(G) = dom(G)$. Moreover, $d_k\text{-}dom(G) \leq dom(G)$.

Example 7.2.5: The solid vertices in the graph G of Figure 7.2.4 form a minimum distance-2 dominating set. Thus, d_2-$dom(G) = 2$. Notice that $dom(G) = 4$.

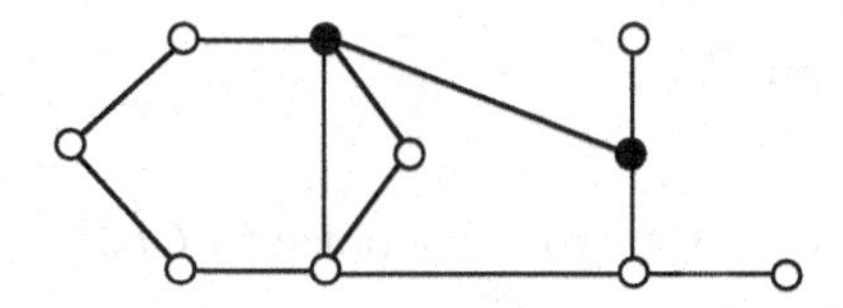

Figure 7.2.4 A minimum distance-2 dominating set.

EXERCISES for Section 7.2

7.2.1 Find all minimum dominating sets of the following graph and argue why there are no smaller ones.

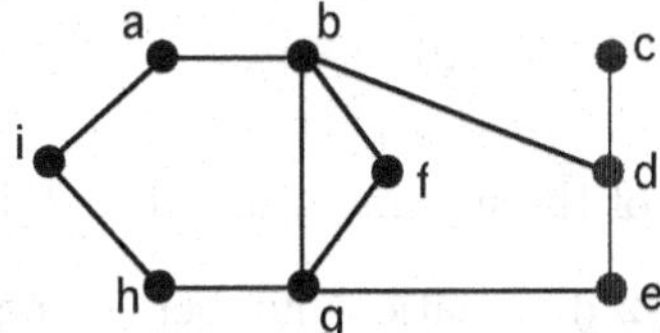

7.2.2 Find upper and lower bounds for the size of a minimum dominating set of an n-vertex graph. Then draw three 8-vertex graphs: one that achieves the lower bound, one that achieves the upper bound, and one that achieves neither.

7.2.3 Determine the domination number $dom(G)$ of the given graph, and justify your answer.

(a) The n-vertex path graph P_n.

(b) The n-vertex cycle graph C_n.

(c) Circular ladder graph CL_6. Can you generalize to CL_n?

(d) The 3-dimensional hypercube Q_3. Can you generalize to Q_n?

(e) The Petersen graph.

7.2.4 Find the domination number of each of the following circulant graphs, and justify your answer.

(a) $circ(5 : 1, 2)$ (b) $circ(6 : 1, 2)$ (c) $circ(8 : 1, 2)$

7.2.5 Determine the minimum number of queens that can be placed on a chessboard so that each square is either occupied by a queen or can be reached in one move by a queen.

7.2.6 Determine the minimum number of knights that can be placed on a chessboard so that each square is either occupied by a knight or can be reached in one move by a knight.

7.2.7 Determine the independent domination number and the connected domination number of the given graph, and justify your answer.

(a) The n-vertex path graph P_n.

(b) The n-vertex cycle graph C_n.

(c) The Petersen graph.

(d) Circular ladder graph CL_6. Can you generalize to CL_n?

(e) The 3-dimensional hypercube Q_3.

(f) The 4-dimensional hypercube Q_4.

7.2.8 Determine the minimum number of non-attacking queens that can be placed on a chessboard so that each square is either occupied by a queen or can be reached in one move by a queen.

7.2.9 Find the independent domination number and the connected domination number of each of the following circulant graphs, and justify your answer.

(a) $circ(5 : 1, 2)$ (b) $circ(6 : 1, 2)$ (c) $circ(8 : 1, 2)$

7.2.10 In Figure 7.2.4, find all of the minimum distance-2 dominating sets of the graph.

7.2.11 Determine the distance-2 domination number of the given graph, and justify your answer.

(a) The n-vertex path graph P_n

(b) The n-vertex cycle graph C_n

(c) The Petersen graph

(d) The circular ladder graph CL_6. Can you generalize to CL_n?

(e) The 3-dimensional hypercube Q_3

7.2.12 Draw a graph G such that $d_k\text{-}dom(G) = 4 - k$, for $k = 1, 2$, and 3.

7.2.13 For any integer $m \geq 3$, describe how to construct a graph G such that $d_k\text{-}dom(G) = m + 1 - k$, for $k = 1, 2, \ldots, m$.

7.2.14 Let G be a graph with $dom(G) \geq 3$. Prove that the diameter of the edge-complement graph $\overline{G}$ is at most 2.

7.2.15 Prove Corollary 7.2.4.

7.2.16 Prove Theorem 7.2.5.

7.2.17 Prove Theorem 7.2.6.

7.2.18 Prove Proposition 7.2.10.

7.3 BANDWIDTH

In this section, *bandwidth* is described in terms of vertex labelings and adjacency matrices. In §7.6, bandwidth is reinterpreted from the perspective of graph homomorphisms.

Since the addition of self-loops and multi-edges does not change the bandwidth of a graph, we assume that all graphs in this section are simple graphs.

TERMINOLOGY: A ***standard (1-based) vertex-labeling***(or simply, ***numbering***) of an n-vertex graph G is a bijection $f : V_G \to \{1, 2, \ldots, n\}$.

TERMINOLOGY: The ***induced numbering*** $\hat{f}_H$ on a subgraph H of a graph G with numbering f is obtained by placing the numbers of the set $f(V_H)$ in ascending order, say, $a_1, a_2, \ldots$, and then, for each vertex $v \in V_H$, defining $\hat{f}_H(v) = s$, if $f(v) = a_s$.

DEFINITION: The ***bandwidth of a numbering*** f of a graph G is given by

$$bw_f(G) = \max\{|f(u) - f(v)| \mid uv \in E_G\}$$

DEFINITION: The ***bandwidth*** of a graph G is given by

$$bw(G) = \min\{bw_f(G) | f \text{ is a numbering of } G\}$$

DEFINITION: A ***bandwidth numbering*** of a graph G is a numbering f that achieves $bw(G)$, i.e., $bw_f(G) = bw(G)$.

Example 7.3.1: Two numberings, f and g, of the 3-dimensional hypercube Q_3 are shown in Figure 7.3.1. One can show that g is a bandwidth numbering of Q_3, that is, $bw(Q_3) = 4$ (see Exercises).

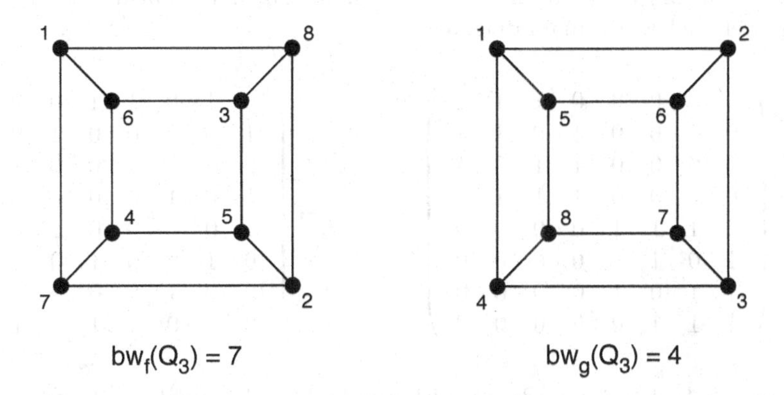

Figure 7.3.1 Two different numberings of the hypercube Q_3.

Application 7.3.1: *Electrical Circuit Design* We can model an electrical circuit as a graph, where adjacent vertices in the graph correspond to a pair of nodes in the circuit that are connected by a single wire. If the nodes of the circuit are to be equally spaced along a single row, then the left-to-right ordering of the nodes corresponds to a numbering of the graph, and the bandwidth of that numbering is the length of the longest wire needed. It follows that the bandwidth of the graph is the smallest possible length of the longest wire needed among all left-to-right orderings of the nodes.

Figure 7.3.2 shows the two left-to-right orderings for the two numberings of Q_3 shown in Figure 7.3.1.

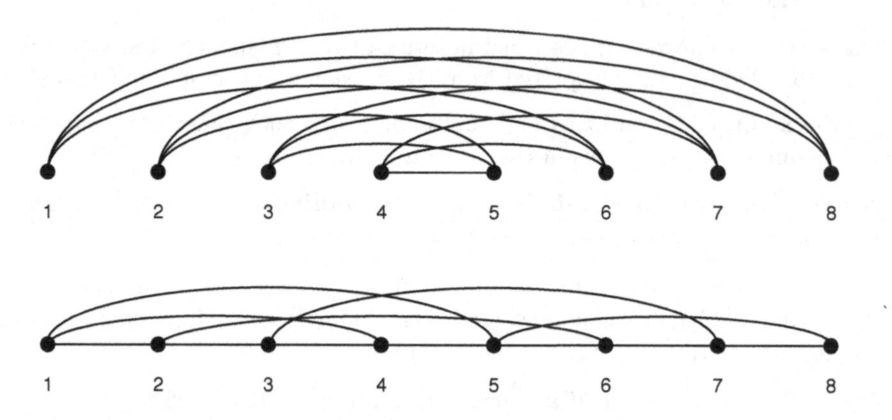

Figure 7.3.2 Two different designs of the hypercube Q_3 circuit.

Interpreting Bandwidth in Terms of Adjacency Matrices

The term *bandwidth* originates from its description in terms of adjacency matrices.

NOTATION: For a given numbering f of a graph G, the adjacency matrix of G whose rows and columns are ordered according to f is denoted M_f.

Example 7.3.1 continued: The adjacency matrices M_f and M_g for the numberings f and g from Figure 7.3.1 are shown below. Notice that the 1s in M_f are spread over the seven diagonals immediately above and the seven immediately below the main diagonal. In contrast, the 1s in M_g are concentrated in the four diagonals immediately above and the four immediately below the main diagonal.

$$M_f = \begin{pmatrix} 0 & 0 & 0 & 0 & 0 & 1 & 1 & 1 \\ 0 & 0 & 0 & 0 & 1 & 0 & 1 & 1 \\ 0 & 0 & 0 & 0 & 1 & 1 & 0 & 1 \\ 0 & 0 & 0 & 0 & 1 & 1 & 1 & 0 \\ 0 & 1 & 1 & 1 & 0 & 0 & 0 & 0 \\ 1 & 0 & 1 & 1 & 0 & 0 & 0 & 0 \\ 1 & 1 & 0 & 1 & 0 & 0 & 0 & 0 \\ 1 & 1 & 1 & 0 & 0 & 0 & 0 & 0 \end{pmatrix} \qquad M_g = \begin{pmatrix} 0 & 1 & 0 & 1 & 1 & 0 & 0 & 0 \\ 1 & 0 & 1 & 0 & 0 & 1 & 0 & 0 \\ 0 & 1 & 0 & 1 & 0 & 0 & 1 & 0 \\ 1 & 0 & 1 & 0 & 0 & 0 & 0 & 1 \\ 1 & 0 & 0 & 0 & 0 & 1 & 0 & 1 \\ 0 & 1 & 0 & 0 & 1 & 0 & 1 & 0 \\ 0 & 0 & 1 & 0 & 0 & 1 & 0 & 1 \\ 0 & 0 & 0 & 1 & 1 & 0 & 1 & 0 \end{pmatrix}$$

Proposition 7.3.1: *Let f be a bandwidth numbering of a graph G. Then every 1 in the corresponding adjacency matrix M_f lies in the* **band** *containing the $bw_f(G)$ diagonals above and the $bw_f(G)$ diagonals below that are closest to the main diagonal.*

Proof: The result follows immediately from the observation that the (i, j)-entry in any matrix is in the $|i-j|^{\text{th}}$ diagonal above the main diagonal if $i < j$ or the $|i-j|^{\text{th}}$ below if $i > j$. ◇

Application 7.3.2: *Sparse Matrices* A sparse matrix is one whose elements are mostly zeros. Much work has been done to find ways of storing such matrices so that the storage as well as the performance of matrix operations are efficient. One strategy is to store only the "band" of diagonals around the main diagonal containing the nonzero elements.

Some Basic Results and Examples

Proposition 7.3.2: *For any positive integer n,*

(i) $bw(P_n) = 1$.
(ii) $bw(K_n) = n - 1$.
(iii) $bw(C_n) = 2$. ◊ *(Exercises)*

Example 7.3.2: We claim that $bw(K_{4,3}) = 4$. The numbering shown in Figure 7.3.3 has bandwidth 4, so it suffices to show that $bw_f(K_{4,3}) \geq 4$ for any numbering f of $K_{4,3}$. If the numbers 1, 6, and 7 are not all assigned to the same side of the bipartition, then $bw_f(K_{4,3}) \geq 5$. Moreover, if 2 is not assigned to the same side as 7, then $bw_f(K_{4,3}) \geq 5$. Thus, every numbering different from the one shown in Figure 7.3.3 has bandwidth greater than 4.

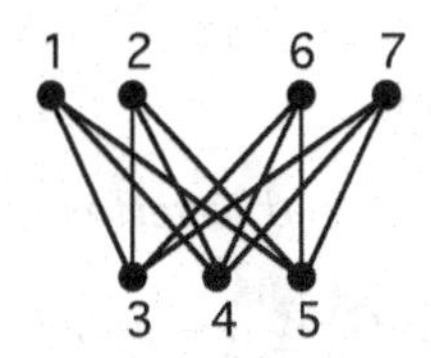

Figure 7.3.3 A bandwidth numbering of $K_{4,3}$.

Proposition 7.3.3: *Let H be a subgraph of G. Then $bw(H) \leq bw(G)$.*

Proof: Let f be a bandwidth numbering on G and $\hat{f}_H$ the induced numbering on the subgraph H. Then $bw(H)$ is less than or equal to the bandwidth of the numbering $\hat{f}_H$, which is no larger than the bandwidth of f, which equals $bw(G)$. Symbolically,

$$bw(H) \leq bw_{\hat{f}}(H) \leq bw_f(G) = bw(G)$$ ◊

Characterizing Bandwidth via Powers of the Path Graph

DEFINITION: The mth **power of graph** G, denoted G^m, is the graph having vertex-set $V_{G^m} = V_G$ and edge-set $E_{G^m} = \{uv \mid d_G(u, v) \leq m\}$.

OBSERVATION Let $\langle v_1, v_2, \ldots, v_n \rangle$ be the vertex sequence of the path graph P_n. Then for $1 \leq m \leq n-1$, the adjacency matrix M_f of P_n^m, where f is the natural numbering $f(v_i) = i$, has *all* 1s on its first m diagonals (above and below the main diagonal) and 0s everywhere else.

Proposition 7.3.4: *For the path graph P_n, $bw(P_n^m) = m$ for $1 \leq m \leq n - 1$.*

Proof: Proposition 7.3.1 and the observation above imply that the natural numbering corresponding to the vertex sequence of P_n is a bandwidth numbering of P_n. ◊

Proposition 7.3.5: *Let G be an n-vertex graph. Then $bw(G) \leq m$ if and only if G is a subgraph of P_n^m.*

Proof: The result follows immediately from Propositions 7.3.1, 7.3.3, and 7.3.4. ◊

Corollary 7.3.6: *Let G be an n-vertex graph. Then $bw(G) = m$ if and only if m is the smallest integer such that G is a subgraph of P_n^m.*

Some Bounds on the Bandwidth

We close the section with four results, due to Dewdney [ChDeGiKo75], that establish bounds on the bandwidth, in terms of maximum degree, diameter, chromatic number, and vertex-connectivity.

Proposition 7.3.7: *Let G be a graph with maximum degree $\delta_{\max}(G) = \Delta$. Then*

$$bw(G) \geq \left\lceil \frac{\Delta}{2} \right\rceil$$

Proof: Let f be a bandwidth numbering of G, and let v be a vertex with neighborhood $N(v) = \{w_1,\ w_2, \ldots, w_\Delta\}$. We may assume that $f(w_1) < f(w_2) < \cdots < f(w_\Delta)$. Then

$$bw_f(G) \geq \max\left\{|f(v) - f(w_1)|\,,\ |f(v) - f(w_\Delta)|\right\} \geq \left\lceil \frac{\Delta}{2} \right\rceil$$

whether $f(w_1) < f(v) < f(w_\Delta)$ or not. ♢

Proposition 7.3.8: *Let G be a connected n-vertex graph. Then*

$$bw(G) \leq n - diam(G)$$

Proof: Let $s, t \in V_G$ such that $d(s,t) = diam(G) = l$, and let $\langle u_0, u_1, \ldots, u_l\rangle$, where $s = u_0$ and $t = u_l$, be the vertex sequence of a shortest s-t path P in G. For $0 \leq k \leq l$, let V_k be the set of vertices in G whose distance from s equals k. Then the sets $V_0, V_1, \ldots V_l$ form a partition of V_G, and for each k, $0 \leq k \leq l$, $u_k \in V_k$.

Let f be a numbering that assigns s the number 1, the vertices in V_1 the numbers from 2 to $|V_1|+1$, the vertices in V_2 the numbers from $|V_1|+2$ to $|V_1|+|V_2|+1$, and so on. Figure 7.3.4 illustrates such a numbering.

Observe that if xy is an edge in G, then $|d(s,x) - d(s,y)| \leq 1$, and hence, $x, y \in V_k \cup V_{k+1}$, for some k, $0 \leq k \leq l$. The numbering f satisfies $|f(x) - f(y)| \leq |V_k \cup V_{k+1}| - 1$, and since $V_k \cup V_{k+1}$ contains only two of the $l+1$ vertices of the path P, $|V_k \cup V_{k+1}| \leq n - (l+1-2) = n - l + 1$. Therefore $|f(x) - f(y)| \leq n - l = n - diam(G)$. ♢

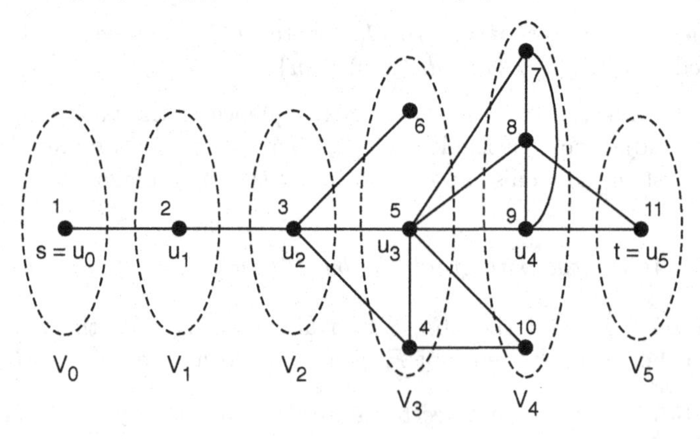

Figure 7.3.4 A numbering based on distance from s.

The next bound involves the *chromatic number* of a graph.

REVIEW FROM §6.1:

- The **(vertex) chromatic number** of graph G, denoted $\chi(G)$, is the smallest number k such that there is a function $g : V(G) \to \{1, 2, \ldots, k\}$ with the property that, if $uv \in E_G$, then $g(u) \neq g(v)$.
- If H is a subgraph of graph G, then $\chi(H) \leq \chi(G)$.

Proposition 7.3.9: *For the path graph* $P_n = \langle v_1, v_2, \ldots, v_n \rangle$, $\chi(P_n^m) = m+1$ *for* $1 \leq m \leq n-1$.

Proof: We have $\chi(P_n^m) \geq m+1$, since the complete graph K_{m+1} is a subgraph of P_n^m. To establish the reverse inequality, observe that in P_n^m, vertex v_1 is not adjacent to v_{m+2}, v_2 is not adjacent to v_{m+3}, and so on. Thus, the function g defined by $g(v_i) = i(\text{mod } m+1)$ is a proper $(m+1)$-coloring of P_n^m. ♢

Corollary 7.3.10: *Let G be an n-vertex graph. Then*

$$bw(G) \geq \chi(G) - 1$$

Proof: Let $bw(G) = m$. Then G is a subgraph of P_n^m, by Proposition 7.3.5. It follows that $\chi(G) \leq \chi(P_n^m) = m+1$. ♢

Remark: Brigham and Dutton [BrDu85] showed that $bw(G) \geq \chi'(G)/2$, where $\chi'(G)$ is the *edge-chromatic number* of G (§6.4).

Our last result shows that the *vertex-connectivity* is a lower bound on the bandwidth.

REVIEW FROM §1.3 and §3.4:

- A vertex subset S is a **vertex-cut** of a connected graph G if $G - S$ is non-connected.
- The **vertex-connectivity** of a connected graph G, denoted $\kappa_v(G)$, is the minimum number of vertices whose removal can either disconnect G or reduce it to a 1-vertex graph.

Lemma 7.3.11: *For the path graph* $P_n = \langle v_1, v_2, \ldots, v_n \rangle$, *the bound* $\kappa_v(P_n^m) \leq m$ *holds for* $1 \leq m \leq n-1$.

Proof: Since v_1 is not adjacent to v_{m+2} in P_n^m, the vertex subset $\{v_2, v_3, \ldots, v_{m+1}\}$ is a vertex-cut in P_n^m. ♢

Proposition 7.3.12: *For any graph G,* $bw(G) \geq \kappa_v(G)$. ♢ *(Exercises)*

Remark: For comprehensive surveys on bandwidth, each with extensive bibliographies, see [Br04a], [ChChDeGi82], and [LaWi99]. Further results can be found in [Ch88] and [Mi91].

EXERCISES for Section 7.3

7.3.1 Calculate the bandwidth of the given graph.

(a) Circular ladder CL_3. (b) $K_n \times K_2$. (c) $K_{m,n}$, for $m, n \geq 2$.

7.3.2 Prove Proposition 7.3.2.

7.3.3 Prove that the bandwidth of $K_{1,n}$ equals $\lceil \frac{n}{2} \rceil$.

7.3.4 Prove that the bandwidth of $K_{2,4}$ equals 3.

7.3.5 Show that $bw(Q_3) = 4$. (Hint: make use of the symmetry in arguing that there is no numbering f with $bw_f(Q_3) < 4$.)

7.3.6 Prove that the $2n$-vertex wheel W_{2n-1} has bandwidth equal to n.

7.3.7 Prove Proposition 7.3.12. (Hint: Use Lemma 7.3.11.)

7.4 INTERSECTION GRAPHS

Suppose that each vertex of a graph is associated with a subset of some set. We can define two vertices to be adjacent if the cardinality of their intersection exceeds some predetermined *tolerance* threshold. The threshold value of 1 corresponds to the family of *intersection graphs.*

CONVENTION FOR THIS SECTION: All graphs in this section are assumed to be simple graphs, even when the modifier *simple* does not appear in the assertions.

Intersection Graphs

DEFINITION: Let $\mathcal{F} = \{S_1, S_2, \ldots, S_n\}$ be a family of subsets of a set. The ***intersection graph of*** $\mathcal{F}$, denoted $\Omega(\mathcal{F})$, is the graph whose vertex- and edge-sets are given by

$$V_{\Omega(\mathcal{F})} = \{S_1, S_2, \ldots, S_n\}$$
$$E_{\Omega(\mathcal{F})} = \{S_iS_j \mid i \neq j \text{ and } S_i \cap S_j \neq \emptyset\}$$

DEFINITION: Let G be a graph with $V_G = \{v_1, v_2, \ldots, v_n\}$. Then G is an ***intersection graph*** if there exists a family of sets $\mathcal{F} = \{S_1, S_2, \ldots, S_n\}$ such that $G \cong \Omega(\mathcal{F})$ with the natural correspondence $f(v_i) = S_i$, $i = 1, 2, \ldots, n$. The family $\mathcal{F}$ is called a ***set representation*** of graph G.

NOTATION: If G is an intersection graph with set representation $\mathcal{F} = \{S_1, S_2, \ldots, S_n\}$, then we typically refer to the vertex-set as $\{S_1, S_2, \ldots, S_n\}$.

Example 7.4.1: Figure 7.4.1 depicts the cycle graph C_4 as an intersection graph with a set representation given by the family $\mathcal{F} = \{\{1,4\}, \{1,2\}, \{2,3\}, \{3,4\}\}$.

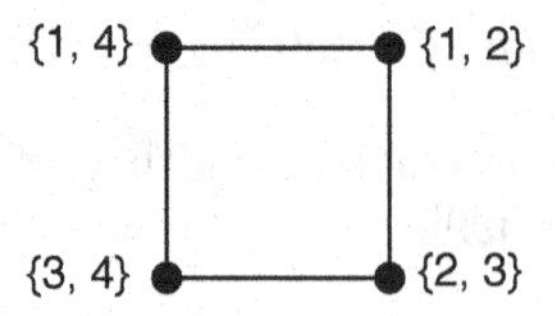

Figure 7.4.1 The cycle graph C_4 depicted as an intersection graph.

Proposition 7.4.1: [Ma45] *Every simple graph G is an intersection graph.*

Proof: For each vertex $v \in V_G$, let E_v be the subset of edges incident on v. Then the family of subsets $\{E_v\}_{v \in V_G}$ is a set representation of G. ◇

Some Special Kinds of Intersection Graphs

DEFINITION: A graph G is an ***interval graph*** if it is an intersection graph corresponding to a family of intervals on the real line.

Example 7.4.2: The assignment of intervals to the vertices of the graph in Figure 7.4.2 shows that it is an interval graph.

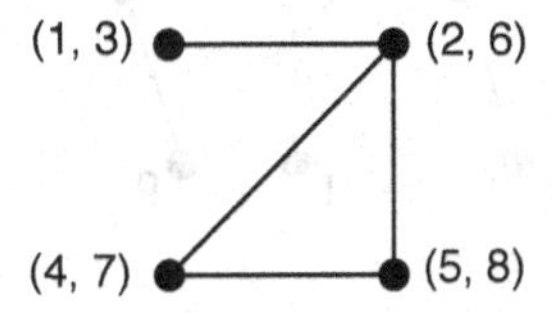

Figure 7.4.2 An interval graph.

Remark: The class of interval graphs is a *proper* subclass of intersection graphs. For instance, one can show that the cycle graph C_n for all $n \geq 4$ is not an interval graph (see Exercises).

REVIEW FROM §1.1: The ***line graph*** $L(G)$ of a graph G has a vertex for each edge of G, and two vertices in $L(G)$ are adjacent if and only if the corresponding edges in G have a vertex in common.

Thus, the line graph $L(G)$ is the intersection graph corresponding to the endpoint sets of the edges of G.

Example 7.4.3: Figure 7.4.3 shows a graph G and its line graph $L(G)$.

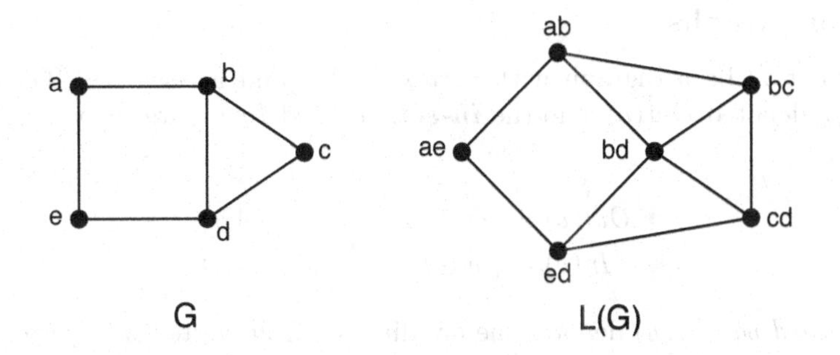

Figure 7.4.3 A graph and its line graph.

Subtree Graphs

DEFINITION: A graph G is a ***subtree graph*** if G is the intersection graph of a family $\mathcal{F} = \{T_1, T_2, \ldots, T_n\}$ of subtrees of a tree T, i.e., the vertex-sets of the T_is form a set representation of G. The tree T and the family $\mathcal{F}$ are called a ***tree representation*** of G.

If T is a path (and the T_is are subpaths), then the tree representation is called a ***path representation***, and G is said to be an intersection graph of a family of subpaths of a path.

Example 7.4.4: A subtree graph G and one of its tree representations are shown in Figure 7.4.4. The vertex-sets of the subtrees $T_1, T_2, \ldots, T_7$ of tree T are listed at the right.

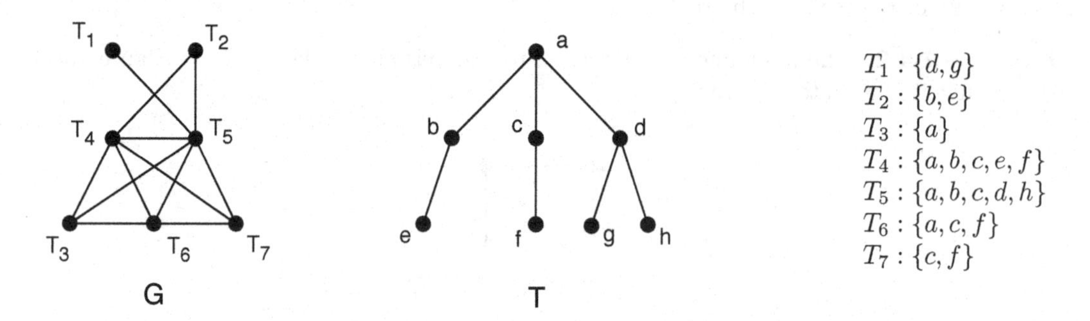

Figure 7.4.4 A subtree graph G and one of its tree representations.

Proposition 7.4.2: *A graph is an interval graph if and only if it is an intersection graph of a family of subpaths of a path.* ♢ *(Exercises)*

DEFINITION: A graph is a ***chordal graph*** if, for all $n \geq 4$, it does not contain an n-vertex cycle graph C_n as an induced subgraph.

Example 7.4.4, continued: The graph G in Figure 7.4.4 is a chordal graph.

Proposition 7.4.3: *[Bu74, Ga74, Wa78] A graph is a chordal graph if and only if it is a subtree graph.* ♢ *(See, e.g., [McMc99], Theorem 2.4.)*

Competition Graphs

DEFINITION: Let D be a digraph with vertex-set V_D and arc-set A_D. The ***out-set*** of a vertex $v \in V_D$, denoted $Out(v)$, and the ***in-set***, denoted $In(v)$, are the vertex subsets given by

$$Out(v) = \{w \in V_D \mid (v, w) \in A_D\}$$
$$In(v) = \{w \in V_D \mid (w, v) \in A_D\}$$

where the *ordered pair* (x, y) denotes the arc directed from vertex x to vertex y.

DEFINITION: The ***competition graph of a digraph*** D is the intersection graph of the family of out-sets of the vertices of D.

DEFINITION: A graph G is a ***competition graph*** if there is a digraph D such that G is (isomorphic to) the competition graph of D.

Example 7.4.5: A digraph D and its competition graph G are shown in Figure 7.4.5.

Application 7.4.1: *Food Webs* [Co78] Suppose that $a_1, a_2, \ldots, a_n$ are n animals in an ecosystem. A ***food web*** is a digraph model D of the predator-prey relationship among the a_is, where the vertex- and arc-sets of D are given by

$$V_D = \{a_1, a_2, \ldots, a_n\}$$
$$A_D = \{(a_i,\ a_j) \mid a_i \text{ preys on } a_j\}$$

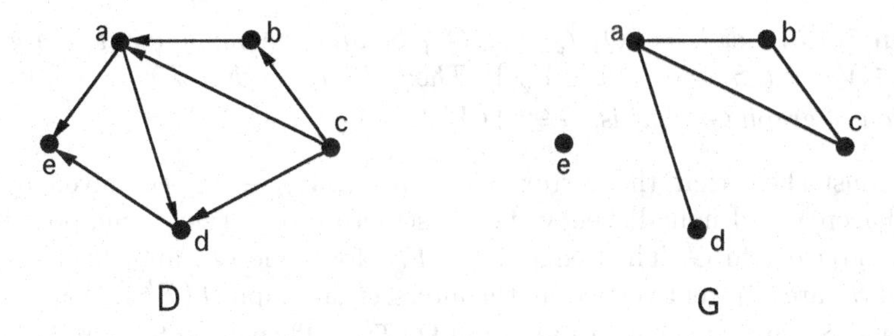

Figure 7.4.5 A digraph D and its competition graph G.

Then two animals a_i and a_j compete for the same prey if and only if a_i and a_j are adjacent in the competition graph of D. In the competition graph of the food web shown in Figure 7.4.6, the yellow warbler is adjacent to the frog, since they both prey on the spider, the yellow warbler is adjacent to the spider since they both prey on the flea beetle, and the frog and the spider are adjacent since they both prey on the sawfly.

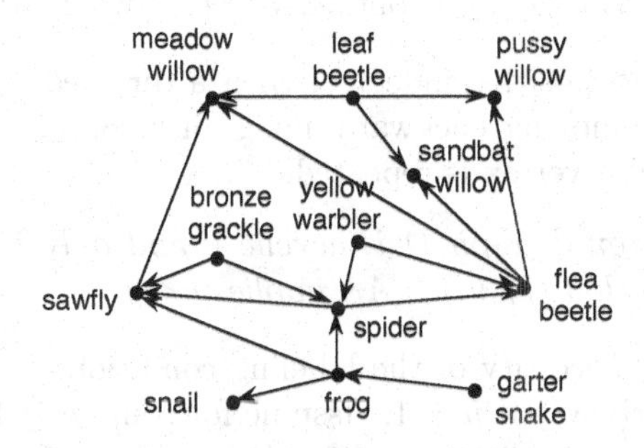

Figure 7.4.6 The food web in a Canadian willow forest.

Edge Clique Covers and Intersection Graphs

Edge clique covers and set representations for intersection graphs are closely related.

DEFINITION: An ***edge clique cover*** of a graph G is a family $\mathcal{K} = \{Q_1, Q_2, \ldots, Q_t\}$ of complete subgraphs of G such that every edge of G is an edge of at least one of the Q_is, that is,

$$E_G = E_{Q_1} \cup E_{Q_2} \cup \ldots E_{Q_t}$$

TERMINOLOGY NOTE: Elsewhere in this book, a *clique* is a *maximal* complete subgraph, but to be consistent with the terminology used in intersection graph theory, we do *not* require that the complete subgraphs in an edge clique cover be maximal.

Proposition 7.4.4: *Let $\mathcal{F} = \{S_1, S_2, \ldots, S_n\}$ be a set representation for an intersection graph G, and let $S = \bigcup_{i=1}^{n} S_i$. For each $x \in S$, let Q_x be the subgraph induced on the vertex subset $V_x = \{S_i \mid x \in S_i\}$. Then the family $\mathcal{K}^{\mathcal{F}} = \{Q_x\}_{x \in S}$ is an edge clique cover of graph G.*

Proof: To show that each Q_x is a complete subgraph of G, suppose that vertices $S_i, S_j \in V_x$ with $i \neq j$. Then $x \in S_i \cap S_j$, and, hence, S_i and S_j are adjacent in G. Next, let $e = S_k S_l$ be any edge of G. Then, since S_k and S_l are adjacent, there is at least one $x \in S_k \cap S_l$, i.e., $S_k, S_l \in V_x$. Thus, $e \in E_{Q_x}$, which shows that $\mathcal{K}^{\mathcal{F}}$ is an edge clique cover of G. ♢

Proposition 7.4.5: *Let $\mathcal{K} = \{Q_1, Q_2, \ldots, Q_t\}$ be an edge clique cover of a graph G, and for each $v \in V_G$, let $S_v = \{i \mid v \in V_{Q_i}\}$. Then the collection $\mathcal{F}^{\mathcal{K}} = \{S_v\}_{v \in V_G}$ is a set representation of graph G, that is, $G \cong \Omega\left(\mathcal{F}^{\mathcal{K}}\right)$.*

Proof: We must show that the vertex bijection $f : V_G \to V_{\Omega(\mathcal{F}^{\mathcal{K}})}$ given by $f(v) = S_v$ preserves adjacency and nonadjacency (i.e., is structure-preserving). Suppose that vertices v and w are adjacent in G. Then edge $vw \in E_{Q_i}$ for some Q_i, and, hence, $i \in S_v \cap S_w$. Thus, S_v and S_w are adjacent vertices in the intersection graph $\Omega\left(\mathcal{F}^{\mathcal{K}}\right)$. Conversely, suppose vertices S_v and S_w are adjacent in the graph $\Omega\left(\mathcal{F}^{\mathcal{K}}\right)$. Then $S_v \cap S_w \neq \emptyset$, i.e., $i \in S_v \cap S_w$ for some i. It follows that $v, w \in V_{Q_i}$, and, since Q_i is a complete graph, v is adjacent to w in graph G. ◇

Characterizing Competition Graphs of Acyclic Digraphs

Initial investigations of competition graphs focused on acyclic digraphs, because food webs are typically acyclic.

Lemma 7.4.6: *Every acyclic digraph contains a vertex with indegree 0.*

Proof: If every vertex had positive indegree, then a directed cycle could be constructed by starting at any vertex, moving backward along an incoming arc to its tail vertex, and repeating the process until a vertex is repeated. ◇

Lemma 7.4.7: *An n-vertex digraph D is acyclic if and only if its vertex-set V_D can be labeled $\{v_1, v_2, \ldots, v_n\}$ so that $(v_i, v_j) \in A_D$ implies $i < j$.*

Proof: ($\Rightarrow$) To prove the necessity of the labeling condition, we use induction on n. The labeling condition holds trivially if $n = 1$. Assume for some $n \geq 1$ that all n-vertex acyclic digraphs satisfy the labeling condition, and let D be an acyclic digraph on $n+1$ vertices. By Lemma 7.4.6, D has a vertex w with indegree 0. By the induction hypothesis, the n vertices of the digraph $\hat{D} = D - v$ can be labeled $v_1, v_2, \ldots, v_n$ so that $(v_i, v_j) \in A_{\hat{D}}$ implies $i < j$. Then by adding 1 to each subscript of the v_is and labeling vertex w as v_1, we obtain a labeling of the vertices of D that satisfies the condition.

($\Leftarrow$) To prove the sufficiency of the labeling condition, suppose that the vertices of digraph D are $v_1, v_2, \ldots, v_n$ and that $(v_i, v_j) \in A_{\hat{D}}$ implies $i < j$. If there were a vertex sequence $\langle v_{i_1}, v_{i_2}, \ldots, v_{i_s}, v_{i_1} \rangle$ that represented a directed cycle, then the condition would imply that $i_1 < i_2 < \cdots < i_s < i_1$, a contradiction. Thus, D must be acyclic. ◇

Theorem 7.4.8: [DuBr83, LuMa83] *An n-vertex graph G is a competition graph of an acyclic digraph if and only if V_G can be labeled $\{v_1, v_2, \ldots, v_n\}$ and G has an edge clique cover $\mathcal{K} = \{Q_1, Q_2, \ldots, Q_n\}$ such that $v_i \in Q_j$ implies $i < j$.*

Proof: ($\Rightarrow$) Suppose that G is the competition graph of an acyclic digraph D. By Lemma 7.4.7, we may assume that $V_G = V_D = \{v_1, v_2, \ldots, v_n\}$, where $(v_i, v_j) \in A_D$ implies $i < j$. Let $\mathcal{K} = \{Q_1, Q_2, \ldots, Q_n\}$ be the family of in-sets in D, i.e., $Q_i = In(v_i)$, $i = 1, 2, \ldots, n$. Then $\mathcal{K}$ is an edge clique cover of G such that $v_i \in Q_j$ implies $i < j$ (see Exercises).

($\Leftarrow$) Suppose that $V_G = \{v_1, v_2, \ldots, v_n\}$ and $\mathcal{K} = \{Q_1, Q_2, \ldots, Q_n\}$ is an edge clique cover of G such that $v_i \in Q_j$ implies $i < j$. Define a digraph D with $V_D = V_G$ and $A_D = \{(v_i, v_j) \mid v_i \in Q_j\}$. By Lemma 7.4.7, D is acyclic. Moreover, G is the competition graph of D (see Exercises). ◇

Remark: Competition graphs of arbitrary digraphs and of loopless digraphs are characterized in [DuBr83].

Duality Between Edge Clique Covers and Set Representations

DEFINITION: The edge clique cover $\mathcal{K}^{\mathcal{F}} = \{Q_x\}_{x \in S}$, defined in Proposition 7.4.4, corresponding to a set representation $\mathcal{F}$ of the intersection graph G is called the ***dual edge clique cover of*** $\mathcal{F}$.

DEFINITION: The set representation $\mathcal{F}^{\mathcal{K}} = \{S_v\}_{v \in V_G}$, defined in Proposition 7.4.5, corresponding to an edge clique cover $\mathcal{K}$ of a graph G is called the ***dual set representation of*** $\mathcal{K}$.

NOTATION: Let G be an intersection graph with set representation $\mathcal{F}$ and edge clique cover $\mathcal{K}$. Then $\mathcal{F}^{(\mathcal{K}^{\mathcal{F}})}$ denotes the dual set presentation of the dual edge clique cover of $\mathcal{F}$, and $\mathcal{F}^{(\mathcal{K}^{\mathcal{F}})}$ denotes the dual edge clique cover of the dual set representation of $\mathcal{K}$.

Proposition 7.4.9: *Let G be an intersection graph with set representation $\mathcal{F}$ and edge clique cover $\mathcal{K}$. Then both of the following are true.*

(i) There is a one-to-one correspondence between the sets in $\mathcal{F}$ and the sets in $\mathcal{F}^{(\mathcal{K}^{\mathcal{F}})}$.

(ii) There is a one-to-one correspondence between the subgraphs in $\mathcal{K}$ and the subgraphs in $\mathcal{K}^{(\mathcal{F}^{\mathcal{K}})}$. ♢ *(Exercises)*

Intersection Number of a Graph

DEFINITION: The ***intersection number*** of an n-vertex graph G, denoted $int(G)$, is the minimum cardinality of a set S such that G is the intersection graph of a family $\mathcal{F} = \{S_1, S_2, \ldots, S_n\}$ of subsets of S.

DEFINITION: The ***edge-clique-cover number*** of a graph G, denoted $\theta_e(G)$, is the smallest t for which there exists an edge clique cover $\mathcal{K} = \{Q_1, Q_2, \ldots, Q_t\}$.

Theorem 7.4.10: [ErGoPo66] *For every graph G, $int(G) = \theta_e(G)$.*

Proof: Let S be a smallest set such that $S = \bigcup_{i=1}^{n} S_i$ and $\mathcal{F} = \{S_1, S_2, \ldots, S_n\}$ is a set representation of G (i.e., $|S| = int(G)$). Then by Proposition 7.4.4, $\mathcal{K}^{\mathcal{F}}$ is an edge clique cover of G of size $|S|$, which shows that $int(G) \geq \theta_e(G)$. The reverse inequality follows similarly from Proposition 7.4.5. ♢

p-Intersection Graphs

DEFINITION: The ***p-intersection graph*** of a family of subsets $\mathcal{F} = \{S_1, S_2, \ldots, S_n\}$ of a finite set S, denoted $\Omega_p(\mathcal{F})$, is the graph whose vertex- and edge-sets are given by

$$V_{\Omega_p(\mathcal{F})} = \{S_1,\ S_2, \ldots, S_n\}$$
$$E_{\Omega_p(\mathcal{F})} = \{S_iS_j \mid i \neq j \text{ and } |S_i \cap S_j| \geq p\}$$

DEFINITION: A graph G is a ***p-intersection graph*** if there exists a family of sets $\mathcal{F} = \{S_1, S_2, \ldots, S_n\}$ such that $G \cong \Omega_p(\mathcal{F})$.

Thus, the 1-intersection graphs are the ordinary intersection graphs.

Example 7.4.6: The subsets of $\{1, 2, \ldots, 7\}$ that label the vertices of the graph in Figure 7.4.7 show that it is a 2-intersection graph.

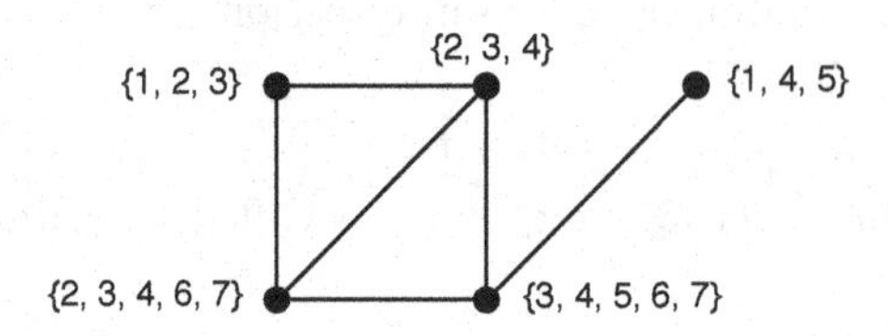

Figure 7.4.7 A 2-intersection graph.

Proposition 7.4.11: *Every graph is a p-intersection graph for all integers $p \geq 1$.*

Proof: Proposition 7.4.1 establishes the base of an inductive proof on p. Assume that a graph G is a p-intersection graph for some $p \geq 1$, and let $\mathcal{F} = \{S_1, S_2, \ldots, S_n\}$ be a family of sets such that $G \cong \Omega_p(\mathcal{F})$. Let $x \notin \bigcup_{i=1}^{n} S_i$. Then

$$\hat{\mathcal{F}} = \{S_1 \cup \{x\}, S_2 \cup \{x\}, \ldots, S_n \cup \{x\}\}$$

is a family of sets such that $G \cong \Omega_{p+1}(\hat{\mathcal{F}})$. ◇

Tolerance Graphs

Intersection graphs and their variations discussed in this section have been generalized to *tolerance graphs*. The most general version of the definition was introduced in [JaMcMu91] and [JaMcSc91].

DEFINITION: Let $\mathcal{F} = \{S_1,\ S_2, \ldots, S_n\}$ be a family of subsets of a finite set S; let $\phi : R^+ \times R^+ \to R^{\geq 0}$ be a symmetric function taking pairs of positive real numbers to nonnegative reals; let $\mu : \mathcal{P}(S) \to R^{\geq 0}$ be a function taking subsets of S to nonnegative reals; and let t_i be a positive real number, called a ***tolerance***, assigned to subset S_i, $i = 1, 2, \ldots, n$. The ***ϕ-tolerance intersection graph*** (or simply ***tolerance graph***) G of the family F with respect to ϕ, μ, and the tolerances t_i has vertex- and edge-sets given by

$$V_G = \{S_1,\ S_2, \ldots, S_n\}$$
$$E_G = \{S_i S_j \mid i \neq j \text{ and } \mu(S_i \cap S_j) \geq \phi(t_i,\ t_j)\}$$

DEFINITION: An n-vertex graph G is a ***ϕ-tolerance intersection graph*** if there exists a family $\mathcal{F}$, functions ϕ and μ, and tolerances t_i, $i = 1, 2, \ldots, n$, as defined above, such that G is isomorphic to the ϕ-tolerance intersection graph of $\mathcal{F}$ with respect to ϕ, μ, and the t_is.

Some Special Classes of Tolerance Graphs

For many of the classes of tolerance graphs under recent investigation, the function μ is taken to be the cardinality of a set or the length of an interval, and sometimes, the tolerances t_i are constant or are set equal to $\mu(S_i)$. Some of the choices for ϕ include the constant, minimum, maximum, sum, and absolute difference functions.

Example 7.4.7: A p-intersection graph of a family $\mathcal{F}$ is a ϕ-tolerance intersection graph, where $\mu(S') = |S'|$ for each $S' \subseteq S$, tolerance $t_i = \mu(S_i)$ for $i = 1, 2, \ldots, n$, and $\phi(t_i, t_j) = p$ for all pairs of tolerances.

DEFINITION: A ***min-tolerance interval graph*** of a family $\mathcal{F} = \{I_1, I_2, \ldots, I_n\}$ of intervals on the real line is a ϕ-tolerance intersection graph, where t_i is a positive real number assigned to interval I_i, $i = 1, 2, \ldots, n$, $\phi(t_i, t_j) = \min(t_i, t_j)$, and for any interval J, $\mu(J)$ is its length.

Proposition 7.4.12: *Every interval graph is a min-tolerance interval graph.*

◊ *(Exercises)*

Remark: The ***max-tolerance interval graph*** and ***sum-tolerance interval graph*** are analogously defined.

TERMINOLOGY NOTE: The term *tolerance* was introduced in [GoMo82, GoMoTr84] in the context of min-tolerance interval graphs, and several authors refer to these simply as *tolerance graphs*.

Example 7.4.8: Figure 7.4.8 depicts the cycle graph C_4 as a min-tolerance interval graph, which shows that the interval graphs are a proper subclass of the min-tolerance interval graphs.

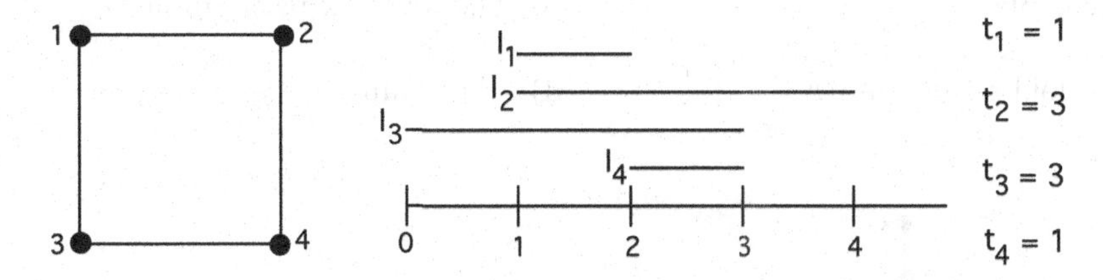

Figure 7.4.8 The cycle graph C_4 as a min-tolerance interval graph.

DEFINITION: Let $\mathcal{F} = \{S_1, S_2, \ldots, S_n\}$ be a family of subsets of a finite set S. An ***abdiff-tolerance intersection graph*** of $\mathcal{F}$ is a ϕ-tolerance intersection graph, where $\mu(S') = |S'|$ for each $S' \subseteq S$, t_i is a positive real number assigned to subset S_i, $i = 1, 2, \ldots, n$, and $\phi(t_i, t_j) = |t_i - t_j|$.

Example 7.4.9: Figure 7.4.9 depicts the cycle graph C_4 as an abdiff-tolerance graph. Next to each vertex is its corresponding subset and tolerance.

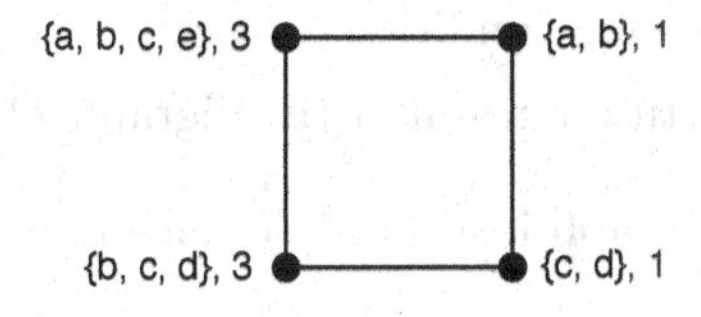

Figure 7.4.9 The cycle graph C_4 as an abdiff-tolerance intersection graph.

DEFINITION: Let ϕ be a symmetric function mapping pairs of nonnegative integers to nonnegative integers. A graph is a ***ϕ-tolerance competition graph*** if it is the ϕ-tolerance intersection graph of the family of out-sets of the vertices of some digraph.

Remark: For examples and results on ϕ-tolerance competition graphs, see [AnLaLuMcMe94], [BrMcVi95], [BrMcVi96], and [BrCaVi100]. Some of the functions ϕ considered in those papers include $\phi(t_i, t_j) = min(t_i, t_j)$, $max(t_i, t_j)$, $t_i + t_j$ and $|t_i - t_j|$.

Remark: For several other results, examples, and applications of intersection graphs and tolerance graphs, see the monograph by McKee and McMorris [McMc99]. Also, Golumbic and Trenk have written the first book [GoTr03] devoted entirely to tolerance graphs.

EXERCISES for Section 7.4

7.4.1 Draw the interval graph for the intervals $(0, 2), (3, 8), (1, 4)(3, 4), (2, 5), (7, 9)$.

7.4.2 Show that the complete graph K_n is an interval graph for all $n \geq 1$.

7.4.3 Show that for all $n \geq 4$, the cycle graph C_n is not an interval graph.

7.4.4 Determine the intersection number of the given graph, and justify your answer.

(a) The complete graph K_n

(b) The 3-vertex path graph P_3

(c) The n-fold self-union nK_2

(d) The complete graph $K_{2,3}$

(e)

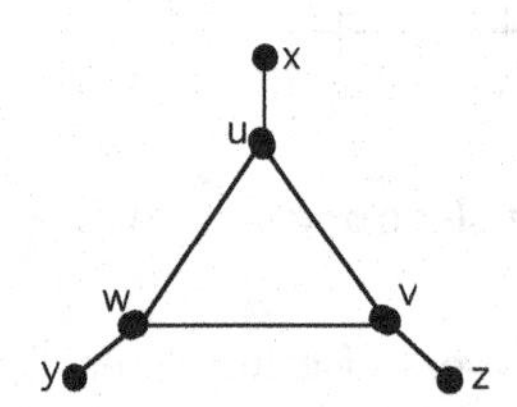

7.4.5 Prove Proposition 7.4.2.

7.4.6 Prove Proposition 7.4.9.

7.4.7 This problem refers to the proof of Theorem 7.4.8.

(a) Show that the family $\mathcal{K}$ of in-sets defined in the first part of the proof is an edge clique cover that satisfies the labeling condition.

(b) Show that G is the competition graph of the digraph D defined in the second part.

7.4.8 Show that every interval graph is a min-tolerance interval graph.

7.5 Graph Homomorphisms

Recall from §2.1 that an isomorphism between two simple graphs is a mapping that preserves adjacency and nonadjacency. The more general mapping, a graph homomorphism, preserves adjacency, but not necessarily nonadjacency. This section explores this second type of mapping.

REVIEW FROM §2.1:

DEFINITION: A ***graph homomorphism*** $f : G \to H$ from a graph G to a graph H is a pair of functions $(f_V : V_G \to V_H, f_E : E_G \to E_H)$ between their vertex- and edge-sets, respectively, such that for every edge $e \in E_G$, the function f_V maps the endpoints of e to the endpoints of edge $f_E(e)$. When there is a graph homomorphism $f : G \to H$, we say that G is ***homomorphic*** to H.

DEFINITION: A ***graph isomorphism*** $f : G \to H$ from a graph G to a graph H is a graph homomorphism $(f_V : V_G \to V_H, f_E : E_G \to E_H)$ for which f_V and f_E are bijections. When there is a graph isomorphism $f : G \to H$, we say that G is ***isomorphic*** to H, denoted $G \cong H$.

Theorem 2.1.1 *Let G and H be simple graphs. If (f_V, f_E) is a graph homomorphism from G to H, then f_V preserves adjacency. Conversely, if $f_V : V_G \to V_H$ preserves adjacency, then there exists a unique function $f_E : E_G \to E_H$, such that (f_V, f_E) is a graph homomorphism from G to H.*

NOTATION: The subscripts V and E on f are sometimes omitted when no ambiguity can result.

Example 7.5.1: Let $v_0, v_1, \ldots, v_n$ be the successive vertices along a path graph P_{n+1} of length n, and let $w_0, w_1, \ldots, w_{n-1}$ be the successive vertices along an n-vertex cycle graph C_n. The vertex mapping $f_V : V_{P_{n+1}} \to V_{C_n}$ given by $f(v_0) = f(v_n) = w_0$ and $f(v_i) = w_i$ for $i = 1, 2, \ldots, n-1$ is adjacency-preserving. Thus, if $f_E : E_{P_{n+1}} \to E_{C_n}$ is its induced edge mapping, then $f = (f_V, f_E)$ is a graph homomorphism from the path graph of length n to the n-vertex cycle graph. Figure 7.5.1 illustrates the mapping for $n = 4$.

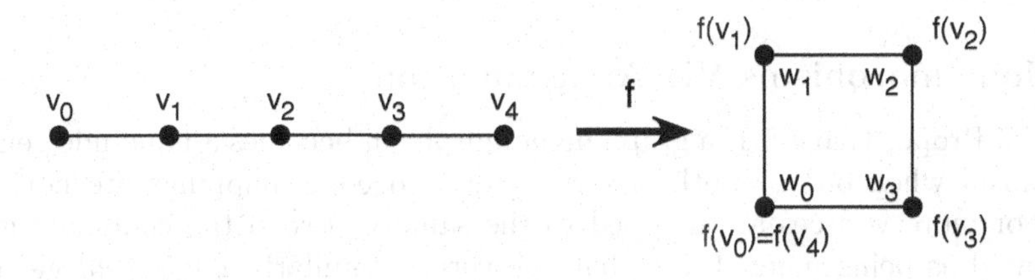

Figure 7.5.1 A graph homomorphism from P_5 to C_4.

Example 7.5.2: A graph homomorphism $f : Q_3 \to P_4$ is represented in Figure 7.5.2 by the vertical projection of Q_3 downward onto P_4. The shading of the vertices and the features of the edges further reinforce the imagery.

DEFINITION: The ***image of a graph*** G under a graph homomorphism $f : G \to H$ is the subgraph of H whose vertex-set is $f(V_G)$ and whose edge-set is $f(E_G)$.

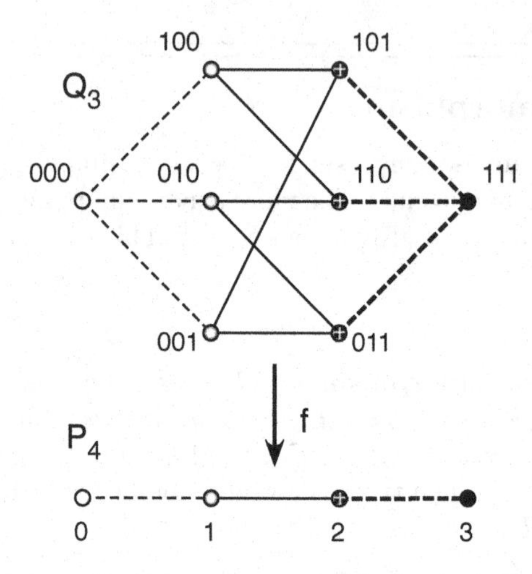

Figure 7.5.2 Mapping the cube graph Q_3 to a path of length 3.

Example 7.5.3: Figure 7.5.3 shows a graph homomorphism f from a simple graph to a non-simple graph. The vertex and edge labels on the codomain graph indicate the preimages of the functions f_V and f_E.

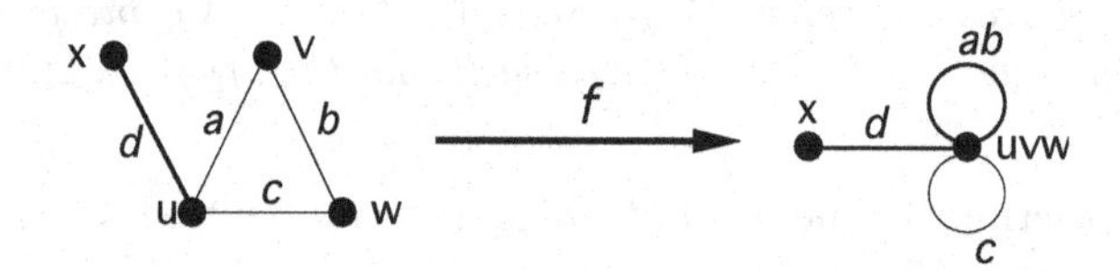

Figure 7.5.3 Mapping a simple graph to a non-simple graph.

Proposition 7.5.1: *If both the vertex function and the edge function of a graph homomorphism $G \to H$ are one-to-one (but not necessarily onto), then the domain graph G is isomorphic to its image in the codomain graph H.*

Proof: The functions $f_V : V_G \to f(V_G)$ and $f_E : E_G \to f(E_G)$ are clearly onto. ♢

Graph Homomorphisms Via Amalgamation

In light of Proposition 7.5.1, a graph homomorphism becomes a bona fide generalization of isomorphism when one or both of the vertex and edge mappings are not one-to-one. When two or more vertices are mapped to the same vertex in the codomain graph, they may be viewed as being merged into that one vertex. Similarly, edges that get mapped to the same edge in the codomain are merged into that one edge. These observations motivate the following definitions of two kinds of amalgamation that provide a different perspective for graph homomorphisms.

DEFINITION: Let G be any graph, and let $\pi = \{V_1, V_2, \ldots, V_t\}$ be a partition of its vertex-set V_G. The ***vertex-partition amalgamation*** of graph G corresponding to partition π is the transformation of G into a graph G/π that results from merging (amalgamating) all of the vertices in each cell of the partition.† That is, for each $i = 1, 2, \ldots, t$, the vertices in subset V_i are merged into a single vertex, generically denoted V_i. Thus, the vertex-set

†This notation is analogous to the one used for quotient groups in group theory.

$V_{G/\pi} = \{V_1, V_2, \ldots, V_t\}$, and the edge-set $E_{G/\pi} = E_G$, except that any edge in G that had $u \in V_i$ as an endpoint now has, in G/π, the amalgamated vertex V_i as an endpoint instead.

Example 7.5.4: Figure 7.5.4 shows how the cycle graph C_4 can be obtained by a vertex-partition amalgamation of the path graph $P_5 = \langle v_0, v_1, v_2, v_3, v_4 \rangle$ corresponding to the partition $\pi = \{\{v_0,\ v_4\}, \{v_1\}, \{v_2\}, \{v_3\}\}$. The amalgamated vertex is labeled by juxtaposing the names of the vertices that were merged.

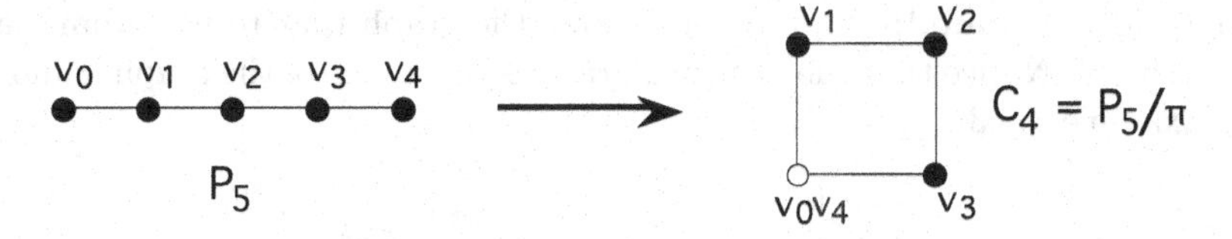

Figure 7.5.4 Cycle graph C_4 obtained from a vertex-amalgamation of P_5.

Notice that this transformation is identical to the graph homomorphism of Example 7.5.1.

Example 7.5.5: Figure 7.5.5 shows the vertex-partition amalgamation of the cube graph Q_3 corresponding to the partition of the vertices according to the number of 1s in their label.

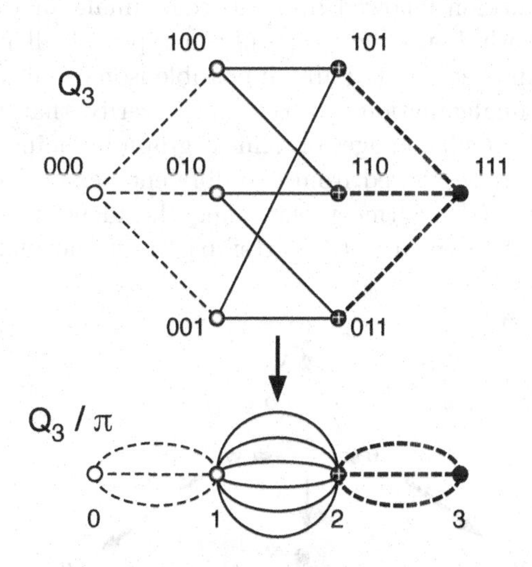

Figure 7.5.5 A vertex-partition amalgamation of the cube graph Q_3.

Example 7.5.6: Figure 7.5.6 shows the vertex-partition amalgamation of the domain graph from Example 7.5.3, where the corresponding partition of the vertex-set is $\{\{x\}, \{u, v, w\}\}$.

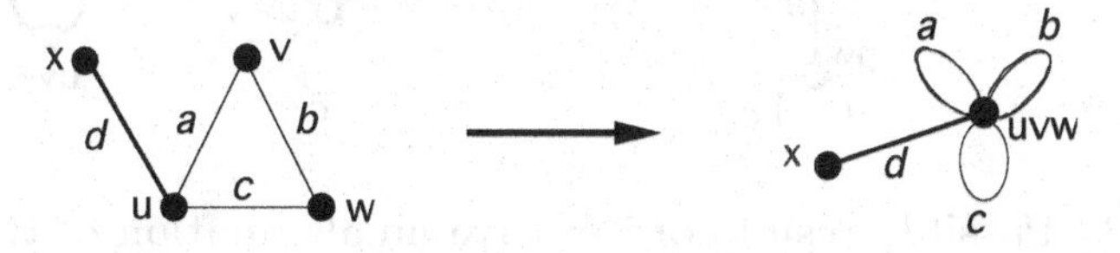

Figure 7.5.6 A vertex-partition amalgamation based on $\{\{x\}, \{u, v, w\}\}$.

Multi-Edge Amalgamation

DEFINITION: Let G be any graph, and let $A = \{e_1, e_2, \ldots, e_s\} \subseteq E_G$ be any subset of edges having the same endpoints. The ***multi-edge amalgamation*** corresponding to A is the graph G_A that results from merging (amalgamating) all of the edges in A into a single edge generically denoted $e_1 e_2 \cdots e_s$. Thus,

$$V_{G_A} = V_G \quad \text{and} \quad E_{G_A} = (E_G - A) \cup \{e_1 e_2 \cdots e_s\}$$

Example 7.5.6 continued: Figure 7.5.7 shows the graph that results from merging the edges in $A = \{a, b\}$. Notice that the resulting graph is the image of the graph homomorphism defined in Example 7.5.3

Figure 7.5.7 A multi-edge amalgamation of $A = \{a, b\}$.

Example 7.5.7: The six isomorphism types of graphs derivable from the cycle graph C_3 by vertex-partition amalgamation followed by multi-edge amalgamation are shown in Figure 7.5.8. The graphs at depth 1 are the isomorphism types of all possible vertex-partition amalgamations. The graphs at depth 2 are all possible isomorphism types that result from subsequent multi-edge amalgamations. It is easy to verify that these six types are the isomorphism types of all possible images of a linear graph mapping of C_3. Observe that the vertex and edge labels can be viewed from two different perspectives, either as preimages of graph homomorphisms (the leftmost one being the identity isomorphism), or as the vertices and edges that are merged in the vertex-partition and multi-edge amalgamations, respectively.

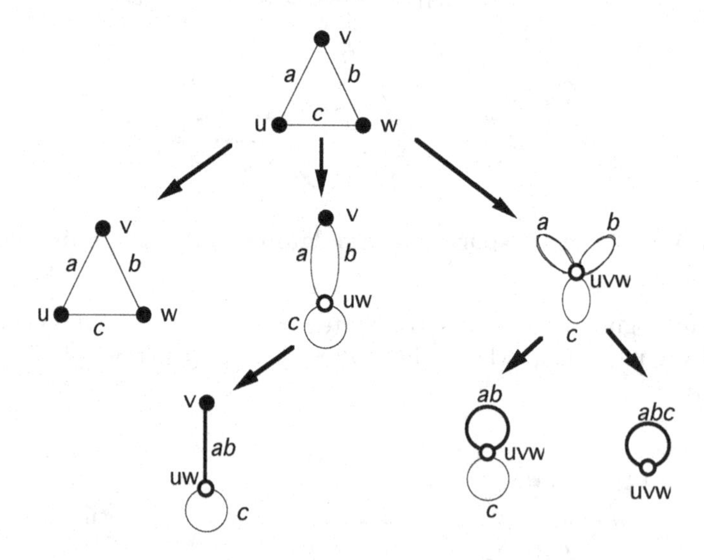

Figure 7.5.8 Possible results of iterative amalgamation of the graph C_3.

Complete Multi-Edge Amalgamation

DEFINITION: The ***complete multi-edge amalgamation*** of any graph G is the graph that results from taking each multi-edge and merging its edges into a single edge.

Example 7.5.8: The complete multi-edge amalgamation of the graph Q_3/π (from Example 7.5.5) results in the path graph P_4, as shown in Figure 7.5.9. Recall that P_4 was the image of the graph homomorphism given in Example 7.5.2.

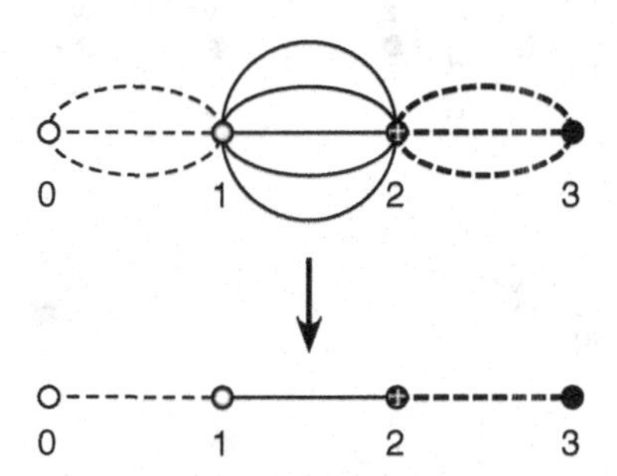

Figure 7.5.9 The complete multi-edge amalgamation of the graph Q_3/π.

Remark: The last three examples showed how a graph homomorphism can be realized by vertex-partition amalgamation followed by iterative (or complete) multi-edge amalgamation. In fact, every graph homomorphism that is onto (both f_V and f_E) can be realized using these two kinds of amalgamations, and vice versa. We state and prove the result for simple graphs.

Proposition 7.5.2: *Let $f : G \to H$ be a graph homomorphism from a simple graph G* **onto** *a simple graph H. Then H can be obtained from G by a vertex-partition amalgamation G/π of G followed by a complete multi-edge amalgamation of G/π. Conversely, any vertex-partition amalgamation of a simple graph G followed by a complete multi-edge amalgamation that results in a graph H corresponds to a graph homomorphism from G onto H, where H has no multi-edges.*

Proof: Let $f^{-1}(\{w\})$ be the preimage of vertex w for each $w \in V_H$, and let $f^{-1}(\{e\})$ be the preimage of edge e for each $e \in E_H$. Then $\pi = \{f^{-1}(\{w\}) \mid w \in V_H\}$ is a partition of V_G, and each $f^{-1}(\{e\})$ corresponds to a multi-edge in the vertex-partition amalgamation G/π (since f preserves adjacency). It follows that H is obtained from the complete multi-edge amalgamation of G/π. The converse assertion is an immediate consequence of the definitions of the vertex-partition and complete multi-edge amalgamations. ◇

Mapping by Merging Two Ordinary Edges

Merging any two edges of a graph G requires amalgamating their endpoint sets. If each of the other vertices and edges of G is mapped to itself, then the resulting transformation is a graph homomorphism from G onto the transformed graph. The next two examples illustrate four instances of this kind of transformation.

Example 7.5.9: The first and last edge of the path graph P_5 could be merged in two different ways (without merging the other two edges), according to how their endpoints are merged. This is determined by the vertex-partition amalgamation that is used. The transformation shown on the left in Figure 7.5.10 results from the vertex-partition amalgamation corresponding to the partition $\{\{v,z\},\{w,y\},\{x\}\}$. The one on the right uses the partition $\{\{v,y\},\{w,z\},\{x\}\}$.

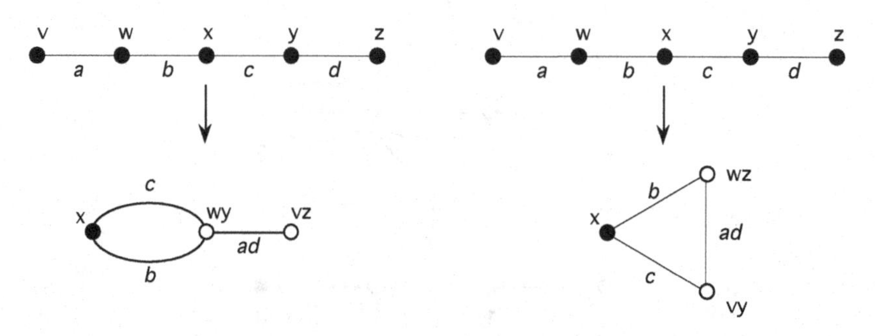

Figure 7.5.10 Two ways to merge the first and last edges of P_5.

Example 7.5.10: The second and third edges of the path graph P_5 could be merged in two different ways (without merging the other two edges). The transformation shown on the left in Figure 7.5.11 results from the vertex-partition amalgamation corresponding to the partition $\{\{w,y\},\{v\},\{x\},\{z\}\}$. The one on the right uses the partition $\{\{v\},\{z\},\{w,x,y\}\}$.

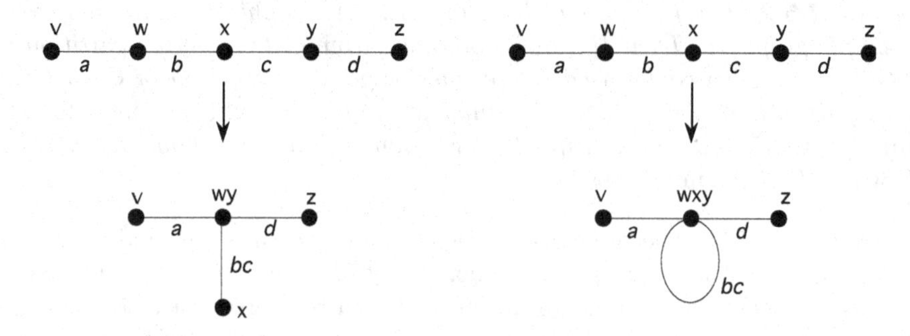

Figure 7.5.11 Two ways to merge the second and third edges of P_5.

Graph Homomorphisms of Paths

A walk of length n in a graph can be regarded as the image of a path of length n under a graph homomorphism.

Example 7.5.11: Figure 7.5.12 shows a path P of length 6 being mapped to a walk W of length 6. Each vertex label along walk W indicates the preimage or preimages of the vertex.

Proposition 7.5.3: *Every connected graph is the image of a path.*

Proof: In a connected graph, there is a walk that traverses every edge. According to the remark immediately above, there is a path whose image is that walk. ◇

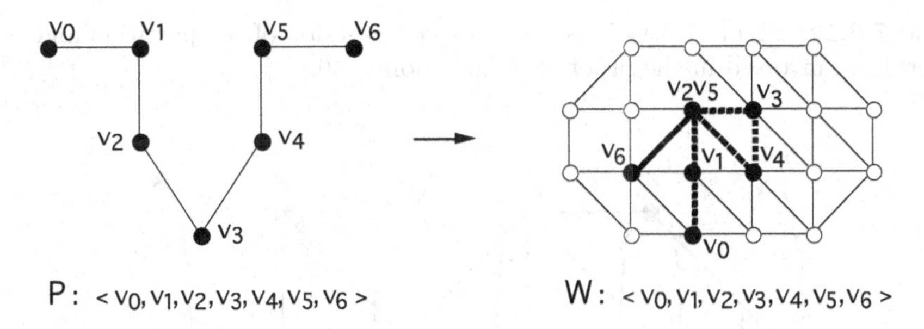

Figure 7.5.12 Mapping a path of length 6 to a walk of length 6.

Proposition 7.5.4: *Let $f : G \to H$ be a graph homomorphism of graphs. Then G is bipartite whenever $f(G)$ is bipartite.*

Proof: Suppose that $f(G)$ is bipartite, and that (U, V) is the corresponding bipartition of $V_{f(G)}$. If there were an edge e joining two vertices of $f^{-1}(U)$, then its endpoints would both be in U, which would imply (by the definition of a graph homomorphism) that its image $f(e)$ would join two vertices of U. An edge joining two vertices of $f^{-1}(V)$ would lead to the same contradiction. ♢

Corollary 7.5.5: *A graph can be linearly mapped onto a path if and only if it is bipartite.*

Proof: It follows from Proposition 7.5.4 that if a graph G can be linearly mapped onto a path, which is a bipartite graph, then G itself must be bipartite.

Conversely, consider a bipartite graph G on vertex parts X and Y. Graph G can be linearly mapped onto the path P_2 of length 1, by mapping all the vertices of part X to one endpoint of P_2 and all the vertices of part Y to the other. All the edges are mapped to the only edge of P_2. ♢

Mapping Graphs Onto Pseudopaths

DEFINITION: A ***pseudopath of length*** n consists of a path of length n with one or more self-loops at one or more of its vertices. (See Figure 7.5.13.)

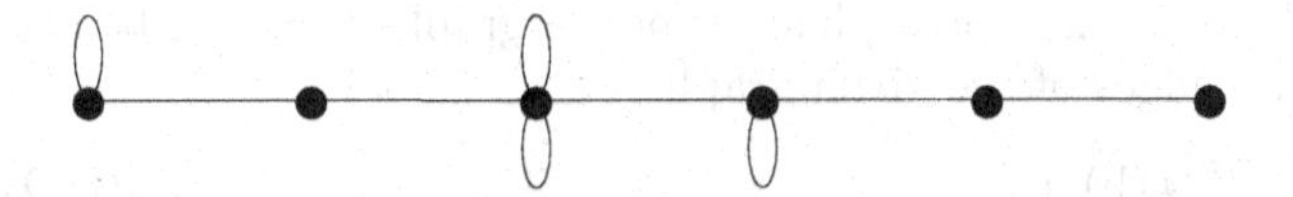

Figure 7.5.13 A pseudopath of length 5.

Proposition 7.5.6: *Let G be a connected graph with diameter d. There exists a graph homomorphism of G onto a pseudopath of length d.*

Proof: Let v be a vertex with maximum eccentricity d, and let $W_i = \{x \in V_G \mid d(v, x) = i\}$, $i = 0, \ldots d$. Then the collection of subsets $\{W_0, W_1, \ldots W_d\}$ is a partition of V_G. Moreover, any two adjacent vertices in G are either both in one W_j, $1 \leq j \leq d$, or separately in a W_j and W_{j+1}, $0 \leq j \leq d-1$. Let P be the pseudo path $\langle v_0, v_1, \ldots v_d \rangle$ of length d having a self-loop on vertex v_i if there is an edge joining two vertices in W_i, and consider the mapping $f : V_G \to V_P$ given by $f(x) = v_i$ if $x \in W_i$. It follows from the properties of the partition that f is a graph homomorphism from G onto P. ♢

Example 7.5.12: Figure 7.5.14 illustrates the construction of the partition and the graph homomorphism involved in the proof of Proposition 7.5.6.

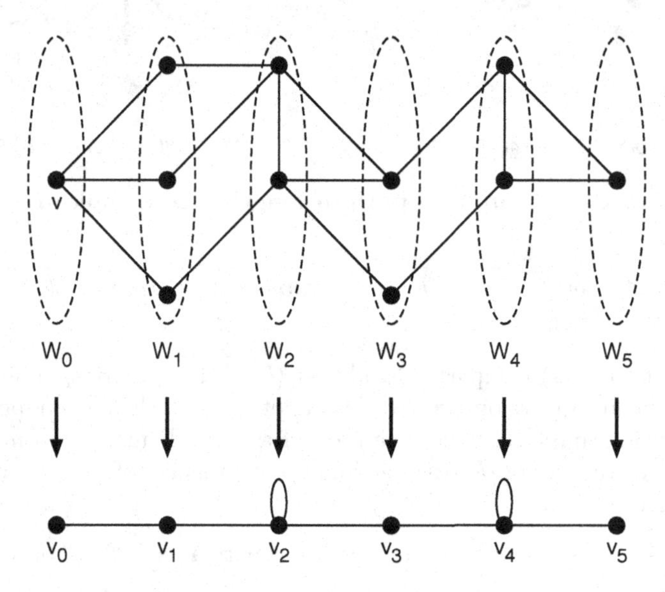

Figure 7.5.14 Mapping a graph to a pseudopath.

EXERCISES for Section 7.5

7.5.1 Draw all the different isomorphism types of graphs that can be obtained by amalgamating exactly two vertices of the given graph.

(a) P_3 (b) C_5 (c) W_5 (d) Q_3

(e) $K_4 - K_2$ (f) W_6 (g) C_6 (h) $K_{3,3}$

7.5.2 Draw all the different isomorphism types of graphs that can be obtained by amalgamating exactly two edges of the given graph.

(a) P_3 (b) C_5 (c) W_5 (d) Q_3

(e) $K_4 - K_2$ (f) W_6 (g) C_6 (h) $K_{3,3}$

7.5.3 Draw all the different isomorphism types of graphs that can be obtained as images of the given graph under a graph homomorphism.

(a) P_3 (b) P_4 (c) $K_{1,3}$ (d) $2K_2$

7.5.4 The graph at the tail of the arrow is to be the homomorphic image of the graph at the head (i.e., via an onto graph homomorphism). Construct a table with all the vertices and edges of the domain graph in the first row. In the second row, below each vertex or edge, write its image in the codomain graph.

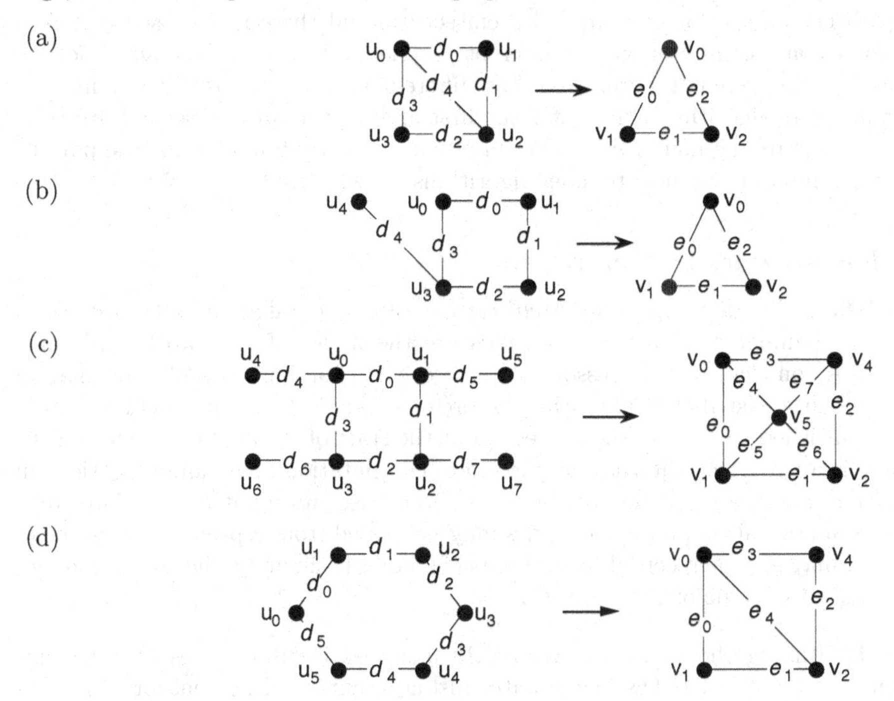

7.5.5 Describe a graph homomorphism from the designated domain onto the designated codomain.

(a) From the cube graph Q_3 onto the complete graph K_4.

(b) From the cycle graph C_{10} onto the complete graph K_5.

(c) From the cycle graph C_{2n} onto the path graph P_{n+1}, for $n \geq 1$.

(d) From the path graph P_n onto an arbitrary path graph of shorter length.

(e) From the cycle graph C_{kn} onto the cycle graph C_n, for $k, n \geq 1$.

7.5.6 Prove that there is no graph homomorphism from the designated domain onto the designated codomain.

(a) From the cycle graph C_3 onto the path graph P_4.

(b) From the cycle graph C_3 onto the path graph P_3.

(c) From $K_4 - K_2$ onto the path graph P_6.

(d) From $K_4 - K_2$ onto the path graph P_5.

(e) From $K_4 - K_2$ onto the cycle graph C_4.

(f) From the cycle graph C_{2n+1} onto the cycle graph C_{2n}, for $n \geq 1$.

(g) From the cycle graph C_{2n} onto the cycle graph C_{2n-1}, for $n \geq 3$.

7.5.7 Give an example of a graph homomorphism of a connected graph *onto* a connected graph such that the cycle rank (see §3.2) of the codomain is larger than the cycle rank of the domain.

7.5.8 Give an example of a graph homomorphism of connected graphs such that the cycle rank of the codomain is smaller than the cycle rank of the domain.

7.6 MODELING NETWORK EMULATION

Sometimes a computational algorithm for some fixed process, such as matrix inversion or sorting, has been written for one parallel architecture, and there is a subsequent need to run that process on a computer with a different parallel architecture. Instead of designing, coding, and testing a new program for that different parallel architecture, it may be simpler to *emulate* the algorithm written for the first architecture on the second architecture, by means of a graph homomorphism. We briefly discuss graph models for the parallel architecture of a computer and how parallel algorithms are executed.

Graph Models for Parallel Computers

Application 7.6.1: *Parallel Computer Architectures* In a parallel architecture for a computer, a large number of identical processors are the nodes of a network, and some miniaturized wires connect each processor to a few others. For purposes of coordinating all the processors in a distributed algorithm, interconnection network models are highly symmetrical. Input is fed to some of the processors at the start of the execution. During the execution, there is a two-phase alternating pattern of computation and communication. In the computational phase, each of the identical processors uses its input to calculate some output. In the communication phase, some bits may be passed from a processor to some of its neighbors and may also be received from some neighbors. Ultimately, the correct answer is collected as output of some of the processors.

Example 7.6.1: Finding the minimum among 2^n unsorted items requires $O(2^n)$ steps, since each item must be considered as a candidate. Just as a tennis tournament for 2^n players can find a winner in n elimination rounds, a parallel computer can find a minimum among 2^n unsorted items in n computational steps. A parallel computer for finding the minimum of 2^n items can be modeled as a hypercube of processors labeled by binary numerals, as shown in Figure 7.6.1.

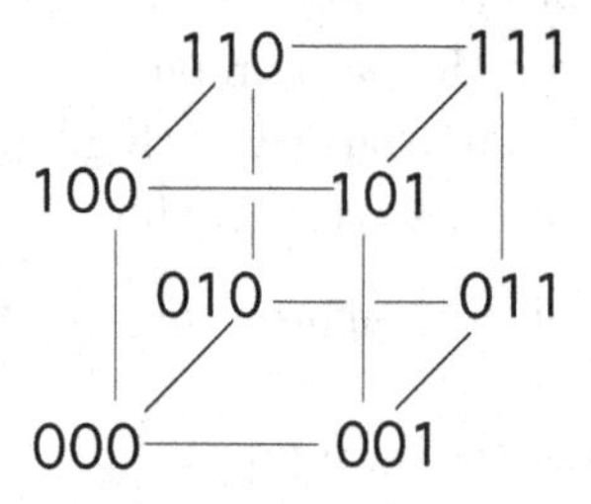

Figure 7.6.1 Small-scale model of a hypercube parallel computer.

On the first step, the item in each processor is compared with the item in the processor whose bitstring label differs only in the first bit (reading right to left), and they are swapped, if necessary, so that the smaller item moves to the processor indexed by the smaller numeral. Thus, in Figure 7.6.1, all the winners of the first round would be in the left plane of the hypercube.

On the second step, the item in each processor is compared to the item in the processor whose bitstring label differs only in the second bit, and they are swapped, if necessary, so that, once again, the smaller item moves to the processor indexed by the smaller numeral. Thus, in Figure 7.6.1, all the winners of the second round would be in the front plane of the hypercube, and so on.

In general, within n steps, the minimum item is in processor $00\cdots 0$. The execution time of this parallel algorithm is only $O(n)$, a vast improvement over $O(2^n)$ strictly sequential comparison steps. Figure 7.6.2 illustrates the application of this algorithm to a list of eight items.

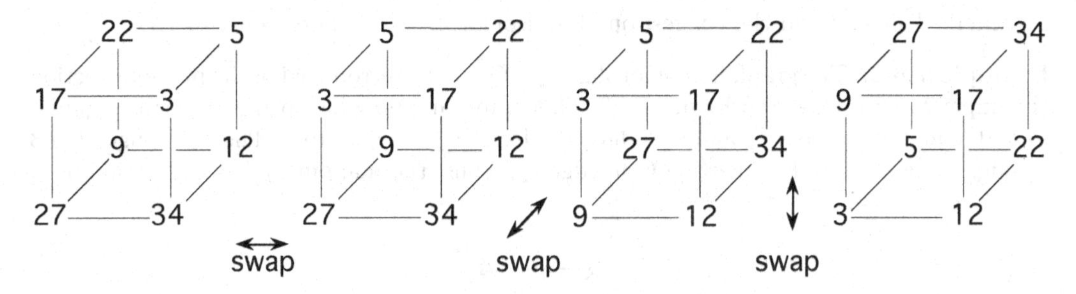

Figure 7.6.2 Finding the minimum on a hypercube.

A Model for Porting Parallel Algorithms

Application 7.6.2: *Emulating an Interconnection Network* It is sometimes necessary to execute a distributed algorithm on a parallel computer whose processors and network configuration differ from those on which the distributed algorithm is based, via a process called ***emulation***. In an emulation, the network model for which the distributed algorithm was written is called the *guest*, and the network model on which it is to be executed is called the *host*. The process of *porting* the algorithm from the guest parallel computer to the host parallel computer is modeled by a graph mapping from the guest network to the host network. This terminology has been adopted also by the graph-theoretic model.

TERMINOLOGY: For a graph mapping $f : G \to H$ in the context of network emulation, the domain graph G is called the ***guest***, and the codomain graph H is called the ***host***.

Sometimes the network emulation model requires a more general graph mapping, called a *semilinear* graph mapping, which we define later in this section. The terminology introduced in the next subsection applies to this more general graph mapping as well.

Load and Congestion of a Graph Mapping

Although a host processor can use its memory to do the job of more than one guest processor, it emulates just one of them at a time. The number of guest processors a host processor must emulate is called its *load*. Load delays a computation, compared to how long it would have taken on the guest network itself.

DEFINITION: The ***load at a vertex*** v of the host H of a graph mapping $f : G \to H$ is the number of vertices in the preimage $f_V^{-1}(\{v\})$.

DEFINITION: The ***load of a graph mapping*** $f : G \to H$ is its maximum load at any vertex, taken over all vertices of the host H.

Moreover, if two neighboring host processors are emulating more than one guest processor apiece, then only one message at a time can traverse the wire between those processors, lest two messages interfere. While emptying the message queues for all the competing messages, more delay occurs. The number of guest wires that must be emulated by a given host wire is called its *congestion*.

DEFINITION: The ***congestion on an edge*** e of the host H of a graph mapping $f : G \to H$ is the number of edges in the preimage $f_E^{-1}(\{e\})$.

DEFINITION: The ***congestion of a graph mapping*** $f : G \to H$ is its maximum congestion at any edge, taken over all edges of the host H.

Remark: The load and the congestion of an isomorphism are both equal to 1.

Example 7.6.2: The graph homomorphism $f : Q_3 \to P_4$ introduced in the previous section (Example 7.5.2) is shown in Figure 7.6.3. This graph mapping has load 1 at vertices v_0 and v_3 and load 3 at vertices v_1 and v_2. Thus, the load of f is equal to 3. There is congestion 3 on edges e_1 and e_3 and congestion 6 on edge e_2. Thus, the mapping f has congestion 6.

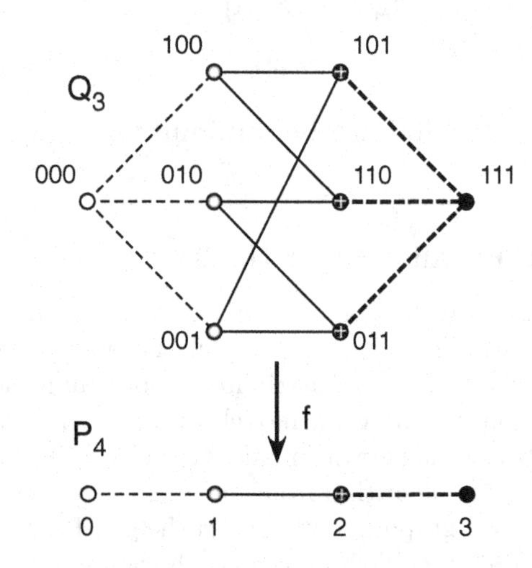

Figure 7.6.3 A graph homomorphism with load 3 and congestion 6.

Example 7.6.3: There are many possible mappings of the cube graph Q_3 to the complete graph K_4. The vertex labels on K_4 in Figure 7.6.4 specify the vertex images of one such mapping. Each edge e of Q_3 is mapped to the edge of K_4 whose endpoints are the images of the endpoints of e.

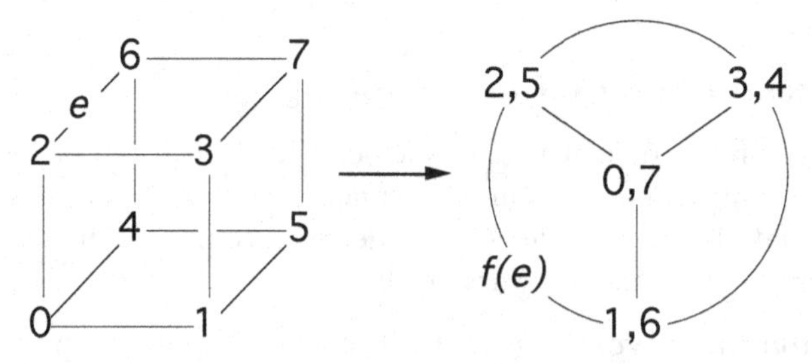

Figure 7.6.4 A graph homomorphism from Q_3 onto K_4.

Thus, the load of this graph homomorphism from Q_3 to K_4 equals 2, since two vertices of the guest Q_3 are mapped to every vertex of the host K_4. Also, it has congestion equal to 2, since two edges of the guest are mapped to every edge of the host.

Proposition 7.6.1: *Let $f : G \to H$ be any graph homomorphism from a guest graph G to a host graph H. Then the load of f is at least $\lceil |V_G|/|V_H| \rceil$, and the congestion of f is at least $\lceil |E_G|/|E_H| \rceil$.*

Proof: In the terminology of the generalized pigeonhole principle, the vertex function f_V has $|V_G|$ pigeons and $|V_H|$ pigeonholes, and the edge function f_E has $|E_G|$ pigeons and $|E_H|$ pigeonholes. ◊

Proposition 7.6.2: *Let $f : G \to H$ be a graph homomorphism, and let e be a proper edge of guest G such that both endpoints of e are mapped to the same vertex of the host H. Then $f(e)$ is a self-loop at that host vertex.*

Proof: This is an immediate consequence of the incidence-preservation property of a graph homomorphism. ◊

Corollary 7.6.3: *Let $f : K_{m,n} \to H$ be a graph homomorphism to a graph with no self-loops, and let u and v be vertices in different parts of the bipartition of $K_{m,n}$. Then $f(u) \neq f(v)$.*

Proof: This follows from the contrapositive of Proposition 7.6.2. ◊

Example 7.6.4: *Minimizing Load and Congestion* Consider the bipartite graph $K_{3,3}$, with 6 vertices and 9 edges, as guest and the bipartite graph $K_{2,3}$, with 5 vertices and 6 edges, as host. Then the load of any graph homomorphism is at least $\lceil 6/5 \rceil = 2$, and the congestion is at least $\lceil 9/6 \rceil = 2$, by Proposition 7.6.1. Figure 7.6.5 depicts a graph homomorphism that achieves these lower bounds. The number of strikes on an edge of the host indicates its congestion.

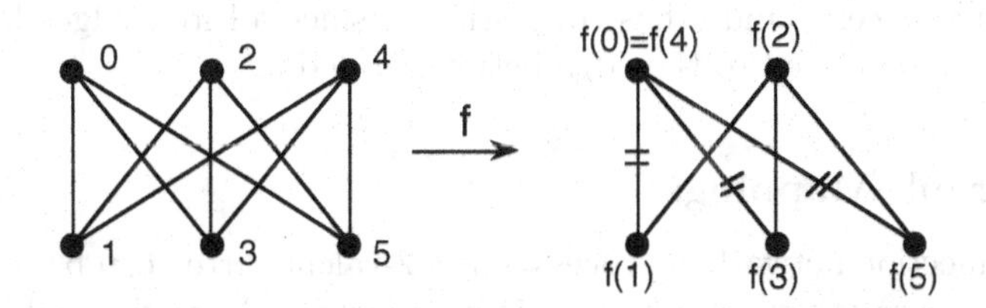

Figure 7.6.5 A homomorphism $f : K_{3,3} \to K_{2,3}$ with load 2 and congestion 2.

Example 7.6.5: *Minimizing Load and Congestion* Consider the bipartite graph $K_{3,3}$, with 6 vertices and 9 edges, as guest and the complete graph K_4, with 4 vertices and 6 edges, as host. Then the load of any graph homomorphism is at least $\lceil 6/4 \rceil = 2$, and the congestion is at least $\lceil 9/6 \rceil = 2$. The lower bound on load is realized by the mapping f in Figure 7.6.6.

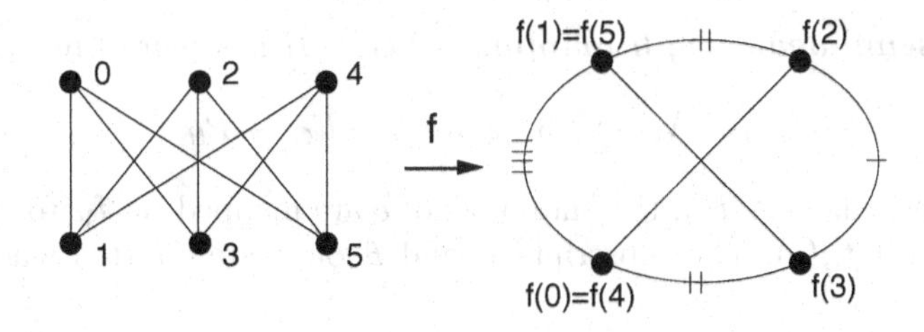

Figure 7.6.6 A homomorphism $f : K_{3,3} \to K_4$ with load 2 and congestion 4.

Figure 7.6.7 shows that congestion 3 is achievable by a linear graph mapping g with load 3.

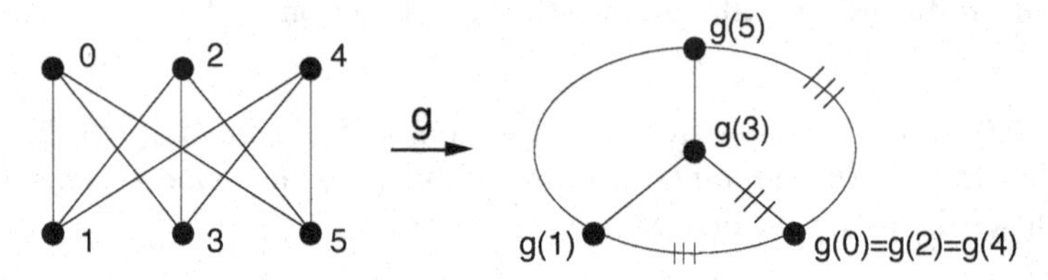

Figure 7.6.7 A homomorphism $g : K_{3,3} \to K_4$ with load 3 and congestion 3.

The following proposition shows that 3 is the minimum possible congestion for a graph homomorphism from $K_{3,3}$ to K_4.

Proposition 7.6.4: *The congestion of a graph homomorphism $f : K_{3,3} \to K_4$ is at least 3.*

Proof: Suppose that all three vertices in one part of the bipartition of $K_{3,3}$ are mapped to the same vertex v of K_4. Then, by incidence preservation, all nine edges of $K_{3,3}$ are mapped to the three edges of K_4 terminating at v. Thus, at least one of these three edges of K_4 carries congestion at least 3.

If neither part of the bipartition of $K_{3,3}$ is mapped to a single vertex of K_4, then Corollary 7.6.3 implies that one vertex b of K_4 is the image of two vertices u, v from one part of $K_{3,3}$, and another vertex c is the image of two vertices x, y from the other part. This implies that the edge between vertices b and c has congestion 4, since all four edges between the pairs u, v and x, y must be mapped to the edge between a and b. $\diamond$

Semilinear Graph Mappings

In an interconnection network, a processor at a 2-valent vertex can be regarded as a relay station, since whatever other computation it may have performed could just as well have been accomplished by the next processor. This insight motivated the introduction of a more general form of graph mapping, under which, in effect, an edge of the guest (domain) graph can be mapped to a path in the host (codomain). In a network emulation, the intermediate vertices along that path simply relay the information to the next node. This *dilating* of an edge into a path introduces an additional source of delay in the emulation, but it sometimes compensates for this cost by facilitating the construction of mappings that would otherwise be impossible.

NOTATION: P_H denotes the set of all paths in a graph H.

DEFINITION: A ***semilinear graph mapping*** $f : G \to H$ is a pair of functions

$$f_V : V_G \to V_H \quad \text{and} \quad f_E : E_G \to P_H$$

such that for every edge $e \in E_G$, the endpoints of e are mapped by f_V to the first and last vertices of the path $f_E(e)$. The subscripts V and E on f are omitted when no ambiguity can result.

Example 7.6.6: Figure 7.6.8 illustrates a semilinear graph mapping from P_2 to P_4, in which the only edge of P_2 is mapped to a path that spans P_4. The two vertices of P_2 are mapped to the first and last vertices of P_4. One may think of the single edge of the domain graph P_2 as being stretched into a path of length 3 in the codomain P_4.

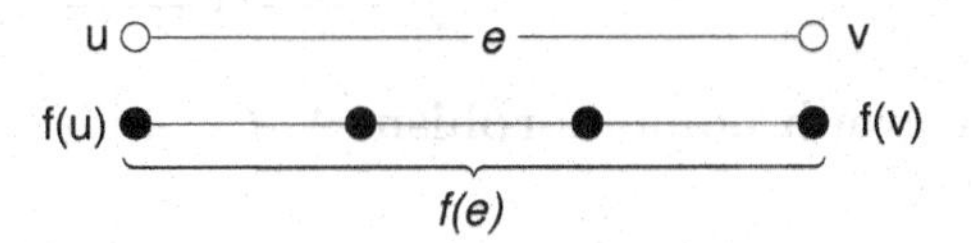

Figure 7.6.8 A length-1 path semilinearly mapped to a length-3 path.

DEFINITION: The ***dilation of an edge*** in the guest graph G under a semilinear graph mapping $f : G \to H$ is the length of the path onto which it is mapped.

DEFINITION: The ***dilation of a semilinear graph mapping*** $f : G \to H$ is the maximum dilation of any edge in the guest graph G.

Proposition 7.6.5: *A graph homomorphism is a semilinear graph mapping with dilation 1.*

Proof: This is an immediate consequence of the definitions. ◇

The following example demonstrates that dilation of edges permits the existence of semilinear graph mappings between graphs where no graph homomorphisms are possible.

Example 7.6.7: There is no graph homomorphism $C_5 \to C_4$, since the host C_4 is bipartite but the guest C_5 is not. Having a semilinear mapping, as illustrated in Figure 7.6.9, is regarded as much better than having no mapping. Each of the edges a, b, c, and d is mapped to one edge of the cycle. Edge e is stretched so that it wraps around the entire 4-cycle, exactly as shown. Thus, edge e has dilation 4, and the semilinear graph mapping f has dilation 4.

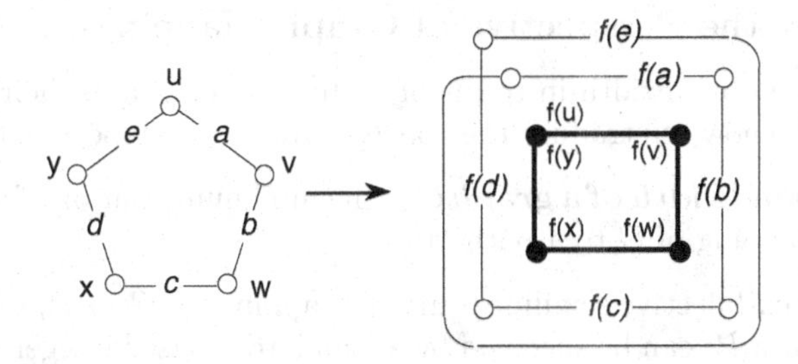

Figure 7.6.9 A 5-cycle (with one dilated edge) doubly wrapped on a 4-cycle.

Sometimes, dilating edges permits the congestion of the resulting semilinear mapping to be lower than the least possible congestion of a graph homomorphism.

Example 7.6.8: In any graph homomorphism $f : C_5 \to C_3$, there is a 3-path in the domain C_5 whose three edges are mapped to different edges of the codomain C_3. Moreover, the other two edges of C_5 are mapped to the same edge of C_3, as illustrated in Figure 7.6.10. (Observe that the more natural wraparound, with $f(d) = f(a)$ and $f(e) = f(b)$, is not linear.) Thus, the smallest possible congestion for a graph homomorphism $f : C_5 \to C_3$ is 3.

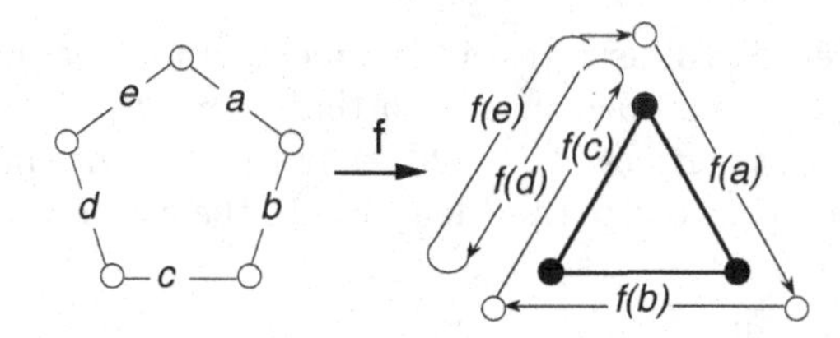

Figure 7.6.10 A graph homomorphism $f : C_5 \to C_3$ with congestion 3.

However, this congestion can be reduced to 2 with a semilinear graph mapping $h : C_5 \to C_3$, illustrated in Figure 7.6.11, in which one of the edges is dilated so that C_5 can be wrapped twice around C_3.

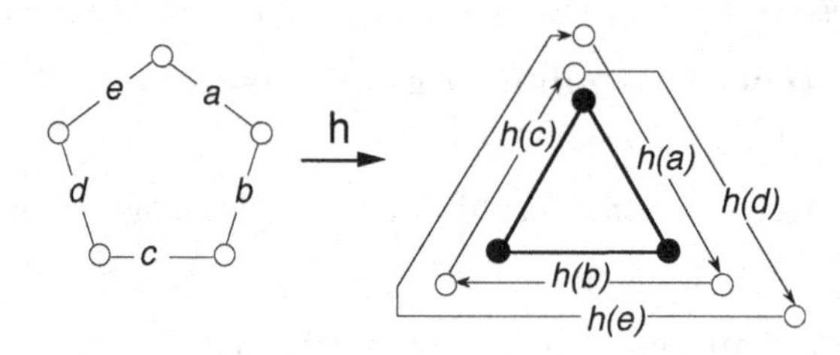

Figure 7.6.11 A semilinear graph mapping $h : C_5 \to C_3$ with congestion 2.

Proposition 7.6.6: *Given any graph G and any connected graph H, there is a semilinear graph mapping $G \to H$.*

Proof: Start with any vertex function $f : V_G \to V_H$. Given an edge $e \in E_G$ with endpoints u and v, define $f(e)$ to be any path between $f(u)$ and $f(v)$ in H. At least one such path exists, because H is connected. ◇

Bandwidth from the Perspective of Graph Mappings

In §7.3, we defined bandwidth in terms of vertex labelings and their corresponding adjacency matrices. We now reintroduce the concept from the perspective of graph mappings.

DEFINITION: The ***bandwidth of a graph*** G is the minimum dilation of any vertex-bijective semilinear graph mapping of G to a path graph.

Remark: Any vertex-bijective semilinear graph mapping $f : G \to P_n$ of an n-vertex graph G to the n-vertex path P_n can be specified by a standard 0-based integer labeling of V_G. The dilation of the mapping $f : G \to P_n$ will then be equal to the maximum difference between the labels on the endpoints of an edge, taken over all edges. Thus, the graph-mapping and the numbering-of-a-graph perspectives of bandwidth do indeed coincide.

This suggests the following approach to calculating the bandwidth of a graph with a high level of symmetry. For each of the essentially different (up to symmetry) standard vertex labelings of the given graph, calculate the dilation, as described above. The minimum of these dilations is the bandwidth of the given graph.

Example 7.6.9: [*Bandwidth of the complete graph* K_n] No matter how the labels $0, \ldots, n-1$ are assigned to the vertices of K_n, the numbers 0 and $n-1$ will be adjacent. Thus, the bandwidth is $n-1$.

Example 7.6.10: [*Bandwidth of the path* P_n] Assigning the labels $0, \ldots, n$ in sequence to the vertices of P_n corresponds to a mapping of dilation 1. Thus, the bandwidth of P_n is equal to 1.

Example 7.6.11: [*Bandwidth of the cycle* C_n] Since the cycle C_n has n edges, and since there are only $n-1$ consecutive pairs among the numbers $0, \ldots, n-1$, it follows that some adjacent pair of vertices of C_n must be labeled with non-consecutive numbers. Hence, the bandwidth of C_n is at least 2. This lower bound is realizable, as illustrated in Figure 7.6.12, by first assigning ascending even integers to adjacent vertices, starting from 0 anywhere on C_n, and then finishing with descending odd numbers.

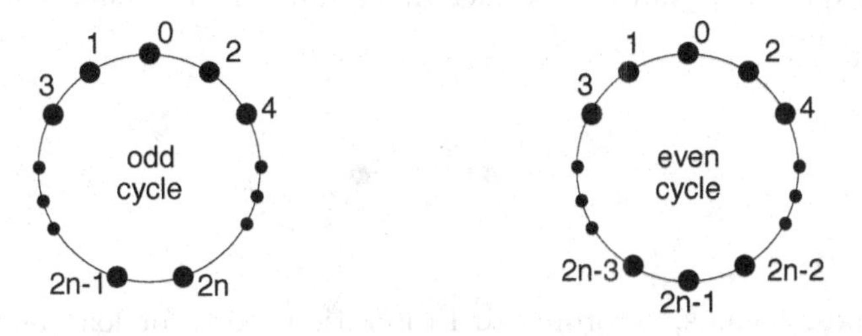

Figure 7.6.12 A vertex-bijective labeling of C_n with dilation 2.

EXERCISES for Section 7.6

7.6.1 Specify a graph homomorphism from the graph G to the graph H that has the prescribed load and congestion.

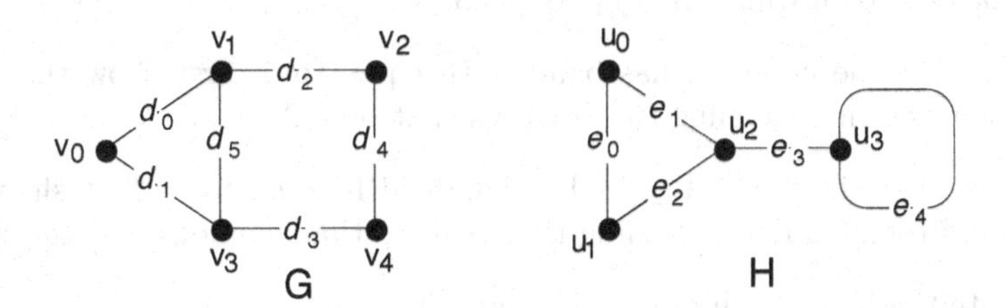

(a) Load 2 and congestion 2. (b) Load 2 and congestion 3.

(c) Load 2 and congestion 4. (d) Load 4 and congestion 4.

7.6.2 Specify a graph homomorphism from the given domain graph to the given codomain graph with the prescribed load and congestion.

(a) From Q_3 to C_4 with load equal to 2 and congestion equal to 3.

(b) From Q_3 to the dipole D_4 with load equal to 4 and congestion equal to 3.

(c) From Q_3 to K_4 with load equal to 2 and congestion equal to 2.

(d) From CL_6 to $K_{3,3}$ with load equal to 2 and congestion equal to 2.

7.6.3 Specify a graph homomorphism from the given domain graph to the given codomain graph with the minimum possible load.

(a) $K_{8,4} \to D_4$.
(b) $K_{m,n} \to C_4$.
(c) $C_6 \to K_4 - K_2$.
(d) $CL_4 \to C_4$.

7.6.4 Specify a graph homomorphism from the given domain graph to the given codomain graph with the minimum possible congestion.

(a) $K_{8,4} \to D_4$.
(b) $K_{m,n} \to C_4$.
(c) $C_6 \to K_4 - K_2$.
(d) $CL_4 \to C_4$.

7.6.5 The ***dumbbell graph*** has two vertices and one proper edge joining them, plus a self-loop at each vertex, as shown. Consider the graph homomorphisms from the complete graph K_4.

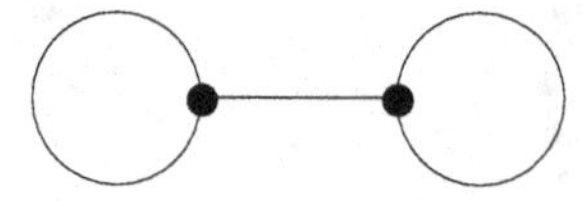

What are the lower bounds, according to Proposition 7.6.1, for load and for congestion? Prove that the lower bound for congestion cannot be achieved. (Hint: The vertex mapping completely determines the edge mapping.)

7.6.6 Do the previous exercise for the general case in which the guest is K_n, with $n \geq 3$.

NOTE: Some of the bandwidth exercises from §7.3 are repeated below for convenience.

7.6.7 Prove that the bandwidth of $K_{1,n}$ equals $\lceil \frac{n}{2} \rceil$.

7.6.8 Prove that the bandwidth of $K_{2,4}$ equals 3.

7.6.9 To prove that the cube Q_3 has bandwidth equal to 4, first show that 4 is a lower bound, and then exhibit a labeling that realizes that bound.

7.6.10 To prove that the wheel W_{2n-1} has bandwidth equal to n, first show that n is a lower bound, and then exhibit a labeling that realizes that bound. (See Exercise 7.3.6.)

7.6.11 Calculate the bandwidth of the given graph.

(a) Circular ladder CL_3.
(b) $K_n \times K_2$.
(c) $K_{m,n}$, for m, $n \geq 2$.
(d) $K_n \times K_m$.

7.6.12 Calculate the bandwidth of the graph obtained by amalgamating two 3-cycles at a vertex.

7.7 SUPPLEMENTARY EXERCISES

7.7.1 Construct a graph whose center and median subgraph have no vertices in common, or argue why no such graph exists.

7.7.2 Construct a graph whose periphery and median subgraph have a vertex in common, or argue why no such graph exists.

7.7.3 Let G be a connected graph and s an integer such that $rad(G) \leq s \leq diam(G)$. Prove that there is a vertex v with eccentricity s.

7.7.4 Let s and n be any two integers such that $1 \leq s \leq \frac{n}{2}$. Show that there exists an n-vertex connected graph G with $dom(G) = s$.

7.7.5 Let G be a graph with at least two components. Prove that the edge-complement graph $\overline{G}$ has $dom(\overline{G}) \leq 2$.

7.7.6 Find a largest minimal dominating set of the 10-vertex path graph P_{10}

7.7.7 Prove or disprove: For every tree, the domination number equals the cardinality of a minimum vertex cover.

7.7.8 Determine the bandwidth of the tripartite graph $K_{2,3,5}$.

7.7.9 Let $G = G_1 \cup G_2 \cup \cdots \cup G_t$ be the disjoint union of t graphs. Express the bandwidth $bw(G)$ in terms of the bandwidths $bw(G_1), bw(G_2), \ldots, bw(G_t)$.

7.7.10 Show that for any positive integer m, there exists a graph whose intersection number equals m.

7.7.11 Draw a 6-vertex, 12-edge simple graph with bandwidth = 3, or prove that none exists.

7.7.12 Draw a 3-regular, 8-vertex simple graph with bandwidth = 3.

7.7.13 Calculate the bandwidth of the Petersen graph.

7.7.14 Prove that 5 is the bandwidth of the wheel W_9.

7.7.15 Calculate $bw(K_{4,8})$. (Hint: consider diameter.)

7.7.16 Calculate the bandwidth of the given graph.

(a) CL_4 (b) CL_5 (c) CL_{2n} (d) CL_{2n+1}

(e) ML_4 (f) ML_5 (g) ML_{2n} (h) ML_{2n+1}

(i) W_4 (j) W_5 (k) W_{2n} (l) W_{2n+1}

7.7.17 Prove that any graph mapping from $K_{3,3}$ to P_3 has congestion at least 5.

7.7.18 Let $f : G \to H$ be a graph homomorphism onto graph H, and let $u, v \in V_G$. Prove that the distance between $f(u)$ and $f(v)$ is less than or equal to the distance between u and v.

7.7.19 Let $f : G \to H$ be a graph homomorphism onto graph H. Prove that the diameter of H is less than or equal to the diameter of G.

7.7.20 Label the vertices of $K_{4,8}$ with numbers 0 to 7 so as to indicate a graph homomorphism to C_8. This provides an upper bound on congestion of such a mapping. Calculate the congestion of your mapping.

7.7.21 Calculate a lower bound on the congestion of a graph homomorphism from $K_{4,8}$ to C_8. (Hint: consider diameter.)

7.7.22 Label each vertex of the graph G in Figure 7.7.1 with the name of the vertex of the graph H to which it is linearly mapped, so that the resulting mapping has load = 2. Prove that all graph homomorphisms from G to H have load = 2.

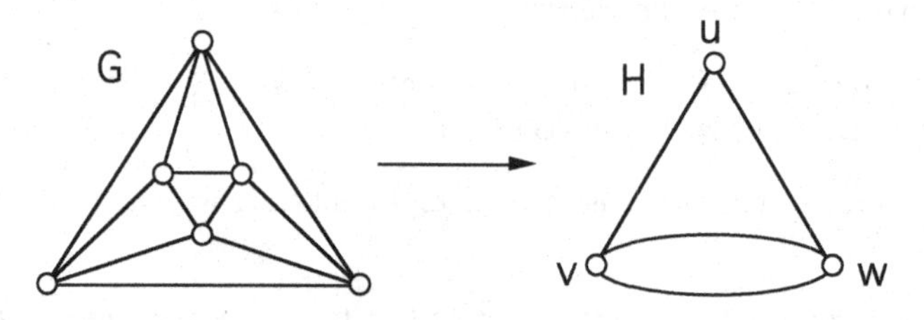

Figure 7.7.1

7.7.23 Calculate the minimum possible load for a graph homomorphism from graph G to graph H in Figure 7.7.2.

7.7.24 Calculate the minimum possible congestion for a graph homomorphism from graph G to graph H in Figure 7.7.2.

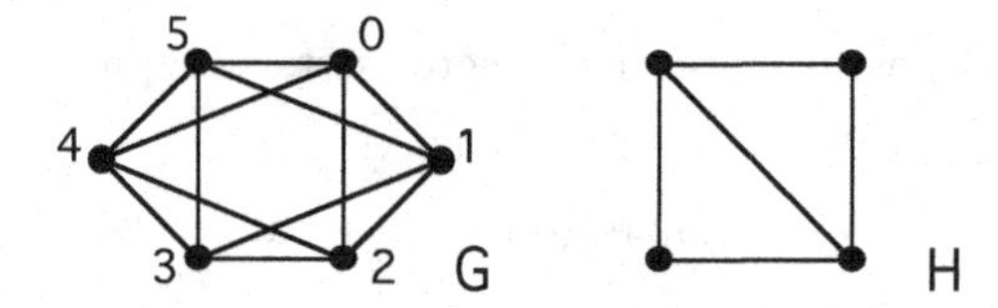

Figure 7.7.2

Glossary

adjacency-preserving vertex mapping: see graph homomorphism between two simple graphs.

bandwidth of a numbering f of a graph G, denoted $bw_f(G)$: the quantity given by $bw_f(G) = \max\{|f(u) - f(v)| \mid uv \in E_G\}$.

bandwidth numbering of a graph G**:** a numbering f that achieves $bw(G)$, i.e., $bw_f(G) = bw(G)$.

bandwidth of a graph G , denoted $bw(G)$: the quantity prescribed by the rule $bw(G) = \min\{bw_f(G) \mid f \text{ is a numbering of } G\}$; the notations $ban(G)$ and $B(G)$ are also used for the bandwidth of G.

bandwidth of a graph from the perspective of graph mappings: the minimum dilation of any vertex-bijective semilinear graph mapping of that graph to a path graph; equivalent to the number-of-a-graph perspective given above.

block of a graph G**:** a maximal connected subgraph that contains no cut-vertices of G.

center of a graph G, denoted $Z(G)$: the subgraph induced on the set of central vertices of G.

central vertex v of a graph G: a vertex with minimum eccentricity, i.e., such that $ecc(v) = rad(G)$.

chordal graph: a graph G such that for all $n \geq 4$, G does not contain an n-vertex cycle graph C_n as an induced subgraph.

chromatic number of graph G , denoted $\chi(G)$: the smallest number k such that there is a function $g : V(G) \to \{1, 2, \ldots, k\}$ with the property that, if $uv \in E_G$, then $g(u) \neq g(v)$.

circumference of a graph G with at least one cycle: the length of a longest cycle in G; denoted $circum(G)$. The circumference of an acyclic graph is undefined.

closed interval for two vertices u, v of a connected graph G, denoted $I[u, v]$: the union of the sets of vertices of G that lie on u-v geodesics.

closed interval for a vertex subset S of a connected graph G, denoted $I[S]$: the union $\bigcup_{u,v \in S} I[u, v]$.

competition graph of a digraph D**:** the intersection graph of the family of out-sets of the vertices of D.

competition graph: a graph G that is (isomorphic to) the competition graph of some digraph D.

—, ϕ**-tolerance:** a special kind of *tolerance graph.*

congestion on an edge of the codomain of a graph mapping: the number of edges mapped to that edge.

congestion of a graph mapping: the maximum congestion on any edge of the codomain.

convex hull of a vertex subset S in a graph G: the smallest convex set that contains the subset S.

convex vertex subset S in a graph G: a vertex subset S such that $I[S] = S$.

diameter of a graph G, denoted $diam(G)$: the maximum of the vertex eccentricities in G or, equivalently, the maximum distance between two vertices in G, i.e., $diam(G) = \max_{x \in V_G} ecc(x) = \max_{x,y \in V_G} d(x, y)$.

dilation of an edge in the domain of a semilinear graph mapping: the length of the path onto which that edge is mapped.

dilation of a semilinear graph mapping: the maximum dilation of any edge in the domain graph.

distance $d(s, t)$ from a vertex s to a vertex t in a graph G: the length of a shortest s-t path if one exists; otherwise, $d(s, t) = \infty$.

dominates: what a vertex does to itself and to each of its neighbors.

dominating set in a graph G: a vertex subset $D \subseteq V_G$ such that every vertex of G is in D or is adjacent to at least one vertex in D.

—, **connected:** a dominating set D such that the subgraph induced on D is connected.

—, **distance-k** for an integer $k \geq 1$: vertex subset D such that for all $v \in V_G - D$, there exists $x \in D$ such that $d(v, x) \leq k$.

—, **independent:** an independent set of vertices that is also a dominating set of G.

—, **minimal:** a dominating set such that every proper subset is non-dominating.

domination number of a graph G: the cardinality of a minimum dominating set of G; denoted $dom(G)$ (elsewhere, often $\gamma(G)$).

—, **connected,** denoted $c\text{-}dom(G)$; the cardinality of a minimum connected dominating set of G.

—, **distance-k,** denoted $d_k\text{-}dom(G)$: the cardinality of a minimum distance-k dominating set of G.

—, **independent,** denoted $i\text{-}dom(G)$: the cardinality of a minimum independent dominating set of G.

dual edge clique cover of a set representation $\mathcal{F}$ for an intersection graph G: the edge clique cover $\mathcal{K}^{\mathcal{F}}$ defined in Proposition 7.4.4.

dual set representation of an edge clique cover $\mathcal{K}$ of a graph G: the set representation $\mathcal{F}^{\mathcal{K}}$ defined in Proposition 7.4.5.

dumbbell graph: the graph that has two vertices and one proper edge joining them, plus a self-loop at each vertex.

eccentricity of a vertex v in a graph G, denoted $ecc(v)$: the distance from v to a vertex farthest from v, i.e., $ecc(v) = \max\limits_{x \in V_G} d(v, x)$.

edge clique cover of a graph G: a family $\mathcal{K} = \{Q_1, Q_2, \ldots, Q_t\}$ of complete subgraphs of G such that every edge of G is an edge of at least one of the Q_is, i.e., $E_G = E_{Q_1} \cup E_{Q_2} \cup \cdots \cup E_{Q_t}$.

food web a digraph model D of the predator-prey relationship among the n animals $a_1, a_2, \ldots, a_n$ in an ecosystem, where the vertex- and arc-sets of D are given by

$$V_D = \{a_1,\ a_2, \ldots, a_n\}$$
$$A_D = \{(a_i, a_j) \mid a_i \text{ preys on } a_j\}$$

geodesic between vertices u and v: a u-v path of minimum length.

girth of a graph G with at least one cycle: the length of a shortest cycle in G; denoted $girth(G)$. The girth of an acylic graph is undefined.

graph homomorphism $f : G \to H$ from a graph G to a graph H: a pair of functions $(f_V : V_G \to V_H, f_E : E_G \to E_H)$ between their vertex- and edge-sets, respectively, such that for every edge $e \in E_G$, the function f_V maps the endpoints of e to the endpoints of edge $f_E(e)$.

guest, in the context of network emulation: the domain graph of the graph mapping that models the emulation.

host, in the context of network emulation: the codomain graph of the graph mapping that models the emulation.

image of a graph G under a graph homomorphism $f : G \to H$: the subgraph of H whose vertex-set is $f(V_G)$ and whose edge-set is $f(E_G)$.

incidence-preservation property: see *graph homomorphism.*

independence number of a graph G, denoted $\alpha(G)$: the cardinality of a largest independent set of G.

independent (vertex) set in a graph G: a vertex subset S such that no two vertices in S are adjacent.

induced numbering $\hat{f}_H$ on a subgraph H of a standard labeled graph G: the result of placing the numbers of the set $f(V_H)$ in ascending order, say $a_1, a_2, \ldots$, and then, for each vertex $v \in V_H$, defining $\hat{f}_H(v) = s$, if $f(v) = a_s$.

in-set $In(v)$ **of a vertex** v in a digraph D with vertex-set V_D and arc-set A_D: the vertex subset $In(v) = \{w \in V_D \mid (w, v) \in A_D\}$.

intersection graph of a family of subsets $\mathcal{F} = \{S_1, S_2, \ldots, S_n\}$, denoted $\Omega(\mathcal{F})$: the graph whose vertex- and edge-sets are given by

$$V_{\Omega(\mathcal{F})} = \{S_1, S_2, \ldots, S_n\}$$
$$E_{\Omega(\mathcal{F})} = \{S_iS_j \mid i \neq j \text{ and } S_i \cap S_j \neq \emptyset\}$$

—, *p*- of a family of subsets $\mathcal{F} = \{S_1, S_2, \ldots, S_n\}$ of a finite set S, denoted $\Omega_p(\mathcal{F})$: the graph whose vertex- and edge-sets are given by

$$V_{\Omega_p(\mathcal{F})} = \{S_1, S_2, \ldots, S_n\}$$
$$E_{\Omega_p(\mathcal{F})} = \{S_iS_j \mid i \neq j \text{ and } |S_i \cap S_j| \geq p\}$$

intersection graph: a graph G with vertex-set $V_G = \{v_1, v_2, \ldots, v_n\}$ such that there exists a family of sets $\mathcal{F} = \{S_1, S_2, \ldots, S_n\}$ such that $G \cong \Omega(\mathcal{F})$ with the natural correspondence $f(v_i) = S_i$, $i = 1, 2, \ldots, n$.

—, **abdiff-tolerance:** a special kind of *tolerance graph.*

—, **min-tolerance:** a special kind of *tolerance graph.*

—, ***p*-:** a graph G such that there exists a family of sets $\mathcal{F} = \{S_1, S_2, \ldots, S_n\}$ such that $G \cong \Omega_p(\mathcal{F})$.

intersection number of a graph G, denoted $int(G)$: the minimum cardinality of a set S such that G is the intersection graph of a family of subsets of S.

interval graph: a graph G that is an intersection graph of a family of intervals on the real line.

—, **min-tolerance:** a generalization of an interval graph.

isomorphism between two simple graphs G and H: a structure-preserving vertex bijection $f: V_G \to V_H$.

line graph of a graph G: the graph $L(G)$ that has a vertex for each edge of G, and such that two vertices in $L(G)$ are adjacent if and only if the corresponding edges in G have a vertex in common.

load of a graph mapping: the maximum load on any vertex of the codomain.

load on a vertex of the codomain of a graph mapping: the number of vertices mapped to that vertex.

median vertex in a connected graph: a vertex whose total distance is minimum.

median subgraph $M(G)$ of a connected graph G: the subgraph induced on the median vertices of G.

metric on the vertex-set V_G of a graph G: a mapping $d: V_G \times V_G \to \mathbb{R}$ that satisfies the following four conditions:

- $d(x,y) \geq 0$ for all $x, y \in V_G$.
- $d(x,y) = 0$ if and only if $x = y$.
- $d(x,y) = d(y,x)$ for all $x, y \in V_G$. (*symmetry property*)
- $d(x,y) + d(y,z) \geq d(x,z)$ for all $x, y, z \in V_G$. (*triangle inequality*)

multi-edge amalgamation corresponding to a subset $A = \{e_1, e_2, \ldots, e_s\} \subseteq E_G$ of edges having the same endpoints: the graph G_A that results from merging (amalgamating) all of the edges in A into a single edge generically denoted $e_1 e_2 \cdots e_s$.

—, **complete:** the graph that results from taking each multi-edge and merging its edges into a single edge.

neighborhood of a vertex v in a graph G:

—, **open,** denoted $N(v)$: the set of all the neighbors of vertex v.

—, **closed,** denoted $N[v]$: the set containing vertex v and all its neighbors, i.e., $N[v] = N(v) \cup \{v\}$.

numbering: see *vertex-labeling*.

out-set $Out(v)$ **of a vertex** v in a digraph D with vertex-set V_D and arc-set A_D: the vertex subset $Out(v) = \{w \in V_D \mid (v,w) \in A_D\}$.

path representation of a graph G: a family $\mathcal{F} = \{P_1, P_2, \ldots, P_n\}$ of subpaths of a path P such that G is the intersection graph of $\mathcal{F}$.

peripheral vertex v of a graph G: a vertex with maximum eccentricity, i.e., $ecc(v) = diam(G)$.

periphery of a graph G, denoted $per(G)$: the subgraph induced on the set of peripheral vertices of G.

m^{th} **power of graph** G: the graph G^m having vertex-set $V_{G^m} = V_G$ and edge-set $E_{G^m} = \{uv \mid d_G(u,v) \leq m\}$.

private neighbor of a vertex v relative to a dominating set D of a graph G: a vertex $w \in V_G$ such that v is the only vertex in D that dominates w. That is, $N[w] \cap D = \{v\}$.

pseudopath of length n: a spanning path of length n with one or more self-loops at one or more of its vertices.

radius of a graph G, denoted $rad(G)$: the minimum of the vertex eccentricities, i.e., $rad(G) = \min_{x \in V_G} ecc(x)$.

semilinear graph mapping $f : G \to H$: a pair of functions $f_V : V_G \to V_H$ and $f_E : E_G \to P_H$, such that for every edge $e \in E_G$, the endpoints of e are mapped by f_V to the first and last vertices of the path $f_E(e)$.

set representation of a graph G: A family of subsets $\mathcal{F}$ such that G is the intersection graph of $\mathcal{F}$.

Steiner distance of a vertex subset U in a connected graph G: the number of edges in a Steiner tree for U; denoted $sd(U)$.

Steiner tree for a vertex subset U in a connected graph G: a smallest tree subgraph of G that contains all the vertices of U.

structure-preserving vertex bijection between two **simple** graphs G and H: a vertex bijection $f : V_G \to V_H$ that preserves adjacency and nonadjacency, i.e., for every pair of vertices in G,

$$u \text{ and } v \text{ are adjacent in } G \Leftrightarrow f(u) \text{ and } f(v) \text{ are adjacent in } H$$

subtree graph: a graph G that is the intersection graph of a family $\mathcal{F}$ of subtrees $\{T_1, T_2, \ldots, T_n\}$ of a tree T, i.e., the vertex-sets of the T_is form a set representation of G. The tree T and the family $\mathcal{F}$ are called a **tree representation** of G.

symmetry property of a metric on the vertex-set of a graph: see *metric*.

total distance of a vertex v in a connected graph G, denoted $td(v)$: the quantity $td(v) = \sum_{w \in V_G} d(v, w)$.

tolerance graph: a generalization of intersection graphs.

tree representation of a graph G: a family $\mathcal{F} = \{T_1, T_2, \ldots, T_n\}$ of subtrees of a tree T such that G is the **intersection graph** of $\mathcal{F}$.

triangle inequality of a metric on the vertex-set of a graph: see *metric*.

vertex-connectivity of a connected graph G, denoted $\kappa_v(G)$: the minimum number of vertices whose removal can either disconnect G or reduce it to a 1-vertex graph.

vertex-cut of a connected graph G: a vertex subset S such that $G - S$ is non-connected.

vertex-labeling, (1-based) standard, of an n-vertex graph G: a bijection $f : V_G \to \{1, 2, \ldots, n\}$.

vertex-partition amalgamation of a graph G corresponding to a partition $\pi = \{V_1, V_2, \ldots, V_t\}$ of its vertex-set V_G: the transformation of G into a graph G/π that results from merging (amalgamating) all of the vertices in each cell of the partition.

Chapter 8

ANALYTIC GRAPH THEORY

INTRODUCTION

Several areas of graph theory are concerned with the likelihood or certainty of the presence in a graph of various subgraphs or, more generally, of graph properties that emerge as the number of vertices and/or the number of edges increases. Collectively they are grouped as *analytic graph theory.*

The foundational results in analytic graph theory were obtained decades ago, and there has been extensive subsequent development. Even though many such results can be stated easily, they can be quite difficult to prove. As an introduction to this deep and well-developed area, this chapter concentrates on the basic results whose proofs are the most readily accessible and illuminating.

8.1 RAMSEY THEORY

A theorem of the British logician Frank Ramsey in 1930 launched what is now called Ramsey theory. The classical Ramsey puzzle, a simple case of that theorem, is to prove that in any group of six persons, there are either three mutual acquaintances or three mutual non-acquaintances.

Theorem 8.1.1: *In any simple graph G with six vertices, either there are three mutually adjacent vertices or there are three mutually nonadjacent vertices. Equivalently, either the graph G or its edge-complement $\overline{G}$ contains a K_3-subgraph.*

Proof: Let $u \in V_G$. Then either $\deg_G(u) \geq 3$ or $deg_{\overline{G}}(u) \geq 3$, since $\deg_G(u) + deg_{\overline{G}}(u) = 5$. We may assume without loss of generality that $\deg_G(u) \geq 3$, because of the symmetry of the theorem statement with respect to G and $\overline{G}$. Let x, y, and z be three of the neighbors of u.
If any two of vertices x, y, and z are adjacent, then those two, together with vertex u, form a K_3-subgraph of graph G. Alternatively, if x, y, and z are mutually nonadjacent, then they form a K_3-subgraph in graph $\overline{G}$. ♢

Ramsey Numbers

The Ramsey puzzle is readily generalized to a vertex-extremal problem of calculating a minimum number of vertices in a simple graph that guarantees the occurrence of a property.

DEFINITION: Let s and t be positive integers. The ***Ramsey number*** $r(s,t)$ is the minimum positive number n such that for every simple n-vertex graph G, either G contains a K_s-subgraph or its edge-complement graph $\overline{G}$ contains a K_t-subgraph.

Remark: Thus, Theorem 8.1.1 and the fact that C_5 contains neither K_3 nor three independent vertices imply that $r(3,3) = 6$.

ALTERNATIVE DEFINITION: The ***Ramsey number*** $r(s,t)$ is the minimum number n such that for every edge-coloring of K_n with colors red and blue, there exists either a red K_s-subgraph or a blue K_t-subgraph.

This version of the definition generalizes to arbitrarily many edge colors and to arbitrary subgraphs in each color.

Fundamental Properties of Ramsey Numbers

Proposition 8.1.2: $r(s,t) = r(t,s)$, *for all* $s, t \geq 1$.

Proof: The alternative definition is clearly symmetric in the roles of s and t. ♢

Proposition 8.1.3: $r(s,1) = 1$, *for all* $s \geq 1$.

Proof: The edge-complement of a nontrivial graph always contains K_1. ♢

Proposition 8.1.4: $r(s,2) = s$, *for all* $s \geq 1$.

Proof: If an s-vertex simple graph G is not complete, then its edge-complement $\overline{G}$ contains K_2. Thus, $r(s,2) \leq s$. The reverse inequality also holds, since K_{s-1} obviously does not contain a K_s-subgraph, and $\overline{K_{s-1}}$ contains no edges. ♢

Theorem 8.1.5: [***Erdös and Szekeres, 1935***]

(a) For all $s, t \geq 2$,

$$r(s,t) \leq r(s-1,t) + r(s,t-1)$$

(b) Moreover, if $r(s-1,t)$ *and* $r(s,t-1)$ *are both even, then*

$$r(s,t) < r(s-1,t) + r(s,t-1)$$

Proof: This proof of part (a) generalizes the proof given for Theorem 8.1.1.

(a) Let G be a simple graph on $r(s-1,t)+r(s,t-1)$ vertices, and let $u \in V_G$. Then either $\deg_G(u) \geq r(s-1,t)$ or $\deg_{\overline{G}}(u) \geq r(s,t-1) = r(t-1,s)$, since $\deg_G(u) + deg_{\overline{G}}(u) = r(s-1,t)+r(s,t-1)-1$. If $\deg_G(u) \geq r(s-1,t)$ then the subgraph H of G induced on the open neighborhood of u has at least $r(s-1,t)$ vertices. Thus, either subgraph H contains a K_{s-1}-subgraph, which together with vertex u forms a K_s-subgraph of G, or $\overline{H}$ (and hence $\overline{G}$) contains a K_t-subgraph. If $\deg_{\overline{G}}(u) \geq r(t-1,s)$, then the result follows by applying the same argument to $\overline{G}$ with the roles of s and t reversed.

(b) Now suppose that $r(s-1,t)$ and $r(s,t-1)$ are both even, and that graph G has $r(s-1,t)+r(s,t-1)-1$ vertices. Since $|V_G|$ is odd, there is some vertex u of G with even degree, by Corollary 1.1.3. Then $\deg_G(u)+\deg_{\overline{G}}(u) = r(s-1,t)+r(s,t-1)-2$. If $\deg_G(u) \geq r(s-1,t)$, then we are done, as in part (a) above. If $\deg_G(u) < r(s-1,t)$, then $\deg_G(u) \leq r(s-1,t)-2$ and, hence, $\deg_{\overline{G}}(u) \geq r(t-1,s)$. The result again follows, as in part (a).

Remark: The next result shows that Ramsey numbers exist for all pairs of positive integers.

Corollary 8.1.6: [***Erdös and Szekeres, 1935***] *For all positive integers* s *and* t,

$$r(s,t) \leq \binom{s+t-2}{s-1}$$

Proof: We use induction on the sum $s+t$. The assertion is trivially true for $s+t=2$, by Proposition 8.1.3. Assume for some $k \geq 3$ that the inequality is true for all positive integers s and t whose sum is smaller than k, and suppose that $s+t=k$. Then

$$\begin{aligned}
r(s,t) &\leq r(s-1,t) + r(s,t-1) \quad \text{(by Theorem 8.1.5(a))} \\
&\leq \binom{(s-1)+t-2}{(s-1)-1} + \binom{s+(t-1)-2}{s-1} \quad \text{(by the induction hypothesis)} \\
&= \binom{s+t-3}{s-2} + \binom{s+t-3}{s-1} \\
&= \binom{s+t-2}{s-1} \quad \text{(by Pascal's recursion)}
\end{aligned}$$

◇

Ramsey Number Calculations

Determining values of the Ramsey numbers $r(s,t)$ for $3 \leq s \leq t$ has proved to be quite difficult, except for the first few. In fact, the Ramsey numbers for only nine such pairs are known; they are shown in Table 8.1.1. Theorem 8.1.1 established $r(3,3)=6$. The next two results calculate $r(3,4)$ and $r(3,5)$, and a third result produces an upper bound for $r(4,4)$.

	$t=3$	4	5	6	7	8	9
$s=3$	6	9	14	18	23	28	36
4		18	25				

Table 8.1.1: The known Ramsey numbers $r(s,t)$ for $3 \leq s \leq t$.

Proposition 8.1.7: $r(3,4) = 9$.

Proof: Since the Ramsey numbers $r(2,4) = 4$ and $r(3,3) = 6$ are both even, Theorem 8.1.5(b) implies the upper bound

$$r(3,4) \leq r(2,4) + r(3,3) - 1 = 9$$

For the reverse inequality, observe that the 8-vertex circulant graph $G = circ(8;1,4)$, shown in Figure 8.1.1, has neither a K_3-subgraph nor an independent set of four vertices (i.e., $\overline{G}$ has no K_4-subgraph). ♢

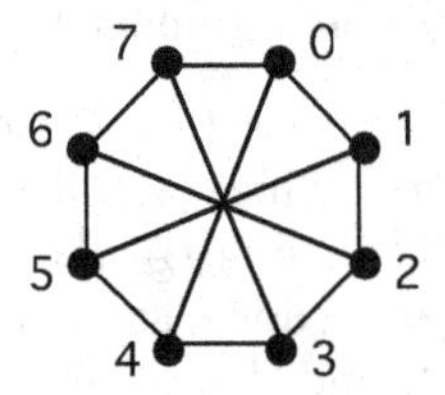

Figure 8.1.1 The circulant graph $circ(8;1,4)$.

Proposition 8.1.8: $r(3,5) = 14$.

Proof: The upper bound for $r(3,5)$ follows from Theorem 8.1.5(a) and Propositions 8.1.4 and 8.1.7. In particular,

$$r(3,5) \leq r(2,5) + r(3,4) = 5 + 9 = 14$$

For the reverse inequality, consider the 13-vertex circulant graph $G = circ(13;1,5)$, shown in Figure 8.1.2 below. G clearly has no K_3-subgraph. If it had a set S of five mutually nonadjacent vertices, then two of them would have to be within distance two of each other. By symmetry, we may take these two vertices to be 0 and 2. This excludes vertices 1, 3, 5, 7, 8, 10, and 12, leaving only 4, 6, 9, and 11 as possible members of S. But at most one of the vertices 4 and 9 can be in S, because they are adjacent. Likewise, at most one of 6 and 11 can be in S. ♢

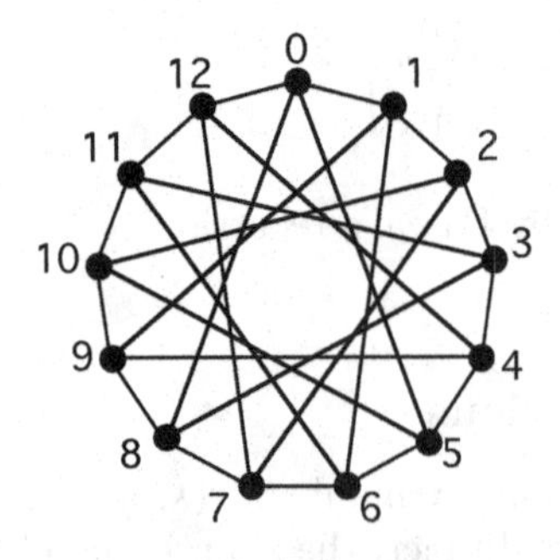

Figure 8.1.2 The circulant graph $circ(13;1,5)$.

Proposition 8.1.9: $r(4,4) \leq 18$.

Proof: By Theorem 8.1.5(a) and Propositions 8.1.2 and 8.1.7, we have

$$r(4,4) \leq r(3,4) + r(4,3) = 9 + 9 = 18$$

Ramsey Numbers for Arbitrary Subgraphs

DEFINITION: For arbitrary simple graphs H_1 and H_2, the ***Ramsey number*** $r(H_1, H_2)$ is the minimum number n such that for every n-vertex simple graph G, either H_1 is a subgraph of G or H_2 is a subgraph of $\overline{G}$.

Proposition 8.1.10: *The Ramsey number $r(H_1, H_2)$ exists for any two simple graphs H_1 and H_2.*

Proof: Let H_1 and H_2 be simple graphs with m and n vertices, respectively. Since H_1 is a subgraph of K_m and H_2 is a subgraph of K_n, we have $r(H_1, H_2) \leq r(m, n)$. ◇

Proposition 8.1.11:

$$r(K_{1,m}, K_{1,n}) = \begin{cases} m+n & \text{if } m, n \text{ not both even} \\ m+n-1 & \text{if } m, n \text{ both even} \end{cases}$$

Proof:

Case 1: m and n are not both even. We may assume without loss of generality that m is odd. Let G be the $(m+n-1)$-vertex, $(m-1)$-regular circulant graph $G = circ(m+n-1; 1, 2, \ldots, \frac{m-1}{2})$. Then G does not contain $K_{1,m}$, and its $(n-1)$-regular edge-complement $\overline{G}$ does not contain $K_{1,n}$. Therefore, $r(K_{1,m}, K_{1,n}) \geq m+n$. For the reverse inequality, suppose that G is an $(m+n)$-vertex graph that does not contain $K_{1,m}$. Then its maximum degree $\delta_{\max}(G) \leq m-1$, and, hence, its edge-complement $\overline{G}$ has maximum degree $\delta_{\max}(\overline{G}) \geq n$. Thus, $K_{1,n}$ is a subgraph of $\overline{G}$, which shows that $r(K_{1,m}, K_{1,n}) \leq m+n$.

Case 2: m and n are both even. The $(m+n-2)$-vertex, $(m-1)$-regular circulant graph $G = circ\left(m+n-2; 1, 2, \ldots, \frac{m-2}{2}, \frac{m+n-2}{2}\right)$ does not contain $K_{1,m}$, and $\overline{G}$ does not contain $K_{1,n}$. Thus, $r(K_{1,m}, K_{1,n}) \geq m+n-1$. Now suppose that G is an $(m+n-1)$-vertex graph that does not contain $K_{1,m}$. Then, as in Case 1, $\delta_{\max}(G) \leq m-1$. But G cannot be $(m-1)$-regular since $m+n-1$ and $m-1$ are both odd and the degree sum must be even. Therefore, there is at least one vertex v with $\deg_G(v) < m-1$. Then $\deg_{\overline{G}}(v) \geq n$, and hence, $K_{1,n}$ is a subgraph of $\overline{G}$. Thus, $r(K_{1,m}, K_{1,n}) \leq m+n-1$, which completes the proof. ◇

Theorem 8.1.12: [***Chvátal, 1977***] *Let T be a tree with t vertices. Then*

$$r(K_s, T) = (s-1)(t-1) + 1$$

Proof: The assertion is trivially true for $s = 1$ and $s = 2$. Suppose that $s \geq 3$. We observe that the complete $(s-1)$-partite graph $G = K_{t-1,\ldots,t-1}$ does not contain K_s, and that each component of $\overline{G} = (s-1)K_{t-1}$ has only $t-1$ vertices, so it does not contain T. Thus, we have the lower bound

$$r(K_s, T) \geq (s-1)(t-1) + 1$$

To establish the upper bound, let G be any graph with $(s-1)(t-1)+1$ vertices, such that G does not contain K_s. Then its edge-complement $\overline{G}$ does not contain a subset of s mutually nonadjacent vertices, or, equivalently, the independence number $\alpha\left(\overline{G}\right) \leq s-1$. By Proposition 6.1.5, we have

$$\chi\left(\overline{G}\right) \geq \left\lceil \frac{(s-1)(t-1)+1}{\alpha(\overline{G})} \right\rceil \geq \left\lceil \frac{(s-1)(t-1)+1}{s-1} \right\rceil = t$$

Thus, $\overline{G}$ is not $(t-1)$-colorable. It follows that $\overline{G}$ must contain a critically t-chromatic subgraph H (see §6.1). By Theorem 6.1.22, we have $\delta_{\min}(H) \geq t-1$, and hence, by Theorem 3.1.9, it follows that H contains tree T. Therefore, $\overline{G}$ contains T, which establishes the upper bound. ♢

Generalization to Arbitrarily Many Edge-Partition Classes

DEFINITION: The ***(generalized) Ramsey number*** $r(G_1, G_2, \ldots, G_k)$ for any collection of simple graphs $G_1, G_2, \ldots, G_k$ is the minimum number n such that in any edge k-coloring of a complete graph K_n, at least one of the subgraphs G_j has all of its edges in color class j.

In 1955, Greenwood and Gleason established the following result, which is still, as of this writing, the only nontrivial generalized Ramsey number known for the case in which all the graphs $G_1, \ldots, G_k$ are complete. A proof may be found in [GrGl55].

Theorem 8.1.13: [***Greenwood and Gleason, 1955***] $r(3,3,3) = 17$.

Remark: For other results, conjectures, variations, and references in Ramsey theory, see [Fa04] and [GrRoSp90].

EXERCISES for Section 8.1

8.1.**1** Use Theorem 8.1.5 and Corollary 8.1.6 to calculate upper bounds for $r(3,6)$. Which gives the tighter bound?

8.1.**2** Prove that $r(3,6) \geq 17$ by considering *circ*$(16; 1, 5, 8)$.

8.1.**3** Use Exercise 8.1.1 to calculate an upper bound for $r(3,7)$.

8.1.**4** Prove that $r(3,7) \geq 21$ by considering *circ*$(21; 1, 5, 9)$.

8.1.**5** Use Theorem 8.1.5 and Corollary 8.1.6 to calculate upper bounds for $r(4,5)$. Which gives the tighter bound?

8.1.**6** Prove that $r(3,t) \leq \frac{t^2+t}{2}$, by using Corollary 8.1.6.

8.1.**7** Prove that $r(3,t) \leq \frac{t^2+t}{2}$, for $t \geq 3$, by using Theorem 8.1.5 inductively.

8.1.**8** Prove that $r(3,t) \leq \frac{t^2+3}{2}$, for $t \geq 3$ and t odd, by using Exercise 8.1.7.

8.1.**9** Prove that $r(3,t) \leq \frac{t^2+3}{2}$, for $t \geq 3$ and t odd, by using Theorem 8.1.5(b) and Exercise 8.1.8.

8.1.**10** Prove that $r(3,3,3) \leq 17$.

8.2 EXTREMAL GRAPH THEORY

Extremal graph theory is primarily concerned with determining the maximum number of edges an n-vertex simple graph may contain without having a given property. This is an area of graph theory in which many of the proofs are long and computationally involved. We are focusing here on easily proved results.

DEFINITION: Within a given domain of graphs, a property $\mathcal{P}$ is said to hold for every n-vertex simple graph with ***sufficiently many edges*** if there is a number M such that every n-vertex graph in that domain with at least M edges has property $\mathcal{P}$.

DEFINITION: Within a given domain of graphs, a graph property $\mathcal{P}$ that holds for every n-vertex simple graph with sufficiently many edges, for every sufficiently large number n of vertices, is said to hold for every ***sufficiently large graph*** in the domain.

Extremal Graphs and Extremal Functions

DEFINITION: Within a given domain of simple graphs, an n-vertex simple graph with m edges is an ***extremal graph*** for a property $\mathcal{P}$ if it does not have property $\mathcal{P}$, but every n-vertex simple graph with more than m edges does have property $\mathcal{P}$.

DEFINITION: Within a given domain of simple graphs, an ***extremal function*** for a property $\mathcal{P}$ is the function $ex(n, \mathcal{P})$ whose value is the number of edges in an extremal graph for that property.

Example 8.2.1: Consider the property $\mathcal{P}$ of having at least one cycle. An n-vertex graph with $n-1$ edges has no cycles if and only if it is a tree. Moreover, every n-vertex graph with at least n edges has a cycle. Thus, the extremal graphs for having a cycle are the trees, and the extremal function is $ex(n, \mathcal{P}) = n - 1$.

Example 8.2.2: Consider the property $\mathcal{P}$ of non-planarity. An n-vertex graph with at least $3n-5$ edges is non-planar (see §4.5). To construct a planar n-vertex graph with $3n-6$ edges, join an $(n-2)$-cycle to two new vertices, as shown in Figure 8.2.1. Thus, the extremal function is $ex(n, \mathcal{P}) = 3n - 6$.

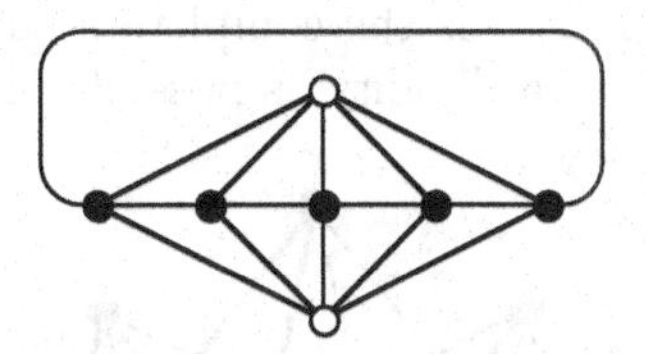

Figure 8.2.1 The graph $2K_1 + C_5$ is extremal for non-planarity.

Example 8.2.3: The property of Eulerian traversability (§1.2) does not hold for sufficiently large graphs. In particular, it does not hold for a complete graph K_{2n} on evenly many vertices.

Turán's Theorem

The prototypical problem of extremal graph theory and its solution are due to Turán [Tu41]: what is the maximum number of edges that an n-vertex simple graph could have and still contain no subgraph isomorpic to K_3?

TERMINOLOGY: A graph is said to be ***F-free*** if it contains no subgraph isomorphic to a given graph F. In particular, Turan's problem is about largest K_3-free graphs. More generally, if $\mathcal{F}$ is a family of graphs, then we may describe a graph as $\mathcal{F}$-free, if it is free of every member of $\mathcal{F}$.

NOTATION: For a given family $\mathcal{F}$ of graphs, $ex(n, \mathcal{F})$ denotes the maximum number of edges possible in an n-vertex graph that is $\mathcal{F}$-free. If $\mathcal{F}$ contains only one graph F, then we commonly write $ex(n, F)$.

Turán showed that $ex(n, K_3) = \left\lfloor \frac{n^2}{4} \right\rfloor$. Theorems 8.2.1 and 8.2.2 establish the upper and lower bounds, respectively.

Theorem 8.2.1: [***Turán, 1941***] *Let G be a simple graph with n vertices and at least $\left\lfloor \frac{n^2}{4} \right\rfloor + 1$ edges. Then G contains a subgraph isomorphic to K_3.*

Proof: Using $n = 3$ and $n = 4$ as base cases for an induction, we have $\left\lfloor \frac{n^2}{4} \right\rfloor + 1 = 3$ and 5, respectively. Thus, graph G is isomorphic to K_3 or to $K_4 - K_2$, so in both cases, G contains K_3. Assume that the assertion is true for all graphs with k vertices, with $3 \leq k < n$ and $n \geq 5$, and let graph G have n vertices and at least $\left\lfloor \frac{n^2}{4} \right\rfloor + 1$ edges.

Let u and v be adjacent vertices in G. Then the subgraph $H = G - \{u, v\}$ has $n - 2$ vertices. If there is some vertex w of $G - \{u, v\}$ that is adjacent to both u and v, then the subgraph of G induced on $\{u, v, w\}$ is isomorphic to K_3 if there is no such vertex w, then at most $n - 2$ edges join u and v to H, and hence, the number of edges in H is at least.

$$\left\lfloor \frac{n^2}{4} \right\rfloor + 1 - (n - 1) = \left\lfloor \frac{n^2 - 4n + 4}{4} \right\rfloor + 1 = \left\lfloor \frac{(n-2)^2}{4} \right\rfloor + 1$$

By the induction hypothesis, H contains a K_3 subgraph. ◊

REVIEW FROM §6.1: An ***r-partite graph*** is a loopless graph whose vertices can be partitioned into r independent sets, which are sometimes called the ***partite sets*** of the partition.

DEFINITION: A ***complete*** r-partite graph has an edge between two vertices if and only if they are in different partite sets.

DEFINITION: The ***Turán graph*** $T_{n,r}$ is the complete r-partite graph on n vertices whose partite sets are as nearly equal in cardinality as possible.

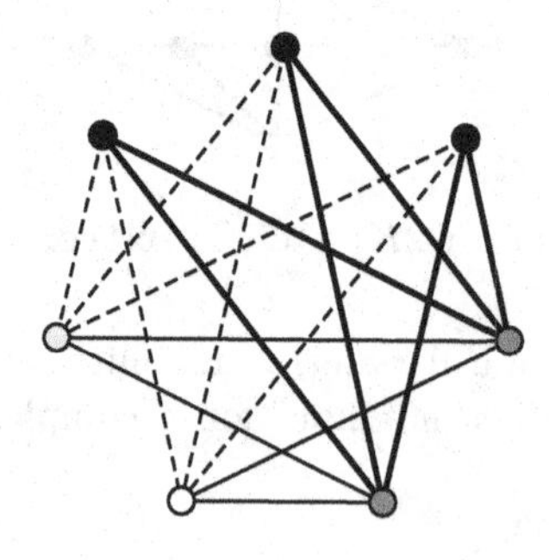

Figure 8.2.2 The Turán graph $T_{7,3}$.

Theorem 8.2.2: [**Turán, 1941**] *The bipartite graph $T_{n,2}$ is an extremal graph for $ex(n, K_3)$.*

Proof: Since $T_{n,2}$ is bipartite, it is K_3-free. If n is even, then $T_{n,2} \cong K_{\frac{n}{2},\frac{n}{2}}$ has $\frac{n^2}{4}$ edges. If n is odd, then $T_{n,2} \cong K_{\frac{n+1}{2},\frac{n-1}{2}}$ has $\frac{n^2-1}{4}$ edges. In either case, the graph $T_{n,2}$ has $\left\lfloor \frac{n^2}{4} \right\rfloor$ edges. ♢

Remark: The methods above for deriving $ex(n, K_3)$ generalize readily to calculating $ex(n, K_r)$ for arbitrary $r \geq 3$. In general, the Turán graph $T_{n,r}$ is an extremal graph for this problem.

Vertex Degree

A graph is extremal for the property of having a vertex of degree d if and only if it is extremal for not being $K_{1,d}$-free.

Theorem 8.2.3: $ex(n, K_{1,d}) = \left\lfloor \frac{n(d-1)}{2} \right\rfloor$.

Proof: Let G be a graph with $\left\lfloor \frac{n(d-1)}{2} \right\rfloor + 1$ edges. Then

$$2\,|E_G| = 2\left(\left\lfloor \frac{n(d-1)}{2} \right\rfloor + 1\right) > 2\frac{n(d-1)}{2} = n(d-1)$$

Therefore,

$$\delta_{avg}(G) = \frac{2\,|E_G|}{n} > d-1$$

It follows that some vertex has degree at least d, which shows that

$$ex(n, K_{1,d}) \leq \left\lfloor \frac{n(d-1)}{2} \right\rfloor$$

If d is odd or n is even, then a $(d-1)$-regular graph on n vertices is $K_{1,d}$-free and has $\left\lfloor \frac{n(d-1)}{2} \right\rfloor$ edges. Otherwise, add a maximum matching to a $(d-2)$-regular n-vertex graph to obtain a $K_{1,d}$-free graph whose number of edges is

$$\frac{n(d-2)}{2} + \frac{n-1}{2} = \frac{nd-n-1}{2} = \left\lfloor \frac{nd-n}{2} \right\rfloor$$

which establishes the lower bound and completes the proof. ♢

Connectedness

REVIEW FROM §1.3: A ***vertex-cut*** in a graph G is a vertex-set U such that $G - U$ has more components than G.

REVIEW FROM §3.4: A graph G is ***k*-*connected*** if G is connected and every vertex-cut has at least k vertices. Thus, "1-connected" and "connected" are synonymous.

Proposition 8.2.4: *In the domain of simple graphs, the graph $K_{n-1} \cup K_1$ is an extremal graph for the property of connectedness.*

Proof: Let G be an n-vertex extremal graph for connectedness. Then G must have at least two components, or it would be connected. There cannot be more than two components, since otherwise, two components could be joined by an edge without yielding a connected graph. Hence, there are exactly two components. Both components must be complete, since otherwise an edge could be added without yielding a connected graph. Thus, $G \cong K_r \cup K_{n-r}$. It follows that

$$|E_G| = \binom{r}{2} + \binom{n-r}{2} = \frac{r^2 - r}{2} + \frac{(n-r)^2 - (n-r)}{2} = \frac{n^2 - 2nr + 2r^2 - n}{2}$$

which has a maximum on the interval $[1, n-1]$ when $r = 1$ and $r = n - 1$. ◊

The graph in Figure 8.2.3 illustrates Proposition 8.2.4 with an extremal graph on six vertices for connectedness. The next proposition, generalizes the construction to give extremal graphs for k-connectedness. Figure 8.2.4 gives extremal graphs on six vertices for 2-connectedness and 3-connectedness.

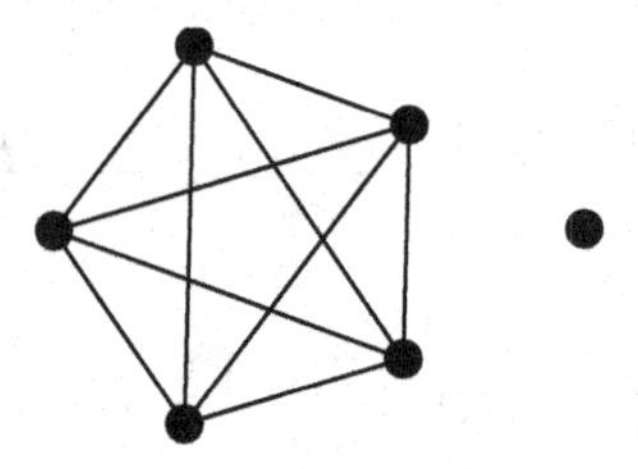

Figure 8.2.3 A six-vertex extremal graph for connectedness.

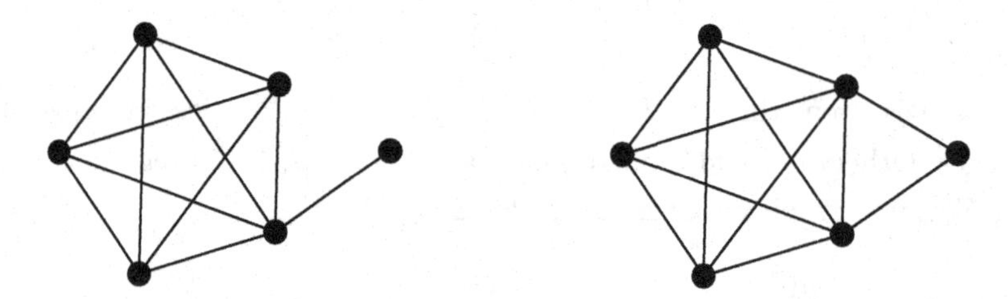

Figure 8.2.4 Six-vertex extremal graphs for 2- and 3-connectedness.

Proposition 8.2.5: *In the domain of simple graphs, the graph G that results from joining one new vertex to $k - 1$ vertices of K_{n-1}, where $n \geq 2$ and $1 \leq k \leq n - 1$, is an extremal graph for the property of k-connectedness.*

Proof: Since the neighbors of the new vertex form a vertex-cut of size $k - 1$, G is not k-connected. Thus, it suffices to prove that any n-vertex simple graph with one more edge is k-connected.We use induction on n.

The base cases are:

- For $n = 2$ and $k = 1$, we get a 2-vertex graph with 0 edges, which is extremal for 1-connectedness, since the 2-vertex graph with 1 edge *is* 1-connected.

- For $n = 3$ and $k = 1$, we get a 3-vertex graph with 1 edge, which is extremal for 1-connectedness, since the 3-vertex graph with 2 edges *is* 1-connected.

- For $n = 3$ and $k = 2$, we get a 3-vertex graph with 2 edges, which is extremal for 2-connectedness, since the 3-vertex graph with 3 edges *is* 2-connected.

Assume for some $n \geq 3$, a simple graph on n vertices with $\binom{n-1}{2} + k$ edges, is k-connected, for any k, $1 \leq k \leq n-1$. Let H be a $(n+1)$-vertex simple graph with $\binom{n}{2} + k$ edges, $1 \leq k \leq n$. If $k = 1$, H is 1-connected by Proposition 8.2.4. If $k = n$, then H is a complete graph on $n + 1 = k + 1$ vertices, and so is k-connected. Suppose $1 < k < n$, and let v be a vertex of minimum degree in graph H. Since $k < n$, E_H is not a complete graph. Thus, $\deg_H(v) \leq n - 1$. Also, $\deg_H(v) \geq k$, since

$$\binom{n}{2} \geq |E_{H-v}| = |E_H| - \deg_H(v) = \binom{n}{2} + k - \deg_H(v)$$

Therefore,

$$\begin{aligned}
|E_{H-v}| &= |E_H| - \deg_H(v) \\
&\geq \binom{n}{2} + k - (n-1) \\
&= \binom{n}{2} - \binom{n-1}{1} + k \\
&= \binom{n-1}{2} + k \qquad \text{(by Pascal's Identity)}
\end{aligned}$$

This implies, by the induction hypothesis, that the n-vertex graph $H - v$ is k-connected. Therefore, since $\deg_H(v) \geq k$, graph H must also be k-connected, which completes the proof. ◇

Traversability

REVIEW FROM §1.2

- A cycle that includes every vertex of a graph is call a ***Hamiltonian cycle***.

- A ***Hamiltonian graph*** is a graph that has a Hamiltonian cycle.

DEFINITION: A path that includes every vertex of a graph is call a ***Hamiltonian path***.

Theorem 8.2.6: [***Ore, 1960***] *Let G be a simple n-vertex graph, where $n \geq 3$, such that $\deg(x) + \deg(y) \geq n$ for each pair of nonadjacent vertices x and y. Then G is Hamiltonian.*

Proof: By way of contradiction, assume that the theorem is false, and let G be a maximal counterexample. That is, G is non-Hamiltonian and satisfies the conditions of the theorem, and the addition of any edge joining two nonadjacent vertices of G results in a Hamiltonian graph.

Let x and y be two nonadjacent vertices of G (G is not complete, since $n \geq 3$). To reach a contradiction, it suffices to show that $\deg(x) + \deg(y) \leq n - 1$.

Since the graph $G + xy$ contains a Hamiltonian cycle, G contains a Hamiltonian path whose endpoints are x and y. Let $\langle x = v_1, v_2, \ldots v_n = y\rangle$ be such a path, as shown below on the left.

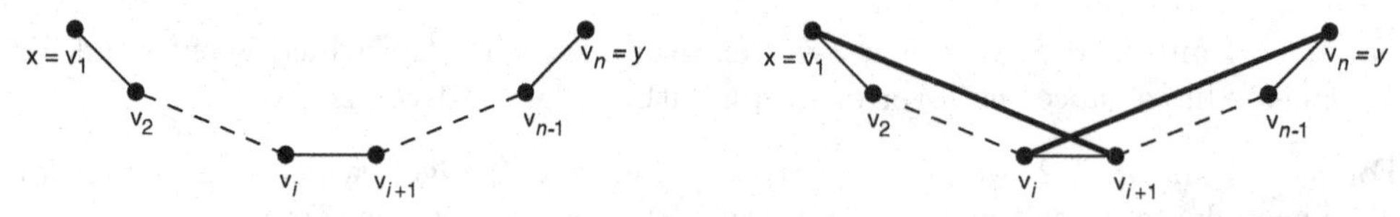

For each $i = 2, \ldots, n-1$, at least one of the pairs v_1, v_{i+1} and v_i, v_n is nonadjacent, since otherwise, $\langle v_1, v_2, \ldots, v_i, v_n, v_{n-1}, \ldots, v_{i+1}, v_1 \rangle$ would be a Hamiltonian cycle in G, as shown on the right. This means that if $(a_{i,j})$ is the adjacency matrix for G, then $a_{1,i+1} + a_{i,n} \leq 1$, for $i = 2, \ldots, n-2$.
Thus,

$$\begin{aligned}
\deg(x) + \deg(y) &= \sum_{i=2}^{n-1} a_{1,i} + \sum_{i=2}^{n-1} a_{i,n} \\
&= a_{1,2} + \sum_{i=3}^{n-1} a_{1,i} + \sum_{i=2}^{n-2} a_{i,n} + a_{n-1,n} \\
&= 1 + \sum_{i=2}^{n-2} a_{1,i+1} + \sum_{i=2}^{n-2} a_{i,n} + 1 \\
&= 2 + \sum_{i=2}^{n-2} (a_{1,i+1} + a_{i,n}) \\
&\leq 2 + n - 3 = n - 1
\end{aligned}$$

which establishes the desired contradiction. ◊

Proposition 8.2.7: *A simple n-vertex graph G with at least $\binom{n-1}{2}+2$ edges is Hamiltonian.*

Proof: Let u and v be two nonadjacent vertices of G. Then,

$$\begin{aligned}
\deg(u) + \deg(v) &= |E_G| - \left|E_{G-\{u,v\}}\right| \\
&\geq \left(\binom{n-1}{2} + 2\right) - \binom{n-2}{2} \\
&= \binom{n-2}{1} + 2 \qquad \text{by Pascal's Identity} \\
&= n
\end{aligned}$$

By Ore's Theorem (Theorem 8.2.6), graph G is Hamiltonian. ◊

Corollary 8.2.8: *The graph G obtained by joining a new vertex to one vertex of a complete graph K_{n-1} is extremal for Hamiltonian traversability. (See Figure 8.2.5.)*

Proof: The graph G is non-Hamiltonian, and it has $\binom{n-1}{2} + 1$ edges. By Proposition 8.2.7, the result of adding one more edge to G would be a Hamiltonian graph. ◊

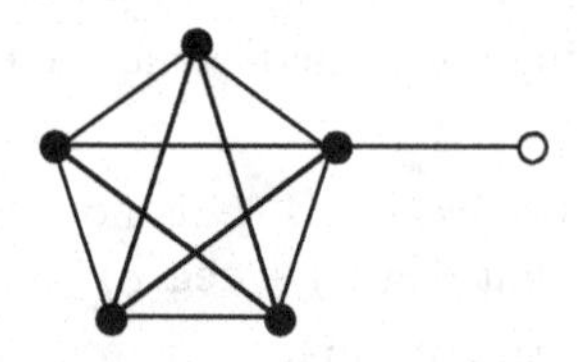

Figure 8.2.5 A 6-vertex extremal graph for Hamiltonian traversability.

Corollary 8.2.9: *A simple n-vertex graph with at least* $\binom{n-1}{2}+1$ *edges contains an open Hamiltonian path.*

Proof: Choose any two nonadjacent vertices u and v. By Proposition 8.2.7, the result of adding an edge between vertices u and v would be a Hamiltonian graph. Thus, there must be a Hamiltonian path from u to v. ◇

Corollary 8.2.10: $ex(n, P_n) = \binom{n-1}{2}$.

Proof: In view of Corollary 8.2.9, the graph $K_1 \cup K_{n-1}$ is extremal. ◇

Chromatic Number

It is trivial to calculate extremal numbers for some extreme values of chromaticity. It is easy to calculate them for nearly extreme values.

Proposition 8.2.11: *Let* P_k *be the property that a graph has chromatic number k.*

(i) $ex(n, P_2) = 0$

(ii) $ex(n, P_n) = \binom{n}{2} - 1$

(iii) $ex(n, P_3) = \left\lfloor \frac{n^2}{4} \right\rfloor$

Proof: Parts (*i*) and (*ii*) are obvious.
(*iii*) The complete bipartite graph $K_{\lfloor n/2 \rfloor, \lceil n/2 \rceil}$ has chromatic number 2. It is extremal for chromatic number 3, by Turán's theorem, since its number of edges is

$$\left\lfloor \frac{n}{2} \right\rfloor \cdot \left\lceil \frac{n}{2} \right\rceil = \left\lfloor \frac{n^2}{4} \right\rfloor$$

◇

EXERCISES for Section 8.2

8.2.1 List all 5-vertex graphs that are extremal for non-planarity.

8.2.2 List all 6-vertex graphs that are extremal for non-planarity.

8.2.3 Prove that your list in Exercise 8.2.2 is complete.

8.2.4 Calculate $ex(n, K_4)$ and $ex(n, K_5)$, by using the fact that the Turán graphs are extremal.

8.2.5 Calculate $ex(4, 2K_2)$, and draw the extremal graphs.

8.2.6 Calculate $ex(5, 2K_2)$, and draw the extremal graphs.

8.2.7 Calculate $ex(6, 2K_2)$, and draw the extremal graphs.

8.2.8 Calculate $ex(6, 3K_2)$, and draw the extremal graphs.

8.2.9 Calculate the extremal number for the property of having radius 1.

8.2.10 Calculate the extremal number for the property of having diameter 2.

8.3 RANDOM GRAPHS

A *random graph* is a graph in which the number of vertices is specified and the adjacencies between vertices are determined in some random way. The vertices are usually considered to have the *standard labeling* $1, 2, \ldots, n$. The predominant concern is large graphs, with many vertices.

The *theory of random graphs* uses probabilistic methods to establish the existence of certain kinds of graphs and to determine some properties of "almost all" graphs in various families. Whereas we were concerned in §8.2 with *extremal numbers* for various properties, here we focus mainly on what happens *on average*, i.e., on what the *expected* properties are.

It is assumed throughout this section that the reader is reasonably familiar with the basic concepts of elementary probability, including the uniform and binomial distributions, expectation, and conditional probability.

NOTATION: We use $pr(E)$ to denote the probability of an event E.

Labeled Graph Isomorphism

DEFINITION: An isomorphism $f : G \to H$ of two labeled graphs is a ***label-preserving isomorphism*** if for every vertex $v \in V_G$, the vertex $f(v) \in V_H$ has the same label as v.

In the theory of random graphs, we usually distinguish between two standard labeled graphs, unless there is a label-preserving isomorphism between them.

The Probabilistic Method of Proof

As an illustration of the probabilistic method, we consider the problem of calculating a lower bound on ***diagonal Ramsey numbers***, that is, Ramsey numbers of the form $r(s, s)$. We recall from §8.1 that $r(3, 3) = 6$ and that $r(4, 4) = 18$.

Theorem 8.3.1: *For every integer $s \geq 3$, we have $r(s, s) > \lfloor 2^{s/2} \rfloor$.*

Proof: Let $n = \lfloor 2^{s/2} \rfloor$. The complete graph K_n has $\binom{n}{2}$ edges. Thus, the space of all red-blue edge-colorings of a standard-labeled K_n contains $2^{\binom{n}{2}}$ different red-blue graphs. Our objective is to show that at least one of them contains no monochromatic K_s, that is, neither a red K_s nor a blue K_s. It is sufficent to show that the probability that a random red-blue graph in the space has no monochromatic K_s is less than 1.

Toward that objective, suppose that each edge ij is red with probability $1/2$, and that we take the edge colors to be mutually independent. Then each of the red-blue graphs occurs with probability $2^{-\binom{n}{2}}$.

For any fixed subset $L \subset \{1, \ldots, n\}$ of cardinality s, let E_L be the event that the complete subgraph induced by the vertex-set L is either a red K_s or a blue K_s. Then

$$pr(E_L) = \left(\frac{1}{2}\right)^{\binom{s}{2}} + \left(\frac{1}{2}\right)^{\binom{s}{2}} = 2\left(\frac{1}{2}\right)^{\binom{s}{2}} = 2^{1-\binom{s}{2}}$$

Let E_s denote the union of all such events E_L where L ranges over all subsets of cardinality s. Since there are exactly $\binom{n}{s}$ such subsets, it follows that

$$\begin{aligned} pr(E_s) \le \sum pr(E_L) &= \binom{n}{s} \cdot 2^{1-\binom{s}{2}} \\ &= \frac{n!}{(n-s)! \cdot s!} \cdot 2^{1-\frac{s^2}{2}+\frac{s}{2}} \\ &< \frac{n^s}{s!} \cdot 2^{1-\frac{s^2}{2}+\frac{s}{2}} \\ &\le \frac{2^{\frac{s^2}{2}}}{s!} \cdot 2^{1-\frac{s^2}{2}+\frac{s}{2}} \quad (\text{since } n \le 2^{\frac{s}{2}}) \\ &= \frac{1}{s!} \cdot 2^{1+\frac{s}{2}} \\ &< 1 \quad (\text{since } s \ge 3) \end{aligned}$$

◊

As a second illustration of a probabilistic proof, we consider a problem on directed graphs.

DEFINITION: A ***tournament*** is a complete graph in which each edge is assigned a direction.

Theorem 8.3.2: [*Szele, 1943*] *For every positive integer n, there is an n-vertex tournament with at least $n!2^{-(n-1)}$ (directed) Hamiltonian paths.*

Proof: Let D be an n-vertex tournament whose edge-directions are randomly assigned, with probability $1/2$ in each direction. For each ordering τ of the vertices $1, \ldots, n$, let

$$h_\tau(D) = \begin{cases} 1 & \text{if } \tau \text{ represents a Hamiltonian path in D} \\ 0 & \text{otherwise} \end{cases}$$

Thus, the sum over all orderings τ,

$$h(D) = \sum_\tau h_\tau(D)$$

is the number of Hamiltonian paths in tournament D.

The probability that a given ordering τ corresponds to a Hamiltonian path in D equals $2^{-(n-1)}$, and, hence, the expected value $E(h_\tau(D)$ equals $2^{-(n-1)}$. It follows that the expected number of Hamitonian paths in a random tournament D is

$$E(h(D)) = E\left(\sum_\tau h_\tau(D)\right) = \sum_\tau E(h_\tau(D)) = n!2^{-(n-1)}$$

Since $n!2^{-(n-1)}$ is the average number of Hamiltonian paths in an n-vertex tournament, there must be at least one tournament whose number of Hamiltonian paths is at least $n!2^{-(n-1)}$. ◊

Models for Random Graphs

DEFINITION: A ***model for random graphs*** is a probability space whose elements are graphs.

TERMINOLOGY: The phrase *random graph* can refer to the model for random graphs itself or to one of the elements in that probability space.

The two most common models — the only ones we consider here — are the *uniform random graph*, in which the number m of edges is specified and all labeled graphs with m edges are equally likely, and the *Bernoulli random graph*, in which the number of edges is not fixed and each edge is equally likely to occur.

The Erdös-Rényi Random Graph

DEFINITION: For $0 \leq m \leq \binom{n}{2}$, the uniform random graph $\mathcal{G}(n,m)$, also called the ***Erdös-Rényi random graph***, is the equiprobable space of all n-vertex, m-edge, labeled simple graphs, each having probability $\binom{\binom{n}{2}}{m}^{-1}$.

Example 8.3.1: There are $\binom{\binom{5}{2}}{3} = 120$ random graphs in $\mathcal{G}(5,3)$. Figure 8.3.1 shows the four possible isomorphism types.

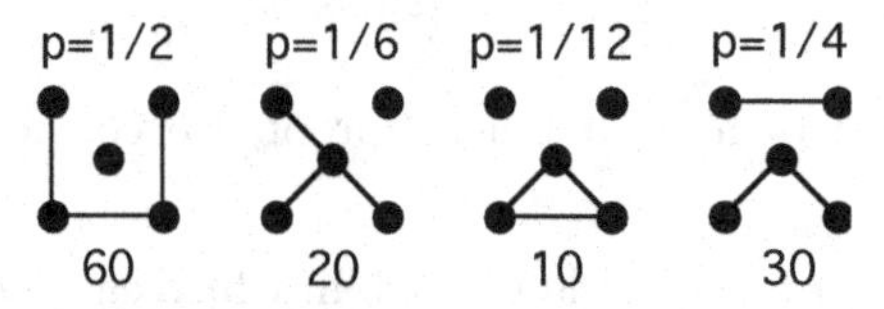

Figure 8.3.1 Probabilities of isomorphism types of graphs in $\mathcal{G}(5,3)$.

Below each type is the number of labeled graphs corresponding to that type. Above each is the probability that a random $\mathcal{G}(5,3)$ graph will be of that type. We see that although each graph is equally likely, some isomorphism types are more likely than others. Notice also that the probabilities for a $\mathcal{G}(5,3)$ graph having 0, 1, and 2 isolated vertices are

$$\frac{30}{120} = \frac{1}{4}, \quad \frac{80}{120} = \frac{2}{3}, \quad \text{and} \quad \frac{10}{120} = \frac{1}{12}$$

respectively. Thus, the expected number of isolated vertices is

$$0 \cdot \frac{1}{4} + 1 \cdot \frac{2}{3} + 2 \cdot \frac{1}{12} = \frac{5}{6}$$

Expected Number of Copies of K_s in $\mathcal{G}(n,m)$

Calculating the expected number of copies of the complete subgraph K_3 in a random graph $\mathcal{G}(n,m)$ in the proof of Theorem 8.3.3 involves all of the steps needed for the general complete subgraph K_s. The following definition simplifies the representation of summations over arbitrary sets and is used in that calculation.

DEFINITION: The ***Iverson truth function*** is

$$true(A) = \begin{cases} 1 & \text{if assertion } A \text{ is true} \\ 0 & \text{if assertion } A \text{ is false} \end{cases}$$

Example 8.3.2: Since there are $\binom{\binom{4}{2}}{4} = 15$ random graphs in $\mathcal{G}(4,4)$, each has probability $\frac{1}{15}$. The number of copies of K_3 in a given graph G is

$$\sum_{u,v,w \in V_G} true(uv, uw, vw \in E_G)$$

Thus, the expected number of copies of K_3 in a $\mathcal{G}(4,4)$ graph equals

$$\frac{1}{15} \sum_{G \in \mathcal{G}(4,4)} \sum_{u,v,w \in V_G} true(uv, uw, vw \in E_G)$$

Figure 8.3.2 shows the two isomorphism types of 4-vertex, 4-edge graphs. There are three $\mathcal{G}(4,4)$ graphs of type A and 12 of type B. Graphs of type A have no K_3 and graphs of type B each have one. Thus, the value of the double sum above is 12, and, hence, the expected number of copies of K_3 in a $\mathcal{G}(4,4)$ graph is $\frac{12}{15} = \frac{4}{5}$.

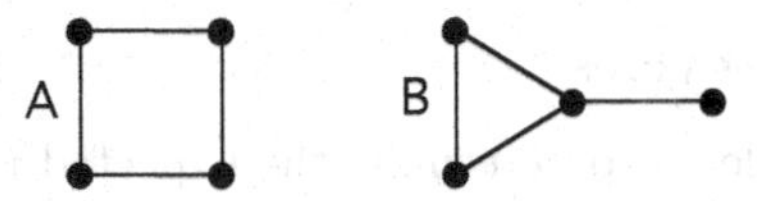

Figure 8.3.2 The two isomorphism types of 4-vertex, 4-edge graphs.

Remark: Example 8.3.2 underscores the distinction between labeled graphs, which are the elements of the probability space $\mathcal{G}(4,4)$, and the isomorphism types of unlabeled graphs. Each of the 15 elements of $\mathcal{G}(4,4)$ is equally likely, but a graph of isomorphism type B is 4 times as likely as the one of type A. If both were equally likely, then the expected number of copies of K_3 would have been $\frac{1}{2}$.

The next result establishes a general formula for the expected number of copies of K_3 in the random graph $\mathcal{G}(n,m)$. The proof is based on a simple interchange of the order of summation.

Theorem 8.3.3: *The expected number of copies of the complete graph K_3 in a random graph in $\mathcal{G}(n,m)$ is $\binom{n}{3} \cdot \binom{m}{3} \cdot \binom{\binom{n}{2}}{3}^{-1}$.*

Proof: Arguing as in Example 8.3.2, we calculate that the expected number of copies of K_3 in a random graph in $\mathcal{G}(n,m)$ equals

$$\binom{\binom{n}{2}}{m}^{-1} \sum_{G \in \mathcal{G}(4,4)} \sum_{u,v,w \in V_G} true(uv, uw, vw \in E_G)$$

Interchanging the order of summation, we obtain

$$\binom{\binom{n}{2}}{m}^{-1} \sum_{u,v,w \in \{1,\ldots,n\}} \sum_{G \in \mathcal{G}(4,4)} true(uv, uw, vw \in E_G)$$

For each of the $\binom{n}{3}$ triples $u, v, w \in \{1, \ldots, n\}$, the inner sum has constant value

$$\binom{\binom{n}{2} - 3}{m-3}$$

It follows that the expected number of copies is

$$\begin{aligned} \binom{n}{3} \cdot \binom{\binom{n}{2} - 3}{m-3} \cdot \binom{\binom{n}{2}}{3}^{-1} &= \binom{n}{3} \cdot \frac{\left[\binom{n}{2} - 3\right]!}{(m-3)!\left[\binom{n}{2} - m\right]!} \cdot \frac{m!\left[\binom{n}{2} - m\right]!}{\binom{n}{2}!} \\ &= \binom{n}{3} \cdot \frac{m!}{(m-3)!} \cdot \frac{\left[\binom{n}{2} - 3\right]!}{\binom{n}{2}!} \\ &= \binom{n}{3} \cdot \binom{m}{3} \cdot \binom{\binom{n}{2}}{3}^{-1} \end{aligned}$$

◇

Remark: If we evaluate the formula of Theorem 8.3.3 for $m \geq n+1$, we see that the expected number of copies of K_3 is greater than 1.

The proof of Theorem 8.3.3 has an immediate generalization.

Theorem 8.3.4: *The expected number of copies of the complete graph K_s in a random graph $G \in \mathcal{G}(n,m)$ is*

$$\binom{n}{s} \cdot \binom{m}{\binom{s}{2}} \cdot \binom{\binom{n}{2}}{\binom{s}{2}}^{-1}$$

◇ *(Exercises)*

Expected Number of k-cycles

Similar calculations enable us to determine the expected number of copies of any fixed graph H in a random graph $G \in \mathcal{G}(n,m)$. As an illustration, we consider the case in which H is a k-cycle. We count *unoriented* k-cycles, since a vertex sequence and its reverse represent the same k-cycle subgraph. The proof uses the elementary combinatorial result that the number of circular permutations of a k-element set is $(k-1)!$.

Theorem 8.3.5: *The expected number of k-cycles in a random graph $G \in \mathcal{G}(n,m)$ is*

$$\frac{(k-1)!}{2} \cdot \binom{n}{k} \cdot \binom{m}{k} \cdot \binom{\binom{n}{2}}{k}^{-1}$$

Proof: Any given labeled k-cycle C uses k edges. Thus, the number of graphs in $\mathcal{G}(n,m)$ that contain C is

$$\binom{\binom{n}{2}-k}{m-k}$$

The number of different labeled k-cycles in K_n is

$$\frac{1}{2}\binom{n}{k}(k-1)! = \frac{(k-1)!}{2}\binom{n}{k}$$

where the division by 2 results from counting unoriented cycles. It follows that the total number of k-cycles occurring among the graphs in $\mathcal{G}(n,m)$ is

$$\frac{(k-1)!}{2}\binom{n}{k} \cdot \binom{\binom{n}{2}-k}{m-k}$$

Thus, the expected number of k-cycles in a random graph in $\mathcal{G}(n,m)$ is

$$\begin{aligned}\frac{(k-1)!}{2}\binom{n}{k}\binom{\binom{n}{2}-k}{m-k}\binom{\binom{n}{2}}{m}^{-1} &= \frac{(k-1)!}{2}\binom{n}{k} \cdot \frac{\left[\binom{n}{2}-k\right]!}{(m-k)!\left[\binom{n}{2}-m\right]!} \cdot \frac{m!\left[\binom{n}{2}-m\right]!}{\binom{n}{2}!} \\ &= \frac{(k-1)!}{2}\binom{n}{k} \cdot \frac{m!}{(m-k)!} \cdot \frac{\left[\binom{n}{2}-k\right]!}{\binom{n}{2}!} \\ &= \frac{(k-1)!}{2} \cdot \binom{n}{k} \cdot \binom{m}{k} \cdot \binom{\binom{n}{2}}{k}^{-1}\end{aligned}$$

◇

The Bernoulli Random Graph

DEFINITION: The ***Bernoulli random graph*** $\mathcal{G}(n,p)$ is the probability space of all labeled simple graphs on n vertices, where each of the $\binom{n}{2}$ possible edges occurs with probability p. Thus, the probability of each m-edge graph is $p^m(1-p)^{\binom{n}{2}-m}$.

Example 8.3.3: Since there are $\binom{\binom{5}{3}}{3}$ possible 5-vertex, 3-edge graphs, the probability that a random graph in $\mathcal{G}(5,p)$ has three edges is

$$\binom{10}{3}p^3(1-p)^7$$

For $p = \frac{1}{2}$, this probability equals $\frac{120}{1024} = \frac{15}{128}$.

Bernoulli Space as a Union of Erdös-Rényi Spaces

The Bernoulli space $\mathcal{G}(n,p)$ of $2^{\binom{n}{2}}$ labeled simple n-vertex graphs is partitioned into $\binom{n}{2} + 1$ different Erdös-Rényi spaces

$$\left\{\mathcal{G}(n,m) \mid m = 0, \ldots, \binom{n}{2}\right\}$$

where each $\mathcal{G}(n,m)$ has $\binom{\binom{n}{2}}{m}$ labeled simple graphs. Here, the probability distribution in $\mathcal{G}(n,m)$ is the conditional distribution of $\mathcal{G}\left(n, \frac{1}{2}\right)$ restricted to the space of n-vertex, m-edge graphs.

Example 8.3.4: The probability that a random graph is a complete graph in $\mathcal{G}(n,p)$ is $p^{\binom{n}{2}}$. Of course, in the Erdös-Rényi space $\mathcal{G}\left(n, \binom{n}{2}\right)$, the probability is 1 (since the complete graph is the only element).

Example 8.3.5: To calculate the probability in $\mathcal{G}(4,p)$ that a random graph is a 3-edge path graph P_4, we observe that there are $\frac{4!}{2} = 12$ P_4s in K_4. Since it has 3 edges, each possible P_4 occurs with probability $p^3(1-p)^3$. Thus, the net probability is $12p^3(1-p)^3$. In the Erdös-Rényi space $\mathcal{G}(4,3)$, the $\binom{6}{3} = 20$ graphs fall into three isomorphism types. There are four cases of $K_3 \cup K_1$, four cases of $K_{1,3}$, and 12 cases of P_4. Thus, the probability of P_4 is $12/20 = 3/5$.

Some Properties of Almost Every Graph

DEFINITION: A graph property P is said to hold for ***almost every graph*** if the probability that a random graph $G \in \mathcal{G}(n,p)$ has property P has the limiting value of 1 as $n \to \infty$, for $p > 0$.

Theorem 8.3.6: *Almost every graph contains K_3.*

Proof: In a random graph $G \in \mathcal{G}(n,p)$, select $\lfloor \frac{n}{3} \rfloor$ disjoint triples (i.e., size-3 subsets) of V_G. For any particular one of these triples $\{u, v, w\}$, the probability that all of the edges uv, uw, vw are present is p^3. Thus, the probability that not all of them are present is $1 - p^3$. Since the presence of edges in G is probabilistically independent, it follows that the probability that none of these triples is spanned by a K_3 is

$$(1-p^3)^{\lfloor n/3 \rfloor}$$

whose limit is 0 as $n \to \infty$ for $p > 0$, because $1 - p^3 < 1$. Consequently, the probability that at least one of the triples is spanned by K_3 has limiting value 1 as $n \to \infty$ for $p > 0$. ◇

The proof of Theorem 8.3.6 has an immediate generalization.

Theorem 8.3.7: *For any integer $s \geq 1$, almost every graph contains K_s.*

◇ *(Exercises)*

Corollary 8.3.8: *Let H be any fixed simple graph. Then almost every graph contains a copy of H.*

Proof: This follows immediately from Theorem 8.3.7, because the graph H is a subgraph of some complete graph K_s. ◊

Proposition 8.3.9: *In almost every graph, every pair of vertices is joined by a path of length 2.*

Proof: Let $G \in \mathcal{G}(n,p)$, let $u, v \in V_G$, and let $X_{\{u,v\}}$ be the event that there is no path of length 2 between u and v. For each vertex $w \in V_G - \{u,v\}$, the probability that not both of the edges uw and vw are present is $1 - p^2$. Thus, the probability that there is no path of length 2 joining u and v is

$$pr(X_{\{u,v\}}) = (1-p^2)^{n-2}$$

The probability that there is at least one pair of vertices with no path of length 2 between them is

$$pr\left(\bigcup_{\{u,v\}} X_{\{u,v\}}\right) \leq \sum_{\{u,v\}} pr(X_{\{u,v\}}) = \binom{n}{2}(1-p^2)^{n-2}$$

Applying L'Hôpital's Rule, we see that for $p > 0$,

$$\lim_{n\to\infty} \binom{n}{2}(1-p^2)^{n-2} = 0$$

which finishes our proof. ◊

Theorem 8.3.10: *Almost every graph is connected.*

Proof: This is an immediate consequence of Proposition 8.3.9. ◊

The following generalization of Proposition 8.3.9 enables us to strengthen Theorem 8.3.10 to higher orders of connectedness.

Proposition 8.3.11: *Let r, s, and t be nonnegative integers. In almost every graph, for any r-subset $U = \{u_1, \ldots, u_r\}$ of vertices and any s-subset $V = \{v_1, \ldots, v_s\}$ disjoint from U, there is a t-subset $W = \{w_1, \ldots, w_t\}$, disjoint from both U and V, such that every vertex $w_j \in W$ is adjacent to every vertex of U and nonadjacent to every vertex of V.*

Proof: The probability that an arbitrary subset of t vertices disjoint from both U and V has this adjacency-nonadjacency property is $(p^r(1-p)^s)^t$. The $n - r - s$ vertices in $\overline{U \cup V}$ can be partitioned into $\lfloor \frac{n-r-s}{t} \rfloor$ t-subsets of vertices that are pairwise disjoint and disjoint from both U and V, with perhaps a few vertices left over. The probability that none of these t-subsets has the adjacency-nonadjacency property is

$$\left(1 - (p^r(1-p)^s)^t\right)^{\lfloor \frac{n-r-s}{t} \rfloor}$$

whose limit is 0 as $n \to \infty$ for $p > 0$. ◊

Theorem 8.3.12: *For $k > 0$, almost every graph is k-connected.*

Proof: Let u and v be an arbitrary pair of vertices and let $X_{\{u,v\}}$ be the event that there are not k internally disjoint paths of length 2 between u and v. It follows from Proposition 8.3.11, with $r = 2$, $s = 0$, and $t = k$, that

$$pr(X_{\{u,v\}}) = (1 - p^{2k})^{\lfloor \frac{n-2}{k} \rfloor}$$

The probability that there is at least one pair of vertices that do not have k internally disjoint paths of length 2 between them is

$$pr\left(\bigcup_{\{u,v\}} X_{\{u,v\}}\right) \leq \sum_{\{u,v\}} pr(X_{\{u,v\}}) = \binom{n}{2}(1-p^2)^{n-2}$$

The result follows, since this limit is 0 as $n \to \infty$ or $p > 0$. ◇

Some Statistical Properties of Random Graphs in $\mathcal{G}(n,p)$

Theorem 8.3.13: *For a random graph $G \in \mathcal{G}(n,p)$, the number of edges has mean μ and standard deviation σ given by*

$$u(|E_G|) = p \cdot \binom{n}{2} \quad \text{and} \quad \sigma(|E_G|) = \sqrt{p(1-p) \cdot \binom{n}{2}}$$

Proof: By definition of the Bernoulli space $\mathcal{G}(n,p)$, the random variable $|E_G|$ has the binomial distribution with $\cdot\binom{n}{2}$ trials and success rate p, and, hence, has the stated mean and standard deviation. ◇

Corollary 8.3.14: *For a random graph $G \in \mathcal{G}(n,p)$, $\sigma(|E_G|) \leq \frac{n}{2\sqrt{2}}$.*

Proof: For p such that $0 \leq p \leq 1$, we have $p(1-p) \leq \frac{1}{4}$. Thus, by Theorem 8.3.13,

$$\sigma(|E_G|) = \sqrt{p(1-p) \cdot \binom{n}{2}} \leq \sqrt{\frac{n^2}{8}} = \frac{n}{2\sqrt{2}}$$

EXERCISES for Section 8.3

8.3.1 Draw all six isomorphism types of 5-vertex, 4-edge simple graphs. Calculate the probability of each type in $\mathcal{G}(5,4)$.

8.3.2 Using Exercise 8.3.1, calculate the probability that a random graph in $\mathcal{G}(5,4)$ contains a K_3.

8.3.3 Use Theorem 8.3.3 to calculate the expected number of copies of K_3 in a random graph in $\mathcal{G}(5,4)$. (Compare to Exercise 8.3.2.)

8.3.4 Calculate the expected number of copies of the bipartite graph $K_{3,3}$ in a random graph $G \in \mathcal{G}(n,m)$.

8.3.5 Calculate the expected number of vertices of degree 3 or more in a random graph in $\mathcal{G}(5,4)$. (Hint: This is the same as the expected number of copies of $K_{1,3}$.)

8.3.6 Calculate the probability that a random graph in $\mathcal{G}(5,5)$ is the 5-cycle C_5.

8.3.7 Calculate the probability that a random graph in $\mathcal{G}(5,\frac{1}{2})$ is the 5-cycle C_5.

8.3.8 Prove the remark immediately after Theorem 8.3.3.

8.3.9 Prove that almost every graph is non-planar.

8.3.10 Prove that for every k, the chromatic number of almost every graph is larger than k.

8.3.11 Prove Theorem 8.3.4, that the expected number of copies of the complete graph K_s in a random graph $G \in \mathcal{G}(n, m)$ is

$$\binom{n}{s} \cdot \binom{m}{\binom{s}{2}} \cdot \binom{\binom{n}{2}}{\binom{s}{2}}^{-1}$$

8.3.12 Prove Theorem 8.3.7, that almost every graph contains K_s.

8.3.13 Use Proposition 8.3.11 to prove that for any $d > 0$, almost every graph has minimum degree at least d.

8.4 SUPPLEMENTARY EXERCISES

8.4.1 Prove that $r(K_3, K_{1,3}) > 6$.

8.4.2 Prove that $r(K_{1,3}, C_4) = 6$.

8.4.3 Calculate $r(C_4, C_4)$.

8.4.4 Use Theorems 8.2.1 and 8.2.3 to prove that $r(K_3, K_{1,3}) \leq 7$.

8.4.5 The proof of Theorem 8.2.3 calls for a $(d-1)$-regular graph on n vertices, when d is odd or n is even. Specify such a graph.

8.4.6 Prove that if m and n are both even, then $r(K_{1,n}, K_{1,m}) \leq m + n - 1$.

8.4.7 Draw a 6-vertex graph that does not contain K_3 and whose complement does not contain the 3-edge path P_4.

8.4.8 Calculate as tight an upper bound as you can for the Ramsey number $r(5, 5)$.

8.4.9 Prove that $ex(5, C_4) = 6$.

8.4.10 Prove that $ex(5, C_5) = 8$.

8.4.11 Let G be a 6-vertex, 8-edge simple graph. Prove that G contains a 4-edge path P4, for these cases: (a) $\delta_{\max}(G) = 5$; (b) $\delta_{\max}(G) = 4$; (c) $\delta_{\max}(G) = 3$. Conclude that $ex(6, P_4) = 7$. (Hint: Let vertex u have maximum degree in G. If any two of its neighbors are joined, then at each neighbor, there starts a 3-edge-path within the closed neighborhood $N(u)$.)

8.4.12 For $n \geq 3$, let $t(n)$ be the smallest number such that there exists a K_3-free, simple, n-vertex, $t(n)$-edge graph $G(n)$ such that joining any two nonadjacent vertices of $G(n)$ creates a K_3. Establish a formula for the function $t(n)$.

8.4.13 Calculate the expected number of copies of $K_{2,3}$ in a random graph in $\mathcal{G}(n, m)$.

8.4.14 Calculate the expected number of copies of CL_3 in a random graph in $\mathcal{G}(n, m)$.

8.4.15 Calculate the expected number copies of the circulant graph $circ(8 : 1, 4)$ in a random graph in $\mathcal{G}(n, m)$.

8.4.16 A simple graph is constructed with 12 vertices and 36 randomly selected edges. Calculate the expected number of triangles.

8.4.17 A simple graph is constructed with eight vertices and seven randomly selected edges. Calculate the probability that it is connected.

Glossary

almost every graph, property that holds for: a property P such that the probability that a random graph $G \in \mathcal{G}(n,p)$ has property P has the limiting value of 1 as $n \to \infty$, for $p > 0$.

Bernoulli random graph $\mathcal{G}(n,p)$: the probability space of all labeled simple graphs on n vertices, where each of the $\binom{n}{2}$ possible edges occurs with probability p; thus, the probability of each m-edge graph in $\mathcal{G}(n,p)$ is $p^m(1-p)^{\binom{n}{2}-m}$.

Erdös-Rényi random graph $\mathcal{G}(n,m)$: the equiprobable space of all n-vertex, m-edge, labeled simple graphs, each having probability $\binom{\binom{n}{2}}{m}^{-1}$; also called the *uniform random graph.*

extremal function *ex(n, P)* for a property P: the function whose value is the number of edges in an n-vertex extremal graph for that property.

extremal n-vertex graph for a property P: a largest n-vertex simple graph that does not have property P, where "largest" means having the maximum number of edges.

Iverson truth function: the function

$$true(A) = \begin{cases} 1 & \text{if assertion } A \text{ is true} \\ 0 & \text{if assertion } A \text{ is false} \end{cases}$$

label-preserving isomorphism; an isomorphism $f : G \to H$ of two labeled graphs such that for every vertex $v \in V_G$, the vertex $f(v) \in V_H$ has the same label as v.

model for random graphs: a probability space whose elements are graphs.

multipartite graph: a loopless graph whose vertices can be partitioned into k independent sets, which are sometimes called the *partite sets*, is said to be k-partite.

partite sets: see *multipartite graph.*

Ramsey number $r(s,t)$ for positive integers s and t: the minimum positive number n such that for every simple n-vertex graph G, either G contains a K_s-subgraph or its edge-complement graph $\overline{G}$ contains a K_t-subgraph.

—**, diagonal:** a Ramsey number $r(s,s)$.

—**, for arbitrary subgraphs H_1 and H_2,** denoted $r(H_1,H_2)$: the minimum number n such that for every n-vertex simple graph G, either H_1 is a subgraph of G or H_2 is a subgraph of $\overline{G}$.

—, **generalized,** denoted $r(G_1, G_2, \ldots, Gk)$: for any collection of simple graphs $G_1, G_2, \ldots, G_k$, the minimum number n such that in any edge k-coloring of a complete graph K_n, at least one of the subgraphs G_j has all of its edges in one color class j.

random graph: either the *model for random graphs* itself or one of the elements in that probability space.

sufficiently large graph, with respect to a graph property P**:** a graph whose number of edges, relative to its number of vertices, is large enough that every graph with as many edges must have the specified property.

tournament: a complete graph in which each edge is assigned a direction.

Turán graph $T_{n,r}$**:** the complete r-partite graph on n vertices whose partite sets are as nearly equal in cardinality as possible.

uniform random graph $\mathcal{G}(n, m)$**:** the equiprobable space of all n-vertex, m-edge, labeled simple graphs, each having probability $\binom{\binom{n}{2}}{m}^{-1}$; also called the *Erdös-Rényi random graph.*

Chapter 9

GRAPH COLORINGS AND SYMMETRY

INTRODUCTION

This chapter explores the interplay between a graph's symmetry and the number of different colorings of that graph. For example, if one coloring of a graph can be obtained from another coloring by a simple rotation of the graph, then the two colorings are equivalent. The symmetries of a graph are precisely defined using the graph's automorphism group, along with its vertex- and edge-permutations. The equivalence classes under these group actions determine when colorings are equivalent. We develop the basic mathematical concepts and tools to enumerate these equivalence classes.

Although we restrict our attention mainly to simple graphs, the methods studied in this chapter are applicable to general graphs.

9.1 AUTOMORPHISMS OF SIMPLE GRAPHS

The concept of graph automorphism was introduced in §2.2. For convenience, its definition is presented below.

REVIEW FROM §2.1 AND §2.2:

- A ***graph homomorphism*** $f : G \to H$ from a graph G to a graph H is a pair of functions $(f_V : V_G \to V_H, f_E : E_G \to E_H)$ between their vertex- and edge-sets, respectively, such that for every edge $e \in E_G$, the function f_V maps the endpoints of e to the endpoints of edge $f_E(e)$. That is, if u and v are adjacent vertices, $f(uv) = f(u)f(v)$.
- A ***graph isomorphism*** $f : G \to H$ from a graph G to a graph H is a graph homomorphism $(f_V : V_G \to V_H, f_E : E_G \to E_H)$ for which f_V and f_E are bijections. When there is a graph isomorphism $f : G \to H$, we say that G is ***isomorphic*** to H, denoted $G \cong H$.
- Let G and H be simple graphs. A bijection $f : V_G \to V_H$ gives a graph isomorphism if and only if f preserves both adjacency and nonadjacency (that is, if f is ***structure-preserving***).
- An isomorphism from a graph G to itself is called an ***automorphism***.

NOTATION: The set of automorphisms of a graph G is denoted $\mathcal{A}ut(G)$, and the corresponding sets of vertex-permutations and of edge-permutations are denoted $\mathcal{A}ut_V(G)$ and $\mathcal{A}ut_E(G)$, respectively.

The Sets $\mathcal{A}ut_V(G)$ and $\mathcal{A}ut_E(G)$ are Permutation Groups

DEFINITION: A collection P of permutations on the same set of objects is ***closed under composition*** if for every pair $\pi_1, \pi_2 \in P$, the composition $\pi_1\pi_2$ is in P.

DEFINITION: Let P be a non-empty collection of permutations of the finite set of objects Y such that P is closed under composition. Then the set P together with set Y is called a ***permutation group*** and is denoted $[P : Y]$.

TERMINOLOGY: The permutation group is said to ***act on the set*** Y.

DEFINITION: The ***full symmetric group*** Σ_Y on a set Y is the collection of *all* permutations on Y.

Theorem 9.1.1: *The set $\mathcal{A}ut(G)$ of all automorphisms of a simple graph G acts as a permutation group on V_G and on E_G.*

Proof: It is straightforward to verify that:

1. the composition of two structure-preserving vertex-permutations is a structure-preserving vertex-permutation, and
2. the inverse of a structure-preserving vertex-permutation is a structure-preserving vertex-permutation.

Thus, $\mathcal{A}ut_V(G)$ and $\mathcal{A}ut_E(G)$ are permutation groups. (see Exercises) ◇

DEFINITION: Let $[P : Y]$ be a permutation group. The ***orbit*** of an object $y \in Y$ is the set of all objects $z \in Y$ such that $\pi(y) = z$ for some permutation $\pi \in P$.

REVIEW FROM §2.2

- The action of the automorphism group $\mathcal{A}ut(G)$ on a graph G partitions V_G into ***vertex orbits***. That is, the vertices u and v are in the same orbit if there exists an automorphism π such that $\pi(u) = v$. Similarly, $\mathcal{A}ut(G)$ partitions E_G into ***edge orbits***.
- A graph G is ***vertex-transitive*** if all the vertices are in a single orbit. Similarly, G is ***edge-transitive*** if all the edges are in a single orbit.

Remark: In this chapter, permutations are represented in *disjoint-cycle form* (introduced in §2.2). Permutation groups and abstract groups are discussed in Appendix A.4.

Example 9.1.1: The graph $K_{1,3}$, shown below, has six automorphisms, and the vertex- and edge-permutation groups corresponding to $\mathcal{A}ut(K_{1,3})$ are given by

$$\mathcal{A}ut_V(K_{1,3}) = \{(u)(v)(w)(x),\ (x)(u\ v\ w),\ (x)(u\ w\ v),\ (x)(u)(v\ w),\ (x)(v)(u\ w),\ (x)(w)(u\ v)\}$$

$$\mathcal{A}ut_E(K_{1,3}) = \{(a)(b)(c),\ (a\ b\ c),\ (a\ c\ b),\ (a)(b\ c),\ (b)(a\ c),\ (c)(a\ b)\}$$

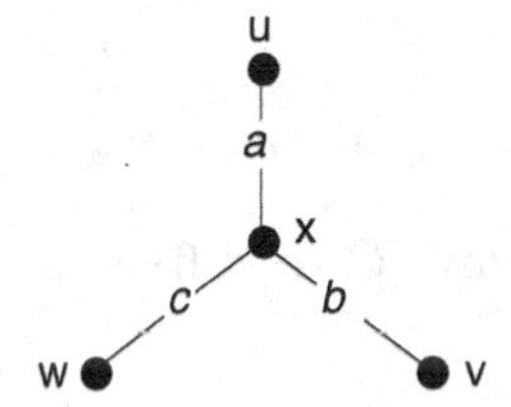

Figure 9.1.1 The graph $K_{1,3}$.

The vertex-orbits are $\{u, v, w\}$ and $\{x\}$. The only edge-orbit is $\{a, b, c\}$. Therefore, $K_{1,3}$ is edge-transitive, but not vertex-transitive.

Remark: Figure 9.1.1 gives a drawing of $K_{1,3}$ whose geometric symmetry captures all of the automorphisms of that graph. However, this is not always possible. The Petersen graph has no such drawing, as shown in Example 2.2.4.

Automorphism Groups of Some Other Simple Graphs

The methods used above to determine $\mathcal{A}ut(K_{1,3})$ are now applied to calculating the automorphism groups of some other graphs.

Example 9.1.2: $\mathcal{A}ut(K_n)$ Each of the $n!$ permutations on the vertex-set of K_n is structure-preserving, because every pair of vertices is joined by an edge. As illustrated in Figure 9.1.2, a permutation of vertices determines a permutation of edges. Thus, every vertex-permutation specifies a different automorphism of K_n.

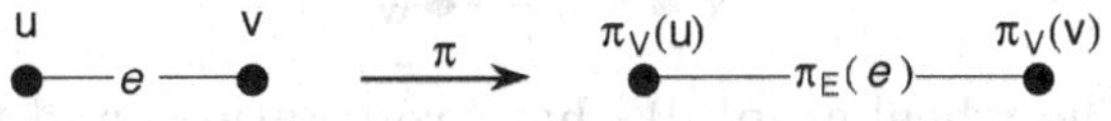

Figure 9.1.2 Automorphism action on an edge and its endpoints.

The complete graph K_4 can be represented as the 1-skeleton of a regular tetrahedron in Euclidean 3-space. All 24 automorphisms of K_4 can be realized by rotations and reflections of a regular tetrahedron. To generalize this geometric viewpoint to larger values of n, represent the complete graph K_n as the 1-skeleton of a regular n-simplex in Euclidean n-space. Then all $n!$ automorphisms can be realized by combinations of rotations and reflections of a regular n-simplex.

Example 9.1.3: $Aut(C_n)$ The cycle graph C_n can be represented as the 1-skeleton of a regular n-gon in the plane, as illustrated in Figure 9.1.3 for $n = 5$. All of its automorphisms can be realized as rotations and reflections of a regular n-gon. For instance, rotation 72° clockwise corresponds to the automorphism of C_5 whose vertex-permutation is $(u\ v\ w\ x\ y)$ and whose edge-permutation is $(a\ b\ c\ d\ e)$. Also, reflection through the vertical axis of symmetry corresponds to the automorphism whose vertex-permutation is $(u)(v\ y)(w\ x)$ and whose edge-permutation is $(a\ e)(b\ d)(c)$. In general, a regular n-gon has n rotations and n reflections, which gives the graph C_n a total of $2n$ automorphisms.

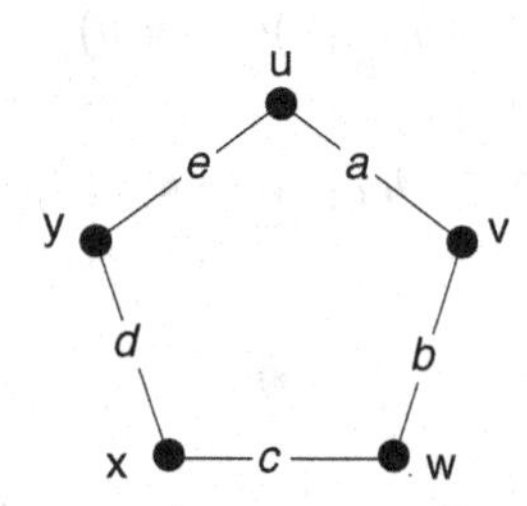

Figure 9.1.3 The cycle graph C_5 has five rotations and five reflections.

Example 9.1.4: $Aut(W_n)$ A geometric representation of the n-spoked wheel graph W_n can be constructed by placing a vertex at the center of a regular n-gon in the plane, and joining it to every vertex of the n-gon, as illustrated for $n = 5$ in Figure 9.1.4. All of its automorphisms can be realized as rotations and reflections of the n-gon. For instance, rotation 72° clockwise corresponds to the automorphism of W_5 whose vertex-permutation is $(t)(u\ v\ w\ x\ y)$ and whose edge-permutation is $(a\ b\ c\ d\ e)(f\ g\ h\ i\ j)$. Also, reflection through the vertical axis of symmetry corresponds to the automorphism whose vertex-permutation is $(t)(u)(v\ y)(w\ x)$ and whose edge-permutation is $(a\ e)(b\ d)(c)(f\ i)(g\ h)(j)$. Thus, the wheel graph W_n has n rotations and n reflections, for a total of $2n$ automorphisms, except when $n = 3$. In the special case $W_3 \cong K_4$, in which the hub vertex has the same degree as the rim vertices, there are 24 automorphisms.

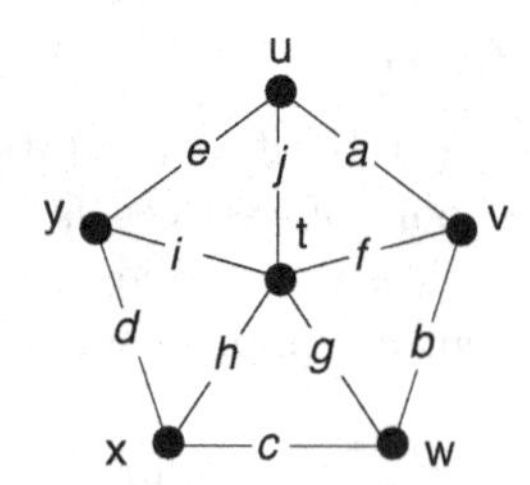

Figure 9.1.4 The wheel graph W_5 has five rotations and five reflections.

Example 9.1.5: $\mathcal{A}ut(K_{m,n})$ Let $K_{m,n}$ have vertex bipartition $R \cup S$, with $m = |R|$ and $n = |S|$, and let $\tau \in \mathcal{A}ut(K_{m,n})$. Then, since τ preserves nonadjacency, two vertices on one side of the bipartition must map to two vertices on that same side or to two vertices on the other side. It follows that either $\tau(R) = R$ and $\tau(S) = S$, or else the bijection τ swaps R and S (i.e., $\tau(R) = S$ and $\tau(S) = R$).

If $m \neq n$, then $\tau(R) = R$ and $\tau(S) = S$ is the only possibility. Thus, τ permutes the elements of set R and permutes the elements of set S. That is, $\tau = \rho \oplus \sigma$, where $\rho \in \Sigma_R$ and $\sigma \in \Sigma_S$, and

$$(\rho \oplus \sigma)(u) = \begin{cases} \rho(u), & \text{if } u \in R \\ \sigma(u), & \text{if } u \in S \end{cases}$$

Conversely, for each pair of permutations $\rho \in \Sigma_R$ and $\sigma \in \Sigma_S$, $\rho \oplus \sigma$ is a structure-preserving vertex-permutation of $K_{m,n}$. Therefore, if $m \neq n$, then $|\mathcal{A}ut_V(K_{m,n})| = m!n!$.

If $m = n$, then $\tau \in \mathcal{A}ut(K_{n,n})$ is either a permutation of the form $\rho \oplus \sigma$ or is a vertex bijection that swaps R and S. Moreover, each of the $(n!)^2$ vertex bijections that swap R and S is structure-preserving, since every vertex in R is adjacent to every vertex in S. Thus, $|\mathcal{A}ut(K_{n,n})| = 2(n!)^2$.

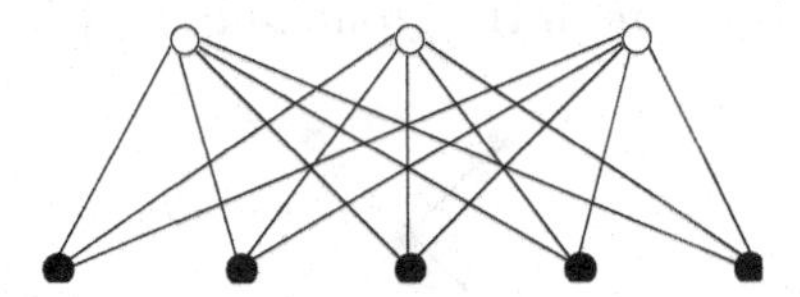

Figure 9.1.5 The bipartite graph $K_{3,5}$ has $3!5! = 720$ automorphisms.

EXERCISES for Section 9.1

9.1.1 Write the vertex-permutation and the edge-permutation for every automorphism of the graph shown. Also list the vertex-orbits and the edge-orbits.

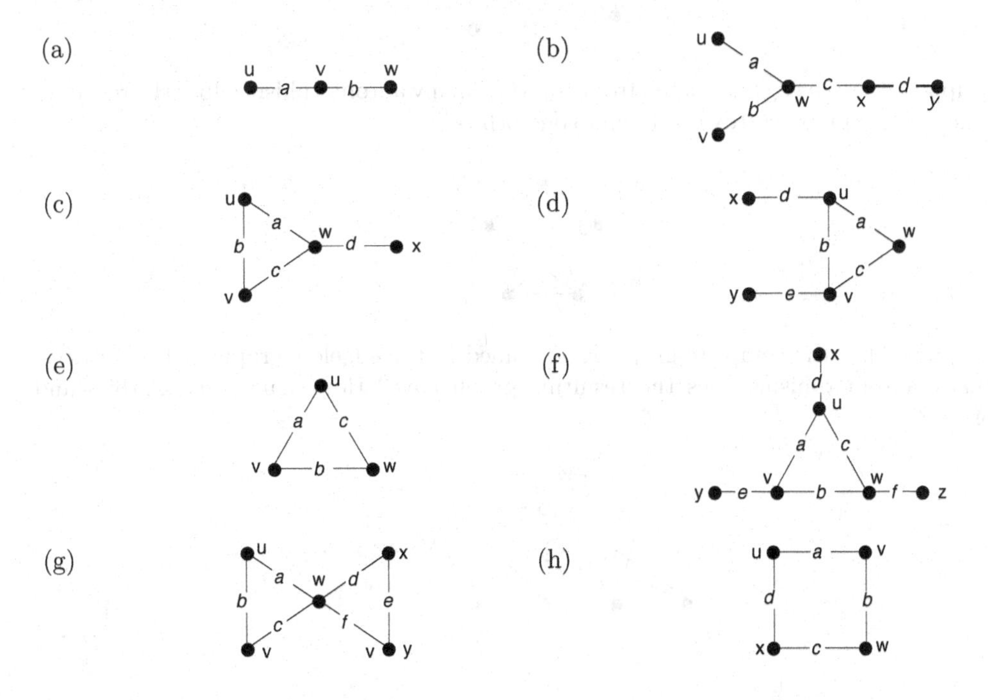

9.1.2 Redraw the graph so that as many automorphisms as possible are represented by symmetries of the drawing.

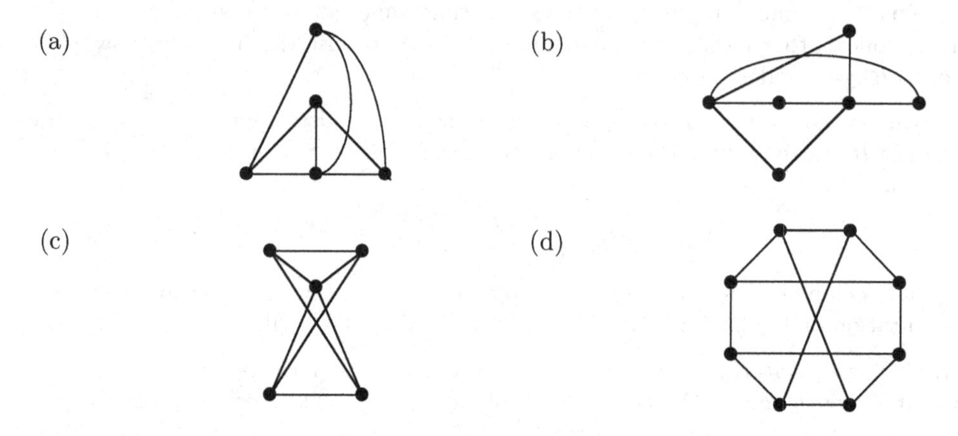

9.1.3 Prove Theorem 9.1.1.

9.1.4 How many automorphisms are in the group $\mathcal{A}ut(K_{1,4})$?

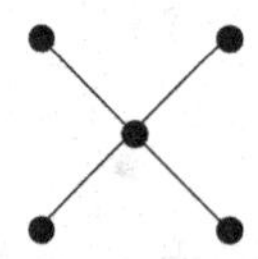

9.1.5 Suppose that K_3 and K_4 are amalgamated at a vertex. How many automorphisms does the resulting graph have? How many vertex-orbits and edge-orbits?

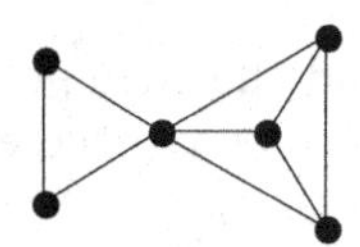

9.1.6 Suppose that an edge is deleted from K_5. How many automorphisms does the resulting graph have? How many vertex-orbits and edge-orbits?

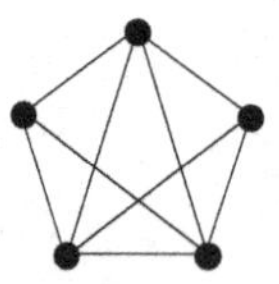

9.1.7 Suppose that the complete graph K_3 is joined to the edgeless graph on five vertices. How many automorphisms does the resulting graph have? How many vertex-orbits and edge-orbits?

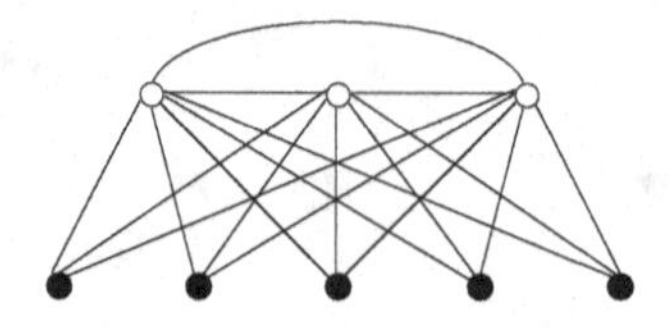

9.1.8 How many automorphisms are in the group $\mathcal{A}ut(K_{2,3,4})$, where $K_{2,3,4}$ is the *complete tripartite graph* with tripartition consisting of a 2-vertex, 3-vertex, and 4-vertex subset? How many vertex-orbits and edge-orbits?

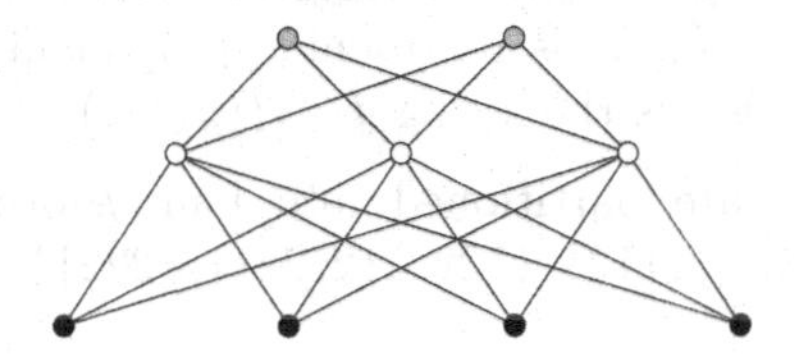

9.1.9 How many automorphisms are in the group $\mathcal{A}ut(CL_n)$, where CL_n is the n-vertex circular ladder graph (§1.1)? How many vertex-orbits and edge-orbits?

9.2 EQUIVALENCE CLASSES OF COLORINGS

The symmetry of a graph G has a substantial effect on how we determine the number of *different* vertex-colorings and edge-colorings. What we actually count is the number of equivalence classes of its vertex- and edge-colorings, induced by the automorphism group $\mathcal{A}ut(G)$, and this section gives a precise description of these equivalence classes.

Coloring a Set Subject to the Action of a Permutation Group

Counting equivalence classes of graph colorings lies within the general context of counting equivalent colorings of a set acted upon by a permutation group.

DEFINITION: A ***k*-coloring** of a set Y is a mapping f from Y *onto* the set $\{1, 2, \ldots, k\}$, in which the value $f(y)$ is called the ***color*** of y. Any k-coloring of Y is also called a ***coloring***.

DEFINITION: A **($\leq k$)-*coloring*** of a set Y is a coloring that uses k or *fewer* colors, formally a mapping f from Y onto any set $\{1, 2, \ldots, t\}$ with $t \leq k$.

NOTATION: The set of all ($\leq k$)-colorings of the elements of a set Y is denoted $Col_k(Y)$.

Proposition 9.2.1: *Let Y be a set. Then $|Col_k(Y)| = k^{|Y|}$.*

Proof: This is a direct application of the Rule of Product, a familiar counting principle in elementary discrete mathematics (see Appendix A.3). ♢

DEFINITION: Let $\mathcal{P} = [P : Y]$ be a permutation group acting on a set Y, and let f and g be ($\leq k$)-colorings of the objects in Y. Then the coloring f is ***$\mathcal{P}$-equivalent*** to the coloring g if there is a permutation $\pi \in P$ such that $g = f\pi$, that is, if for every object $y \in Y$, the color $g(y)$ is the same as the color $f(\pi(y))$.

TERMINOLOGY: The $\mathcal{P}$-equivalence classes are also called ***coloring classes***, or ***$\mathcal{P}$-orbits***. That is, any two $\mathcal{P}$-equivalent colorings are in the same $\mathcal{P}$-orbit.

NOTATION: The set of $\mathcal{P}$-orbits of $Col_k(Y)$ is denoted $\{Col_k(Y)\}_{\mathcal{P}}$. Thus, $|\{Col_k(Y)\}_{\mathcal{P}}|$ equals the number of non-equivalent ($\leq k$)-colorings.

NOTATION: The identity permutation on a set Y is often denoted ε.

- For any $y \in Y$, $\varepsilon(y) = y$.
- For any permutation in $\pi \in \Sigma_Y$, $\varepsilon\pi = \pi = \pi\varepsilon$.

Example 9.2.1: Let $Y = \{x, y, z\}$ be the vertex-set V_{C_3} of the 3-cycle graph C_3, shown on the left in Figure 9.2.1, and let $P = \{\varepsilon, (x\ y\ z), (x\ z\ y)\}$ be the group of vertex-permutations corresponding, respectively, to the 0°, 120°, and 240° clockwise rotations of C_3. The set $\text{Col}_2(V_{C_3})$, which consists of eight (≤ 2)-colorings, is shown on the right in Figure 9.2.1. Each (≤ 2)-coloring $f : V_{C_3} \to \{1, 2\}$ is represented graphically using white (color 1) and black (color 2) vertices, and also as the string $f(x)f(y)f(z)$.

These eight vertex-colorings are partitioned into four $\mathcal{P}$-orbits of $Col_2(V_{C_3})$ given by $[Col_2(V_{C_3})]_{\mathcal{P}} = \{\{111\}, \{211, 121, 112\}, \{122, 212, 221\}, \{222\}\}$.

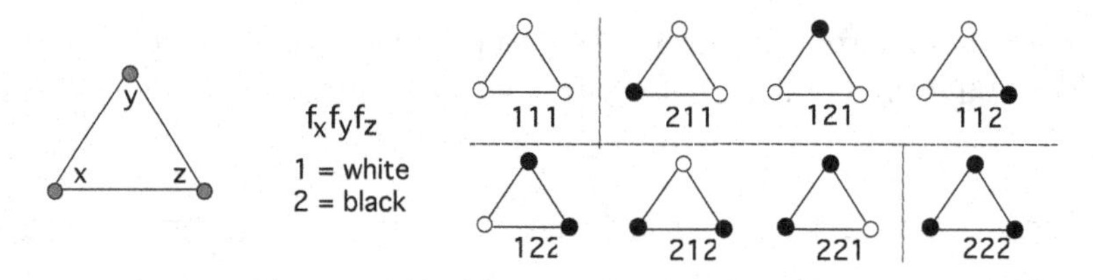

Figure 9.2.1 The four coloring classes of V_{C_3} under rotational equivalence.

Remark: Intuitively, the reason why the colorings in one class, e.g., $\{211, 121, 112\}$, are equivalent is because rotating a drawing of the graph superimposes one of these colorings automorphically onto another with correct color matching. This intuitive notion is what is formally represented by the condition $g = f\pi$.

Induced Permutation Actions

In Example 9.2.1, we saw how the permutations acting on the three vertices of C_3 *induced* another action on the set of eight vertex-(≤ 2)-colorings. More generally, a permutation group acting on a set of objects also acts on the set of colorings of those objects, as we now show.

Proposition 9.2.2: *Let $\mathcal{P} = [P : Y]$ be a permutation group acting on a set Y. Let $f \in Col_k(Y)$ be a $(\leq k)$-coloring of Y, and let $\pi \in P$ be a permutation in $\mathcal{P}$. Then the composition $f\pi$ of permutation π followed by coloring f is a coloring in $Col_k(Y)$.*

Proof: The composition $f\pi$ is a coloring of Y because it assigns to each object $y \in Y$ whatever color f assigns to the object $\pi(y)$. ◇

Corollary 9.2.3: *Let $\mathcal{P} = [P : Y]$ be a permutation group acting on a set Y, and let $\pi \in P$. Then the mapping $\pi_{YC} : Col_k(Y) \to Col_k(Y)$ defined by*

$$\pi_{YC} : (f) = f\pi, \text{ for every coloring } f \in Col_k(Y)$$

is a permutation on the set $Col_k(Y)$.

Proof: The mapping π_{YC} is a bijection, because it has an inverse, namely, the rule $g \mapsto g\pi^{-1}$. ◇

DEFINITION: The mapping $\pi_{YC} : Col(Y) \to Col_k(Y)$ defined (in Corollary 9.2.3) by the rule $f \mapsto f\pi$ is called the ***induced permutation action*** of π on $Col_k(Y)$.

NOTATION: To distinguish between the action of a permutation π on a set Y and its induced action on the set $Col_k(Y)$ of colorings of Y, we let π_Y denote its action on Y and π_{YC} its action on $Col_k(Y)$. When there is no risk of confusion, the subscripts Y and YC may both be omitted.

DEFINITION: Let $\mathcal{P} = [P : Y]$ be a permutation group acting on a set Y. The collection $\mathcal{P}_{YC} = [P : Col_k(Y)]$ of induced permutations on $Col_k(Y)$ is called the ***induced permutation group***.

NOTATION: When it is necessary to distinguish between the group that acts on the set Y and the group that acts on the set $Col_k(Y)$, the respective notations $\mathcal{P}_Y$ and $\mathcal{P}_{YC}$ are used.

Example 9.2.1, continued: Figure 9.2.2 below depicts the disjoint-cycle form of each permutation π_Y and also of the corresponding induced permutation π_{YC} acting on the set $Col_2(Y) = \{111, 112, 121, 122, 211, 212, 221, 222\}$.

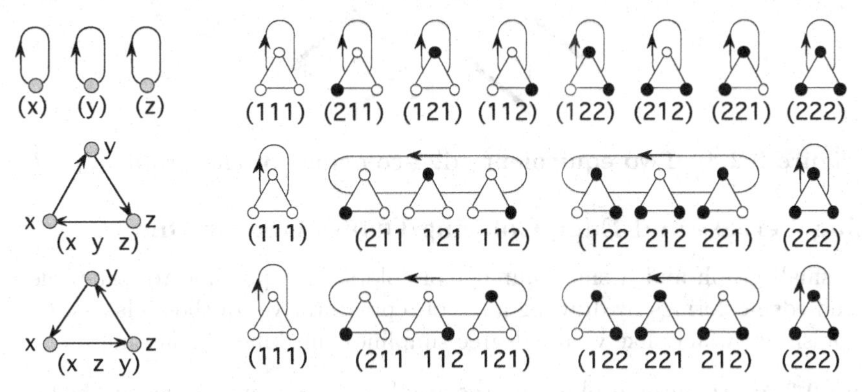

Figure 9.2.2 Correspondence of permutations and induced permutations.

Notice that two colorings are $\mathcal{P}$-equivalent (i.e., in the same $\mathcal{P}$-orbit) if the two colorings appear in a cycle of at least one induced permutation.

Equivalent Colorings of a Graph G under $\mathcal{A}ut(G)$

If the symmetries of a graph are ignored, then counting vertex- or edge-colorings is trivial; using k or fewer colors, there are $k^{|V_G|}$ vertex-colorings and $k^{|E_G|}$ edge-colorings (by Proposition 9.2.1). However, taking symmetry into account, by counting orbits of colorings, is more complicated.

TERMINOLOGY: $Col_k(V_G)$ is sometimes called the ***full set of vertex-($\leq k$)-colorings*** of a graph G, and $Col_k(E_G)$ the ***full set of edge-($\leq k$)-colorings***.

DEFINITION: Let G be a graph with automorphism group $\mathcal{A}ut(G)$. ***Equivalent vertex-colorings*** are vertex-colorings that are $\mathcal{A}ut_V(G)$-equivalent, and ***equivalent edge-colorings*** are edge-colorings that are $\mathcal{A}ut_E(G)$-equivalent.

Thus, equivalent vertex-colorings are in the same $\mathcal{A}ut_V(G)$-orbit, and equivalent edge-colorings are in the same $\mathcal{A}ut_E(G)$-orbit.

NOTATION: The set of all $\mathcal{A}ut_V(G)$-orbits (coloring classes of $Col_k(V_G)$) is denoted $\{\text{Col}_k(V_G)\}_{\mathcal{A}ut_V(G)}$. Similarly, $\{Col_k(E_G)\}_{\mathcal{A}ut_E(G)}$ denotes the set of all $\mathcal{A}ut_E(G)$-orbits of $Col_k(E_G)$.

Example 9.2.2: The two vertex-colorings in Figure 9.2.3 are equivalent, because a 180°-rotation or a reflection of one graph drawing through its horizontal axis corresponds to a graph automorphism.

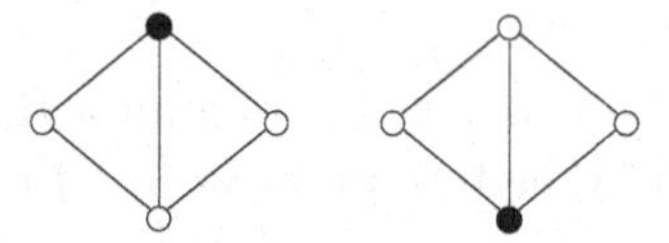

Figure 9.2.3 Two equivalent vertex-colorings of the graph $K_4 - K_2$.

Example 9.2.3: The two edge-colorings in Figure 9.2.4 are equivalent. Reflection of the graph through its vertical or horizontal axis corresponds to an automorphism that maps one coloring onto the other.

Figure 9.2.4 Two equivalent edge-colorings of the graph $K_4 - K_2$.

Counting Vertex- and Edge-Coloring Orbits One by One

For a small graph and a small number of colors, it is possible to count the orbits of vertex- and edge-colorings by drawing a list of representatives of those classes. Using graph automorphism invariants like vertex degree simplifies this kind of counting.

Example 9.2.4: An automorphism on the graph $K_4 - K_2$ must either fix the two 2-valent vertices or swap them, and it must either fix the two 3-valent vertices or swap them. Thus, $\mathcal{A}ut(K_4 - K_2)$ has the following representations as a group of vertex-permutations and as a group of edge-permutations.†

Symmetry	$\pi \in \mathcal{A}ut_V(G)$	$\pi \in \mathcal{A}ut_E(G)$
identity	$(u)(v)(w)(x)$	$(a)(b)(c)(d)(e)$
refl. thru vert. axis	$(u)(w)(v\ x)$	$(e)(a\ b)(c\ d)$
refl. thru horiz. axis	$(v)(x)(u\ w)$	$(e)(a\ b)(b\ d)$
180° rotation	$(v\ w)(v\ x)$	$(e)(a\ d)(b\ c)$

Figure 9.2.5 shows how the (full) set $Col_2(V_{K_4-K_2})$ of 16 vertex-($\leq$2)-colorings is partitioned into nine $\mathcal{A}ut_V(K_4 - K_2)$-orbits.

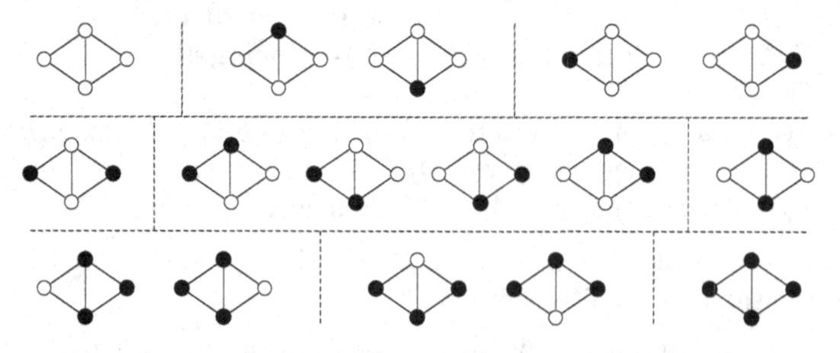

Figure 9.2.5 The $\mathcal{A}ut_V(K_4 - K_2)$-orbits of $Col_2(V_{K_4-K_2})$.

†The reader familiar with group theory will notice that the group $\mathcal{A}ut(K_4 - K_2)$ is abstractly isomorphic to $Z_4 \times Z_2$.

Example 9.2.4, continued: Since the graph $K_4 - K_2$ has five edges, there are, ignoring equivalences, $2^5 = 32$ edge-colorings that use two or fewer colors. Figure 9.2.6 shows a representative of each of the 14 $Aut_E(K_4 - K_2)$-orbits of $Col_2(E_{K_4-K_2})$. Next to each representative edge-($\leq$2)-coloring of $K_4 - K_2$ in the figure is the number of colorings in its orbit.

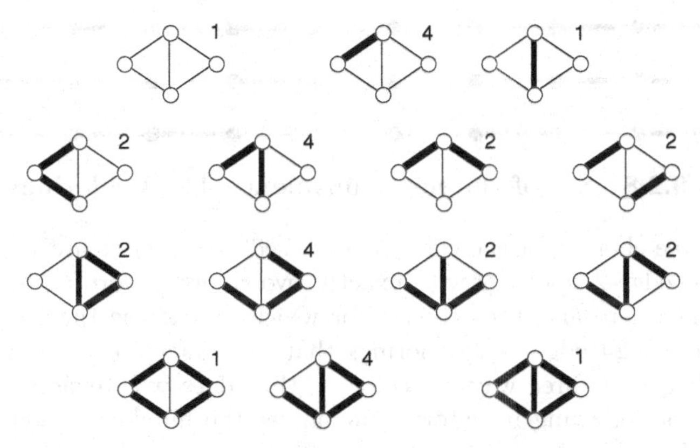

Figure 9.2.6 Representatives and sizes of the 14 orbits of $Col_2(E_{K_4-K_2})$.

Elementary Application of Symmetries in Itemizing Colorings

As the size of the graph and the number of colors increase, it becomes progressively less practical to count orbits by ordinary itemization. However, systematic exploitation of graph symmetries often reduces the work.

Example 9.2.5: A systematic approach to counting the $Aut_V(P_5)$-orbits (coloring classes) of vertex-($\leq$2)-colorings of the path graph P_5 may begin with the observation that there is only one vertex-($\leq$2)-coloring with all vertices white. There are three coloring classes with four white vertices and one black, depending on whether the black vertex is at an end, next to an end, or in the middle. There are six coloring classes with three white and two black, as shown in Figure 9.2.7.

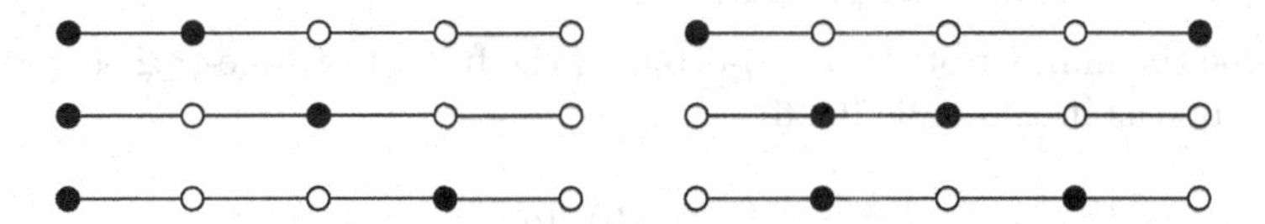

Figure 9.2.7 The six vertex-colorings of P_5 with three white and two black.

By symmetry between the colors white and black, there are also six coloring classes with three black vertices and two white, three coloring classes with four black and one white, and one class with all five vertices black, for a total of 20 $Aut_V(P_5)$-orbits of colorings. In other words, there are 20 non-equivalent vertex-($\leq$2)-colorings.

Example 9.2.6: The number of coloring classes of edge-($\leq$2)-colorings of P_5 is 10, according to the following inventory:

$$10 \text{ total} = \begin{cases} 1 & \text{with all four edges light} \\ 1 & \text{with all four edges dark} \\ 2 & \text{with three edges light and one edge dark} \\ 2 & \text{with one edge light and three edges dark} \\ 4 & \text{with two edges light and two edges dark} \end{cases}$$

Example 9.2.7: The number of (non-equivalent) edge-($\leq$3)-colorings of P_5 can be derived from the number of edge-($\leq$2)-colorings. We begin by calculating the number of edge-3-colorings. The three different edge colors are depicted by *bold*, *plain*, and *dashed* edges. Figure 9.2.8 shows the six kinds of edge-3-colorings of P_5 with the color *bold* used twice.

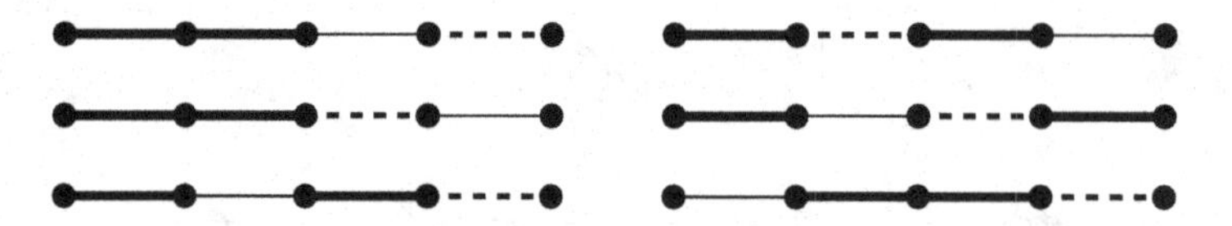

Figure 9.2.8 Six of the non-equivalent edge-3-colorings of P_5.

Since there are three choices for the color that is used twice, there are 18 edge-3-colorings. Since P_5 has eight edge-2-colorings with exactly two colors, according to the inventory in Example 9.2.6, and there are three choices for which two of the three colors are used, it follows that there are 24 edge-($\leq$3)-colorings that use exactly two of the three available colors. Finally, there are three ways to color all the edges with exactly one of the three available colors. The following inventory summarizes the number of non-equivalent edge-($\leq$3)-colorings of P_5.

$$45 \text{ total} = \begin{cases} 18 & \text{using all three colors} \\ 24 & \text{using exactly two of the three colors} \\ 3 & \text{using only one color} \end{cases}$$

If equivalences were ignored, the total would be $3^4 = 81$.

EXERCISES for Section 9.2

9.2.1 For the given graph G:

i. Group the full set of vertex-($\leq$2)-colorings into $\mathcal{A}ut_V(G)$-orbits, as in Figure 9.2.5.
ii. Determine the number of $\mathcal{A}ut_V(G)$-orbits of the full set of vertex-($\leq$3)-colorings. Itemize by inventory as in Example 9.2.7.
iii. Determine the number of $\mathcal{A}ut_E(G)$-orbits of the full set of edge-($\leq$2)-colorings. Itemize by inventory as in Example 9.2.6.

(a) K_3

(b) P_4

(c) $K_{1,3}$

(d) $K_{2,3}$

(e) $K_5 - K_2$

(f) CL_3

(g)

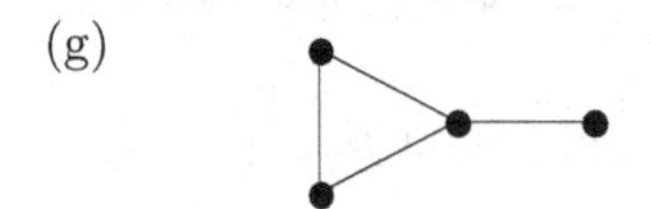

9.3 BURNSIDE'S LEMMA

There is a mathematical principle whose application permits quick calculation of the number of orbits under the action of a permutation group. It is commonly called Burnside's lemma, although it was first published by Frobenius. To state and prove Burnside's lemma, some additional terminology is helpful.

We illustrate the definitions and results in this section using the basic actions of $\mathcal{A}ut(G)$ on V_G and E_G. However, we will apply Burnside's lemma (in §9.5) to the *induced* actions of $\mathcal{A}ut(G)$ on the sets $Col_k(V_G)$ and $Col_k(E_G)$.

Stabilizer of an Object

DEFINITION: Let $\mathcal{P} = [P : Y]$ be a permutation group, and let $y \in Y$. The ***stabilizer*** of y is the subgroup $\mathcal{S}tab(y) = \{\pi \in P \mid \pi(y) = y\}$.

Thus, the stabilizer of an object y is simply the subgroup comprising all the permutations whose disjoint-cycle form contains the 1-cycle (y).

Example 9.3.1: The analysis of $K_4 - K_2$ and $\mathcal{A}ut(K_4 - K_2)$ continues.

Symmetry	$\pi \in \mathcal{A}ut_V(G)$
identity	$\varepsilon = (u)(v)(w)(x)$
refl. thru vert. axis	$\pi_1 = (u)(w)(v\ x)$
refl. thru horiz. axis	$\pi_2 = (v)(x)(u\ w)$
180° rotation	$\pi_3 = (u\ w)(v\ x)$

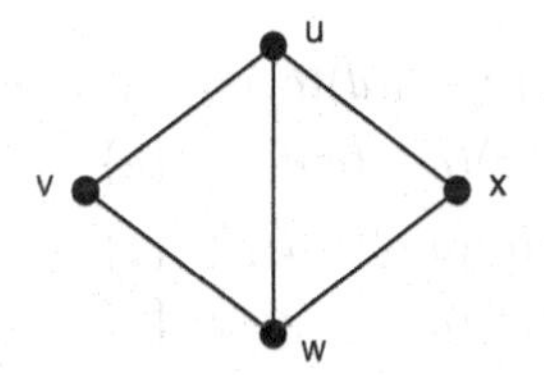

In $\mathcal{A}ut_V(K_4 - K_2)$, the stabilizers are as follows:

$$\mathcal{S}tab(u) = \{\varepsilon, \pi_1\}$$
$$\mathcal{S}tab(v) = \{\varepsilon, \pi_2\}$$
$$\mathcal{S}tab(w) = \{\varepsilon, \pi_1\}$$
$$\mathcal{S}tab(x) = \{\varepsilon, \pi_2\}$$

Example 9.3.2: That stabilizers of different objects need not have the same number of permutations is illustrated by $\mathcal{A}ut_E(K_4 - K_2)$.

Symmetry	$\pi \in \mathcal{A}ut_E(G)$
identity	$\varepsilon = (a)(b)(c)(d)(e)$
refl. thru vert. axis	$\pi_1 = (e)(a\ b)(c\ d)$
refl. thru horiz. axis	$\pi_2 = (e)(a\ c)(b\ d)$
180° rotation	$\pi_3 = (e)(a\ d)(b\ c)$

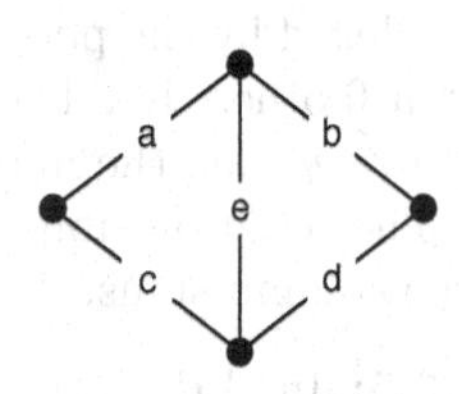

In $\mathcal{A}ut_E(K_4 - K_2)$, the stabilizers are as follows:

$$\mathcal{S}tab(a) = \mathcal{S}tab(b) = \mathcal{S}tab(c) = \mathcal{S}tab(d) = \{\varepsilon\}$$
$$\mathcal{S}tab(e) = \{\varepsilon, \pi_1, \pi_2, \pi_3\}$$

Fixed-Point Set of a Permutation

DEFINITION: Let $\mathcal{P} = [P : Y]$ be a permutation group, and let $\pi \in P$. The ***fixed-point set*** of the permutation π is the subset $Fix(\pi) = \{y \in Y \mid \pi(y) = y\}$.

Thus, $Fix(\pi)$ consists of the objects of Y appearing as 1-cycles in the disjoint-cycle form of π.

TERMINOLOGY: For a given automorphism π on a graph, the ***fixed-vertex-set*** and ***fixed-edge-set*** are the fixed-point sets of π_V and π_E, respectively.

Example 9.3.3: The vertex-permutations in $\mathcal{A}ut_E(K_4 - K_2)$ have the following fixed-point sets

$$Fix((u)(v)(w)(x)) = \{u, v, w, x\}$$
$$Fix((u)(w)(v\ x)) = \{u, w\}$$
$$Fix((v)(x)(u\ w)) = \{v, x\}$$
$$Fix((u\ w)(v\ x)) = \emptyset$$

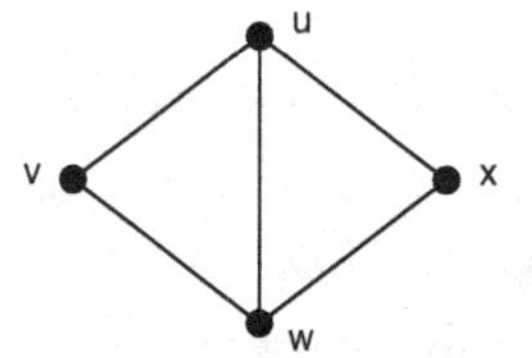

Example 9.3.4: The edge-permutations in $\mathcal{A}ut_E(K_4 - K_2)$ have the following fixed-point sets

$$Fix((a)(b)(c)(d)(e)) = \{a,\ b,\ c,\ d,\ e\}$$
$$Fix((e)(a\ b)(c\ d)) = \{e\}$$
$$Fix((e)(a\ c)(b\ d)) = \{e\}$$
$$Fix((e)(a\ d)(b\ c)) = \{e\}$$

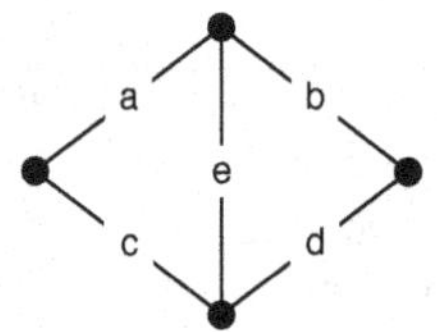

Relationships Involving Stabilizers

The next three lemmas are needed for the proof of Burnside's lemma.

Lemma 9.3.1: *Let $\mathcal{P} = [P : Y]$ be a permutation group. Then*

$$\sum_{y \in Y} |Stab(y)| = \sum_{\pi \in P} |Fix(\pi)|$$

Proof: Consider a matrix whose rows are indexed by the objects of the set Y and whose columns are indexed by the permutations in P, and whose entry in row y and column π is 1 if $\pi(y) = y$, but 0 otherwise. Then for each y, the summand on the left side of the equation is the sum of row y (i.e., the number of 1s), and the summand on the right side is the sum of column π. The equation simply asserts that the sum of the row sums of the matrix equals the sum of the column sums. ◊

Example 9.3.5: In $\mathcal{A}ut_V(K_4 - K_2)$, the sum of the stabilizer sizes (from Example 9.3.1) is

$$\sum_{y \in \mathcal{A}ut_E(K_4 - K_2)} |Stab(y)| = 2 + 2 + 2 + 2 = 8$$

and the sum of the sizes of the fixed-vertex-sets (from Example 9.3.3) is

$$\sum_{\pi \in P} |Fix(\pi)| = 4 + 2 + 2 + 0 = 8$$

Example 9.3.6: In $Aut_E(K_4 - K_2)$, the sum of the stabilizer sizes (from Example 9.3.2) is,

$$\sum_{y \in Aut_E(K_4-K_2)} |Stab(y)| = 1+1+1+1+4 = 8;$$

and the sum of the sizes of the fixed-edge-sets (from Example 9.3.4) is

$$\sum_{\pi \in P} |Fix(\pi)| = 5+1+1+1 = 8$$

Lemma 9.3.2: *Let $\mathcal{P} = [P : Y]$ be a permutation group and $y \in Y$. Then*

$$|Stab(y)| = \frac{|P|}{|orbit(y)|}$$

Proof: Suppose that $orbit(y) = \{y = y_1, y_2, \ldots, y_n\}$ and that, for $j = 1, \ldots, n$, P_j is the subset of permutations of P that map object y to object y_j. Then the subsets $P_1, P_2, \ldots, P_n$ partition the permutation group P, and $P_1 = Stab(y)$.

For $j = 1, \ldots, n$, let π_j be any permutation such that $\pi_j(y) = y_j$. Then the rule $\pi \mapsto \pi_j \circ \pi$ (composition with π_j) is a bijection from P_1 to P_j, which implies that $|P_j| = |P_1| = |Stab(y)|$, for $j = 1, \ldots, n$.

Since each of the n partition cells $P_1, P_2, \ldots, P_n$ of group P has cardinality $|Stab(y)|$, it follows that $n \cdot |Stab(y)| = |P|$. But $n = |orbit(y)|$, which completes the proof. ♢

Example 9.3.7: Analysis of $Aut_V(K_4 - K_2)$ continues.

Symmetry	$\pi \in Aut_V(G)$
identity	$\varepsilon = (u)(v)(w)(x)$
refl. thru vert. axis	$\pi_1 = (u)(w)(v\,x)$
refl. thru horiz. axis	$\pi_2 = (v)(x)(u\,w)$
180° rotation	$\pi_3 = (u\,w)(v\,x)$

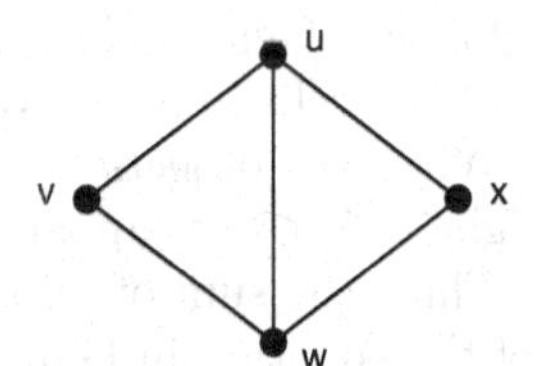

The orbits of $Aut_V(K_4 - K_2)$ are $\{u, w\}$ and $\{v, x\}$, both of cardinality 2. All the stabilizers are of cardinality 2, as determined in Example 9.3.1. The cardinality of the group $Aut_V(K_4 - K_2)$ is 4. Thus, for each vertex, the equation $|Stab(y)| = \frac{|P|}{|orbit(y)|}$ takes the form $2 = \frac{4}{2}$.

Example 9.3.8: Analysis of $Aut_E(K_4 - K_2)$ continues.

Symmetry	$\pi \in Aut_E(G)$
identity	$\varepsilon = (a)(b)(c)(d)(e)$
refl. thru vert. axis	$\pi_1 = (e)(a\,b)(c\,d)$
refl. thru horiz. axis	$\pi_2 = (e)(a\,c)(b\,d)$
180° rotation	$\pi_3 = (e)(a\,d)(b\,c)$

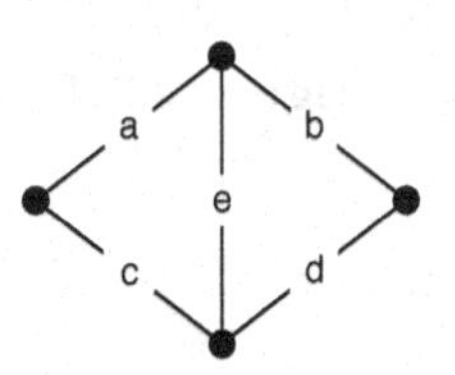

The orbits of $Aut_E(K_4 - K_2)$ are $\{e\}$ and $\{a, b, c, d\}$. For each of the edges a, b, c, and d, the equation $|Stab(y)| = \frac{|P|}{|orbit(y)|}$ becomes $1 = \frac{4}{4}$. For edge e, it becomes $4 = \frac{4}{1}$.

Remark: Although the next lemma is expressed in terms of the orbits of permutation groups, it is really a fact about set partitions.

Lemma 9.3.3: *Let $\mathcal{P} = [P : Y]$ be a permutation group with n orbits. Then*

$$\sum_{y \in Y} \frac{1}{|orbit(y)|} = n$$

Proof: Suppose that $Y_1, \ldots, Y_n$ are the orbits. Then $Y = Y_1 \cup \cdots \cup Y_n$. It follows that

$$\begin{aligned}\sum_{y \in Y} \frac{1}{|orbit(y)|} &= \sum_{j=1}^{n} \sum_{y \in Y_j} \frac{1}{|orbit(y)|} \\ &= \sum_{j=1}^{n} \sum_{y \in Y_j} \frac{1}{|Y_j|} \\ &= \sum_{j=1}^{n} \left[\frac{1}{|Y_j|} \sum_{y \in Y_j} 1 \right] \\ &= \sum_{j=1}^{n} \frac{1}{|Y_j|} |Y_j| \\ &= \sum_{j=1}^{n} 1 \\ &= n\end{aligned}$$

$\diamondsuit$

Example 9.3.9: The left side of the equation in Lemma 9.3.3 is the sum of the reciprocals of the sizes of the cells in a partition. The partition depicted at the left of Figure 9.3.1 has three cells. When the objects in the cells are converted to the reciprocals of the cell sizes, as on the right, it becomes apparent that the sum of the fractions within each cell must be equal to 1. Thus, the sum of all the reciprocals must equal the number of cells, as on the right side of the equation in Lemma 9.3.3.

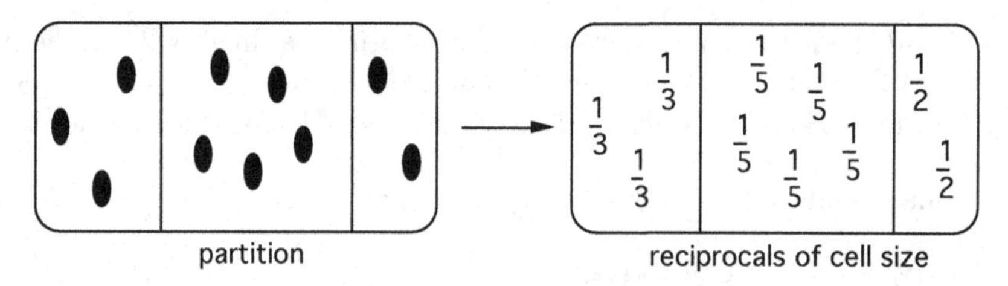

Figure 9.3.1 Reciprocals of cell sizes of a partition.

Proof of Burnside's Lemma

Theorem 9.3.4: [***Burnside's Theorem***] *Let $\mathcal{P} = [P : Y]$ be a permutation group with n orbits. Then*

$$n = \frac{1}{|P|} \sum_{\pi \in P} |Fix(\pi)|$$

Proof: Lemmas 9.3.1, 9.3.2, and 9.3.3 establish the following chain of equalities, which proves Burnside's lemma.

$$\begin{aligned}
\frac{1}{|P|} \sum_{\pi \in P} |Fix(\pi)| &= \frac{1}{|P|} \sum_{y \in Y} |Stab(y)| && \text{(by Lemma 9.3.1)} \\
&= \frac{1}{|P|} \sum_{y \in Y} \frac{|P|}{|orbit(y)|} && \text{(by Lemma 9.3.2)} \\
&= \frac{1}{|P|} |P| \sum_{y \in Y} \frac{1}{|orbit(y)|} \\
&= \sum_{y \in Y} \frac{1}{|orbit(y)|} \\
&= n && \text{(by Lemma 9.3.3)}
\end{aligned}$$

Direct Application of Burnside's Lemma

The most powerful applications of Burnside's lemma are not to counting vertex orbits or edge orbits, but rather, to counting induced equivalence classes, for which purpose they require the use of auxiliary results and techniques, which are the focus of the rest of the chapter. However, a direct orbit-counting application of Burnside's lemma to a permutation group $\mathcal{P} = [P : Y]$ would proceed as follows:

1. the values of $|Fix(\pi)|$ are added over all $\pi \in P$;
2. the resulting sum is divided by $|P|$.

The following two examples apply the direct method to counting vertex-orbits and edge-orbits.

Example 9.3.10: $\mathcal{A}ut_V(K_4 - K_2)$ has four automorphisms. The sum of the sizes of their fixed-point sets, previously calculated in Example 9.3.5, is

$$\sum_{y \in P} |Fix(\pi)| = 4 + 2 + 2 + 0 = 8$$

The orbits are $\{u, w\}$ and $\{v, x\}$. Since $|\mathcal{A}ut_V(K_4 - K_2)| = 4$, Burnside's lemma implies correctly that the number of orbits is $2 = 8/4$.

Example 9.3.11: $\mathcal{A}ut_E(K_4 - K_2)$ has four automorphisms. The sum of the sizes of their fixed-point sets, previously calculated in Example 9.3.6, is

$$\sum_{\pi \in P} |Fix(\pi)| = 5 + 1 + 1 + 1 = 8$$

The orbits are $\{a, b, c, d\}$ and $\{e\}$. Since $|\mathcal{A}ut_V(K_4 - K_2)| = 4$, Burnside's lemma implies correctly that the number of orbits is $2 = 8/4$.

EXERCISES for Section 9.3

9.3.1 For the vertex-permutation group of the indicated graph:

i. Determine the stabilizers of all objects.
ii. Determine the fixed points of all permutations.
iii. Determine the number of orbits.
iv. Confirm Lemma 9.3.1.
v. Confirm Lemma 9.3.2.
vi. Confirm Lemma 9.3.3.
vii. Confirm Burnside's lemma.

(a) The graph of Exercise 9.1.1(a)
(b) The graph of Exercise 9.1.1(b)
(c) The graph of Exercise 9.1.1(c)
(d) The graph of Exercise 9.1.1(d)
(e) The graph of Exercise 9.1.1(e)
(f) The graph of Exercise 9.1.1(f)
(g) The graph of Exercise 9.1.1(g)
(h) The graph of Exercise 9.1.1(h)
(i) The graph of Exercise 9.1.2(a)
(j) The graph of Exercise 9.1.2(b)
(k) The graph of Exercise 9.1.2(c)
(l) The graph of Exercise 9.1.2(d)
(m) P_4
(n) P_4+K_1
(o) The graph of Exercise 9.1.5
(p) The graph of Exercise 9.1.6
(q) $P_4 + K_2$

9.3.2 For the edge-permutation group of the indicated graph:

i. Determine the stabilizers of all objects.
ii. Determine the fixed points of all permutations.
iii. Determine the number of orbits.
iv. Confirm Lemma 9.3.1.
v. Confirm Lemma 9.3.2.
vi. Confirm Lemma 9.3.3.
vii. Confirm Burnside's lemma.

(a) The graph of Exercise 9.1.1(a)
(b) The graph of Exercise 9.1.1(b)
(c) The graph of Exercise 9.1.1(c)
(d) The graph of Exercise 9.1.1(d)
(e) The graph of Exercise 9.1.1(e)
(f) The graph of Exercise 9.1.1(f)
(g) The graph of Exercise 9.1.1(g)
(h) The graph of Exercise 9.1.1(h)
(i) The graph of Exercise 9.1.2(a)
(j) The graph of Exercise 9.1.2(b)
(k) The graph of Exercise 9.1.2(c)
(l) The graph of Exercise 9.1.2(d)
(m) P_4
(n) P_4+K_1
(o) The graph of Exercise 9.1.5
(p) The graph of Exercise 9.1.6
(q) $P_4 + K_2$

9.4 CYCLE-INDEX POLYNOMIAL OF A PERMUTATION GROUP

The *cycle-index polynomial* is a polynomial that displays the cycle structure of a permutation group. What is commonly considered to be the most important principle in enumerative graph theory is that substituting the value k into the cycle-index polynomial yields the number of equivalence classes of ($\leq k$)-colorings. This section examines some examples of applications of this substitution principle and then proves its correctness.

Cycle-Structure Monomial of a Permutation

TERMINOLOGY: The ***cycle structure of a permutation*** is the number of cycles of each length in its disjoint-cycle form.

TERMINOLOGY: A ***monomial*** is a polynomial with only one term.

DEFINITION: Let $\mathcal{P} = [P : Y]$ be a permutation group on a set Y of n objects, and let $\pi \in P$. The ***cycle-structure monomial*** of π is the n-variable monomial

$$\zeta(\pi) = \prod_{k=1}^{n} z_k^{r_k} = z_1^{r_1} z_2^{r_2} \cdots z_n^{r_n}$$

where z_k is a formal variable and r_k is the number of k-cycles in the disjoint-cycle form of π.

Example 9.4.1: The permutation $\pi = (1\ 7\ 9\ 3)(2\ 4\ 8\ 6)(5)$ has cycle-structure monomial $\zeta(\pi) = z_1 z_4^2$, because π has one 1-cycle and two 4-cycles.

Example 9.4.2: The cycle structure of all the vertex-permutations in $\mathcal{A}ut_V(K_4 - K_2)$ is given in the following table.

Symmetry	$\pi \in \mathcal{A}ut_V(G)$	Cycle structure
identity	$(u)(v)(w)(x)$	z_1^4
refl. thru vert. axis	$(u)(w)(v\ x)$	$z_1^2 z_2$
refl. thru horiz. axis	$(v)(x)(u\ w)$	$z_1^2 z_2$
180° rotation	$(u\ w)(v\ x)$	z_2^2

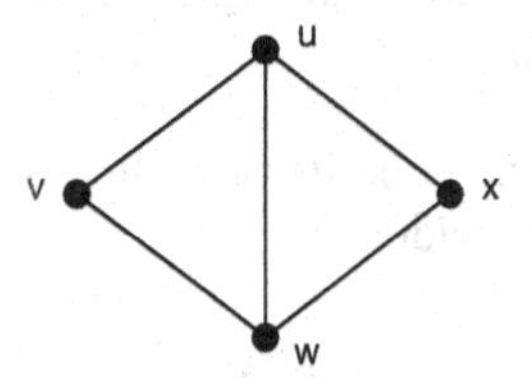

Example 9.4.3: The cycle structure of all the edge-permutations in $\mathcal{A}ut_V(K_4 - K_2)$ is given in the following table.

Symmetry	$\pi \in \mathcal{A}ut_E(G)$	Cycle structure
identity	$(a)(b)(c)(d)(e)$	z_1^5
refl. thru vert. axis	$(e)(a\ b)(c\ d)$	$z_1 z_2^2$
refl. thru horiz. axis	$(e)(a\ c)(b\ d)$	$z_1 z_2^2$
180° rotation	$(e)(a\ d)(b\ c)$	$z_1 z_2^2$

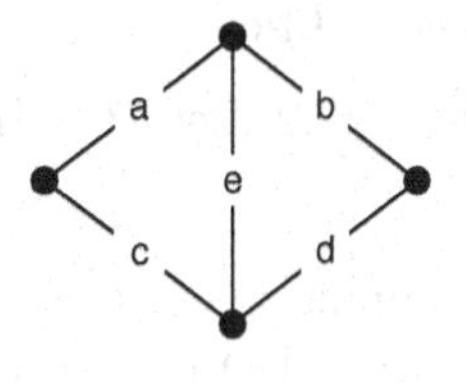

Cycle-Index Polynomial

The last two examples used the actions of $\mathcal{A}ut(G)$ on V_G and E_G, but the remaining examples use the *induced* actions of $\mathcal{A}ut(G)$ on V_G on the sets $Col_k(V_G)$ and $Col_k(E_G)$.

DEFINITION: Let $\mathcal{P} = [P : Y]$ be a permutation group on a set of n objects. Then the ***cycle-index polynomial*** of $\mathcal{P}$ is the polynomial

$$\mathcal{Z}_P(z_1, \ldots, z_n) = \frac{1}{|P|} \sum_{\pi \in P} \zeta(\pi)$$

where $\zeta(\pi)$ is the cycle-structure monomial of the permutation π.

Example 9.4.4: From Example 9.4.2, it follows that the cycle-index polynomial of the permutation group $\mathcal{A}ut_V(K_4 - K_2)$ is

$$\mathcal{Z}_{\mathcal{A}ut_V(K_4-K_2)}(z_1, z_2) = \frac{1}{4}\left(z_1^4 + 2z_1^2 z_2 + z_2^2\right)$$

Substituting 2, the number of colors, for both of the variables z_1 and z_2 yields the number

$$\mathcal{Z}_{\mathcal{A}ut_V(K_4-K_2)}(2, 2) = \frac{1}{4}\left(2^4 + 2 \cdot 2^2 \cdot 2 + 2^2\right) = 9$$

which was shown in Figure 9.2.5 to be the number of $\mathcal{A}ut_V(K_4 - K_2)$-orbits of vertex-($\leq$2)-colorings of the graph $K_4 - K_2$.

Example 9.4.5: From Example 9.4.3, it follows that the cycle-index polynomial of the permutation group $\mathcal{A}ut_E(K_4 - K_2)$ is

$$\mathcal{Z}_{\mathcal{A}ut_E(K_4-K_2)}(z_1, z_2) = \frac{1}{4}\left(z_1^5 + 3z_1 z_2^2\right)$$

Again, substituting 2 for both variables, we obtain

$$\mathcal{Z}_{\mathcal{A}ut_E(K_4-K_2)}(2, 2) = \frac{1}{4}\left(2^5 + 3 \cdot 2 \cdot 2^2\right) = 14$$

which was shown in Figure 9.2.6 to be the number of $\mathcal{A}ut_E(K_4 - K_2)$-orbits of edge($\leq$2)-colorings.

That the substitutions in the last two examples yielded the number of coloring classes is not a coincidence.

Correctness of Substituting into the Cycle-Index Polynomial

We now show as a consequence of Burnside's lemma (Theorem 9.3.4) that substituting the value k into the cycle-index polynomial *always* yields the number of coloring classes (orbits) of the ($\leq k$)-colorings.

REVIEW FROM §9.2: Let $\mathcal{P} = [P : Y]$ be a permutation group acting on a set Y, and let $\pi \in P$.

- The mapping $\pi_{YC} : Col_k(Y) \to Col_k(Y)$ defined by $\pi_{YC}(f) = f\pi$, for every coloring $f \in Col_k(Y)$ is a permutation on the set $Col_k(Y)$, called the ***induced permutation action*** of π on $Col_k(Y)$.
- To distinguish between the action of a permutation π on a set Y and its induced action on the set $Col_k(Y)$ of colorings of Y, we let π_Y denote its action on Y and π_{YC} its action on $Col_k(Y)$. When there is no risk of confusion, the subscripts Y and YC may both be omitted.

- The collection $\mathcal{P}_{YC} = [P : Col_k(Y)]$ of induced permutations on $Col_k(Y)$ is called the ***induced permutation group.***

- The set of orbits (coloring classes) of the induced permutation group on $Col_k(Y)$ is denoted $\{Col_k(Y)\}_{\mathcal{P}}$.

- When it is necessary to distinguish between the group that acts on the set Y and the group that acts on the set $Col_k(Y)$, the respective notations $\mathcal{P}_Y$ and $\mathcal{P}_{YC}$ are used.

NOTATION: If $p(x_1, \ldots, x_n)$ is a multivariate polynomial, then $p(k, \ldots, k)$ denotes the result of substituting the value k for every variable x_j.

Lemma 9.4.1: *Let $\mathcal{P} = [P : Y]$ be a permutation group, and let $\pi \in P$, with induced action π_{YC} on $Col_k(Y)$. Then the number of ($\leq k$)-colorings of Y that are fixed by π_{YC} is given by*

$$|Fix(\pi_{YC})| = \zeta(\pi_Y)(k, \ldots, k)$$

Proof: A ($\leq k$)-coloring c is fixed by π_{YC} if and only if within each cycle of π_Y, all the objects are assigned the same color by c. Thus, there are k independent choices possible for each cycle of π_Y. Therefore, $|Fix(\pi_{YC})| = k^n$, where n is the number of cycles in π_Y. But k^n is precisely the value of $\zeta(\pi_Y)(k, \ldots, k)$. ♢

Example 9.4.6: Recall from Figure 9.2.2 the induced permutations from the action of $\mathcal{A}ut(C_3)$ on $Col_2(C_3)$. The following table illustrates the application of Lemma 9.4.1.

$\pi_Y \in P_Y$	$\pi_{YC} \in P_{YC}$	$\zeta(\pi_Y)$	$\|Fix(\pi_{YC})\|$
ϵ_Y	ϵ_{YC}	z_1^3	8
$(x\ y\ z)$	(111)(211 121 112)(122 212 221)(222)	z_3	2
$(x\ z\ y)$	(111)(211 112 121)(122 221 212)(222)	z_3	2

Theorem 9.4.2: *Let $\mathcal{P} = [P : Y]$ be a permutation group. Then*

$$|\{Col_k(Y)\}_{\mathcal{P}}| = \mathcal{Z}_{\mathcal{P}}(k, \ldots, k)$$

Proof: Applying Burnside's lemma to the induced permutation group $\mathcal{P}_{YC}$ yields the equation

$$\begin{aligned}
|\{Col_k(Y)\}_{\mathcal{P}}| &= \frac{1}{|P_{YC}|} \sum_{\pi_{YC} \in P_{YC}} |Fix(\pi_{YC})| \\
&= \frac{1}{|P_{YC}|} \sum_{\pi_{YC} \in P_{YC}} \zeta(\pi_Y)(k, \ldots, k) \qquad \text{by Lemma 9.4.1} \\
&= \frac{1}{|P_Y|} \sum_{\pi_Y \in P_Y} \zeta(\pi_Y)(k, \ldots, k) \\
&= \mathcal{Z}_{\mathcal{P}}(k, \ldots, k)
\end{aligned}$$

♢

Example 9.4.7: $\mathcal{A}ut(P_5)$ has two automorphisms, the identity and the reflection, and the cycle-index polynomial for $\mathcal{A}ut_V(P_5)$ is given by

$$\mathcal{Z}_{\mathcal{A}ut_V(P_5)}(z_1, z_2) = \frac{1}{2}(z_1^5 + z_1 z_2^2)$$

Substituting 2 for both the variables yields

$$\mathcal{Z}_{Aut_V(P_5)}(2,2) = \frac{1}{2}(2^5 + 2 \cdot 2^2) = 20$$

which is the number of vertex-($\leq$2)-colorings calculated in Example 9.2.5.

Similarly, the cycle-index polynomial for $Aut_E(P_5)$ is

$$\mathcal{Z}_{Aut_E(P_5)}(z_1, z_2) = \frac{1}{2}(z_1^4 + z_2^2)$$

and, hence,

$$\mathcal{Z}_{Aut_E(P_5)}(2,2) = \frac{1}{2}(2^4 + 2^2) = 10$$

which is the number of edge-($\leq$2)-colorings calculated in Example 9.2.6.

Moreover, substituting 3 for both the variables yields

$$\mathcal{Z}_{Aut_E(P_5)}(3,3) = \frac{1}{2}(3^4 + 3^2) = 45$$

which was calculated in Example 9.2.7 to be the number of edge-($\leq$3)-colorings of P_5.

EXERCISES for Section 9.4

9.4.1 For the given graph:

i. Calculate the cycle-index polynomial of $Aut_V(G)$, and evaluate the result of substituting the number 2 for all the variables.
ii. Calculate the cycle-index polynomial of $Aut_E(G)$, and evaluate the result of substituting the number 2 for all the variables.
iii. Use drawings and/or description to account for all the non-equivalent vertex-($\leq$2)-colorings, and make sure that their number agrees with your answer to (i).
iv. Use drawings and/or description to account for all the non-equivalent edge-($\leq$2)-colorings, and make sure that their number agrees with your answer to (ii).

(a) The graph of Exercise 9.1.1(a)
(b) The graph of Exercise 9.1.1(b)
(c) The graph of Exercise 9.1.1(c)
(d) The graph of Exercise 9.1.1(d)
(e) The graph of Exercise 9.1.1(e)
(f) The graph of Exercise 9.1.1(f)
(g) The graph of Exercise 9.1.1(g)
(h) The graph of Exercise 9.1.1(h)
(i) The graph of Exercise 9.1.2(a)
(j) The graph of Exercise 9.1.2(b)
(k) The graph of Exercise 9.1.2(c)
(l) The graph of Exercise 9.1.2(d)
(m) P_4
(n) P_4+K_1
(o) The graph of Exercise 9.1.5
(p) The graph of Exercise 9.1.6
(q) $P_4 + K_2$

9.4.2 Make a table, as in the continuation of Example 9.2.1, for the automorphism group $\mathcal{A}ut(G)$ of the given graph. In the first column, list all the automorphisms (as vertex-permutations), and in the second column, list the corresponding induced permutations on $Col_2(V_G)$. In the strings that code the colorings, the left-to-right order of the symbols should correspond to lexicographic order of the vertex-names.

(a) The graph of Exercise 9.1.1(a)
(b) The graph of Exercise 9.1.1(b)
(c) The graph of Exercise 9.1.1(c)
(d) The graph of Exercise 9.1.1(d)
(e) The graph of Exercise 9.1.1(e)

9.4.3 Make a table, as in the continuation of Example 9.2.1, for the automorphism group $\mathcal{A}ut(G)$ of a given graph. In the first column, list all the automorphisms (as edge-permutations), and in the second column, list the corresponding induced permutations on $Col_2(E_G)$. In the strings that code the colorings, the left-to-right order of the symbols should correspond to lexicographic order of the edge-names.

(a) The graph of Exercise 9.1.1(a)
(b) The graph of Exercise 9.1.1(b)
(c) The graph of Exercise 9.1.1(c)
(d) The graph of Exercise 9.1.1(d)
(e) The graph of Exercise 9.1.1(e)

9.5 MORE COUNTING, INCLUDING SIMPLE GRAPHS

In this section, we illustrate further applications of Burnside's lemma and the substitution principle (Theorem 9.4.2). The section ends by showing how the problem of counting the number of isomorphism classes of simple graphs can be reduced to the problem of counting the non-equivalent edge-($\leq$2)-colorings of the complete graph.

Counting Necklaces

Example 9.5.1: Consider the set of 4-beaded necklaces that can be constructed using only black beads and white beads. This set can be modeled as the coloring classes of vertex-($\leq$2)-colorings of the cycle graph C_4 (i.e., the $\mathcal{A}ut_V(C_4)$-orbits of $Col_2(V_{C_4})$). Figure 9.5.1 shows all the possible kinds of colorings.

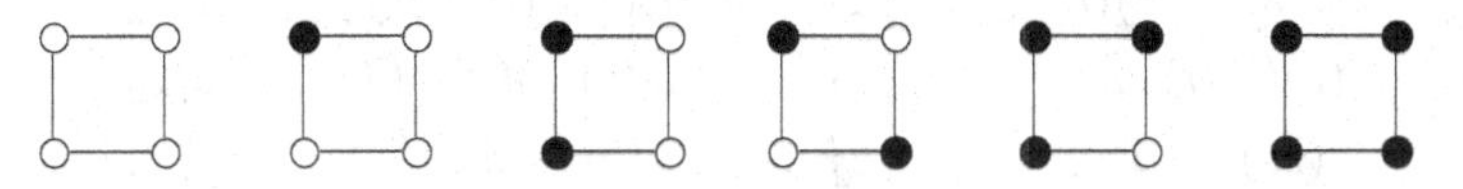

Figure 9.5.1 The six non-equivalent vertex-($\leq$2) colorings of C_4.

It is straightforward to verify that the cycle-index polynomial for $\mathcal{A}ut_V(C_4)$ is given by

$$\mathcal{Z}_{\mathcal{A}ut_V(C_4)}(z_1, z_2, z_3, z_4) = \frac{1}{8}\left(z_1^4 + 2z_1^2 z_2 + 3z_2^2 + 2z_4\right)$$

Applying Theorem 9.4.2, we have

$$\mathcal{Z}_{\mathcal{A}ut_V(C_4)}(2, 2, 2, 2) = \frac{1}{8}\left(2^4 + 2 \cdot 2^2 \cdot 2 + 3 \cdot 2^2 + 2 \cdot 2\right) = 6$$

Example 9.5.2: Now consider vertex-($\leq$3)-colorings of C_4. By Theorem 9.4.2, the number of non-equivalent vertex-($\leq$3)-colorings can be calculated as follows:

$$Z_{Aut_V(C_4)}(3,3,3,3) = \frac{1}{8}\left(3^4 + 2 \cdot 3^2 \cdot 3 + 3 \cdot 3^2 + 2 \cdot 3\right) = 21$$

Direct counting gives three colorings that use only one of the three colors. Since there are four ways to vertex-color C_4 with exactly two colors and three ways to choose two colors from a set of three, there are $3 \cdot 4 = 12$ colorings of C_4 that use exactly two of the three colors. Figure 9.5.2 shows that when exactly three colors are used, there are six colorings.

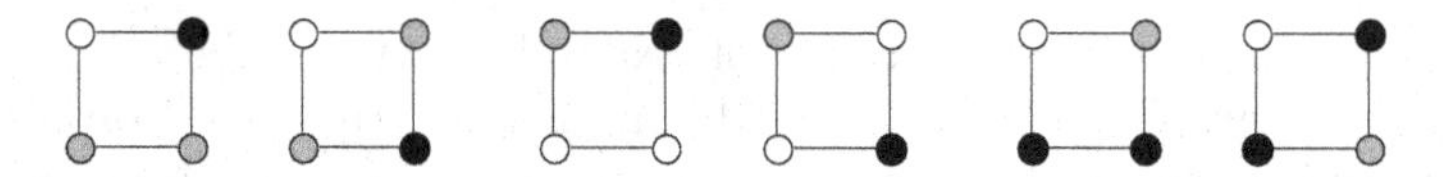

Figure 9.5.2 The six vertex-colorings of C_4 with three colors.

Thus, in all, there are $3 + 12 + 6 = 21$ vertex-($\leq$3)-colorings of C_4, confirming the value obtained by Theorem 9.4.2.

Counting Vertex-Colorings of a Bowtie

Example 9.5.3: Figure 9.5.3 shows the *bowtie BT*, the graph that results from the vertex-amalgamation of two copies of K_3.

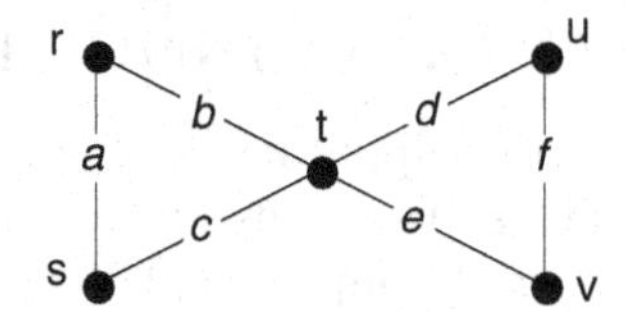

Figure 9.5.3 The bowtie graph *BT*.

The bowtie BT has eight automorphisms. The corresponding vertex- and edge-permutations, along with their cycle structures, are given in the table below.

$\pi \in Aut_V(BT)$	Cycle structure	$\pi \in Aut_E(BT)$	Cycle structure
$(r)(s)(t)(u)(v)$	z_1^5	$(a)(b)(c)(d)(e)(f)$	z_1^6
$(r\ s)(t)(u)(v)$	$z_1^3 z_2$	$(a)(b\ c)(d)(e)(f)$	$z_1^4 z_2$
$(r)(s)(t)(u\ v)$	$z_1^3 z_2$	$(a)(b)(c)(d\ e)(f)$	$z_1^4 z_2$
$(rs)(t)(u\ v)$	$z_1 z_2^2$	$(a)(b\ c)(d\ e)(f)$	$z_1^2 z_2^2$
$(r\ u)(t)(s\ v)$	$z_1 z_2^2$	$(a\ f)(b\ d)(c\ e)$	z_2^3
$(r\ v\ s\ u)(t)$	$z_1 z_4$	$(a\ f)(b\ e\ c\ d)$	$z_2 z_4$
$(r\ u\ s\ v)(t)$	$z_1 z_4$	$(a\ f)(b\ d\ e\ c)$	$z_2 z_4$
$(r\ v)(t)(s\ u)$	$z_1 z_2^2$	$(a\ f)(b\ e)(c\ d)$	z_2^3

Thus, the cycle-index polynomial for $Aut_V(BT)$ is

$$Z_{Aut_V(BT)}(z_1, z_2, z_3, z_4) = \frac{1}{8}\left(z_1^5 + 2z_1^3 z_2 + 3z_1 z_2^2 + 2z_1 z_4\right)$$

and the cycle-index polynomial for $\mathcal{A}ut_E(BT)$ is

$$\mathcal{Z}_{\mathcal{A}ut_E(BT)}(z_1, z_2, z_3, z_4) = \frac{1}{8}\left(z_1^6 + 2z_1^4 z_2 + z_1^2 z_2^2 + 2z_2^3 + 2z_2 z_4\right)$$

By Theorem 9.4.2, the number of non-equivalent vertex-($\leq$2)-colorings of BT is

$$\mathcal{Z}_{\mathcal{A}ut_V(BT)}(2,2,2,2) = \frac{1}{8}(2^5 + 2 \cdot 2^3 \cdot 2 + 3 \cdot 2 \cdot 2^2 + 2 \cdot 2 \cdot 2) = 12$$

Figure 9.5.4 confirms this calculation. Since there are six white-black vertex-colorings with at most two black, there must also be another six with at most two white.

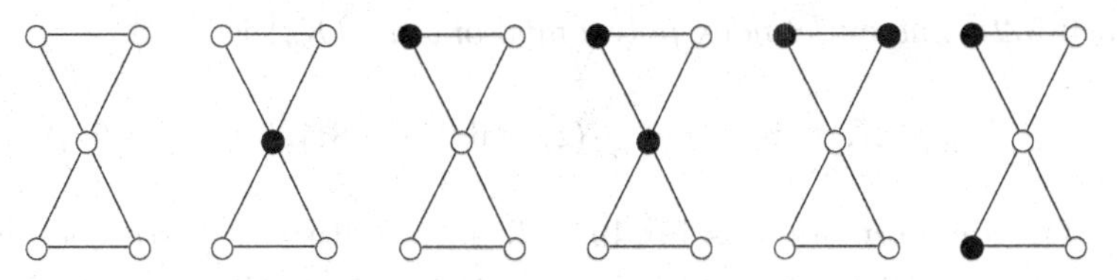

Figure 9.5.4 White-black vertex-colorings of BT with at most two black.

Example 9.5.3, continued: Similarly, by Theorem 9.4.2, the number of edge($\leq$ 2)-colorings of BT is obtained as follows:

$$\mathcal{Z}_{\mathcal{A}ut_E(BT)}(2,2,2,2) = \frac{1}{8}(2^6 + 2 \cdot 2^4 \cdot 2 + 2^2 \cdot 2^2 + 2 \cdot 2^3 + 2 \cdot 2 \cdot 2) = 21$$

Confirming this calculation by drawings is left as an exercise.

Counting Simple Graphs

Theorem 9.4.2 can be used to count the isomorphism types (classes) of n-vertex simple graphs by modeling the types as the coloring classes ($\mathcal{A}ut_E(K_n)$-orbits) of the edge-($\leq$2)-colorings of the complete graph K_n. The following proposition formalizes this connection.

Proposition 9.5.1: *The number of isomorphism types of the n-vertex simple graphs equals the number of coloring classes of the edge-($\leq$ 2)-colorings of K_n, that is,* $\left|\{Col_2(E_{K_n})\}_{\mathcal{A}ut_E(K_n)}\right|$.

Proof: If the two colors used for coloring the edges of K_n are regarded as present and absent, then the full set, of $2^{\binom{n}{2}}$ edge-($\leq$2)-colorings is in one-to-one correspondence with the set of $2^{\binom{n}{2}}$ n-vertex simple graphs, and hence, the coloring classes correspond to the isomorphism types. ♢

Proposition 9.5.1 suggests the following strategy for counting the isomorphism types of the n-vertex simple graphs. The strategy is illustrated by a series of propositions for counting the isomorphism types of simple graphs with four and five vertices.

General Strategy for Counting n-Vertex Simple Graphs

Calculate the cycle-index polynomial of $\mathcal{A}ut_V(K_n)$.

Step 1. *Calculate the cycle-index polynomial of $\mathcal{A}ut_V(K_n)$.*

Since knowing the cycle-index polynomial is sufficient for algebraic counting, writing out all the permutations in a large permutation group can be avoided.

Step 2. *Calculate the cycle-index polynomial of $\mathcal{A}ut_E(K_n)$.*

The cycle-index polynomial of $\mathcal{A}ut_E(K_n)$ is obtained by considering each cycle size and each pair of cycle sizes in the cycle-index polynomial of $\mathcal{A}ut_V(K_n)$.

Step 3. *Apply Theorem 9.4.2.*

This final step in counting the isomorphism types of graphs with n vertices is simply to substitute 2 for every variable in the cycle-index polynomial $\mathcal{Z}_{\mathcal{A}ut_E(K_n)}$.

Counting Simple Graphs on Four Vertices

Proposition 9.5.2: *The cycle-index polynomial of $\mathcal{A}ut_V(K_4)$ is*

$$\mathcal{Z}_{\mathcal{A}ut_V(K_4)}(z_1, z_2, z_3, z_4) = \frac{1}{24}(z_1^4 + 6z_1^2z_2 + 8z_1z_3 + 3z_2^2 + 6z_4)$$

Proof: The 24 vertex-permutations in $\mathcal{A}ut_V(K_4)$ are naturally partitioned according to the five possible cycle structures: z_1^4, $z_1^2z_2$, z_1z_3, z_2^2, and z_4. Each cell in this partition is to be counted.

z_1^4: 1 automorphism.
Only the identity permutation has this cycle structure.

$z_1^2z_2$: 6 automorphisms.
There are $\binom{4}{2} = 6$ ways to choose two vertices for the 2-cycle.

z_1z_3: 8 automorphisms.
There are $\binom{4}{3} = 4$ ways to choose three vertices for the 3-cycle and $(3-1)! = 2$ ways to arrange them in a cycle.

z_2^2: 3 automorphisms.
There are three ways to group four objects into two cycles, when it does not matter which cycle is written first.

z_4: 6 automorphisms.
They correspond to the $(4-1)! = 6$ ways that four objects can be arranged in a cycle. ◇

Proposition 9.5.3: *The cycle-index polynomial of $\mathcal{A}ut_E(K_4)$ is*

$$\mathcal{Z}_{\mathcal{A}ut_E(K_4)}(z_1, z_2, z_3, z_4) = \frac{1}{24}(z_1^6 + 9z_1^2z_2^2 + 8z_3^2 + 6z_2z_4)$$

Proof: The size of the cycle to which an edge belongs is determined by the cycles to which its endpoints belong. Thus, for every automorphism $\pi \in \mathcal{A}ut_E(K_n)$, the cycle structure $\zeta(\pi_E)$ of the edge-permutation is determined by the cycle structure $\zeta(\pi_V)$ of the vertex-permutation.

Case 1. If $\zeta(\pi_V) = z_1^4$, then $\zeta(\pi_E) = z_1^6$.

If both endpoints of an edge e are in a 1-cycle of the vertex-permutation, then they are both fixed points. In a simple graph, the corresponding edge-permutation must map that edge to itself.

Case 2. If $\zeta(\pi_V) = z_1^2 z_2$, then $\zeta(\pi_E) = z_1^2 z_2^2$.

An edge of K_4 is mapped to itself if both its endpoints are in a 2-cycle or if each endpoint is in a 1-cycle. Thus, two edges of K_4 are fixed by π_V. Each of the other four edges has one endpoint in a 1-cycle, which is fixed by π_V, and the other in a 2-cycle of π_V, which is mapped by π_V to the other vertex in that 2-cycle. It follows that such an edge lies in a 2-cycle of π_E.

Case 3. If $\zeta(\pi_V) = z_1 z_3$, then $\zeta(\pi_E) = z_3^2$.

The three edges of K_4 that have both their ends in the 3-cycle of π_V lie in a 3-cycle of π_E. The three edges of K_4 that have one endpoint in a 1-cycle of π_V and the other endpoint in a 3-cycle all lie in another 3-cycle of π_E.

Case 4. If $\zeta(\pi_V) = z_2^2$, then $\zeta(\pi_E) = z_1^2 z_2^2$.

The two edges that have both endpoints in the same 2-cycle of π_V are both fixed by π_E. If an edge has one endpoint in one 2-cycle of π_V and the other endpoint in another 2-cycle of π_V, then that edge lies in a 2-cycle of π_E with the edge whose respective endpoints are the other vertices of those 2-cycles of π_V.

Case 5. If $\zeta(\pi_V) = z_4$, then $\zeta(\pi_E) = z_2 z_4$.

The four edges whose endpoints are consecutive vertices in the 4-cycle of π_V form a cycle of π_E. The two edges whose endpoints are spaced 2 apart in the 4-cycle of π_V form a 2-cycle of π_E. ◊

Corollary 9.5.4: *There are exactly 11 isomorphism types of simple graphs with 4 vertices.*

Proof: Using Proposition 9.5.3, we have

$$\mathcal{Z}_{Aut_E(K_4)}(2,2,2,2) = \frac{1}{24}(2^6 + 9 \cdot 2^2 \cdot 2^2 + 8 \cdot 2^2 + 6 \cdot 2 \cdot 2) = 11$$

The 11 graphs promised by this calculation are shown in Figure 9.5.5. ◊

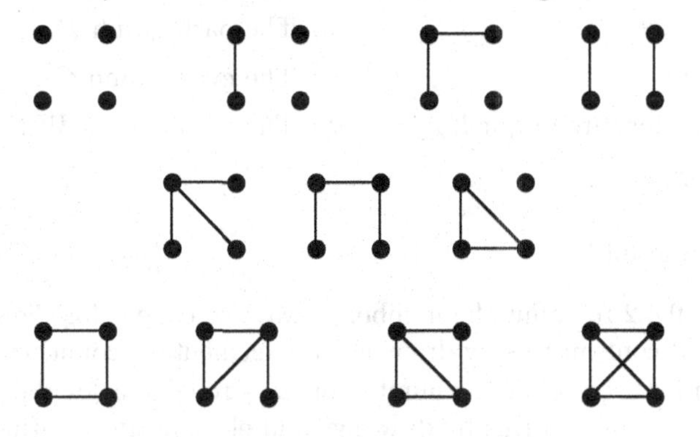

Figure 9.5.5 The 11 simple 4-vertex graphs.

Remark: The full set of ($\leq$2)-colorings of E_{K_4} has cardinality 64 (each of the six edges can be colored *present* or *absent*), and each of the 11 graphs in Figure 9.5.5 represents an isomorphism class. For instance, the second graph in the top row represents a class with six graphs, since there are six ways to choose the two endpoints of the *present* edge. The sizes of these 11 isomorphism classes must, of course, sum to 64 (see Exercises).

Simple Graphs on Five Vertices

Proposition 9.5.5: *There are exactly 34 isomorphism types of simple graphs with 5 vertices.*

Proof: By using the same approach as in the calculation of the number of 4-vertex simple graphs, it can be shown that

$$\mathcal{Z}_{Aut_V(K_5)}(z_1,\ldots,z_5) = \frac{1}{120}(z_1^5 + 10z_1^3z_2 + 15z_1z_2^2 + 20z_1^2z_3 + 30z_1z_4 + 20z_2z_3 + 24z_5)$$

Therefore,

$$\mathcal{Z}_{Aut_E(K_5)}(z_1,\ldots,z_6) = \frac{1}{120}\left(z_1^{10} + 10z_1^4z_2^3 + 15z_1^2z_2^4 + 20z_1z_3^3 + 30z_2z_4^2 + 20z_1z_3z_6 + 24z_5^2\right)$$

and, accordingly,

$$\begin{aligned}\mathcal{Z}_{Aut_E(K_5)}(2,\ldots,2) &= \frac{1}{120}\left(2^{10} + 10\cdot 2^7 + 15\cdot 2^6 + 20\cdot 2^4 + 30\cdot 2^3 + 20\cdot 2^3 + 24\cdot 2^2\right)\\ &= 34\end{aligned}$$

◇

Remark: The same process can be repeated for any number of vertices.

EXERCISES for Section 9.5

9.5.1 For the given graph:

i. Use Theorem 9.4.2 to count the number of ways to vertex-color the graph with a set of two colors, and confirm this by drawings and elementary counting methods.
ii. Use Theorem 9.4.2 to count the number of ways to vertex-color the graph with a set of three colors, and confirm this by drawings and elementary counting methods.

(a) The path graph P_4.
(b) The path graph P_5.
(c) The cycle graph C_5.
(d) The cycle graph C_6.
(e) The complete bipartite graph $K_{2,3}$.
(f) The wheel graph W_5.
(g) The graph $K_5 - K_2$.

9.5.**2** For the given graph:

i. Use Theorem 9.4.2 to count the number of ways to edge-color the graph with a set of two colors, and confirm this by drawings and elementary counting methods.
ii. Use Theorem 9.4.2 to count the number of ways to edge-color the graph with a set of three colors, and confirm this by drawings and elementary counting methods.

(a) The path graph P_4.
(b) The path graph P_5.
(c) The cycle graph C_5.
(d) The cycle graph C_6.
(e) The complete bipartite graph $K_{2,3}$.
(f) The wheel graph W_5.
(g) The graph $K_5 - K_2$.

9.5.**3** In the bowtie graph BT:

i. Draw the white-black edge-colorings, using at most two black edges.
ii. Draw the white-black edge-colorings, using exactly three black edges.
iii. Add twice the number from part (i) to the number from part (ii). Compare your answer with the one obtained in Example 9.5.3.

9.5.**4** For each graph in Figure 9.5.5, state how many graphs are in its isomorphism class in the *present-absent* coloring of K_4.

9.5.**5** Apply the algebraic enumeration techniques of this section to counting the number of isomorphism types of simple graphs with three vertices.

9.5.**6** Draw all the isomorphism types of simple graphs with three vertices.

9.5.**7** Draw all the isomorphism types of 3-vertex graphs with no self-loops and with at most two edges between any two vertices.

9.5.**8** Substitute the number 3 into $\mathcal{Z}_{\mathcal{A}ut_E(K_3)}$ and evaluate. Why should this agree with the result from Exercise 9.5.7?

9.5.**9** Determine the number of isomorphism types of simple graphs with six vertices.

9.6 PÖLYA-BURNSIDE ENUMERATION

This section describes Pölya's ingenious extension of the enumeration strategy presented in §9.3, §9.4, and §9.5. With this extension, we are not only able to count the number of coloring classes, but we can also categorize those classes according to the number of objects that are assigned each color. For the application to counting the non-isomorphism types of the n-vertex simple graphs, we can determine the number of non-isomorphism types having each possible number of edges.

Inventory of the Coloring Classes

DEFINITION: Let $\mathcal{P} = [P : Y]$ be a permutation group acting on a set Y of n objects. The ***inventory*** of the coloring classes, $\{Col_k(Y)\}_{\mathcal{P}}$, indicates for each possible combination of how many times each color is used, the number of coloring classes that use that combination of colors.

Example 9.6.1: According to Figure 9.2.5, the inventory of coloring classes of vertex-($\leq$2)-colorings of the graph $K_4 - K_2$ is given by

1	4 white 0 black
2	3 white 1 black
3	2 white 2 black
2	1 white 3 black
1	0 white 4 black

Example 9.6.2: According to Figure 9.2.6, the inventory of coloring classes of edge-($\leq$2)-colorings of the graph $K_4 - K_2$ is

1 5 light 0 dark
2 4 light 1 dark
4 3 light 2 dark
4 2 light 3 dark
2 1 light 4 dark
1 0 light 5 dark

Pölya Substitution

Pölya ([Po37]) sharpened the application of Theorem 9.4.2 to enumeration by devising a special way of substituting a k-variate polynomial (instead of the number k) into the cycle-index polynomial.

DEFINITION: Let $\mathcal{Z}_p(z_1, \ldots, z_n)$ be the cycle-index polynomial for a permutation group $\mathcal{P} = [P : Y]$ on a set Y of n objects. The ***Pölya substitute of order*** k for the cycle-index variable z_j is the k-variate polynomial

$$x_1^j + x_2^j + \cdots + x_k^j$$

DEFINITION: Let $\mathcal{Z}_p(z_1, \ldots, z_n)$ be the cycle-index polynomial for a permutation group $\mathcal{P} = [P : Y]$ on a set Y of n objects. The ***Pölya-inventory polynomial of order*** k is the polynomial obtained by replacing each cycle-index variable z_j by its Pölya substitute of order k. It is denoted $\mathcal{Z}_{\mathcal{P}}(x_1 + \cdots + x_k)$.

Pölya's Enumeration Theorem

The culminating result of this chapter, the *Pölya Enumeration Theorem*, provides an elegant method for determining the number of non-equivalent colorings for a specified number of occurrences of each color. For a proof, see, for example, [Bo00] or [Br04b].

Theorem 9.6.1: **[*Pölya Enumeration Theorem*]** *Let $\mathcal{P} = [P : Y]$ be a permutation group on a set Y of n objects, and let $\mathcal{Z}_{\mathcal{P}}(x_1 + \cdots + x_k)$ be the Pölya-inventory polynomial of order k for the coloring classes of $Col_k(P)$. Then the coefficient of the term $x_1^{j_1} x_2^{j_2} \cdots x_k^{j_k}$ in the expansion of $\mathcal{Z}_{\mathcal{P}}(x_1 + \cdots + x_k)$ is the number of coloring classes that use color i exactly j_i times, $i = 1, 2, \ldots, k$.*

Example 9.6.3: Returning to Example 9.4.4, the cycle-index polynomial for $\mathcal{A}ut_V(K_4 - K_2)$ is

$$\mathcal{Z}_{\mathcal{A}ut_V(K_4-K_2)}(z_1, z_2) = \frac{1}{4}\left(z_1^4 + 2z_1^2 z_2 + z_2^2\right)$$

Thus, the Pölya-inventory polynomial of order 2 is

$$\begin{aligned}
\mathcal{Z}_{\mathcal{A}ut_V(K_4-K_2)}(x_1 + x_2) &= \frac{1}{4}\left((x_1 + x_2)^4 + 2(x_1 + x_2)^2(x_1^2 + x_2^2) + (x_1^2 + x_2^2)^2\right) \\
&= \frac{1}{4}(x_1^4 + 4x_1^3 x_2 + 6x_1^2 x_2^2 + 4x_1 x_2^3 + x_2^4 \\
&\quad + 2x_1^4 + 4x_1^3 x_2 + 4x_1^2 x_2^2 + 4x_1 x_2^3 + 2x_2^4 \\
&\quad + x_1^4 + 2x_1^2 x_2^2 + x_2^4) \\
&= x_1^4 + 2x_1^3 x_2 + 3x_1^2 x_2^2 + 2x_1 x_3^3 + x_2^4
\end{aligned}$$

In this Pölya-inventory polynomial, the variable x_1 stands for the color white and x_2 for the color black. Thus, its coefficients agree exactly with the inventory given in Example 9.6.1.

Example 9.6.4: Returning to Example 9.4.5, the cycle-index polynomial for $\mathcal{A}ut_E(K_4 - K_2)$ is

$$\mathcal{Z}_{\mathcal{A}ut_E(K_4-K_2)}(z_1, z_2) = \frac{1}{2}\left(z_1^5 + 3z_1z_2^2\right)$$

Thus, the Pölya-inventory polynomial of order 2 is

$$\begin{aligned}\mathcal{Z}_{\mathcal{A}ut_E(K_4-K_2)}(x_1 + x_2) &= \frac{1}{4}\left((x_1 + x_2)^5 + 3(x_1 + x_2)(x_1^2 + x_2^2)^2\right)\\ &= \frac{1}{4}(x_1^5 + 5x_1^4x_2 + 10x_1^3x_2^2 + 10x_1^2x_2^3 + 5x_1x_2^4 + x_2^5\\ &\quad + 3x_1^5 + 3x_1^4x_2 + 6x_1^3x_2^2 + 6x_1^2x_2^3 + 3x_1x_2^4 + 3x_2^5)\\ &= x_1^5 + 2x_1^4x_2 + 4x_1^3x_2^2 + 4x_1^2x_2^3 + 2x_1x_2^4 + x_2^5\end{aligned}$$

This represents the detailed inventory given in Example 9.6.2.

Inventory of n-Vertex Simple Graphs

The Pölya-inventory polynomial of order 2 for $\mathcal{A}ut_E(K_n)$ gives an inventory of the n-vertex simple graphs according to their number of edges. A term of the form $c_kx_1^jx_2^k$ means that there are c_k configurations with k edges present and j edges absent.

Example 9.6.5: According to Proposition 9.5.3, the cycle-index polynomial for $\mathcal{A}ut_E(K_4)$ is

$$\mathcal{Z}_{\mathcal{A}ut_E(K_4)} = \frac{1}{24}\left(z_1^6 + 9z_1^2z_2^2 + 8z_3^2 + 6z_2z_4\right)$$

This leads to the following calculation of $\mathcal{Z}_{\mathcal{A}ut_E(K_4)}(x_1 + x_2)$.

$$\frac{1}{24}\left((x_1 + x_2)^6 + 9(x_1 + x_2)^2(x_1^2 + x_2^2)^2 + 8(x_1^3 + x_2^3)^2 + 6(x_1^2 + x_2^2)(x_1^4 + x_2^4)\right)$$

$$\begin{aligned}&= \frac{1}{24}\left(x_1^6 + 6x_1^5x_2 + 15x_1^4x_2^2 + 20x_1^3x_2^3 + 15x_1^2x_2^4 + 6x_1x_2^5 + x_2^6\right.\\ &\quad + 9x_1^6 + 18x_1^5x_2 + 27x_1^4x_2^2 + 36x_1^3x_2^3 + 27x_1^2x_2^4 + 18x_1x_2^5 + 9x_2^6\\ &\quad + 8x_1^6 + 16x_1^3x_2^3 + 8x_2^6\\ &\quad \left. + 6x_1^6 + 6x_1^4x_2^2 + 6x_1^2x_2^4 + 6x_2^6\right)\\ &= x_1^6 + x_1^5x_2 + 2x_1^4x_2^2 + 3x_1^3x_2^3 + 2x_1^2x_2^4 + x_1x_2^5 + x_2^6\end{aligned}$$

In complete agreement with Figure 9.5.5, this means that the inventory of 4-vertex simple graphs is as follows:

1 graph with 0 edges
1 graph with 1 edge
2 graphs with 2 edges
3 graphs with 3 edges
2 graphs with 4 edges
1 graph with 5 edges
1 graph with 6 edges

EXERCISES for Section 9.6

9.6.**1** Construct the Pölya-inventory polynomial for 3-vertex simple graphs.

9.6.**2** Draw all the different isomorphism types of 5-vertex simple graphs with three or fewer edges.

9.6.**3** Draw all the different isomorphism types of 5-vertex simple graphs with four edges.

9.6.**4** Draw all the different isomorphism types of 5-vertex simple graphs with five edges.

9.6.**5** Add the number of graphs in Exercise 9.6.4 to twice the number of graphs in Exercise 9.6.3 and twice the number of graphs in Exercise 9.6.2. Why should this number equal 34?

9.6.**6** Construct the Pölya-inventory polynomial for 5-vertex simple graphs.

9.6.**7** Construct the Pölya-inventory polynomial for 6-vertex simple graphs.

9.6.**8** Substitute the Pölya-inventory polynomial of order 3 into $\mathcal{Z}_{Aut_E}(K_3)$, and compare the result with your answer for Exercise 9.5.20.

9.7 SUPPLEMENTARY EXERCISES

9.7.**1** For the given graph:

i. List all the vertex automorphisms.
ii. Write the cycle index of the vertex automorphism group.
iii. Use Burnside enumeration to count the essentially different ways to color the vertices with at most two colors.
iv. Calculate a Pölya inventory of the colorings from part (iii).
v. Draw all the different ways from part (iii).
vi. List all the edge automorphisms.
vii. Write the cycle index of the edge automorphism group.
viii. Use Burnside enumeration to count the essentially different ways to color the edges with at most two colors.
ix. Calculate a Pölya inventory of the colorings from part (iii).
x. Draw all the different ways from part (viii).

(a) The octahedral graph $\mathcal{O}_3$.
(b) The result of adding an edge joining two nonadjacent vertices of $K_{3,3}$.
(c) The result of a vertex-amalgamation of two copies of K_4.
(d) The result of an edge-amalgamation of two copies of K_4.
(e) The Mobius ladder ML_4.
(f) The Cartesian product $C_3 \times C_3$.

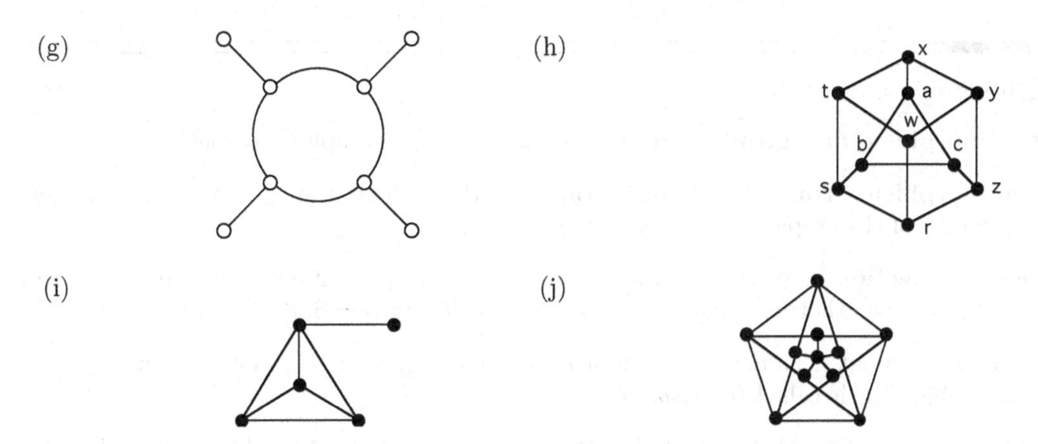

9.7.2 Write the vertex orbits and edge orbits of the graph below. (Use the notation uv for an edge between vertices u and v.) Calculate the cycle index of the vertex automorphism group.

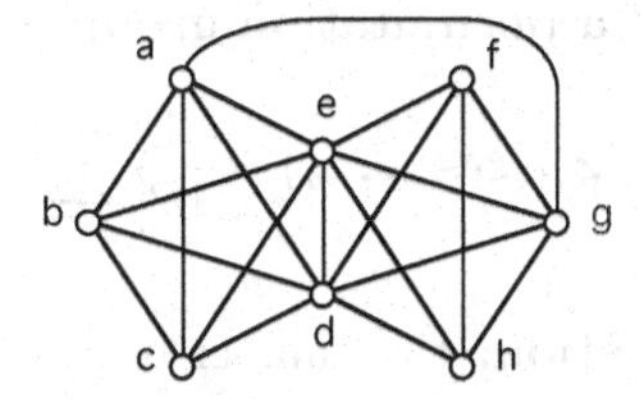

9.7.3 List all six automorphisms that preserve directions on the arcs of the digraph below, and write the cycle index for this automorphism group. Use it to count the possible isomorphism types of simple digraphs on three vertices. Verify your answer with drawings.

9.7.4 In the graph below:

i. Find the only vertex that does not lie on a 3-cycle, and thus, lies in an orbit by itself.
ii. Partition the vertices according to distance from this special vertex.
   Suggestion: redraw the graph.
iii. List the vertex orbits.

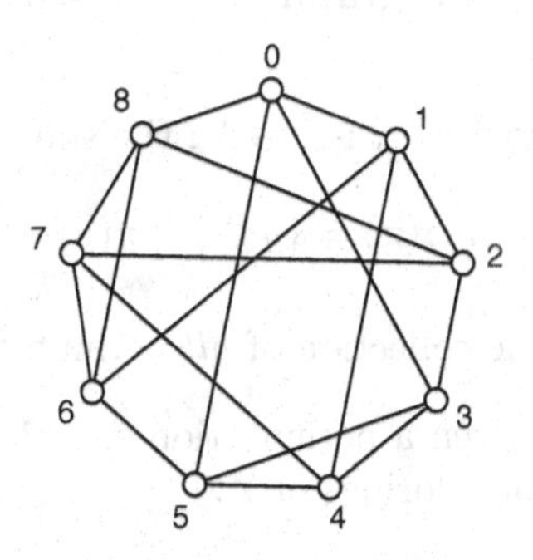

Glossary

automorphism of a graph G: an isomorphism from the graph G to itself.

automorphism group $\mathcal{A}ut(G)$ **of a graph** G: the permutation group of all automorphisms of the graph G.

closed collection of permutations: a collection P of permutations on the same set of objects, such that for every pair $\pi_1, \pi_2 \in P$, the composition $\pi_1\pi_2$ is in P.

coloring of a set Y: a mapping f from Y *onto* a set $\{1, 2, \ldots, k\}$ of integers, in which the number $f(y)$ is called the *color* of y.

($\leq k$)**-coloring of a set** Y: a coloring that uses k or *fewer* colors. The set of all ($\leq k$)-colorings of Y is denoted $Col_k(Y)$.

coloring class for a set with a coloring: a subset of all objects of like color.

cycle-index (polynomial) of a permutation group $\mathcal{P}$: the polynomial

$$\mathcal{Z}_P(z_1, \ldots, z_n) = \frac{1}{|P|} \sum_{\pi \in P} \zeta(\pi)$$

cycle structure of a permutation: the number of cycles of each length in its disjoint-cycle form.

cycle-structure monomial of a permutation π: the multivariate monomial $\zeta(\pi) = \prod_{k=1}^{n} z_j^{r_j}$ where z_k is a formal variable and where r_k is the number of k-cycles in the disjoint cycle form of π.

edge-automorphism group $\mathcal{A}ut_E(G)$ **of a graph** G: the permutation group whose object set is E_G and whose permutations are the edge functions of the automorphisms of the graph G.

$\mathcal{P}$**-equivalent colorings of a set** Y **under a permutation group** $\mathcal{P} = [P : Y]$: colorings f and g for which there is a permutation $\pi \in P$ such that $g = f\pi$. That is, for every object $y \in Y$, the color $g(y)$ is the same as the color $f(\pi(y))$. The set of $\mathcal{P}$-equivalence classes of $Col_k(Y)$ is denoted $\{Col_k(Y)\}_P$.

equivalent edge-colorings on a graph G: edge-colorings f_1 and f_2 on G that are $\mathcal{A}ut_E(G)$-equivalent.

equivalent vertex-colorings on a graph G: vertex-colorings f_1 and f_2 on G that are $\mathcal{A}ut_V(G)$-equivalent.

fixed-point set of a permutation $\pi : Y \to Y$: the subset

$$Fix(\pi) = \{y \in Y \mid \pi(y) = y\}$$

full symmetric group Σ_Y: the collection of *all* permutations on a set Y.

induced permutation π_{YC}: given a permutation $\pi \in \mathcal{P} = [P : Y]$, the rule $f \mapsto f\pi$ that permutes the set $Col_k(Y)$ of colorings of Y.

induced permutation group $\mathcal{P}_k$: given a permutation group $\mathcal{P} = [P : Y]$, the group of induced permutations on $Col_k(Y)$.

inventory of the coloring classes: for a permutation group acting on a set Y of n objects, specification for each possible combination of how many times each color is used, the number of coloring classes that use that combination of colors.

involution: a permutation π such that $\pi = \pi^{-1}$.

monomial: a polynomial with only one term.

orbit of an object $y \in Y$ **under a permutation group** P: the set $\{\pi(y) | \pi \in P\}$.

permutation group $[P : Y]$: a mathematical structure such that P is a non-empty closed collection of permutations on the same finite set of objects Y.

Pölya-counting polynomial: the polynomial that gives detailed inventories, obtained by substituting the Pölya substitutes into the cycle-index polynomial.

Pölya substitute of order k **for the cycle-index variable** z_j: the k-variate polynomial $x_1^j + x_2^j + \cdots + x_k^j$ that replaces the cycle-index variable z_j in the cycle-index polynomial $\mathcal{Z}_P(z_1, \ldots, z_n)$.

stabilizer of an object y: in a permutation group $\mathcal{P} = [P : Y]$, the subgroup $Stab(y) = \{\pi \in P \mid \pi(y) = y\}$.

symmetric group Σ_Y **on a set** Y: the collection of *all* permutations on Y.

vertex-automorphism group $\mathcal{A}ut_V(G)$ **of a graph** G: the permutation group whose object set is V_G and whose permutations are the vertex functions of the automorphisms of the graph G.

Chapter 10

ALGEBRAIC SPECIFICATION OF GRAPHS

INTRODUCTION

Exploiting the algebraic symmetries of a large network can be crucial to realizing its most efficient computational and communications capabilities. In a specification by incidence table of a graph with, say, a million vertices, algebraic symmetries and other global features can be obscured by the sheer mass of details of the specification. Using a *voltage graph* to specify a network puts the symmetries into focus. A voltage graph is a vertex-labeled digraph whose edges are labeled by algebraic elements. If G is a large graph with n-fold symmetry, then the vertex-set and edge-set of the voltage graph used to specify G are smaller than those of G by a factor of n. Many interesting large graphs are specifiable by a voltage graph with only one vertex. The general idea is that incipient patterns designed into the voltage graph are developed and replicated throughout the larger graph it specifies. Voltage graphs were first developed for application to the construction of surface imbeddings of graphs with geometric symmetry. Application of voltage graphs to algebraic specification of parallel-processor networks is more recent.

Some algebraic results used in this chapter are summarized in Appendix A.4.

10.1 CYCLIC VOLTAGES

Useful ideas commonly evolve from prototype solutions, in mathematics as well as in engineering. The prototype form of voltage graph described in this section is adapted to cyclic symmetry. In Sections 10.4 and 10.5, more general forms of voltage graphs are developed.

Symmetry and Specification Reduction

To understand how voltage graphs might help with a large problem, this section begins with a small example that illustrates how a graph can be specified by a smaller graph. Definitions, terminology, and precise descriptions of the notational conventions follow this example. The underlying idea of a voltage graph is that it is used to generate patterns in the graph that it specifies.

Example 10.1.1: The vertices and edges of the directed 8-cycle graph on the left in Figure 10.1.1 are labeled with subscripts 0, 1, 2, and 3. One pattern in this labeled 8-cycle is that the u-vertices and the v-vertices alternate, as do the d-arcs and the e-arcs. Notice that every d-arc is directed from a u-vertex to a v-vertex, and every e-arc is directed from a v-vertex to a u-vertex. This pattern is represented by the directed 2-cycle on the right side of Figure 10.1.1.

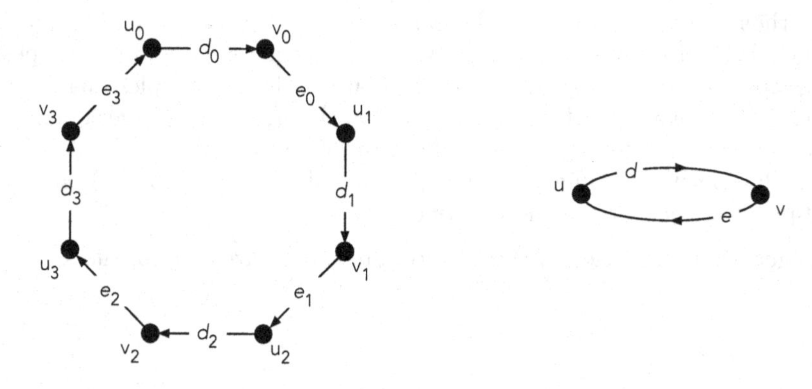

Figure 10.1.1 A directed 8-cycle and a directed 2-cycle.

The subscript patterns of the 8-cycle are yet to be specified. Observe that the subscript on the head vertex of each d-arc of the 8-cycle is the same as the subscript on the tail vertex, and that the subscript on the head of each e-arc is obtained by incrementing the subscript on the tail vertex by 1 (mod 4). Labeling the 2-cycle arcs d and e by 0 and 1, respectively, as shown in Figure 10.1.2, is used to specify when and how the subscripts change. The labels 0 and 1 are referred to as ***voltages***.

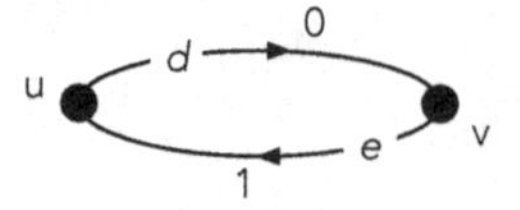

Figure 10.1.2 A directed 2-cycle with voltages mod 4 on its arcs.

Using a smaller digraph to specify the directed 8-cycle in this example is facilitated by the following three features of the labeling of the directed 8-cycle:

1. All the subscripts on vertex- and edge-names in the 8-cycle are selected from the same algebraic structure; in particular, they are all integers modulo 4.
2. The number of u-vertices, the number of v-vertices, the number of d-edges, and the number of e-edges are all exactly the same; in particular, there are four of each.
3. Every d-arc in the 8-cycle has a u-tail and a v-head, while every e-arc has a v-tail and a u-head. The vertex subscripts of the e-arcs increment by 1 (mod 4), while the vertex subscripts of the d-arcs remain the same.

Voltage Assignments

When applying *voltage graphs* to the problem of writing a convenient-sized specification for a large graph or digraph, one starts by imagining a system of vertex labels and edge labels (for the large graph object) that has the three kinds of features described in Example 10.1.1, and one then tries to design the appropriate voltage graph. In contrast, the definitions start with a *voltage graph* and use it to give details of the large graph that it specifies. In Section 10.4, $\mathcal{B}$-voltage graphs will be defined for any group $\mathcal{B}$. In this section, we restrict our attention to the cyclic groups. For any positive integer $n \geq 2$, the cyclic group $\mathbb{Z}_n$ is the set $\{0, 1, 2, \ldots, n-1\}$ such that for any $i, j \in \mathbb{Z}_n$, $i + j$ means addition mod n, 0 is the additive identity, and the additive inverse of an element i is $-i = n - i$, since $i + (n - i) = 0$ (mod n).

DEFINITION: A $\mathbb{Z}_n$-***voltage*** on an arc of a digraph $\vec{G}$ is a label that is an element in $\mathbb{Z}_n$.

DEFINITION: A $\mathbb{Z}_n$-***voltage assignment*** on a digraph $\vec{G}$ is a function $\alpha : E_{\vec{G}} \to \mathbb{Z}_n$ that labels every arc $e \in E_{\vec{G}}$ with a $\mathbb{Z}_n$-voltage.

DEFINITION: A $\mathbb{Z}_n$-***voltage graph*** is a pair $\left\langle \vec{G},\ \alpha : E_{\vec{G}} \to \mathbb{Z}_n \right\rangle$ such that $\vec{G}$ is a digraph and α is a $\mathbb{Z}_n$-voltage assignment on $\vec{G}$. A $\mathbb{Z}_n$-voltage graph is also called a ***cyclic voltage graph***.

TERMINOLOGY: One might be inclined to use the term "voltage digraph" since a voltage graph is a labeled digraph, but "voltage graph" is the standard term.

DEFINITION: The ***voltage group*** of a voltage graph $\left\langle \vec{G}, \alpha :\ E_{\vec{G}} \to \mathbb{Z}_n \right\rangle$ is the group $\mathbb{Z}_n$ from which the voltages are assigned.

Example 10.1.1, continued: The labeled 2-cycle in Figure 10.1.2 is the $\mathbb{Z}_4$-voltage graph, $\left\langle \vec{G}, \alpha :\ E_{\vec{G}} \to \mathbb{Z}_4 \right\rangle$, where $\vec{G}$ is the directed 2-cycle, and α is defined by $\alpha(d) = 0$ and $\alpha(e) = 1$. The 8-cycle in Figure 10.1.1 is the digraph it specifies. The labels on the vertices and edges in the 8-cycle are formed from the vertex- and edge-labels in the 2-cycle, with the elements of the voltage group $\mathbb{Z}_4$ as subscripts.

DEFINITION: The ***covering digraph*** of the $\mathbb{Z}_n$-voltage graph $\left\langle \vec{G}, \alpha : E_{\vec{G}} \to \mathbb{Z}_n \right\rangle$ is the graph $\vec{G}^\alpha = (V^\alpha, E^\alpha)$ with

$$V^\alpha = \{v_j \mid v \in V_{\vec{G}} \text{ and } j \in \mathbb{Z}_n\} \qquad E^\alpha = \{e_j \mid e \in E_{\vec{G}} \text{ and } j \in \mathbb{Z}_n\}$$

such that if arc e in the voltage graph is directed from vertex u to vertex v, then the arc e_j in the covering digraph is directed from vertex u_j to vertex $v_{j+\alpha(e)}$ (addition in the subscript is mod n).

Example 10.1.2: A $\mathbb{Z}_3$-voltage graph and its covering digraph are illustrated in Figure 10.1.3. Observe that the drawing must specify that the voltages are in $\mathbb{Z}_3$. When a voltage graph is specified by a drawing, the voltage group is indicated in the drawing itself, as shown. In the covering digraph, each label from the $\mathbb{Z}_3$-voltage graph appears 3 times, with each of the elements of $\mathbb{Z}_3$ as a subscript. Each b arc is directed from a u vertex to a v vertex having the same subscript. Each c arc is also directed from a u vertex to a v vertex, but the subscript of the u vertex is incremented by 1 (mod 3) to get the subscript of the v vertex. Similarly, the subscripts of the vertices of the d arcs increment by 2 (mod 3).

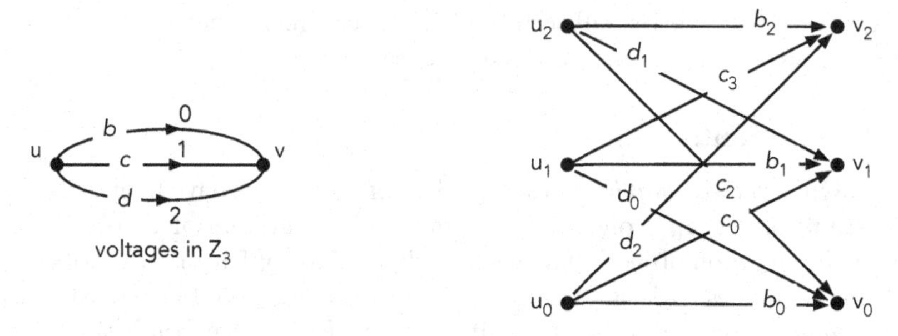

Figure 10.1.3 A $\mathbb{Z}_3$-voltage graph and its covering digraph.

DEFINITION: The ***covering graph*** of the $\mathbb{Z}_n$-voltage graph $\left\langle \vec{G}, \alpha : E_{\vec{G}} \to \mathbb{Z}_n \right\rangle$ is the *underlying graph* of the covering digraph $\vec{G}^\alpha$. It is denoted G^α.

Example 10.1.2, continued: For the $\mathbb{Z}_3$-voltage graph in Figure 10.1.3, the covering graph is $K_{3,3}$ with vertex-set V^α and edge-set E^α, where

$$V^\alpha = \{u_0, u_1, u_2, v_0, v_1, v_2\} \quad \text{and} \quad E^\alpha = \{b_0, b_1, b_2, c_0, c_1, c_2, d_0, d_1, d_2\}$$

Vertex Fibers and Edge Fibers in the Covering Graph

The vertex-set of the covering graph has a natural partition, in which the vertices that arise from a single vertex of the voltage graph are grouped into a cell. The edge-set of the covering graph is similarly partitioned. The terminology and notation for these partitions are borrowed from algebraic topology.

DEFINITION: Let $\left\langle \vec{G}, \alpha : E_{\vec{G}} \to \mathbb{Z}_n \right\rangle$ be a $\mathbb{Z}_n$-voltage graph.

1. For each vertex $v \in V_{\vec{G}}$, the vertex-set $\{v_j \mid j \in \mathbb{Z}_n\}$ in the covering graph G^α is called the ***(vertex) fiber*** over v and is denoted $\tilde{v}$.

2. For each edge $e \in E_G$, the edge-set $\{e_j | j \in \mathbb{Z}_n\}$ in the covering graph G^α is called the ***(edge) fiber*** over e and is denoted $\tilde{e}$.

Example 10.1.2, continued: The vertex fibers are

$$\tilde{u} = \{u_0, u_1, u_2\} \quad \textit{and} \quad \tilde{v} = \{v_0, v_1, v_2\}$$

and the edge fibers are

$$\tilde{b} = \{b_0, b_1, b_2\}, \quad \tilde{c} = \{c_0, c_1, c_2\} \quad \textit{and} \quad \tilde{d} = \{d_0, d_1, d_2\}$$

Example 10.1.3: Now suppose that the voltages in the voltage graph of Figure 10.1.2 are regarded as elements of $\mathbb{Z}_3$, rather than $\mathbb{Z}_4$. Then the covering vertex-set and covering edge-set are

$$V^\alpha = \{u_0, u_1, u_2, v_0, v_1, v_2\} \quad \text{and} \quad E^\alpha = \{d_0, d_1, d_2, e_0, e_1, e_2\}$$

and the covering digraph is a 6-cycle, as shown in Figure 10.1.4.

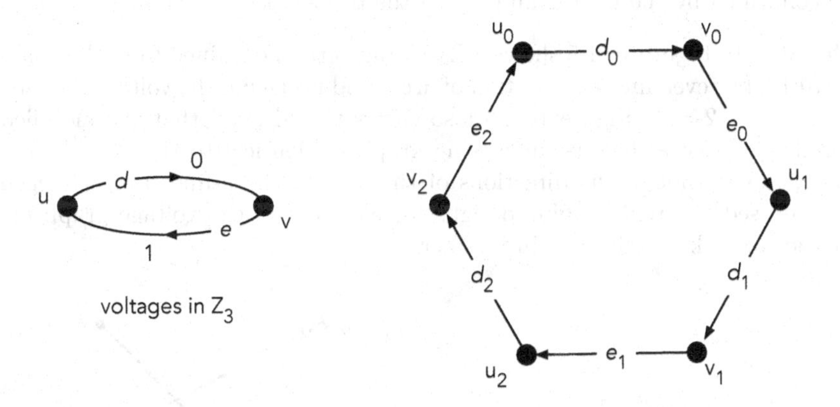

Figure 10.1.4 A 2-cycle with $\mathbb{Z}_3$-voltages and the 6-cycle it specifies.

The vertex and edge fibers are

$$\tilde{u} = \{u_0, u_1, u_2\}, \tilde{v} = \{v_0, v_1, v_2\}, \tilde{d} = \{d_0, d_1, d_2\}, \text{ and } \tilde{e} = \{e_0, e_1, e_2\}$$

If the voltages were regarded as elements of $\mathbb{Z}_5$ (making it a $\mathbb{Z}_5$-voltage graph), then the covering digraph would be a 10-cycle. Moreover, each vertex fiber and each edge fiber would have five elements.

NOTATION: When it is clear what voltage group is intended, the voltage graph may be denoted $\left\langle \vec{G}, \alpha \right\rangle$.

Proposition 10.1.1: *Let $\left\langle \vec{G}, \alpha \right\rangle$ be a $\mathbb{Z}_n$-voltage graph. Then there are n times as many vertices and edges in the covering digraph $\vec{G}^\alpha$ as in $\vec{G}$.*

Proof: There are n vertices v_j and n edges e_j, respectively, in each vertex fiber $\tilde{v}$ and edge fiber $\tilde{e}$ of $\vec{G}^\alpha$ for each vertex $v \in V_{\vec{G}}$ and for each edge $e \in E_{\vec{G}}$, respectively. ◇

Proposition 10.1.2: *Let $\left\langle \vec{G}, \alpha \right\rangle$ be a $\mathbb{Z}_n$-voltage graph, and let v be any vertex of $\vec{G}$. Then the outdegree and indegree of each vertex v_j in the fiber $\tilde{v}$ are equal to the outdegree and indegree, respectively, of vertex v.*

Proof: For each arc e directed from v in $\vec{G}$, there is exactly one arc e_j in the fiber $\tilde{e}$ directed from v_j in the covering digraph. For each arc d directed from some vertex w to v, there is exactly one arc d_k directed from w_k to v_j, where $k + \alpha(d) = j$. That is, there is exactly one arc $d_{j-\alpha(d)}$ in $\tilde{d}$ directed to v_j. ◇

Artificiality of the Arc Directions

A voltage graph is defined formally to be a digraph, and the mathematical object derived from a voltage graph is also defined formally to be a digraph. However, the arc directions in this context are usually regarded simply as an ingredient of the construction (and not as "onewayness"). The result of reversing an arc direction in a voltage graph and replacing its voltage by the inverse voltage is a new voltage graph that still specifies the same undirected graph, as confirmed by the next example and the proposition that follows it.

Example 10.1.4: Figure 10.1.5 shows a $\mathbb{Z}_3$-voltage graph obtained from the voltage graph of Figure 10.1.3 by reversing the direction of arc d and replacing the voltage 2 by its additive inverse, $-2 = 3 - 2 = 1$. Figure 10.1.5 also shows the digraph that the modified voltage graph specifies. Observe that its underlying graph is identical to the underlying graph of Figure 10.1.3, even though the directions of the arcs in the d-fiber have changed. Notice that we have used different graphic designs for each arc in the voltage graph in order to make it easier to pick out the arcs in its fiber.

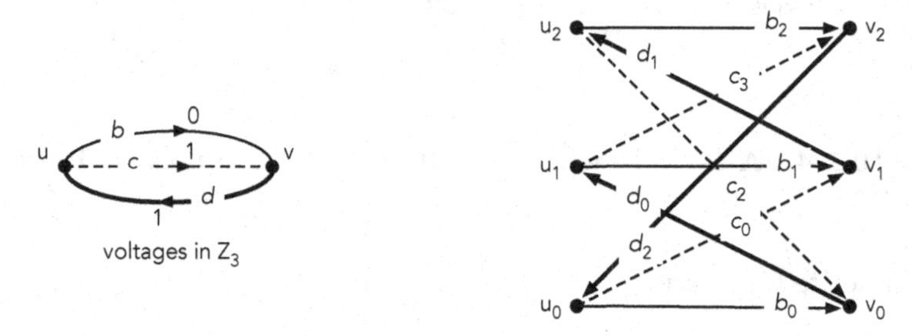

Figure 10.1.5 A $\mathbb{Z}_3$-voltage graph and the digraph it specifies.

Proposition 10.1.3: *Let $\left\langle \vec{G}, \alpha \right\rangle$ be a $\mathbb{Z}_n$-voltage graph. If $(\vec{H}, \beta)$ is the $\mathbb{Z}_n$-voltage graph obtained from $(\vec{G}, \alpha)$ by simply reversing the direction of an arc in $\left\langle \vec{G}, \alpha \right\rangle$ and negating the voltage on that arc, then the covering graph H^β is identical to the covering graph G^α.*

Proof: Let d be an arc in $\left\langle \vec{G}, \alpha \right\rangle$, directed from u to v and suppose $\left\langle \vec{H}, \beta \right\rangle$ is the $\mathbb{Z}_n$-voltage graph obtained by reversing the direction of d and assigning it the voltage $\beta(d) = -\alpha(d)$. This alteration in the voltage graph only affects the arcs in the d-fiber, and its effect on those arcs is to change the names and directions, without changing the vertex pairs involved. That is, for $i = 0, \ldots, n-1$, an arc d_i from u_i to $v_{i+\alpha(d)}$ in $\vec{G}^\alpha$, corresponds to an arc $d_{i+\alpha(d)}$ in $\vec{H}^\beta$ from $v_{i+\alpha(d)}$ to $u_{i+\alpha(d)+\beta(d)} = u_i$ that still joins the same two vertices. Thus, the underlying graphs are identical. ◇

The following proposition and its proof further illustrate the usefulness of de-emphasizing edge directions in voltage graphs.

Proposition 10.1.4: *If $\langle \vec{G}, \alpha \rangle$ is a $\mathbb{Z}_n$-voltage graph such that G is bipartite, then the covering graph G^α is bipartite.*

Proof: Let (U, W) be the bipartition of G. Then every edge in the covering graph joins some vertex u_i with $u \in U$ to a vertex w_j with $w \in W$. Thus, the vertices of the covering graph are bipartitioned, so that one part is the union of the vertex fibers over vertices in U and the other part is the union of the vertex fibers over vertices in W. ◇

Remark: In this chapter, discussion often focuses on the underlying graph of a digraph. However, including arc directions in the drawing makes it easier to verify its correctness.

Voltage Graphs with Self-Loops

The same specification rules apply when there is a self-loop in the voltage graph. If e is a self-loop with endpoint v in a voltage graph, and if the voltage assigned to edge e is $\alpha(e)$, then the edge e_j in the covering digraph has tail v_j and head $v_{j+\alpha(e)}$.

Example 10.1.5: Figure 10.1.6 shows that adding a self-loop with voltage 1 at both vertices of the voltage graph of Figure 10.1.3 transforms it into a specification of the complete graph K_6.

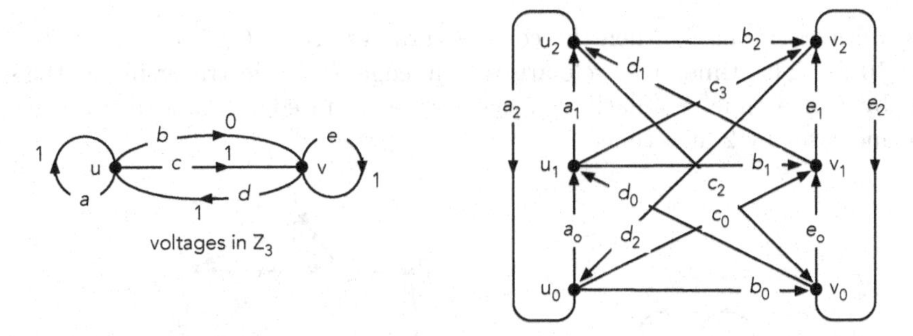

Figure 10.1.6 A $\mathbb{Z}_3$-voltage graph for K_6 and its covering digraph.

REVIEW FROM §1.1. For any set $S \subseteq \{1, 2, \ldots, n-1\}$, the ***circulant graph*** $circ(n : S)$ has vertex-set Z_n, and two vertices i and j are adjacent if and only if there is a number $s \in S$ such that $i + s = j \mod n$ or $j + s = i \mod n$.

Remark: Every circulant graph can be specified using only integers between 1 and $(n+1)/2$ in the set S, due to the fact that $i + s = j \mod n$ if and only if $j + (n - s) = i \mod n$. Thus, the integers s and $n - s$ in S create exactly the same edges in the graph.

Sometimes a voltage graph has only one vertex, so that all the arcs are self-loops. These digraphs are called *bouquets*, denoted B_r, if there are r self-loops. When it is possible to specify a graph or digraph with a one-vertex voltage graph, this is optimal, in the sense that it is the smallest possible specification.

In Example 10.1.3, we saw how changing the voltage group changes the covering graph. The following example is another illustration.

Example 10.1.6: Suppose voltages of 1 and 2 are assigned to the two arcs in the bouquet B_2, as shown in Figure 10.1.7. If the voltage group is $\mathbb{Z}_5$, then this $\mathbb{Z}_5$-voltage graph specifies the complete graph $K_5 \cong circ(5 : 1, 2)$. Observe that self-loop d has voltage 1, and that in traversing the 5-cycle d_0, d_1, d_2, d_3, d_4 arising from edge d, the subscript increases by 1 at a time. Similarly, the self-loop e has voltage 2, and in traversing the 5-cycle e_0, e_2, e_4, e_1, e_3, arising from edge e, the subscript increases by 2 at a time.

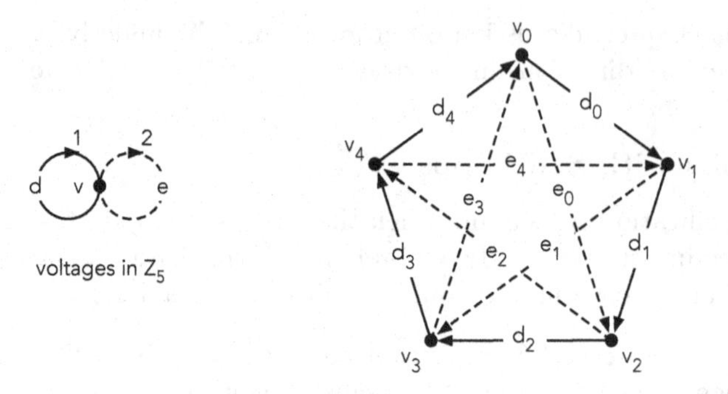

Figure 10.1.7 $\mathbb{Z}_5$-voltages on B_2 and the covering graph, $K_5 \cong circ(5:1,2)$.

If the voltages are in $\mathbb{Z}_7$, then the covering graph is $K_7 - C_7 \cong circ(7:1,2)$, as shown in Figure 10.1.8. This time, a 7-cycle arises from edge d, and in traversing it, the subscript increases by 1 at a time. Similarly, a 7-cycle arises from edge e, and in its traversal, the subscript increases by 2 at a time.

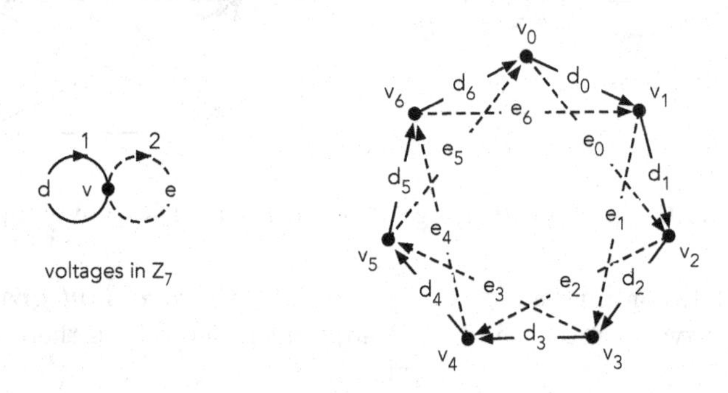

Figure 10.1.8 $\mathbb{Z}_7$-voltages on B_2 and the covering graph, $K_7 - C_7 \cong circ(7:1,2)$.

When the voltages are in $\mathbb{Z}_6$, then the covering graph is $K_6 - 3K_2 \cong circ(6:1,2)$, as shown in Figure 10.1.9. This time, a 6-cycle arises from edge d, and in traversing it, the subscript increases by 1 at a time. However, two 3-cycles arise from edge e, and in the traversal of either of them, the subscript increases by 2 at a time.

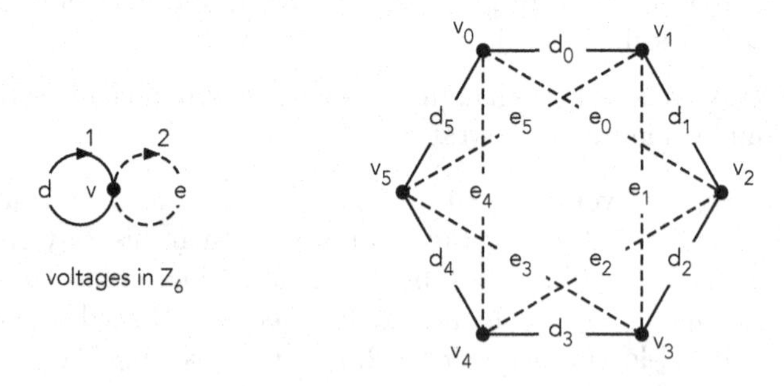

Figure 10.1.9 $\mathbb{Z}_6$-voltages on B_2 and the covering graph, $K_6 - 3K_2 \cong circ(6:1,2)$.

The next proposition generalizes the cycle generation that occurs in Example 10.1.6. Its proof is left as an exercise.

Proposition 10.1.5: *In a $\mathbb{Z}_n$-voltage graph $\left\langle \vec{G}, \alpha \right\rangle$, let e be a self-loop at vertex v with voltage k. Then the subgraph induced on the edges $e_0, e_1, \ldots, e_{n-1}$ in the covering graph G^α is the disjoint union of $\frac{n}{\gcd(n,k)}$-cycles.*

By Proposition 10.1.5, a voltage n on a self-loop in a $\mathbb{Z}_{2n}$-voltage graph would create 2-cycles in the covering digraph, causing doubled edges to appear in the covering graph. Collapsing such doubled edges to a single edge enhances the usefulness of voltage graphs in algebraic specification.

DEFINITION: ***Bidirected-arc convention***: Suppose e is a self-loop with voltage n on a vertex v in a $\mathbb{Z}_{2n}$-voltage graph. The pair of arcs that would otherwise join vertex v_j to vertex v_{j+n} and vertex v_{j+n} to vertex v_j in the covering digraph is replaced by a single bidirected arc between v_j and v_{j+n}. The ***underlying graph of a digraph with bidirected arcs*** has a single edge for each bidirected arc.

Example 10.1.7: In Figure 10.1.10, a $\mathbb{Z}_6$-voltage graph is shown where one of the self-loops has voltage 3. The covering digraph has oppositely directed arcs between pairs of vertices, each of these can be collapsed into a single bidirected arc. The resulting covering graph, using this convention, would have vertices of degree 3. In Figure 10.1.10, the edges are not labeled; instead, solid and dashed lines are used to distinguish the two edge fibers.

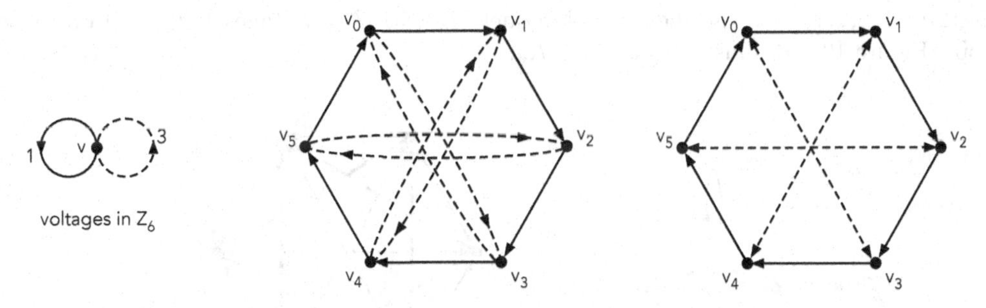

Figure 10.1.10 Replacing 2-cycles in a covering digraph with bidirected arcs.

CONVENTION: In the last few examples, the arcs in the voltage graph were distinguished by using different graphic designs. This made it easier to distinguish the arc fibers in the covering graph. With this graphic-design convention, it is not necessary to label the arcs in the voltage graph or its covering digraph, though it is sometimes convenient to use both.

Circulant Graphs as Covering Graphs of $\mathbb{Z}_n$-Voltage Graphs

In Example 10.1.6, three circulant graphs were obtained as the covering graphs of bouquets with cyclic voltages. In fact, all circulant graphs may be obtained this way, as stated in the next theorem.

CONVENTION: If a $\mathbb{Z}_n$-voltage graph has only one vertex v, then every vertex in the covering graph is labeled v_i, for some element $i \in \mathbb{Z}_n$. It is common to simplify the covering graph diagram by labeling each vertex i instead of v_i.

Proposition 10.1.6: *Let the distinct positive integers $x_1, \ldots, x_r$, each less than $\frac{n}{2}$, be assigned as $\mathbb{Z}_n$-voltages to the self-loops of the bouquet B_r. Then the covering graph is isomorphic to the circulant graph $circ(n : x_1, \ldots, x_r)$.*

Proof: This is an immediate consequence of the definitions of circulant graph and $\mathbb{Z}_n$-voltage graph. Figure 10.1.11 illustrates the proposition for $circ(13 : 1, 5)$.

◇

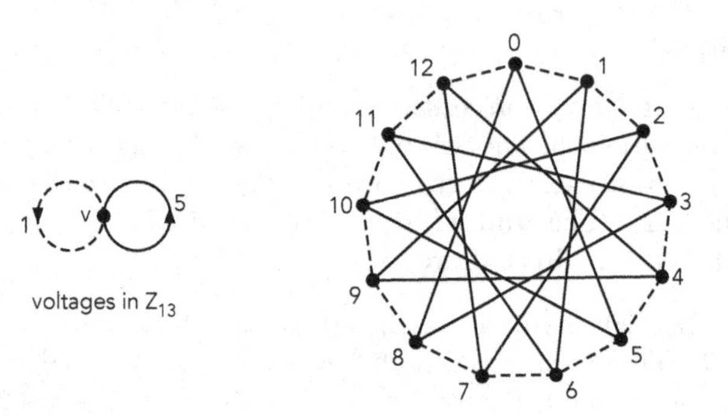

Figure 10.1.11 *circ*$(13 : 1, 5)$ **as a covering graph of** B_2 **with** $\mathbb{Z}_{13}$**-voltages.**

Example 10.1.8: With the bidirected-arc convention in effect, the complete graph $K_{2n} \cong circ(2n : 1, 2, \ldots, n)$ is specified by a bouquet B_n with $\mathbb{Z}_{2n}$-voltages $1, 2, \ldots, n$ on its self-loops. Figure 10.1.12 illustrates this for K_6.

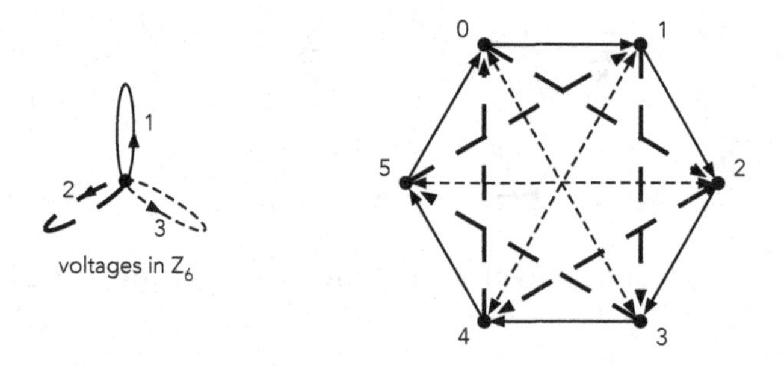

Figure 10.1.12 A K_6-digraph with bidirected arcs.

Self-loops and multi-arcs are common in voltage graphs, and do not necessarily generate self-loops and multi-edges in the covering graph. The next proposition establishes precisely when they do.

Proposition 10.1.7: *Let $\vec{G}^\alpha$ be the covering digraph of a $\mathbb{Z}_n$-voltage graph $\left\langle \vec{G}, \alpha \right\rangle$.*

1. *$\vec{G}^\alpha$ has a self-loop if and only if $\vec{G}$ has a self-loop e with voltage 0.*
2. *$\vec{G}^\alpha$ has a multi-arc if and only if $\vec{G}$ has a multi-arc for which at least two of the arcs have the same voltage.*
3. *$\vec{G}^\alpha$ has two oppositely directed arcs if and only if one or both of the following conditions hold:*
   (a) $\vec{G}$ has two oppositely directed arcs whose voltages are additive inverses.
   (b) n is even and some self-loop in $\vec{G}$ has voltage $\frac{n}{2}$.

Proof:

1. See Exercise 10.1.4.

2. Suppose $\vec{G}^\alpha$ has two arcs d_i and e_i, directed from v_i to w_j, for some $i, j \in \mathbb{Z}_n$. Since, $j = i + \alpha(d)$ and $j = i + \alpha(e)$, $\alpha(d) = \alpha(e)$.

   Conversely, two arcs d and e directed from v to w in $\vec{G}$, with $\alpha(d) = \alpha(e)$, generate arcs d_0 and e_0 in $\vec{G}^\alpha$, both directed from v_0 to $w_{\alpha(d)} = w_{\alpha(e)}$.

3. Suppose d_i and e_j are distinct arcs in $\vec{G}^\alpha$, where d_i is directed from v_i to w_j and e_j is directed from w_j to v_i, for some $i, j \in \mathbb{Z}_n$. Then in $\vec{G}$, d is an arc from v to w and e is an arc from w to v with $\alpha(e) = i - j = -(j - i) = -\alpha(d)$. Since d_i and e_j are different arcs, either (i) $d \neq e$ or (ii) $d = e$ and $i \neq j$.

   If $d \neq e$, then they are oppositely directed arcs whose voltages are additive inverses. In the second case, since $d = e$, d is a self-loop and $\alpha(d) = -\alpha(d)$, which implies $\alpha(d) = 0$ or $n/2$. But $\alpha(d) = j - i \neq 0$, and hence, $\alpha(d) = n/2$. Therefore, d is a self-loop with voltage $n/2$.

   Suppose condition (a) holds. Let d be an arc directed from v to w in $\vec{G}$, and let e be an arc directed from w to v with $\alpha(d) = -\alpha(e)$. Arcs d and e generate arcs d_0 and $e_{\alpha(d)}$ in $\vec{G}^\alpha$, where d_0 is directed from v_0 to $w_{\alpha(d)}$ and $e_{\alpha(d)}$ is directed from $w_{\alpha(d)}$ to $v_{\alpha(d)+\alpha(e)} = v_0$.

   Now suppose condition (b) holds. If n is even and e is a self-loop on v in $\vec{G}$ with $\alpha(e) = \frac{n}{2}$, then e_0 is an arc in $\vec{G}^\alpha$ directed from v_0 to $v_{n/2}$ and $e_{n/2}$ is directed from $v_{n/2}$ to $v_{n/2+n/2} = v_0$.

◇

DEFINITION: A $\mathbb{Z}_n$-voltage graph $\langle G, \alpha \rangle$ is ***simple*** if G has no self-loops with voltage 0, no multi-arcs with the same voltage, and no oppositely directed arcs whose voltages are additive inverses.

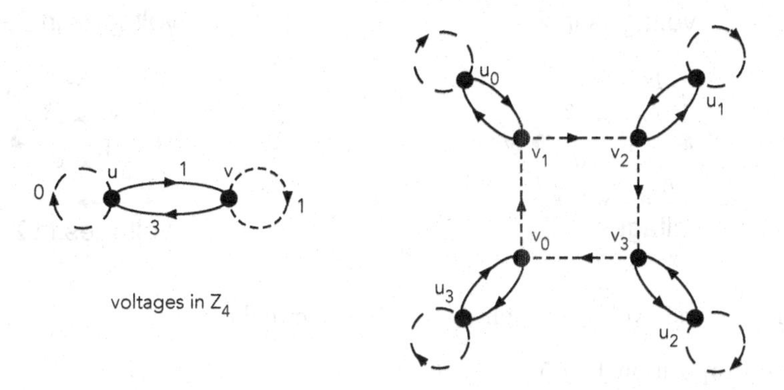

Figure 10.1.13 A non-simple $\mathbb{Z}_4$-voltage graph and its covering digraph.

Corollary 10.1.8: *Using the bidirected-arc convention, the covering graph G^α is a simple graph if and only if the $\mathbb{Z}_n$-voltage graph $\langle G, \alpha \rangle$ is simple.*

Proof: By Proposition 10.1.7, the only case we need to consider is when n is even and $\vec{G}$ has a self-loop with voltage $\frac{n}{2}$. In this case, the oppositely directed arcs in $\vec{G}^\alpha$ that it generates are combined into one edge by the bidirected arc convention. ◇

EXERCISES for Section 10.1

10.1.1 Draw the covering digraph for the given voltage graph.

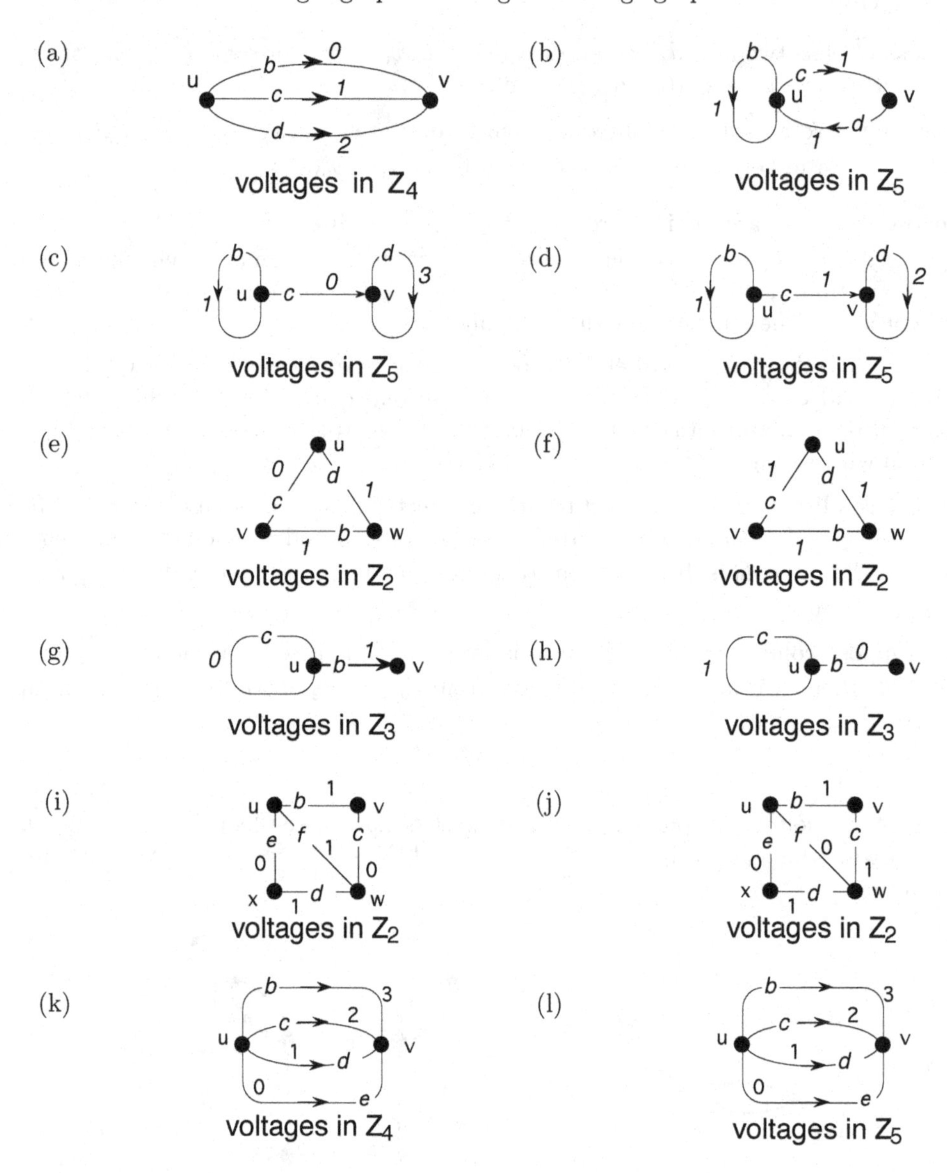

10.1.2 Design a cyclic-voltage graph to specify the circular ladder CL_n.

10.1.3 Prove Proposition 10.1.5.

10.1.4 Prove that the covering graph for a cyclic-voltage graph has no self-loops, unless the voltage graph has a self-loop with voltage 0.

10.1.5 How can the voltage graph specifying K_{2n-1} be modified so that it specifies the octahedral graph O_n?

10.2 SPECIFYING CONNECTED GRAPHS

REVIEW FROM Section §1.2 A graph is ***connected*** if for every pair of vertices u and v, there is a walk from u to v. A digraph is ***connected*** if its underlying graph is connected.

We begin the section with an example of a covering digraph having multiple components. Our goal in this section is to characterize those $\mathbb{Z}_n$-voltage graphs whose covering graphs are connected.

Example 10.2.1: The covering digraph in Figure 10.2.1, has three components: two 6-cycles and a 12-cycle. It may not be surprising that covering graph is not connected, given that the voltage graph is not connected. However, the upper component of the $\mathbb{Z}_4$-voltage graph generates two disjoint cycles in the covering graph, whereas the lower component generates only one.

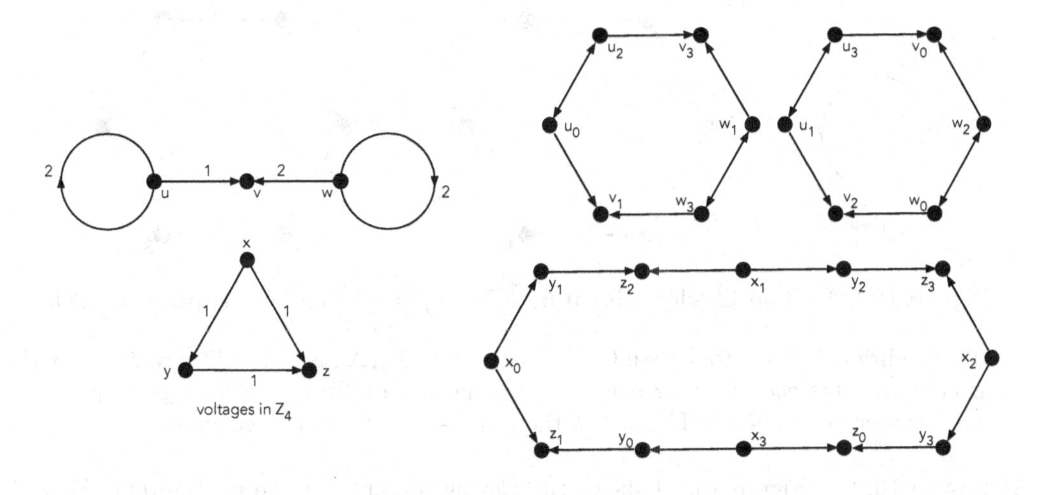

Figure 10.2.1 A $\mathbb{Z}_4$-voltage and its non-connected covering digraph.

Cayley Graphs

The simplest $\mathbb{Z}_n$-voltage graphs have a single vertex, i.e., they are bouquets. Remember that in this case, it is common not to label that vertex and to simply label the vertices in the covering graph with the elements of $\mathbb{Z}_n$.

DEFINITION: A subset X of the cyclic group $\mathbb{Z}_n$ is a ***generating set*** if every element of $\mathbb{Z}_n$ is a sum of elements of X. We say X *generates* $\mathbb{Z}_n$.

Example 10.2.2: Suppose $X = \{x_1, x_2, \ldots, x_r\}$ is a subset of $\mathbb{Z}_n$. We can also view X as the set of nonnegative integers less than n, and in this context, compute the greatest common divisor $d = \gcd(n, x_1, x_2, \ldots, x_r)$. X is a generating set of $\mathbb{Z}_n$, if and only if $d = 1$. This observation follows from the fact that the greatest common divisor of a set of numbers equals the smallest positive number that can be formed by taking sums and differences of numbers in the set. Since $\gcd(6, 2, 3) = 1$, the set $\{2, 3\}$ generates $\mathbb{Z}_6$. Since $\gcd(6, 2) = 2$, the set $\{2\}$ does not generate $\mathbb{Z}_6$; only 0, 2, and 4 can be obtained as a sum of 2s.

DEFINITION: Let $X = \{x_1, x_2, \ldots, x_r\}$ be a generating set of $\mathbb{Z}_n$. The ***Cayley digraph*** $\vec{C}(\mathbb{Z}_n, X)$ is the covering digraph for the voltage graph $\langle B_r, \alpha \rangle$, where B_r is the bouquet with r self-loops, and α assigns each voltage in X to one of the self-loops. $\vec{C}(\mathbb{Z}_n, X)$ has as its vertex-set and arc-set

$$V_{\vec{C}(\mathbb{Z}_n,X)} = \mathbb{Z}_n \quad \text{and} \quad E_{\vec{C}(\mathbb{Z}_n,X)} = \{x_i \mid x \in X, i \in \mathbb{Z}_n\}$$

respectively. Arc x_i joins vertex i to vertex $i + x$. (Bidirected arcs are sometimes used for the generator $\frac{n}{2}$, when n is even, as described in §10.1.)

The ***Cayley graph*** $C(\mathbb{Z}_n, X)$ is the *underlying graph* of the Cayley digraph $\vec{C}(\mathbb{Z}_n, X)$.

Example 10.2.3: Figure 10.2.2 shows the Cayley digraph for the cyclic group Z_6 with generating set $\{1\}$ and the corresponding Cayley graph, which is clearly isomorphic to the circulant graph *circ*$(6 : 1)$.

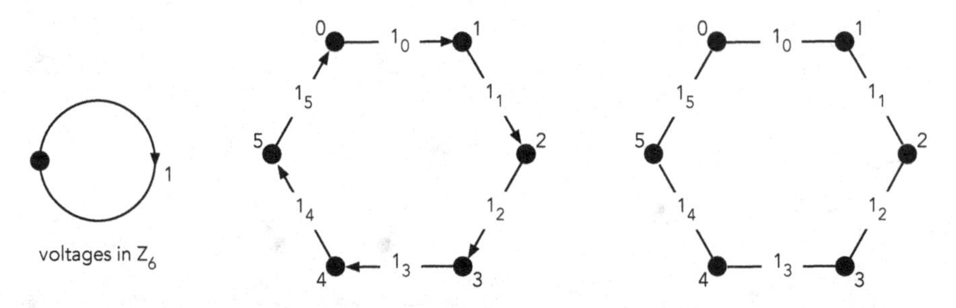

Figure 10.2.2 The Cayley digraph $\vec{C}(\mathbb{Z}_6, \{1\})$ and Cayley graph $C(\mathbb{Z}_6, \{1\})$.

The traditional way is to draw a Cayley digraph $\vec{C}(\mathbb{Z}_n, X)$ is to label the vertices by the elements of $\mathbb{Z}_n$. Instead of giving edges distinct names, a different color or graphic feature is used for each edge fiber $\tilde{x}$. This led to the terminology *Cayley color graph.*

Example 10.2.4: Figure 10.2.3 shows the Cayley digraph for the cyclic group Z_6 with generating set $\{2, 3\}$ and the corresponding Cayley graph, which is isomorphic to the circulant graph *circ*$(6 : 2, 3)$. This illustrates that a non-minimum generating set for a group may be used in a Cayley-graph specification of a graph.

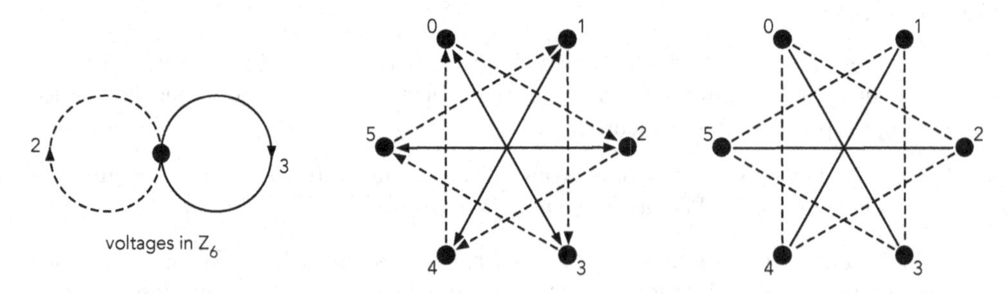

Figure 10.2.3 The Cayley digraph $\vec{C}(\mathbb{Z}_6, \{2, 3\})$ and Cayley graph $C(\mathbb{Z}_6, \{2, 3\})$.

Example 10.2.5: Figure 10.2.4 shows a covering graph, which is isomorphic to the circulant graph *circ*$(6:2)$. It is not a Cayley graph since $\{2\}$ does not generate $\mathbb{Z}_6$. Notice that this graph is not connected. Since X is a generating set of $\mathbb{Z}_n$, a Cayley graph $C(\mathbb{Z}_n, X)$ is always a connected graph, as will be shown in the next subsection.

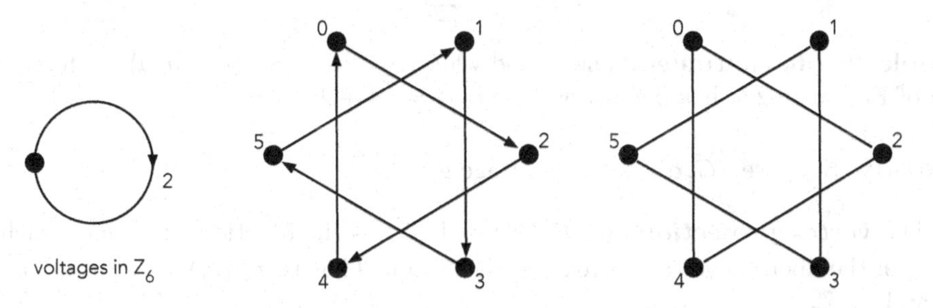

Figure 10.2.4 The circulant graph *circ*$(6:2)$ is not a Cayley graph.

Signed Walks and Connectivity

DEFINITION: A ***signed walk*** in a digraph is an alternating sequence

$$W = \langle v_1, \delta_1 e_1, v_2, \delta_2 e_2, \ldots v_{n-1}, \delta_{n-1} e_{n-1}, v_n \rangle$$

of vertices and arcs, such that for each i, $\delta_i \in \{1, -1\}$, and

- if $\delta_i = 1$, then e_i is directed from v_i to v_{i+1} and
- if $\delta_i = -1$, then e_i is directed from v_{i+1} to v_i.

Example 10.2.6: Consider the $\mathbb{Z}_4$-voltage graph $\left\langle \vec{G}, \alpha \right\rangle$ and its covering digraph $\vec{G}^\alpha$ in Figure 10.2.5. Using the bidirected-arc convention, the self-loop e on vertex v with

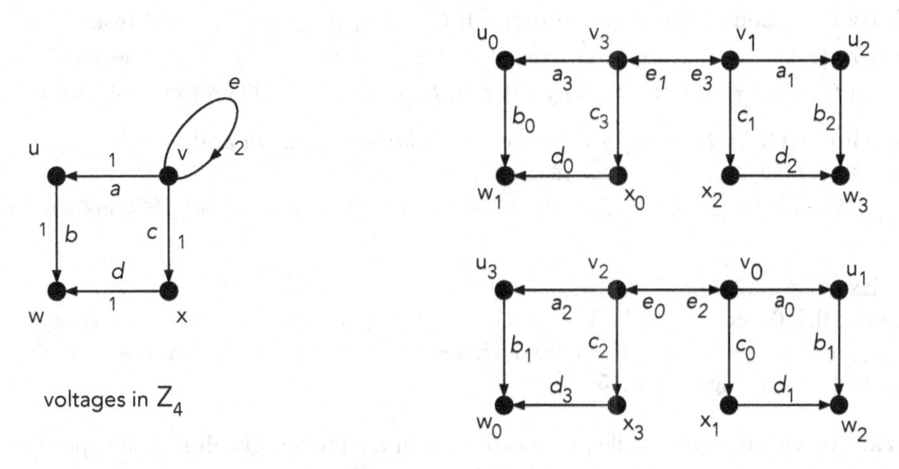

Figure 10.2.5 A $\mathbb{Z}_4$-voltage graph and its covering digraph.

voltage 2, generates two bidirected arcs e_1/e_3 and e_0/e_2. The signed walk $W_1 = \langle u_0, -a_3, v_3, e_3, v_1, a_1, u_2 \rangle$ in $\vec{G}^\alpha$ traverses the arcs: a_3, e_3, and a_1 in the digraph $\vec{G}^\alpha$, where arc a_3 is traversed in the opposite direction. The signed walk $W_2 = \langle u, -a, v, e, v, a, u \rangle$ in the $\mathbb{Z}_4$-voltage graph traverses arc a in the opposite direction, followed by arcs e and a in the normal direction.

DEFINITION: If $W = \langle v_1, \delta_1 e_1, v_2, \delta_2 e_2, \ldots v_{n-1}, \delta_{n-1} e_{n-1}, v_n \rangle$ is a signed walk in a $\mathbb{Z}_n$-voltage graph $\left\langle \vec{G}, \alpha \right\rangle$, then the ***net voltage*** of W, denoted $nv(W)$, is

$$nv(W) = \sum_{i=1}^{n-1} \delta_i \alpha(e_i)$$

Example 10.2.6, continued: The signed walk $W_2 = \langle u, -a, v, e, v, a, u \rangle$ in the $\mathbb{Z}_4$-voltage graph of Figure 10.2.5, has net voltage $nv(W_2) = -1 + 2 + 1 = 2$.

DEFINITION: Suppose $\left\langle \vec{G}, \alpha \right\rangle$ is a $\mathbb{Z}_n$-voltage graph.

- The ***vertex projection*** $\pi_V : V(\vec{G}^\alpha) \to V(\vec{G})$, is the function, that maps each vertex v_i in the fiber above the vertex $v \in V(\vec{G})$ to v. That is, $\pi_V(v_i) = v$, for any $v \in V(\vec{G})$ and $i \in \mathbb{Z}_n$.
- The ***edge projection*** $\pi_E : E(\vec{G}^\alpha) \to E(\vec{G})$ maps each arc e_i in the fiber above the arc $e \in E(\vec{G})$ to e. That is, $\pi_E(e_i) = e$, for any $e \in E(\vec{G})$ and $i \in \mathbb{Z}_n$.
- The ***natural projection*** $\pi : \vec{G}^\alpha \to \vec{G}$ is the pair $\pi = (\pi_V, \pi_E)$.

DEFINITION: A ***digraph homomorphism*** $f : \vec{G} \to \vec{H}$ from digraph $\vec{G}$ to digraph $\vec{H}$ is a pair of functions

$$f_V : V(\vec{G}) \to V(\vec{H}) \text{ and } f_E : E(\vec{G}) \to E(\vec{H})$$

such that if e is an arc in $\vec{G}$ directed from v to w, then $f_E(e)$ is an arc in $\vec{H}$ directed from $f_V(v)$ to $f_V(w)$.

DEFINITION: A digraph homomorphism $f = (f_V, f_E)$ is ***onto*** if both f_V and f_E are onto.

Proposition 10.2.1: *For any $\mathbb{Z}_n$-voltage graph $\left\langle \vec{G}, \alpha \right\rangle$, the natural projection π is a digraph homomorphism from $\vec{G}^\alpha$ onto $\vec{G}$.*

Proof: By definition of the covering digraph $\vec{G}^\alpha$, if e_i is an arc directed from v_i to w_j in $\vec{G}^\alpha$, then $\pi_E(e_i) = e$ is an arc in $\vec{G}$ from $v = \pi_V(v_i)$ to $w = \pi_V(w_j)$. Furthermore, for any vertex v in $\vec{G}$, $v = \pi_V(v_0)$ and for any arc e in $\vec{G}$, $e = \pi_E(e_0)$. Therefore, π is onto. ◇

Proposition 10.2.2: *If $f : \vec{G} \to \vec{H}$ is a digraph homomorphism and $W = \langle v_1, \delta_1 e_1, v_2, \delta_2 e_2, \ldots v_{n-1}, \delta_{n-1} e_{n-1}, v_n \rangle$ is a signed walk in $\vec{G}$, then $f(W) = \langle f(v_1), \delta_1 f(e_1), f(v_2), \delta_2 f(e_2), \ldots f(v_{n-1}), \delta_{n-1} f(e_{n-1}), f(v_n) \rangle$ is a signed walk in $\vec{H}$.*

Proof: Exercise 10.2.2. ◇

Example 10.2.6, continued: The signed walk $W_1 = \langle u_0, -a_3, v_3, e_3, v_1, a_1, u_2 \rangle$ in the covering graph projects onto the closed signed walk $\pi(W_1) = \langle u, -a, v, e, v, a, u \rangle$ in the $\mathbb{Z}_4$-voltage graph of Figure 10.2.5.

Remark: As with directed walks, a signed walk in a simple digraph may be specified as a vertex sequence.

Example 10.2.7: In $\vec{G}_1^\alpha$ from Figure 10.2.6, the u_0-u_1 signed walk $\langle u_0, v_2, v_3, u_1 \rangle$ projects onto the signed walk $\langle u, -f, v, e, v, f, u \rangle$, which has net voltage $-2 + 1 + 2 = 1$; the u_0-v_1 signed walk $\langle u_0, v_2, v_1 \rangle$ projects onto the signed walk $\langle u, -f, v, -e, v \rangle$, which has net voltage $-2 - 1 = -3 = 1$; and the v_0-v_1 signed walk $\langle v_0, v_1 \rangle$ projects onto the signed walk $\langle v, e, v \rangle$, which has net voltage 1.

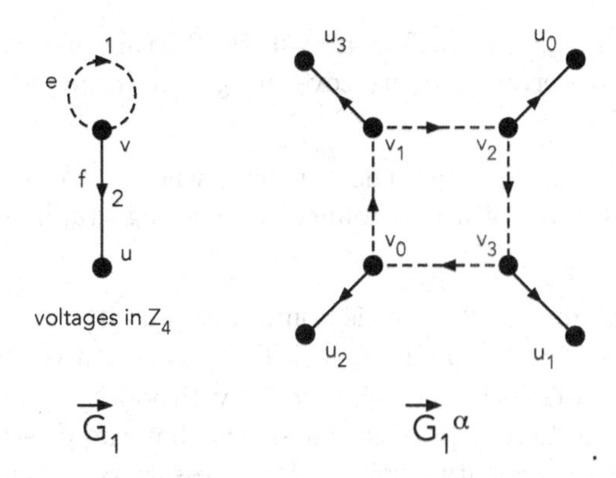

Figure 10.2.6 A $\mathbb{Z}_4$-voltage graph with a connected covering digraph.

Proposition 10.2.3: *If $\left\langle \vec{G}, \alpha \right\rangle$ is a $\mathbb{Z}_n$-voltage graph and W is a v_i-w_j signed walk in $\vec{G}^{\alpha}$ then $\pi(W)$ is a v-w signed walk in $\left\langle \vec{G}, \alpha \right\rangle$ with net voltage $nv(\pi(W)) = j - i$.*

Proof: By Proposition 10.2.2, $\pi(W)$ is a signed walk from $v = \pi(v_i)$ to $w = \pi(w_j)$.

To show $nv(\pi(W)) = j - i$, we use induction on the length of the signed walk. If W is a signed walk in $\vec{G}^{\alpha}$ of length 1, then $\pi(W) = \langle u, (+1)e, w \rangle$ or $\pi(W) = \langle u, (-1)e, w \rangle$, for some arc e. In the first case, e is directed from u to w, and so $W = \langle v_i, (+1)e_i, w_j \rangle$ with e_i directed from v_i to w_j. Therefore, $j = i + \alpha(e)$, or $nv(W) = \alpha(e) = j - i$. In the second case, e is directed from w to v, and so $W = \langle v_i, (-1)e_j, w_j \rangle$ with e_j directed from w_j to v_i. Therefore, $i = j + \alpha(e)$, or $nv(W) = -\alpha(e) = -(i - j) = j - i$.

Suppose the proposition holds for any signed walk of length $n - 1$. To show it holds for any signed walk of length n, let $W = W_1 \circ W_2$ be a v_i-u_k signed walk of length $n - 1$ followed by a u_k-w_j signed walk of length 1 in $\vec{G}^{\alpha}$. By the definition of signed walk and net voltage, $\pi(W) = \pi(W_1) \circ \pi(W_2)$ and $nv(\pi(W)) = nv(\pi(W_1)) + nv(\pi(W_2))$. By the induction hypothesis, $nv(\pi(W_1)) = k - i$ and by the base case, $nv(W_2) = j - k$. Therefore, $nv(\pi(W)) = (k - i) + (j - k) = j - i$. ♢

Example 10.2.6, continued: In Figure 10.2.5, the signed walk $W = \langle u, -a, v, e, v, a, u \rangle$ in the voltage graph generates four signed walks in the covering graph: $\langle u_0, -a_3, v_3, e_3, v_1, a_1, u_2 \rangle$, $\langle u_1, -a_0, v_0, e_0, v_2, a_2, u_3 \rangle$, $\langle u_2, -a_1, v_1, e_1, v_3, a_3, u_0 \rangle$, and $\langle u_3, -a_2, v_2, e_2, v_0, a_0, u_1 \rangle$. Notice that each of these signed walks is a u_i-u_j walk with $j - i = 2 = nv(W)$.

Proposition 10.2.4: *Suppose W is a signed walk from v in a $\mathbb{Z}_n$-voltage graph $\left\langle \vec{G}, \alpha \right\rangle$. For each i in $\mathbb{Z}_n$, there is a directed walk $\tilde{W}$ from v_i in $\vec{G}^{\alpha}$ such that $\pi(\tilde{W}) = W$.*

Proof: Again, we use induction on the length of the signed walk W. If the length is 1, then $W = \langle v, \delta e, w \rangle$, where $\delta = \pm 1$. By Proposition 10.2.3, if $\delta = 1$, then $\tilde{W} = \langle v_i, e_i, w_j \rangle$, where $j = i + \alpha(e)$ and if $\delta = -1$, then $\tilde{W} = \langle v_i, -e_j, w_j \rangle$, where $j = i - \alpha(e)$, which establishes the base.

Suppose that the result holds for signed walks of length $n - 1$. Any signed walk W of length n in $\left\langle \vec{G}, \alpha \right\rangle$ is the concatenation $W = W_1 \circ W_2$ of a signed v-u walk W_1 of length $n-1$ with a signed u-w walk W_2 of length 1. By the induction hypothesis, there is a signed walk $\tilde{W}_1$ from v_i such that $\pi(\tilde{W}_1) = W_1$ and by Proposition 10.2.3, $\tilde{W}_1$ ends at $u_{i+nv(W_1)}$. By the base case, there is a signed walk $\tilde{W}_2$ from $u_{i+nv(W_1)}$ such that $\pi(\tilde{W}_2) = W_2$. Therefore, $\pi(\tilde{W}_1 \circ \tilde{W}_2) = W$. ♢

Remark: For each i in $\mathbb{Z}_n$, the "lift" of a walk W starting at a vertex v in a $\mathbb{Z}_n$-voltage to a walk $\tilde{W}$ starting at vertex v_i in its covering graph is unique. We leave the proof as Exercise 10.2.**3**.

Before stating a proposition that characterizes when a covering graph is connected, we look at one more example of a non-connected covering graph specified by a connected $\mathbb{Z}_n$-voltage graph.

Example 10.2.8: In Figure 10.2.7, $\vec{G}_2$ is connected, but its covering graph is not, since there is no signed walk in $\vec{G}_2^\alpha$ from v_0 to v_1. The projection of such a walk would be a closed signed v-v walk in $\vec{G}_2$ with net voltage 1 by Proposition 10.2.3. However, no such walk exists in $\vec{G}_2$. The following proposition shows that the existence of a closed signed walk with net voltage 1 in the voltage graph is a necessary condition for the connectedness of a covering graph.

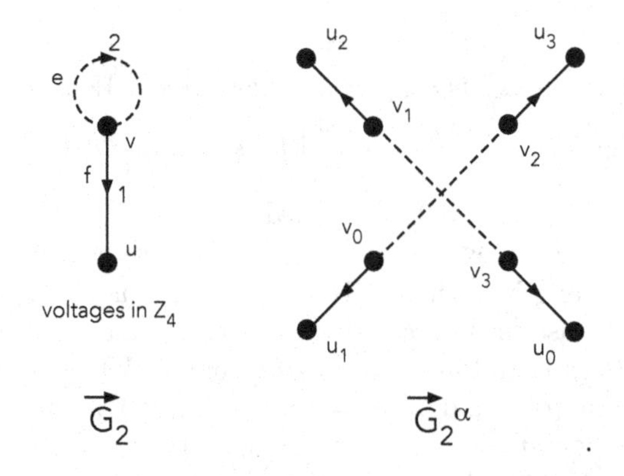

Figure 10.2.7 A $\mathbb{Z}_4$-voltage graph with a non-connected covering digraph.

Proposition 10.2.5: *Let $\left\langle \vec{G}, \alpha \right\rangle$ be a $\mathbb{Z}_n$-voltage graph. The covering digraph $\vec{G}^\alpha$ is connected if and only if $\vec{G}$ is connected and there is a closed signed walk in $\vec{G}$ with net voltage 1.*

Proof: Suppose $\vec{G}^\alpha$ is connected and let v and w be any two vertices in $\vec{G}$. Since $\vec{G}^\alpha$ is connected, there is a signed v_0-w_0 walk in $\vec{G}^\alpha$, which projects onto a signed v-w walk in $\vec{G}$, by Proposition 10.2.4. Therefore, $\vec{G}$ is connected. Also, there is a signed walk in $\vec{G}^\alpha$ from v_0 to v_1, which projects onto a closed signed walk from v in $\vec{G}$ with net voltage 1, by Proposition 10.2.3.

Now suppose $\left\langle \vec{G}, \alpha \right\rangle$ is connected and has a closed signed walk W_1 from a vertex v with net voltage 1. Let v_i and w_j be two vertices in $\vec{G}^\alpha$. Since $\vec{G}$ is connected, there is a v-w signed walk W_2 in $\vec{G}$. Let $t = nv(W_2)$. Let W be the concatenation of $(j-i-t)$ copies of W_1 with W_2. By Proposition 10.2.4, there is a signed walk $\tilde{W}$ in $\vec{G}_\alpha$ from v_i such that $\pi(\tilde{W}) = W$. Moreover, $nv(W) = (j-i-t)\cdot 1 + t = j-i$, and hence, $\tilde{W}$ is a signed walk in $\vec{G}^\alpha$ from v_i to w_j, by Proposition 10.2.2. Since, this is true for any two vertices v_i and w_j in $\vec{G}^\alpha$, the covering digraph is connected. ♢

Example 10.2.1, continued: The $\mathbb{Z}_4$-voltage graph from Figure 10.2.1, reprinted in Figure 10.2.8, has two components and therefore we expect at least two components in its covering graph. The closed walk $\langle x, y, z, x\rangle$ has net voltage 1, and so the lower component of the $\mathbb{Z}_4$-voltage graph generates a single component of the covering graph. However, there is no closed signed walk on the upper component of the $\mathbb{Z}_4$-voltage graph, and so the upper component generates more than one component in the covering graph.

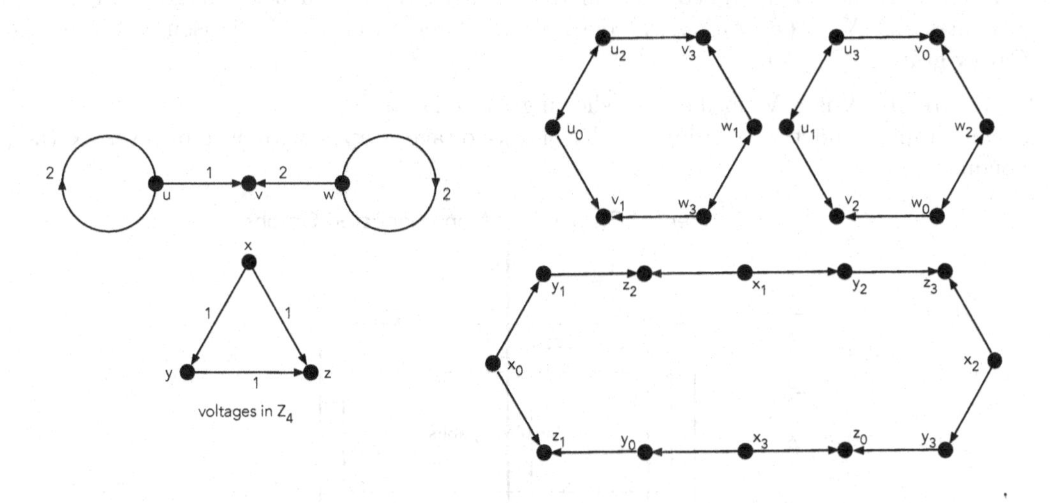

Figure 10.2.8 A $\mathbb{Z}_4$-voltage graph with a 3-component covering digraph.

By definition, every Cayley graph is the covering graph of a bouquet, whose voltages generate $\mathbb{Z}_n$. Showing that the conditions of Proposition 10.2.5 are met would prove the following corollary, and it is left as an exercise.

Corollary 10.2.6: *Every Cayley graph is connected.*

Proof: Exercise 10.2.4. ♢

Relationships among Cayley, Circulant, and Covering Graphs

The preceding examples have shown that Cayley graphs, circulant graphs, and the covering graphs of $\mathbb{Z}_n$-voltage graphs are related. By definition, every Cayley graph is the covering graph of a $\mathbb{Z}_n$-voltage graph $\langle B_r, \alpha\rangle$ and by Proposition 10.1.6, the same is true of circulant graphs. The following two propositions encapsulate these relationships.

Proposition 10.2.7: *A Cayley graph is a circulant graph if and only if it is simple.*

Proof: If a Cayley graph is a circulant graph, then it is simple, by definition of circulant graph. Conversely, suppose the Cayley graph $C(\mathbb{Z}_n, X)$ is simple. By Proposition 10.1.7, $0 \notin X$, and thus, the circulant graph $circ(n, X)$ is defined. Both $C(\mathbb{Z}_n, X)$ and $circ(n, X)$ are simple graphs on the same vertex-set, and they have the same adjacencies. In particular, vertices i and j are adjacent in both $C(\mathbb{Z}_n, X)$ and $circ(n, X)$ if and only if $i - j \in X$ or $j - i \in X$. Hence, the two graphs are isomorphic. ♢

Proposition 10.2.8: *A circulant graph is a Cayley graph if and only if it is connected.*

Proof: If a circulant graph is a Cayley graph, then it must be connected, since every Cayley graph is connected (by Corollary 10.2.6). Conversely, suppose a circulant graph $circ(n : x_1, x_2, \ldots, x_r)$ is connected. By definition, $circ(n : x_1, x_2, \ldots, x_r)$ is the covering graph of a $\mathbb{Z}_n$-voltage graph $\langle B_r, \alpha \rangle$, where α assigns each voltage x_i to one of the self-loops in B_r. Also, we may assume $1 \leq x_1 < x_2 < \cdots < x_r \leq (n+1)/2$. Since, $circ(n : x_1, \ldots, x_r)$ is connected, there is a signed walk in the voltage graph with net voltage 1. Since $\{1\}$ generates $\mathbb{Z}_n$, $X = \{x_1, x_2, \ldots, x_r\}$ does also. Therefore, $circ(n, X)$ is isomorphic to the Cayley graph $C(\mathbb{Z}_n, X)$. ♢

Figure 10.2.9 is a Venn diagram showing the relationships among Cayley graphs, circulant graphs, and the covering graphs of $\mathbb{Z}_n$-voltage graphs with just one vertex (i.e., bouquets).

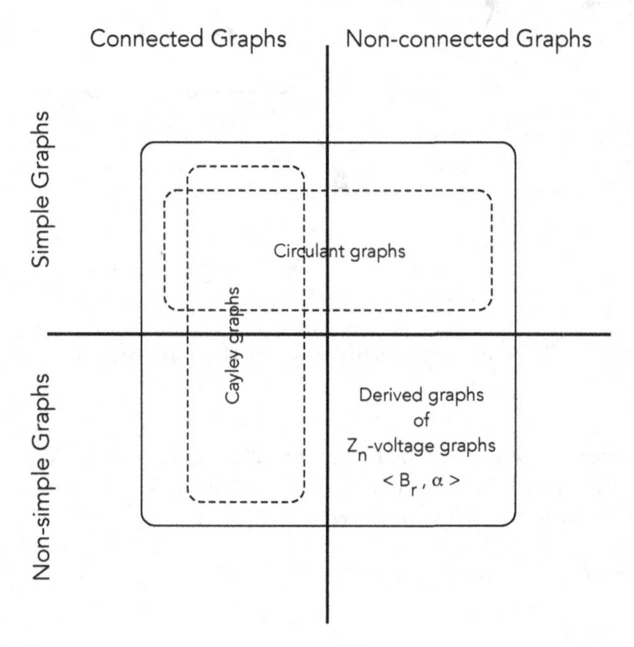

Figure 10.2.9 Cayley, circulant, and covering graphs.

EXERCISES for Section 10.2

10.2.**1** Draw the Cayley graph for the given group and generating set.

(a) The cyclic group $\mathbb{Z}_9$ with generating set $\{1, 3\}$.

(b) The cyclic group $\mathbb{Z}_{10}$ with generating set $\{2, 5\}$.

10.2.**2** Prove Proposition 10.2.2.

10.2.**3** Show that for each i in $\mathbb{Z}_n$, the "lift" of a walk W starting at a vertex v in a $\mathbb{Z}_n$-voltage to a walk $\tilde{W}$ starting at vertex v_i in its covering graph is unique. That is, show that there is a unique walk starting at vertex v_i in the covering graph, that projects onto W.

10.2.**4** Give the details of a proof for Corollary 10.2.6.

10.3 $\mathbb{Z}_n$-VOLTAGE GRAPHS AND GRAPH COLORINGS

There are two natural ways of coloring the vertices and/or edges of a covering graph. The first introduced in this section uses a coloring of the voltage graph to induce a coloring of the covering graph, and the second uses the voltage-group as the color set to color vertices in the covering graph.

REVIEW FROM §6.1 The ***vertex chromatic number*** of a graph G, denoted $\chi(G)$, is the minimum number of colors required for a proper vertex-coloring of G.

REVIEW FROM §6.4 The ***edge chromatic number*** of a graph G, denoted $\chi'(G)$, is the minimum number of colors required for a proper edge-coloring of G.

Induced Colorings

DEFINITION: Suppose $\left\langle \vec{G}, \alpha \right\rangle$ is a $\mathbb{Z}_n$-voltage graph and π is the projection from $\vec{G}^\alpha$ to $\vec{G}$. If $f : V(G) \to C$ is a vertex-coloring of G, the ***induced vertex-coloring*** of $\vec{G}^\alpha$ is $f \circ \pi_V : V(G^\alpha) \to C$. If $g : E(G) \to C'$ is an edge-coloring of G, then the ***induced edge-coloring*** of $\vec{G}^\alpha$ is $g \circ \pi_E : E(G^\alpha) \to C'$.

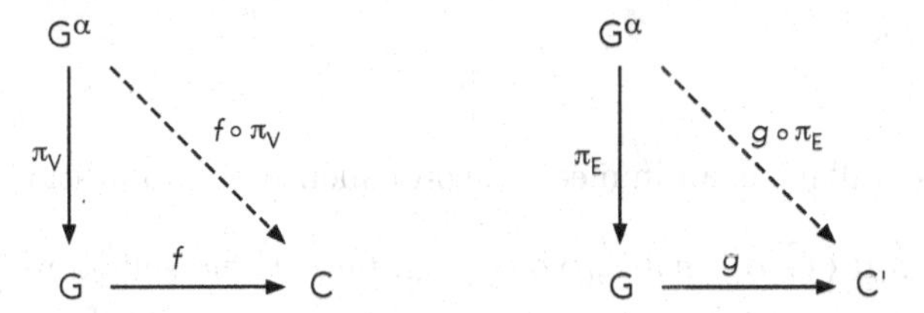

Figure 10.3.1 Induced vertex- and edge-colorings are compositions.

Example 10.3.1: The underlying graph of the $\mathbb{Z}_3$-voltage graph in Figure 10.3.2 is a complete graph on three vertices. The (black, grey, white) vertex 3-coloring of G induces a proper vertex 3-coloring of the covering graph G^α.

The proper (solid, long-dash, short-dash) edge 3-coloring of G induces a proper edge 3-coloring of G^α.

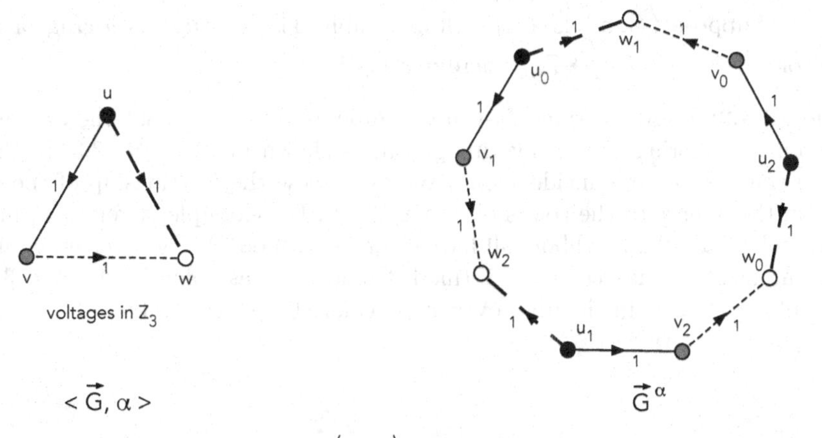

Figure 10.3.2 Colorings of $\left\langle \vec{G}, \alpha \right\rangle$ and their induced colorings of $\vec{G}^\alpha$.

Proposition 10.3.1: *Suppose $\langle G, \alpha \rangle$ is a $\mathbb{Z}_n$-voltage graph, $f : V(G) \to C$ is a vertex-coloring of G, and $g : V(G) \to C$ is an edge-coloring of G.*

1. *f is a proper vertex-coloring of G*

   if and only if $f \circ \pi_V$ is a proper vertex-coloring of G^α.

2. *g is a proper edge-coloring of G*

   if and only if $g \circ \pi_E$ is a proper edge-coloring of G^α.

Proof:

1. Suppose f is a proper vertex-coloring of G. If e_i is an arc in $\vec{G}^\alpha$ from v_i to w_j, then its image, $\pi_E(e_i) = e$, is an arc in G from $\pi_V(v_i) = v$ to $\pi_V(w_j) = w$. Since f is a proper vertex-coloring, $f(v) \neq f(w)$. Therefore, $(f \circ \pi_V)(v_i) = f(v) \neq f(w) = (f \circ \pi_V)(w_j)$. Since this is true for every arc in $\vec{G}^\alpha$, the induced vertex-coloring $f \circ \pi_V$ is a proper vertex-coloring of G^α.

   If f is not a proper vertex-coloring of G, then there is an arc e in $\vec{G}$ from v to w such that $f(v) = f(w)$. Then, for any $i \in \mathbb{Z}_n$, e_i is an arc in $\vec{G}^\alpha$ from v_i to $w_{i+\alpha(e)}$. Therefore, the induced vertex-coloring $f \circ \pi_V$ is not a proper vertex-coloring of G^α, since $(f \circ \pi_V)(v_i) = f(v) = f(w) = (f \circ \pi_V)(w_{i+\alpha(e)})$.

2. Exercise 10.3.8.

$\diamondsuit$

The following corollary is an immediate consequence of Proposition 10.3.1.

Corollary 10.3.2: *If $\left\langle \vec{G}, \alpha \right\rangle$ is a $\mathbb{Z}_n$-voltage graph with no self-loops, then*

- $\chi(G^\alpha) \leq \chi(G)$ *and*
- $\chi'(G^\alpha) \leq \chi'(G)$

Voltage-colorings

If $\left\langle \vec{G}, \alpha \right\rangle$ is a $\mathbb{Z}_n$-voltage graph, then we can obtain a vertex n-coloring of the covering graph by using $\mathbb{Z}_n$ as the color-set.

DEFINITION: Suppose $\left\langle \vec{G}, \alpha \right\rangle$ is a $\mathbb{Z}_n$-voltage graph. The ***voltage-coloring*** of G^α is the vertex n-coloring $\tilde{\alpha} : V(G^\alpha) \to \mathbb{Z}_n$, where $\tilde{\alpha}(v_i) = i$.

Example 10.3.2: Using the same $\mathbb{Z}_3$-voltage graph from Figure 10.3.2, the voltage-coloring gives a proper 3-coloring of the covering graph, as shown in Figure 10.3.3. In the voltage graph, the voltages on arcs incident on a vertex x show the relationship of the color of a vertex x_i in the fiber $\tilde{x}$ to the colors of its neighbors. For example, a vertex u_i in the fiber over u has color i and its neighbors all have color $i + 1 \pmod 3$. For a vertex v_i in the fiber above v, one neighbor has color $i - 1 \pmod 3$ and one has color $i + 1 \pmod 3$. All the neighbors of a vertex w_i in the fiber over w are colored $i - 1 \pmod 3$.

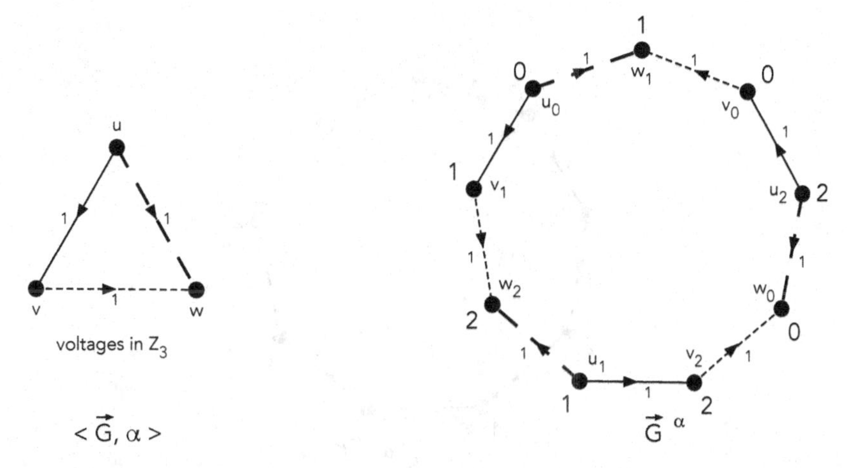

Figure 10.3.3 A $\mathbb{Z}_3$-voltage graph $\left\langle \vec{G}, \alpha \right\rangle$ and the voltage-coloring of $\vec{G}^\alpha$.

Proposition 10.3.3: *Suppose $\left\langle \vec{G}, \alpha \right\rangle$ is a $\mathbb{Z}_n$-voltage graph. The voltage-coloring, $\tilde{\alpha}(v_i) = i$, is a proper vertex-coloring of G^α if and only if $\vec{G}$ has no voltage assignments equal to 0.*

Proof: If e is an arc in $\vec{G}$ directed from v to w with voltage equal to 0, then for any $i \in \mathbb{Z}_n$, e_i is an arc in $\vec{G}^\alpha$ from v_i to $w_{i+\alpha(e)} = w_i$. Since $\tilde{\alpha}(v_i) = i = \tilde{\alpha}(w_i)$, $\tilde{\alpha}$ is not a proper vertex-coloring.

Conversely, suppose no arc in $\vec{G}$ has voltage equal to 0, and let e_i be an arbitrary arc in $\vec{G}^\alpha$ from v_i to $w_{i+\alpha(e)}$. Since $\alpha(e) \neq 0$, $i \neq i+\alpha(e)$, and so, $\tilde{\alpha}(v_i) = i \neq i+\alpha(e) = \tilde{\alpha}(w_{i+\alpha(e)})$. Therefore, $\tilde{\alpha}$ is a proper vertex-coloring of $\vec{G}^\alpha$, and hence, G^α. ♢

Corollary 10.3.4: *If $\left\langle \vec{G}, \alpha \right\rangle$ is a $\mathbb{Z}_n$-voltage graph and $\vec{G}$ has no voltage assignments equal to 0, then $\chi(G^\alpha) \leq n$.*

Irregular Colorings

A proper coloring of a graph allows one to distinguish a vertex from each of its neighbors by its color. An *irregular coloring* is a proper coloring that allows one to distinguish all vertices by looking at their closed neighborhoods.

DEFINITION: Given a vertex k-coloring $f : V_G \to \{0, \ldots, k-1\}$ of a graph G, the ***color code of a vertex*** v is the k-tuple $f_v = (f_v^0, f_v^1, \ldots, f_v^{k-1})$, where f_v^i, $0 \leq i \leq k-1$, is the number of neighbors of v assigned the color i.

DEFINITION: A proper vertex-coloring f of a graph is an ***irregular coloring*** if for any two distinct vertices $v, w \in V_G$, $f(v) \neq f(w)$ or $f_v \neq f_w$.

Example 10.3.2: (continued) The voltage-coloring $\tilde{\alpha}$ from Figure 10.3.3 and shown below in Figure 10.3.4 with color codes is an irregular 3-coloring. Since every vertex v has degree 2, if $\tilde{\alpha}(v) = i$, then the ith coordinate of $\tilde{\alpha}_v$ must be 0, and the other two coordinates are 0 and 2, 1 and 1, or 2 and 0. This example achieves every possible color code for vertices of every color, and thus has the maximum number of vertices of degree 2 possible for an irregular 3-coloring. This was actually evident from the voltage assignments in G as we will show.

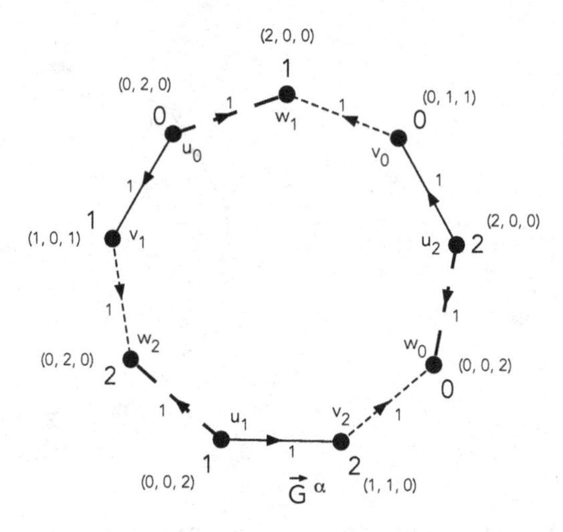

Figure 10.3.4 An irregular 3-coloring with color codes.

DEFINITION: Suppose $\left\langle \vec{G}, \alpha \right\rangle$ is a $\mathbb{Z}_n$-voltage graph. The ***voltage code*** of a vertex v in $\vec{G}$ is an n-tuple $\alpha_v = (\alpha_v^0, \alpha_v^1, \ldots, \alpha_v^{n-1})$, defined as follows. If n is even and $i = n/2$, then α_v^i is the number of arcs with voltage i incident on v. Otherwise, α_v^i is the number of arcs with voltage i directed from v plus the number of arcs with voltage $n - i$ directed to v. The ***voltage degree*** of a vertex v is the sum of the coordinates in α_v.

Example 10.3.3: A voltage graph and its covering graph are shown in Figure 10.3.5. The self-loop on vertex u has voltage 1, and it contributes 2 to the voltage degree of u, since it adds 1 to both α_u^1 and α_u^3. The self-loop on vertex v has a voltage of $2 = 4/2$, and it contributes 1 to the voltage degree, adding 1 only to α_v^2. The voltage degree of u is 3 and the voltage degree of v is 2, corresponding to the degrees of the vertices in the fibers over u and v, respectively.

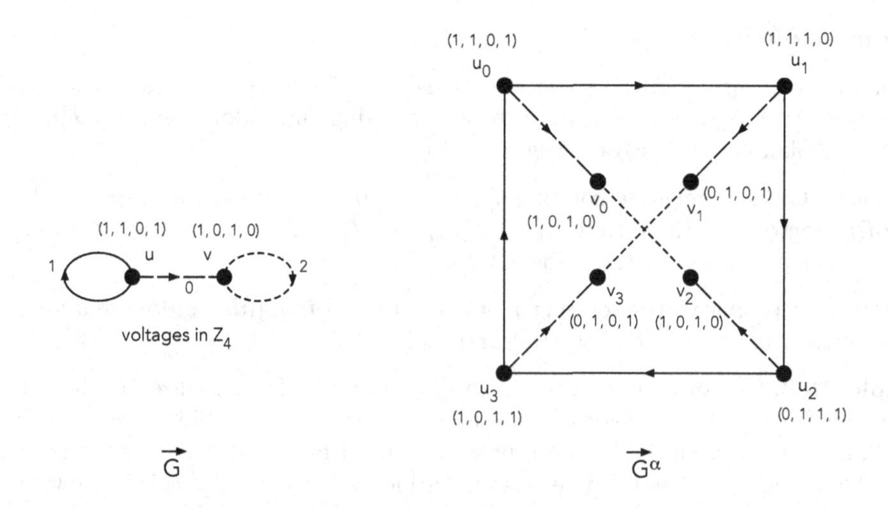

Figure 10.3.5 A $\mathbb{Z}_4$-voltage graph with voltage codes and a voltage-coloring with color codes.

DEFINITION: Given a voltage graph $\left\langle \vec{G}, \alpha \right\rangle$, the voltage assignment α is ***irregular*** if no two vertices have the same voltage code.

Proposition 10.3.5: *Suppose $\left\langle \vec{G}, \alpha \right\rangle$ is a $\mathbb{Z}_n$-voltage graph and $\tilde{\alpha}$ is the voltage-coloring of G^α. For all $i, j \in \mathbb{Z}_n$ and $u \in V_G$, $\tilde{\alpha}^i_{u_j} = \alpha^{i-j}_u$.*

Proof: If n is even, and $i - j = n/2$, then α^{i-j}_u is the number of arcs in the voltage graph incident on u with voltage $i - j$. Each of these arcs that is a self-loop corresponds to a bidirected arc in $\vec{G}^\alpha$ between u_j and u_i, and thus, adds one to the value of $\tilde{\alpha}^i_{u_j}$. Each arc that is not a self-loop corresponds to an arc in $\vec{G}^\alpha$ incident on u_j and another vertex with index i, also adding one to the value of $\tilde{\alpha}^i_{u_j}$. Thus, in this case, $\tilde{\alpha}^i_{u_j} = \alpha^{i-j}_u$.

If $i - j \neq n/2$, then α^{i-j}_u is the number of arcs in the voltage graph directed from u with voltage $i - j$ plus the number of arcs in the voltage graph directed to u with voltage $j - i$. In the first case, the arc corresponds to an arc in $\vec{G}^\alpha$ from u_j to some vertex with index i and in the second case, it corresponds to an arc directed from a vertex with index i to u_j. In both cases, the arc in $\vec{G}^\alpha$ contributes one to the value of $\tilde{\alpha}^i_{u_j}$, and hence $\tilde{\alpha}^i_{u_j} = \alpha^{i-j}_u$. ◊

Example 10.3.3, continued: Moreover, the color codes of u_0 and v_0 are identical to the voltage codes of u and v, respectively, and the color codes of u_i and v_i, $i = 1, 2, 3$, are $(\alpha^{-i}_u, \alpha^{1-i}_u, \alpha^{2-i}_u, \alpha^{3-i}_u)$ and $(\alpha^{-i}_v, \alpha^{1-i}_v, \alpha^{2-i}_v, \alpha^{3-i}_v)$, respectively. That is, the color codes of u_i and v_i are cyclic permutations of the voltage codes of u and v.

Theorem 10.3.6: *Suppose $\left\langle \vec{G}, \alpha \right\rangle$ is a simple $\mathbb{Z}_n$-voltage graph with no voltages equal to 0, and let $\tilde{\alpha}$ be the voltage-coloring of G^α.*

The voltage-coloring $\tilde{\alpha}$ is irregular if and only if the voltage assignment α is irregular.

Proof: Since no voltage is equal to 0, $\tilde{\alpha}$ is a proper coloring of G^α, by Proposition 10.3.3.

If α is not irregular, then there are two vertices v and w in G for which $\alpha_v = \alpha_w$. In this case, $\tilde{\alpha}(v_0) = 0 = \tilde{\alpha}(w_0)$ and $\tilde{\alpha}_{v_0} = \tilde{\alpha}_{w_0}$. Therefore, $\tilde{\alpha}$ is not irregular.

Conversely, if $\tilde{\alpha}$ is not irregular, then G^α has two vertices v_j and w_j, for which $\tilde{\alpha}_{v_j} = \tilde{\alpha}_{w_j}$. Therefore, by Proposition 10.3.5, for each $i \in \mathbb{Z}_n$, $\alpha^{i-j}_v = \alpha^{i-j}_w$. But this means that $\alpha_v = \alpha_w$, and hence, α is not irregular. ◊

The following upper bound for the number of vertices of degree r was established in [RaZh07].

Theorem 10.3.7: *If a graph has an irregular k-coloring, where $k \geq 2$, then the number of vertices of degree r is at most $k\binom{r+k-2}{r}$.*

Proof: Suppose f is an irregular k-coloring of a graph G with $f(v) = 0$ for some vertex v of degree r. Let $f_v = (0, f^1_v, f^2_v, \ldots, f^{k-1}_v)$ be one of the possible color codes for vertex v. This color code corresponds uniquely to a binary string (shown below) of length $r + k - 2$ having exactly $k - 2$ zeros. For $i = 1, \ldots, k - 2$, f^i_v is the number of consecutive ones that immediately precede the ith zero, and f^{k-1}_v is the number of ones following the last zero.

$$\underbrace{1\cdots1}_{f^1_v \text{ ones}}\ 0\ \underbrace{1\cdots1}_{f^2_v \text{ ones}}\ 0\cdots0\ \underbrace{1\cdots1}_{f^{k-1}_v \text{ ones}}$$

Clearly, there are $\binom{r+k-2}{k-2}$ such binary strings, and hence, there are $\binom{r+k-2}{k-2}$ color codes for a vertex with color 0. Since, the number of color codes for a vertex of any other color is the same, the total number of color codes is $k\binom{r+k-2}{k-2} = k\binom{r+k-2}{r}$. ◇

The following results are immediate.

Corollary 10.3.8: *If f is an irregular k-coloring of a cycle C_n, then $n \leq k\binom{k}{2}$.*

Corollary 10.3.9: *If f is an irregular 3-coloring of an r-regular, n-vertex graph, then $n \leq 3(r+1)$.*

Irregular Colorings of r-regular graphs

DeBruijn graphs can be used to create irregular k-colorings of n-cycles with $n = k\binom{k}{2}$, as shown in [AnBaBr09]. In [AnViYe12], $\mathbb{Z}_k$-voltage graphs were used to create irregular k-colorings of r-regular graphs, that achieved the bound in Proposition 10.3.7. The final examples of this section use $\mathbb{Z}_k$-voltage graphs to achieve the bounds for stated in Corollaries 10.3.8 and 10.3.9.

Example 10.3.4: The voltage assignment for the $\mathbb{Z}_3$-voltage graph in Figure 10.3.6 is irregular and has no voltages equal to 0. Thus, the voltage-coloring of the covering graph is an irregular 3-coloring of a 2-regular graph with $3 \cdot 3 = 9$ vertices, by Theorem 10.3.6. Since the voltage graph is connected and has a closed signed walk with net voltage 1, the

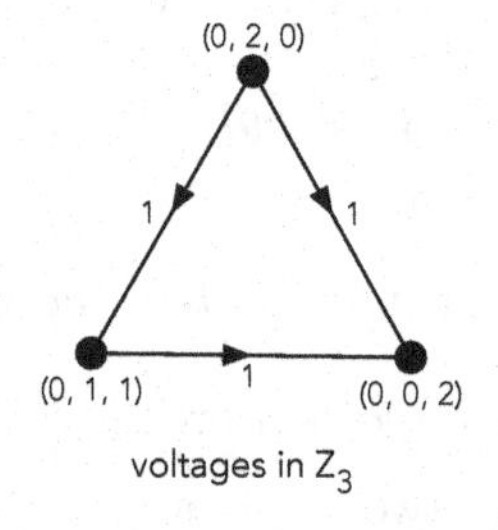

Figure 10.3.6 A $\mathbb{Z}_3$-voltage graph with an irregular voltage assignment.

covering graph is connected, and thus, a 9-cycle, as shown in Figure 10.3.3. By Corollary 10.3.8, this is the maximum possible size for a cycle with an irregular 3-coloring.

Example 10.3.5: The voltage assignment for the $\mathbb{Z}_4$-voltage graph in Figure 10.3.7 is irregular and has no voltages equal to 0. Thus, the voltage-coloring of the covering graph is an irregular 4-coloring of a 2-regular graph with $4 \cdot 6 = 24$ vertices, by Theorem 10.3.6.

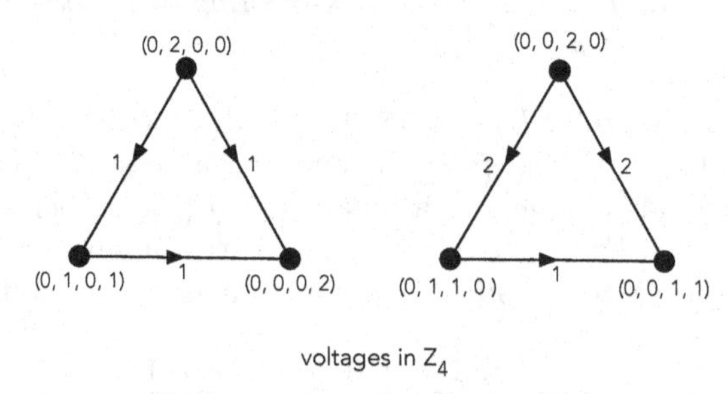

Figure 10.3.7 A $\mathbb{Z}_4$-voltage graph with an irregular voltage assignment.

The voltage graph is not connected, and therefore the covering graph is not connected. However, the net voltage of the walk going once around (counterclockwise) either of the 3-cycles is 1. Therefore, the covering graph of each component is connected and the covering graph of the whole voltage graph consists of two 12-cycles, as shown in Figure 10.3.8. We

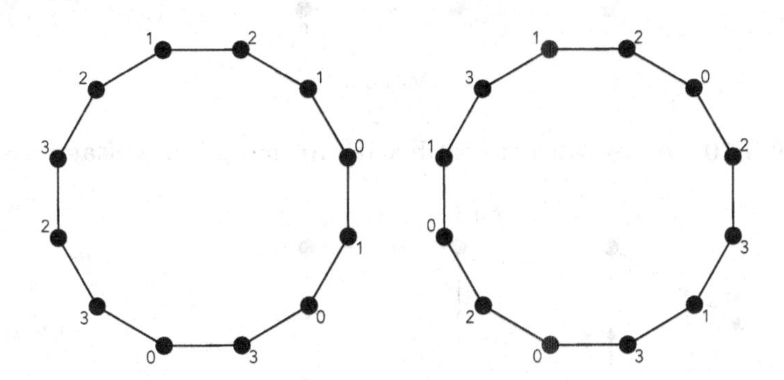

Figure 10.3.8 A $\mathbb{Z}_4$-voltage graph with an irregular voltage assignment.

can adjust the covering graph to be a cycle, rather than having two components, by noticing that both cycles have an edge from a vertex with color 0 to a vertex with color 1. If these edges are replaced by edges joining the two cycles, the color codes remain the same, as shown in Figure 10.3.9. This gives us an irregular 4-coloring of a cycle with 24 vertices, which is the bound given in Corollary 10.3.8.

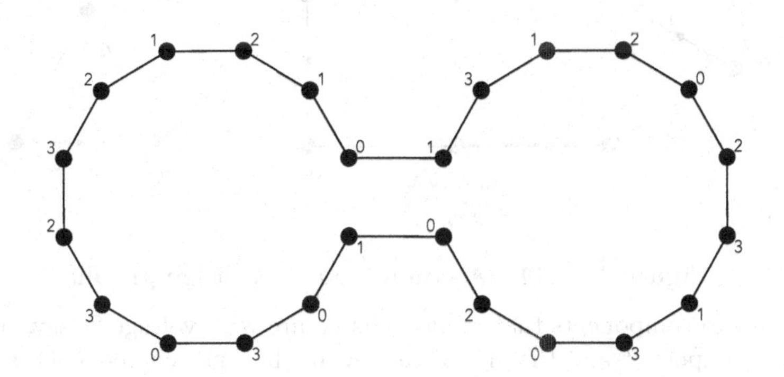

Figure 10.3.9 A $\mathbb{Z}_4$-voltage graph with an irregular voltage assignment.

Example 10.3.6: The voltage assignment for the $\mathbb{Z}_5$-voltage graph in Figure 10.3.10 is irregular and has no voltages equal to 0. Thus, the voltage-coloring of the covering graph is an irregular 5-coloring of a 2-regular graph with $5 \cdot 10 = 50$ vertices, by Theorem 10.3.6. However, since it is not connected, the covering graph is not connected, and so is a union of disjoint cycles. Since the two components on the left both have an arc with voltage one, it is possible to create one component with vertices having the same voltage codes, by exchanging the heads of the two arcs. The two components on the left side both have arcs with voltage 1. If we switch the head vertices of these two arcs, then this joins the two components, as shown in Figure 10.3.11.

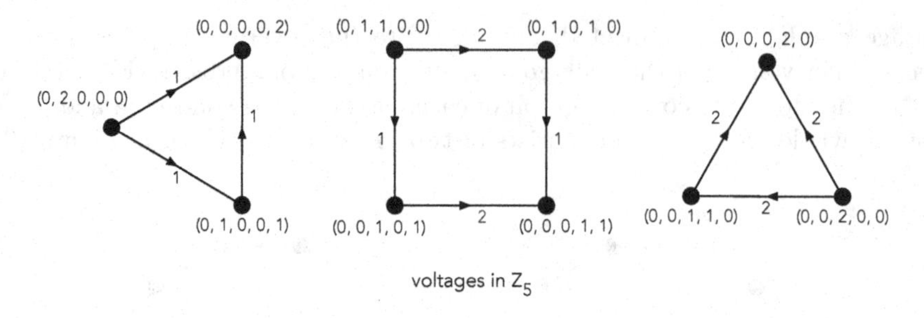

Figure 10.3.10 A $\mathbb{Z}_5$-voltage graph with an irregular voltage assignment.

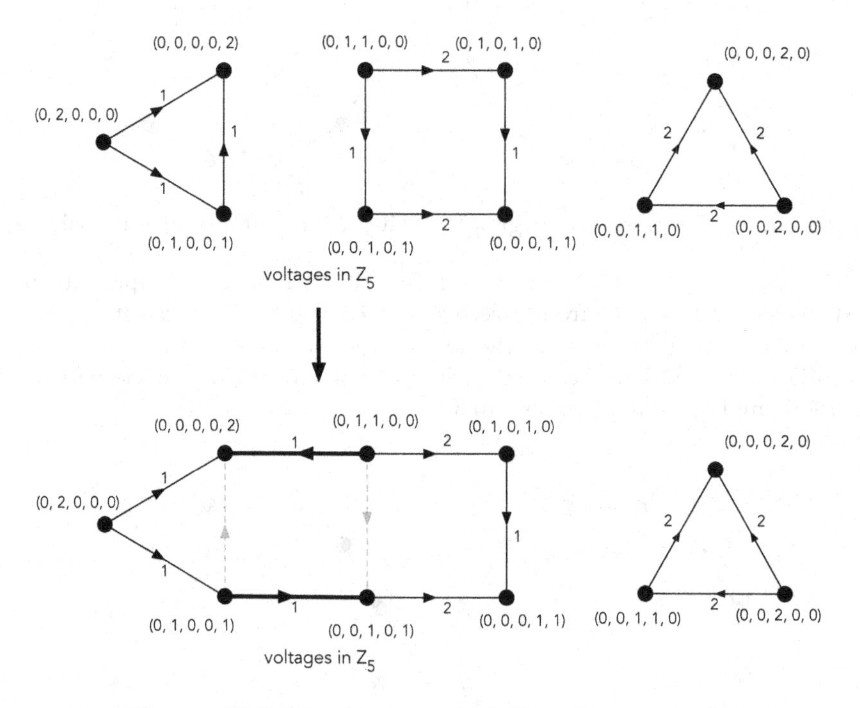

Figure 10.3.11 A connected $\mathbb{Z}_5$-voltage graph.

Each of the two components that remain has an arc with voltage 2 (say, the arc on the top of the left component and the arc on the left of the right component), and the heads of these two arcs can also be switched. In Figure 10.3.12, the resulting voltage graph is

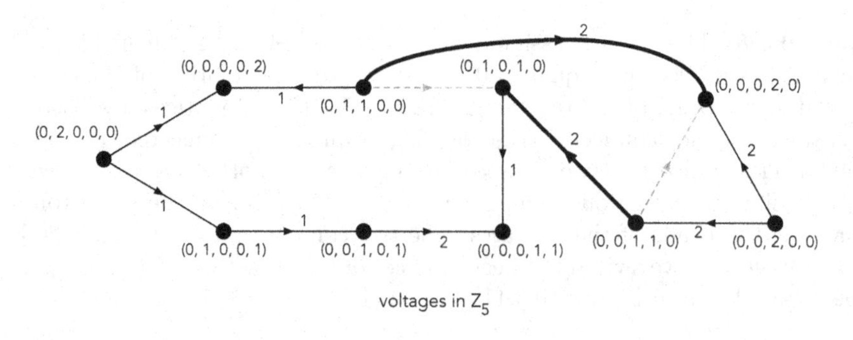

Figure 10.3.12 A connected $\mathbb{Z}_5$-voltage graph.

connected, and the signed walk going once around the cycle (clockwise) has net voltage 1. Therefore, the covering graph of this new $\mathbb{Z}_5$-voltage graph is a 50-cycle. The voltage-coloring is an irregular 5-coloring (by Theorem 10.3.6) on a cycle with as many vertices as possible (by Corollary 10.3.8).

Example 10.3.7: The voltage assignment for the $\mathbb{Z}_3$-voltage graph in Figure 10.3.13 is irregular and every vertex has voltage degree 3. Since it is connected and has a closed signed walk with net voltage 1, the covering graph is a connected 3-regular graph on 12 vertices with an irregular 3-coloring. This satisfies the bound in Corollary 10.3.9.

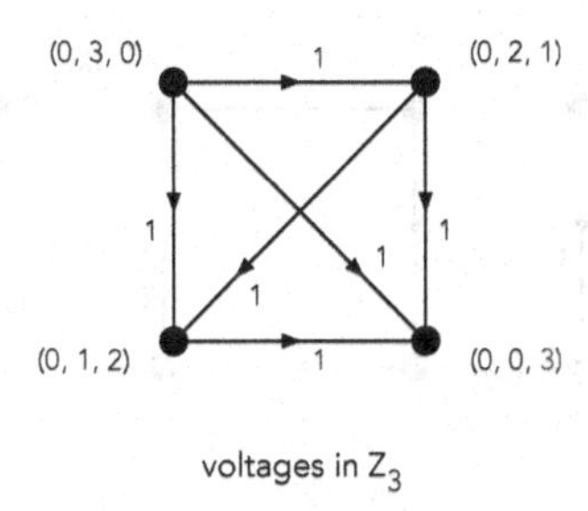

Figure 10.3.13 A $\mathbb{Z}_3$-voltage graph.

EXERCISES for Section 10.3

10.3.1 Specify a 4-regular graph with an irregular 3-coloring, by assigning directions and voltages to the edges of the complete graph K_5.

10.3.**2** Generalize the results of Examples 10.3.2 and 10.3.7, and Exercise 10.3.1, to specify an r-regular graph with an irregular 3-coloring, having the maximum number of vertices possible.

10.3.**3** What is the covering graph for the $\mathbb{Z}_2$-voltage graph B_2, where both self-loops have voltage 1? What can you say about the covering graph and its covering coloring?

10.3.**4** Find the voltage codes for each vertex in the $\mathbb{Z}_6$-voltage graph shown below. What can you say about the covering graph and its derived coloring? Is there a connected graph that has the same properties?

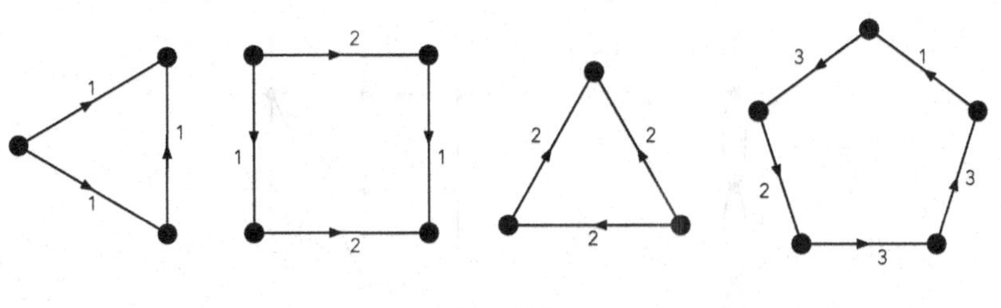

10.3.**5** Find the voltage codes for each vertex in the $\mathbb{Z}_7$-voltage graph shown below. What can you say about the covering graph and its derived coloring? Is there a connected graph that has the same properties?

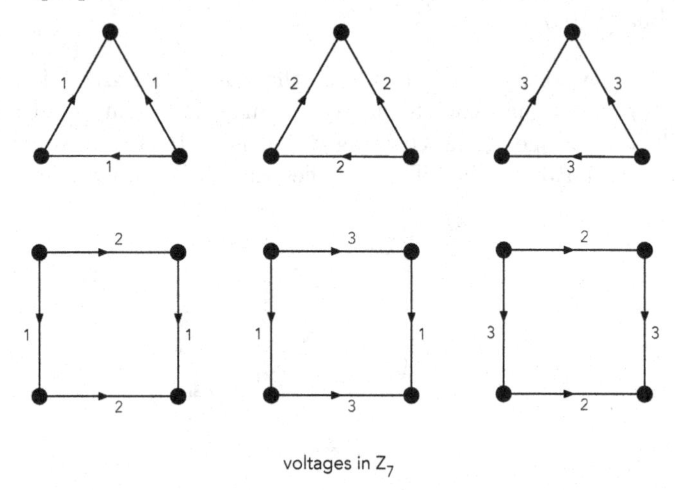

10.3.**6** Generalize the results of Examples 10.3.5 and Exercises 10.3.3 and 10.3.4.

10.3.**7** Generalize the results of Examples 10.3.4 and 10.3.6 and Exercise 10.3.5.

10.3.**8** Prove Part 2 of Proposition 10.3.1.

10.3.**9** Label the graph on the left with voltages in $\mathbb{Z}_3$ so that the covering graph is isomorphic to $C_3 \times C_3$, which has been drawn in two different ways on the right. Conclude that $\chi(C_3 \times C_3) = 3$. Note that each graph is drawn on a flat torus, so that edges going off the top of the picture, enter on the bottom, and edges leaving from the right re-enter on the left.

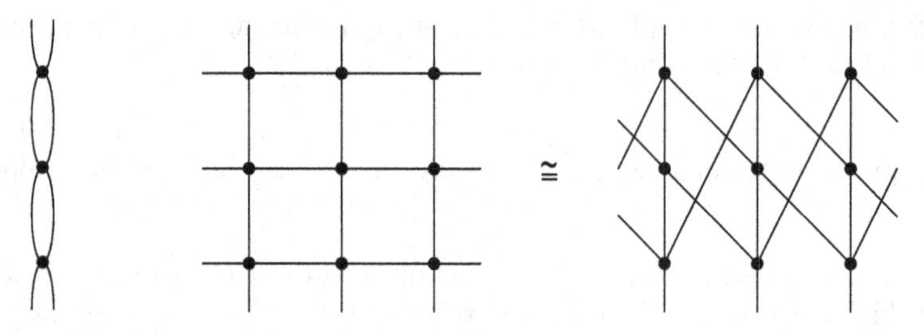

10.3.**10** Find the chromatic number of the graph on the right. Conclude that it is not a covering graph for any $\mathbb{Z}_3$-voltage graph defined on the graph on the left. Note that each graph is drawn on a flat torus, so that edges going off the top of the picture enter on the bottom, and edges leaving from the right, re-enter on the left.

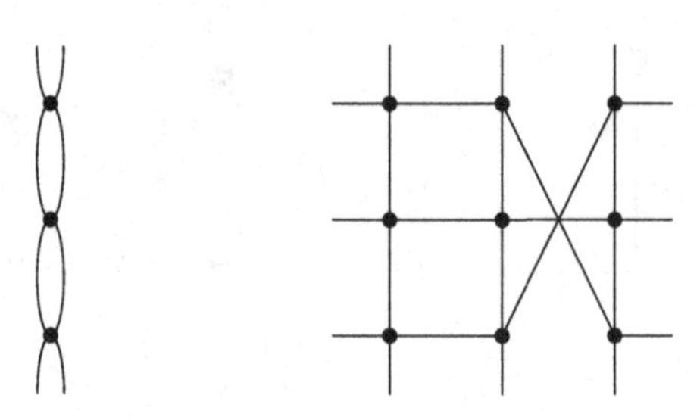

10.4 GENERAL VOLTAGE GRAPHS

The power of voltage graphs is increased when the algebraic structure from which the voltages are selected is permitted to be an arbitrary group (see Appendix A.4), possibly non-Abelian. When the group is Abelian, additive notation is used for the group operation, but in general, multiplicative notation is used and ε is used for the *identity element* of the group. The *order* of a group element b is the smallest nonnegative integer k for which $b^k = \varepsilon$.

Remark: The following definitions are natural extensions of those for $\mathbb{Z}_n$-voltage graphs.

DEFINITION: Let $\vec{G} = (V, E)$ be a digraph, and let $\mathcal{B}$ be a group. A ***$\mathcal{B}$*-voltage assignment** on $\vec{G}$ is a function α that assigns to every arc $e \in E$ an element $\alpha(e) \in \mathcal{B}$. The element $\alpha(e) \in \mathcal{B}$ is called the **voltage** on e.

DEFINITION: A ***$\mathcal{B}$*-voltage graph** is a pair $\left\langle \vec{G}, \alpha \right\rangle$ such that $\vec{G} = (V, E)$ is a digraph, $\mathcal{B}$ is a group, and α is a $\mathcal{B}$-voltage assignment on $\vec{G}$.

DEFINITION: The **voltage group** of a voltage graph $\left\langle \vec{G}, \alpha \right\rangle$ is the group $\mathcal{B}$ from which the voltages are assigned.

DEFINITION: The ***covering digraph*** of the regular-voltage graph $\left\langle \vec{G}, \alpha \right\rangle$ is the directed graph $\vec{G}^\alpha = (V^\alpha, E^\alpha)$, with

$$V^\alpha = \{v_b \mid v \in V(G) \text{ and } b \in \mathcal{B}\} \quad \text{and} \quad E^\alpha = \{e_b \mid e \in E(G) \text{ and } b \in \mathcal{B}\}$$

such that if an arc e in a voltage graph is directed from vertex u to vertex v, then the arc e_b is directed from vertex u_b to vertex $v_{b\alpha(e)}$. The ***covering graph*** is the underlying graph of the covering digraph $\vec{G}^\alpha$. It is denoted G^α.

DEFINITION: Let X be a generating set of size r of a group $\mathcal{B}$. The ***Cayley digraph*** $\vec{C}(\mathcal{B}, X)$ is the covering digraph for the voltage graph $\langle B_r, \alpha \rangle$, where B_r is the bouquet with r self-loops, and α assigns each voltage in X to one of the self-loops. The vertex-set and arc-set of $\vec{C}(\mathcal{B}, X)$ are

$$V_{\vec{C}(\mathcal{B},X)} = \mathcal{B} \quad \text{and} \quad E_{\vec{C}(\mathcal{B},X)} = \{x_b \mid x \in X, b \in \mathcal{B}\}$$

respectively. Arc x_b joins vertex b to vertex bx.
The ***Cayley graph*** $C(\mathcal{B}, X)$ is the *underlying graph* of the Cayley digraph $\vec{C}(\mathcal{B}, X)$.

DEFINITION: Let $\left\langle \vec{G}, \alpha \right\rangle$ be a $\mathcal{B}$-voltage graph.

1. For each vertex $v \in V_{\vec{G}}$, the vertex-set $\{v_b \mid b \in \mathcal{B}\}$ in the covering graph G^α is called the ***(vertex) fiber*** over v and is denoted $\tilde{v}$.

2. For each edge $e \in E_G$, the edge-set $\{e_b | b \in \mathcal{B}\}$ in the covering graph G^α is called the ***(edge) fiber*** over e and is denoted $\tilde{e}$.

DEFINITION: ***Bidirected-arc convention***: Let e be an arc directed from vertex u to vertex v in a $\mathcal{B}$-voltage graph $\left\langle \vec{G}, \alpha \right\rangle$. If the voltage y on arc e has order 2, then the pair of arcs in the covering graph directed from vertex v_b to vertex v_{by} and from vertex v_{by} to vertex v_b is replaced by a single bidirected arc between v_b and v_{by}. The ***underlying graph of a digraph with bidirected arcs*** has a single edge for each bidirected arc.

Example 10.4.1: Figure 10.4.1 illustrates that the 3×4 wraparound mesh is a Cayley graph for the group $\mathbb{Z}_3 \times \mathbb{Z}_4$.

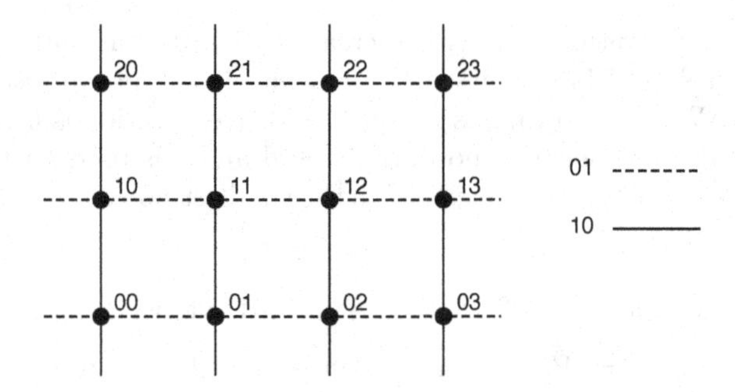

Figure 10.4.1 The Cayley graph $C(\mathbb{Z}_3 \times \mathbb{Z}_4, \{01, 10\})$.

NOTATION: The notation used here for an element of a direct sum of k small cyclic groups is a string of k digits, rather than a k-tuple. For instance, 20 stands for the element $(2, 0)$. This convention avoids cluttering the drawings with parentheses and commas.

Example 10.4.2: The voltage 10 in the group $\mathbb{Z}_2 \times \mathbb{Z}_4$ has order 2, and thus produces bidirected arcs in the Cayley digraph $\vec{C}(\mathbb{Z}_2 \times \mathbb{Z}_4, \{10, 01\})$.

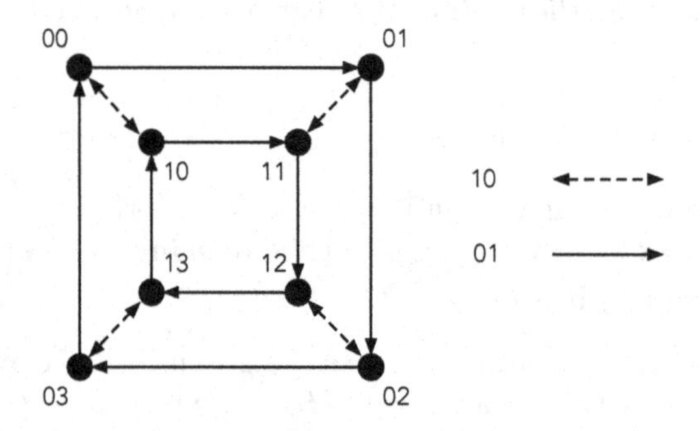

Figure 10.4.2 The Cayley digraph $\vec{C}(\mathbb{Z}_2 \times \mathbb{Z}_4, \{10, 01\})$.

Example 10.4.3: The dihedral group $\mathcal{D}_4$ is describable as the group of rigid-body motions on the unit square. Let r denote a 90° clockwise rotation and let s denote a reflection through a vertical axis. Then the elements of $\mathcal{D}_4$ are $\{\varepsilon, r, r^2, r^3, s, rs, r^2s, r^3s\}$. Figure 10.4.3 shows a Cayley digraph for $\mathcal{D}_4$ with generating set $\{r, s\}$.

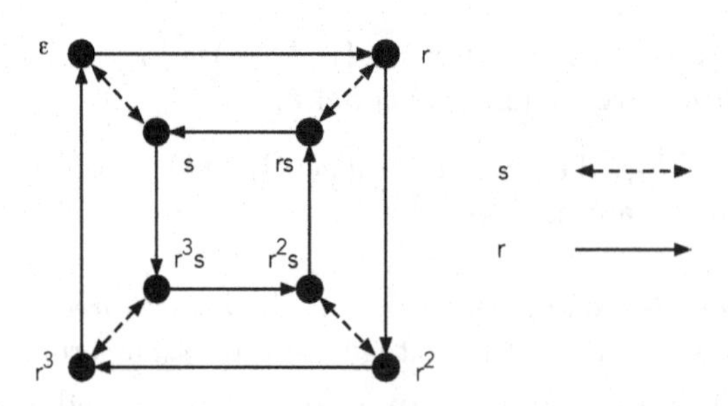

Figure 10.4.3 A Cayley digraph for the dihedral group $\mathcal{D}_4$.

Example 10.4.4: The hypercube graph Q_n can be specified as a Cayley graph for the group $(\mathbb{Z}_2)^n$, where the generating set contains all bitstrings with exactly one 1. Since every voltage in the generating set has order 2, every edge is a bidirected arc in the Cayley digraph, as shown in Figure 10.4.4.

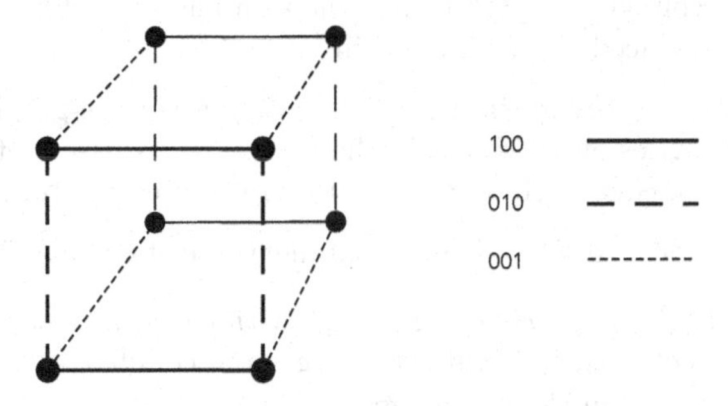

Figure 10.4.4 The Cayley graph $C(\mathbb{Z}_2 \times \mathbb{Z}_2 \times \mathbb{Z}_2, \{100, 010, 001\})$.

Remark: Examples 10.4.2, 10.4.3, and 10.4.4 specify Q_3 as Cayley graph for the groups $\mathbb{Z}_2 \times \mathbb{Z}_4$, $\mathcal{D}_4$, and $(\mathbb{Z}_2)^3$, respectively. Finding a polynomial-time algorithm to decide whether a given regular graph is a Cayley graph, or proving that this decision problem is NP-complete, is a research problem. Accordingly, finding all the different ways in which a given regular graph can be specified as a Cayley graph is often regarded as a difficult problem.

Example 10.4.5: The Petersen graph is a 3-regular graph. We leave it as an exercise to show that it cannot be specified as a Cayley graph. It can be specified as a $\mathbb{Z}_5$-voltage graph as shown in Figure 10.4.5.

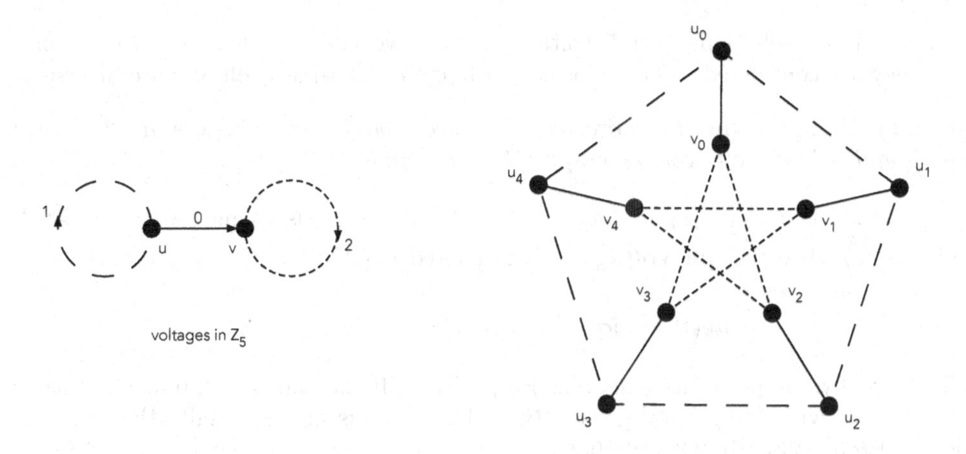

Figure 10.4.5 A $\mathbb{Z}_5$-voltage-graph specification of the Petersen graph.

Properties of Covering Graphs

All of the definitions and propositions about cyclic voltage graphs from the first three sections of this chapter hold for general voltage graphs, with little or no modification in either the statements or their proofs, other than generalizing the operation. In particular, the definition for net voltage is copied below: the sum has been changed to a product, and additive inverses are replaced by generalized inverses.

DEFINITION: The ***natural projection*** $\pi : \vec{G}^\alpha \to \vec{G}$ of a voltage-graph construction is the graph mapping that carries every vertex in the fiber $\tilde{v}$ to its base vertex v and every edge in the fiber $\tilde{e}$ to its base edge e. (In effect, the natural projection "erases" the subscripts.)

The proofs of Theorems 10.4.1 and 10.4.2 parallel those of Propositions 10.1.1 and 10.1.2.

Theorem 10.4.1: *If $\langle G, \alpha \rangle$ is a $\mathcal{B}$-voltage graph with $|\mathcal{B}| = n$. Then there are n times as many vertices and directed edges (bidirected edges are counted as two directed edges) in the covering digraph $\vec{G}^\alpha$ as in the voltage graph $\vec{G}$.*

Remark: Another way to express Theorem 10.4.1 is to say that the natural projection is n-to-1, for both vertices and edges.

Theorem 10.4.2: *Let $\left\langle \vec{G}, \alpha \right\rangle$ be a $\mathcal{B}$-voltage graph, and let v be any vertex of G. Then the outdegree and indegree of each vertex v_j in the fiber $\tilde{v}$ are equal to the outdegree and indegree, respectively, of vertex v.*

Remark: Another way to express Theorem 10.4.2 is to say that the natural projection preserves degree.

DEFINITION: A $\mathcal{B}$-voltage graph $\langle G, \alpha \rangle$ is ***simple*** if G has no self-loops with voltage the identity, no multi-arcs with the same voltage, and no oppositely directed arcs whose voltages are inverses.

Remark: If two self-loops in a $\mathcal{B}$-voltage graph have voltages that are inverses of each other, they are considered to be "oppositely directed arcs whose voltages are inverses."

Corollary 10.4.3: *Using the bidirected-arc convention, the covering graph G^α is a simple graph if and only if the $\mathcal{B}$-voltage graph $\langle G, \alpha \rangle$ is simple.*

DEFINITION: If $W = \langle v_1, \delta_1 e_1, v_2, \delta_2 e_2, \ldots v_{n-1}, \delta_{n-1} e_{n-1}, v_n \rangle$ is a signed walk in a $\mathcal{B}$-voltage graph $\left\langle \vec{G}, \alpha \right\rangle$, then the **net voltage** of W, denoted $nv(W)$, is

$$nv(W) = [\alpha(e_1)]^{\delta_1} [\alpha(e_2)]^{\delta_2} \cdots [\alpha(e_{n-1})]^{\delta_{n-1}}$$

The next two propositions generalize Propositions 10.2.3 and 10.2.4, replacing the cyclic group $\langle \mathbb{Z}_n, + \rangle$ with the general group $\langle \mathcal{B}, \cdot \rangle$. Their proofs are essentially the same as the proofs of Propositions 10.2.3 and 10.2.4.

Proposition 10.4.4: *If $\left\langle \vec{G}, \alpha \right\rangle$ is a $\mathcal{B}$-voltage graph and W is a v_b-w_d signed walk in $\vec{G}^\alpha$, then $\pi(W)$ is a v-w signed walk in $\left\langle \vec{G}, \alpha \right\rangle$ with net voltage $nv(\pi(W)) = b^{-1}d$.*

Proposition 10.4.5: *Suppose W is a signed walk from v in a $\mathcal{B}$-voltage graph $\left\langle \vec{G}, \alpha \right\rangle$. For each b in $\mathcal{B}$, there is a directed walk $\tilde{W}$ from v_b in $\vec{G}^\alpha$ of the same length as W such that $\pi(\tilde{W}) = W$.*

Corollary 10.4.6: *Suppose $\left\langle \vec{G}, \alpha \right\rangle$ is a $\mathcal{B}$-voltage graph. If W is a closed signed walk in $\vec{G}$ with net voltage equal to the identity, then $\vec{G}^\alpha$ has a closed walk $\tilde{W}$ of the same length as W.*

Proof: Suppose W is a closed signed walk in $\vec{G}$ from vertex v with $nv(W) = \varepsilon$. By Proposition 10.4.5, there is a walk $\tilde{W}$ of the same length as W from v_ε. By Proposition 10.2.2, $\tilde{W}$ ends at a vertex in the fiber $\tilde{v}$ and by 10.4.4, this terminal vertex is v_ε. Therefore, $\tilde{W}$ is a closed walk. ◇

REVIEW FROM §1.2. The ***girth*** of a graph is the length of its shortest cycle.

DEFINITION: A nontrivial closed signed walk in a simple $\mathcal{B}$-voltage graph is ***terminal*** if its net voltage is the identity. A signed walk in a simple $\mathcal{B}$-voltage graph is ***minimal*** if it contains no terminal subwalks.

If a covering graph G^α is simple (using the bidirected arc convention), then a cycle $\langle v_1, e_1, v_2, e_2, \ldots, e_{n-1}, v_n \rangle$ (directions of edges ignored) corresponds uniquely to the closed signed walk $\langle v_1, \delta_1 e_1, v_2, \delta_2 e_2, \ldots, \delta_{n-1} e_{n-1}, v_n \rangle$, where $\delta_i = 1$ if e_i is directed from v_i to v_{i+1} and $\delta_i = -1$ if e_i is directed from v_{i+1} to v_i.

Corollary 10.4.7: *Suppose $\left\langle \vec{G}, \alpha \right\rangle$ is a simple $\mathcal{B}$-voltage graph. The girth of the covering graph is the minimum length of all terminal walks in $\vec{G}$ of length 3 or greater.*

Proof: Since G^α is a simple graph, $girth(G) \geq 3$. Suppose W is a closed signed walk in $\vec{G}^\alpha$ which corresponds to a cycle in G^α with length equal to $girth(G)$. Its projection, $\pi(W)$, has length equal to $girth(G)$ and $nv(\pi(W)) = \varepsilon$, by Proposition 10.4.4. Therefore, the minimum length of all terminal walks in $\vec{G}$ of length 3 or greater is less than or equal to the girth of G.

Conversely, suppose W is a terminal walk in $\vec{G}$ with length at least 3. Then, by Corollary 10.4.6, there is a closed walk $\tilde{W}$ in $\vec{G}^\alpha$ of the same length as W. Therefore, $girth(G)$ is less than or equal to the minimum length of all terminal walks in $\vec{G}$ of length 3 or greater. ◇

Proposition 10.4.8: *Suppose G is a Cayley graph $C(\mathcal{B}, X)$ with $|X| \geq 2$. If $\mathcal{B}$ is Abelian, then $girth(G) \leq 4$.*

Proof: Suppose G is a Cayley graph $C(\mathcal{B}, X)$ with $|X| \geq 2$. Let $x \neq y \in X$, and let the d and e be the self-loops in the $\mathcal{B}$-voltage graph with these voltages. If $\mathcal{B}$ is Abelian, then the closed walk $\langle d, e, -d, -e \rangle$ is a terminal walk of length 4 in the $\mathcal{B}$-voltage graph. The result follows by Corollary 10.4.7. ◇

Example 10.4.5, continued: The Petersen graph has girth 5. Therefore, by Proposition 10.4.8, it cannot be specified as a Cayley graph $C(\mathbb{Z}_{10}, X)$. The only other group (up to isomorphism) with 10 elements is the dihedral group D_5. If a Cayley graph $C(D_5, X)$ specifies a 3-regular graph, then either X contains one element s of order 2 (a reflection) and one element r of order 5 (a rotation), or X contains three elements s_1, s_2, s_3 of order 2 (three reflections). In the first case, rs is a reflection and so has order 2. Since $rsrs = \varepsilon$, the D_5-voltage graph has a minimal, terminal walk of length 4 and so the Cayley graph has a cycle of length 4. In the second case, the Cayley graph has no odd cycles, since the product of an odd number of reflections is a reflection (not the identity). Therefore, the Petersen graph is not a Cayley graph.

Example 10.4.6: The website *https://www.jaapsch.net/puzzles/cayley.htm* specifies Platonic and Archimedean solids, prisms, and anti-prisms as Cayley graphs (where possible). The *icosahedron* has twelve vertices of degree 5 and twenty 3-sided faces. The site specifies this Platonic solid as the Cayley graph $C(A_4, \{(123), (234), (13)(24)\})$, where A_4 is the alternating group on four elements, as illustrated in Figure 10.4.6. If the three arcs in the A_4-voltage graph B_3 are labeled a, b, c with $\alpha(a) = (123)$, $\alpha(b) = (234)$, and $\alpha(c) = (13)(24)$, then the 3-cycles in the Cayley graph are generated by three terminal walks of length 3 in the A_4-voltage graph:

$$
\begin{aligned}
W_1 &= \langle a, a, a \rangle \\
&\qquad nv(W_1) = (123) \cdot (123) \cdot (123) = \varepsilon, \\
W_2 &= \langle b, b, b \rangle \\
&\qquad nv(W_2) = (234) \cdot (234) \cdot (234) = \varepsilon, \\
W_3 &= \langle -a, c, -b \rangle \\
&\qquad nv(W_3) = (123)^{-1} \cdot (13)(24) \cdot (234)^{-1} = (132)(13)(24)(243) = \varepsilon.
\end{aligned}
$$

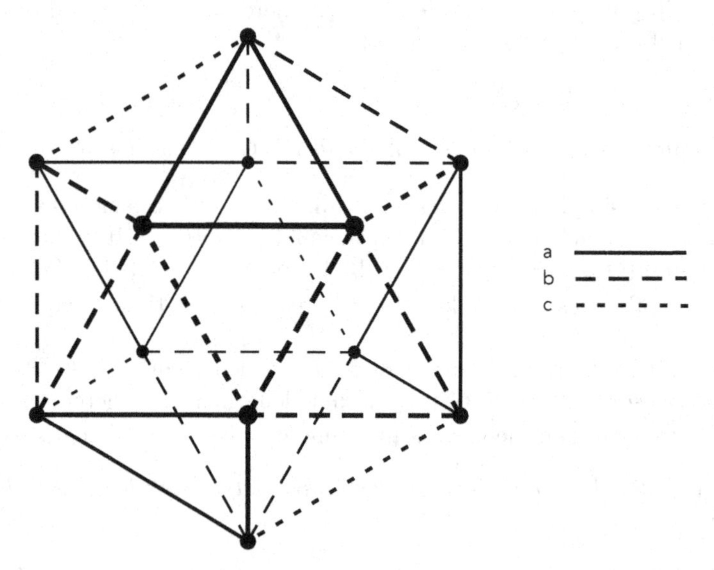

Figure 10.4.6 $C(A_4, \{(123), (234), (13)(24)\})$ **is the icosahedron graph.**

Example 10.4.7: The *truncated icosahedron* is created by "slicing the corners off the icosahedron," creating pentagonal faces at each corner and turning the original triangular faces into hexagons. This Archimedean solid has vertices of degree 3. Since this graph has girth 5, it cannot be specified by a Cayley graph using an Abelian group. The site specifies the truncated icosahedron as the Cayley graph $C(A_5, \{(12345), (23)(45)\})$. It is illustrated in Figure 10.4.7. If the two arcs in the A_5-voltage graph B_2 are labeled d, e with $\alpha(d) = (12345)$ and $\alpha(e) = (23)(45)$, then the two terminal walks in the voltage graph that generate the faces are:

$$
\begin{aligned}
W_1 &= \langle d, d, d, d, d \rangle \\
&\qquad nv(W_1) = (12345)^5 = \varepsilon, \\
W_2 &= \langle -d, e, -d, e, -d, e \rangle \\
&\qquad nv(W_2) = [(12345)^{-1} \cdot (23)(45)]^3 = [(15432) \cdot (23)(45)]^3 = (142)^3 = \varepsilon.
\end{aligned}
$$

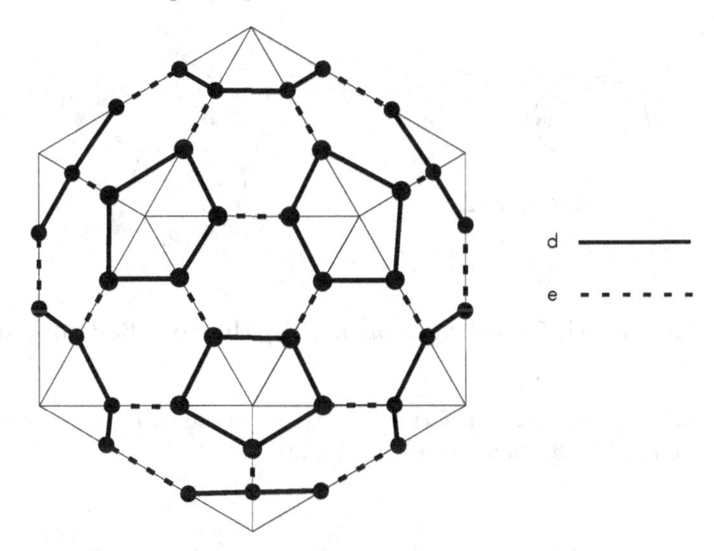

Figure 10.4.7 The truncated icosahedron graph is a Cayley graph.

Natural Transformation and Vertex-Transitivity

DEFINITION: Let $\langle G, \alpha \rangle$ be a $\mathcal{B}$-voltage graph, and let $a \in \mathcal{B}$. The ***natural transformation*** φ_a on the covering graph G^α is given by the rules: $\varphi_a(v_b) = v_{ab}$ and $\varphi_a(e_b) = e_{ab}$, for all $b \in B$.

Remark: When the voltage group is non-Abelian, it is important to observe that the automorphism φ_a acts by multiplying on the left. When the voltage group is Abelian, as in most of our examples, the natural transformation is given by addition to the subscripts.

Proposition 10.4.9: *Let $\langle G, \alpha \rangle$ be a $\mathcal{B}$-voltage graph, and let $a \in \mathcal{B}$. Then the natural transformation $\varphi_a : G^\alpha \to G^\alpha$ is an automorphism.*

Proof: Let arc $e \in E_G$ have tail u and head v. Then the endpoints of the edge e_b in G^α are u_b and $v_{b\alpha(e)}$. The definition of the natural transformation asserts that $\varphi_a(e_b) = e_{ab}$, and the definition of the voltage-graph construction prescribed that the endpoints of the edge e_{ab} are u_{ab} and $v_{ab\alpha(e)}$. The definition of the natural transformation asserts that $\varphi_a(u_b) = u_{ab}$ and $\varphi_a(v_{b\alpha(e)}) = v_{ab\alpha(e)}$. Thus, an arc e is directed from u to v if and only if $\varphi_a(e)$ is directed from $\varphi_a(u)$ to $\varphi_a(v)$, and hence, φ_a is a digraph homomorphism. Since φ_a has an inverse, namely $\varphi_{a^{-1}}$, it follows that φ_a is an automorphism. ♢

Example 10.4.8: The natural transformation φ_1 on the specified copy of K_5 in Figure 10.4.8 (which is identical to Figure 10.1.7) corresponds to rotation of $\frac{2\pi}{5}$ radians clockwise. More generally, the natural transformation φ_m corresponds to a rotation of $\frac{2m\pi}{5}$ radians clockwise.

Corollary 10.4.10: *Let $\langle G, \alpha \rangle$ be a $\mathcal{B}$-voltage graph, and let $v \in V(G)$. Then any two components of the covering graph G^α that contain a vertex from the vertex fiber $\tilde{v}$ are isomorphic.*

Proof: Let $v_a, v_b \in \tilde{v}$. Then $\varphi_{ba^{-1}}$ is an isomorphism from the component of v_a to the component of v_b. ♢

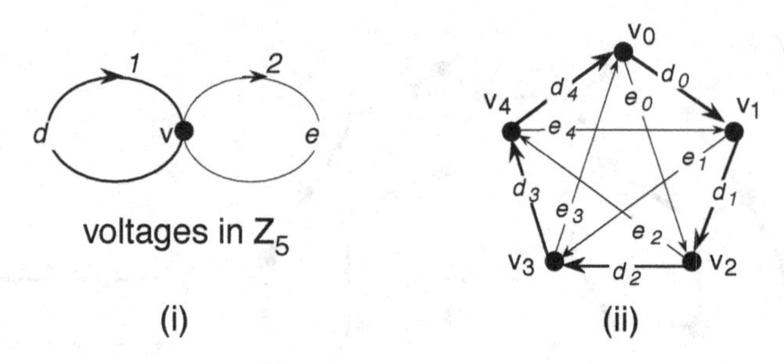

Figure 10.4.8 (i) $\mathbb{Z}_5$-voltages on B_2; (ii) the specified copy of K_5.

Example 10.4.9: The natural transformation φ_3 swaps the two components in the covering graph of Example 10.2.6, shown in Figure 10.4.9.

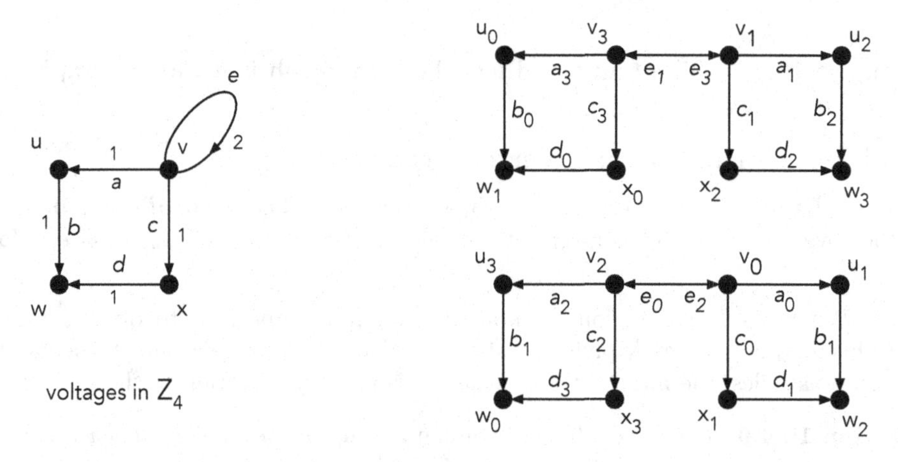

Figure 10.4.9 A $\mathbb{Z}_4$-voltage graph and its covering digraph.

REVIEW FROM §2.2: A graph G is ***vertex-transitive*** if for all vertex pairs $u, v \in V_G$, there is an automorphism of G that maps u to v.

Proposition 10.4.11: *For any group $\mathcal{B}$, the Cayley digraph $\vec{C}(\mathcal{B}, X)$ is vertex-transitive.*

Proof: Let $b, c \in \mathcal{B}$. Then the natural transformation $\varphi_{cb^{-1}}$ maps vertex b to vertex c in $\vec{C} \to (\mathcal{B}, X)$. ◊

EXERCISES for Section 10.4

10.4.1 Draw the Cayley graph for the given group and generating set.

(a) The cyclic group $\mathbb{Z}_9$ with generating set $\{1, 3\}$.

(b) The cyclic group $\mathbb{Z}_{10}$ with generating set $\{2, 5\}$.

(c) The group $\mathbb{Z}_2 \times \mathbb{Z}_3$ with generating set $\{10, 01\}$.

(d) The group $\mathbb{Z}_2 \times \mathbb{Z}_3 \times \mathbb{Z}_4$ with generating set $\{110, 101, 011\}$.

10.4.2 Draw the Cayley graph for the given group and generating set, and draw its voltage-graph specification.

(a) The group $\mathbb{Z}_2 \times \mathbb{Z}_3$ with generating set $\{11, 12\}$.

(b) The dihedral group $\mathcal{D}_5$ of rigid-body motions on a regular pentagon, using a 72° rotation r and a reflection s through the vertical axis as generators.

(c) The symmetric group Σ_4 of permutations on the set $\{1, 2, 3, 4\}$, with generating set $\{(1\ 2\ 3\ 4), (1\ 2)(3)(4)\}$.

(d) The *alternating group* $\mathcal{A}_4$ of permutations on the set $\{1, 2, 3, 4\}$, with generating set $\{(1\ 2\ 3)(4), (1\ 2\ 4)(3)\}$.

10.4.3 Draw the covering graph for the given regular-voltage graph.

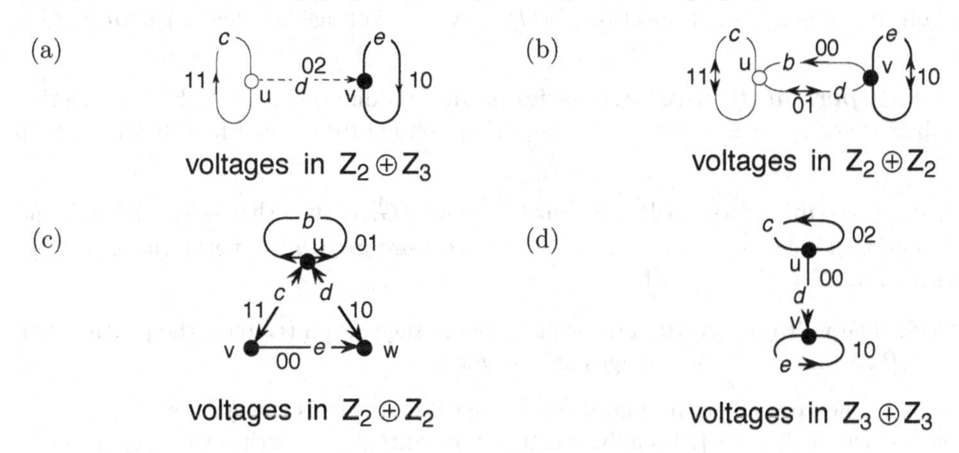

10.4.4 Give a geometric description of the effect produced by the given natural transformation on the covering graph of the given voltage graph.

(a) Voltage graph of Exercise 10.4.1(a) and transformation φ_2

(b) Voltage graph of Exercise 10.4.1(b) and transformation φ_3.

(c) Voltage graph of Exercise 10.4.2(b) and transformation φ_s.

10.4.5 For the covering graph of the given voltage graph, write the vertex-permutation and the edge-permutation of the given natural transformation.

(a) Voltage graph of Exercise 10.4.3(a) and transformation φ_{11}.

(b) Voltage graph of Exercise 10.4.3(b) and transformation φ_{10}.

(c) Voltage graph of Exercise 10.4.3(c) and transformation φ_{01}.

10.4.6 Let $\langle G, \alpha \rangle$ be a regular-voltage graph. Prove that the composition $\varphi_a \circ \varphi_b$ is the natural automorphism φ_{ab}.

10.5 PERMUTATION VOLTAGES

Modifying the covering graph so that subscripts may be the objects of any permutation group is a powerful way to expand the class of graphs that can be specified by voltage graphs.

Permutation-Voltage Specifications

The terminology for each detail of a permutation-voltage graph is similar to the terminology for the corresponding detail of a $\mathcal{B}$-voltage graph.

DEFINITION: Let $\mathcal{P} = [P : Y]$ be a permutation group. A ***$\mathcal{P}$-voltage*** on an edge of a digraph is a label on that edge by a permutation $\pi \in P$. Any such voltage is called a ***permutation voltage***.

DEFINITION: A ***permutation-voltage assignment*** on a digraph $\vec{G} = (V, E)$ is a function α that labels every arc $e \in E$ with a permutation voltage from some permutation group $\mathcal{P} = [P : Y]$.

DEFINITION: A ***permutation-voltage graph*** is a pair $(\vec{G}, \alpha)$ such that $\vec{G}$ is a digraph and α is a permutation-voltage assignment on $\vec{G}$. A ***$\mathcal{P}$-voltage graph*** is a permutation-voltage graph with voltages in $\mathcal{P} = [P : Y]$.

DEFINITION: The ***voltage group*** of a permutation-voltage graph $(\vec{G}, \alpha)$ is the permutation group $\mathcal{P} = [P : Y]$ in which the voltages are assigned.

Most often, the group for a permutation-voltage assignment is the symmetric group Σ_n acting on the set $\{1, 2, \ldots, n\}$. Usually, permutation voltages are written in disjoint cycle form.

Example 10.5.1: Figure 10.5.1 shows (i) a permutation-voltage graph and (ii) the digraph it specifies (i.e., the *covering digraph*).

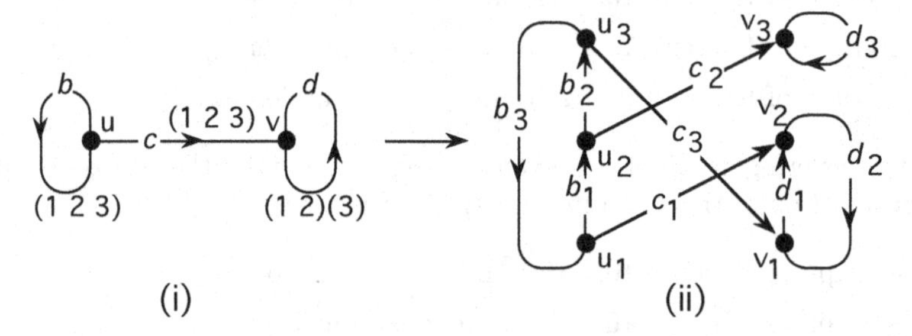

Figure 10.5.1 (i) A Σ_3-voltage graph; (ii) the covering digraph.

Observe in the voltage graph that arc c joins vertex u to vertex v, and that the voltage on arc c is the permutation (1 2 3). Observe in the covering digraph that arc c_1 joins vertex u_1 to vertex v_2, since the permutation (1 2 3) maps the object 1 to the object 2. Similarly, arc c_2 joins vertex u_2 to vertex v_3, since (1 2 3) maps the object 2 to the object 3, and arc c_3 joins vertex u_3 to vertex v_1, since (1 2 3) maps the object 3 to the object 1.

DEFINITION: The ***covering digraph*** of the $\mathcal{P}$-voltage graph $\left\langle \vec{G}, \alpha : E(\vec{G}) \to \mathcal{P} \right\rangle$ with voltages in $\mathcal{P} = [P : Y]$ is the digraph $\vec{G}^\alpha = (V^\alpha, E^\alpha)$ with

$$V^\alpha = \{v_j \mid v \in V(\vec{G}) \text{ and } j \in Y\}$$
$$E^\alpha = \{e_j | e \in E(\vec{G}) \text{ and } j \in Y\}$$

such that if an arc e in the voltage graph is directed from vertex u to vertex v, then the arc e_j in the covering graph is directed from vertex u_j to vertex $v_{\alpha(e)(j)}$. The ***covering graph***, denoted G^α, is the underlying graph of the covering digraph $\vec{G}^\alpha$.

TERMINOLOGY: The terminology ***vertex fiber*** and ***edge fiber*** is used for permutation voltage graphs, as for $\mathcal{B}$-voltage graphs.

CONVENTIONS: Permutation-voltage graphs may also use the bidirected-arc convention.

Remark: The preceding definition specifies that the subscript of the tail vertex u_j always agrees with the subscript of its arc e_j. It also specifies that the subscript of the head vertex $v_{\alpha(e)(j)}$ is the object $a(e)(j)$ to which object j is permuted by the voltage $\alpha(e)$ assigned to edge e, as in Example 10.5.1.

Remark: Every regular $\mathcal{B}$-voltage graph is equivalent to a permutation-voltage graph under the regular representation of the group $\mathcal{B}$ as a permutation group on itself. In this sense, permutation-voltage graphs generalize regular-voltage graphs.

Drawing Graphs Specified by Permutation Voltages

There is no fixed rule to determine the best way to position the vertices in drawing the covering graph from a voltage graph. If the voltage graph has only one vertex, one way is to arrange all the vertices in a circle, and another is to group the vertices according to the partition imposed by the disjoint-cycle structure of the permutation voltage on one of the self-loops of the voltage graph.

Example 10.5.2: Figure 10.5.2(i) shows Σ_6-voltages on the bouquet B_2. Figure 10.5.2(ii) shows the covering graph with its vertices arranged around the regular hexagon generated by self-loop d. Figure 10.5.2(iii) shows the covering graph with the vertices grouped into vertical columns according to the partition imposed by the disjoint-cycle structure of the permutation voltage on self-loop e.

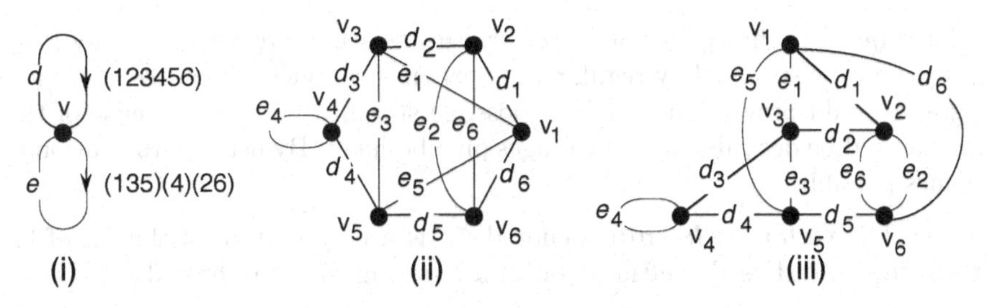

Figure 10.5.2 A Σ_6-voltage graph and two views of its covering graph.

When the voltage graph has more than one vertex, sometimes the covering graph is drawn so that its vertices are organized into columns according to their vertex fibers.

Example 10.5.3: Figure 10.5.3(i) shows a Σ_4-voltage graph. Figure 10.5.3(ii) shows the covering graph with its vertices in fiber columns. Figure 10.5.3(iii) shows the covering graph with its vertices arranged around a regular octagon. This covering graph is isomorphic to the cube graph Q_3 (see Exercises).

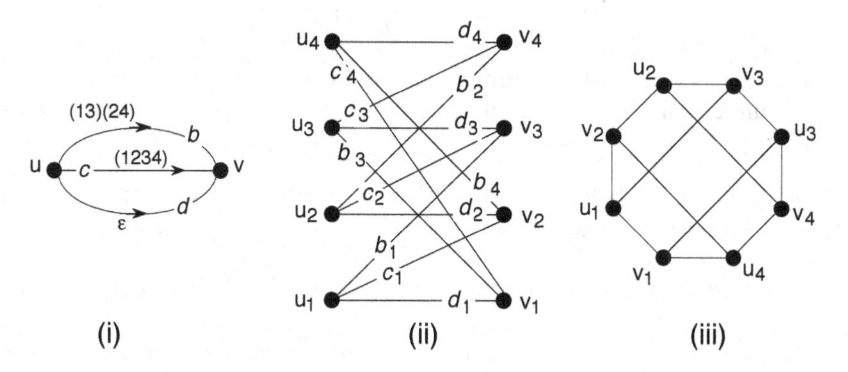

Figure 10.5.3 A Σ_4-voltage graph and two views of its covering graph.

Specifying the deBruijn Graphs

DEFINITION: The $(2, n)$-***deBruijn digraph*** $D_{2,n}$ consists of 2^{n-1} vertices, labeled by the bitstrings of length $n-1$, and 2^n arcs, labeled by the bitstrings of length n. The arc $b_1b_2 \ldots b_{n-1}b_n$ is directed from vertex $b_1b_2 \ldots b_{n-1}$ to vertex $b_2 \ldots b_{n-1}b_n$. Figure 10.5.4 shows $D_{2,4}$.

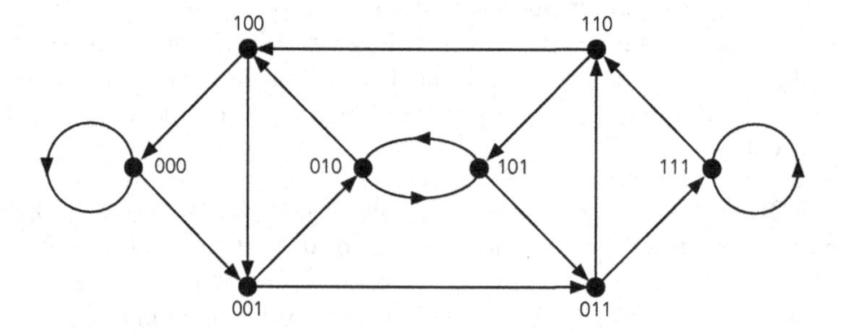

Figure 10.5.4 The $(2, 4)$-deBruijn digraph $D_{2,4}$.

The deBruijn digraph $D_{2,n}$ is not vertex-transitive (see Exercises). Thus, by Proposition 10.4.11, it cannot be specified by regular voltages on a bouquet. Of course, since a bouquet voltage graph would provide the smallest possible specification (in the sense of the fewest vertices), it would be desirable to use voltages on a bouquet. By using permutation voltages, this becomes possible.

DEFINITION: The ***(left) cycle-shift***, denoted ζ_n, is a permutation of the set of bitstrings of length n, that transfers the leftmost bit of a bitstring to the right end.

$$\zeta_n(b_1b_2 \ldots b_{n-1}b_n) = b_2 \ldots b_{n-1}b_nb_1$$

When the context is clear, it is simply denoted ζ.

DEFINITION: The ***deBruijn permutation***, denoted δ_n, is a permutation of the set of bit-strings of length n that consists of a left cycle-shift followed by "flipping" the rightmost bit.

$$\zeta_n(b_1 b_2 \ldots b_{n-1} b_n) = b_2 \ldots b_{n-1} b_n (1 - b_1)$$

When the context is clear, it is simply denoted δ.

Example 10.5.4: Consider the set $Y = \{000, 001, 010, 011, 100, 101, 110, 111\}$ of all bit-strings of length 3. The disjoint-cycle representation of the cycle-shift ζ_3 is

$$\zeta_3 = (000)\ (001\ 010\ 100)\ (011\ 110\ 101)\ (111)$$

and the disjoint-cycle representation of the deBruijn permutation δ_3 is

$$\delta_3 = (000\ 001\ 011\ 111\ 110\ 100)\ (101\ 010)$$

Let $\mathcal{P} = [P : Y]$ be the permutation group generated by ζ_3 and δ_3. (The group $\mathcal{P}$ has 24 elements, as explained below, but it is not necessary to know this.) Figure 10.5.5(i) shows a $\mathcal{P}$-voltage graph that specifies the deBruijn graph $D_{2,4}$. Figure 10.5.5(ii) shows the covering graph, using the clarifying-graphic conventions. This construction generalizes to all deBruijn graphs.

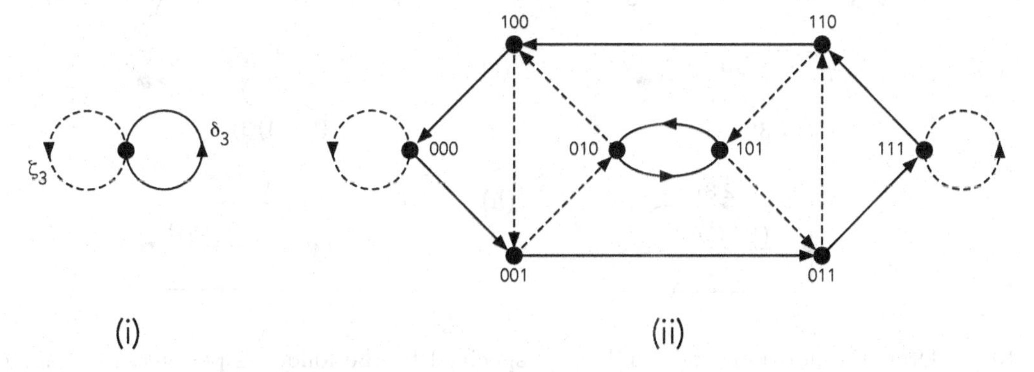

Figure 10.5.5 Specifying the deBruijn digraph $D_{2,4}$ with permutation voltages.

Wreath Product $\mathbb{Z}_n \otimes_{wr} \mathbb{Z}_2$ †

The permutalion group generated by ζ_n and δ_n is known as the *wreath product* $\mathbb{Z}_n \otimes_{wr} \mathbb{Z}_2$. Each of its elements can be denoted by a pair $\langle k, b_0 b_1 \ldots b_{n-1} \rangle$ containing an integer $k \in \{0, 1 \ldots, n-1\}$ and a bitstring $b_0 b_1 \ldots b_{n-1}$ of length n. It permutes a bitstring of length n by first cycle-shifting its bits k positions leftward and then adding the bitstring $b_0 b_1 \ldots b_{n-1}$ to the result. In this notation,

$$\zeta_3 = \langle 1, 000 \rangle \quad \text{and} \quad \delta_3 = \langle 1, 001 \rangle$$

For instance,

$$\langle 1, 001 \rangle\,(011) = 110 + 001 = 111$$

and

$$\langle 1, 000 \rangle\,(011) = 110 + 000 = 110$$

† Prior acquaintance with wreath products is probably necessary to understand this subsection.

Remark: The graph G^α specified by any voltage-graph construction is like a set of layers of G in which some edges cross-connect layers, instead of staying in their own layer. Voltage graphs are used extensively in the construction of surface imbeddings, as described in Chapter 11. The paper [GrTu77] in which Gross and Tucker introduce permutation-voltage graphs shows that they are sufficiently powerful to construct every possible instance of what algebraic topologists call a *covering space* of a graph.

EXERCISES for Section 10.5

10.5.**1** Draw the covering digraph for the given permutation-voltage graph.

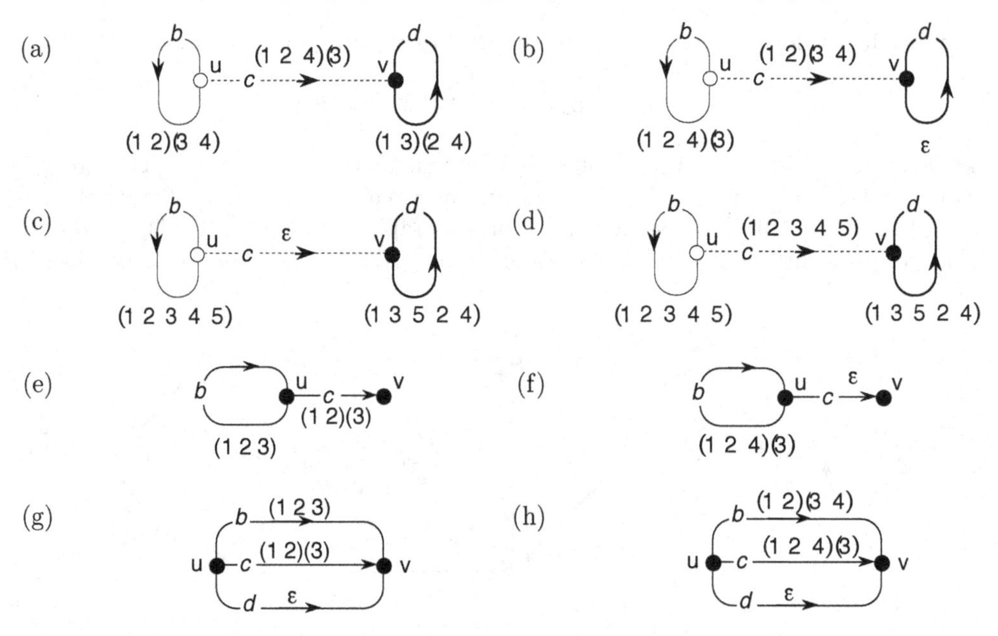

10.5.**2** Draw the deBruijn digraph $D_{2,4}$ as specified by the following permutation-voltage graph.

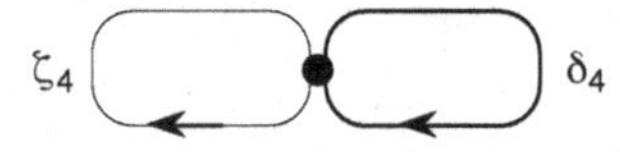

10.5.**3** Prove that the covering graph of Figure 10.5.3 is isomorphic to the cube graph Q_3.

10.5.**4** Prove that the deBruijn graph $D_{2,3}$ is not vertex-transitive.

10.5.**5** Write the disjoint-cycle decomposition of the product $\delta_3\delta_3$.

10.5.**6** Write the product $\delta_3\delta_3$ as a pair $\langle k, b_0b_1b_3\rangle$.

10.5.**7** Calculate the image of the given permutation from the wreath product $\mathbb{Z}_3 \otimes_{wr} \mathbb{Z}_2$ on the given bitstring.

(a) $\langle 1, 101\rangle(010)$

(b) $\langle 2, 010\rangle(100)$

(c) $\langle 2, 101\rangle(101)$

10.6 SYMMETRIC GRAPHS AND PARALLEL ARCHITECTURES

When there is a need to write a brief specification of a given fixed graph G, it is natural to turn to voltage-graph methods to try to design such a specification. Historically, voltage graphs were developed to specify network layouts on surfaces. Use of voltage graphs to specify networks of processors for parallel-computation architectures began more recently.

Theoretical Criteria for Designing Base Graphs

DEFINITION: When voltages are assigned to the edges of a digraph G, that digraph is called the ***base digraph*** of the construction.

DEFINITION: The ***base graph*** of a voltage-graph construction is the underlying graph of the base digraph.

DEFINITION: The ***natural projection*** of a voltage-graph construction is the graph mapping that carries every vertex fiber $\tilde{v}$ to its base vertex v and every edge fiber $\tilde{e}$ to its base edge e. (In effect, the natural projection "erases" the subscripts.)

Theorem 10.6.1 and Theorem 10.6.2 are fundamental to every application of voltage graphs, since they prescribe properties of the base graph that narrow the candidates to a tractably small number of possibilities. The importance of Theorem 10.6.3 is its implication that the base graph can be developed first and the directions assigned afterward. Theorems 10.6.1 through 10.6.4 and their proofs parallel those of Propositions 10.1.1 through 10.1.4.

Theorem 10.6.1: *Let $\langle G, \alpha : E_G \to \mathcal{A}\rangle$ be either a regular-voltage graph with $|\mathcal{A}| = n$ or a permutation-voltage graph with n objects in the permuted set. Then there are n times as many vertices and edges in the covering digraph G^α as in the base graph G.*

Theorem 10.6.2: *Let $\langle G, \alpha\rangle$ be any kind of voltage graph, and let v be any vertex of G. Then the outdegree and indegree of each vertex v_j in the fiber $\tilde{v}$ are equal to the outdegree and indegree, respectively, of vertex v.*

Remark: Theorem 10.6.1 and Theorem 10.6.2 say that the natural projection is n-to-1 and preserves degree.

Theorem 10.6.3: *Let $\langle G, \alpha\rangle$ be any kind of voltage graph, and let d be an arc of G from u to v. Let $\langle H, \beta\rangle$ be the voltage graph obtained from $\langle G, \alpha\rangle$ by reversing the direction of arc d and changing its voltage to the inverse of $\alpha(d)$ in the voltage group. Then the covering graph H^β is isomorphic to the covering graph G^α.*

Theorem 10.6.4: *Let $\langle G, \alpha\rangle$ be any kind of voltage graph, such that the base graph G is bipartite. Then the covering graph G^α is bipartite.*

Specification Problems with Elementary Solutions

Given a fixed graph $\tilde{G} = (\tilde{V}, \tilde{E})$, a solution to the fundamental specification problem has three steps:

(1) Design a plausible base digraph $G = (V, E)$.
(2) Select a plausible voltage group.
(3) Design a suitable voltage assignment.

Quite often, the objective is to specify every member of a graph family, not just a single graph.

DEFINITION: A ***unified specification*** for an entire family of graphs (typically) consists of a single base graph and a single voltage assignment formula that specifies every graph in the family.

In accordance with Theorem 10.6.1, the first step in designing a specification for a given graph $\tilde{G}$ (or family of graphs) begins by calculating $\gcd(|\tilde{V}|, |\tilde{E}|)$ (or a formula for the whole family). Ideally, there will be a base digraph $G = (V, E)$ that satisfies the conditions

$$|V| = \frac{|\tilde{V}|}{\gcd(|\tilde{V}|, |\tilde{E}|)} \quad \text{and} \quad |E| = \frac{|\tilde{E}|}{\gcd(|\tilde{V}|, |\tilde{E}|)}$$

With the aid of Theorem 10.6.2, it is often easy to restrict the supply of plausible candidates for the base digraph to a reasonable number.

Example 10.6.1: Consider the problem of specifying the circular ladder CL_n. Since

$$|\tilde{V}| = |V_{CL_n}| = 2n \quad \text{and} \quad |\tilde{E}| = |E_{CL_n}| = 3n$$

it follows that

$$\gcd(|\tilde{V}|, |\tilde{E}|) = \gcd(2n, 3n) = n$$

Thus, the first base graphs $G = (V, E)$ to consider have

$$|V| = \frac{|\tilde{V}|}{n} = 2 \quad \text{and} \quad |E| = \frac{|\tilde{E}|}{n} = 3$$

Since every vertex of CL_n has degree 3, it follows from Theorem 10.6.2 that the vertices of the base digraph must have degree 3. The only two 3-regular graphs with $|V| = 2$ and $|E| = 3$ are the dipole D_3 and the dumbbell graph. For odd n, the circular ladder CL_n is not bipartite, so, by Theorem 10.6.4, the dipole, which is bipartite, cannot be the base graph. Thus, the dumbbell is the only plausible candidate for a unified specification.
The cyclic group $\mathbb{Z}_n$ is a plausible candidate for the voltage group, since it has n elements. Figure 10.6.1 shows a $\mathbb{Z}_n$-voltage assignment on the dumbbell that specifies CL_n.

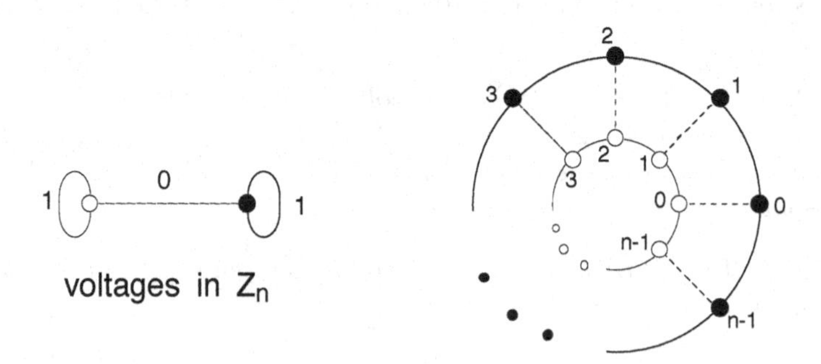

Figure 10.6.1 Using the dumbbell to specify the circular ladders CL_n.

Example 10.6.2: The complete bipartite graph $K_{n,n}$ has

$$|\tilde{V}| = |V(K_{n,n})| = 2n \quad \text{and} \quad |\tilde{E}| = |E(K_{n,n})| = n^2$$

so that

$$\gcd(|\tilde{V}|, |\tilde{E}|) = \gcd(2n, n^2) = \begin{cases} n & \text{if } n \text{ is odd} \\ 2n & \text{if } n \text{ is even} \end{cases}$$

Thus, for n odd, the first base graphs $G = (V, E)$ to consider have

$$|V| = \frac{|\tilde{V}|}{n} = 2 \quad \text{and} \quad |E| = \frac{|\tilde{E}|}{n} = n$$

Since every vertex of $K_{n,n}$ has degree n, it follows from Theorem 10.6.2 that the vertices of the base digraph must have degree n. Thus, the plausible candidates for the base digraph have two vertices, $2k+1$ arcs joining them, and $\frac{n-2k-1}{2}$ self-loops at each vertex.

The most attractive candidate among them (it is likely to be the easiest to work with) is the dipole D_n. By Theorem 10.6.3, all the arcs can go in the same direction from one vertex to the other. By Theorem 10.6.4, with this base graph, it is certain in advance that every voltage assignment specifies a bipartite graph.

A logical choice for the voltage group is the cyclic group $\mathbb{Z}_n$. In this case, the only plausible voltage assignment is to use a different voltage on every edge, as shown in Figure 10.6.2, since otherwise, there would be multi-edges. Observe that this solution is a generalization of Example 10.1.2.

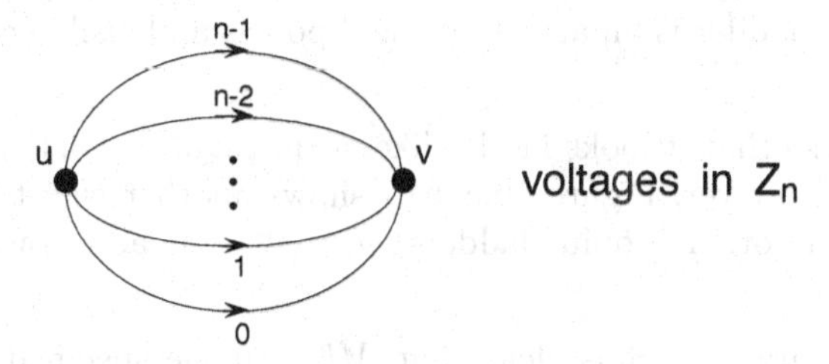

Figure 10.6.2 Specifying the complete bipartite graph $K_{n,n}$.

It is easy to see that this solution for odd n also works for even n. However, the theoretical criteria suggest that for even n, there might be a base graph with $|V| = 1$ and $|E| = \frac{n}{2}$ that is, the base graph $G = B_{n/2}$. Assigning the $\frac{n}{2}$ odd voltages in $\mathbb{Z}_{2n}$ to the n self-loops forces the endpoints of every arc from a vertex in the specified digraph to have opposite parity. The net effect is that the specified graph is $K_{n,n}$, as illustrated in Figure 10.6.3 for the case $n = 4$.

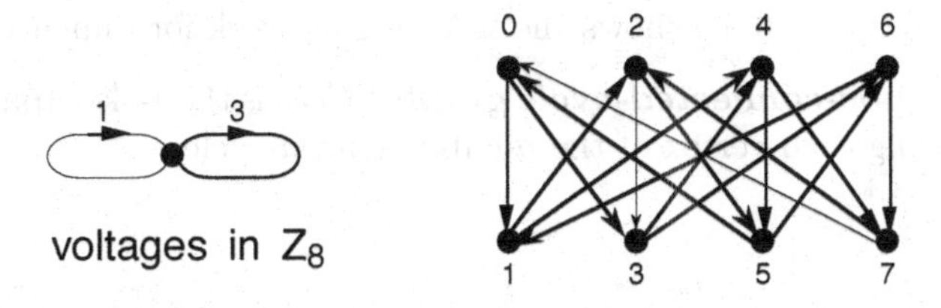

Figure 10.6.3 Using B_2 to specify the complete bipartite graph $K_{4,4}$.

Example 10.6.3: The Möbius ladder ML_n (depicted for $n = 4$ in Figure 10.6.4) has

$$\left|\tilde{V}\right| = |V_{ML_n}| = 2n \quad \text{and} \quad \left|\tilde{E}\right| = |E_{ML_n}| = 3n$$

Thus, the circular ladder CL_n, we have $\gcd(|\tilde{V}|, |\tilde{E}|) = \gcd(2n, 3n) = n$.

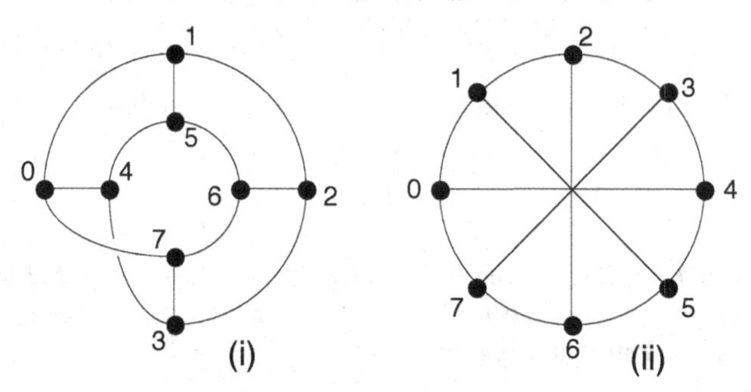

Figure 10.6.4 Two drawings of the Möbius ladder ML_4.

Once again, the first base graphs $G = (V, E)$ to consider have

$$|V| = \frac{|\tilde{V}|}{n} = 2 \quad \text{and} \quad |E| = \frac{|\tilde{E}|}{n} = 3$$

and as for CL_n, every vertex of ML_n has degree 3. Once again, the only two 3-regular graphs with $|V| = 2$ and $|E| = 3$ are the dipole D_3 and the dumbbell graph. Unlike some circular ladders, every Möbius ladder is bipartite, so the dipole is a plausible candidate for a unified specification.

When ML_n is drawn so that it looks ladder-like, as in Figure 10.6.4(i), some of the symmetry remains hidden. However, Figure 10.6.4(ii) shows another way to visualize ML_4 (and, through generalization, other Möbius ladders) so that some additional symmetry becomes apparent.

From the symmetric drawing, it is clear that ML_n can be specified as the Cayley graph $C(\mathbb{Z}_{2n}, \{1, n\})$, while adopting the bidirected-arc convention.

Specification Problems with Complicated Solutions

For various computational problems, it would be desirable to perform a distributed computation on an n-dimensional hypercube network. However, there is a physical limit to the number of communications links that can converge on a microscopic-size processor. To overcome this problem, the processor at each corner of the hypercube is replaced by an n-cycle of processors, with one edge of the hypercube linking each processor in the n-cycle to another corner. Figure 10.6.5 shows the resulting network for dimension 3.

DEFINITION: The ***cube-connected-cycle graph*** CCC_n is the 1-skeleton of the polyhedron obtained by replacing each vertex of the n-cube by an n-cycle.

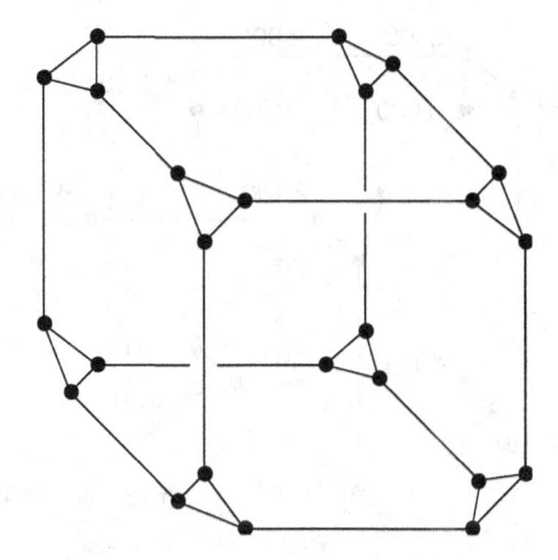

Figure 10.6.5 The cube-connected-cycle graph CCC_3

Example 10.6.4: The n-dimensional hypercube has 2^n corners and $n2^{n-1}$ edges. The cube-connected-cycle graph CCC_n replaces each vertex of the hypercube by n vertices and n edges. Thus,

$$|\tilde{V}| = |V(CCC_n)| = n2^n \quad \text{and} \quad |\tilde{E}| = |E(CCC_n)| = n2^{n-1} + n2^n = 3n2^{n-1}$$

so that

$$\gcd(|\tilde{V}|, |\tilde{E}|) = \gcd(n2^n, 3n2^{n-1}) = n2^{n-1}$$

which suggests that $|V| = 2$ and $|E| = 3$. Since CCC_n is 3-regular, the dumbbell and the dipole D_3 seem to be the two best candidates for a base graph. However, since CCC_n is not bipartite when n is odd, the best hope for a unified solution is to use the dumbbell as a base graph. Although not obvious, it turns out that a solution is the regular-($\mathbb{Z}_n \otimes_{wr} \mathbb{Z}_2$)-voltage graph in Figure 10.6.6, whose covering graph is two isomorphic copies of CCC_n.

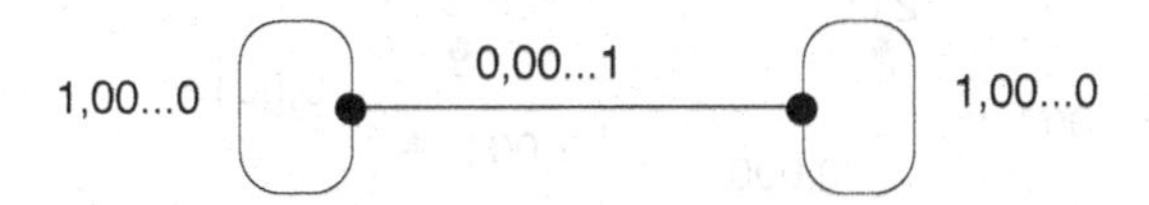

Figure 10.6.6 A ($\mathbb{Z}_n \otimes_{wr} \mathbb{Z}_2$)-voltage graph specifying CCC_n.

The specified graph for $n = 3$ is shown in Figure 10.6.7.

The next two examples show that when the voltages on a graph are from a permutation group, it is necessary to declare explicitly whether they are to be regarded as regular voltages or as permutation voltages.

Example 10.6.5: If Figure 10.6.6 were interpreted as a permutation-voltage assignment, then the resulting specified graph would have only 2^n vertices, rather than $n2^n$, as is clear from the definitions of the two different types of voltage-graph constructions of specified graphs.

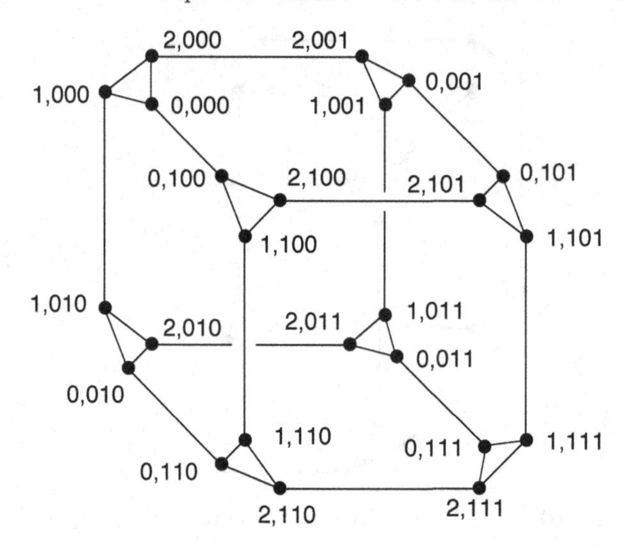

Figure 10.6.7 The graph CC_3, as specified by Figure 10.6.6.

Example 10.6.6: The permutation-voltage graph of Figure 10.5.5, which specified a de-Bruijn graph, is reproduced here as Figure 10.6.8.

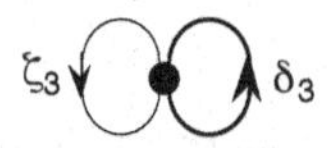

Figure 10.6.8 A regular-voltage graph specifying WBF_3**.**

Suppose it is now reinterpreted as a regular-($\mathbb{Z}_3 \otimes_{wr} \mathbb{Z}_2$)-voltage graph, instead of a permutation-($\mathbb{Z}_3 \otimes_{wr} \mathbb{Z}_2$)-voltage graph. Then the result, shown in Figure 10.6.9, is called the ***3-dimensional wrapped-butterfly graph*** WBF_3. Higher dimensional wrapped-butterfly graphs are used in parallel computation of fast Fourier transforms.

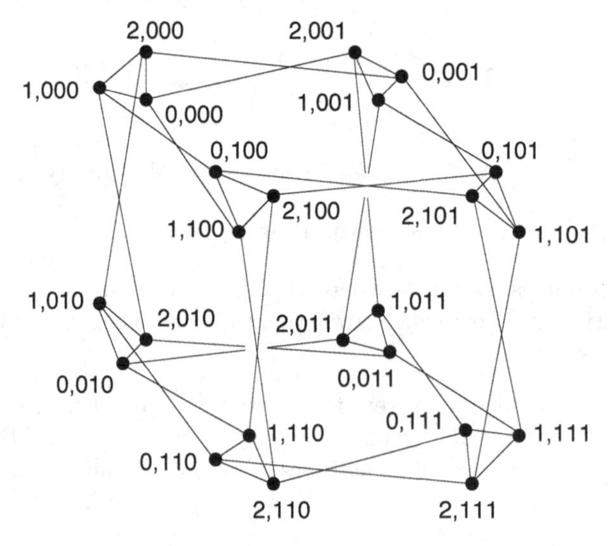

Figure 10.6.9 The wrapped-butterfly graph WBF_3**.**

EXERCISES for Section 10.6

10.6.1 Design a voltage graph to specify the given graph or family of graphs.

(a) The octahedral graph Q_4.

(b) The family O_n of octahedral graphs.

(c) The graph $Q_3 \times C_3$.

(d) The family $Q_3 \times C_n$.

(e) The graph $K_5 \times C_5$.

(f) The family$K_5 \times C_n$.

(g) The graph $K_5 \times K_5$.

(h) The graph $K_5 + C_5$.

DEFINITION: The ***shuffle-exchange digraph*** $\vec{SE}_n$ has as its vertex-set the length-n bitstrings. There is an arc from each bitstring v to the bitstring $\zeta_n(v)$ obtainable by cycle-shifting leftward one position. There is also a bidirected arc between each bitstring and the bitstring that can be obtained by changing its rightmost bit.

10.6.2 Construct a permutation-voltage graph that specifies the shuffle-exchange graph.

10.6.3 Give a counterexample to the converse of Theorem 10.6.4.

10.7 SUPPLEMENTARY EXERCISES

10.7.1 Draw the covering graph specified by this voltage graph with cyclic voltages in the group Z_5. Suppose the voltages were from an arbitrary cyclic group Z_n. Why is the covering graph 2-factorable (see §6.5)?

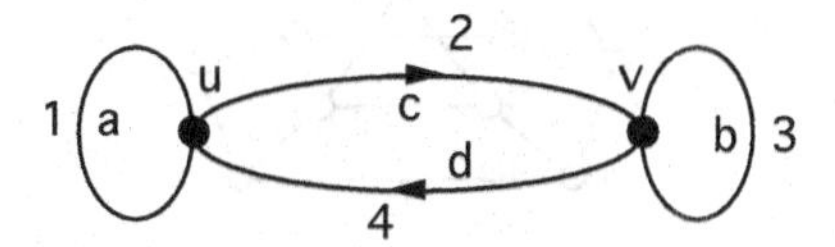

10.7.2 Give a regular voltage assignment on $K_{3,3}$ and a labeling of the vertices of CL_6 as the covering graph specified by that voltage graph.

10.7.3 Construct the covering graph for the following permutation voltage graph.

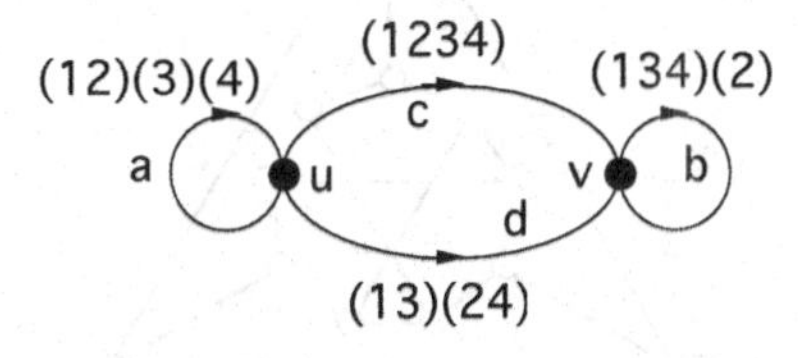

10.7.4 Give a permutation voltage assignment on D_3 that specifies CL_6 as the covering graph.

10.7.5 Label the dumbbell with cyclic voltages in Z_6 so that the covering graph is isomorphic to the graph at the right.

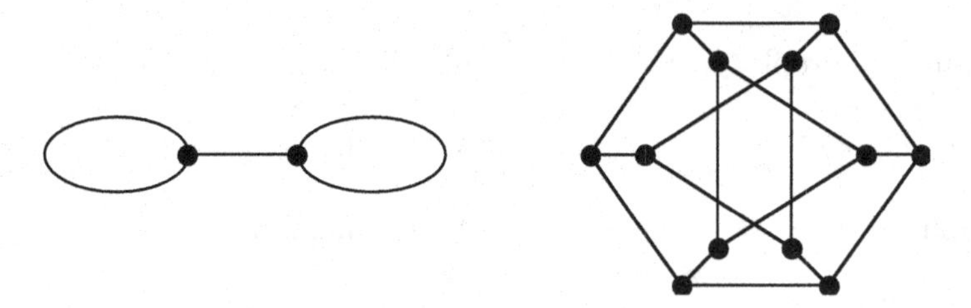

10.7.6 Label the bouquet with cyclic voltages in Z_6 so that the covering graph is isomorphic to the graph at the right.

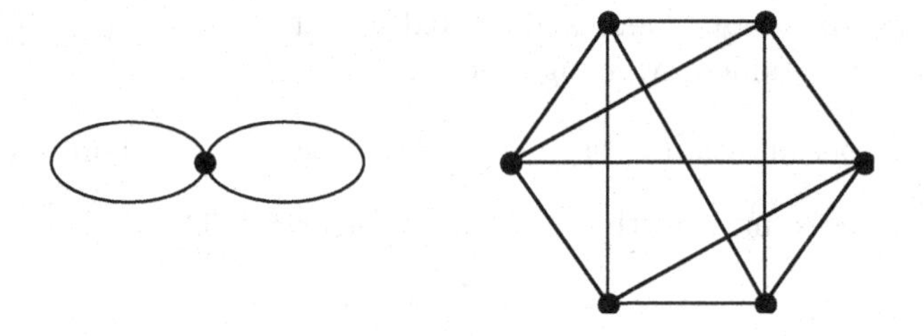

10.7.7 The graph below is called ***Franklin's graph***. Draw a 4-vertex voltage graph for it with vertex labels w, x, y, and z, and label the vertices of Franklin's graph as the specified covering graph.

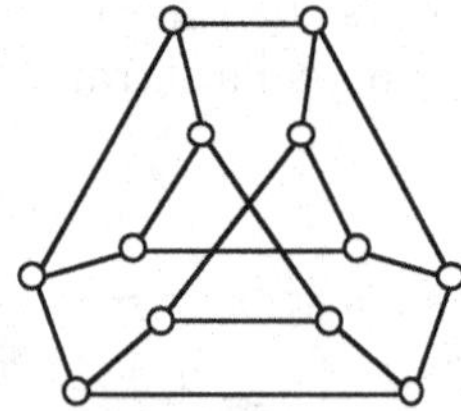

10.7.8 The graph below is called the ***Tietze graph***. Draw a 4-vertex cyclic voltage graph for it, and label the vertices w, x, y, and z. Label the vertices of the Tietze graph as the specified covering graph.

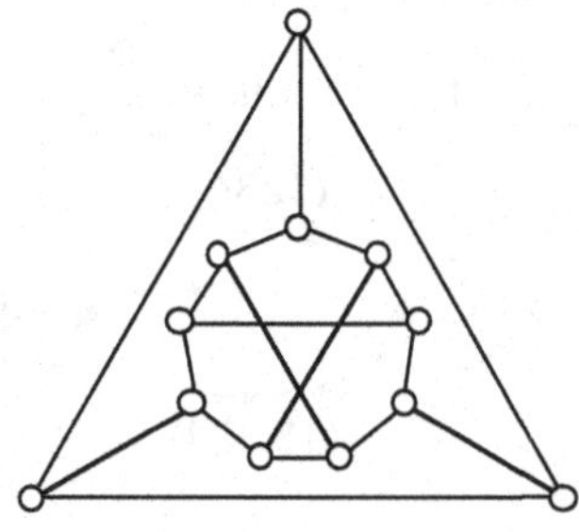

10.7.9 Draw three mutually non-isomorphic 3-regular, 10-vertex simple graphs, each vertex transitive. (Optional hint: use cyclic voltage graphs.) Sketch proofs that your three graphs are mutually non-isomorphic.

10.7.10 (a) Draw a 2-vertex Z_8-voltage graph whose covering graph is isomorphic to the graph below.

(b) Draw a 1-vertex $Z_2 \times Z_8$-voltage graph whose covering graph is isomorphic to the graph below.

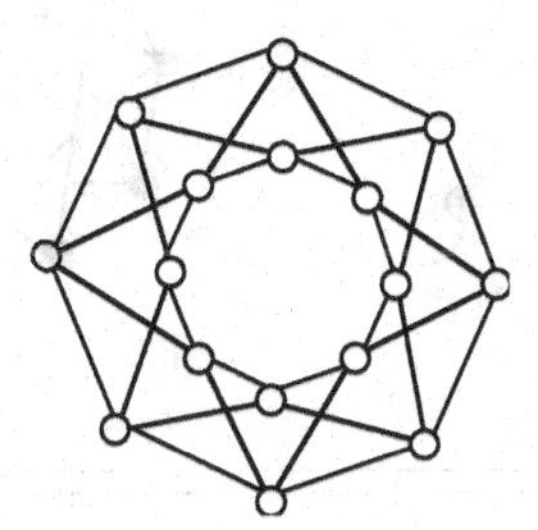

10.7.11 Consider the covering graphs for the three voltage graphs below with voltages in Z_{11}. Decide whether some pair of them is isomorphic or whether they are mutually non-isomorphic.

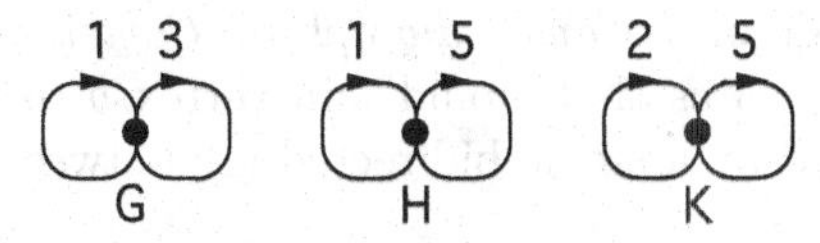

10.7.12 a. Draw the covering graph for this voltage graph. b. Then draw a 1-vertex voltage graph whose covering graph is isomorphic to that covering graph of part a.

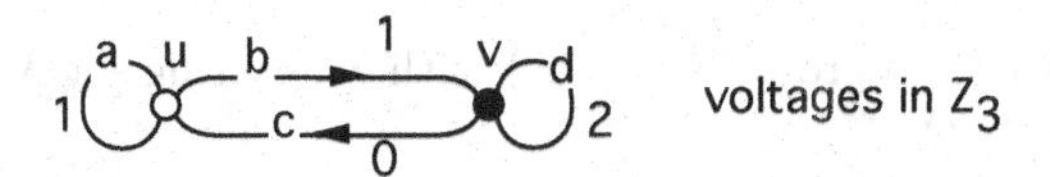

10.7.13 Assign cyclic voltages in Z_4 to the graph at the left so that the graph at the right is the covering graph.

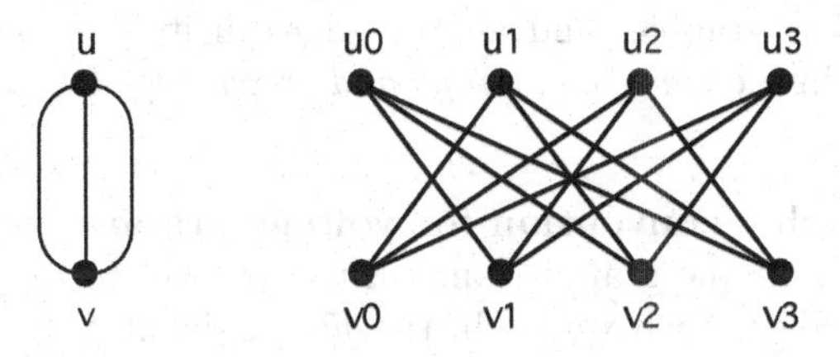

10.7.14 Assign cyclic voltages in Z_9 to the graph at the left so that the graph at the right is the covering graph.

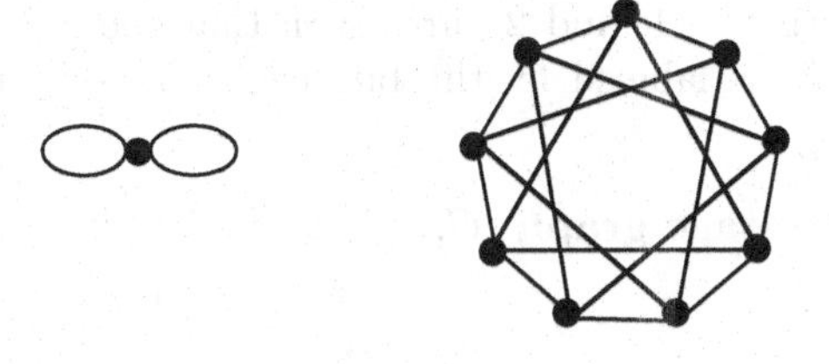

10.7.15 Assign cyclic voltages in Z_3 to the graph at the left so that the graph at the right is the covering graph.

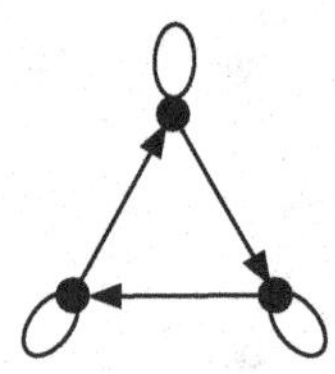
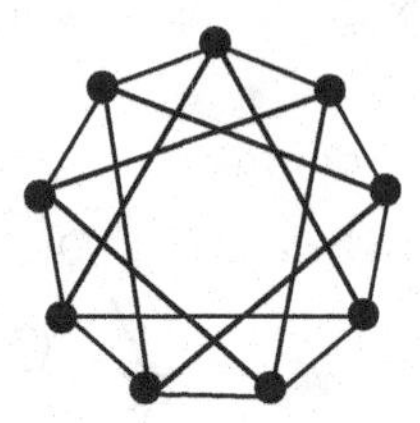

Glossary

base digraph of a voltage graph $\langle G, \alpha \rangle$: the digraph G.

base graph of a voltage graph $\langle G, \alpha \rangle$: the underlying graph of the *base digraph.*

bidirected-arc convention in a *Cayley digraph* or *Cayley graph*: if y is a generator of order 2, then the pair of arcs that would join vertex b to vertex by and vertex by to vertex b may be replaced by a single bidirected arc between b and by.

Cayley digraph $\vec{C}(\mathcal{B}, X)$ for a group $\mathcal{B} = \langle B, \cdot \rangle$ with generating set X: the digraph with vertex-set $V_{\vec{C}(\mathcal{B},X)} = B$ and arc-set $E_{\vec{C}(\mathcal{B},X)} = \{x_b \mid x \in X, b \in B\}$, such that arc x_b joins vertex b to vertex bx. (*Bidirected arcs* are sometimes used for generators of order 2.)

Cayley graph $C(\mathcal{B}, X)$ for a group $\mathcal{B} = \langle B, \cdot \rangle$ with generating set X: the underlying graph of the *Cayley digraph* $\vec{C}(\mathcal{B}, X)$.

Cayley graph: any graph G such that there exists a group $\mathcal{B}$ and a generating set X such that G is isomorphic to the *Cayley graph* $C(\mathcal{B}, X)$.

clarifying-edge-graphic convention for voltage graphs: each edge of the *base graph* is drawn in a unique graphic, and all the edges in its *fiber* are drawn in that same graphic; moreover, in the *covering digraph and graph*, edge labels are commonly omitted altogether.

clarifying-vertex-graphic convention for voltage graphs: each vertex of the *base graph* is drawn in a unique graphic, and all the vertices in its *fiber* are drawn in that same graphic; moreover, each vertex in the *fiber* is labeled by its subscript only, with its *mainscript* omitted.

covering digraph for a voltage graph $\langle G, \alpha \rangle$: the graph $G^\alpha = (V^\alpha, E^\alpha)$.

deBruijn digraph $D_{2,n}$: the digraph consisting of 2^{n-1} vertices, each labeled by a different bitstring of length $n-1$, and 2^n arcs such that the arc from vertex $b_1 b_2 \cdots b_{n-1}$ to vertex $b_2 \cdots b_{n-1} b_n$ is labeled by the bitstring $b_1 b_2 \cdots b_n$ referred to as the $(2, n)$-deBruijn digraph.

covering graph for a voltage graph $\langle G, \alpha \rangle$: the underlying graph of the *covering digraph* G^α.

cube-connected-cycle graph CCC_n: the vertex-set is $\{(k,b) \mid k \in \mathbb{Z}_n, b \in \mathbb{Z}_2^n\}$, and two vertices (k,b) and (m,c) are adjacent if and only if $k = m$ while b and c differ in only one bit or $b = c$ and $k = m \pm 1$; isomorphic to the 1-skeleton of the polyhedron obtained by chopping the corners off an n-dimensional hypercube.

(left) cycle-shift ζ_n: a permutation of the set of bitstrings of length n that transfers the leftmost bit of a bitstring to the right end. When the context is clear, it is simply denoted ζ.

cyclic voltage (or $\mathbb{Z}_n$-voltage): a label on an arc by a number $j(\text{mod } n)$ (i.e., by an element of the cyclic group $\mathbb{Z}_n$.

deBruijn permutation δ_n: a permutation of the set of bitstrings of length n that consists of a left cycle-shift followed by adding 1 (mod 2) to the rightmost bit. When the context is clear, it is simply denoted δ.

diameter $diam(G)$ of a graph G: the maximum of the distances $d(u,v)$, taken over all pairs of vertices of G.

fiber $\tilde{e}$ over an edge e of a voltage graph: (1) for *regular voltages* in a group $\mathcal{B} = \langle B, \cdot \rangle$, the edge-set $\{e_b \mid b \in B\}$ in the covering graph G^α; (2) for *permutation voltages* in a permutation group $\mathcal{P} = [P : Y]$, the edge-set $\{e_y \mid y \in Y\}$ in G^α.

fiber $\tilde{v}$ over a vertex v of a voltage graph: (1) for *regular voltages* in a group $\mathcal{B} = \langle B, \cdot \rangle$, the vertex-set $\{v_b \mid b \in B\}$ in the covering graph G^α; (2) for *permutation voltages* in a permutation group $\mathcal{P} = [P : Y]$, the vertex-set $\{v_y \mid y \in Y\}$ in G^α.

mainscript of a subscripted variable X_j: the variable name X, i.e., the part without the subscript.

mean pair-distance $d_{avg}(G)$ of a graph G: the average of the distances $d(u,v)$, taken over all pairs of distinct vertices in G.

natural projection of a voltage-graph construction: the "subscript-erasing" graph mapping that carries every vertex fiber $\tilde{v}$ to its base vertex v and every edge fiber $\tilde{e}$ to its base edge e.

natural transformation φ_a on a voltage-specified covering graph G^α: the graph automorphism with vertex function $v_b \mapsto v_{ab}$ and edge function $e_b \mapsto e_{ab}$, for all $y \in B$.

permutation voltage: a label on an arc by an element of a permutation group $\mathcal{P} = [P : Y]$.

regular voltage: a label on an arc by an element of a group $\mathcal{B}$.

shuffle-exchange digraph $\vec{SE}_n$: the vertex-set is the length-n bitstrings; there is an arc from each bitstring v to the bitstring $\zeta_n(v)$ obtainable by cycle-shifting leftward one position. There is also a bidirected arc between each bitstring and the bitstring that can be obtained by changing its rightmost bit.

unified specification for a family of graphs or digraphs: usually, a single *base graph* and a single *voltage-assignment* formula that specifies every graph or digraph in the family.

vertex-transitive graph: a graph such that for all vertex pairs u, v, there is a graph automorphism that maps u to v.

voltage on an arc of a digraph: a label by an element of a group $\mathcal{B}$ or of a permutation group $\mathcal{P} = [P : Y]$.

voltage assignment on a digraph: a function α that labels every arc with a voltage from a group $\mathcal{B}$ or from a permutation group $\mathcal{P} = [P : Y]$.

voltage graph: a pair $\langle G, \alpha \rangle$ such that G is a digraph and α is a *voltage assignment* on G.

—, **$\mathcal{B}$-voltage graph:** a voltage graph with voltage group $\mathcal{B}$.

—, **$\mathbb{Z}_n$-voltage graph:** a voltage graph with voltage group $\mathbb{Z}_n$.

voltage group: the group from which the *voltages* are assigned.

Chapter 11

NON-PLANAR LAYOUTS

INTRODUCTION

The most important quantification of non-planarity of a graph is the number of handles or crosscaps one must add to the plane to eliminate all the crossings. There are two elementary ways to represent an arbitrary graph drawing on an arbitrary surface. When specifying imbeddings devoid of adequate symmetry, one resorts to these elementary representations. One way is to draw the graph on a flat polygon representation (§5.5) of the surface. Another way, completely combinatorial, is to write the cycle of edge-ends incident on each vertex, as described in §11.1. The list of such cycles, called a *rotation system*, completely specifies an imbedding. The set of all such rotation systems for a graph corresponds to the set of all imbeddings, thereby yielding information about the distribution of the imbeddings according to the genus of the imbedding surface.

When a graph is sufficiently rich in symmetry that it can be specified by a small voltage graph, there is an extension of the voltage-graph construction. The idea is that a voltage graph drawn on a surface can specify a symmetric imbedding of the covering graph. Voltage graphs have been used in problem solutions. They are used frequently to find an imbedding of a graph into a surface with the smallest possible number of handles. Solving this imbedding-minimization problem for complete graphs was the main part of the solution by Ringel and Youngs of the Heawood problem of finding the chromatic number of all the higher-order closed surfaces.

11.1 REPRESENTING IMBEDDINGS BY ROTATIONS

We have seen that a graph imbedding can be represented by a drawing on a flat polygon representation of a surface. It is helpful also to have a precise combinatorial form of represention of a graph imbedding that does not depend on drawing pictures.

Review of Closed Surfaces and Flat Polygon Drawings

REVIEW FROM §5.2:

Theorem 5.2.2 *Classification of Closed Orientable Surfaces. Every closed orientable surface is topologically equivalent to exactly one of the surfaces in the infinite sequence* $S_0, S_1, S_2, \ldots$.

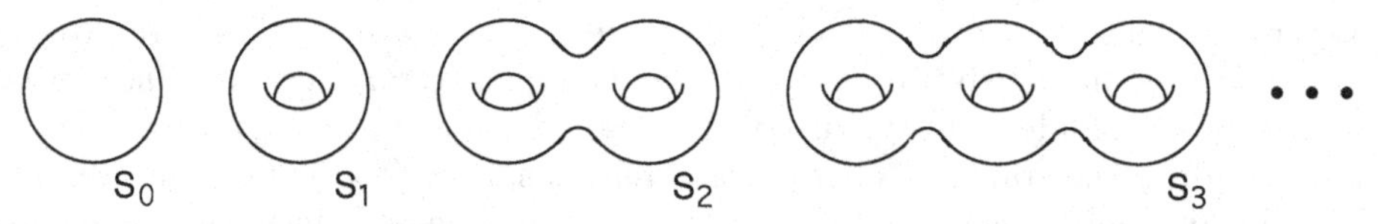

Figure 5.2.6 The sequence of closed orientable surfaces.

REVIEW FROM §5.2: A ***crosscap*** is a portion of a surface that is topologically equivalent to a Möbius band. In a drawing of a non-orientable surface, a crosscap is commonly represented by a circle with an inscribed crossmark.

Theorem 5.2.3 *Classification of Closed Non-Orientable Surfaces. Every closed non-orientable surface is topologically equivalent to exactly one of the surfaces in the infinite sequence* $N_1, N_2, N_3, \ldots$.

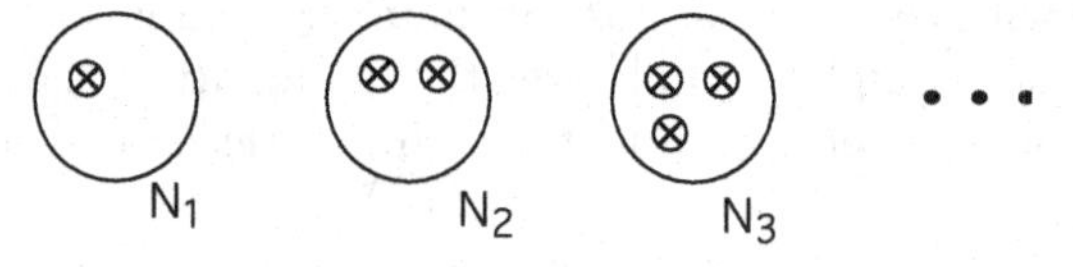

Figure 5.2.7 The sequence of closed non-orientable surfaces.

REVIEW FROM §5.5: A ***flat polygon representation*** of a surface S is a drawing of a polygon with markings to match its sides in pairs, such that when the sides are pasted together as the markings indicate, the resulting surface obtained is topologically equivalent to S.

Parametrization

It was explained in §5.1 how each edge in a graph embedding can be modeled as a space curve, $f : [0,1] \to \mathbb{R}^3$. This parametrization of each edge of a graph by the unit interval [0, 1] creates a distinction between its *0-end* and its *1-end.* Such a distinction facilitates the combinatorial description of a graph imbedding.

NOTATION: Let e be an edge of a graph. Then e^+ denotes the 0-end of edge e, and e^- denotes the 1-end. The mnemonic for this notation is that starting at e^+ means going in the positive direction from 0 to 1. The usual rules of sign composition apply, so that

$$e^{-\sigma} = \begin{cases} e^+ & \text{if } \sigma = - \\ e^- & \text{if } \sigma = + \end{cases}$$

The set of edge-ends of a graph G is denoted $E_G^{\pm}$, or sometimes, simply $E^{\pm}$.

NOTATION: In a ***drawing*** of an undirected parametrized graph, the *parametrization arrow* points to the 1-end.

Remark: A directional arrow on an edge, if it exists, is independent of the parametrization arrow; that is, both arrows may point the same way or opposite ways. The distinction in meaning is that a parametrization arrow gives different names to the two possible senses in which an edge can be traversed, whereas a direction arrow forbids passage in one of those two senses. It is not necessary here to display both arrows simultaneously. However, in circumstances where it is necessary to display both arrows simultaneously, it is helpful to use graphically different types of arrowhead.

Rotations and Rotation Systems

DEFINITION: A ***rotation at a vertex*** v of a graph G is a cyclic permutation of the edge-ends incident on v.

DEFINITION: A ***rotation system*** for a graph G is an assignment of a rotation to every vertex.

TERMINOLOGY: A rotation system is also regarded as a permutation on the set $E_G^{\pm}$ of edge-ends. This permutation is also called a *rotation system*.

Remark: It is assumed here that the orientable surfaces we are discussing are subsets of Euclidean 3-space. The intent of this assumption is to impose a fixed distinction between *clockwise* and *counterclockwise* on the surface.

TERMINOLOGY: An orientable surface is ***oriented*** if one of the orientations *clockwise* or *counterclockwise* is designated as *preferred*.

DEFINITION: Let $h : G \to S_G$ be a graph imbedding in an oriented surface. The ***induced rotation at*** v is the cyclic permutation $\rho_h(v)$ of edge-ends incident on v in the order in which they are encountered in a traversal around v in the preferred orientation.

DEFINITION: Let $h : G \to S_G$ be a graph imbedding in an oriented surface. The ***induced rotation system*** ρ_h is the function that assigns to each vertex $v \in V_G$ the rotation $\rho_h(v)$.

DEFINITION: Let ρ be a rotation system for a graph G. The ***rotation table for*** ρ, denoted T_ρ, has one row for each vertex of G. The content of each row of the table is the name of a vertex, followed by a complete list of the edge-ends incident on that vertex, in an order consistent with $\rho(v)$. (It is a common custom to write each row in lexicographic order.)

Example 11.1.1: Figure 11.1.1 shows an imbedding of K_4 in the sphere and a table representing the rotation system it induces. The direction arc around vertex v_1 indicates that the preferred orientation is counterclockwise (globally).

Proposition 11.1.1: *Let $h : G \to S$ be an oriented graph imbedding, and let ρ_h be the induced rotation. Then $\prod_{v \in V_G} \rho_h(v)$ is a permutation on the edge-ends of G.*

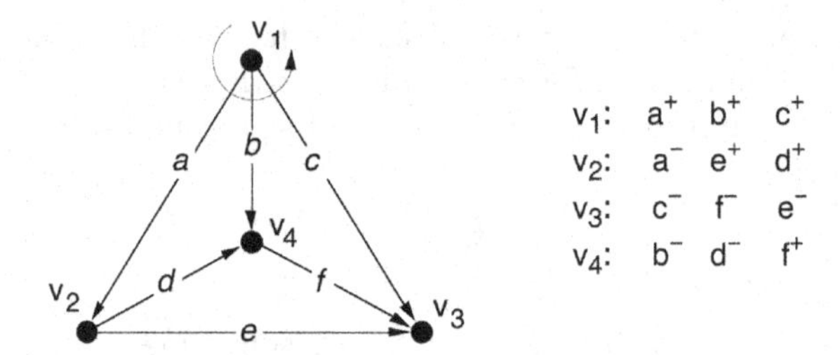

Figure 11.1.1 An imbedding $g : K_4 \to S_0$ and its induced rotation system ρ_g.

Proof: Every edge-end occurs at exactly one vertex. Thus, the product of the rotations over all vertices is the disjoint-cycle form of a permutation. ◊

Example 11.1.2: Figure 11.1.2 shows an imbedding of K_4 in the torus and its rotation system. The only difference from the rotation system induced by the imbedding $g : K_4 \to S_0$ is that the cyclic orders of the v_2-row and of the v_4-row are reversed; yet, despite this seemingly small difference, one imbedding is in the sphere and the other in the torus.

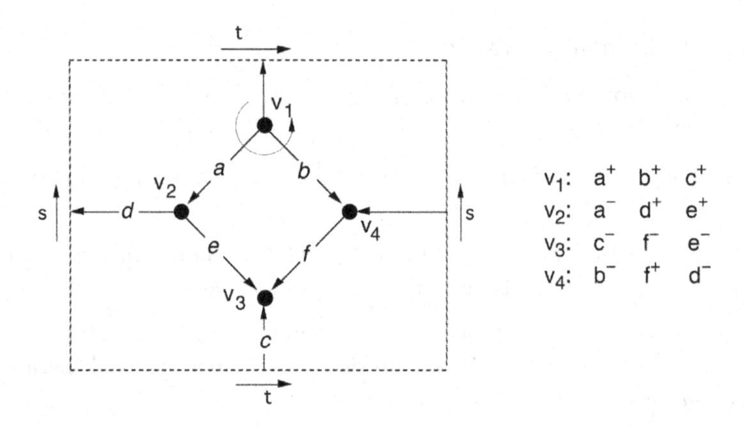

Figure 11.1.2 An imbedding $h : K_4 \to S_1$ and its induced rotation system ρ_h.

Dual Rotations

DEFINITION: Let $\rho : E_G^{\pm} \to E_G^{\pm}$ be a rotation system on a graph G. The ***dual rotation*** (or ***circulation***) ρ^* is the permutation $\rho^* : E_G^{\pm} \to E_G^{\pm}$ given by the rule $e^{\sigma} \mapsto \rho\,(e^{-\sigma})$.

Invoking the following algorithm is the practical way to use a rotation table T_ρ to determine the edge-end $\rho^*\,(e^{\sigma})$.

Algorithm 11.1.1: Circulation
Input: a rotation table T_ρ, an edge-end e^{σ}
Output: the edge-end $\rho^*\,(e^{\sigma})$
Find the (only) row v in table T_ρ that contains the edge-end $e^{-\sigma}$.
Return (whatever edge-end follows $e^{-\sigma}$ in row v).

Example 11.1.3: In the imbedding $g : K_4 \to S_0$ of Example 11.1.1,

$$\rho^* = (a^+\ e^+\ c^-)\ (a^-\ b^+\ d^-)\ (b^-\ c^+\ f^-)\ (d^+\ f^+\ e^-)$$

Example 11.1.4: In the imbedding $h : K_4 \to S_1$ of Example 11.1.2,

$$\rho^* = (a^- \; b^+ \; f^+ \; e^-) \; (a^+ \; d^+ \; b^- \; c^+ \; f^- \; d^- \; e^+ \; c^-)$$

Observe that the four cycles of ρ^* in Example 11.1.3 correspond to the boundary walks of the four faces in Figure 11.1.1 and that the two cycles of ρ^* in Example 11.1.4 correspond to the boundary walks of the two faces in Figure 11.1.2. The Heffter-Edmonds theorem below asserts that this phenomenon always occurs.

TERMINOLOGY: A sequence of edge-ends is said to *coincide* with a walk in a graph if the order of the edge-ends is the order in which they occur on a traversal of the walk.

Theorem 11.1.2: [***Heffter-Edmonds Theorem***] [He1891] [Ed60] *In a cellular imbedding $h : G \to S_g$, each cycle of the induced dual rotation ρ_h^* coincides with a face-boundary walk of the imbedding h.*

Proof: The edge $\rho_h^*(e^\sigma)$ is defined to be $\rho(e^{-\sigma})$, which is precisely the next edge-end after $e^{-\sigma}$ in the boundary walk that traverses edge e from e^σ to $e^{-\sigma}$. ◇

TERMINOLOGY: In view of Theorem 11.1.2, the construction from a rotation system ρ of the cycles of its induced circulation ρ^* is commonly called ***face-tracing***.

Algorithm 11.1.2: Face-Tracing
Input: edge-end list $E^\pm$, rotation table T_ρ
Output: list of all cycles of the circulation $\rho^*(e^\sigma)$
{*Initialize*} mark all edge-ends *unused*
While any unused edge-ends remain
Choose next (lex order) unused edge-end e_1^σ from $E^\pm$
Start new cycle by writing left parenthesis "("
$e := e_1^\sigma$
Repeat
Write e next in current cycle
$e = \rho^*(e)$ {call Circulation Algorithm}
Until $e = e_1^\sigma$
Close current cycle by writing right parenthesis ")"
Return

The Induced Surface of a Rotation System

NOTATION: Let z be a cyclic permutation. Then $length(z)$ denotes its length.

DEFINITION: Let ρ be a rotation system for a graph G. For each cycle z of the dual rotation ρ^*, let p_z be a polygon with $length(z)$ sides, labeled consecutively by the edges of the cycle in graph G that corresponds to the permutation cycle z. The set of all such polygons is called the ***polygon set*** for rotation ρ.

Proposition 11.1.3: *Let ρ be a rotation system for a graph G, and let $\{p_z\}$ be the polygon set for ρ. Then the list of all boundary walks of all the polygons in $\{p_z\}$ mentions each edge of graph G exactly twice.*

Proof: This follows immediately from the fact that rotation ρ is a permutation of the set of edge-ends. ◇

DEFINITION: Let ρ be a rotation system for a graph G. The ***surface induced by*** ρ is the topological space $S(\rho)$ formed from the graph G and the polygon set for ρ by fitting the boundary walk of each polygon to the corresponding cycle of G.

Example 11.1.5: Figure 11.1.3 shows a rotation table for K_4 and the set of labeled polygons that arise when the induced surface is constructed from that table. Observe that this rotation table differs from the table in Example 11.1.2 only in the bottom row.

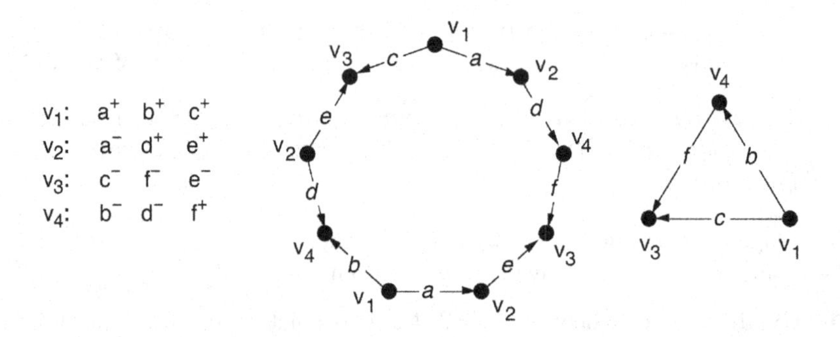

Figure 11.1.3 A rotation system for K_4 and its labeled polygons.

Whereas the imbedding of K_4 in Example 11.1.2 has as its faces a 4-gon and an 8-gon, the faces of this imbedding are a 3-gon and 9-gon. Figure 11.1.4 below shows the resulting imbedding in the torus.

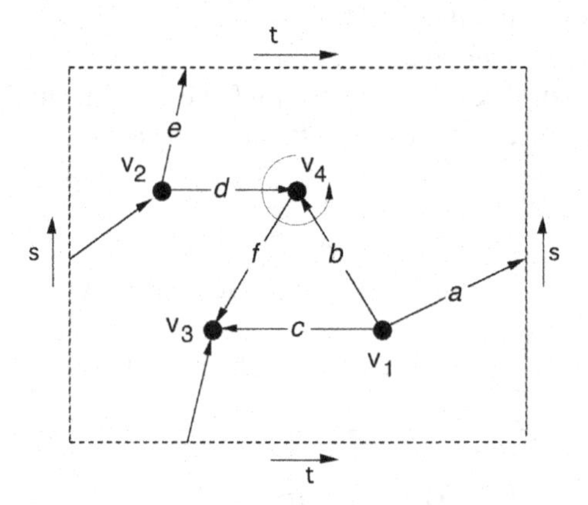

Figure 11.1.4 An imbedding $h : K_4 \to S_1$ and its induced rotation system ρ_h.

DEFINITION: Let ρ be a rotation system for a graph G. The ***imbedding induced by*** ρ is the oriented imbedding of G into the induced surface $S(\rho)$ whose face-boundary walks coincide with the cycles of the circulation ρ^*. It is denoted ι_ρ.

Counting the Imbeddings of a Graph

TERMINOLOGY: An imbedding on an oriented surface is called an *oriented imbedding.*

DEFINITION: Let $h : G \to S$ and $h' : G \to S'$ be two oriented imbeddings of a graph G. They are ***equivalent imbeddings*** if they have exactly the same set of face-boundary walks.

Proposition 11.1.4: *Let $h : G \to S$ and $h' : G \to S'$ be two oriented imbeddings such that $\rho_h = \rho'_h$. Then h and h' are equivalent imbeddings.*

Proof: It follows from Theorem 11.1.2 that oriented imbeddings with the same induced rotation system have the same set of face-boundary walks. ♢

NOTATION: $\gamma_g(G)$ is the number of rotation systems ρ such that the induced surface has genus g.

Theorem 11.1.5: *Let G be a connected graph. The correspondence between equivalence classes of oriented imbeddings of G and rotation systems of G is a bijection.*

Proof: Proposition 11.1.4 establishes that the correspondence is one-to-one. Since every rotation system ρ on a connected graph G induces a surface $S(\rho)$ and an imbedding $\iota_\rho : G \to S(\rho)$ whose induced rotation system is ρ, the correspondence is onto. ♢

Remark: Given a rotation system ρ, the genus of the induced imbedding surface $S(\rho)$ can be deduced from the formula $|V| - |E| + |F| = 2 - 2g$, by substituting the number of cycles of the circulation ρ^* for $|F|$. Thus, it is not necessary to construct the surface $S(\rho)$ to determine its genus.

Theorem 11.1.6: *Let G be a connected graph. Then*

$$\sum_{g=0}^{\infty} \gamma_g(G) = \prod_{v \in V_G} [\deg(v) - 1]!$$

Proof: The left side is the total number of rotation systems of graph G. The rotations at each vertex v are in one-to-one correspondence with the ways of arranging the edge-ends at v into a cycle, so there are $[\deg(v) - 1]!$ rotations at v. Thus, the number of rotation systems of G equals the value of the product on the right side. ♢

DEFINITION: The ***genus distribution*** of a graph G is the function that assigns to each nonnegative integer n the number $\gamma_n(G)$.

Example 11.1.6: By Theorem 11.1.6, it follows that K_4 has 16 inequivalent imbeddings. All of them can be obtained by systematically listing the rotation systems for K_4. Two of them are imbeddings in S_0 with four 3-sided faces, six of them are imbeddings in S_1 with a 4-sided face and an 8-sided face, and eight of them are imbeddings in S_1 with a 3-sided face and a 9-sided face. Rather than exhaustively applying the Face-Tracing Algorithm to all 16 rotation systems, it is possible to use the symmetry of the graph and Examples 11.1.1, 11.1.2, and 11.1.5 to complete this calculation.

EXERCISES for Section 11.1

11.1.1 Write the rotation system for the given imbeddings in the sphere S_0.

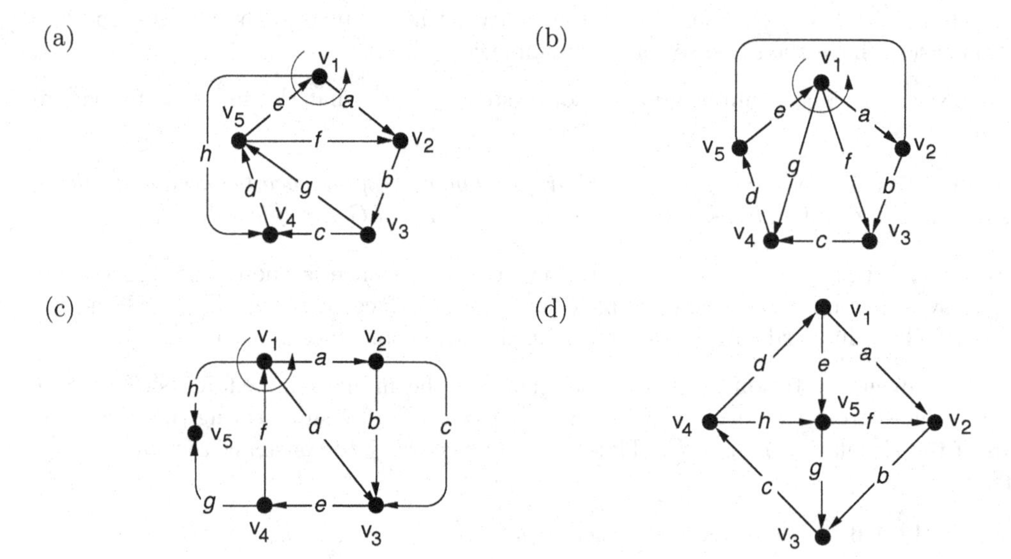

11.1.2 Write the rotation system for the given imbeddings in the torus S_1.

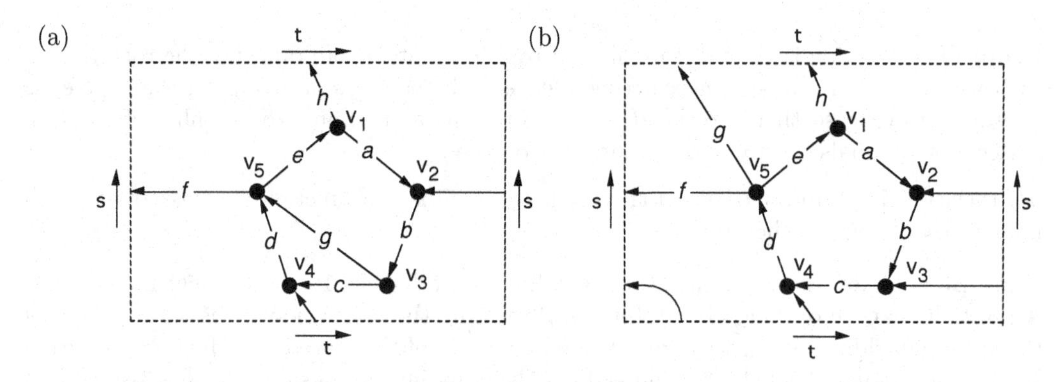

11.1.3 For the given rotation system, (*a*) draw the graph, and (*b*) write the circulation.

(a)

$$\begin{array}{llll} u: & a^+ & b^+ & d^+ \\ v: & b^- & c^- & e^+ \\ w: & a^- & f^+ & c^+ \\ x: & d^- & g^- & i^+ \\ y: & e^- & h^+ & g^+ \\ z: & f^- & i^- & h^- \end{array}$$

(b)

$$\begin{array}{llll} u: & a^+ & b^+ & d^+ \\ v: & b^- & e^+ & c^- \\ w: & a^- & f^+ & c^+ \\ x: & d^- & g^- & i^+ \\ y: & e^- & h^+ & g^+ \\ z: & f^- & i^- & h^- \end{array}$$

(c)

$$\begin{array}{llll} u: & a^+ & b^+ & d^+ \\ v: & b^- & e^+ & c^- \\ w: & a^- & f^+ & c^+ \\ x: & d^- & g^- & i^+ \\ y: & e^- & g^+ & h^+ \\ z: & f^- & i^- & h^- \end{array}$$

(d)

$$\begin{array}{lllll} v: & a^+ & c^- & b^+ & \\ w: & a^- & d^+ & e^+ & \\ x: & e^- & g^+ & h^+ & \\ y: & c^+ & h^- & f^+ & \\ z: & b^- & f^- & g^- & d^- \end{array}$$

(e)

$$\begin{array}{lllll} v: & a^+ & c^- & b^+ & \\ w: & a^- & d^+ & e^+ & \\ x: & e^- & g^+ & h^+ & \\ y: & c^+ & h^- & f^+ & \\ z: & b^- & f^- & d^- & g^- \end{array}$$

(f)

$$\begin{array}{lllll} v: & a^+ & c^- & b^+ & \\ w: & a^- & d^+ & e^+ & \\ x: & e^- & h^+ & g^+ & \\ y: & c^+ & h^- & f^+ & \\ z: & b^- & f^- & d^- & g^- \end{array}$$

11.1.4 For the given rotation system, (*i*) draw the labeled polygons, and (*ii*) draw the imbedding.

(a) The rotation system of Exercise 11.1.3(a)
(b) The rotation system of Exercise 11.1.3(b)
(c) The rotation system of Exercise 11.1.3(c)
(d) The rotation system of Exercise 11.1.3(d)
(e) The rotation system of Exercise 11.1.3(e)
(f) The rotation system of Exercise 11.1.3(f)

11.1.5 Determine the number of inequivalent imbeddings for the given graph.

(a) K_4.
(b) $K_5 - K_2$.
(c) $K_6 - 3K_2$.
(d) W_n.
(e) Q_n.
(f) K_n.
(g) $Q_3 + K_1$.
(h) $K_{5,5}$.
(i) $K_2 \times K_4$.
(j) $P_4 + P_4$.

11.2 GENUS DISTRIBUTION OF A GRAPH

The difficult problem of minimizing the number of handles needed to eliminate all the edge-crossings from a drawing of a complete graph arose in the 19th century. Gerhard Ringel worked from 1950 to 1968 to solve this problem, which he completed in 1968, with the aid of J.W.T. Youngs. The present perspective incorporates a much wider range of problems, involving algebraic enumeration and algorithmics.

Minimum Genus

DEFINITION: The ***minimum genus*** of a connected graph G is the number

$$\gamma_{\min}(G) = \min\{g \mid \gamma_g(G) > 0\}$$

that is, the minimum genus of any orientable surface on which G can be imbedded.

Example 11.2.1: If a graph is planar, then its minimum genus is zero.

Example 11.2.2: If a non-planar graph G has an edge e such that $G - e$ is planar, then $\gamma_{\min}(G) = 1$. Starting with a planar drawing of $G - e$, using surgery to add a handle that joins the regions incident on the endpoints of e, and then drawing edge e on that handle yields an imbedding $G \to S_1$.

TERMINOLOGY NOTE: Before 1970, the phrase *genus of a graph* was used universally for what we now call its *minimum genus*. When there is no danger of ambiguity, we sometimes use the old phrase.

DEFINITION: A ***minimum (orientable) surface*** for a graph G is a surface whose genus is $\gamma_{\min}(G)$.

DEFINITION: Any imbedding of a graph G on a minimum (orientable) surface is called a ***minimum (orientable) imbedding***.

REVIEW FROM §1.2: The ***girth of a graph*** G is the length of a smallest cycle in G.

Theorem 11.2.1: *Let G be a connected graph. Then*

$$\gamma_{\min}(G) \geq \left\lceil \frac{|E|(girth(G)-2)}{2 \cdot girth(G)} - \frac{|V|}{2} + 1 \right\rceil$$

Proof: Given a minimum imbedding $G \to S$, the Euler polyhedral equation is

$$|V| - |E| + |F| = 2 - 2\gamma_{\min}(G)$$

The Edge-Face Inequality (see §4.5 and §5.5) implies that

$$|F| \leq \frac{2|E|}{girth(G)}$$

which implies that

$$|V| - \frac{E|(girth(G)-2)}{girth}(G) \geq 2 - 2\gamma_{\min}(G)$$

and, in turn, that

$$\gamma_{\min}(G) \geq \frac{|E|(girth(G)-2)}{2 \cdot girth(G)} - \frac{|V|}{2} + 1$$

Since $\gamma_{\min}(G)$ is integer-valued, the conclusion follows. ◇

Corollary 11.2.2: *Let G be a simple connected graph. Then*

$$\gamma_{\min}(G) \geq \left\lceil \frac{|E|}{6} - \frac{|V|}{2} + 1 \right\rceil$$

Proof: Since G is simple, it follows that $girth(G) \geq 3$. Thus,

$$\frac{|E|(girth(G)-2)}{2 \cdot girth(G)} = \frac{|E|}{2} - \frac{|E|}{girth(G)} \geq \frac{|E|}{6}$$

and the result follows immediately from Theorem 11.2.1. ◇

Example 11.2.3: Since $girth(K_8) = 3$, it follows from Corollary 11.2.2 that

$$\begin{aligned} \gamma_{\min}(K_8) &\geq \left\lceil \frac{|E|}{6} - \frac{|V|}{2} + 1 \right\rceil \\ &= \left\lceil \frac{28}{6} - \frac{8}{2} + 1 \right\rceil = \left\lceil \frac{5}{3} \right\rceil = 2 \end{aligned}$$

Maximum Genus

The study of the surface of largest genus on which a graph can be cellularly imbedded began with [NoStWh71], followed by [NoRiStWh72].

DEFINITION: The ***maximum genus*** of a connected graph G is the number

$$\gamma_{\max}(G) = \max\{g \mid \gamma_g(G) > 0\}$$

that is, the maximum genus of any orientable surface on which G can be cellularly imbedded.

DEFINITION: A ***maximum (orientable) surface*** for a graph G is a surface whose genus is $\gamma_{\max}(G)$.

DEFINITION: Any imbedding of a graph on a maximum (orientable) surface is called a ***maximum (orientable) imbedding***.

REVIEW FROM §3.2: The ***cycle rank*** (or ***Betti number***) of a connected graph G is

$$\beta(G) = |E| - |V| + 1$$

Theorem 11.2.3: *Let G be a connected graph. Then*

$$\gamma_{\max}(G) \leq \left\lfloor \frac{\beta(G)}{2} \right\rfloor$$

Proof: Given a maximum imbedding $G \to S$, the Euler polyhedral equation is

$$|V| - |E| + |F| = 2 - 2\gamma_{\max}(G)$$

Rearranging the Euler equation yields

$$2\gamma_{\max}(G) + |F| = |E| - |V| + 2 = \beta(G) + 1$$

which implies (since $|F| \geq 1$) that

$$\gamma_{\max}(G) \leq \frac{\beta(G)}{2}$$

Since $\gamma_{\max}$ is integer-valued, the conclusion follows. ◇

Example 11.2.4: The maximum genus of a tree T is 0, since $\beta(T) = 0$.

Example 11.2.5: Example 11.1.2 contains an imbedding of K_4 in S_1. Therefore, $\gamma_{\max}(K_4) \geq 1$. Moreover, by Theorem 11.2.3,

$$\gamma_{\max}(K_4) \leq \left\lfloor \frac{\beta(K_4)}{2} \right\rfloor = \left\lfloor \frac{3}{2} \right\rfloor = 1$$

Combining these two bounds yields the value $\gamma_{\max}(K_4) = 1$.

Genus Range

Our immediate objective is to prove that a graph can be cellularly imbedded on any surface whose genus lies between its minimum genus and its maximum genus.

DEFINITION: The ***genus range*** of a graph G is the set

$$\{g \mid G \text{ has a rotation system } \rho \text{ such that } g \text{ is the genus of } S(\rho)\}$$

DEFINITION: Two rotation systems are ***adjacent*** if one can be obtained from the other by the operation of transposing two consecutive edge-ends in the rotation at one vertex.

Theorem 11.2.4: *Let ρ and λ be adjacent rotation systems for a graph G. Then the genus of the induced surface $S(\rho)$ differs from the genus of the induced surface $S(\lambda)$ by at most 1.*

Proof: Suppose that transposing edge-ends d^σ and e^τ at vertex v transforms rotation system ρ into rotation system λ. Then deleting edge e from the respective induced imbeddings ι_ρ and ι_λ (which eliminates e^τ and $e^{-\tau}$ from their rotation systems) transforms both ρ and λ into the same rotation system η. From an alternative viewpoint, each of the imbeddings

$$\iota_\rho : G \to S(\rho) \quad \text{and} \quad \iota_\lambda : G \to S(\lambda)$$

can be obtained by inserting edge e into the same imbedding $\iota_\eta : G - e \to S(\eta)$. Inserting one edge into an imbedding cannot decrease the genus of the surface, and it increases the genus by at most 1, since it can be drawn, if necessary, on a single new handle. This yields the inequalities

$$\gamma(S(\eta)) \leq \gamma(S(\rho)) \leq \gamma(S(\eta)) + 1 \quad \text{and} \quad \gamma(S(\eta)) \leq \gamma(S(\lambda)) \leq \gamma(S(\eta)) + 1$$

Thus, $|\gamma(S(\rho)) - \gamma(S(\lambda))| \leq 1$. ♢

Corollary 11.2.5: **[*Interpolation Theorem*]** [Du66] *Let G be a connected graph, and let $\gamma_{\min}(G) \leq g \leq \gamma_{\max}(G)$. Then $\gamma_g(G) \geq 1$.*

Proof: Any rotation system can be obtained from any other by a sequence of transpositions of edge-ends. Thus, there is a sequence of adjacent rotation systems from some minimum-genus imbedding to a maximum-genus imbedding. By Theorem 11.2.4, the genus of the imbedding surface varies by at most 1 as one progresses through this sequence. Thus, some intermediate imbedding surface has genus g. ♢

Remark: Thus, Corollary 11.2.5 asserts that the genus range of a graph G is the integer interval $[\gamma_{\min}(G), \gamma_{\max}(G)]$.

Example 11.2.6: The complete graph K_8 has $(7!)^8$ different imbeddings, by Theorem 11.1.6. By Theorems 11.2.1 and 11.2.3, they all lie in the genus range $[2, 10]$.

COMPUTATIONAL NOTE: Since there are super-exponentially many rotation systems to consider, trying to determine the minimum genus or the maximum genus of a given graph, by an exhaustive algorithm, is computationally infeasible. The upper bound of Theorem 11.2.3 is not necessarily the maximum genus. There is a polynomial-time algorithm [FuGrMc88] to calculate the maximum genus. The lower bound of Theorem 11.2.1 is not necessarily the minimum genus. Deciding whether a given integer is the minimum genus of a graph is an NP-complete problem [Th89].

Non-Orientable Imbeddings

The study of non-orientable imbeddings has the significant complication that a graph may be imbedded so that traversing some particular edges, in effect, reverses the rotation order at the destination vertex. For this reason, the coverage here is limited to two basic definitions and two basic theorems.

DEFINITION: The ***minimum crosscap number*** of a connected graph G is the smallest crosscap number of any surface N_k on which G can be imbedded. It is denoted $\overline{\gamma}_{\min}(G)$. If G is planar, then $\overline{\gamma}_{\min}(G) = 0$.

DEFINITION: The ***maximum crosscap number*** of a connected graph G is the largest crosscap number of any surface N_k on which G can be cellularly imbedded. It is denoted $\overline{\gamma}_{\max}(G)$. If G is planar, then $\overline{\gamma}_{\max}(G) = 0$.

Theorem 11.2.6: *Let G be a connected simple graph. Then*

$$\overline{\gamma}_{\min}(G) \geq \left\lceil \frac{|E|}{3} - |V| + 2 \right\rceil$$

Proof: This proof is completely analogous to the proof of Corollary 11.2.2. Details are left as an exercise. ♢

Theorem 11.2.7: *Let G be a connected graph. Then*

$$\overline{\gamma}_{\max}(G) = \beta(G)$$

where $\beta(G) = |E| - |V| + 1$.

Proof: That $\beta(G)$ is an upper bound for $\overline{\gamma}_{\max}(G)$ is provable by reasoning parallel to that for Theorem 11.2.3. Proof of equality was first published by [Ed65c]. ◊

EXERCISES for Section 11.2

11.2.1 Use Corollary 11.2.2 to calculate a lower bound for the minimum genus of the given graph.

(a) K_9. (b) $K_9 - C_9$.
(c) $K_{10} - C_3$. (d) $K_{10} - P_5$.
(e) K_8. (f) $K_8 - K_{2,3}$.
(g) $C_3 \times C_3 \times C_3$. (h) $C_7 + C_8$.
(i) $K_{6,6}$. (j) Q_6.
(k) The result of amalgamating two copies of K_5 at a vertex.
(l) The result of amalgamating two copies of K_5 on an edge.

11.2.2 Apply Theorem 11.2.1 to derive a lower bound for minimum genus that is sharper than the bound of Corollary 11.2.2, for graphs with girth 4.

11.2.3 Apply the result of Exercise 11.2.2 to the graphs of Exercises 11.2.1(i) and 11.2.1(j).

11.2.4 Use Theorem 11.2.3 to calculate an upper bound for the maximum genus of the given graph.

(a) K_9. (b) $K_9 - C_9$.
(c) $K_{10} - C_3$. (d) $K_{10} - P_5$.
(e) K_8. (f) $K_8 - K_{2,3}$.
(g) $C_3 \times C_3 \times C_3$. (h) $C_7 + C_8$.
(i) W_6. (j) CL_3.

11.2.5 Prove Theorem 11.2.6.

11.3 VOLTAGE-GRAPH SPECIFICATION OF GRAPH LAYOUTS

The application that inspired the invention of voltage graphs originally was the specification of graph imbeddings. This section explains how a cellular drawing of a voltage graph on a surface generates an imbedding for the covering graph. In this section, it is assumed, for the sake of simplicity, that the voltages are regular and that the group is additive.

Signed Walks

The concept of signed walks illustrates emphatically that in voltage graphs the edge-directions are *not* to be regarded as traversal restrictions.

DEFINITION: A ***signed walk*** in a digraph is an alternating sequence of vertices and edge-ends

$$v_0, e_1^{\sigma_1}, v_1, e_2^{\sigma_2}, v_2, \ldots, v_{n-1}, e_n^{\sigma_n}, v_n$$

such that the result of deleting signs and ignoring directions is a walk

$$v_0, e_1, v_1, e_2, v_2, \ldots, v_{n-1}, e_n, v_n$$

(i.e., in the underlying graph), and such that reversing the direction on every minus-signed edge and then deleting the signs would yield a directed walk. The fundamental idea is not difficult: in a signed walk, a minus-signed edge-end represents an edge traversed against its designated direction.

DEFINITION: Let

$$W = \langle v_0, e_1^{\sigma_1}, v_1, e_2^{\sigma_2}, v_2, \ldots, v_{n-1}, e_n^{\sigma_n}, v_n \rangle$$

be a signed walk in a voltage graph $\langle G, \alpha : E_G \to B \rangle$. The ***net voltage*** on the walk W is the sum

$$\alpha(W) = \sum_{j=1}^{n} \sigma_j \alpha(e_j)$$

That is, the net voltage is the sum of the voltages on the plus-signed edges minus the sum of the voltages on the minus-signed edges.

Example 11.3.1: Figure 11.3.1 shows a specification of the Petersen graph. The signed walk

$$W = \langle v, d^+, v, c^+, u, e^-, u \rangle$$

starts at vertex v, traverses self-loop d in forward direction back to vertex v, next traverses edge c to vertex u, and then traverses self-loop e in reverse direction back to vertex u. Thus, its net voltage is

$$\alpha(W) = \alpha(d) + \alpha(c) - \alpha(e) = 1 + 0 - 2 = -1 (\text{mod } 5)$$

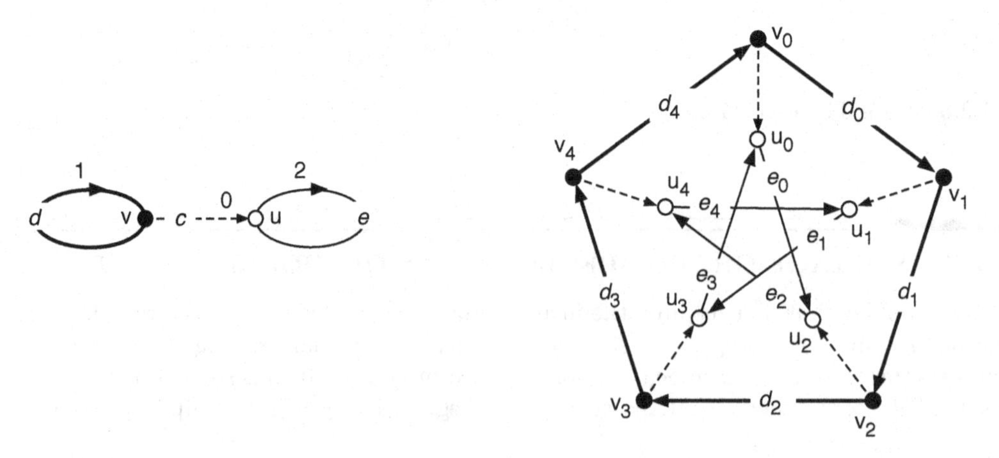

Figure 11.3.1 A voltage graph specifying the Petersen graph.

Theorem 11.3.1: Unique Walk Lifting *Let $\langle G, \alpha : E_G \to A\rangle$ be a voltage graph in which there is a signed walk*

$$W = \langle v_0,\ e_1^{\sigma_1},\ v_1,\ e_2^{\sigma_2},\ v_2,\ \ldots,\ v_{n-1},\ e_n^{\sigma_n},\ v_n\rangle$$

and let $a \in A$. Then there is a unique alternating sequence

$$W_a = v_{0,a},\ e_{1,a}^{\sigma_1},\ v_{1,k_1},\ e_{2,k_1}^{\sigma_1},\ v_{2,k_2},\ \ldots,\ v_{n-1,k_{n-1}},\ e_{n,k_{n-1}}^{\sigma_n},\ v_{n,k_n}$$

that forms a signed walk in the covering digraph G^α.

Proof: For $j = 1, \ldots, n$, choose

$$k_j = a + \sum_{\ell=1}^{j-1} \alpha(e_\ell)$$

This formula supports an inductive argument that each successive edge choice e_{j,k_j} is the only choice of an edge in the fiber of e_j that extends the signed walk. ♢

DEFINITION: Let W be a signed walk in a voltage graph with initial vertex v. The unique signed walk W_a in the covering graph that starts at vertex v_a and projects onto walk W (as in Theorem 11.3.1) is called a ***lift of walk*** W.

Example 11.3.2: *The walk $W = \langle v, d^+, v, c^+, u, e^-, u\rangle$ in Example 11.3.1 has these five lifts:*

$$W_0 = \langle v_0, d_0^+, v_1, c_1^+, u_1, e_4^-, u_4\rangle$$
$$W_1 = \langle v_1, d_1^+, v_2, c_2^+, u_2, e_0^-, u_0\rangle$$
$$W_2 = \langle v_2, d_2^+, v_3, c_3^+, u_3, e_1^-, u_1\rangle$$
$$W_3 = \langle v_3, d_3^+, v_4, c_4^+, u_4, e_2^-, u_2\rangle$$
$$W_4 = \langle v_4, d_4^+, v_0, c_0^+, u_0, e_3^-, u_3\rangle$$

Proposition 11.3.2: *Let $\langle G,\ \alpha : E_G \to A\rangle$ be a voltage graph, and let W be a signed walk in G from u to v. Then the lift W_b is a signed walk in the covering graph G^α from the vertex u_b to the vertex $v_{b+\alpha(W)}$.*

Proof: This follows by induction on the length of the walk. ♢

Example 11.3.2, continued: The net voltage on walk W from vertex v to vertex u is 4 (mod 5). Observe that each walk W_j begins at v_j and ends at u_{j+4}.

Kirchoff's Voltage Law

DEFINITION: A closed walk of an imbedded voltage graph satisfies the ***Kirchhoff's voltage law (KVL)*** if its net voltage is the identity element of the voltage group.

Theorem 11.3.3: *Let W be a signed closed walk in a voltage graph that satisfies KVL. Then every lift W_b is a closed walk in the covering graph G^α.*

Proof: This is an immediate consequence of Proposition 11.3.2. ♢

Theorem 11.3.4: *Let $\langle G,\ \alpha : E_G \to A\rangle$ be a voltage graph, and let W be a signed closed walk in G, such that $\alpha(W) = b$ is of order m in the voltage group A. Then the concatenation*

$$W_a W_{a+b} \cdots W_{a+(m-1)b}$$

is a closed walk in G^α.

Proof: Suppose that walk W starts at vertex v. By Proposition 11.3.2, the lift $W_{a+\ell b}$ is a walk from $v_{a+\ell b}$ to $v_{a+(\ell+1)b}$, for $\ell = 0, \ldots, m-1$. It follows that the iterated concatenation $W_a W_{a+b} \cdots W_{a+(m-1)b}$ is a walk from v_a to $v_{a+mb} = v_a$, that is, a closed walk. ◇

Remark: A cyclic permutation of the constituents of the concatenation

$$W_a W_{a+b} \cdots W_{a+(m-1)b}$$

may start and stop at some vertex $v_{a+\ell b}$ different from v_a, but it traverses exactly the same cyclic sequence of vertices and edges. In the context of imbedded voltage graphs, it is called a *cyclically equivalent concatenation.*

KVL-Imbedded Voltage Graphs

DEFINITION: An ***imbedded voltage graph*** is a pair $\langle \iota : G \to S,\ \ \alpha : E_G \to A \rangle$. The first component is a cellular imbedding of a graph G in a surface S, called the **base imbedding**. The second component is a voltage assignment on graph G.

DEFINITION: A face of an imbedded voltage graph is said to satisfy the KVL if the net voltage on the face boundary walk is 0.

DEFINITION: A ***KVL-imbedded voltage graph*** is an imbedded voltage graph in which every face satisfies KVL.

Remark: In the preceding definition, notice the difference between the classical Kirchoff's voltage law of physics and its topological namesake. Whereas the physical law is that a net voltage gain of 0 occurs in the traversal of every closed walk, the topological law imposes this requirement only on face-boundary walks. (See Exercises.)

DEFINITION: In a KVL-imbedded voltage graph with voltage group A, let f be a k-sided face with boundary walk W. Then the ***fiber over*** f, denoted $\hat{f}$, is the set $\{f_a \mid a \in A\}$ of k-sided polygons, called the ***covering faces***. The boundary walk of the covering face f_a is labeled by the lifted walk W_a of the covering graph.

DEFINITION: Let $\langle \iota : G \to S,\ \ \alpha : E_G \to A \rangle$ be a KVL-imbedded voltage graph. Fitting each covering face f_a, for $f \in F_\iota$ and $a \in A$, in accordance with its labeling to the corresponding lift of a boundary walk of f forms the ***covering surface*** S^α and the ***covering imbedding*** $\iota^\alpha : G^\alpha \to S^\alpha$.

Example 11.3.3: Figure 11.3.2 illustrates a one-face KVL imbedding $B_2 \to S_1$ with voltages in $\mathbb{Z}_5$ and the covering imbedding $K_5 \to S_1$ that it specifies.

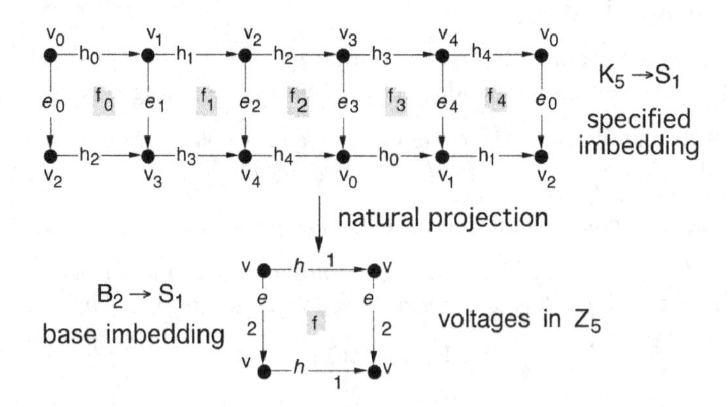

Figure 11.3.2 A KVL voltage graph and its covering imbedding.

Example 11.3.4: Often, it helps to construct the covering graph alone before constructing the covering imbedding. For instance, Figure 11.3.3 shows how to assign voltages from $\mathbb{Z}_7$ to the bouquet B_3 so that the covering graph is the complete graph K_7, with seven vertices and 21 edges.

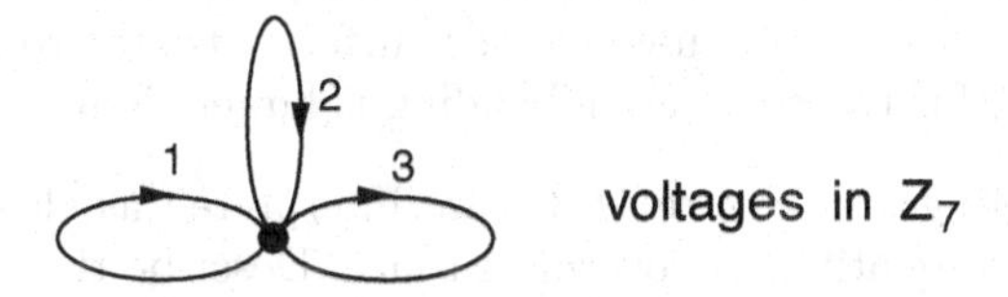

Figure 11.3.3 Specifying K_7 by $\mathbb{Z}_7$-voltages on B_3.

Figure 11.3.4 illustrates a two-face KVL imbedding of that voltage graph on S_1 and the covering imbedding. Since the covering imbedding has 14 faces, the covering surface must be the surface S_g whose Euler characteristic is $7 - 21 + 14 = 0 = 2 - 2g$, that is, the surface S_1. Shading is used to group the faces of the covering imbedding into fibers.

TERMINOLOGY: The imbedded-voltage-graph construction extends the *natural projection* $p : G^\alpha \to G$, so that it maps each face f_a in the fiber $\tilde{f}$ to the face f, and thereby becomes a mapping $p : S^\alpha \to S$ of surfaces.

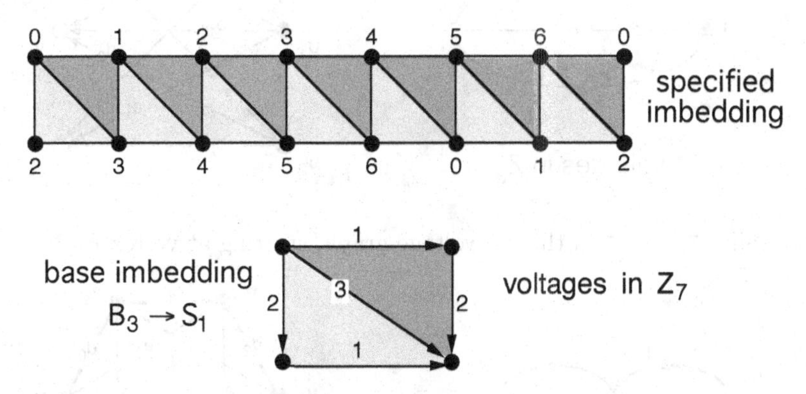

Figure 11.3.4 Specification of an imbedding $K_7 \to S_1$.

EXERCISES for Section 11.3

11.3.1 Draw the covering imbedding generated by the given KVL voltage graph in S_1.

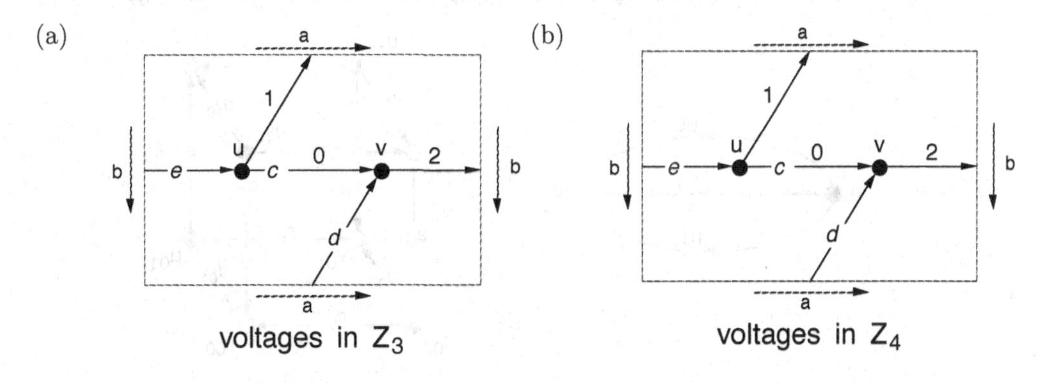

11.3.2 Prove that a KVL-imbedded voltage graph in S_1 specifies an imbedding such that every component of the imbedding surface is a torus. (Hint: Prove that the value of the Euler polyhedral formula $|V| - |E| + |F|$ for the covering imbedding is 0.)

11.3.3 Prove that a KVL-imbedded voltage graph in S_0 with nontrivial voltage group specifies a non-connected graph and a non-connected surface. (Hint: Prove that the value of the formula $|V| - |E| + |F|$ for the covering imbedding is larger than 2.)

11.3.4 Suppose the voltages in Figure 11.3.2 are interpreted as elements of $\mathbb{Z}_6$. Draw the covering imbedding and identify the covering surface. Describe the covering graph.

11.3.5 Suppose the voltages in Figure 11.3.4 are interpreted as elements of $\mathbb{Z}_8$. Draw the covering imbedding and identify the covering surface. Describe the covering graph.

11.3.6 Construct the designated lift of the given signed walk.

(a) Signed walk c^+, d^-, c^+, b^- in this $\mathbb{Z}_3$-voltage graph, starting at vertex u_2.

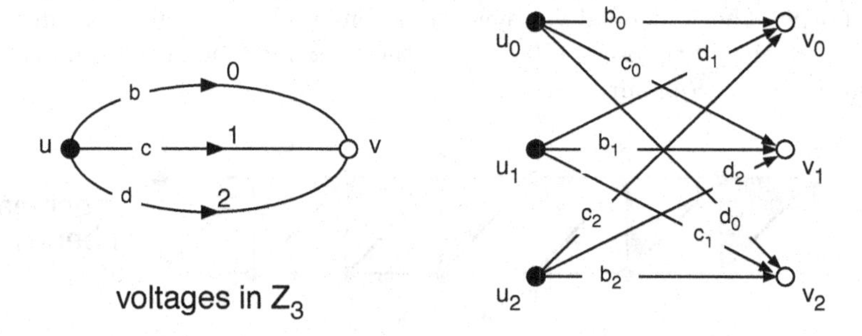

(b) Signed walk e^-, e^-, d^+ in this $\mathbb{Z}_6$-voltage graph, starting at vertex v_4.

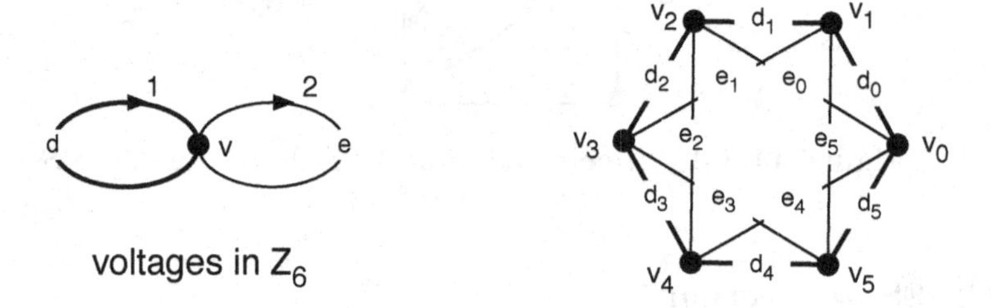

(c) Signed walk c^+, d^-, e^+ in this $\mathbb{Z}_2 \times \mathbb{Z}_2$-voltage graph, starting at vertex u_{01}.

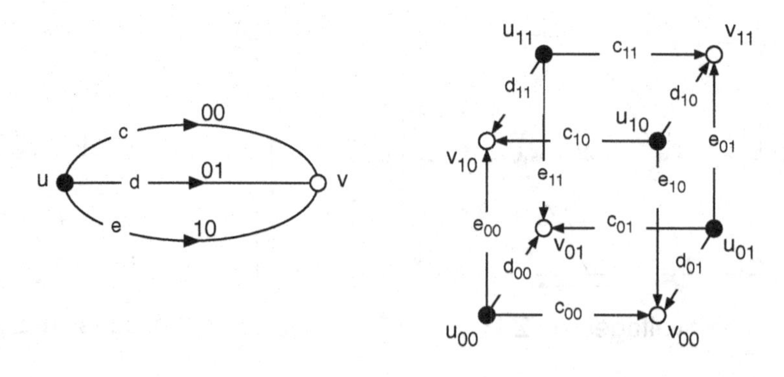

(d) Signed walk c^+, d^-, e^+ in this $\mathbb{Z}_2 \times \mathbb{Z}_2 \times \mathbb{Z}_2$-voltage graph, starting at vertex u_{101}.

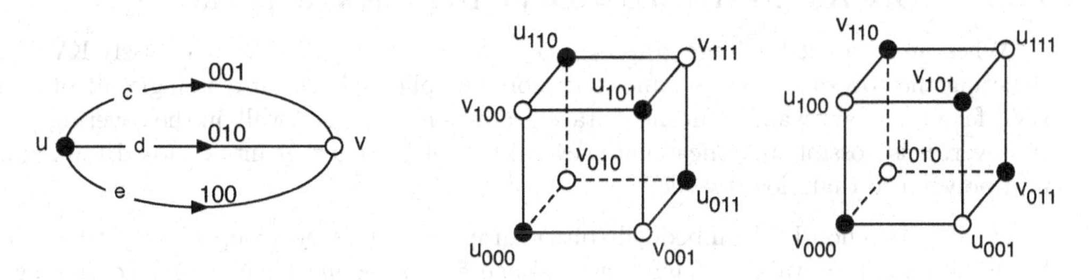

(e) Signed walk b^+, c^+, d^- in this permutation-Σ_3-voltage graph, starting at vertex u_3.

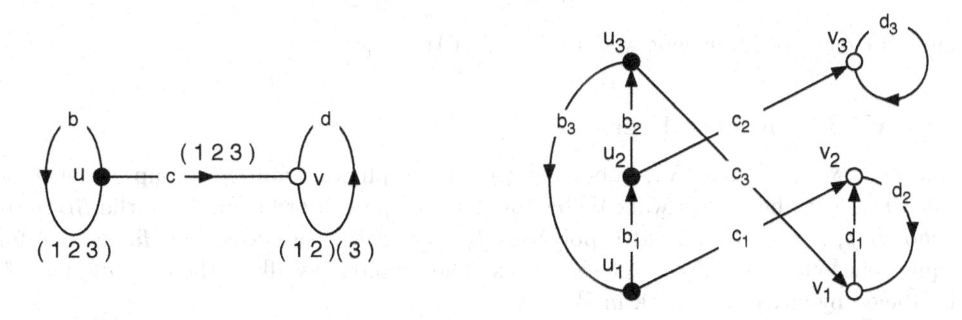

DEFINITION: Let $\langle G, \alpha : E_G \to \mathcal{B}\rangle$ be a regular-voltage graph, and let $v \in V_G$. The ***local group*** at v is the subgroup of $\mathcal{B}$ comprising the net voltages on all closed walks based at v. (This concept concerns components of the covering graph.)

11.3.7 Let $\langle G, \alpha : E_G \to \mathcal{B}\rangle$ be a voltage graph, and let $u, v \in V_G$. Prove that vertex v_j is in the same component of the covering graph as vertex u_i if and only if there is a signed walk from u to v in G whose net voltage equals $j - i$.

11.3.8 Let $\langle G, \alpha : E_G \to \mathcal{B}\rangle$ be a voltage graph. Use Exercise 11.3.7 to prove that the components of the covering graph are in one-to-one correspondence with the cosets of the local group in the voltage group $\mathcal{B}$ [AlGr76].

11.3.9 Let G be a bipartite graph with voltage $1 \in \mathbb{Z}_2$ assigned to every edge. Prove that the Kirchoff's voltage law holds for every closed walk. Prove also that the covering graph consists of two disjoint copies of G. (This illustrates what happens when KVL holds on every closed walk, rather than only on face-boundary walks.)

11.4 NON-KVL-IMBEDDED VOLTAGE GRAPHS

When an imbedded voltage graph $\langle \iota : G \to S, \;\; \alpha : E_G \to A \rangle$ does not satisfy KVL, constructing the covering faces is somewhat more complicated, because a single lift of a non-KVL face-boundary walk W in the voltage graph is not a closed walk in the covering graph. However, it is possible to concatenate several lifts of W together into a closed walk and to fit a polygon to that closed walk.

NOTATION: In a non-KVL-imbedded voltage graph with voltage group A, let f be a k-sided face with boundary walk W having net voltage b of order m. Then the set containing the walk

$$W_a W_{a+b} \cdots W_{a+(m-1)b}$$

and all cyclically equivalent walks is denoted $W_{a+<b>}$.

Non-KVL Covering Faces

DEFINITION: In a non-KVL-imbedded voltage graph with voltage group A, let f be a k-sided face with boundary walk W having net voltage b of order m. Then the ***fiber over*** f, denoted $\tilde{f}$, is a set of mk-sided polygons $f_{a+<b>}$, called the ***covering faces***, one for each equivalence class $W_{a+<b>}$ of closed walks. The boundary walk of the covering face $f_{a+<b>}$ is labeled by any closed walk in $W_{a+<b>}$.

DEFINITION: Let $\langle \iota : G \to S, \;\; \alpha : E_G \to A \rangle$ be a non-KVL-imbedded voltage graph. Fitting each covering face $f_{a+<b>}$, for $f \in F_\iota$ and $a \in A$, in accordance with its labeling to the corresponding closed lift of a boundary walk of f forms the ***covering surface*** S^α and the ***covering imbedding*** $\iota^\alpha : G^\alpha \to S^\alpha$.

Example 11.4.1: The imbedded $(Z_2 \times Z_2)$-voltage graph in Figure 11.4.1 has three faces, all 2-sided. The net voltage on each face boundary has order 2. Thus, the 12 boundary walk lifts combine to form six face boundaries in the covering graph. They fit together, as shown, to form the surface of a cube. In particular, the upper digon of the imbedded voltage graph specifies the top and bottom faces of the cube; the lower digon specifies the left and right faces; and the exterior digon specifies the front and back faces.

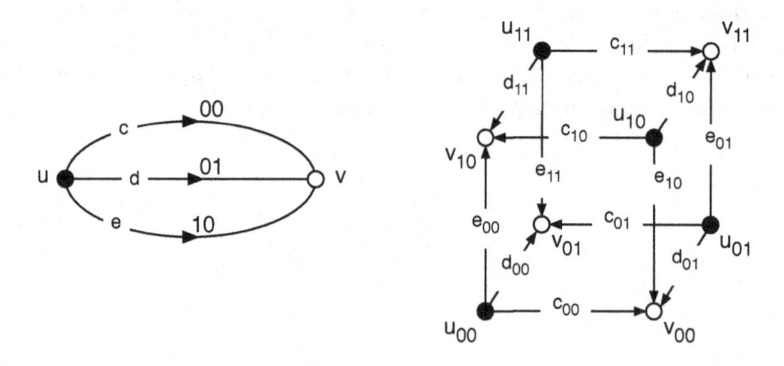

Figure 11.4.1 Specification of an imbedding $Q_3 \to S_0$.

Minimum-Genus Formula for Hypercube Graphs

A standard approach to deriving a minimum-genus formula for a family of graphs is to use algebraic methods to establish a lower-bound formula and a voltage-graph construction of imbeddings that realize that lower bound. The family of hypercubes Q_n illustrates the doubly fortunate circumstance in which no surface surgery is needed, and one simple pattern of imbedded-voltage-graph drawing is enough to specify all the imbeddings.

Proposition 11.4.1: $\gamma_{\min}(Q_n) \geq (n-4) \cdot 2^{n-3} + 1$, *for* $n \geq 2$.

Proof: For $n = 2$ or $n = 3$, the right side is 0. For $n \geq 4$, Theorem 11.2.1 gives the generic lower bound

$$\gamma_{\min}(G) \geq \left\lceil \frac{|E|\,(girth(G) - 2)}{2girth(G)} - \frac{|V|}{2} + 1 \right\rceil$$

Since $|V(Q_n)| = 2^n$, $|E(Q_n)| = n \cdot 2^{n-1}$, and $girth(Q_n) = 4$, this particularizes to

$$\begin{aligned} \gamma_{\min}(Q_n) &\geq \left\lceil \frac{n \cdot 2^{n-1}(4-2)}{2 \cdot 4} - \frac{2^n}{2} + 1 \right\rceil \\ &= \frac{n \cdot 2^{n-1}(4-2)}{2 \cdot 4} - \frac{2^n}{2} + 1 \\ &= n \cdot 2^{n-3} - 2^{n-1} + 1 \\ &= (n-4) \cdot 2^{n-3} + 1 \end{aligned}$$

◇

Proposition 11.4.2: $\gamma_{\min}(Q_n) \leq (n-4) \cdot 2^{n-3} + 1$, *for* $n \geq 2$.

Proof: It suffices to specify an imbedding of Q_n on the surface of genus $(n-4) \cdot 2^{n-3} + 1$. The voltage graph of Figure 11.4.1 for Q_3 generalizes to dimension n, as shown in Figure 11.4.2.

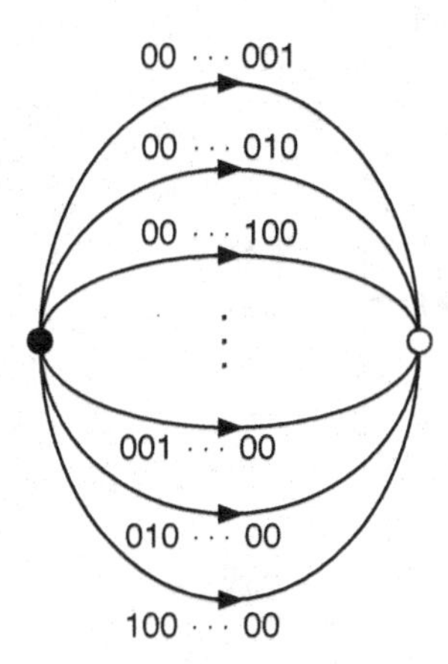

Figure 11.4.2 Voltage graph specifying Q_n.

Since $girth(Q_n) = 4$, every face of an imbedding of Q_n must have at least four sides. The formula provided by Theorem 11.2.1 depends on having as many faces as possible, which implies that the number of sides of almost every face of the imbedding must equal the girth.

With the voltage graph of Figure 11.4.2, it is easy to construct an imbedding in which all the faces are 2-sided and have net voltage of order 2 on their boundary walks. In fact, if that drawing is interpreted as an imbedding in S_0, then each face is a digon with net voltage of order 2 on its boundary walk. ◇

EXERCISES for Section 11.4

11.4.**1** Draw the derived imbedding for the given voltage graph imbedded in S_0.

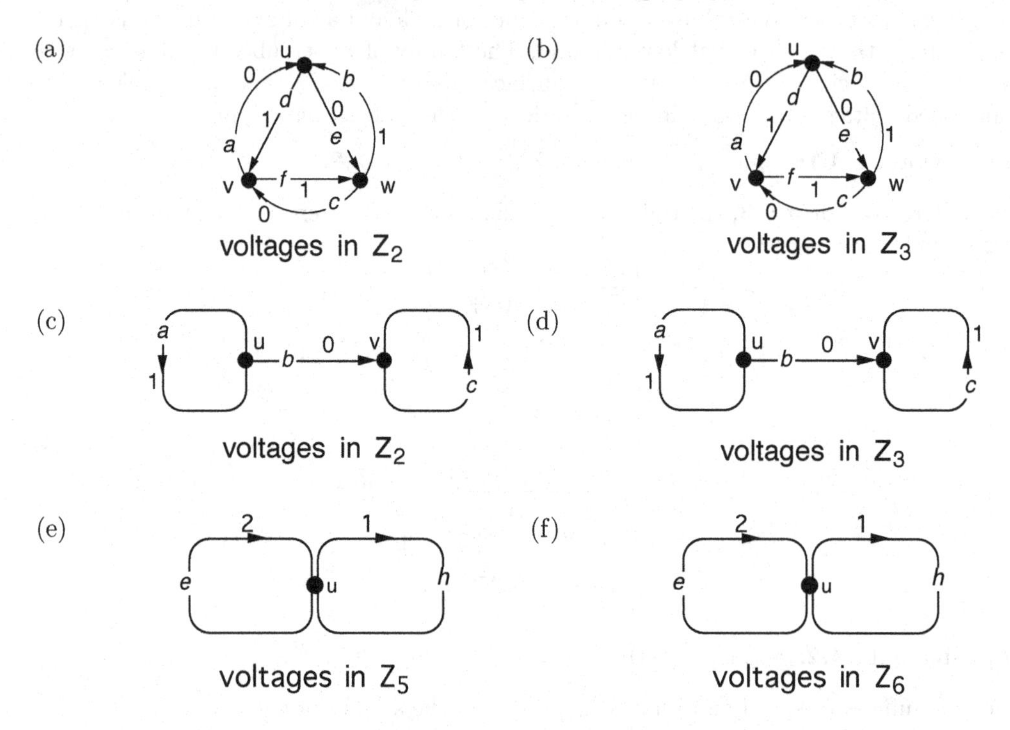

11.4.**2** For the given voltage graph imbedded in S_1, draw the derived faces and determine the genus of the imbedding surface.

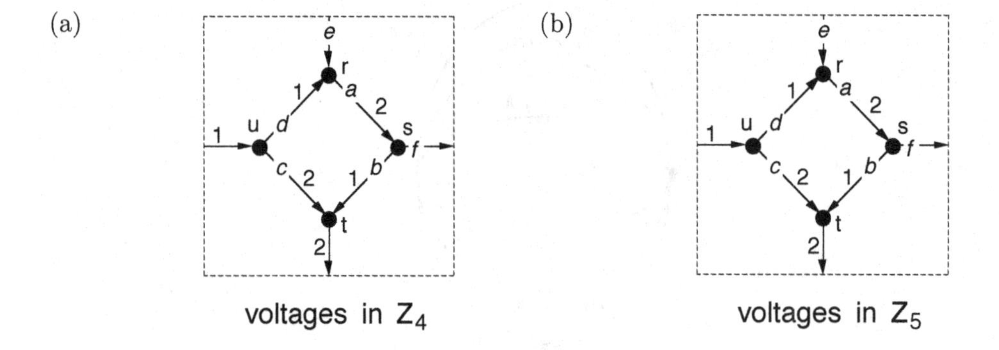

11.5 HEAWOOD MAP-COLORING PROBLEM

The question of sufficiency of four colors (see §6.3) needed for a proper coloring of an arbitrary sphere map was generalized by Percy Heawood. Heawood derived an upper bound for the number of colors needed for an arbitrary map on any closed surface, given as a formula in its Euler characteristic. Heawood's upper bound was ultimately proved to be the exact maximum for all surfaces except the Klein bottle N_2.

DEFINITION: The ***chromatic number of a surface*** S is the maximum of the chromatic numbers of the simple graphs that can be imbedded on S.

REVIEW FROM §5.5: The ***Euler characteristic*** of the surface S_g with g handles is the number $2-2g$.

NOTATION: As explained in §5.5, when colorings arise in topological graph theory, we need to avoid confusion between conflicting uses of the Greek letter χ. Here we denote the chromatic number of a graph or of a surface by $chr(G)$ or $chr(S)$, respectively. The Euler characteristic of a surface is denoted by $ec(S)$.

Heawood Number of a Surface

The following lemmas provide basic upper bounds for the chromatic number.

Lemma 11.5.1: *Let G be a $chr(S)$-colorable chromatically critical graph imbedded on a closed surface S of Euler characteristic $ec(S)$. Then $chr(S) \le \left\lfloor 7 - \frac{6 \cdot ec(S)}{|V_G|} \right\rfloor$.*

Proof: The notations δ_{avg} and $\delta_{\min}$ indicate average degree and minimum degree.

$$
\begin{aligned}
chr(S) &= chr(G) \\
&\le \delta_{\min}(G) + 1 && \text{(by Theorem 6.1.22)} \\
&\le \delta_{avg}(G) + 1 \\
&\le 6 - \frac{6 \cdot ec(S)}{|V_G|} + 1 && \text{(by Theorem 5.5.6)} \\
&= \left\lfloor 7 - \frac{6 \cdot ec(S)}{|V_G|} \right\rfloor && \text{(since } chr(S) \text{ is an integer)}
\end{aligned}
$$

◇

Lemma 11.5.2: *Let S be a closed surface of Euler characteristic $ec(S) \le 0$. Then*

$$chr(S) \le 7 - \frac{6 \cdot ec(S)}{chr(S)}$$

Proof: Let G be a $chr(S)$-colorable chromatically critical graph imbedded on a closed surface S of Euler characteristic $ec(S)$. Since $chr(S) = chr(G) \le |V_G|$ and $ec(S) \le 0$,

$$\frac{6 \cdot ec(S)}{|V_G|} \ge \frac{6 \cdot ec(S)}{chr(S)}$$

Therefore, by Lemma 11.5.1,

$$chr(S) \le 7 - \frac{6 \cdot ec(S)}{|V_G|} \le 7 - \frac{6 \cdot ec(S)}{chr(S)}$$

◇

DEFINITION: The ***Heawood number*** of a closed surface S is the number

$$H(S) = \left\lfloor \frac{7 + \sqrt{49 - 24 \cdot ec(S)}}{2} \right\rfloor$$

Theorem 11.5.3: Heawood, 1890 *Let S be a closed surface, orientable or non-orientable, with Euler characteristic $ec(S) \leq 1$. Then $chr(S) \leq H(S)$.*

Proof: By Poincaré duality, it suffices to show that an arbitrary graph G imbeddable on a surface S has chromatic number less than or equal to the Heawood number of the surface. Moreover, it suffices to assume that G is chromatically critical.

For $ec(S) = 1$, Lemma 11.5.1 yields $chr(N_1) \leq 6 = H(1)$. For $ec(S) \leq 0$, Lemma 11.5.2 implies that

$$chr(S)^2 - 7chr(S) + 6 \cdot ec(S) \leq 0$$

Factoring the quadratic polynomial on the left side yields the inequality

$$\left(chr(S) - \frac{7 - \sqrt{49 - 24 \cdot ec(S)}}{2} \right) \left(chr(S) - \frac{7 + \sqrt{49 - 24 \cdot ec(S)}}{2} \right) \leq 0$$

Since $ec(S) \leq 0$, the value of the radical is larger than 7, from which it follows that the first factor is positive. This implies that the second factor is nonpositive. Since $chr(S)$ is an integer, the conclusion follows. ◇

Remark: Notice that the sphere S_0 satisfies the conclusion of Theorem 11.5.3 ($chr(S_0) = 4 = H(S_0)$), despite not satisfying the hypothesis (since $ec(S_0) = 2$).

Heawood Conjecture and Minimum Genus

Heawood omitted proof that, for each surface, there is a map that realizes the Heawood number. When the gap was noticed, Heawood's assertion was reformulated as a conjecture.

DEFINITION: The ***Heawood conjecture*** is that a surface S has chromatic number $H(S)$.

Proposition 11.5.4: *If $\gamma_{\min}(K_n) \leq \left\lceil \frac{(n-3)(n-4)}{12} \right\rceil$ for all $n \geq 4$, then the Heawood conjecture holds for all orientable surfaces.*

Proof: The surface S_g has Heawood number $H(S_g) = \left\lfloor \frac{7+\sqrt{1+48g}}{2} \right\rfloor$. If

$$\gamma_{\min}(K_n) \leq \left\lceil \frac{(n-3)(n-4)}{12} \right\rceil \quad \text{for all } n \geq 4$$

and if

$$(\clubsuit) \left\lceil \frac{(H(S_g) - 3)\,(H(S_g) - 4)}{12} \right\rceil \leq g$$

then $\gamma_{\min}\left(K_{H(S_g)}\right) \leq g$. Proving inequality ($\clubsuit$) requires routine computation and is left as an exercise. ◇

Philip Franklin proved in 1934 that $chr(N_2) = 6$. Since the Euler characteristic of the Klein bottle (the surface N_2) is 0, and since $H(N_2) = 7$, Franklin's result is a counterexample to the Heawood conjecture. However, the Klein bottle is the only surface whose chromatic number is less than its Heawood number.

Proposition 11.5.5: *If* $\bar{\gamma}_{\min}(K_n) \leq \left\lceil \frac{(n-3)(n-4)}{12} \right\rceil$ *for all* $n \geq 8$, *then the Heawood conjecture holds for all non-orientable surfaces except the Klein bottle.*

Proof: The proof follows the same lines as for the orientable case. ◇

In the 1950s and early 1960, Gerhard Ringel proved the Heawood conjecture completely for the non-orientable surfaces (except N_2, of course), based on Proposition 11.5.5. The denominator of 12 in Proposition 11.5.4 led to the partition of the complete graphs K_n into 12 cases, one for each residue class of n modulo 12. By the early 1960s, Ringel had also constructed minimum imbeddings of K_n for four of the 12 cases.

Ringel described his minimum imbeddings by rotation systems. Each of his rotation systems was developed algebraically from one or more generating rows. Great combinatorial skill was involved in the construction of generating rows. Gustin [Gu63] introduced a graphical *nomogram* called a ***current graph*** as an aid in constructing generating rows of rotation systems. Figure 11.5.1 shows the Gustin nomogram used to construct the generating row for K_7.

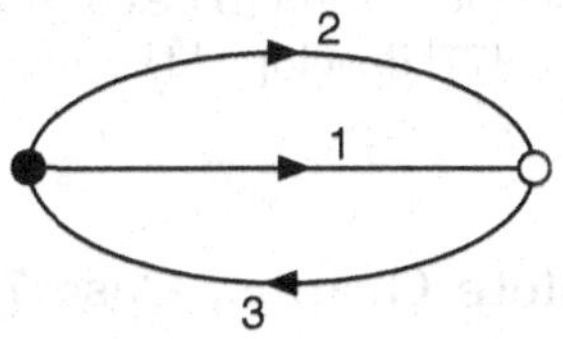

Figure 11.5.1 The Gustin current graph for a toroidal imbedding of K_7.

Gustin called the algebraic elements in a nomogram ***currents***, and he took these currents to be in the group $\mathbb{Z}_7$. The generating row was taken to be the sequence of currents one encountered in a traversal associated with the nomogram, in which a solid vertex (•) indicates clockwise rotation and a hollow vertex (o) indicates counterclockwise rotation.

For instance, if one traverses current 1 to (o), and goes counterclockwise, one next traverses current 3 to (•), where one goes clockwise. One then traverses current 2 to (o), and goes counterclockwise to the edge with current 1, but against the current, so we regard this as current 6, the additive inverse of 1 mod 7. Current 6 terminates at (•), where clockwise rotation leads to current 4, which leads at (o) to current 5, which cycles back to current 1, signifying termination of the tour. We summarize the tour.

0. 1 3 2 6 4 5

This tour is taken as the generating row of a rotation system for K_7, in which each row is generated from the top row by adding the row number modulo 7 to the entries of the top row.

0. 1 3 2 6 4 5
1. 2 4 3 0 5 6
2. 3 5 4 1 6 0
3. 4 6 5 2 0 1
4. 5 0 6 3 1 2
5. 6 1 0 4 2 3
6. 0 2 1 5 3 4

Gustin proved a theorem that if the current graph is 3-regular and if the sum of the currents at each vertex is 0 (a condition called the ***Kirchhoff's current law (KCL)***), then the resulting rotation system specifies a triangulation.

Ringel and Youngs augmented Gustin's method into an extensive family of nomograms, each with its own set of computational rules, and completed their proof of the Heawood conjecture for orientable surfaces in 1968, one of the outstanding mathematical accomplishments of the 20th century. Their complete proof occupies over 300 journal pages.

Gross and Alpert [GrAl73, GrAl74] introduced a topological interpretation of current graphs, which enabled them to generalize the construction greatly and to replace all the nomograms, with their separate rule systems, by a single form of imbedding specification. Topologists recognize the underlying mechanism as a *branched covering*, which generalizes the topological relationship of a Riemann surface to the complex plane. This topological interpretation reduced the length of the proof by about half.

Gross [Gr74] introduced voltage graphs, whose topologically dual perspective simplifies the situation by permitting the specified graph to be considered separately from the specified imbedding. Gross and Tucker [GrTU77] developed the generalization to permutation-voltage graphs (§10.5).

Minimum Genus of Complete Graphs, Case 7

The construction of a minimum imbedding for the complete graph K_n has many details. The least complicated case is for $n = 7 \mod 12$. We recall that Example 11.3.4 gives an imbedded voltage graph that specifies a minimum imbedding of K_7. This smallest instance starts a pattern that continues with minimum imbeddings for K_{19} and for K_{31}.

Example 11.5.1: *The imbedded voltage graph in Figure 11.5.2 specifies a minimum imbedding of* K_{19}.

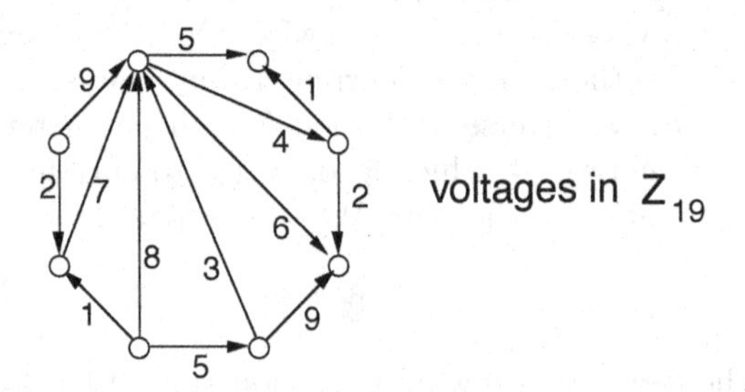

Figure 11.5.2 KVL voltage graph for imbedding $K_{19} \to S_{20}$.

Example 11.5.2: *The imbedded voltage graph in Figure 11.5.3 specifies a minimum imbedding of* K_{31}.

Notice that KVL holds globally. Assigning voltages so that KVL holds globally is the difficult part.

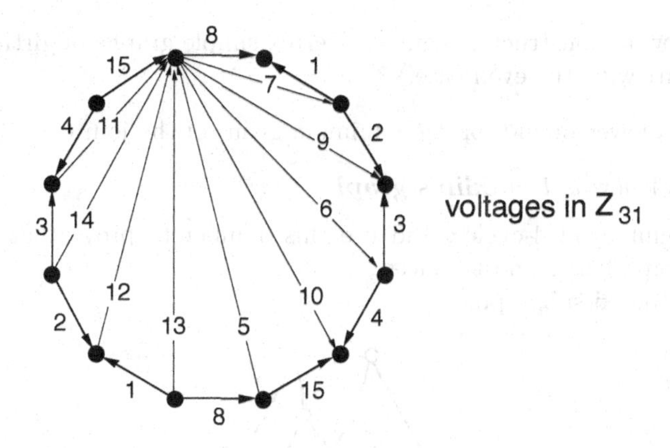

Figure 11.5.3 KVL voltage graph for imbedding of $K_{31} \to S_{63}$.

EXERCISES for Section 11.5

11.5.1 Draw an imbedded voltage graph that specifies a minimum imbedding of the given graph.

(a) K_{43}.
(b) $K_{20} - 10K_2$.
(c) $K_{21} - C_{21}$.
(d) $K_{44} - 22K_2$.
(e) $K_{4,4} - 4K_2$.
(f) $K_{5,5} - C_{10}$.

11.5.2 Verify the inequality (♣) in the proof of Proposition 11.5.4.

11.6 SUPPLEMENTARY EXERCISES

11.6.1 Calculate the face-boundary walks induced by the following rotation system. What is the genus of the induced surface?

$$\begin{array}{llll} v_1: & a^+ & b^+ & c^+ & d^+ \\ v_2: & d^- & e^- & h^+ \\ v_3: & c^- & g^- & h^- \\ v_4: & b^- & f^- & g^+ \\ v_5: & a^- & f^+ & e^+ \end{array}$$

11.6.2 Calculate lower bounds for the minimum genus of the following graphs.

a. K_{13}. b. $K_{13} - K_3$. c. $K_{8,8}$.

11.6.3 Calculate upper bounds for the maximum genus of the following graphs.

a. K_{13}. b. $K_{13} - K_3$. c. $K_{8,8}$.

11.6.4 Prove that minimum genus of $C_4 \times C_4 \times C_4$ is at least 17.

11.6.5 Prove that the minimum genus of a simple, connected graph with 8 vertices and 25 edges is at least two.

11.6.6 Prove that the minimum genus of $K_9 - C_9$ is 1.

11.6.**7** Explain how to construct 6-regular n-vertex simple graphs of girth at least 4 for all $n \geq 12$. (Hint: start with the even case.)

11.6.**8** Calculate a lower bound for the minimum genus of the join $C_7 + Q_3$.

11.6.**9** The graph below is ***Franklin's graph.***

(a) Count the number of 4-cycles, and use this number to prove that an imbedding of Franklin's graph has at most 7 faces.
(b) Find a Kuratowski subgraph.

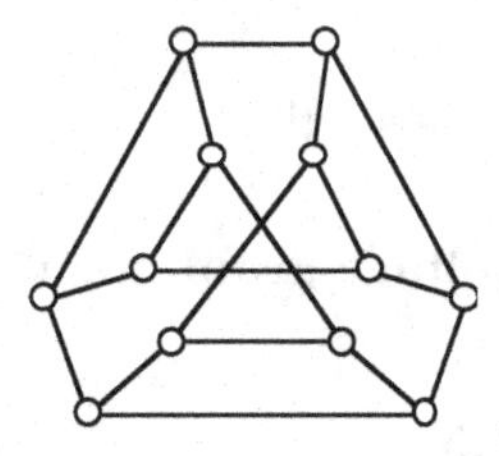

11.6.**10** Prove that for every $n \geq 7$ there is a 6-regular, n-vertex simple graph of genus 1. (Hint: use voltage graphs.)

11.6.**11** Let $G = K_{10} - CL_5$.

(a) Draw a cyclic voltage graph for G, with the minimum possible number of vertices.
(b) To cellularly imbed G into the torus, how many faces would there be?
(c) To cellularly imbed the voltage graph of part (a) in the torus, how many faces would there be?
(d) Draw an imbedded voltage graph for an imbedding of G in the torus.

11.6.**12** The graph shown below is imbedded in the sphere, with voltages in the cyclic group Z_8

(a) How many vertices and edges does the covering graph have?
(b) How many faces are generated by the digon, the triangle, and the exterior face? How many sides do they have?
(c) What is the genus of the covering surface?

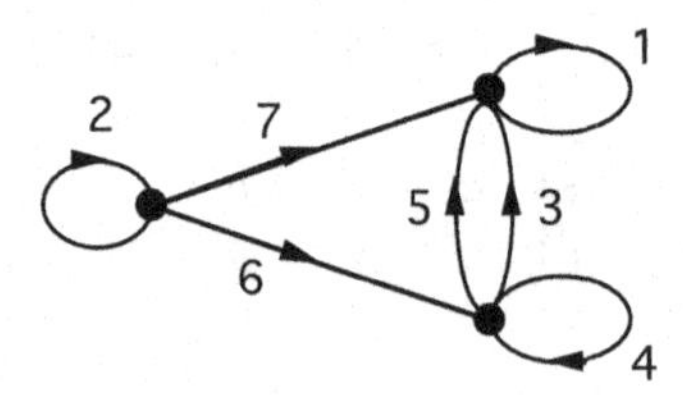

11.6.**13** The voltage graph below is imbedded on the torus.

(a) Calculate the genus of the derived surface when the voltages are in Z_7.
(b) Calculate the genus of the derived surface when the voltages are in Z_8.

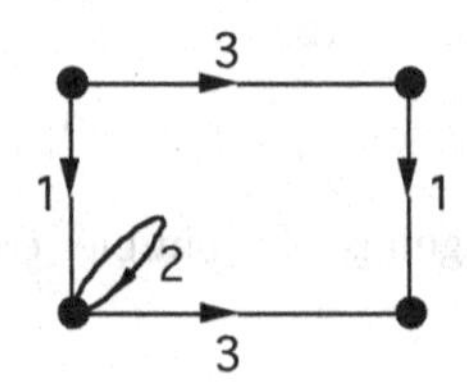

11.6.14 Draw an imbedding of the following graph on the torus.

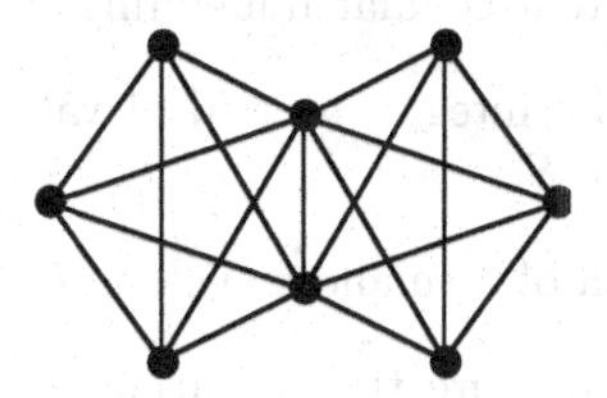

Glossary

base imbedding: the imbedding of the voltage graph in the base surface of the covering construction.

chromatic number of a surface S: the maximum of the chromatic numbers of the simple graphs that can be imbedded on S.

circulation for a rotation system ρ^* (induced): the permutation of edge-ends given by the rule $e^\sigma \mapsto \rho(e^{-\sigma})$.

covering face: a face of the graph imbedding specified by an imbedded voltage graph; a polygon whose boundary walk is matched to a lifted boundary walk or to concatenation of lifted boundary walks of the specified graph.

covering imbedding: the imbedding of the covering graph specified by an imbedded voltage graph.

covering surface: the surface formed by matching the boundaries of the covering faces to the corresponding closed walks in the covering graph.

crosscap on a surface: a subspace that is topologically equivalent to a Möbius band.

current graph: the dual of an imbedded voltage graph.

—, KCL: a current graph in which every vertex satisfies the Kirchoff's current law.

cycle rank of a graph G: the number $\beta(G) = |E| - |V| + 1$.

dual rotation: synonym for circulation.

equivalent imbeddings of a graph: imbeddings that have exactly the same set of face-boundary walks.

Euler characteristic $\chi(S)$ of a surface: the value of the Euler polyhedral formula $|V| - |E| + |F|$ for any cellular imbedding on that surface: $2 - 2g$ for the orientable surface S_g and $2 - k$ for the non-orientable surface N_k.

fiber over a face f of a voltage graph: the set f of covering faces f_j (with f as the mainscript).

flat polygon representation of a surface S: a polygon with its sides marked as pairs, such that when every pair is pasted together, the resulting surface obtained is topologically equivalent to S.

genus distribution of a graph G**:** the function that assigns to each nonnegative integer g the number $\gamma_g(G)$ of different cellular imbeddings in the surface S_g.

genus range of a graph G**:** the integer interval of values of g such that graph G has a cellular imbedding in surface S_g.

girth of a graph G**:** the length of a smallest cycle in G.

Heawood conjecture: the conjecture that a surface of Euler characteristic c has chromatic number $H(c)$, for $c = 1, 0, -1, \ldots$.

Heawood number of a closed surface S**:** $H(S) = \left\lfloor \frac{7+\sqrt{49-24 \cdot ec(S)}}{2} \right\rfloor$.

imbedding induced by rotation system ρ**:** the imbedding ι_ρ whose face-boundary walks coincide with the cycles of the circulation ρ^*.

induced rotation at a vertex v **of a graph imbedding** h**:** the cyclic permutation $\rho_h(v)$ of edge-ends incident at v in the order in which they are encountered in an orientation-consistent traversal around v.

induced rotation system of an oriented graph imbedding h**:** the function that assigns to each vertex its induced rotation.

KCL: stands for *Kirchoff's current law.*

Kirchoff's current law (KCL) at a vertex: in a current graph, the property that the net current into that vertex is the identity element of the voltage group.

Kirchoff's voltage law (KVL) on a face: in an imbedded voltage graph, the property that the net voltage on the boundary walk of that face is the identity element of the voltage group.

Kirchoff's voltage law (KVL) on a closed walk: in a voltage graph, the property that the net voltage on the walk is the identity element of the voltage group.

KVL: stands for *Kirchoff's voltage law.*

lift of a walk W **in a voltage graph:** a walk W_α in the specified graph whose vertices are mapped in sequence onto walk W by the natural projection mapping.

local group at a vertex v **in a voltage graph:** the subgroup comprising the net voltages on all closed walks based at v.

maximum crosscap number $\bar{\gamma}_{\max}(G)$**:** for a connected graph G, the largest crosscap number of any surface N_k on which G can be cellularly imbedded. If G is planar, then $\bar{\gamma}_{\max}(G) = 0$.

maximum genus of a connected graph G**:** the maximum number g such that G has a cellular imbedding on S_g.

maximum (orientable) imbedding: any imbedding of a graph G on a *maximum (orientable) surface.*

maximum (orientable) surface for a graph G: a surface whose genus is $\bar{\gamma}_{\max}(G)$.

minimum crosscap number $\bar{\gamma}_{\max}(G)$ for a connected graph G: the smallest crosscap number of any surface N_k on which G can be imbedded. If G is planar, then $\bar{\gamma}_{\max}(G) = 0$.

minimum genus $\gamma_{\min}(G)$ **of a connected graph** G**:** the smallest number g such that $\gamma_g(G) > 0$.

minimum imbedding: any imbedding of a graph into a *minimum surface.*

minimum (orientable) surface for a graph G: a surface whose genus is $\gamma_{\min}(G)$.

net voltage on a signed walk: the sum of the voltages on the plus-signed edges minus the sum of the voltages on the minus-signed edges.

orientation on a graph imbedding: a choice of either *counterclockwise* or *clockwise* as a preferred traversal direction around the vertices.

oriented graph imbedding: a graph imbedding with a designated orientation.

rotation at a vertex v**:** a cyclic permutation of the edge-ends incident on v.

rotation system for a graph: an assignment of a rotation to every vertex.

—, adjacent: two rotation systems are adjacent if one can be obtained from the other by the operation of transposing two consecutive edge-ends in the rotation at one vertex.

rotation table: representation of a rotation system ρ by a list with one row for each vertex. In each row, the name of a vertex is followed by a complete list of the edge-ends incident on that vertex, in an order consistent with $\rho(v)$. (It is a common custom to write each row in lexicographic order.)

signed walk in a digraph: an undirected walk in which a minus-signed edge-end represents an edge traversed against its designated direction.

surface induced by rotation system ρ**:** the surface $S(\rho)$ that is codomain of the circulation.

voltage graph: a pair $\langle G, \alpha \rangle$ such that G is a digraph and α is a *voltage assignment* on G.

—, imbedded: a cellular imbedding of a graph G with a voltage assignment on G.

—, KVL: an imbedded voltage graph in which every face satisfies KVL.

APPENDIX

A.1 LOGIC FUNDAMENTALS

Propositional Logic

DEFINITION: Let p and q be propositions. The ***conditional proposition*** $p \to q$ is a proposition that is false when p is true and q is false and is true otherwise. It is read "if p then q".

TERMINOLOGY: The following are equivalent statements.

$p \to q$

if p then q

p is a ***sufficient condition*** for q

q is a ***necessary condition*** for p

p only if q

q if p

DEFINITION: The ***converse*** of $p \to q$ is the conditional proposition $q \to p$.

DEFINITION: The ***contrapositive*** of $p \to q$ is the conditional proposition $\neg q \to \neg p$.

Proposition A.1.1: *The proposition $p \to q$ and its contrapositive $\neg q \to \neg p$ are logically equivalent.*

DEFINITION: The statement p ***if and only if*** q means that both conditionals $p \to q$ and $q \to p$ are true.

Types of Proof

TERMINOLOGY: When an assertion is in the form $p \to q$, the proposition p is often referred to as the ***antecedent*** and q the ***consequent***.

DEFINITION: A ***direct proof*** of the conditional $p \to q$ consists of showing that statement q is true under the assumption that statement p is true.

DEFINITION: A ***proof by contrapositive*** of the conditional $p \to q$ is a direct proof of the contrapositive $\neg q \to \neg p$.

Example A.1.1: A proof by contrapositive of the assertion "If n^2 is odd, then n is odd" takes the form: Assume n is even. Then $n = 2k$, for some integer k. Hence, $n^2 = (2k)^2 = 2(2k^2)$, which is even.

DEFINITION: A ***proof by contradiction*** of an assertion S consists of assuming that S is false and deriving some contradiction.

Example A.1.2: Prove: There is no largest prime number.

Proof by contradiction: Assume that P is the largest prime number. Then the product $r = 2 \times 3 \times \cdots \times P + 1$ is not divisible by any prime and, hence, is itself prime, which contradicts the maximality of P.

Mathematical Induction

DEFINITION: ***Axiom of Mathematical Induction – First Form***: Statement $P(n)$ is true for all integers $n \geq b$ if both of the following are true.

Base: $P(b)$

Inductive step: For all $k \geq b, P(k) \rightarrow P(k+1)$

Remark: Typically (but not always), the inductive step is demonstrated by a direct proof of the statement $P(k) \rightarrow P(k+1)$.

Example A.1.3: Prove: For all $n \geq 1, \sum_{i=1}^{n} i = \frac{n}{2}(n+1)$.

Proof: The equation holds for $n = 1$ since its left and right sides are both equal to 1 when $n = 1$. Assume that the equation holds for some $k \geq 1$. Then

$$\sum_{i=1}^{k+1} i = \left(\sum_{i=1}^{k} i\right) + (k+1) = \left(\frac{k}{2}(k+1)\right) + (k+1) = \frac{k+1}{2}(k+2)$$

◇

DEFINITION: ***Axiom of Mathematical Induction — Second Form***: Statement $P(n)$ is true for all integers $n \geq b$ if both of the following are true.

Base: $P(b)$

Inductive step: For all $k \geq b, (P(b) \wedge P(b+1) \wedge \ldots \wedge P(k)) \rightarrow P(k+1)$

Thus, a direct proof of this inductive step is to show that $P(k+1)$ is true under the assumption that the statements $P(b), P(b+1), \ldots, P(k)$ are true.

A.2 RELATIONS AND FUNCTIONS

Relations

DEFINITION: The ***Cartesian product*** of two sets A and B is the set

$$A \times B = \{(a, b) \,|\, a \in A \text{ and } b \in B\}$$

DEFINITION: A ***binary relation*** R from a set S to a set B is a subset of $A \times B$. Alternatively, a binary relation may be regarded as a function from $A \times B$ to the Boolean set $\{true, false\}$.

TERMINOLOGY: Let R be a relation from set A to set B. If $(x, y) \in R$, then x is said to be ***related to*** y (by R). This is often denoted xRy.

DEFINITION: A ***(binary) relation*** on a set A is a relation from A to A.

DEFINITION: A relation R on a set A is ***reflexive*** if for all $x \in A$,

$$xRx \in R$$

DEFINITION: A relation R on a set A is ***symmetric*** if for all $x, y \in A$,

$$xRy \Rightarrow yRx$$

DEFINITION: A relation R on a set A is ***transitive*** if for all $x, y, z \in A$,

$$(xRy \wedge yRz) \Rightarrow xRz$$

DEFINITION: A relation R on a set A is an ***equivalence relation*** if R is reflexive, symmetric, and transitive.

Example A.2.1: Let R be the relation on the set $\mathbb{Z}$ of all integers, given by

$$xRy \Leftrightarrow x - y = 3k, \text{ for some } k \in \mathbb{Z}$$

Then R is an equivalence relation.

DEFINITION: Let R be an equivalence relation on a set A, and let $x \in A$. The ***equivalence class*** of a, denoted $[a]$, is given by

$$[x] = \{a \in A \,|\, aRx\}$$

Example A.2.2: For the equivalence relation defined in Example A.2.1,

$$[0] = \{3k \,|\, k \in \mathbb{Z}\}$$

$$[1] = \{3k + 1 \,|\, k \in \mathbb{Z}\}$$

$$[2] = \{3k + 2 \,|\, k \in \mathbb{Z}\}$$

Observe that the equivalence class of any other element is one of these three classes.

DEFINITION: A collection of distinct non-empty subsets $\{S_1, S_2, \ldots, S_l\}$ of a set A is a ***partition*** of A if both of the following conditions are satisfied.

- $S_i \cap S_j = \emptyset$, for all $1 \leq i < j \leq l$
- $\bigcup_{i=1}^{l} S_i = A$

Proposition A.2.1: *Let R be an equivalence relation on a set A, and let $x, y \in A$. Then the following three statements are logically equivalent.*

1. xRy
2. $[x] = [y]$
3. $[x] \cap [y] \neq \emptyset$

Corollary A.2.2: *Let R be an equivalence relation on a set A. Then the distinct equivalence classes of R partition A.*

Example A.2.3: The three equivalence classes [0], [1], and [2] of Example A.2.2 partition the set $\mathbb{Z}$ of integers.

Partial Orderings and Posets

DEFINITION: A relation R on a set A is ***antisymmetric*** if for all $x, y \in A$,

$$xRy \wedge yRx \Rightarrow x = y$$

DEFINITION: A relation R on a set A is a ***partial ordering*** (or ***partial order***) if R is reflexive, antisymmetric, and transitive.

Example A.2.4: Let R be the *divisibility relation* on the set $\mathbb{Z}^+$ of positive integers, given by $xRy \Leftrightarrow x$ divides y. Then R is a partial ordering.

NOTATION: The symbol $\prec$ is often used to denote a partial ordering.

DEFINITION: Let $\prec$ be a partial ordering on a set A. The pair $\{A, \prec\}$ is called a ***partially ordered set*** (or ***poset***).

Functions

DEFINITION: Let A and B be two sets. A ***function*** f from A to B, denoted $f : A \rightarrow B$, is an assignment of exactly one element of B to each element of A. We also say that f ***maps*** A to B.

ALTERNATIVE DEFINITION: A ***function*** f from A to B, denoted $f : A \rightarrow B$, is a relation from A to B, such that each element of set A appears as a first component in exactly one of the ordered pairs of f.

TERMINOLOGY: Let $f : A \rightarrow B$ be a function from set A to set B. Then A is called the ***domain*** of f, and B is called the ***codomain*** of f.

TERMINOLOGY: Let $f : A \rightarrow B$ be a function from set A to B, and let $a \in A$. The element b assigned to a by f, denoted $b = f(a)$, is called the ***image of*** a ***under*** f, and element a is called a ***preimage of*** b ***under*** f.

DEFINITION: Let $f : A \rightarrow B$ be a function from set A to set B, and let $S \subseteq A$. Then the set $f(S) = \{f(s) \mid s \in S\}$ is the ***image of*** S ***under*** f.

DEFINITION: Let $f : A \to B$ be a function from set A to set B, and let $S \subseteq B$. Then the set $f^{-1}(S) = \{x \in A \mid f(x) \in S\}$ is the ***inverse image of S under f***.

DEFINITION: A function $f : A \to B$ is ***one-to-one*** (or ***injective***) if for all $x, y \in A$, $x \neq y \Rightarrow f(x) \neq f(y)$.

Thus, a function $f : A \to B$ is one-to-one if and only if each element of set B has at most one preimage under f.

DEFINITION: A function $f : A \to B$ is ***onto*** (or ***surjective***) if for all $b \in B$, there exists at least one $a \in A$ such that $f(a) = b$.

Thus, a function $f : A \to B$ is onto if and only if each element of set B has at least one preimage under f.

DEFINITION: A function $f : A \to B$ is a ***bijection*** if f is both injective and surjective.

Example A.2.5: In a standard (0-based) vertex-labeling of an n-vertex graph G, each vertex of G is assigned an integer label from the set $\{0, 1, \ldots, n-1\}$, such that no two vertices get the same label. Thus, the labeling is a bijection

$$f : \{0, 1, \ldots, n-1\} \to V_G$$

It is common to say that the integers from 0 to $n - 1$ ***have been bijectively assigned*** to the vertices of G.

Theorem A.2.3: [***Pigeonhole Principle***] *Let $f : A \to B$ be a function from a finite set A to a finite set B. Let any two of the following three statements be true.*

1. *f is one-to-one.*
2. *f is onto.*
3. *$|A| = |B|$.*

Then the third statement is also true.

The following corollary uses familiar pigeonhole jargon.

Corollary A.2.4: *If each member of a flock of pigeons flies into a pigeonhole, and if there are more pigeons than pigeonholes, then there must be at least one pigeonhole that contains more than one pigeon.*

DEFINITION: Let $f : A \to B$ be a function from a set A to a set B, and let $g : B' \to C$ be a function from a set B' to a set C, where $B \subseteq B'$. Then the **composition of f and g** is the function $g \circ f : A \to C$, given by $(g \circ f)(a) = g(f(a))$.

DEFINITION: Let $f : A \to B$ be a function from a set A onto a set B, and let $g : B \to A$ be a function from a set B to a set A. Then g is the ***inverse function*** of f if for all $a \in A$, $(g \circ f)(a) = a$.

Proposition A.2.5: *Let $f : A \to B$ be a function from a set A onto a set B, and let $g : B \to A$ be a function from a set B onto a set A. Then g is the inverse function f if and only if f is the inverse function of g.*

Proposition A.2.6: *A function $f : A \to B$ has an inverse if and only if f is a bijection.*

A.3 SOME BASIC COMBINATORICS

Proposition A.3.1: [***Rule of Product***] *Let A and B be two finite sets. Then*

$$|A \times B| = |A| \cdot |B|$$

DEFINITION: A ***permutation*** of an n-element set S is a bijection $\pi : S \to S$.

DEFINITION: A ***k*-permutation** of an n-element set S is a permutation of a k-element subset of S.

A consequence of the Rule of Product is the following simple formula for the number of k-permutations.

Proposition A.3.2: *The number of k-permutations of an n-element set is*

$$\frac{n!}{(n-k)!}$$

DEFINITION: A ***k*-combination** of a set S is a k-element subset of S.

NOTATION: The number of k-combinations of an n-element set is denoted $\binom{n}{k}$ or, alternatively, $C(n,k)$.

Proposition A.3.3:

$$\binom{n}{k} = \frac{n!}{k!(n-k)!}$$

Theorem A.3.4: [***Binomial Theorem***] *For every nonnegative integer n and any numbers x and y,*

$$(x+y)^n = \sum_{k=0}^{n} \binom{n}{k} x^{n-k} y^k$$

DEFINITION: Let $\langle a_1, a_2, \ldots, a_k, \ldots \rangle$ be a sequence of real numbers. The ***generating function*** for the sequence $\{a_n\}$ is the power series

$$g(x) = \sum_{k=0}^{\infty} a_k x^k = a_0 + a_1 x + a_2 x^2 + \cdots + a_k x^k + \cdots$$

Example A.3.1: The generating function for the constant sequence $\langle 1, 1, \ldots \rangle$ is

$$\sum_{k=0}^{\infty} x^k = \frac{1}{1-x}$$

Example A.3.2: The generating function for the sequence $\langle 1, 2, 3, \ldots \rangle$ is

$$\sum_{k=0}^{\infty} kx^{k-1} = \sum_{k=0}^{\infty} \frac{d}{dx} x^k = \frac{d}{dx}\left[\sum_{k=0}^{\infty} x^k\right] = \frac{d}{dx}\left(\frac{1}{1-x}\right) = \frac{1}{(1-x)^2}$$

A.4 ALGEBRAIC STRUCTURES

Groups

DEFINITION: A ***binary operation*** $*$ on a non-empty set A is a function $f : A \times A \to A$, given by $f((a,b)) = a * b$.

NOTATION: The set A together with a binary operation $*$ is denoted $(A, *)$.

Example A.4.1: Let $\mathbb{Z}_{10} = \{0, 1, 2, \ldots, 9\}$, and let $\oplus$ and $\odot$ be the operations of addition and multiplication mod 10. For instance, $6 \oplus 6 = 12 \pmod{10} = 2$ and $6 \odot 6 = 36 \pmod{10} = 6$. Then $\oplus$ and $\odot$ are binary operations on the set $\mathbb{Z}_{10}$.

DEFINITION: The binary operation $*$ on set A is ***associative*** if for all $a, b, c \in A$, $(a*b)*c = a*(b*c)$.

DEFINITION: The binary operation $*$ on set A is ***commutative*** if for all $a, b \in A$, $a*b = b*a$.

DEFINITION: Let $*$ be a binary operation on set A. An element $e \in A$, is an ***identity element*** of set A under $*$ if for all $a \in A$,

$$a * e = e * a = a$$

DEFINITION: Let $*$ be a binary operation on set A, and let e be an identity element of $(A, *)$. For $a \in A$, an element $a' \in A$ is an ***inverse*** of a in $(A, *)$ if

$$a * a' = a' * a = e$$

DEFINITION: A ***group*** $\mathcal{G} = (G, *)$ is a non-empty set G and a binary operation $*$ that satisfy the following conditions.

1. The operation $*$ is associative.
2. $\mathcal{G}$ has an identity element.
3. Each $g \in G$ has an inverse in $(G, *)$.

NOTATION: Each element g in a group $\mathcal{G}$ has a *unique* inverse, which is denoted g^{-1}.

DEFINITION: If $\mathcal{G} = (G, *)$ is a group, the set G is the domain (set) of the group $\mathcal{G}$.

NOTATION: When convenient, and when there is no ambiguity, we sometimes write "$g \in \mathcal{G}$" to mean that g is an element of the domain set G.

DEFINITION: An ***Abelian*** group is a group whose operation is commutative.

NOTATION: In an Abelian group, the operation is commonly denoted "+" and called "sum".

TERMINOLOGY: For an abstract group $(G, *)$, the term "product" is used generically to refer to the element $x * y$, and its use does not imply anything about the type of group operation.

Example A.4.2: The group $\mathbb{Z}_n = (\mathbb{Z}_n, +)$, which generalizes $(\mathbb{Z}_{10}, \oplus)$ in Example A.4.1, is Abelian and is denoted $\mathbb{Z}_n$.

Example A.4.3: Observe that $(\mathbb{Z}_n, \odot)$ *is not* a group, since the element 0 has no inverse under $\odot$.

Example A.4.4: Let S be an n-element set, and let P be the set of all permutations of set S. Then $(P, \circ)$, where $\circ$ is the *composition* operation, is a group.

DEFINITION: Let $\mathcal{G} = (G, *)$ be a finite group. A subset X of G is a ***generating set*** for group $\mathcal{G}$ if every element of $\mathcal{G}$ can be expressed as a product of elements of set X.

Example A.4.5: The set $\{2, 5\}$ is a generating set for $\mathbb{Z}_{10}$.

DEFINITION: A ***cyclic group*** is a group that has a 1-element generating set.

NOTATION: If $\{a\}$ is a generating set for a cyclic group $\mathcal{G}$, then we may write $\mathcal{G} = \langle a \rangle$.

Example A.4.6: $\mathbb{Z}_n = (\mathbb{Z}_n, +) = \langle 1 \rangle$.

Example A.4.7: $(\mathbb{Z}_7 - \{0\}, \odot) = \langle 3 \rangle = \langle 5 \rangle$.

DEFINITION: Let $\mathcal{G}_1 = (G_1, *_1)$ and $\mathcal{G}_2 = (G_2, *_2)$ be two groups. The ***direct sum*** (or ***direct product***) $\mathcal{G} \times \mathcal{G}_2$ is the group $(G_1 \times G_2, *)$ whose domain set is the Cartesian product of the domain sets G_1 and G_2 and whose operation is defined by

$$(x_1, x_2) * (y_1, y_2) = (x_1 *_1 y_1, x_2 *_2 y_2)$$

DEFINITION: The n-fold direct sum of groups $\mathcal{G}_1 = (G_1, *_1), \mathcal{G}_2 = (G_2, *_2), \ldots, \mathcal{G}_n = (G_2, *_n)$ is defined recursively as

$$\mathcal{G}_1 \times \mathcal{G}_2 \times \ldots \mathcal{G}_n = (\mathcal{G}_1 \times \mathcal{G}_2 \times \ldots \mathcal{G}_{n-1}) \times \mathcal{G}_n$$

Theorem A.4.1: [***Fundamental Theorem of Finite Abelian Groups***] *Let $\mathcal{G}$ be a finite Abelian group. Then $\mathcal{G}$ can be expressed as the direct sum of cyclic groups.*

Remark: In fact, the cyclic groups in the direct-sum decomposition of an Abelian group have prime-power order, and the direct sum is *essentially* unique, but the above assertion is all that we need in this book.

Order of an Element in a Group

NOTATION: Let x be an element of a group $\mathcal{G} = (G, *)$. The element $x * x$ may be denoted x^2, and the product of r factors of element x may be denoted x^r. In an Abelian group, the element $x + x$ may be denoted $2x$, and the sum of r occurrences of element x may be denoted $r \cdot x$ or sometimes rx.

DEFINITION: Let x be an element of a group with identity e. The ***order*** of x is the smallest positive integer m such that $x^m = e$.

DEFINITION: The ***order*** of a group $\mathcal{G} = (G, *)$ is the cardinality $|G|$ of its domain set.

Proposition A.4.2: *In a finite group, every element has finite order.*

Example A.4.8: For $2 \in \mathbb{Z}_{10}$, we have $5 \cdot 2 = 2 + 2 + 2 + 2 + 2 = 0 \pmod{10}$. Thus, the order of the element $2 \in \mathbb{Z}_{10}$ is 5.

Permutation Groups

DEFINITION: The group of all permutations of a set S with n elements is called the ***full symmetric group*** on S and is denoted Σ_S or Σ_n.

TERMINOLOGY: The elements of the set on which a permutation group acts are called ***objects***.

NOTATION: One way of representing a permutation π of the set $\{1, 2, \ldots, n\}$ is to use a $2 \times n$ matrix, where the first row contains the n integers in order, and the second row contains the images, so that $\pi(j)$ appears below j.

Example A.4.9: The group Σ_3 consists of the six permutations of $S = \{1, 2, 3\}$. They are as follows:

$$\pi_1 = \begin{pmatrix} 1 & 2 & 3 \\ 1 & 2 & 3 \end{pmatrix}; \quad \pi_2 = \begin{pmatrix} 1 & 2 & 3 \\ 1 & 3 & 2 \end{pmatrix}; \quad \pi_3 = \begin{pmatrix} 1 & 2 & 3 \\ 2 & 1 & 3 \end{pmatrix};$$

$$\pi_4 = \begin{pmatrix} 1 & 2 & 3 \\ 3 & 2 & 1 \end{pmatrix}; \quad \pi_5 = \begin{pmatrix} 1 & 2 & 3 \\ 2 & 3 & 1 \end{pmatrix}; \quad \pi_3 = \begin{pmatrix} 1 & 2 & 3 \\ 3 & 1 & 2 \end{pmatrix};$$

The permutation π_1 is the identity element of Σ_3. Remembering that the composition $f \circ g$ is applied *right-to-left*, we have the following sample calculations:

$$\pi_2 \circ \pi_2 = \pi_3 \circ \pi_3 = \pi_4 \circ \pi_4 = \pi_1; \qquad \pi_5 \circ \pi_5 = \pi_6; \qquad \pi_4 \circ \pi_5 = \pi_3; \qquad \pi_5 \circ \pi_4 = \pi_2$$

The last two calculations show that the full symmetric group is *non-Abelian.* Additional calculation shows that the set $\{\pi_4, \pi_5\}$ is a generating set for Σ_3. Moreover, $\pi_5^3 = \pi_6 \circ \pi_5 = \pi_1$. Thus, the order of the permutation π_5 is 3.

DEFINITION: A permutation π is an ***involution*** if $\pi = \pi^{-1}$.

Cycle Notation for a Permutation

Each object of {1, 2, 3} in Example A.4.9 *cycles* back to itself after three applications of the permutation $\pi_5 = \begin{pmatrix} 1 & 2 & 3 \\ 2 & 3 & 1 \end{pmatrix}$. That is,

$$1 \xrightarrow{\pi_5} 2 \xrightarrow{\pi_5} 3 \xrightarrow{\pi_5} 1 \qquad 2 \xrightarrow{\pi_5} 3 \xrightarrow{\pi_5} 1 \xrightarrow{\pi_5} 2 \qquad 3 \xrightarrow{\pi_5} 1 \xrightarrow{\pi_5} 2 \xrightarrow{\pi_5} 3$$

The *cycling* of the images under successive applications of a given permutation is compactly represented by enclosing the chain of images in parentheses so that the image of each element appears to the right of that element, with *wraparound* for the rightmost element.

NOTATION: If S is a set and $a_1, a_2, \ldots, a_k \in S$, then $(a_1 a_2 \ldots a_k)$ denotes the permutation π of S for which

$$\pi(a_i) = \begin{cases} a_{i+1} & \text{for } i = 1, \ldots, k-1 \\ a_1 & \text{for } i = k \end{cases}$$

and

$$\pi(x) = x \quad \text{for all other } x \in S$$

Example A.4.10: The cycle representation of permutation π_5 from Example A.4.9 can be written in any one of three equivalent ways: (1 2 3), (2 3 1), or (3 1 2).

NOTATION: This cycle notation can be extended to any permutation by writing that permutation as a **product of disjoint cycles**. The elements that appear in each cycle form the chain of images under the action of that permutation.

Example A.4.11: Consider the permutation $\alpha \in \Sigma_9$ given by

$$\alpha = \begin{pmatrix} 1 & 2 & 3 & 4 & 5 & 6 & 7 & 8 & 9 \\ 7 & 2 & 6 & 1 & 8 & 3 & 4 & 9 & 5 \end{pmatrix}$$

Then the disjoint cycle representation of α is

$$\alpha = (1\ 7\ 4)(2)(3\ 6)(5\ 8\ 9) \quad \text{or} \quad (1\ 7\ 4)(3\ 6)(5\ 8\ 9)$$

Fields

DEFINITION: A ***field*** $\mathcal{F} = (F, +, \cdot)$ is a set F together with two operations, $+$ and $\cdot$, (called generically *addition* and *multiplication*), such that each of the following conditions is satisfied:

1. $(F, +)$ is an Abelian group.
2. $(F - \{0\}, \cdot)$ is an Abelian group, where 0 is the additive identity.
3. $a \cdot (b + c) = (a \cdot b) + (a \cdot c)$.

Finite Field $GF(2)$

DEFINITION: The ***finite field*** $GF(2)$ consists of the set $\mathbb{Z}_2 = \{0, 1\}$ together with the mod 2 operations $\oplus$ and $\odot$. Thus,

$$0 \oplus 0 = 1 \oplus 1 = 0; \quad 0 \oplus 1 = 1 \oplus 0 = 1; \quad 0 \odot 0 = 1 \odot 0 = 0 \odot 1 = 0; \quad 1 \odot 1 = 1$$

Vector Spaces

DEFINITION: A **vector space** over a field (of **scalars**) $\mathcal{F}$ is a set V (of **vectors**) together with a binary operation $+$ on V and a mapping, called **scalar multiplication**, from the Cartesian product $F \times V$ to $V : (a, v) \mapsto av$, such that the following conditions are satisfied for all scalars $a, b \in F$ and all vectors $v, w \in V$.

1. $(V, +)$ is an Abelian group (the symbol "+" is being used to denote two different operations, addition in the field $\mathcal{F}$ and addition in the set V)
2. $(a \cdot b)v = a(bv)$
3. $(a + b)v = av + bv$
4. $a(v + w) = av + aw$
5. $ev = v$, where e is the multiplicative identity of field $\mathcal{F}$

Example A.4.12: Let $W = \{(\alpha_1, \alpha_2, \alpha_3) \,|\, \alpha_i \in GF(2)\}$, and consider componentwise addition $\oplus$ (mod 2) on the elements of W. For each scalar $\alpha \in GF(2)$ and element $w = (\alpha_1, \alpha_2, \alpha_3)$, let scalar multiplication be defined by

$$\alpha w = (\alpha \odot \alpha_1, \alpha \odot \alpha_2 \alpha \odot \alpha_3)$$

Then it is easy to verify that W with these operations forms a vector space over $GF(2)$.

Remark: The vector space defined in Example A.4.12 is a special case of the vector space $\mathcal{F}^n$ of all n-tuples whose components are elements in a field $\mathcal{F}$.

Independence, Subspaces, and Dimension

The remaining definitions in this section all assume that V is a vector space over a field $\mathcal{F}$. The examples refer to the vector space W defined in Example A.4.12.

DEFINITION: A vector v is a ***linear combination*** of vectors $v_1, v_2, \ldots, v_m$ if $v = \alpha_1 v_1 + \alpha_2 v_2 + \cdots + \alpha_m v_m$ for scalars $\alpha_i \in \mathcal{F}$, $i = 1, \ldots, m$.

Example A.4.13: In vector space W, the vector $(1, 0, 1)$ is a linear combination of vectors $(1, 1, 0)$ and $(0, 1, 1)$, given by $(1, 0, 1) = 1(1, 1, 0) + 1(0, 1, 1)$. This last equation also implies that $(1, 0, 1)$ is a linear combination of any superset of $\{(1, 1, 0), (0, 1, 1)\}$, where the scalar multipliers for the other vectors in the superset are 0.

DEFINITION: A set $\{v_1, v_2, \ldots, v_m\}$ of vectors is ***linearly independent*** if

$$\alpha_1 v_1 + \alpha_2 v_2 + \cdots + \alpha_m v_m = \mathbf{0} \text{ only if } \alpha_1 = \alpha_2 = \cdots = \alpha_m = 0$$

Thus, nonzero vectors $v_1, v_2, \ldots, v_m$ are ***linearly independent*** if none of them is expressible as a linear combination of the remaining ones.

Example A.4.14: It is easy to check that $\{(1,1,0),(0,1,0),(0,1,1)\}$ is a set of linearly independent vectors in vector space W.

DEFINITION: A set S of vectors **spans** V if every vector in V is expressible as a linear combination of vectors in S.

Example A.4.15: The three vectors in Example A.4.14 span W. A more obvious spanning set is $\{(1,0,0),(0,1,0),(0,0,1)\}$.

DEFINITION: A set B is a **basis** for V if the vectors in B span V and are linearly independent.

Proposition A.4.3: *All the bases of a given vector space have the same number of vectors.*

DEFINITION: The ***dimension*** of a vector space V, denoted $\dim(V)$, is the number of vectors in a basis of V.

DEFINITION: A subset U of vector space V is a **subspace** of V if U is itself a vector space with respect to the addition and scalar multiplication of V.

Proposition A.4.4: *Let V be a vector space over a field $\mathcal{F}$, and let U be a subset of V. Then U is a subspace of V if and only if*

1. *$v + w \in U$ for all $v, w \in U$, and*
2. *$\alpha v \in U$ for all $\alpha \in \mathcal{F}$, $v \in U$.*

Example A.4.16: Either of the 3-element sets given in Example A.4.15 shows that W is a 3-dimensional vector space. The set $U = \{(0,0,0),(1,1,0),(0,1,1),(1,0,1)\}$ is a 2-dimensional subspace of W.

Proposition A.4.5: *Let U_1 and U_2 be two subspaces of a vector space V. Then the intersection $U_1 \cap U_2$ is a subspace of V.*

DEFINITION: The **direct sum** $U_1 \oplus U_2$ of two subspaces U_1 and U_2 of V is the set of all vectors of the form $v_1 + v_2$, where $v_1 \in V_1$ and $v_2 \in V_2$.

Proposition A.4.6: *Let U_1 and U_2 be two subspaces of a vector space V. Then the direct sum $U_1 \oplus U_2$ is a subspace of V, and*

$$\dim(U_1 \oplus U_2) = \dim(U_1) + \dim(U_2) - \dim(U_1 \cap U_2)$$

A.5 ALGORITHMIC COMPLEXITY

One approach to the analysis and comparison of algorithms is to consider the *worst-case* performance, ignoring constant factors that are often outside a given programmer's control. These factors include the computer on which the algorithm is run, the operating system, and the machine code translation of the programming language in which the algorithm is written. The objective of this approach is to determine the functional dependence of the running time (or of some other measure) on the number n of inputs (or on some other such variable). A first step toward this goal is to make the notion of "proportional to" mathematically precise.

Big-O Notation

DEFINITION: Let $f : Z^+ \to \mathbb{R}$ and $g : Z^+ \to \mathbb{R}$ be functions from the set of positive integers to the set of real numbers. The function $f(n)$ is in the family $O(g(n))$ (read "big-O" of $g(n)$) if there are constants c and N such that

$$|f(n)| \leq c|g(n)| \text{ for all } n > N$$

TERMINOLOGY: If $f(n)$ is in $O(g(n))$, then $f(n)$ is said to be ***asymptotically dominated*** by $g(n)$. In other words, for sufficiently large n and a sufficiently large constant c, $cg(n)$ is larger than $f(n)$.

TERMINOLOGY NOTE: The usage "f is $O(g(n))$" (omitting the preposition "in") is quite common.

Remark: It follows from the definition that $f(n)$ is in $O(g(n))$ if and only if there is a constant c such that $\lim\limits_{n\to\infty} \frac{f(n)}{g(n)} = c$.

Example A.5.1: The function $f(n) = \sum_{i=1}^{n} i$ is in $O(n^2)$. To see this, observe that for all $n > 0$

$$f(n) = \frac{n^2 + n}{2} < n^2 + n \leq n^2 + n^2 = 2n^2$$

Thus, the condition is satisfied for $c = 2$ and $N = 0$. It is also true that $f(n)$ is in $O(n^r)$ for any $r \geq 2$, but $O(n^2)$ is the *sharpest* among these. In fact, letting $c = 2$ and $N = 0$, the following chain of inequalities shows that n^2 is in $O(f(n))$.

$$n^2 < n^2 + n = 2f(n)$$

The next example illustrates the effect of information-structure representation of a graph on algorithmic computation.

Example A.5.2: Given a simple graph G, find the degree of each vertex $v \in V_G$. For each vertex, we must count the number of edges incident on that vertex. The number of steps required depends on how the input graph is represented in the computer. We consider two of the possibilities. Let n be the number of vertices and m the number of edges.

Representation by adjacency matrix Because the adjacency matrix is symmetric, only the upper (or lower) triangular submatrix (excluding the main diagonal) needs to be examined. There are $\sum_{i+1}^{n-1} i$ cells in this triangular submatrix, and each cell must be checked to see if it is nonzero. Hence, the number of checks is $O(n^2)$.

Representation by adjacency lists of edges For each vertex, the list of edges incident with that vertex is traversed. Thus, the number of add operations is $O(m)$.

DEFINITION: Functions $f(n)$ and $g(n)$ are ***big-O equivalent*** if $f(n)$ is in $O(g(n))$ and $g(n)$ is in $O(f(n))$.

Proposition A.5.1: *Functions $f(n)$ and $g(n)$ are big-O equivalent if and only if $\lim_{n\to\infty} \frac{f(n)}{g(n)} = c$ for some nonzero constant c.*

Proposition A.5.2: *Two polynomial functions are big-O equivalent if and only if they have the same degree.*

DEFINITION: A ***polynomial-time*** algorithm is an algorithm whose running time (as a function on the input size n) is in $O(p(n))$ for some polynomial $p(n)$.

TERMINOLOGY: A *linear-time* algorithm is an algorithm whose running time is in $O(n)$. A *quadratic-time* algorithm is an algorithm whose running time is in $O(n^2)$.

Decision Problems

DEFINITION: A ***decision problem*** is a problem that requires only a *yes* or *no* answer regarding whether some element of its domain has a particular property.

DEFINITION: A decision problem belongs to the ***class P*** if there is a polynomial-time algorithm to solve the problem.

DEFINITION: A decision problem belongs to the ***class NP*** if there is a way to provide evidence of the correctness of a *yes* answer so that it can be confirmed by a polynomial-time algorithm.

Example A.5.3: The decision problem "Is this graph bipartite?" is in class P.

Example A.5.4: It is not known whether the decision problem "Can the vertices of this graph be colored with three colors such that adjacent vertices get different colors?" is in class P. However, a *yes* answer could be supported by supplying a proposed 3-coloring of the vertices. Checking the correctness of this 3-coloring would require counting the number of different colors used on the vertices, which can be accomplished in linear time, and checking for each pair of vertices, whether the two vertices have received the same color and whether they are adjacent, which can be performed in quadratic time. Hence, this decision problem is in class NP.

Remark: The letters NP stand for *nondeterministic polynomial.* This terminology derives from the fact that problems in class NP would be answered in polynomial time if there was a nondeterministic (limitlessly parallel) machine that could check all possible evidence sets simultaneously. This philosophically creative notion provides a meaningful theoretical construct for categorizing problems.

It might appear that the class of problems that could be solved by a nondeterministic computer should "obviously" be larger than those that can be solved by strictly sequential polynomial-time algorithms. However, no one has been able to find a problem in class NP that is provably not in class P. Determining whether such a decision problem exists is among the foremost unsolved problems in theoretical computer science.

The Problem Classes NP-Complete and NP-Hard

There is a large collection of problems (called *NP-complete*), including the vertex-coloring problem and the traveling salesman problem, for which no polynomial-time algorithms have been found, despite considerable effort over the last several decades. It remains open whether there exist such algorithms for these problems. In turns out that if one can find a polynomial-time algorithm for an NP-complete problem, then polynomial-time algorithms can be found for all NP-complete problems. Our goal here is to provide a brief and informal introduction to this topic. A precise, detailed description of this class of problems may be found, for instance, in [Ev79], [GaJo79], or [Wi86].

DEFINITION: A decision problem R is ***polynomially reducible*** to Q if there is a polynomial-time transformation of each instance I_R of problem R to an instance I_Q of problem Q, such that instances I_R and I_Q have the same answer (*yes* or *no*).

DEFINITION: A decision problem is ***NP-hard*** if every problem in class NP is polynomially reducible to it.

DEFINITION: An NP-hard problem R is ***NP-complete*** if R is in class NP.

DEFINITION: Let $X = \{x_1, x_2, \ldots, x_n\}$ be a set of n Boolean variables. The ***satisfiability problem*** is to decide for a propositional form $S(x, \ldots, x_n)$ in the variables $x_i \in X$, whether there is a set of truth values $\{a_1, \ldots, a_n\}$ such that $S(a_1, \ldots, a_n)$ is true.

Cook ([Co71]) showed that the class NP-complete is non-empty by showing that the satisfiability problem is NP-complete. Since that time, the list has grown rapidly.

A.6 SUPPLEMENTARY READING

For discrete mathematics: [PoSt90], [R084], [Ro03], [St93], [Tu95].

For algebra: [Ga90], [MaBi67].

For algorithmics: [AhHoUl83], [Ba88], [Se88], [Wi86].

BIBLIOGRAPHY

A.1 GENERAL READING

The first part of this bibliography is a list of books of general interest on graph theory and on background topics for graph theory, at a level consistent with our present text, grouped according to topic.

ALGEBRA and ENUMERATION

[Ga90] J. A. Gallian, *Contemporary Abstract Algebra*, Second Edition, D. C. Heath, 1990.

[GrKnPa94] R. L. Graham, D. E. Knuth, and O. Patashnik, *Concrete Mathematics: A Foundation for Computer Science*, Second Edition, Addison-Wesley, 1994.

[MaBi67] S. Maclane and G. Birkhoff, *Algebra*, Macmillan, 1967.

ALGORITHMS and COMPUTATION

[AhHoUl83] A. V. Aho, J. E. Hopcroft, and J. D. Ullman, *Data Structures and Algorithms*, Addison-Wesley, 1983.

[Ba83] S. Baase, *Computer Algorithms: Introduction to Design and Analysis*, Addison-Wesley, 1983.

[Ev79] S. Even, *Graph Algorithms*, Computer Science Press, 1979.

[GaJo79] M. R. Garey and D. S. Johnson, *Computers and Intractibility: A Guide to the Theory of NP-Completeness*, W. H. Freeman, 1979.

[Gr97] J. Gruska, *Foundations of Computing*, Thomson, 1997.

[La76] E. L. Lawler, *Combinatorial Optimization: Networks and Matroids*, Holt, Rinehart and Winston, 1976.

[Se88] R. Sedgewick, *Algorithms*, Addison-Wesley, 1988.

[Wi86] H. S. Wilf, *Algorithms and Complexity*, Prentice-Hall, 1986.

COMBINATORIAL MATHEMATICS

[Bo00] K. P. Bogart, *Introductory Combinatorics*, Third Edition, Academic Press, 2000.

[Br04b] R. A. Brualdi, *Introductory Combinatorics*, Fourth Edition, Prentice Hall, 2004.

[Gr08] J. L. Gross, *Combinatorial Methods with Computer Applications*, CRC Press, 2008.

[MiRo91] J. G. Michaels and K. H. Rosen (Eds.), *Applications of Discrete Mathematics*, McGraw-Hill, 1991.

[PoSt90] A. Polimeni and H. J. Straight, *Foundations of Discrete Mathematics*, Brooks/Cole, 1990.

[Ro84] F. S. Roberts, *Applied Combinatorics*, Prentice-Hall, 1984.

[Ro03] K. H. Rosen, *Discrete Mathematics and Its Applications*, Fifth Edition, McGraw-Hill, 2003.

[St93] H. J. Straight, *Combinatorics: An Invitation*, Brooks/Cole, 1993.

[Tu01] A. Tucker, *Applied Combinatorics*, Fourth Edition, John Wiley & Sons, 2001.

GRAPH THEORY

[Be85] C. Berge, *Graphs*, North-Holland, 1985.

[BiLlWi86] N. L. Biggs, E. K. Lloyd, and R. J. Wilson, *Graph Theory* 1736-1936, Oxford, 1986.

[Bo98] B. Bollobás, *Modern Graph Theory*, Springer, 1998.

[BoMu76] J. A. Bondy and U. S. R. Murty, *Graph Theory With Applications*, American Elsevier, 1976.

[ChLe04] G. Chartrand and L. Lesniak, *Graphs & Digraphs*, Fourth Edition, CRC Press, 2004.

[ChZh05] G. Chartrand and P. Zhang, *Introduction to Graph Theory*, McGraw Hill, 2005.

[Di00] R. Diestel, *Graph Theory*, Second Edition, Springer-Verlag, 2000.

[Gi85] A. Gibbons, *Algorithmic Graph Theory*, Cambridge University Press, 1985.

[Go80] M. C. Golumbic, *Algorithmic Graph Theory and Perfect Graphs*, Academic Press, 1980.

[Go88] R. Gould, *Graph Theory*, Benjamin/Cummings, 1988.

[GrYe14] J. L. Gross, J. Yellen, and P. Zhang, eds, *Handbook of Graph Theory*, Second Edition, CRC Press, 2014.

[Ha69] F. Harary, *Graph Theory*, Addison-Wesley, 1969.

[ThSw92] K. Thulasiraman and M. N. S. Swamy, *Graphs: Theory and Algorithms*, John Wiley & Sons, 1992.

[We01] D. B. West, *Introduction to Graph Theory*, Second Edition, Prentice-Hall, 2001.

[Wi96] R. J. Wilson, *Introduction to Graph Theory*, Addison-Wesley Longman, 1996.

[WiWa90] R. J. Wilson and J. J. Watkins, *Graphs: An Introductory Approach*, John Wiley & Sons, 1990.

SURFACES and TOPOLOGICAL GRAPH THEORY

[GrTu87] J. L. Gross and T. W. Tucker, *Topological Graph Theory*, Dover, 2001. (First Edition, Wiley-Interscience, 1987.)

[Ma67] W. S. Massey, *Algebraic Topology: An Introduction*, Harbrace, 1967.

[Ri74] G. Ringel, *Map Color Theorem*, Springer, 1974.

[Wh84] A. T. White, *Graphs, Groups and Surfaces*, Revised Edition, North-Holland, 1984.

A.2 REFERENCES

In addition to the references cited in the text and listed below, the reader may wish to consult the *Handbook of Graph Theory*, which contains many contributed sections with summaries of the most important research results and extensive bibliographies.

Chapter 1: Foundations

[Mc77] B. D. McKay, Computer reconstruction of small graphs, *J. Graph Theory* 1 (1977), 281–283.

[Mc97] B. D. McKay, Small graphs are reconstructible, *Australas. J. Combin.* 15 (1997), 123-126.

[Ni77] A. Nijenhuis, Note on the unique determination of graphs by proper subgraphs, *Notices Amer. Math. Soc.* 24 (1977), A-290.

[Ro84] F. S. Roberts, *Applied Combinatorics*, Prentice-Hall, 1984.

[Wi04] R. J. Wilson, *History of Graph Theory*, §1.3 in *Handbook of Graph Theory*, Second Edition, eds., J. L. Gross, J. Yellen, and P. Zhang, CRC Press, 2014.

[WiWa90] R. J. Wilson and J. J. Watkins, *Graphs: An Introductory Approach*, John Wiley & Sons, 1990.

Chapter 2: Isomorphisms and Symmetry

[Bo00] K. P. Bogart, *Introductory Combinatorics*, Third Edition, Academic Press, 2000.

[Br04b] R. A. Brualdi, *Introductory Combinatorics*, Fourth Edition, Prentice Hall, 2004.

Chapter 3: Trees and Connectivity

[AhHoUl83] A. V. Aho, J. E. Hopcroft, and J. D. Ullman, *Data Structures and Algorithms*, Addison-Wesley, 1983.

[Ba83] S. Baase, *Computer Algorithms: Introduction to Design and Analysis*, Addison-Wesley, 1983.

[FoFu56] L. R. Ford and D. R. Fulkerson, Maximal flow through a network, *Canad. J. Math.* 8 (1956), 399–404.

[Ha62] F. Harary, The maximum connectivity of a graph, *Proc. Natl. Acad. Sci., U.S.* 48 (1962), 1142–1146.

[Ox92] J. G. Oxley, *Matroid Theory*, Oxford, 1992.

[ThSw92] K. Thulasiraman and M. N. S. Swamy, *Graphs: Theory and Algorithms*, John Wiley & Sons, 1992.

[Tu61] W. T. Tutte, A theory of 3-connected graphs, *Indag. Math.* 23 (1961), 441–455.

[We76] D. J. A. Welsh, *Matroid Theory*, Academic Press, 1976.

[Wi96] R. J. Wilson, *Introduction to Graph Theory*, Addison-Wesley Longman, 1996.

Chapter 4: Planarity and Kuratowski's Theorem

[BoMu76] J. A. Bondy and U. S. R. Murty, *Graph Theory With Applications*, American Elsevier, 1976.

[DeMaPe64] G. Demoucron, Y. Malgrange, and R. Pertuiset, Graphes planaires: reconnaissance et construction de representations planaires topologiques, *Rev. Francaise Recherche Operationelle* 8 (1964), 33–47.

[Fa48] I. Fary, On the straight-line representations of planar graphs, *Acta Sci. Math.* 11 (1948), 229–233.

[GrRo79] J. L. Gross and R. H. Rosen, A linear-time planarity algorithm for 2-complexes, *J. Assoc. Comput. Mach.* 20 (1979), 611–617.

[GrRo81] J. L. Gross and R. H. Rosen, A combinatorial characterization of planar 2-complexes, *Colloq. Math.* 44 (1981), 241–247.

[Ha69] F. Harary, *Graph Theory*, Addison-Wesley, 1969.

[HoTa74] J. Hopcroft and R. E. Tarjan, Efficient Planarity Testing, *JACM* 21 (1974), 549–568.

[Ne54] M. H. A. Newman, *Elements of the Topology of Plane Sets of Points*, Cambridge University Press, 1954.

[Th80] C. Thomassen, Planarity and duality of finite and infinite graphs, J. *Combin. Theory Ser. B* 29 (1980), 244–271.

[Th81] C. Thomassen, Kuratowski's theorem, *J. Graph Theory* 5 (1981), 225–241.

[Wa36] K. Wagner, Bemerkungen zum Vierfarbenproblem, *Jber. Deutch. Math. Verien.* 46 (1936), 21–22.

[We01] D. B. West, *Introduction to Graph Theory*, Second Edition, Prentice-Hall, 2001.

[Wi04] R. J. Wilson, *History of Graph Theory*, §1.3 in *Handbook of Graph Theory*, Second Edition, eds., J. L. Gross, J. Yellen, and P. Zhang, CRC Press, 2014.

Chapter 5: Drawing Graphs and Maps

[AgWe88] A. Aggarwal and J. Wein, Computational Geometry, *MIT LCS Research Seminar Series* 3, 1988.

[Ch00] J. Chen, Algorithms and Complexity in Computational Geometry, §13.5 of *Handbook of Discrete and Combinatorial Mathematics* (ed. K. H. Rosen), CRC Press, 2000.

[CoMo72] H. S. M. Coxeter and W. O. J. Moser, *Generators and Relators for Discrete Groups*, Third Edition, Springer-Verlag, 1972.

[EaWh96] P. Eades and S. Whitesides, The realization problem for Euclidean minimum spanning trees is NP-hard, *Algorithmica* 16 (1996), 60–82. (Special issue on Graph Drawing, ed. by G. Di Battista and R. Tamassia.)

[GrTu87] J. L. Gross and T. W. Tucker, *Topological Graph Theory*, Dover, 2001. (First Edition, Wiley-Interscience, 1987.)

[LiDi95] G. Liotta and G. Di Battista, Computing proximity drawings of trees in the 3-dimensional space, pp 239–250 in Proc. 4th Workshop Algorithms Data Struct., *Lecture Notes Comput. Sci.* vol. 955, Springer-Verlag, 1995.

[LiTa04] G. Liotta and R. Tamassia, *Drawings of Graphs*, §10.3 in *Handbook of Graph Theory*, Second Edition, eds., J. L. Gross, J. Yellen, and P. Zhang, CRC Press, 2014.

[Ma67] W. S. Massey, *Algebraic Topology: An Introduction*, Springer-Verlag, 1989. (First Edition: Harcourt Brace & World, 1967.)

[MoSu92] C. Monma and S. Suri, Transitions in geometric minimum spanning trees, *Discrete Comput. Geom.* 8 (1992), 265–293.

Chapter 6: Graph Colorings

[ApHa76] K. I. Appel and W. Haken, Every planar map is four-colorable, *Bull. Amer. Math. Soc.* 82 (1976), 711–712.

[Bä38] F. Bäbler, Über die Zerlegung regulärer Streckencomplexe ungerader Ordnung, *Comment. Math. Helv.* 10 (1938), 275–287.

[Be68] L. W. Beineke, Derived graphs and digraphs, in *Beitrage zur Graphentheorie* Tuebner, 1968.

[Br79] D. Brelaz, New methods to color the vertices of a graph, *Commun. ACM* 22 (1979), 251–256.

[Ca86] M. W. Carter, A survey of practical applications of examination timetabling algorithms, *Oper. Res.* 34 (1986), 193–202.

[GaJo79] M. R. Garey and D. S. Johnson, *Computers and Intractibility: A Guide to the Theory of NP-completeness*, W. H. Freeman, 1979.

[Ho81] I. Holyer, The NP-completeness of edge-coloring, *SIAM J. Computing* 10 (1981), 718–720.

[KiYe92] L. Kiaer and J. Yellen, Weighted graphs and university course timetabling, *Computers and Oper. Res.* 19 (1992), 59–67.

[Kö16] D. König, Über Graphen und ihre Andwendung auf Determinantentheorie und Mengenlehre, *Math. Ann.* 77 (1916), 453–465.

[Lo75] L. Lovász, Three short proofs in graph theory, *J. Combin. Theory Ser. B* 19 (1975), 269–271.

[Ma81] B. Manvel, Coloring large graphs, *Proceedings of the 1981 Southeastern Conference on Graph Theory, Combinatorics and Computer Science*, 1981.

[Pe1891] J. Petersen, Die Theorie der regulären Graphen, *Acta Math.* 15 (1891), 193–220.

[Pl04] M. Plummer, *Factors and Factorization*, §5.4 in *Handbook of Graph Theory*, Second Edition, eds., J. L. Gross, J. Yellen, and P. Zhang, CRC Press, 2014.

[RoSaSeTh97] N. Robertson, D. P. Sanders, P. Seymour, and R. Thomas, The four-colour theorem, *J. Combin. Theory Ser. B* 70 (1997), 166–183.

[WeYe14] A. Wehrer and J. Yellen, The Design and Implementation of an Interactive Course-Timetabling System, *Annals of Oper. Res.* 218, (2014), 327–345.

Chapter 7: Measurement and Mappings

[AnLaLuMcMe94] C. A. Anderson, L. Langley, J. R. Lundgren, P. A. McKenna and S. K. Merz, New classes of p-competition graphs and ϕ-tolerance competition graphs, *Gongr. Numer.* 100 (1994), 97–107.

[Be62] C. Berge, *Theory of Graphs and its Applications*, Methuen, 1962.

[Be73] C. Berge, *Graphs and Hypergraphs*, North-Holland, 1973.

[BiSy83] H. Bielak and M.M. Syslo, Peripheral vertices in graphs, *Studia Sci. Math. Hungar.* 18 (1983), 269–275.

[Br04a] R. C. Brigham, *Bandwidth*, §9.4 in *Handbook of Graph Theory*, Second Edition, eds., J. L. Gross, J. Yellen, and P. Zhang, CRC Press, 2014.

[BrCaVi00] R. C. Brigham, J. R. Carrington and R. P. Vitray, Bipartite graphs and absolute difference tolerances, *Ars Gombin.* 54 (2000), 3–27.

[BrDu85] R. C. Brigham and R. D. Dutton, A compilation of relations between graph invariants, *Networks* 15 (1985), 73–107.

[BrMcVi95] R. C. Brigham F. R. McMorris and R. P. Vitray, Tolerance competition graphs, *Linear Algebra Appl.* 217 (1995), 41–52.

[BrMcVi96] R. C. Brigham, F. R. McMorris and R. P. Vitray, Two-ϕ-tolerance competition graphs, Discrete *Appl. Math* 66 (1996), 101–108.

[Bu74] P. Buneman, A characterisation of rigid circuit graphs, *Discrete Math.* 9 (1974), 205–212.

[BuHa90] F. Buckley and F. Harary, *Distance in Graphs*, Addison-Wesley, 1990.

[Ch88] F. R. K. Chung, Labelings of graphs, *Selected Topics in Graph Theory* 3, Academic, Press Limited, Sail Diego, CA (1988), 151–168.

[ChChDeGi82] P. Z. Chinn, J. Chvátalová, A. K. Dewdney, and N. E. Gibbs, The bandwidth problem for graphs and matrices-a survey, *Journal of Graph Theory* 6 (1982), 223–254.

[ChDeGiKo75] J. Chvátalová, A. K. Dewdney, N. E. Gibbs, and R. R. Korfhage, The bandwidth problem for graphs: a collection of recent results, Research Report 24, Department of Computer Science, University of Western Ontario, (1975).

[ChOeTiZo89] G. Chartrand, O. R. Oellermann, S. Tian, and H. B. Zou, Steiner distance in graphs, *Časopis Pro Pest. Mat.* 114 (1989), 399–410.

[ChZh04] G. Chartrand and P. Zhang, *Distance in Graphs*, §9.1 in *Handbook of Graph Theory*, Second Edition, eds., J. L. Gross, J. Yellen, and P. Zhang, CRC Press, 2014.

[Co78] J. E. Cohen, *Food Webs and Niche Space*, Princeton University Press, 1978.

[DuBr83] R. D. Dutton and R. C. Brigham, A characterization of competition graphs, *Discrete Appl. Math.* 6 (1983), 315–317.

[ErGoPo66] P. Erdös, A. W. Goodman and L. Pósa, The representation of a graph by set intersections, *Canad. J. Math.* 18 (1966), 106–112.

[Ga74] F. Gavril, The intersection graphs of subtrees in trees are exactly the chordal graphs, *Journal of Comb. Theory, Ser. B* 16 (1974), 47–56.

[GoMo82] M. C. Golumbic and C. L. Monma, A generalization of interval graphs with tolerances, *Congr. Numer.* 35 (1982), 321–331.

[GoMoTr84] M. C. Golumbic, C. L. Momna and W. T. Trotter, Tolerance graphs, *Discrete App J. Math.* 9 (1984), 157–170.

[GoTr03] M. C. Golumbic and A. N. Trenk, *Tolerance Graphs*, Cambridge University Press, 2003.

[Ha64] L. H. Harper, Optimal assignment of numbers to vertices, *Journal of SIAM* 12 (1964), 131–135.

[HaHe04] T. W. Haynes and M. A. Henning, *Domination in Graphs*, §9.2 in *Handbook of Graph Theory*, Second Edition, eds., J. L. Gross, J. Yellen, and P. Zhang, CRC Press, 2014.

[HaHeSl98] T. W. Haynes, S. T. Hedetniemi, and P. J. Slater, *Fundamentals of Domination in Graphs*, Marcel Dekker, 1998.

[HaNo53] F. Harary and R. Z. Norman, The dissimilarity characteristic of Husimi trees, *Ann. of Math.* 58 (1953), 134–141.

[HeOeSw90] M. A. Henning, O. R. Oellermann, and H. C. Swart, On Steiner radius and Steiner diameter of a graph, *Ars Combin.* 29 (1990), 13–19.

[HeOeSw91] M. A. Henning, O. R. Oellermann, and H. C. Swart, On vertices with maximum Steiner eccentricity in graphs, *Graph Theory, Gombinatorics, Algorithms and Applications* (eds. Y. Alav, F. R. K. Chung, R. L. Graham, and D. F. Hsu), SIAM Publications (1991), 393–403.

[JaMcMu91] M. S. Jacobson, F. R. McMorris and H. M. Mulder, An introduction to tolerance intersection graphs, in *Graph Theory, Combinatorics and Applications* (Y. Alavi, et al., Eds.) Wiley-Interscience, vol. 2 (1991), pp. 705–723.

[JaMcSc91] M. S. Jacobson, F. R. McMorris and E. R. Scheinerman, General results on tolerance intersection graphs, *J. Graph Theory* 15 (1991), 573–577.

[LaWi99] Y. Lai and K. Williams, A survey of solved problems and applications on bandwidth, edgesum, and profile of graphs, *Journal of Graph Theory* 31 (1999), 75–94.

[LuMa83] J. R. Lundgren and J. S. Maybee, A characterization of graphs of competition number m, *Discrete Appl. Math.* 6 (1983), 319–322.

[Ma45] E. Szpilrajn-Marczewski, Sur deux proprietes des classes d'ensembles, *Fund. Math.* 33 (1945), 303–307.

[McMc99) T. A. McKee and F. R. McMorris, *Topics in Intersection Graph Theory*, SIAM monograph, 1999.

[Mi91] Z. Miller, Graph layouts, *Applications of Discrete Mathematics*, J. G. Michaels and K. H. Rosen (Editors), McGraw-Hill, New York (1991), 365–393.

[Or62] O. Ore, *Theory of graphs, Amer. Math. Soc. Transl.* 38 (1962), 206–212.

[Wa78] J. R. Walter, Representations of chordal graphs as subtrees of a tree, *Journal of Graph Theory* 2 (1978), 265–267.

[WaAcSa79] H. B. Walikar, B. D. Acharya, and E. Sampathkumar, Recent developments in the theory of domination in graphs, Mehta Research Institute, Allahabad, MRI *Lecture Notes in Math*, 1 (1979).

Chapter 8: Analytic Graph Theory

[BoNi04] B. Bollobas and V. Nikiforov, *Extremal Graph Theory*, §8.1 in *Handbook of Graph Theory*, Second Edition, eds., J. L. Gross, J. Yellen, and P. Zhang, CRC Press, 2014.

[Fa04] R. Faudree, *Ramsey Graph Theory*, §8.3 in *Handbook of Graph Theory*, Second Edition, eds., J. L. Gross, J. Yellen, and P. Zhang, CRC Press, 2014.

[GaRoSp90] R. L. Graham, B. 1. Rothschild, and J. H. Spencer, *Ramsey Theory*, John Wiley & Sons, 1990.

[GrGl55] R. E. Greenwood and A. M. Gleason, Combinatorial relations and chromatic graphs, *Canad. J. Math.* 7 (1955), 1–7.

[Tu41] P. Turán, On an extremal problem in graph theory (Hungarian), *Mat. és Fiz. Lapok* 48 (1941), 436–452.

[Wo04] N. Wormald, *Random Graphs*, §8.2 in *Handbook of Graph Theory*, Second Edition, eds., J. L. Gross, J. Yellen, and P. Zhang, CRC Press, 2014.

Chapter 9: Graph Colorings and Symmetry

[Bo00] K. P. Bogart, *Introductory Combinatorics*, Third Edition, Academic Press, 2000.

[Br04b] R. A. Brualdi, *Introductory Combinatorics*, Fourth Edition, Prentice Hall, 2004.

[Gr08] J. L. Gross, *Combinatorial Methods with Computer Applications*, CRC Press, 2008.

[HaPa73] F. Harary and E. M. Palmer, *Graphical Enumeration*, Academic Press, 1973.

[Po37] G. Polya, Kombinatorische Anzahlbestimmungen fur Gruppen, Graphen und Chemische Verbindungen, *Acta Mathematica* 68 (1937), 145–254.

Chapter 10: Algebraic Specification of Graphs

[AnBaBr09] M. Anderson, C. Barrientos, R. Brigham, J. Carrington, M. Kronman, R. Vitray, and J. Yellen, Irregular Colorings of Some Graph Classes, *Bulletin of the Institute of Combinatorics and Its Applications*, (2009).

[AnViYe12] M. Anderson, R. Vitray, J. Yellen, Irregular Colorings of Regular Graphs, *Discrete Mathematics* 312 (2012), 2329-2336.

[GrCh96] J. L. Gross and J. Chen, Algebraic specification of interconnection networks by permutation voltage graph morphisms, *Math. Systems Theory* 29 (1996), 451–470.

[GrTu77] J. L. Gross and T. W. Tucker, Generating all graph coverings by permutation voltage assignments, *Discrete Math.* 18 (1977), 273–283.

[Le92] F. T. Leighton, *Introduction to Parallel Algorithms and Architectures*, Morgan Kaufmann, 1992.

[RaZh07] M. Radcliffe and P. Zhang, Irregular colorings of graphs, *Bull. Inst. Combin. Appl.* 49 (2007) 41-59.

Chapter 11: Non-Planar Layouts

[AlGr76] S. R. Alpert and J. L. Gross, Components of branched coverings of current graphs, *J. Combin. Theory Ser. B* 20 (1976), 283–303.

[Du66] R. A. Duke, The genus, regional number, and Betti number of a graph, *Canad. J. Math.* 18 (1966), 817–822.

[Ed60] J. Edmonds, A combinatorial representation for polyhedral surfaces, *Notices Amer. Math. Soc.* 7 (1960), A-646.

[Ed65c] J. R. Edmonds, On the surface duality of linear graphs, *J. Res. Nat. Bur. Standards Sect. B* (1965), 121–123.

[FuGrMc88] M. Furst, J. L. Gross, and L. McGeoch, Finding a maximum-genus graph imbedding, *JACM* 35 (1988), 523–534.

[Gr74] J. L Gross, Voltage graphs, *Discrete Math.* 9 (1974), 239–246.

[GrAl73] J. L. Gross and S. R. Alpert, Branched coverings of graph imbeddings, *Bull. Amer. Math. Soc.* 79 (1973), 942–945.

[GrAl74] J. L. Gross and S. R. Alpert, The topological theory of current graphs, *J. Combin. Theory Ser. B* 17 (1974), 218–233.

[GrTu77] J. L. Gross and T. W. Tucker, Generating all graph coverings by permutation voltage assignments, *Discrete Math.* 18 (1977), 273–283.

[GrTu87] J. L. Gross and T. W. Tucker, *Topological Graph Theory*, Dover, 2001. (First Edition, Wiley-Interscience, 1987.)

[Gu63] W. Gustin, Orientable embedding of Cayley graphs, *Bull. Amer. Math. Soc.* 69 (1963), 272–275.

[He1891] L. Heffter, Über das Problem del Nachbargebiete, *Math. Annalen* 38 (1891), 477–580.

[NoStWh71] E. A. Nordhaus, B. M. Stewart, and A. T. White, On the maximum genus of a graph, *J. Combin. Theory Ser. B* 11 (1971), 258–267.

[NoRiStWh72] E. A. Nordhaus, R. D. Ringeisen, B. M. Stewart, and A. T. White, A Kuratowski-type theorem for the maximum genus of a graph, *J. Combin. Theory Ser. B* 12 (1972), 260–267.

[Th89] C. Thomassen, The graph genus problem is NP-complete, *J. of Algorithms* 10 (1989), 568–576.

Appendix

[Co71] S. A. Cook, The complexity of theorem proving procedures, *Proc. 3rd Annual ACM Symposium on Theory of Computing (STOC 71)* (1971), 151–158.

[Ev79] S. Even, *Graph Algorithms*, Computer Science Press, 1979.

[GaJo79] M. R. Garey and D. S. Johnson, *Computers and Intractibility: A Guide to the Theory of NP-Completeness*, W. H. Freeman, 1979.

[Wi86] H. S. Wilf, *Algorithms and Complexity*, Prentice-Hall, 1986.

INDEX

Printed in the United States
by Baker & Taylor Publisher Services

Printed in the United States
by Baker & Taylor Publisher Services